Functional Anatomy of the Brain: A View from the Surgeon's Eye

Abhidha Shah • Atul Goel • Yoko Kato
Editors

Functional Anatomy of the Brain: A View from the Surgeon's Eye

Springer

Editors
Abhidha Shah
Department of Neurosurgery
King Edward Memorial Hospital and
Seth G. S. Medical College
Mumbai, Maharashtra, India

Atul Goel
Department of Neurosurgery
Lilavati Hospital and Research Center
Mumbai, Maharashtra, India

Yoko Kato
Department of Neurosurgery
Fujita Health University Hospital
Aichi, Japan

ISBN 978-981-99-3411-9 ISBN 978-981-99-3412-6 (eBook)
https://doi.org/10.1007/978-981-99-3412-6

This Springer imprint is published by the registered company Springer Nature Singapore Pte Ltd.
The registered company address is: 152 Beach Road, #21-01/04 Gateway East, Singapore 189721, Singapore

To my parents Harshad and Ila, Nalin uncle, Aditi and Manan for their constant love, support, understanding, and encouragement.

And

To Prof. Atul Goel for guiding and inspiring me and molding me into the neurosurgeon I am today.

—Abhidha Shah

Dedicated to the companionship, love, and affection of my wife Naina Goel.

—Atul Goel

Dedicated to Prof. Tetsuo Kanno.

—Yoko Kato

Foreword

It is a pleasure for me to write the foreword for the new book *Functional Anatomy of the Brain: A View from the Surgeon's Eye* co-edited by two highly experienced neurosurgeons—Prof. Goel and Prof. Kato.

The idea of the book is to present the general and functional anatomy, which is relevant for surgeries in complex brain regions. Each chapter is devoted to a particular brain area, presenting its structural organization and microsurgical anatomy, related to the most common surgical approaches. It is illustrated with multiple clearly described clinical examples. A special emphasis is placed on the organization of white matter/fiber tracks in the brain and its significance for planning and performing surgery. The reader is led step-by-step through the main steps of the operative approach. Multiple intraoperative photographs enhance the orientation and emphasize potential risks to critical anatomical structures.

A special highlight is the detailed and beautiful illustration of the structural organization of the brainstem and its relevance for surgeries in the area. Not only the nuclei but also all-important fiber tracts and functional systems are depicted in a practically oriented manner rendering it, as well as the whole book, as a very useful teaching manual.

The knowledge of the microanatomy of the brain is a prerequisite for operating complex brain tumors. However, the expanding tumor itself causes unpredictable changes of the normal anatomy and of the location and relation of the tumor to the important surrounding vessels, cranial nerves, or essential brain areas. This "pathological microanatomy" can only be appreciated at surgery under the operating microscope. The clinical experience cannot be supplemented for by studying cadaver anatomy.

My vast experience over the last three to four decades has shown me that there is enormous variability in the course of fiber tracts in individual patients. Additionally, the pathological lesion causes additional changes by either compression or infiltration of the adjacent brain. Slowly growing tumors cause unpredictable functional reorganization of the cortical areas and their connecting fiber tracts. The utilization of brain mapping techniques allows us to demonstrate the functional organization of the brain in each individual case before surgery. As described and nicely illustrated in the current book, this information can be implemented preoperatively for defining the surgical strategy and selecting the safest operative approach. Importantly, the application of functional neuronavigation, which integrates the data from the fMRI and DTI-based tractography, allows us to utilize this information during surgery. Intraoperative MRI allows for the updating of this functional information in order to overcome the effect of brain shift. I hope that future technological advances will allow acquiring more reliable and detailed information before and during surgery, which will further decrease operative risks.

With our increasing understanding of the brain's structure and organization, it becomes obvious that every attempt should be made to preserve all white matter tracts in order to avoid neurological or psychological changes after surgery. The risks of such potential postoperative changes should be considered, but always in relation to the oncological goal of the surgery. The use of transverse callosotomy, for example, allows avoiding larger disruption of the interhemispheric connecting fibers. It could be successfully applied in case of certain small tumors, but for the successful and safe removal of larger ventricular tumors a longitudinal callosotomy is certainly required.

I am confident that the book is a good contribution to the neurosurgical literature. I congratulate the authors and hope very much that the book will help many young neurosurgeons to understand the complex anatomy of the brain and the principles of surgery of complex brain tumors.

INI, Hanover, GmbH Madjid Samii

Preface

Everything we do, every thought we've ever had, is produced by the human brain. But exactly how it operates remains one of the biggest unsolved mysteries, and it seems the more we probe its secrets, the more surprises we find.
—Neil deGrasse Tyson

The brain is the most magnificent and enigmatic organ of the body and that makes neurosurgery one of the most technically challenging and demanding disciplines of medicine.

For a long time the field of neurosurgery was considered as a "no man's land" as the surgical outcomes were often worse than the natural history of the disease itself. The last decade has witnessed the goal of neurosurgery changing from just "saving lives" to achieving a better quality of life that enables the patient to go back to his routine life and occupation.

The human neural network consists of "nodes" (cortical areas) connected by a series of "edges or links" (subcortical network) to create the most fascinating grid of motor, sensory, autonomic, and emotional function. Understanding this neural circuitry is the key to a successful brain surgery. Mastering the three-dimensional architecture of the brain anatomy is the building block to become an accomplished neurosurgeon.

The human brain is an archetypal complex system. This book aims to correlate the exquisite anatomy with the functions of the brain and thus help in building a map of the surgical arena to result in a surgical victory.

Technology has rapidly paved its way into neurosurgery. A modern-day neurosurgical operation theater may very well resemble a space station. A "brain suite" has its microscope, neuronavigation station, CUSA, intraoperative ultrasound, neurophysiological monitoring system, endoscope, and intraoperative CT/MRI. A hybrid brain suite allows for intraoperative cerebral angiography while performing vascular neurosurgery. The neurosurgeon is at the center of this machinery and dictates the use of each. But all this clutter of gadgetry cannot replace exquisite neurosurgical technique. Just as the skill of a pilot is measured by his manual flying skills where the maneuvers are carried out by analysis of raw information obtained from airspeed, altitude, weather, and heading without use of any modern technology, the skill of a neurosurgeon is measured by his neurosurgical technique. Mastery of the technical skills of neurosurgery requires great passion and practice, an amalgamation of heart, mind, and soul till perfection is achieved. This skill defines the caliber of a neurosurgeon much like an artist in any field. Without this mastery of technique no fancy gadgetry will help you perform good neurosurgery.

All in all neurosurgery is an art. Modern-day neurosurgery has advanced to make brain surgery safe. As neurosurgeons we have great responsibility but also have the unique pleasure of seeing a patient become conscious again or walking again or talking again after a good neurosurgical procedure has been performed. There can be no greater joy than this.

> *Only the neurosurgeon dares to improve upon five billion years of evolution in a few hours.*
> *The human brain. A trillion nerve cells storing electrical patterns more numerous than the water molecules of the world's oceans. The soul's tapestry lies woven in the brain's nerve threads. Delicate, inviolate, the brain floats serenely in a bone vault like the crown jewel of biology. What motivated the vast leap in intellectual horse-power between chimp and man? Between tree dweller and moon walker? Is the brain a gift from God, or simply the jackpot of a trillion rolls of DNA dice?*
> *—Frank Vertosick Jr.*

Mumbai, India Atul Goel

Contents

About the Editors

Abhidha Shah is an Associate Professor at the neurosurgery department at King Edward Memorial Hospital and Seth G.S. Medical College. She serves as an associate editor of "Journal of Craniovertebral Junction and Spine" and the "International Journal of Neuroscience." She has more than 160 peer-reviewed international and national publications and has written 20 book chapters. She has been awarded the Vesalius award and the Women in Neurosurgery—Greg Wilkins—Barrick Chair—Visiting International Surgeon award by the American Association of Neurological Surgeons. She has a particular interest in the anatomy of the white fibers of the brain and intra-axial surgery related to anatomy and function. She has been awarded the Indian Society of Neuro-oncology President's award for her clinical research on this subject. She is a member of the International Academy of Neurosurgical Anatomy. She is also on the World Federation of Neurosurgical Societies (WFNS) committee of Neurosurgical Anatomy. Currently she serves as an executive council member of the Neurological Society of India and is the Chair of the Women in Neurosciences section the society. She is also the Secretary of the Women in Neurosurgery Section of the Asian Congress of Neurological Surgeons.

Atul Goel is the Chairman of the neurosurgery department at the Lilavati Hospital and Research Centre. He is Professor Emeritus at the King Edward Memorial Hospital and Seth G.S. Medical College and K.E.M Hospital. Prof. Goel has written over 750 articles finding space in Pubmed. Prof. Goel has vast surgical experience on several complex issues. He has written two books published by the leading medical books publisher (Churchill Livingstone in 1997 and Thieme publishers in 2011). He has published over 180 articles presenting new (small and big) surgical techniques, more than 30 articles introducing philosophical and spiritual issues about neurosurgery. He is currently serving as the president of the cranio-vertebral junction and spine society. He has been awarded the eminent medical teacher, Dr. B.C. Roy National Award, Dandy medal for distinguished neurosurgery services-2018, Outstanding contribution award—Chinese society of Neurosurgery-2019, and International Ambassador award by Congress of Neurological Surgeons, San Francisco 2019. He has been conferred Lifetime Achievement Awards by the Neurospinal Surgeons Association (NSSA) and the Delhi Spine Society. He is a board member of the prestigious International Rhoton Society. He has many US and Indian patents to his name.

Yoko Kato is currently working as Chairman of the neurosurgery department at Fujita Health University Banbuntane Hotokukai Hospital, Japan, where she also holds the position of assistant director of the University. Prof. Kato has received several awards, like the WFNS Medal of Honor award in 2019 and the Honorary Fellowship award in the American College of Surgeons in San Francisco in 2019. She is the editor-in-chief of "Asian Journal of Neurosurgery" and has more than 300 international peer-reviewed publications. She has written numerous chapters in primary neurosurgical textbooks. She is the President of the Asian Congress of Neurological Surgeons.

Anatomy Guided Microneurosurgery of Cerebral Lesions

Guilherme Carvalhal Ribas
and Eduardo Carvalhal Ribas

1 Introduction

The description of a motor language center by Broca [1, 2] and of the motor strip by H. Jackson [2–4] gave rise along the second half of the nineteenth century to the study and description of cranial-cerebral relationships in order to make craniotomies more clinically guided. Broca was the pioneer of these studies [5, 6] having related the eloquent and the main brain gyri and sulci more particularly with the cranial sutures and with the craniometric points he also had described [6], and was soon followed by many other authors [7–11].

More recently, the advent of microneurosurgery started with Yaşargil along the 1970s [12, 13] turned the brain sulci in the major surgical landmarks of the cerebral surface and also made them possible corridors for microneurosurgical procedures [14, 15].

Parallel to the development of microneurosurgery the use of the microscope disclosed a new microanatomical surgical universe which was studied and revealed particularly by Yaşargil himself [13, 15–19] and by Rhoton and his fellows which publications were recently put together in a unique book [20], and it is interesting to point out that the important chapter of cranial-cerebral relationships has not been often revisited along these last four decades. Recent textbooks' mentions of these anatomical relationships are usually very brief and pertinent only to classic descriptions [21–23], what motivated us to restudy this issue in the light of more recent microanatomic knowledge.

Although easily identifiable in standard magnetic resonance images, the brain sulci are usually difficult to be recognized in neurosurgical procedures not only because they are covered by the arachnoid membrane and frequently also by veins, but mainly because most of them are usually interrupted and have diverse anatomical variations [24].

Nevertheless, as already noticed by Broca [5, 6, 25], some specific points of the main brain sulci as their extremities and some gyral sites that underlie particularly prominent cranial points do have a more constant anatomy and, hence, more constant cranial-cerebral relationships. Beyond

G. C. Ribas
University of São Paulo Medical School,
Sao Paulo, Brazil

Albert Einstein Medical School, São Paulo, Brazil

Department of Neurosurgery of the University of Virginia, Charlottesville, VA, USA

Annual Cambridge Lectures in Neurosurgical Anatomy, Cambridge, UK

Hospital Israelita Albert Einstein, São Paulo, Brazil
e-mail: guilherme@neuroribas.com.br

E. C. Ribas (✉)
Discipline of Neurosurgery, Hospital das Clínicas, University of São Paulo Medical School,
São Paulo, Brazil

Neurosurgeon at Hospital Israelita Albert Einstein,
São Paulo, Brazil

A. Shah et al. (eds.), *Functional Anatomy of the Brain: A View from the Surgeon's Eye*,
https://doi.org/10.1007/978-981-99-3412-6_1

the classic sulcal points originally described by Paul Broca and others we also studied other strategic key points like the meeting points of important sulci [26–30], and the regular neural and cranial-cerebral relationships of these specific anatomical points warrant these sites to be considered consistent microneurosurgical sulcal and cortical key-points [28–30]. Together with their corresponding cranial points they constitute a neurosurgical anatomic framework that can help in the understanding of the localization of cortical and subcortical lesions, in the placement of their required craniotomies, and to ease the initial transoperative identification of their related brain sulci and gyri.

According to our findings, the sulcal key points described along this text can be intraoperatively identified within an interval always smaller than 2 cm in relation to their corresponding cranial points, aided by the fact that the sulcal key points are usually visually characterized by a certain degree of enlargement of the subarachnoid space since they generally correspond to an intersection of two sulci. The surgeon's previous knowledge of the usual shape and most frequent anatomic variations of the main brain sulci helps to corroborate the identification of these sulci, and their key points' subaracnoid enlargements favor their initial microsurgical dissections.

Nowadays we fortunately have technology [31] to aid intraoperative localization for neurosurgeons, but unfortunately these extraordinary tools are not homogeneously available throughout our heterogeneous world, and, above all, of course technology should not substitute the anatomic tridimensional knowledge every neurosurgeon should acquire and should continuously develop along his or her practice [30].

On the other hand, it is important to emphasize that the anatomic identification of any eloquent cortical area, even when confirmed by any localizing imaging system, cannot safely substitute the current aid of transoperative neurophysiological testing due to common anatomic functional variations and their possible displacements and/or involvement by the existence of underlying pathology.

The following text is all based on our neurosurgical observations and laboratory findings that were published in four different main articles [24, 28–30], summarized in an important book chapter [32], and more extensively presented and discussed in a whole book that was published in 2018 [26].

Given the cranial-cerebral relationships scope of this chapter, its discussion of brain anatomy features is practically restricted to the main aspects of the brain superolateral surface. For further cranial-cerebral' anatomical details please consult the above references.

For practical purposes, while the anatomy of the brain sulci and gyri is presented here along the text and illustrations, the sulcal and gyral key-points and then cranial relationships are presented and detailed mainly through illustrations and legends.

2 General Features of the Brain Sulci and Gyri Architecture

It is interesting that the general organization of the brain gyri was understood as having a chaotic disposal for many centuries.

Achille Louis Foville depicted the cortical surface properly about 180 years ago [33, 34], and it was only about 160 years ago that Louis Gratiolet, studying comparative anatomy with his teacher Francois Lauret, perceived that the human brain does have an organized arrangement [24, 35, 36] (Figs. 1, 2 and 3).

In order to understand their organization we have initially to remember that the brain sulci were developed along evolution through an infolding process [37], which also takes place along our embryological and fetal developments, and which increases the human brain surface threefold. The generated area of approximately 2200 cm^2 [38] ends up with only one-third exposed and with the other two-thirds buried inside the intrasulcal spaces [39].

Given this infolding process, and since the arachnoid envelopes the whole brain, the sulci are extensions of the subaracnoid space and they have to be understood as anatomical concepts, and not as well defined spaces or compartments. While a few sulci are evident and frequently continuous as the superior frontal and the supe-

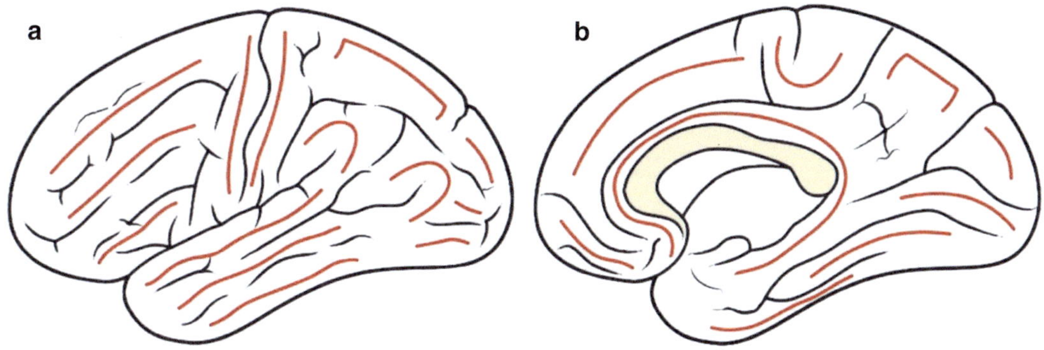

Fig. 1 Basic organization of the brain gyri. (**a**) superolateral surface and (**b**) medial and basal surfaces. Red lines indicate the usual arrangement of the main brain gyri, and the dark lines indicate the usual arrangement of the main brain sulci. (Adapted from [24])

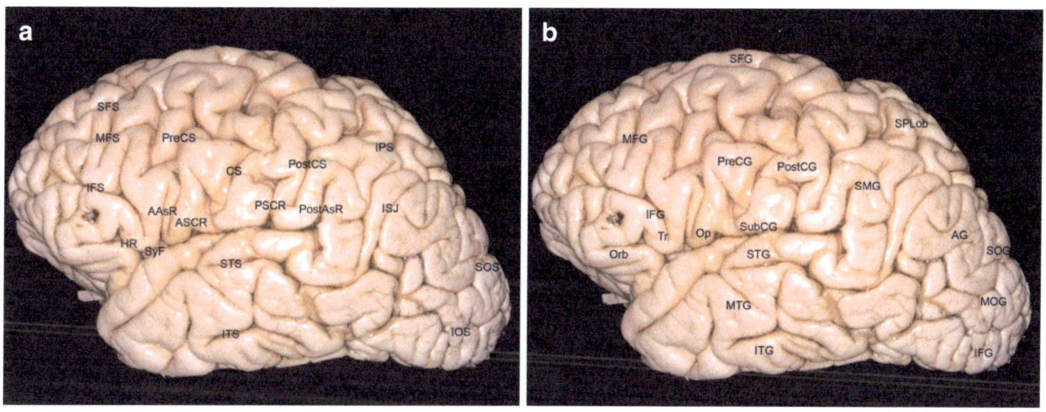

Fig. 2 The main brain sulci (**a**) and gyri (**b**) of the superolateral surface of the brain. *AAsR* Anterior Ascending Ramus, *AG* Angular Gyrus, *ASCR* Anterior Subcentral Ramus of Sylvian Fissure, *CS* Central Sulcus, *HR* Horizontal Ramus, *IFG* Inferior Frontal Gyrus, *IFS* Inferior Frontal Sulcus, *IOS* Inferior Occipital Sulcus, *IPS* Intraparietal Sulcus, *ISJ* Intermediary Sulcus of Jensen, *ITG* Inferior Temporal Gyrus, *ITS* Inferior Temporal Sulcus, *MFG* Middle Frontal Gyrus, *MFS* Middle Frontal Sulcus, *MOG* Middle Occipital Gyrus, *MTG* Middle Temporal Gyrus, *Op* Opercular Part of Inferior Frontal Gyrus, *Orb* Orbital Part of Inferior Frontal Gyrus, *PostAsR* Posterior Ascending Ramus, *PostCG* Postcentral Gyrus, *PostCS* Postcentral Sulcus, *PreCG* Precentral Gyrus, *PreCS* Precentral Sulcus, *PSCR* Posterior Subcentral Ramus of Sylvian Fissure, *SFG* Superior Frontal Gyrus, *SFS* Superior Frontal Sulcus, *SMG* Supramarginal Gyrus, *SOG* Superior Occipital Gyrus, *SOS* Superior Occipital Sulcus, *SPLob* Superior Parietal Lobule, *STG* Superior Temporal Gyrus, *STS* Superior Temporal Sulcus, *SubCG* Subcentral Gyrus, *SyF* Lateral or Sylvian Fissure, *Tr* Triangular Part of Inferior Frontal Gyrus. (Adapted from [24])

rior temporal sulci for example, most of the sulci are usually interrupted as the inferior frontal and inferior temporal for example, and it is the sum of their segments that constitute each of them.

The sulci can then be long or short, continuous or interrupted, straight or with different configurations. When always deep, evident and anatomically constant, they are also referred to as fissures by Broca [6, 24, 35, 36]. The continuous cortical sulci are the lateral or Sylvian Fissure, callosal, calcarine, parieto-occipital, collateral, the central sulcus generally, and frequently the superior frontal and the superior temporal sulci [24, 40].

The main sulci have approximate depths ranging from 1 to 3 cm, and their walls harbor small

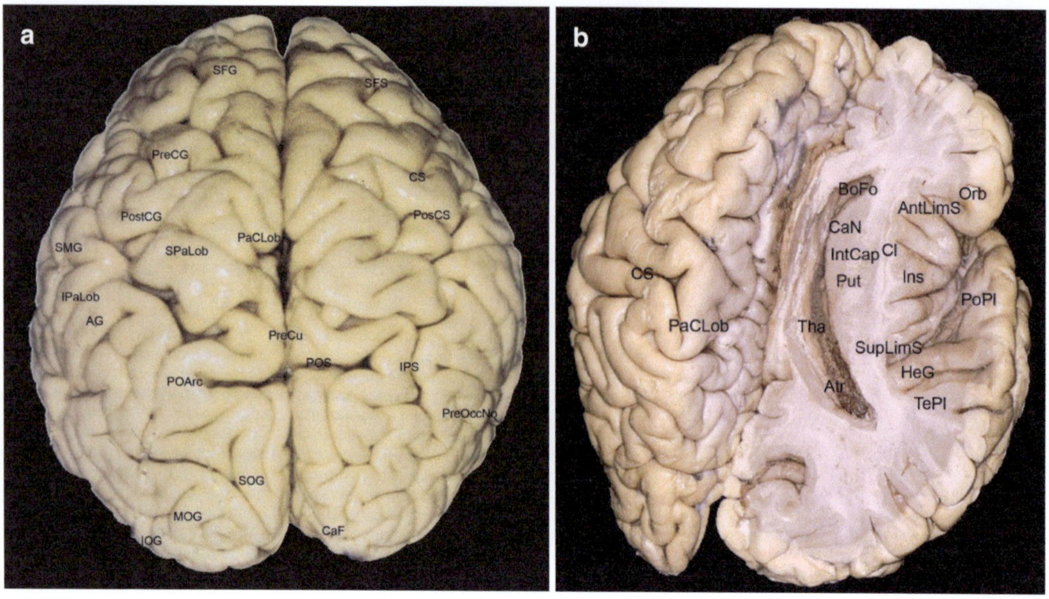

Fig. 3 (**a**) Superior view of the cerebral hemispheres, (**b**) the temporal opercular surface, insula, and the central core of the brain. *AG* Angular Gyrus, *AntLimS* Anterior Limiting Sulcus of insula, *Atr* Atrium of Lateral Ventricle, *BoFo* Body of Fornix, *CaF* Calcarine Fissure, *CaN* Caudate Nucleus, *Cl* Claustrum, *CS* Central Sulcus, *HeG* Heschl's Gyrus, *Ins* Insula, *IntCaps* Internal Capsule fibers, *IOG* Inferior Occipital Gyrus; *IPaLob* Inferior Parietal Lobule, *IPS* Intraparietal Sulcus, *MOG* Middle Occipital Gyrus, *Orb* Orbital part of Inferior Frontal Gyrus, *PaCLob* Paracentral Lobule, *POArc* Parieto-Occipital Arch, *PoPl* Polar Plane of the opercular temporal surface, *POS* Parieto-Occipital Sulcus seen on the superolateral surface, which corresponds to the Parieto-Occipital Incisures within the Parieto-Occipital Arch, *PosCS* Postcentral Sulcus, *PostCG* Postcentral Gyrus, *PreCG* Precentral Gyrus, *PreCu* Precuneus, *PreOccNo* Pre-Occipital Notch, *Put* Putamen, *SFG* Superior Frontal Gyrus, *SFS* Superior Frontal Sulcus, *SMG* Supramarginal Gyrus, *SOG* Superior Occipital Gyrus, *SPaLob* Superior Parietal Lobule, *SupLimS* Superior Limiting Sulcus of Insula, *TePl* Temporal Plane, *Th* Thalamus. (Adapted from [24])

gyri that face and adapt to each other which are generically designated as transverse gyri. The sulci that separate the transverse intrasulcal gyri vary in length and depth and at the surface of the brain they become visible as incisures.

It is noteworthy that the timing of the embryological development of the sulci and their degree of variability [41, 42] define a true morphological hierarchy at the top of which are the fissures and the main sulci. It is equally notable that this structural hierarchy is directly correlated with the functional importance of the areas to which the sulci are related, with the more anatomically constant sulci being those that are topographically related to areas that are more specialized [43, 44].

Due to their frequent variations and connections, the nomenclature of sulci varies among different authors according to their interpretations and designations [11, 24, 40, 45], but the current neurosurgical literature widely adopts the nomenclature proposed by Ono and Yaşargil [13, 15, 17, 40]. For their better understanding, it is important to emphasize that the sulci can vary in size and shape from person to person and these differences can be identified not only among individuals but also between the two hemispheres in the same individual, and, as already mentioned, they should then be understood as concepts and not as well defined spaces or compartments.

Also given the original infolding process the cortical surface is continuous along the sulci depths and around the sulci interruptions and extremities, and the brain gyri constitute then a real continuum which present as a serpentine configuration on the cerebral surface. For this reason the brain gyri, just as the brain sulci,

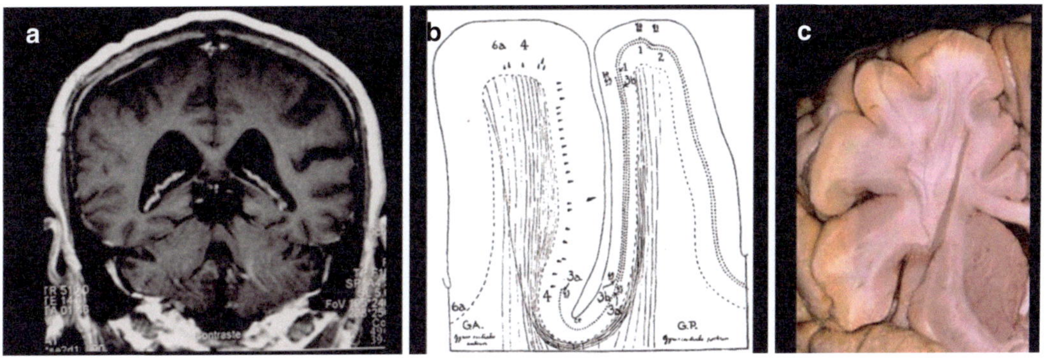

Fig. 4 (**a**) MRI coronal image showing the main sulci of the superolateral and basal brain surfaces point to the nearest ventricular cavity, the sulci of the medial surface are parallel to the corpus callosum; (**b**) classical sketch by Vogt C, Vogt O, 1919, 1928, and (**c**) cadaveric specimen, showing that while the projection fibers arise from the crest of the gyri, the bottom of the sulci are related to U-fibers

should be understood as arbitrary regions, as concepts, and not as well defined anatomical structures [24]. The term convolution usually refers to an anatomic well defined part of a gyrus. The well known orbital, triangular and opercular parts or convolutions together constitute the concept of inferior frontal gyris.

As already mentioned, despite their anatomical variations, the brain gyri do have a basic arrangement (Figs. 1, 2 and 3).

The original infolding process also implies in two important features of surgical interest: one is that all the main sulci of the superolateral and of the basal surfaces of the brain are pointing towards the nearest ventricular cavity, while the sulci of the medial surface are parallel to and dependent on the development of the corpus callosum (Fig. 4a). The other important feature of surgical interest is that while the projection fibers arise from the crest of the gyri, the depths of the sulci are exclusively related with U fibers (Fig. 4b, c). So, a transsulcal approach theoretically spares more projection fibers by putting them apart.

3 The Superolateral Brain Surface and its Sulcal Key-Points

While the cisternal and the intraventricular lesions are usually approached through standard craniotomies and microneurosurgical transfissural routes, the cortical-subcortical and the deep cerebral lesions require initially the understanding of their anatomical localization for the selection of their proper craniotomy placements and of their precise transsulcal, subpial or transgyral approaches (Fig. 5). For the correct approach of these lesions, as well for extrinsic convexity lesions, the understanding and identification of the brain sulci and gyri are mandatory and the framework provided by sulcal key-points can be of significant help (Figs. 6a, b).

In surgery, the only sulcus that can be securely identified is the Sylvian Fissure (SyF) and if we look along the SyF we can practically always identify a small area of a variable enlargement of the subaracnoid space, which corresponds to the Anterior Sylvian Point (ASyP) (Figs. 2, 6 and 7).

The ASyP divides the lateral or SyF in two parts: the anterior, sphenoidal or stem of the SyF, and the lateral or posterior part of the SyF.

From the ASyP always arise two Sylvian rami: the Anterior Ascending Ramus (AAsR) and the Horizontal Ramus (HR) which delineate the Triangular Part (TrP) of the Inferior Frontal Gyrus (IFG), which always harbors inside a small branch from the Inferior Frontal Sulcus (IFS) which comes from superior and anterior to inferior and posterior [28]). Anteriorly to the Triangular Part (TrP) there is the Orbital Part (OrbP), and posteriorly to it there is the Opercular Part (OpP), which is, in our opinion, the most constant U-shaped convolution of the human

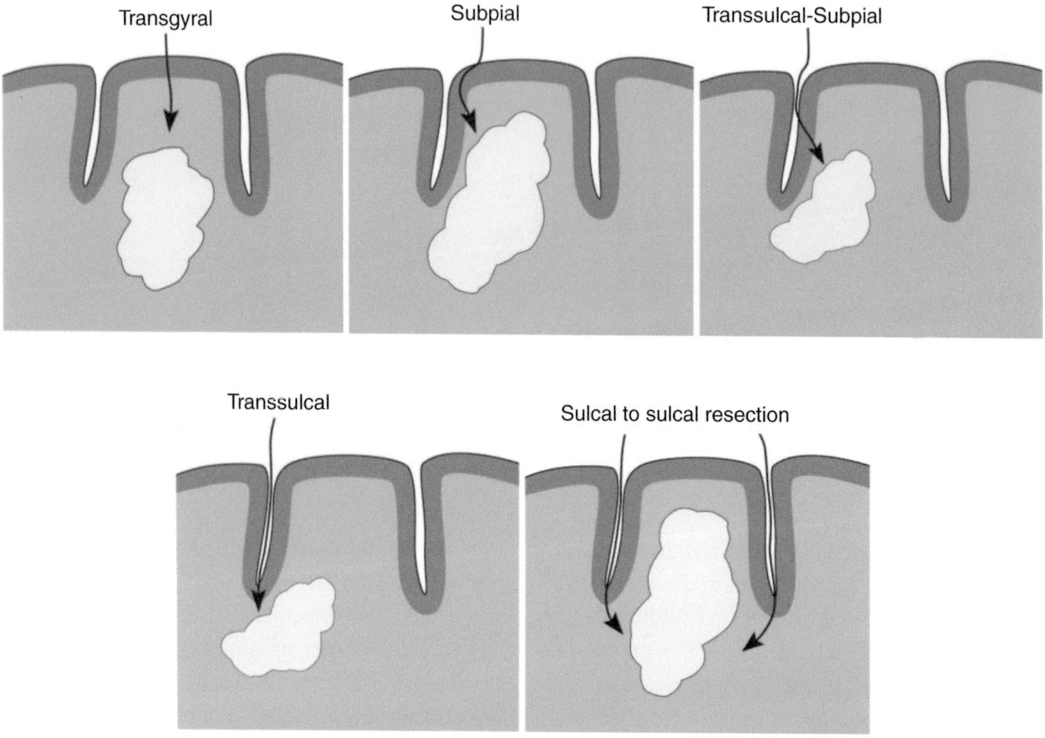

Fig. 5 Possible transcerebral microneurosurgical routes

brain and which always harbor the inferior part of the Precentral Sulcus (PreCS). We usually refer to the opercular part or convolution (OpP) as "the beatiful U".

The ASyP corresponds to an enlargement of the subaracnoid space of the SyF since the Triangular Part (TrP) is usually more retracted within the IFG. The ASyP lies then always at the inferior aspect of the Pars Triangularis (TrP), and in the anterior and basal aspect of the Opercular Part (OpP).

The Orbital, Triangular and Opercular Parts (OrbP, TrP, OpP) together constitute the concept of Inferior Frontal Gyrus (IFG). Its connection with the Precentral Gyrus (PreCG) corresponds to the posterior and ascending half of the U-shaped Opercular Part (OpP) which runs over the Anterior Subcentral Ramus (ASCR).

As already mentioned, the Inferior Frontal Sulcus (IFS) that separates the Inferior Frontal Gyrus (IFG) from the Middle Frontal Gyrus (MFG) is usually an interrupted sulcus formed by two, three or four segments, with the most posterior joining the Precentral Sulcus (PreCS). The so-called Ventral Premotor Area (VPreMA) is roughly located at this level within the Precentral Gyrus (PreCG) (Fig. 2).

The Fronto-Opercular Key Points and their corresponding cranial sites are shown in Fig. 7, and an example of their actual surgical application is shown in Fig. 8.

The Middle Frontal Gyrus (MFG) is the largest of the frontal gyri, and is posteriorly connected to the Precentral Gyrus (PreCG) through a superficial and prominent root responsible by a marked interruption of the Precentral Sulcus (PreCS). The so-called Dorsal Premotor Area (DPreMArea) corresponds roughly to the posterior aspect of the Middle Frontal Gyrus (MFG) (Fig. 2).

The Superior Frontal Gyrus (SFG) is posteriorly connected to the Precentral Gyrus (PreCG) at least through one fold which usually runs medially, already along the Interhemisperic Fissure (IHF). Anteriorly and inferiorly the SFG is continuous with the Rectus Gyrus (RG) already within the orbital frontal brain surface (Fig. 2).

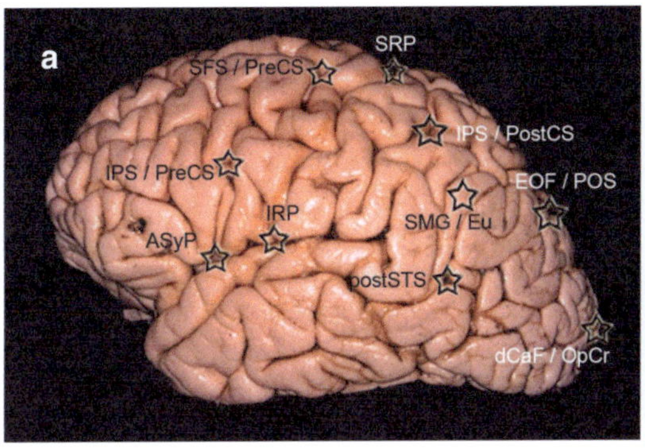

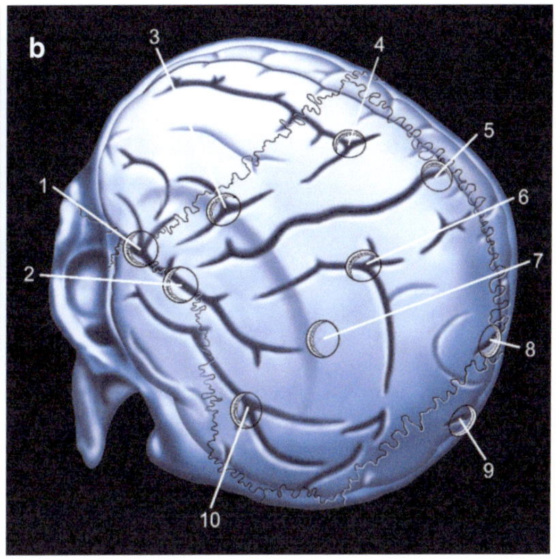

Fig. 6 (**a**) Microneurosurgical Sulcal/Cortical Key Points. The Microneurosurgical Key Points of the brain surface are constituted by real intersections between adjacent sulci or by their prolongations and by gyral and sulcal points located underneath prominent skull points such as the Euryon (center of the parietal tuterosity) and the Opisthocranion (most prominent occipital point). Note that the sulci meeting points are usually characterized by an enlargement of the subarachnoid space. *Eu* Euryon, *OpCr* Opisthocranion, *ASyP* Anterior Sylvian Point, *dCaF/OpCr* Distal Calcarine Fissure Point, underneath the Opisthocranion, *EOF/POS* External Occipital Fissure Medial Point, equivalent to the most superior of the Parieto-Occipital Sulcus in the medial surface of the brain, *IFS/PreCS* Inferior Frontal Sulcus and Precentral Sulcus meeting point, *IPS/PostCS* Intraparietal Sulcus and Postcentral Sulcus transitional or meeting point, *IRP* Inferior Rolandic Point, *postSTS* Superior Temporal Sulcus Posterior segment and extremity, *SFS/PreCS* Superior Frontal Sulcus and Precentral Sulcus meeting pont, *SMG/Eu* Superior aspect of the Supramarginal Gyrus disposed underneath the Euryon, *SRP* Superior Rolandic Point. (**b**) The main Sulcal and Gyral Key-Points and their cranial relationships: (1) the Anterior Sylvian Point (ASyP) lies underneath the most anterior portion of the Squamous suture, (2) the Inferior Rolandic Point (IRP) underneath the most superior portion of the Squamous suture, (3) the intersection of the Inferior Frontal Sulcus (IFS), or of its posterior prolongation, with the Precentral Sulcus (PreCS), lies just posteriorly to the Stephanion (St), craniometric point that corresponds to the intersection of the Superior Temporal Line (STL) with the Coronal Suture (CoSut), (4) the intersection of the Superior Frontal Sulcus (SFS), or of its posterior prolongation, with the Precentral Sulcus (PreCS), lies underneath the cranial point located 1.5 cm posteriorly to the Coronal Suture (CoSut) and 3 cm lateral to the sagittal suture, (5) the Superior Rolandic Point (SRP) lies underneath the cranial point located 5 cm posterior to the Bregma (Br), (6) the intersection of the Intraparietal Sulcus (IPS), or of its anterior prolongation, with the Postcentral Sulcus (PostCS), lies underneath the cranial point located 6 cm anterior to the Lambda (La) and 5 cm lateral to the Sagittal Suture (SagSut), (7) the Euryon (Eu), craniometric point that corresponds to the center of the parietal tuberosity, is located over the superior aspect of the Supramarginal Gyrus (SMG), (8) the superior point of the Parieto-Occipital Sulcus (POS) lies underneath the angle between each Lambdoid (LaSut) and the Sagittal Suture (SagSut), (9) the Opisthocranion (OpCr), that corresponds to the most prominent point of the occipital bossa, is located just superiorly to the Calcarine Fissure Distal Point (dCaF), over the Cuneus (CuG) most prominent aspect, and (10) the Posterior Point of the Superior Temporal Sulcus (STS) lies underneath the cranial point located 3 cm superiorly to the Parietomastoid (PaMaSut) and Squamous (SqSut) sutures meeting point. Measurements with an error <2 cm; adapted from [26–30]

The important Suplementary Motor Area (SMA) corresponds to a not well defined area mostly within the posterior and medial aspect of the SFG, radiologically defined as located between the two lines originated respectively at the anterior and posterior commissures and perpendicular to the line that join these two commissures.

The Superior Frontal Sulcus (SFS) which separates the Superior and the Middle Frontal Gyri (SFG, MFG) is usually a very evident and deep sulcus, parallel to the superomedial margin of the cerebral hemisphere, frequently continuous [40] or with a long continuous segment, and ends posteriorly usually joining the Precentral Sulcus (PreCS) and encroaching the Precentral Gyrus

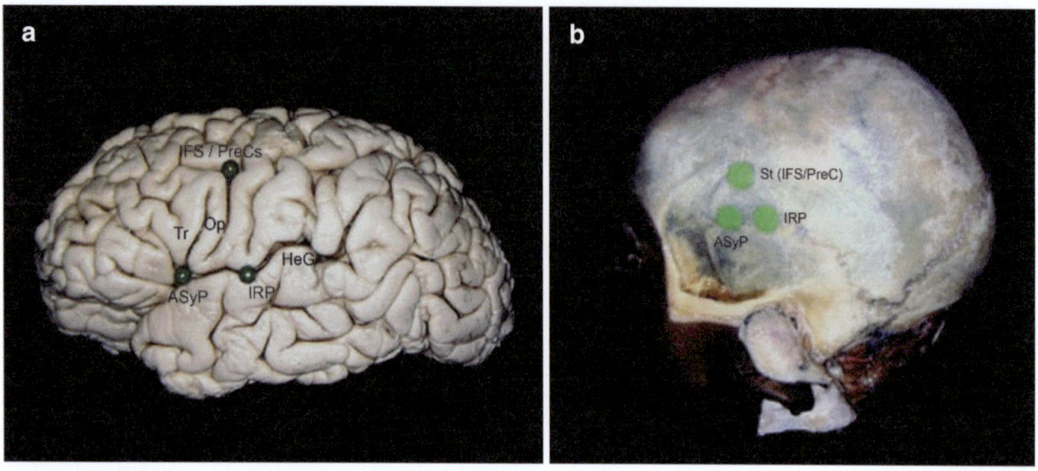

Fig. 7 (**a**) Fronto-Opercular Key Points and (**b**) their corresponding cranial sites: (1) the Anterior Sylvian Point (ASyP) is characterized by enlargement of the Sylvian fissure inferior to the Triangular Part (Tr) and anterior to the Opercular Part (Op) of the Inferior Frontal Gyrus (IFG), and lies underneath the most anterior aspect of the Squamous Suture (SqSut), just posterior to the pterion; (2) the Inferior Rolandic Point (IRP) is located along the Sylvian Fissure (SyF) 2–3 cm posteriorly to the ASyP, just anteriorly to Heschl's Gyrus (HeG), and lies underneath the highest aspect of the Squamous Suture (SqSut), which also corresponds to the intersection of this suture with a vertical line originating at the preauricular depression, about 4 cm high in adults; and (3) the inferior frontal and precentral sulci meeting point (IFS/PreCS), which indicates the superior aspect of the opercular part (Op) of the IFS (up to 3 cm high from the Sylvian fissure), which corresponds to the face motor activation/ventral premotor area (VPM), and which lies underneath the craniometric point known as the Stephanion (St) which corresponds to the site of intersection of the coronal suture with the superior temporal line [28, 30]. *ASyP* Anterior Sylvian Point, *HeG* Heschl's Gyrus, *IFG* Inferior Frontal Gyrus, *IFS/PreCS* Inferior Frontal and Precentral Sulci meeting point, *IFS* Inferior Frontal Sulcus, *IRP* Inferior Rolandic Point, *Op* Opercular Part, *St* Stephanion (meeting point of the coronal suture and superior temporal line), *Tr* Triangular Part of inferior frontal gyrus

(PreCG) exactly at the level of the Omega region which corresponds to the hand motor cortical area [46, 47] (Figs. 2 and 9).

The Superior Frontal Gyrus (SFG) is frequently divided by a secondary and shallow sulcus parallel to Superior Frontal Sulcus (SFS), known as the Medial Frontal Sulcus (MeFS).

The Superior Frontal and Precentral Sulci Meeting Point is then an important sulcal Key Point, located just anteriorly to the Omega region. Its corresponding cranial site is shown in Fig. 9, and its surgical application can be seen in the case shown in Fig. 10.

Anteriorly the three frontal gyri are delineated by the appropriately named Frontomarginal Sulcus (FMaS) which separates the superolateral from the orbital frontal brain surface, and which is more frequently interrupted [45, 48].

Along the Sylvian Fissure (SyF), posteriorly to the anatomically well defined Opercular Part (OpP), there are the Anterior and the Posterior Subcentral Rami (ASCR, PSCR) of the Sylvian Fissure (SyF) which delineate the Subcentral Gyrus (SubCG) that connects inferiorly the oblique Precentral and Postcentral Gyri (PreCG, PostCG) (Fig. 2).

The Precentral (PreCG) and the Postcentral (PostCG) gyri together with the Subcentral Gyrus (SubCG) inferiorly and with the Paracentral Lobule (PaCLob) superiorly and already disposed along the medial surface of the brain hemisphere, constitute all together the so called Central Lobe (CLo) [17, 49–51], which then resembles an ellipse excavated by the Central Sulcus (CS) (Fig. 3).

Anteriorly to the Precentral Gyrus (PreCG) the Precentral Sulcus (PreCS) is almost always very evidently divided by the superficial connection of the Middle Frontal Gyrus (MFG) with the Precentral Gyrus (PreCG) in a superior and a inferior parts. While its superior part ends superiorly at an already medial connection of the

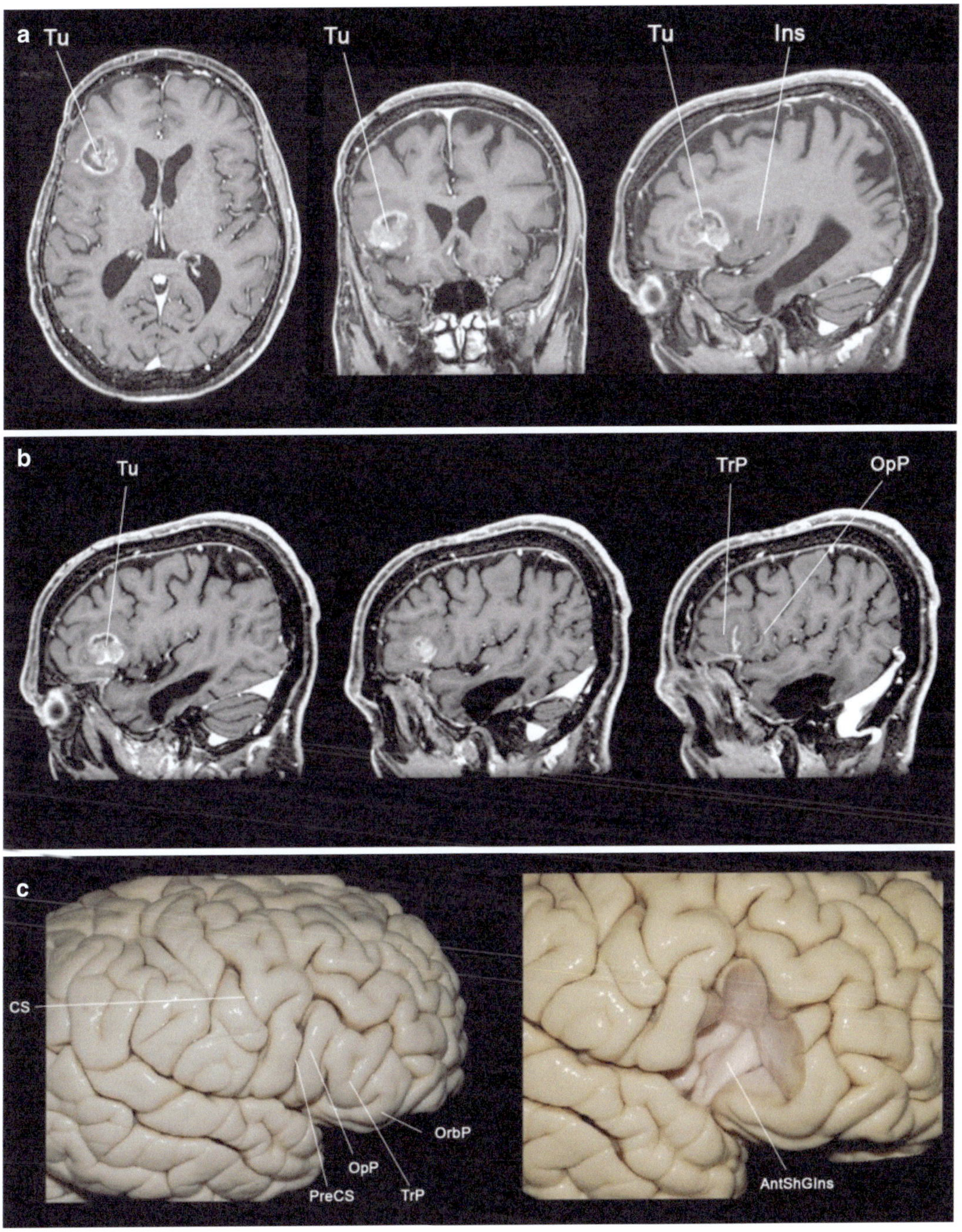

Fig. 8 (**a**) MRI images of a right fronto-opercular high grade glioma, which covers the most anterior aspect of the Insula, (**b**) inside-out images of the tumor showing it lies inside the Triangular Part of the Inferior Frontal Gyrus, (**c**) anatomical selective removal of the Triangular Part of the Inferior Frontal Gyrus exposing the Anterior Short Gyrus of the Insula, (**d**) Fronto-opercular Sulcal Key-points and surgical exposure of the Fronto-parietal Operculum, (**e**) microneurosurgicasl selective removal of the Triangular Part of the Triangular Part of the Inferior Frontal Gyrus, exposing and removing the underlying tumor, (**f**) Post-operative MRI images. *AntShGIns* Anterior Short Gyrus of the Insula, *ASyP* Anterior Sylvian Point, *CS* Central Sulcus, *IFS/PreCS* Inferior Frontal and Precentral Sulci Meeting Point, *Ins* Insula, *IRP* Inferior Rolandic Point, *OpP* Opercular Part of the Inferior Frontal Gyrus, *OrbP* Orbital Part of the Inferior Frontal Gyrus, *PreCS* Precentral Sulcus, *TrP* Triangular Part of the Inferior Frontal Gyrus, *Tu* Tumor

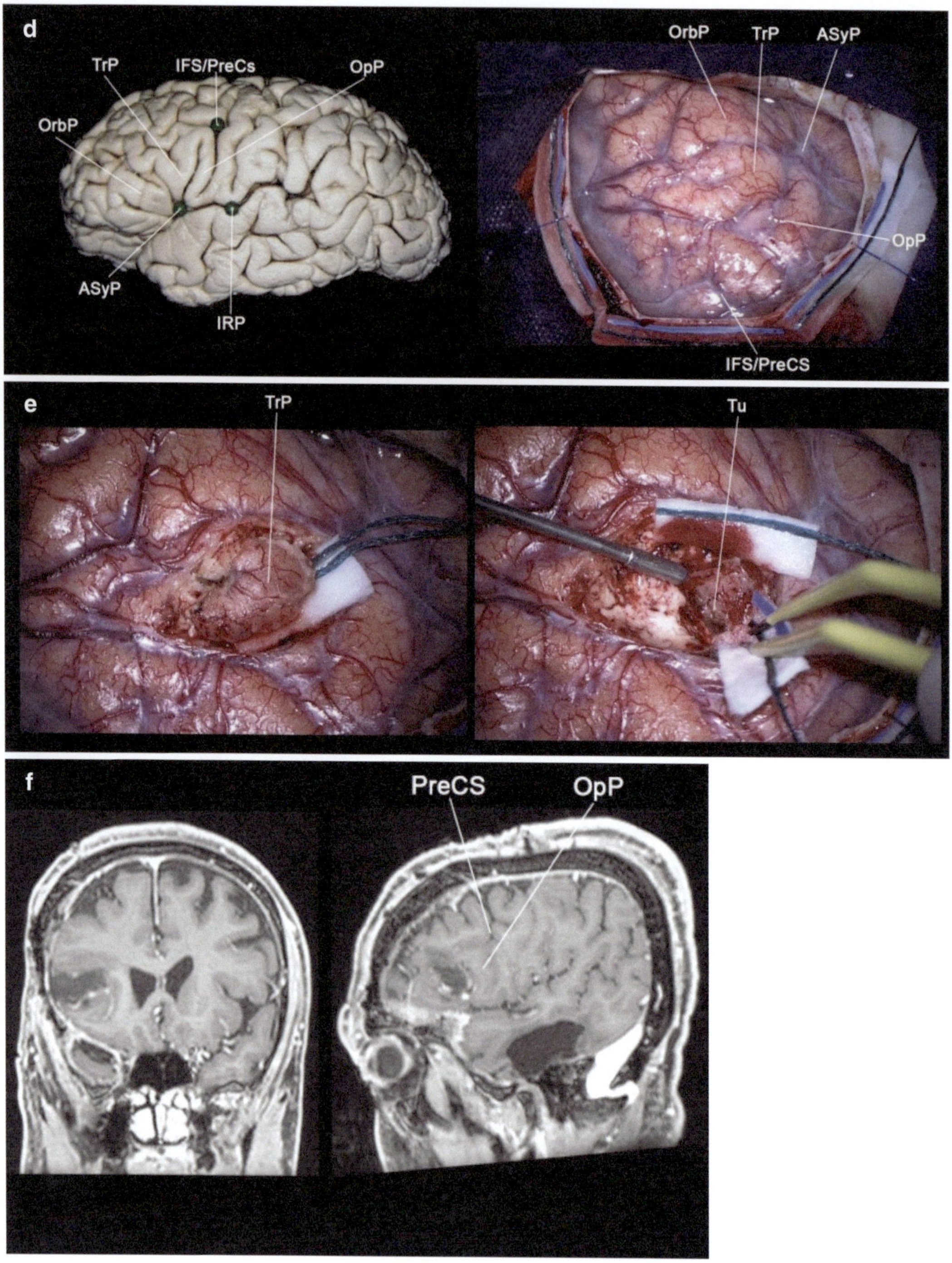

Fig. 8 (continued)

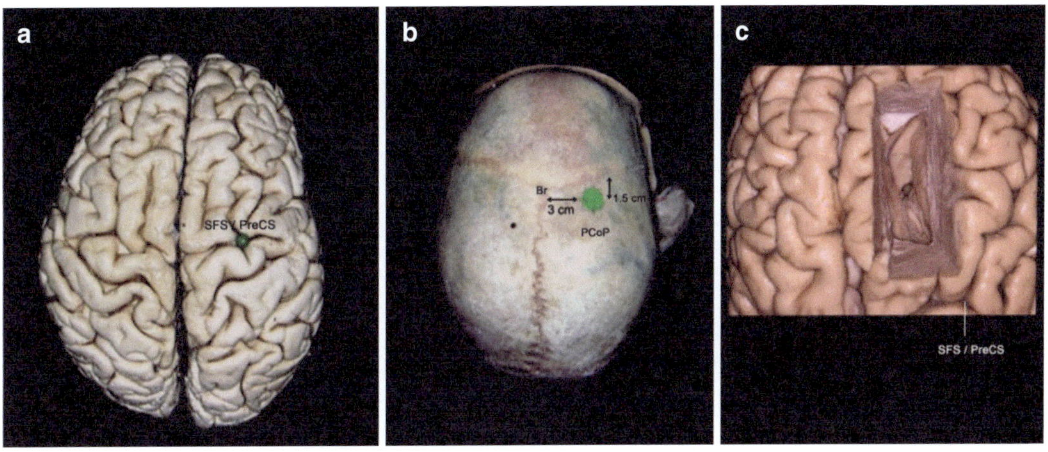

Fig. 9 (**a**) Superior Frontal and Precentral Sulci Meeting Point; (**b**) and its corresponding cranial site: the Superior Frontal and Precentral Sulci Meeting Point (SFS/PreCS) is located underneath the cranial site situated 1.5 cm posterior to the Coronal Suture and 3 cm lateral to the Sagittal Suture, and (**c**) relationship of the superior frontal sulcus with the anterior horn and body of the lateral ventricle. *Br* Bregma, *CS* Central Sulcus, *PreCS* Precentral Sulcus, *SFS/PreCS* Superior Frontal and Precentral Sulci Meeting Point

Superior Frontal Gyrus (SFG) with the Precentral Gyrus (PreCG), the inferior part of the Precentral Sulcus (PreCS) always ends inside the evident U-shaped Opercular Part (OpP) of the Inferior Frontal Gyrus (IFG) [24, 28, 52] (Fig. 2).

The Central Sulcus (CS) separates the primary motor and the somatosensory cortical areas respectively located along the Precentral Gyrus (PreCG) and the narrower Postcentral Gyrus (PostCG). It starts in or near the superomedial border of the hemisphere slightly behind the mid-point between the frontal and occipital poles [39], and runs resembling a lengthened italic S [53] downward and forwards almost always as a continuous sulcus [40], to end usually a little above the Sylvian Fissure (SyF) or, less frequently, inside it (Figa. 2 and 3).

The cranial-cerebral relationships of the Central Sulcus (CS) are shown in Fig. 11.

Posteriorly to the Postcentral Gyrus (PostCG) the Postcentral Sulcus (PostCS) is usually interrupted more frequently due to a connection between the Postcentral Gyrus (PostCG) and the Superior Parietal Lobule (SupPaLob) [40]. Its inferior part is superiorly very frequently continuous with the Intraparietal Sulcus (IPS) [24, 45], and inferiorly it always ends inside a U-shaped connection between the Postcentral Gyrus (PostCG) and the Supramarginal Gyrus (SMG) [24, 29] (Fig. 2).

Looking from above at the superior or opercular surface of the temporal lobe, which lies inside the Sylvian Fissure (SyF), the always evident Transverse Gyrus of Heschl (HeG) divides this surface in two planes. The Polar Plane (PoPl) anteriorly, which is oblique and which covers approximately the inferior half of the insular surface, and the Temporal Plane (TPl) posteriorly which is always flat and triangular with its apex next to the Atrium (Atr) (Fig. 3a).

Given these evident and constant obliquity and flatness of these two planes, a coronal MRI anterior to the Heschl's Gyrus (HeG) will show the Sylvian Fissure (SyF) always oblique while a coronal MRI posterior to it will always disclose the Sylvian Fissure (SyF) flat (Fig. 12c).

Another constant feature of the Heschl's Gyrus (HeG) is that along the Sylvian Fissure (SyF) it always presents as a bulge, a prominence, with the base of the Postcentral Gyrus (PostCG) always lying over this bulge (Fig. 12a, b).

The interesting and close cranial-cerebral relationships of the Squamous Suture (SqSut) and the Sylvian Fissure (SyF), the Heschl's Gyrus (HeG) and the Fronto-Opercular Key Points is shown in Fig. 13.

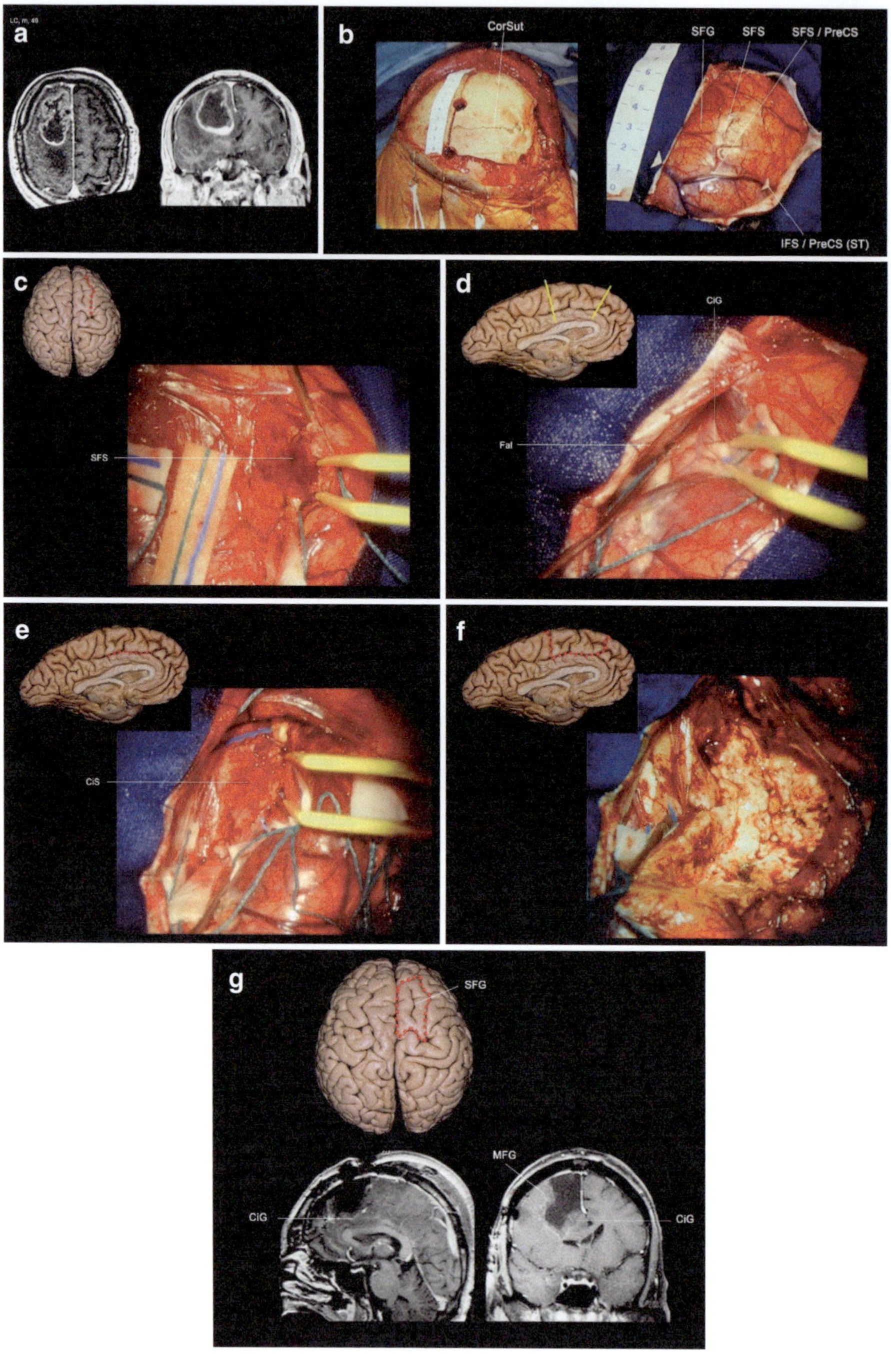

Fig. 10 Sulcal to sulcal resection of a high grade glioma of the right superior Frontal Gyrus. (**a**) MR images of the tumor occupying the Superior Frontal Gyrus; (**b**) craniotomy extending 2–3 cm posteriorly to the Coronal Suture in order to expose the Precentral Sulcus and the Superior Frontal/Precentral Sulci Meeting Point which should correspond to the posterior limit of the resection; operative view and identification of the main anatomical landmarks; (**c**) opening of the Superior Frontal Sulcus; (**d**) identification of the Cingulate Gyrus within the Interhemispheric Fissure; (**e**) opening of the Cingulate Sulcus; (**f**) cavity after en bloc removal of the posterior aspect of the Superior Frontal Gyrus containing the tumor; (**g**) post-operative MR images. *CiG* Cingulated Gyrus, *CiS* Cingulate Sulcus, *CoSut* Coronal Suture, *Fal* Falx, *IFS/PreCS(ST)* Inferior Frontal and Precentral Sulci Meeting Point underneath the Stephanion, *St* Stephanion (Meeting Point of the Coronal Suture and Superior Temporal Line), *MFG* Middle Frontal Gyrus, *SFG* Superior Frontal Gyrus, *SFS/PreCS* Superior Frontal and Precentral Sulci Meeting Point, *SFS* Superior Frontal Sulcus

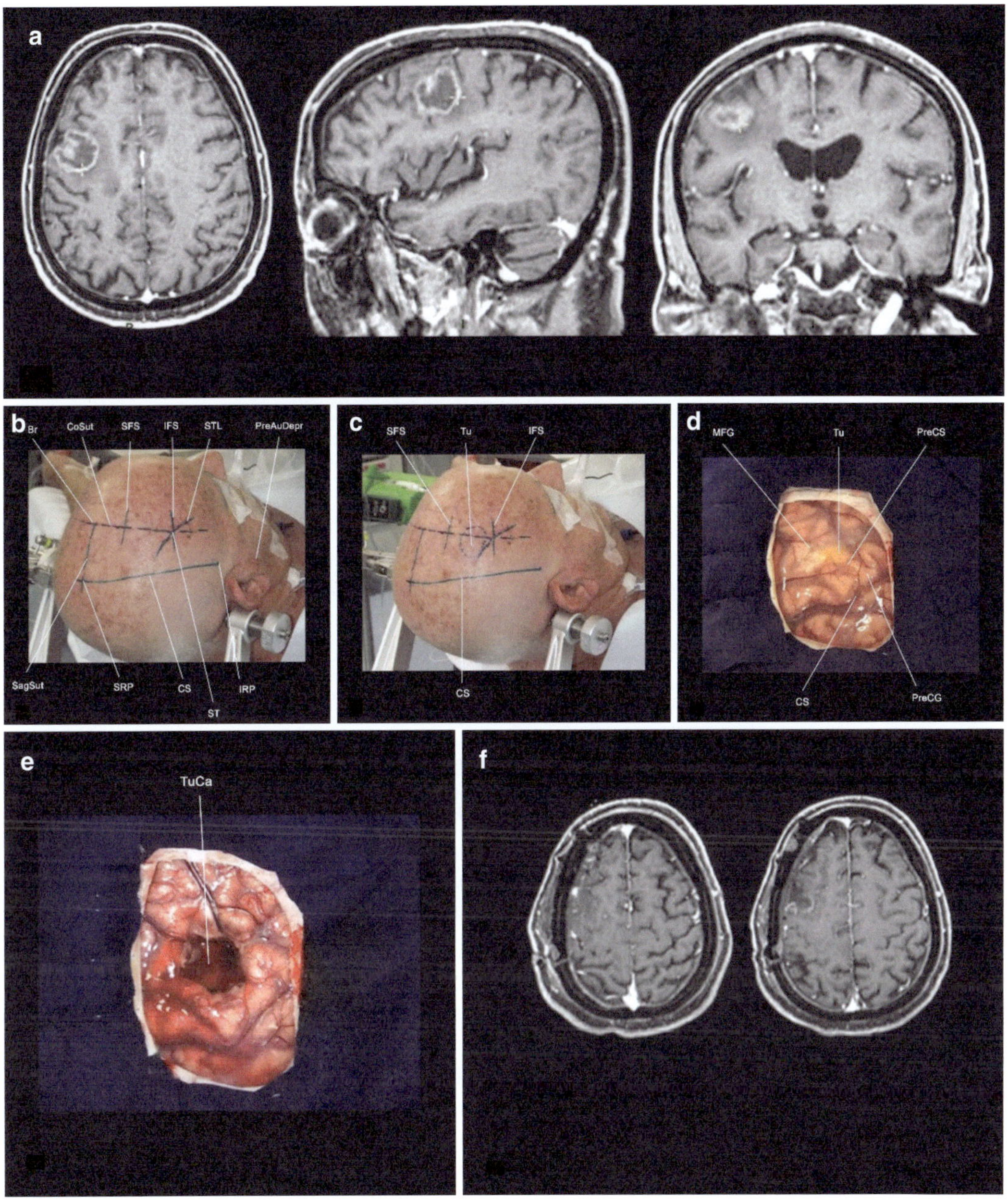

Fig. 11 Resection of a high grade glioma located in the right middle frontal gyrus just anteriorly to the precentral sulcus. (**a**) preoperative MR images; (**b**) main frontal anatomical landmarks and their related sulci: the central sulcus between the Superior Rolandic Point (located 5 cm posterior to the Bregma) and the Inferior Rolandic Point (4 cm superior to the preauricular depression), the Superior Frontal Sulcus which runs approximately 3 cm lateral to the Sagittal Suture, the most posterior aspect of the Inferior Frontal Sulcus underneath the Stephanion (meeting point of the coronal suture with the Superior Temporal Line; (**c**) location of the tumor in the posterior aspect of the right Middle Frontal Gyrus; (**d**) exposure of the brain surface and identification of the sulci and gyri; (**e**) tumor cavity after the resection; (**f**) post operative MR images. *Br* Bregma, *CoSut* Coronal Suture, *CS* Central Sulcus, *IFS* inferior Frontal Sulcus, *IRP* Inferior Rolandic Point, *MFG* Middle Frontal Gyrus, *PreAuDepr* Preauricular Depression, *PreCG* Precentral Gyrus, *PreCS* Precentral Sulcus ,*SagSut* Sagittal Suture, *SFS* Superior Frontal Sulcus, *SRP* Superior Rolandic Point, *St* Stephanion (meeting point of the Coronal Suture and Superior Temporal Line), *STL* superior temporal line, *Tu* tumor, *TuCa* tumor cavity

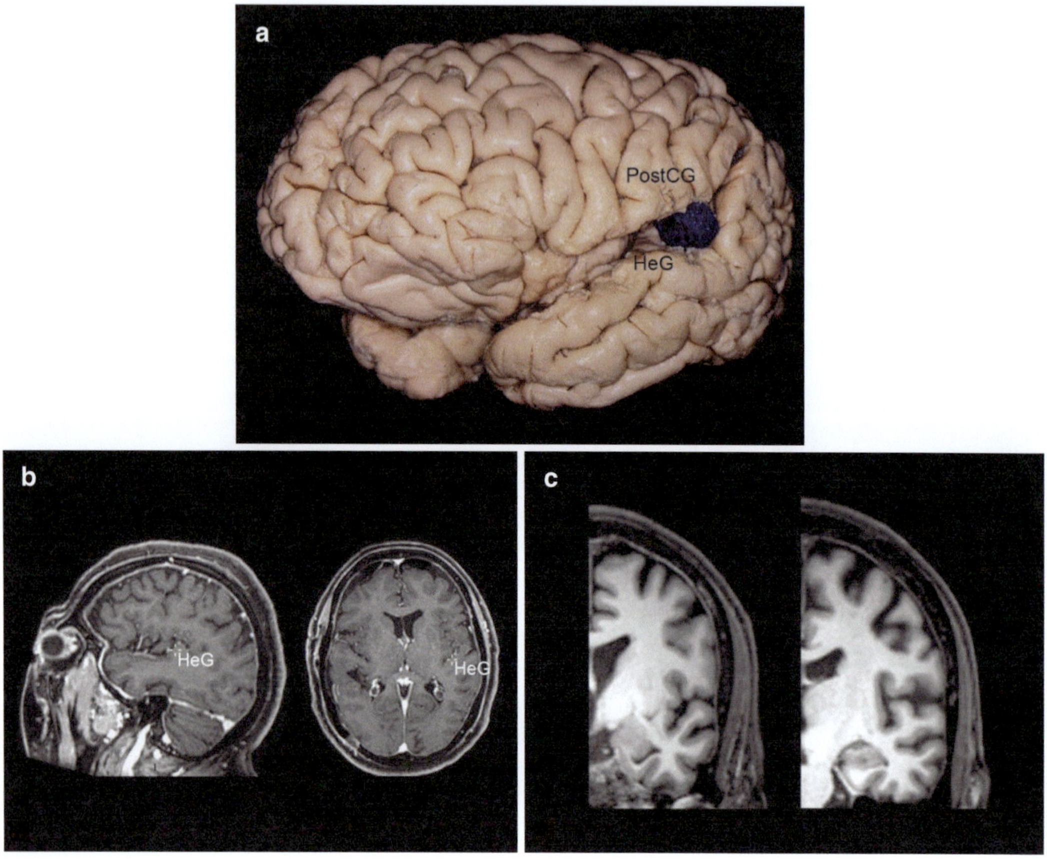

Fig. 12 (**a**, **b**) the Heschl's Gyrus (HeG) forms an evident prominence along the Sylvian Fissure (SyF), and the base of the Postcentral Gyrus (PostCG) always lies over the Heschl's Gyrus (HeG); (**c**) in MRI coronal cuts, the Sylvian Fissure (SyF) is oblique anteriorly to the Heschl's Gyrus (HeG) and is flat posteriorly to it

Operculi in Latin mean covers, curtains and in neuroanatomy correspond to the covers of the Insula (Ins). While the Frontoparietal Operculum (FPOp), which extends from the Horizontal Ramus (HR) to the Posterior Ascending Ramus (PostAsR) of the Sylvian Fissure (SyF) covers approximately the superior half of the Insula (Ins), the Temporal Operculum (TOp) which corresponds to the Superior Temporal Gyrus (STG) and its Polar and Temporal Planes (PoPl, TPl) divided by the Transverse Gyrus of Heschl (HeG), covers approximately the inferior half of the Insula (Ins) [24, 28] (Fig. 3b).

The Insula (Ins) is currently considered one of the cerebral lobes [54–56], and its surface corresponds to the outer surface of the Central Core (CCo) of the brain which is an anatomical very well defined block that harbors part of the Basal Ganglia (BaGa) and the Thalamus (Th) with the Internal Capsule (IntCaps) in between [32, 57]. The Apex (Ap) of the Insula (Ins) lies anteriorly and inferiorly within its surface, is located underneath the Anterior Sylvian Point (ASyP), and covers the Uncinate and its adjoining and posterior Inferior Fronto-Occipital Fascicle (UF, IFOF).

While the anterior half of the Insula (Ins) is related with the anterior half of the Putamen (Put), Anterior Limb of the Internal Capsule (AntLIntCaps), Head of the Caudate Nucleus (HeCa) and with the Frontal Horn of the Lateral Ventricle (FHoLatV), the posterior half of the Insula (Ins) is related with the posterior half of the Putamen (Put), Posterior Limb of the Internal Capsule (PostLIntCaps), Thalamus (Th) and with the IIIrd Ventricle cavity (IIIv) (Fig. 3b).

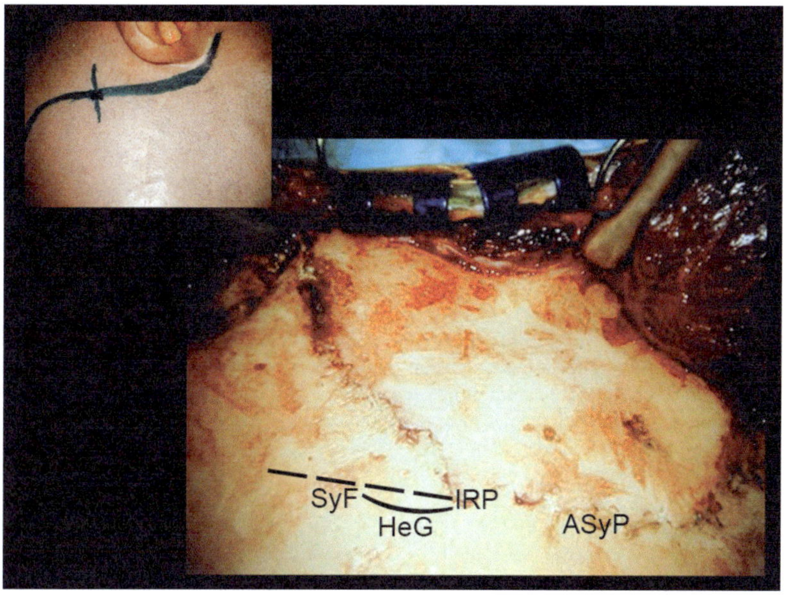

Fig. 13 The Squamous Suture and the Frontoparietal Operculum. While the anterior half of the Squamous Suture (SqSut) runs approximately over the anterior aspect of the Sylvian Fissure (SyF), its posterior half descends while the posterior aspect of the Sylvian Fissure (SyF) ascends. The highest point of the Squamous Suture (SqSut) which correlates with this turning point lies over the Inferior Rolandic Point (IRP) which corresponds to the projection of the Central Sulcus (CS) over the Sylvian Fissure (SyF). The Subcentral Gyrus (SubCG) lies then underneath the mid and upper segment of the Squamous Suture (SqSut). The Anterior Sylvian Point (ASyP) lies 2–2.5 cm [27] anteriorly to the Inferior Rolandic Point (IRP) and along the Sylvian Fissure (SyF), and immediately posterior to the Inferior Rolandic Point (IRP) lies the most lateral aspect of the Heschl's Gyrus (HeG) which in turn lies underneath the base of the Postcentral Gyrus (PostCG)

The body of the Lateral Ventricle (BoLatV) lies above the Thalamus (Th), the Atrium (Atr) lies posterior to the Thalamus (Th), and the Inferior Horn of the Lateral Ventricle (InfHoLatV) lies inferior to the Thalamus (Th), respectively then superior, posterior and inferior to the Central Core (CCo). The Heschl's Gyrus (HeG) runs obliquely and anteriorly to posteriorly towards the Atrium (Atr), hence posteriorly to the Internal Capsule (IntCaps) (Fig. 3b).

Inferiorly to the Sylvian Fissure (SyF) the Superior, Middle and Inferior Temporal Gyri (STG, MTG, ITG) are separated respectively by the Superior and Inferior Temporal Sulci (STS, ITS) (Fig. 2).

While the Superior Temporal Sulcus (STS), just as the Superior Frontal Sulcus (SFS), is usually almost continuous since it usually has at least one long continuous and deep segment, the Inferior Temporal Sulcus (ITS), just as the Inferior Frontal Sulcus (IFS), is always an interrupted sulcus formed by multiple segments. Contrary to the Superior and Middle Temporal Gyri (STG, MTG), the Inferior Temporal Gyrus (ITG) is very short along its lateral aspect, but it is broad along its base which lies over the middle fossa floor and the superior aspect of the Petrous Part of the Temporal bone. Given its low height and non-continuity, the Inferior Temporal Sulcus (ITS) usually is not easily identifiable (Fig. 2).

The parietal surface is divided by the evident Intraparietal Sulcus (IPS) in the quadrangular Superior Parietal Lobule (SPLob) which is continuous with the Precuneus Gyrus (PreCu) along the superomedial border of the cerebral hemisphere, and in the Inferior Parietal Lobule (IPLob) which is formed by the Supramarginal Gyrus (SMG) anteriorly and the Angular Gyrus (AG) posteriorly, separated by the Intermediary Sulcus of Jensen (ISJ) (Fig. 2).

The Intraparietal Sulcus (IPS) is either longitudinal and parallel to the superomedial border or

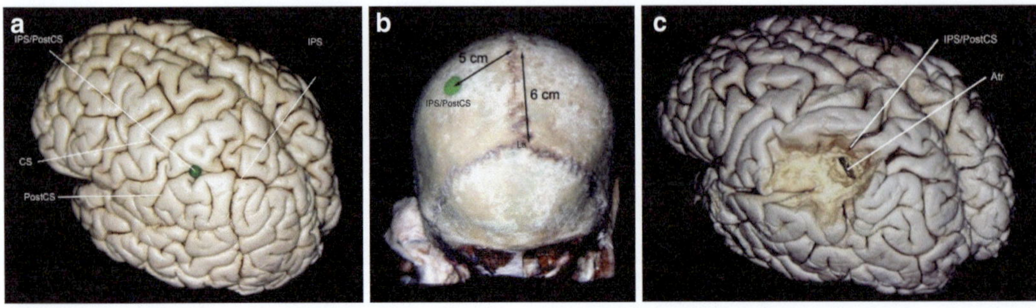

Fig. 14 (**a**) Intraparietal and Postcentral Sulci Meeting Point, or anterior projection of the Intraparietal Sulcus on the Postcentral Sulcus Point when there is an interruption between both sulci; (**b**) its corresponding cranial site; and (**c**) relationship of the Intraparietal and Postcentral sulci point with the Atrium of the Lateral Ventricle. *Atr* Atrium of Lateral Ventricle, *CS* Central Sulcus, *IPS/PostCS* Intraparietal and Postcentral Sulci Meeting Point, *IPS* Intraparietal Sulcus, *La* Lambda, *PostCs* Postcentral Sulcus

is oblique forming an arch with an inferior concavity. It is frequently interrupted and usually continuous with the inferior half of the Postcentral Sulcus (PostCS).

The Intraparietal and Postcentral Sulci Meeting Point (IPS/PostCS) is then an important Sulcal Key Point, and its cranial and deep cerebral (atrial) relationships are shown in Fig. 14.

The Intermediary Sulcus of Jensen (ISJ) always corresponds to a posterior and inferior branch of the Intraparietal Sulcus (IPS) (Fig. 2a). The IPS frequently also has an anterior and superior vertical branch which corresponds to the Transverse Parietal Sulcus of Brissaud (TrPSBr) that penetrates and divides the Superior Parietal Lobule (SupPaLob).

While the Supramarginal Gyrus (SMG) is very constant regarding its quadrangular shape surrounding the superior and distal end of the Sylvian Fissure (SyF), giving its name, the Angular Gyrus (AG) usually is ill-defined anatomically and does not have the classical horseshoe shape generally shown in textbooks (Fig. 2).

The anterior aspect of the Supramarginal Gyrus (SMG) lies underneath the most prominent point of the parietal tuberosity as shown in Fig. 15.

While the Sylvian Fissure (SyF) ends as a bifurcation at the so called Posterior Sylvian Point (PostSyP), giving rise to a very constant Posterior Ascending Ramus (PostAsR) that always penetrates inside the Supramarginal Gyrus (SMG), and a very variable, sometimes absent, descending branch, the Superior Temporal Sulcus (STS) which always ends posteriorly to the Posterior Sylvian Point (PostSyP), as a bifurcation or as a trifurcation, with its more horizontal distal end penetrating inside the Angular Gyrus (AG). Its more superior branch usually corresponds to the Intermediary Sulcus of Jensen (ISJ), and its more inferior branch is very variable (Fig. 2)

The Intermediary Sulcus of Jensen (ISJ) is then a branch of the Intraparietal Sulcus (IPS) and frequently corresponds also to a distal and superior branch of the Superior Temporal Sulcus (STS). When interrupted, its interruption is due to a connection between the Supramarginal (SMG) and Angular (AG) gyri (Fig. 2).

An actual surgical example of the Parietal Sulcal Key Points application is shown in Fig. 16.

Despite being less well defined and less anatomically constant than the gyri of other dorsal cortical areas, the occipital gyri of the superolateral cerebral surface tend to consist of two or three gyri which converge posteriorly to form the occipital pole of each hemisphere. While the Superior and Inferior Occipital Gyri (SOG, IOG) are anatomically more constant, the Middle Occipital Gyrus (MOG) is much more variable regarding its characterization (Fig. 2).

The Inferior Occipital Gyrus (IOG) is always placed horizontally along the inferolateral margin of the cerebral hemisphere, with its base lying over the tentorium. Anteriorly, it is mostly continuous with the Inferior Temporal Gyrus (ITG),

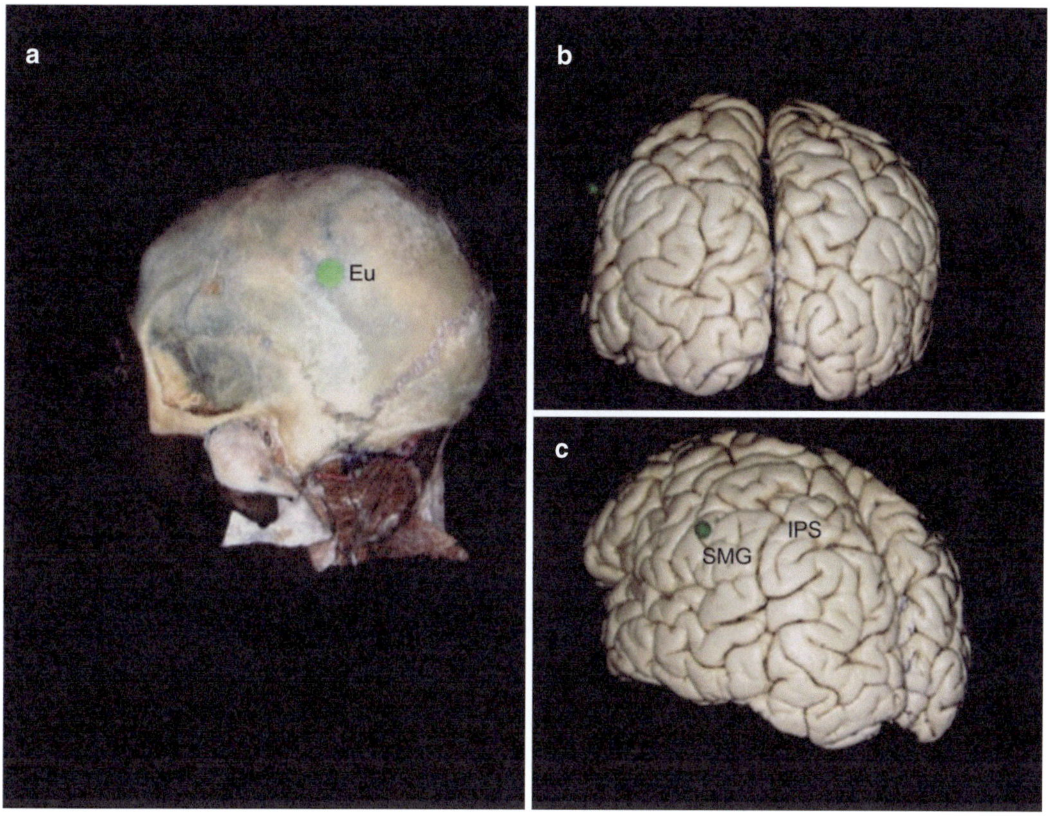

Fig. 15 The Euryon and the Supramarginal Gyrus. (**a**) the Euryon (Eu) corresponds to the most prominent point of the parietal tuberosity or bossa; (**b**, **c**) and lies over the superior aspect of the Supramarginal Gyrus (SMG). *AG* Angular Gyrus, *Eu* Euryon, *IPS* Intraparietal Sulcus, *SMG* Supramarginal Gyrus

and posteriorly it extends medially around the occipital pole becoming continuous with the Lingual Gyrus (LiG) within the medial surface of the hemisphere. Superiorly, the Inferior Occipital Gyrus (IOG) is delimited by the Inferior [11, 40] or Lateral [40, 45] Occipital Sulcus (IOS, LatOS) which frequently is anteriorly directly connected to the Inferior Temporal Sulcus (ITS) [40]. A shorter accessory lateral occipital sulcus frequently runs below the main one [48] (Figs. 2 and 3a).

The Superior Occipital Gyrus (SOG) corresponds usually to a well defined vertical gyrus placed along the Interhemispheric Fissure, which is continuous along the superomedial margin of the hemisphere with the Cuneus Gyrus (CuG) already within the cerebral medial surface. Superiorly, the Superior Occipital Gyrus (SOG) is delimited by the depth of the Parieto-Occipital Sulcus (POS) on the superolateral hemispheric surface which corresponds to the Parieto-Occipital Incisure (POInc), and is superiorly continuous with the Superior Parietal Lobule (SPLob) through the Parieto-Occipital Arcus (POAr) that encircles the Parieto-Occipital Incisure (POInc) and which corresponds to the First or Superior Parieto-Occipital "plis de passage" of Gratiolet (Fig. 3a).

The Intraparietal Sulcus (IPS) systematically extends longitudinally and inferiorly into the occipital lobe [40] becoming then the Intra-Occipital Sulcus (InOS) [11, 24, 45, 58], or Superior Occipital Sulcus (SOS) [11, 45], delimiting the Superior Occipital Gyrus (SOG) laterally. The intra-occipital sulcus can occasionally descend as far as the occipital pole [58] characterizing then a long and vertical sulcus, but most frequently it terminates as a T end which corre-

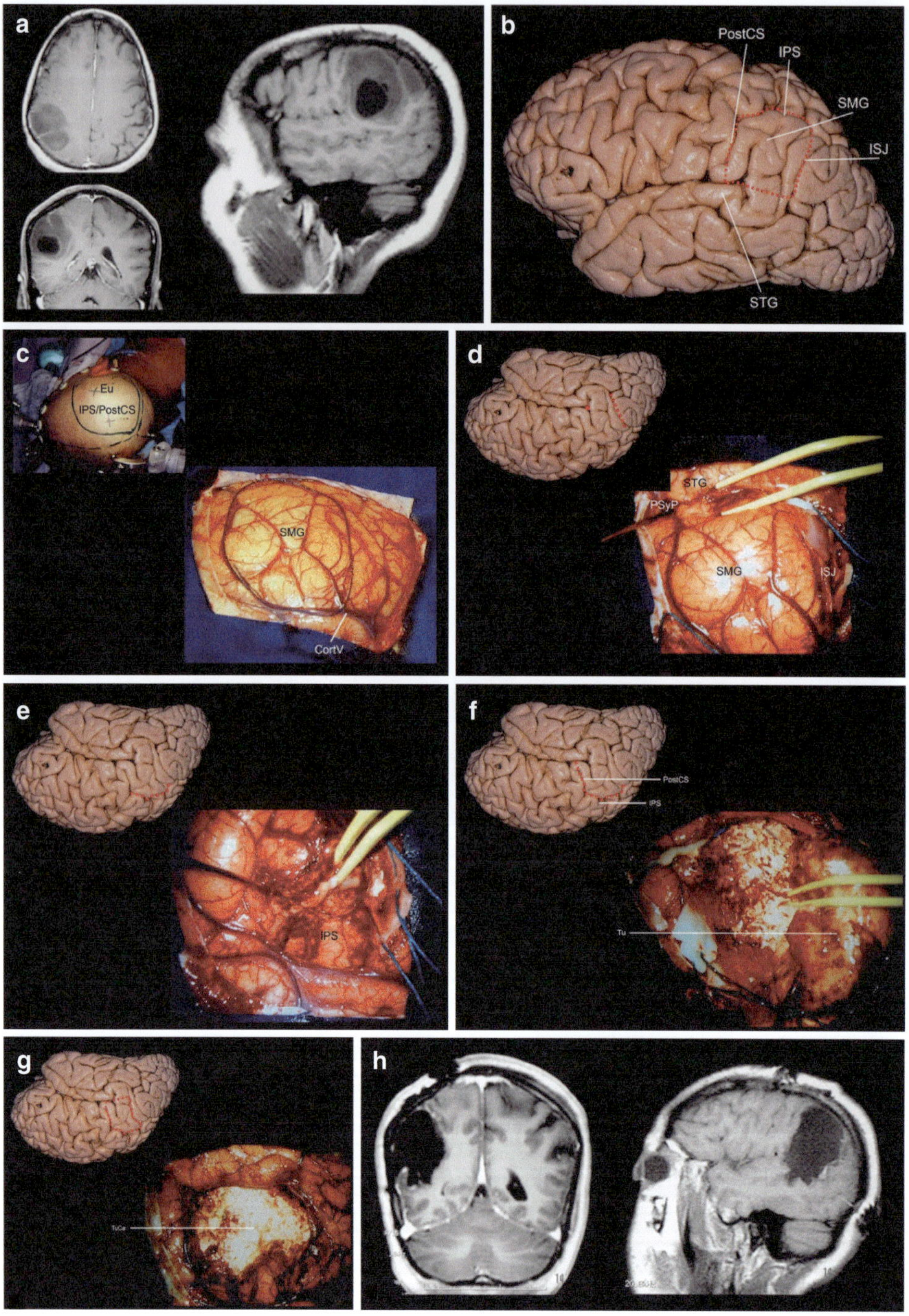

Fig. 16 Anatomical resection of the right Supramarginal Gyrus containing a low grade glioma. (**a**) preoperative MR images; (**b**) anatomical delineation of the Supramarginal Gyrus which is continuous with the Superior Temporal Gyrus along its base; (**c**) planning the craniotomy identifying the Euryon which lies above the Supramarginal Gyrus, a cortical vein running along the Intraparietal Sulcus, and exposure of the Supramarginal Gyrus which harbors the tumor; (**d**) opening of the Intermediary Sulcus of Jensen which separates the Supramarginal from the Angular Gyrus; (**e**) opening of the Intraparietal Sulcus; (**f**) opening of the Postcentral Sulcus which is continuous with the Intraparietal Sulcus, and en bloc removal of the tumor within the Supramarginal Gyrus; (**g**) tumor cavity; and (**h**) post operative MR images. *CortV* cortical vein, *Eu* Euryon, *IPS/PostCS* Intraparietal and Postcentral Sulci Meeting Point, *IPS* Intraparietal Sulcus, *ISJ* Intermediary Sulcus of Jensen, *PostCS* Postcentral Sulcus, *PSyP* Posterior Sylvian Point, *SMG* Supramarginal Gyrus, *STG* Superior Temporal Gyrus, *Tu* tumor, *TuCa* tumor cavity

sponds to the Transverse Occipital Sulcus (TrOS) [40, 58] (Fig. 3a).

The Transverse Occipital Sulcus (TrOS) usually is then an evident T-shaped occipital sulcus that medially penetrates the Superior Occipital Gyrus (SOG) [40, 48, 58], and more rarely it corresponds just to a side branch of the intraoccipital sulcus or to a completely separated sulcus [40].

Since the Inferior [11, 40] or Lateral [40, 45] Occipital Sulcus (IOS, LatOS) is always present, the configuration of the occipital dorsal surface in two gyri (70%) or in three gyri (30%) is more particularly dependent in the lesser or in the greater inferior extension of the Intra-Occipital [11], or Superior Occipital [11, 24, 45] Sulcus (InOS, SOS) [58]. When present, the area of the Middle Occipital Gyrus (MOG) then lies in between the Inferior or Lateral Occipital Sulcus (IOS, LatOS) and the inferior extension of the Intra-Occipital or Superior Occipital Sulcus (InOS, SOS) [58].

In humans, the Lunate Sulcus (LuS) is evident as a well-defined vertical and backward curved sulcus anterior to the occipital pole in less than 40 percent of the brain hemispheres [40, 45, 58]. More frequently it is absent, only vestigial, or may stem from any of the other main superolateral occipital sulci [40, 45].

Posteriorly, the two or three occipital gyri converge to form the occipital pole. Similarly to the temporal pole, the Middle Occipital Gyrus (MOG) is usually shorter and the occipital pole is constituted by the convergence of the Superior and Inferior Occipital Gyri (SOG, IOG) [58] (Fig. 3a).

While the depth of the Parieto-Occipital Sulcus (POS) of the medial surface of the brain corresponds to the Parieto-Occipital Incisure (POInc) on the most medial aspect of superolateral surface of the brain above the occipital pole (Fig. 3a), the Calcarine Fissure (CaF) usually ends branching in a T or Y shape forming the Retrocalcarine Sulcus (RetroCaS) already over the superolateral surface of the cerebral hemisphere within the occipital pole [40, 58] (Fig. 3a).

The Parieto-Occipital Sulcal Key Points are shown in Fig. 17.

Important always to have in mind that the architectural organization of the gyri of the temporo-parieto-occipital surface is such that (Fig. 2b):

1. the Superior Temporal Gyrus (STG) is always continuous with the Supramarginal Gyrus (SMG) around the Posterior Ascending Ramus (PostAsR) of the Sylvian Fissure (SyF),
2. the Middle Temporal Gyrus (MTG) is always continuous with the Angular Gyrus (AG) around the most distal and horizontal aspect of the Superior Temporal Sulcus (STS),
3. the Inferior Temporal Gyrus (ITG) is always continuous with the Inferior Occipital Gyrus (IOG), and
4. around the occipital pole the Inferior Occipital Gyrus (IOG) is medially continuous with the Lingual Gyrus (LiG), which is continuous with the Parahippocampal Gyrus (PaHG) forming an evident anatomical basal gyral ring which, in turn, is continuous with the notorious medial and circular gyral ring of the Limbic lobe, constituted by the parahippocampal and cingulate gyri [56].

The temporo-occipital base is related with two evident Cranial Key Points as shown in Fig. 18.

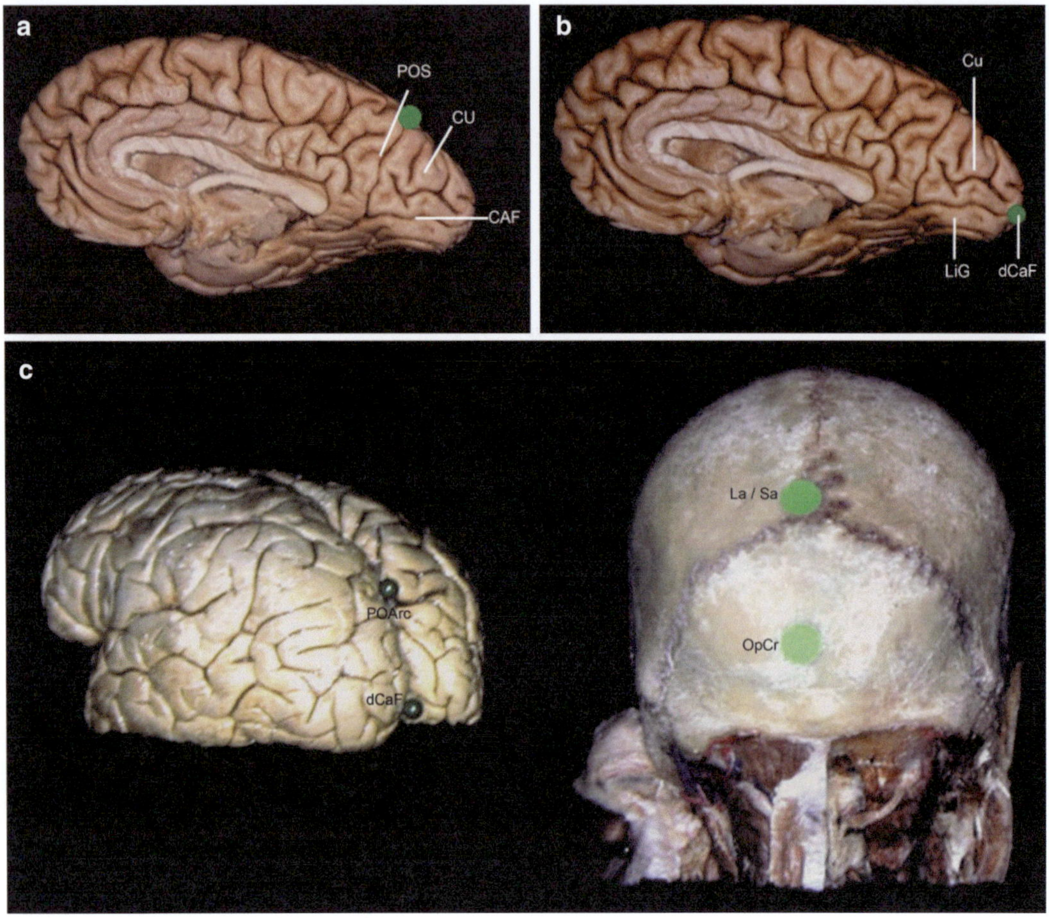

Fig. 17 The Parieto-Occipital Surface and its related Cranial and Sulcal Key-Points: (**a**, **c**) the top of the Parieto-Occipital Sulcus (POS) corresponds to the Parieto-Occipital Incisure (POInc) which is surrounded by the Parieto-Occipital Arc (POArc), and it lies underneath the cranial site given by the angle between the Sagittal (SagSut) and the Lambdoid (LaSut) sutures; (**b**, **c**) the Opisthocranion (OpCr) corresponds to the most prominent point of the occipital bossa, and it lies over the Distal Extremity of the Calcarine Fissure (dCaF)

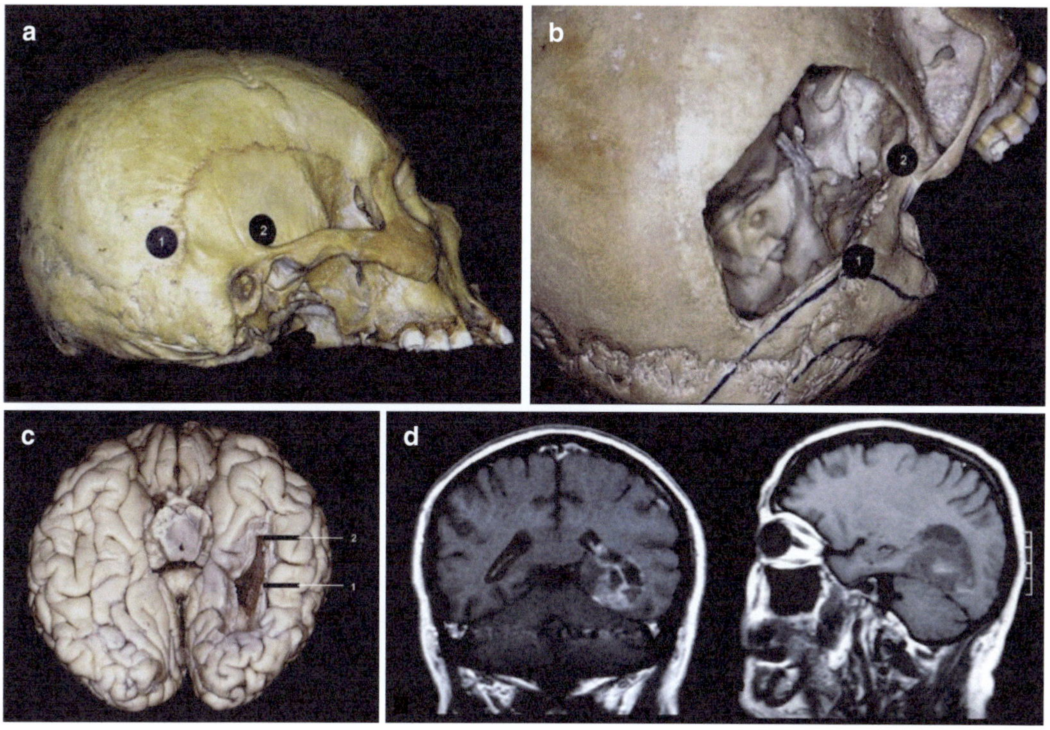

Fig. 18 (**a**) the burr holes made (1) 1 centimeter above the Parietomastoid and Squamous Suture Meeting Point and (2) above the Preauricular Depression (**b**) delimit the external projection of the Petrous Part of the Temporal Bone, (**c**) that lies underneath of the anterior half of the Fusiform Gyrus which constitutes the floor of the Atrium, the tumor protrudes inside this cavity; (**d**) MR coronal and sagittal images a left high grade glioma tumor within the posterior part of the Fusiform Gyrus which lies already over the Tentorium, hence posterior to the Petrous Bone; (**e**) scalpel incision and (**f**) craniotomy site, to expose both the superior surface of the Petrous Bone (between the burr holes made above the Preauricular Depression and above the Parietomastoid and Squamous Sutures Meeting Point) and the upper tentorium surface (posterior half of the craniotomy base); (**g**) subtemporal exposure of the IVth nerve and of the midbrain peduncle after opening of the Ambient Cistern in order to remove CSF; (**h**) basal temporal exposure of the Fusiform Gyrus bulging between the Collateral and Occipitotemporal Sulci; (**i**) post operative CT images showing removal of the basal aspect of the left inferior Temporal Gyrus window and removal of the left Fusiform Gyrus which harbored the tumor. (1) burr hole above the Parietomastoid and Squamous Suture Meeting Point; (2) burr hole above the Preauricular Depression; *FuG* Fusiform Gyrus, *CollS* Collateral Sulcus, *IVn* fourth cranial nerve, *OccTeS* occipitotemporal Sulcus, *Pe* Cerebral Peduncle of the Midbrain, *Tu/FuG* tumor within the Fusiform Gyrus

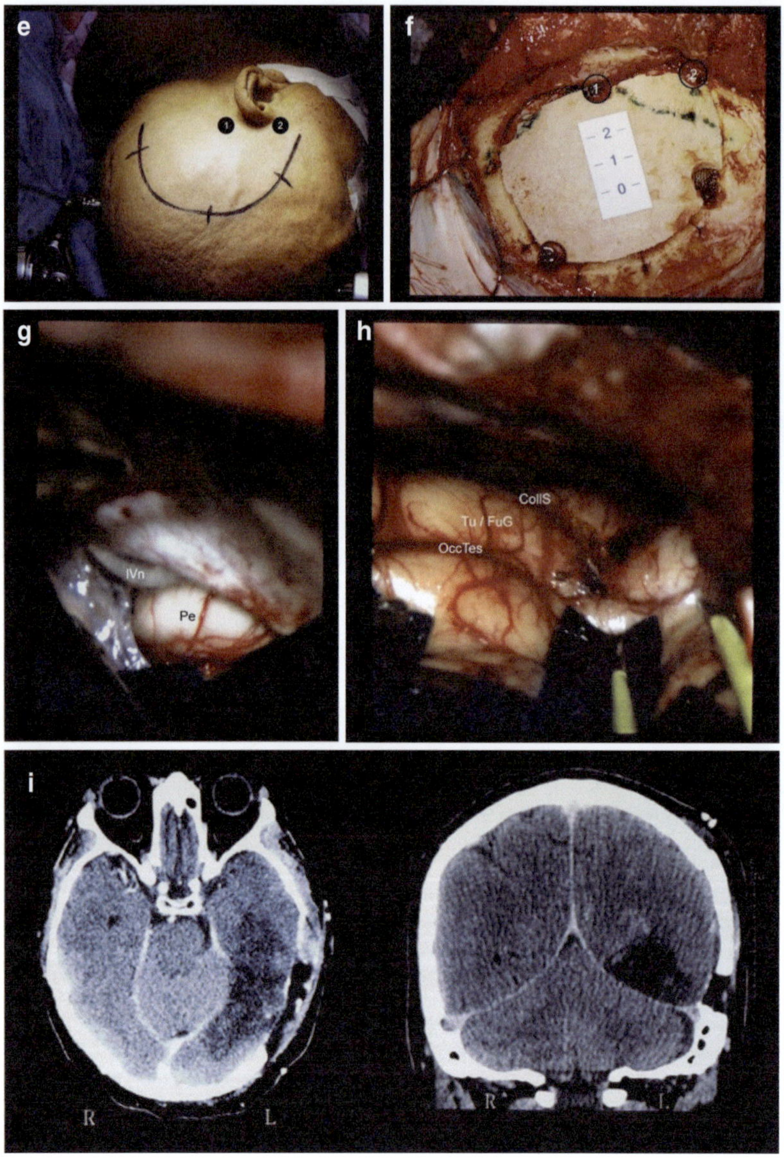

Fig. 18 (continued)

References

1. Broca P. Perte de la parole: ramollissement chronique et destruction partielle du lobe antérieur gauche: sur le siège de la faculté du langage. Bull Soc d'Anth. 1861a;2:235–8 apud Finger S. Origins of neuroscience. New York, NY: Oxford University Press; 1994.
2. Finger S. Origins of neuroscience. New York, NY: Oxford University Press; 1994.
3. Jackson JH. Convulsive spasms of the right hand and arm preceding epileptic seizures. Medical Times and Gazette. 1983a;1:110–1.
4. Jackson JH. Epileptiform seizures – Aura from the thumb – attacks of colored vision. Medical Times and Gazette. 1983b;1:58.
5. Broca P. Sur les rapports anatomiques des divers points de la surface du crâne et des diverses parties des hémisphères cérébraux. Bull Soc d'Anth. 1861b;2:340 apud Gusmão S, Silveira RL, Cabral G. [Broca and the birth of modern surgery.] [Portuguese]. Arq Neuropsiquiatr 2000;58:1149–52.
6. Broca P. Sur la topographie crânie-cerébrale ou sur les rapports anatomiques du crâne et du cerveau. Rev d'Antrop. 1876;5:193–248.
7. Bischoff TW. Die Grosshirnwindungen des Menschen. Munich; 1868 apud Broca P. Sur la topographie crânie-cerébrale ou sur les rapports anatomiques du crâne et du cerveau. Rev d'Antrop 1876;5:193–248.
8. Horsley V. On the topographical relations of the cranium and the surface of the cerebrum. In: Cunningham CJ, editor. Contribution to the surface anatomy of the cerebral hemispheres. Dublin: Academy House; 1892. p. 306–55.
9. Krönlein RU. Topographie cranio cérébrale. V. Bruns' Beiträge zur Keinishen. Chirurgie. 1898;22:364.
10. Taylor EH, Haugton WS. Some recent researches on the topography of the convolutions and fissures of the brain. Trans R Acad (Ireland). 1900;18:511–9.
11. Testut L, Jacob O. Topographic anatomy textbook [in Spanish]. 5th ed. Barcelona. Salvat; 1932.
12. Yaşargil MG. A legacy of microneurosurgery: memoirs, lessons, and axioms. Neurosurgery. 1999;45:1025–91.
13. Yaşargil MG, Kasdaglis K, Jain KK, Weber HP. Anatomical observations of the subarachnoid cisterns of the brain during surgery. J Neurosurg. 1976;44:298–302.
14. Harkey HL, Al-Mefty O, Haines DE, Smith RR. The surgical anatomy of the cerebral sulci. Neurosurgery. 1989;24:651–4.
15. Yaşargil MG, Cravens GF, Roth P. Surgical approaches to "inaccessible" brain tumors. Clin Neurosurg. 1988;34:42–110.
16. Yaşargil MG. Microneurosurgery, vol. v.I. Stuttgart: Georg Thieme; 1984.
17. Yaşargil MG. Microneurosurgery, vol. v.IVa. Stuttgart: Georg Thieme; 1994.
18. Yasargil MG, Fox JL, Ray MW. The operative approach to aneurysms of the anterior communicating artery. In: Krayenbül H, editor. Advances and technical standards in neurosurgery. Wien: Springer; 1975. p. 114–70.
19. Yaşargil MG, Krisht AF, Türe U, Al-Mefty O, Yaşargil DCH. Microsurgery of insular gliomas: part I: surgical anatomy of the Sylvian cistern. Contemp Neurosurg. 2002;24:1–8.
20. Rhoton AL Jr. Cranial anatomy and surgical approaches. Neurosurgery. 2003;53:1–746.
21. Hansebout RR. Surgery of epilepsy, current technique of cortical ressection. In: Schmidek HH, Sweet WH, editors. Operative neurosurgical techniques, vol. 2. New York, NY: Grune and Stratton; 1982. p. 963–79.
22. Pernkoff E. Atlas of topographical and applied human anatomy. Baltimore, MD: Urban & Schwarzenberg; 1980.
23. Rhoton AL Jr. General and micro-operative techniques. In: Youmans JR, editor. Neurological surgery. 4th ed. Philadelphia, PA: W.B. Saunders; 1999. p. 724–66.
24. Ribas GC. The cerebral sulci and gyri. Neurosurg Focus. 2010;28(Suppl 2):E2.
25. Broca P. Instructions Crâniologiques et Crâniométriques de la Société d'Anthropologie, vol. 10. Paris: G. Masson; 1875. p. 377–80.
26. Ribas GC. Applied cranial-cerebral anatomy: brain architecture and anatomically oriented microneurosurgery. Cambridge: Cambridge University Press; 2018.
27. Ribas GC. Study of the anatomic relationships of the lambdoid, occipitomastoid and parietomastoid sutures with the transverse and sigmoid sinuses, and of regional burrholes sites [in Portuguese]. São Paulo, Brazil: Faculdade de Medicina, Universidade de São Paulo; 1991 (thesis doctoral).
28. Ribas GC, Ribas EC, Rodrigues CJ. The anterior sylvian point and the suprasylvian operculum. Neurosurg Focus. 2005;18(Suppl 6b):E1–6.
29. Ribas GC, Rodrigues AJ Jr. The suprapetrosal craniotomy. J Neurosurg. 2007;106:449–54.
30. Ribas GC, Yasuda A, Ribas EC, Nishikuni K, Rodrigues AJ Jr. Surgical anatomy of microneurosurgical sulcal key-points. Neurosurgery. 2006;59 Suppl 4:ONS-177–211.
31. Watanabe E, Watanabe T, Manaka S, Mayanagi Y, Takakura K. Three-dimensional digitizer (neuronavigator): new equipment for computed tomography-guided stereotatic surgery. Surg Neurol. 1987;6:543–7.
32. Ribas EC, Yağmurlu K, Wen HT, Rhoton AL Jr. Microsurgical anatomy of the inferior limiting insular sulcus and the temporal stem. J Neurosurg. 2015;122(Suppl 6):1263–73.
33. Brogna C, Fiengo L, Türe U. Achille Louis Foville's atlas of brain anatomy and the Defoville syndrome. Neurosurgery. 2012;70(Suppl 5):1265–73.
34. Foville AL. Traité complet de l'anatomie, de la physiologie et de la pathologie du système nerveux cérébro-spinal. Paris: Fortin Masson; 1844.
35. Gratiolet LP. Memoire sur les plis cerébraux de l'homme et des primates. Paris: Bertrand; 1854

apud Pearce JM. Luis Pierre Gratiolet (1815-1865): the cerebral lobes and fissures. Eur Neurol 2006;56:262–4.
36. Pearce JM. Luis Pierre Gratiolet (1815–1865): the cerebral lobes and fissures: Eur Neurol. 2006;56:262–4.
37. Sarnat HB, Netsky MG. Evolution of the nervous system. 2nd ed. New York: Oxford University Press; 1981.
38. von Economo C, Koskinas GN. Die Cytoarchitektonik der Hirnrinde des erwachsenen Menschen. Atlas mit 112 mikrophotographischen Tafeln. Wien: J. Springer; 1925 apud Williams PL, Warwick R, editors. Gray's anatomy. 36th ed. Philadelpia: Saunders; 1980.
39. Williams PL, Warwick R, editors. Gray's anatomy. 36th ed. Philadelphia: Saunders; 1980.
40. Ono M, Kubik S, Abernathey CD. Atlas of cerebral sulci. Stuttgart: Thieme; 1990.
41. Chi JG, Dooling EC, Gilles FH. Gyral development of the human brain. Ann Neurol. 1977;1:86–93.
42. Nishikuni K, Ribas GC. Study of the fetal and postnatal morphological development of the brain sulci. J Neurosurg Pediatr. 2013;11(Suppl 1):1–11.
43. Penfield W, Baldwin M. Temporal lobe seizures and technique of subtotal temporal lobectomy. Ann Surg. 1952;136:625–34 apud Hansebout RR. Surgery of epilepsy, current technique of cortical resection. In: Schmidek HH, Sweet WH, editors. Current techniques in operative neurosurgery. New York: Grune and Stratton; 1977. ch. 9, p.101–17.
44. Uematsu S, Lesser R, Fisher RS, Gordon B, Hara K, Krauss GL, Vining EP, Webber RW. Motor and sensory cortex in humans: topography studied with chronic subdural stimulation. Neurosurgery. 1992;31:59–72.
45. Duvernoy HM. The human brain. Wien: Springer; 1991.
46. Yousry TA, Schmid UD, Alkadhi H, Schmidt D, Peraud A, Buettner A, Winkler P. Localization of the motor hand area to a knob on the precentral gyrus a new landmark. Brain. 1997;120:141–57.
47. Yousry TA, Schmid UD, Jassoy AG, Schmidt D, Eisner WE, Reulen HJ, Reiser MF, Lissner J. Topography of the cortical motor hand area: prospective study with functional MR imaging and direct motor mapping at surgery. Radiology. 1995;195:23–9.
48. Petrides M. The human cerebral cortex. London: Elsevier; 2012.
49. Penfield W, Boldrey E. Somatic motor and sensory representation in the cerebral cortex of man as studied by electrical stimulation. Brain 1937;60:389–443 apud Brodal a. neurological anatomy in relation to clinical medicine. 3rd ed. New York: Oxford University Press; 1981
50. Penfield W, Rasmussen T. The cerebral cortex of man. New York: Macmillan; 1950.
51. Rasmussen AT. The extent of recurrent geniculocalcarine fibers (loop of Archambault and Meyer) as demonstrated by gross brain dissection. Anat Rec. 1943;85:277–82.
52. Ribas GC. Cerebral hemispheres. In: Standring S, editor. Gray's anatomy: the anatomical basis of clinical practice. 41st ed. United Kingdom: Elsevier; 2015. p. 373–98.
53. Talairich J, Tornoux P. Referentially oriented cerebral MRI anatomy. New York: Thieme; 1993 apud Yousry TA, Schmid UD, Alkadhi H, Schmidt D, Peraud a, Buettner a, Winkler P Localization of the motor hand area to a knob on the precentral gyrus a new landmark Brain 1997;120:141–157.
54. Excerpta Medica Foundation. Nomina Anatomica. 4th ed. Amsterdam; 1975.
55. Excerpta Medica Foundation. Nomina Anatomica. 6th ed. Amsterdam; 1980.
56. Federative Committee on Anatomical Terminology. International Anatomical Terminology. Stuttgart: Thieme; 1998.
57. Ribas EC, Yağmurlu K, de Oliveira E, Ribas GC, Rhoton A Jr. Microsurgical anatomy of the central core of the brain. J Neurosurg. 2017;22:1–18.
58. Alves RV, Ribas GC, Párraga RG, Oliveira E. The occipital lobe convexity sulci and gyri. J Neurosurg. 2012;166:1014–23.

An Overview of Cortical and Subcortical Anatomy Relevant for Intra-Axial Brain Surgery

Abhidha Shah, Aimee Goel, and Atul Goel

1 Introduction

The human brain is a magnificent masterpiece whose mystery has not been completely unraveled. The last two decades have seen tremendous advancements in understanding the functions of the brain using intra-operative electrophysiology and brain mapping. A key aspect in performing intra-axial brain surgery is having an intricate knowledge of the cortical and subcortical anatomy of the brain and how it relates to function. Traditionally only cortical areas were considered eloquent but now with enhanced understanding the eloquence of the subcortical fiber network has been realized. Whilst damage to a cortical hub can sometimes be compensated by another cortical area, damage to the white fibers is irrevocable. In this chapter we discuss the anatomical landmarks of the major cortical hubs and the architecture of the subcortical network.

The anatomy of the human brain anatomy will be discussed by dividing the brain into 7 lobes. These are the frontal lobe, central lobe, parietal lobe, temporal lobe, occipital lobe, limbic lobe and the insula.

A. Shah (✉)
Department of Neurosurgery, K.E.M. Hospital and Seth, G.S. Medical College, Mumbai, India

A. Goel
Department of Neurosurgery, University Hospitals of North Midlands NHS Trust, Stoke-on-Trent, UK

A. Goel
Department of Neurosurgery, Lilavati Hospital and Research Center, Mumbai, India

Department of Neurosurgery, R.N. Cooper Hospital, Mumbai, India

Bombay Hospital Institute of Medical Sciences and Research Center, Mumbai, India

2 Frontal Lobe

The frontal lobe is the most rostral and largest lobe in each cerebral hemisphere. It extends from the frontal pole anteriorly to the pre- central sulcus posteriorly. Its medial boundary is the interhemispheric fissure and its lateral boundary is the sylvian fissure. The anatomy of the frontal lobe is best understood by studying its three surfaces viz.: the superolateral surface, the medial surface and the basal surface.

2.1 Superolateral Surface of the Frontal Lobe

2.1.1 Cortical Anatomy

The superolateral surface of the frontal lobe is divided into the superior, middle and inferior frontal gyri by two vertical sulci and one horizontal sulcus. (Fig. 1) The sulci are named as the superior and inferior frontal sulci and the precentral sulcus respectively [1]. The superior

A. Shah et al. (eds.), *Functional Anatomy of the Brain: A View from the Surgeon's Eye*,
https://doi.org/10.1007/978-981-99-3412-6_2

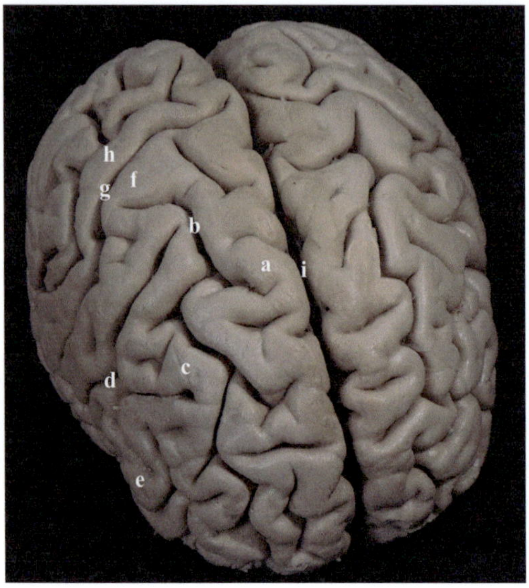

Fig. 1 Superior view of the cerebral hemispheres. a. Superior frontal gyrus, b. Superior frontal sulcus, c. Middle frontal gyrus, d. Inferior frontal sulcus, e. Inferior frontal gyrus, f. Pre-central gyrus, g. Central sulcus, h. Post-central gyrus. i. Interhemispheric fissure

frontal gyrus lies between the superior frontal sulcus and the precentral sulcus. A portion of the superior frontal gyrus extends to the medial surface of the hemisphere. This is usually separated from the dorsolateral portion by a medial frontal sulcus [2, 3]. This portion of the superior frontal gyrus is also called as the medial frontal gyrus. The superior frontal gyrus is continuous anteriorly with the gyrus rectus and connects posteriorly with the precentral gyrus. The middle frontal gyrus is the largest of the three frontal gyri and lies between the superior and inferior frontal sulci. It connects with both the superior and inferior frontal gyri and the precentral gyrus. The inferior frontal gyrus is divided into the pars orbitalis, pars triangularis and the pars opercularis and lies between the inferior frontal sulcus medially and the sylvian fissure laterally. (Fig. 2) The sylvian fissure is divided into an anterior and posterior part by the anterior sylvian point. The anterior sylvian point corresponds to an enlargement of the sylvian fissure recognized intra-operatively as a CSF filled area which lies just beneath the pars triangularis of the inferior frontal gyrus.

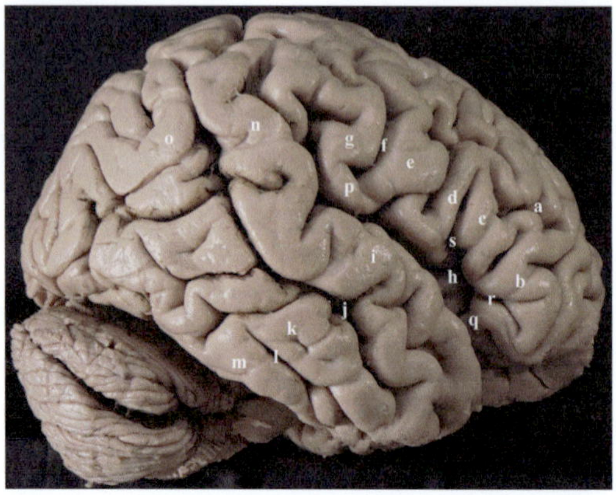

Fig. 2 Lateral view of the cerebral hemisphere. a. Inferior frontal sulcus, b. Pars orbitalis, c. Pars triangularis, d. Pars opercularis, e. Precentral gyrus, f. Central sulcus, g. Postcentral gyrus, h. Anterior sylvian point, i. Superior temporal gyrus, j. Superior temporal sulcus, k. Middle temporal gyrus, l. Inferior temporal sulcus, m. Inferior temporal gyrus, n. Supramarginal gyrus, o. Angular gyrus, p. Subcentral gyrus, q. Sylvian fissure, r. Horizontal ramus of the sylvian fissure, s. Ascending ramus of the sylvian fissure

From the superior aspect of the anterior sylvian point emerge the horizontal and ascending rami of the sylvian fissure demarcating the pars triangularis separating it from the pars orbitalis anteriorly and the pars opercularis posteriorly [2, 3]. The pars orbitalis is the anteriormost part of the inferior frontal gyrus and connects anteriorly with the orbital gyri. The pars opercularis is U shaped due to the inferior termination of the precentral sulcus [2]. The pars opercularis connects prominently with the precentral gyrus just above the sylvian fissure.

The most posterior and medial portion of the superior frontal gyrus just anterior to the precentral gyrus is known as the supplementary motor area (SMA) [2]. The area just anterior to the SMA is the pre-SMA. Both the regions have poorly defined boundaries and are related to performing learned sequences of movements. The most caudal portion of the superior frontal gyrus that lies on the supero-lateral surface in front of the precentral gyrus is the dorsal premotor area, which is involved in the planning of movements. The most caudal portion of the middle frontal gyrus anterior to the precentral gyrus harbors the frontal eye fields responsible for saccadic and voluntary eye movements [2]. In the dominant hemisphere the pars opercularis and the pars triangularis correspond to the Broca's area, which is responsible for motor speech production. The most posterior portion of the pars opercularis near its connection with the precentral gyrus corresponds to the ventral premotor area and intra-operative stimulation of this area causes speech arrest. [2]

2.1.2 Subcortical Anatomy

After removal of the superficial and intersulcal grey matter and the short arcuate fibers the long association fibers in the region of the frontal lobe come into view [1]. The fibers running deep to the middle frontal gyrus are the corona radiata, the superior longitudinal fasciculus and the arcuate fasciculus from medial to lateral [1]. (Figs. 3 and 4) The arcuate fasciculus lies ventrolateral to the superior longitudinal fasciculus in this region. Removal of the inferior frontal gyrus exposes the lateral part of the superior longitudinal fasciculus laterally and superiorly and the arcuate fasciculus medially and inferiorly. Deeper to these fibers lie the fibers of the frontal aslant tract and the fibers of the external capsule joining the internal capsule to form the corona radiata [1].

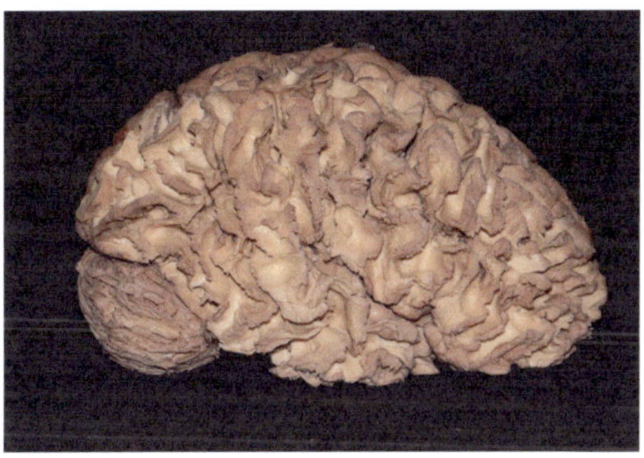

Fig. 3 Lateral view of the hemisphere showing the short arcuate fibers

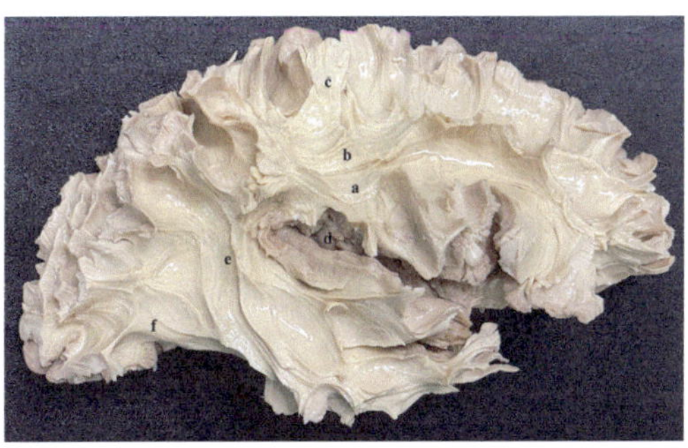

Fig. 4 Subcortical anatomy of the lateral aspect of the cerebral hemisphere. The superficial layer or the short arcuate fibers have been removed to show the middle layer of long association fibers. a. Arcuate fasciculus, b. Lateral part of the superior longitudinal fasciculus, c. Corona radiata, d. Insula, e. Middle longitudinal fasciculus, f. Inferior longitudinal fasciculus

2.2 Medial Surface of Frontal Lobe

2.2.1 Cortical Anatomy

The medial surface of the frontal lobe is composed of the continuation of the medial portion of the superior frontal gyrus superiorly and beneath it the cingulate gyrus. (Fig. 5) Anteriorly at the frontal pole the orbital gyri meet the superior, middle and inferior frontal gyri. The most inferomedial gyri are the paraolfactory and the paraterminal gyri [2, 4]. These gyri are a constituent of the limbic system and will be discussed further with the limbic lobe. Beneath these most inferomedial gyri is the beginning of the cingulum.

2.2.2 Subcortical Anatomy

Beneath the superior frontal gyrus run the fibers of the corona radiata, the callosal fibers and the

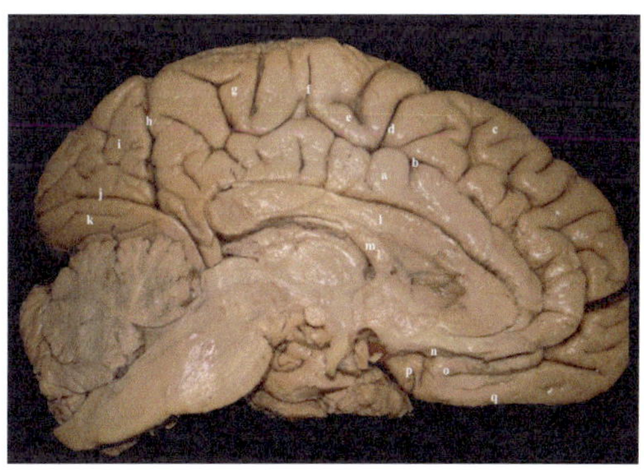

Fig. 5 Medial surface of the cerebral hemisphere. a. Cingulate gyrus, b, Cingulate sulcus, c. Superior frontal gyrus, d. Paracentral sulcus, e. Paracentral lobule, f. Marginal ramus of the cingulate sulcus, g. Precuneus, h. Parieto-occipital sulcus, i. Cuneus, j. Calcarine sulcus, k. Lingual gyrus, l. Corpus callosum, m. Fornix, n. Paraterminal gyrus, o. Paraolfactory gyri, p. Cingulate pole, q. Gyrus rectus

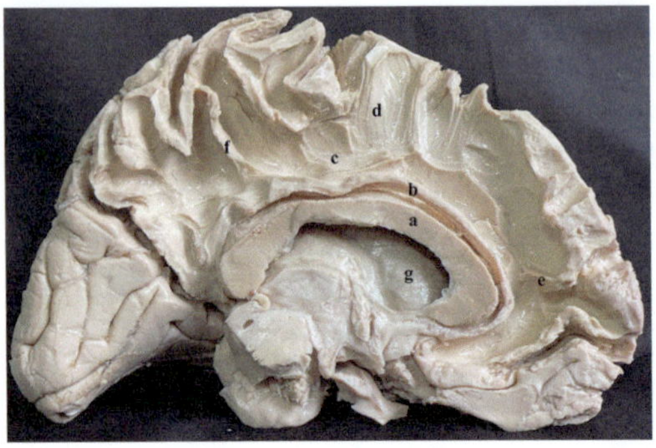

Fig. 6 Subcortical anatomy of the medial aspect of the hemisphere. a. Corpus callosum, b. Cingulum, c. Medial part of the superior longitudinal fasciculus, d. Corona radiata, e. Frontal connection of the cingulum, f. Parietal connection of the cingulum, g. Caudate head

superior longitudinal fasciculus from lateral to medial [1]. (Fig. 6) The fibers of the corona radiata are vertically oriented and the fibers of the superior longitudinal fasciculus are horizontally oriented. Medially deeper to the superior longitudinal fasciculus lies the cingulum and further in the depth the transversely running fibers of the corpus callosum can be seen. The cingulum runs upwards and posteriorly beneath the gray matter of the cingulate gyrus in the medial portion of the fronto-parietal lobe. Removal of the short arcuate fibers of the frontal and parietal lobe shows the connections of the cingulum to the fronto-parietal cortex [1].

2.3 Basal/Orbital Surface of the Frontal Brain

2.3.1 Cortical Anatomy

The orbital surface of the frontal lobe is divided into the medial gyrus rectus and the lateral orbital gyri by the deep olfactory sulcus[1, 4–6] (Figs. 7 and 8). The olfactory sulcus houses the olfactory bulb and the olfactory tract. The orbital gyri make up most of the basal portion of the frontal lobe. An H shaped sulcus known as the cruciform sulcus of Rolando divides the orbital gyri into anterior, posterior, medial and lateral orbital gyri. The posterior orbital gyrus is situated just anterior to the anterior perforated substance and connects medially to the medial orbital gyrus. The medial, anterior and lateral orbital gyri are connected to the superior, middle and inferior frontal gyri along the frontal pole.

The *olfactory tract* runs in the olfactory sulcus on the orbital surface of the frontal lobe. Further

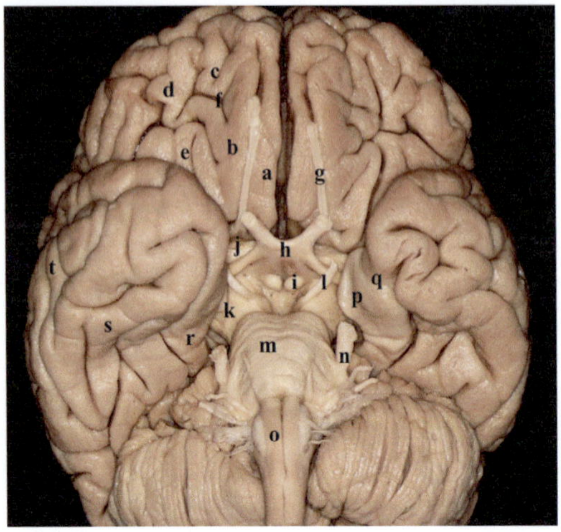

Fig. 7 Inferior view of the cerebral hemispheres showing the basal surfaces of the frontal and temporal lobes. a. Gyrus rectus, b. Medial orbital gyrus, c. Anterior orbital gyrus, d. Lateral orbital gyrus, e. Posterior orbital gyrus, f. Cruciform sulcus of Rolando, g. Olfactory tract, h. Optic chiasm, i. Mammillary body, j. Anterior perforated substance, k. Cerebral peduncle, l. Oculomotor nerve, m. Pons, n. Trigeminal nerve, o. Medullary pyramids, p. Uncus, q. Rhinal sulcus, r. Parahippocampal gyrus, s. Fusiform gyrus, t, Inferior temporal gyrus

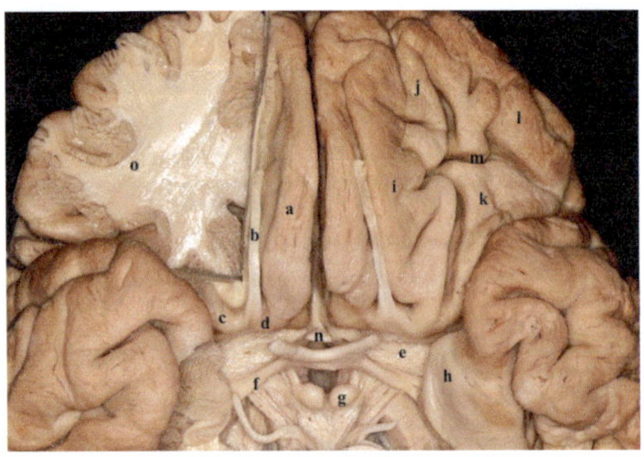

Fig. 8 Closer view of the basal/orbital surface of the frontal lobe. a. Gyrus rectus, b. Olfactory tract, c. Lateral olfactory stria, d. Medial olfactory stria, e. Anterior perforated substance, f. Optic tract, g. Mammillary body, h. Uncus, i. Medial orbital gyrus, j. Anterior orbital gyrus, k. Posterior orbital gyrus, l. Lateral orbital gyrus, m. Cruciform sulcus of Rolando, n. Anterior commissure, o. Short arcuate or U fubers

posteriorly the olfactory tract divides into the *medial and lateral olfactory striae* [1, 4–6]. The medial olfactory stria becomes continuous with the subcallosal (para olfactory) and the paraterminal gyrus on the medial surface of the frontal brain. The lateral olfactory stria passes along the lateral and anterior margin of the anterior perforated substance hugging the posterior margin of the posterior orbital gyri to reach the piriform region in the temporal lobe. The medial striae contain fibers from the anterior olfactory nucleus. These fibers traverse in the anterior commissure to the opposite side. The lateral olfactory stria passes along the lateral margin of the anterior perforated substance giving out some collaterals in the anterior perforated substance to reach the piriform region. These fibers terminate in the prepiriform cortex and the corticomedial part of the amygdaloid nuclear complex. The *prepiriform area* represents the primary olfactory cortex. It corresponds to area 34 of Brodmann and is located on the anterior end of the parahippocampal gyrus and uncus. From here fibers project to the neighboring entorhinal cortex and the medial dorsal nucleus of the thalamus. Fibers from the amgdaloid nuclei connect with the hypothalamus and the septal nuclei [4–6].

2.3.2 Subcortical Anatomy

The first fibers that are visualized after removing the superficial gray matter of the orbital surface of the frontal brain are the arcuate fibers. (Fig. 8) These are short association fibers that interconnect neighboring gyri. Removing the short association fibers in the region of the medial gyrus rectus will reveal the anteriorly curving fibers of the forceps minor [1]. (Figs. 9 and 10) Inferior and lateral to the fibers of the forceps minor are association fiber bundles that connect the orbitofrontal gyri to the supero-medial portion of the thalamus. These fibers run lateral to the gyrus rectus, inferior to the frontal horn and further posteriorly run lateral to the caudate nucleus. Deeper dissection in the region of the medial orbital gyrus will reach the frontal horn. The caudate head lies lateral to the frontal horn and roughly corresponds to the transverse limb of the H shaped cruciform sulcus. Fiber bundles running lateral to the caudate connect the orbitofrontal gyri to the midbrain via the crus cerebri. The medial portion of the orbitofrontal region thus connects to the thalamus and the brainstem. The lateral portion of the orbitofrontal gyri mainly connects with the ipsilateral temporal, parietal and occipital cortices by two major association bundles namely the inferior occipitofrontal fasciculus and the uncinate fasciculus. (Figs. 9 and 10).

Fig. 9 Fiber dissection of the basal surface of the frontal lobe showing the various fibers. a. Genu of corpus callosum, b. Forceps minor, c. Uncinate fasciculus, d. Inferior fronto-occipital fasciculus, e. Insular gyri, f. Mammillary body, g. Optic tract, h. Uncus, j. Short arcuate fibers

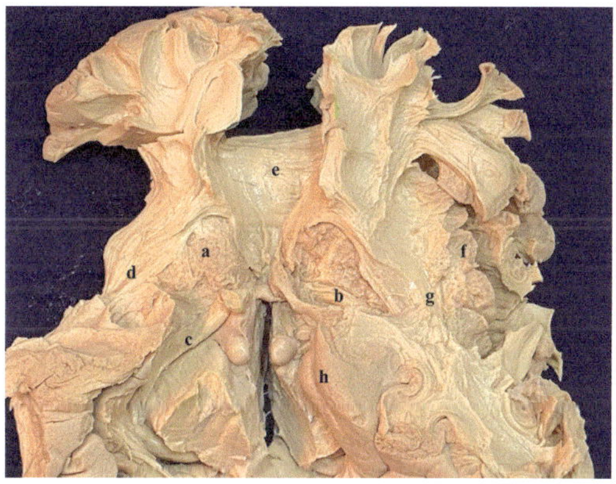

Fig. 10 Deeper dissection in the region of the basal forebrain. a. Substantial innominata, b. Anterior commissure, c. Optic tract, d. Uncinate fasciculus, e. Forceps minor, f. Insular gyri, g. Inferior fronto-occipital fasciculus, h. Uncus

3 Central Lobe

3.1 Cortical Anatomy

The central lobe consists of the precentral and the postcentral gyrus and forms an ellipsoid on the superolateral surface of the hemisphere between the frontal and parietal lobes. (Fig. 11) The precentral gyrus lies between the precentral sulcus anteriorly and the central sulcus posteriorly. The post central gyrus lies between the central sulcus anteriorly and the post central sulcus posteriorly. The central sulcus separates the pre central gyrus and the post-central gyrus. The central sulcus is generally a continuous sulcus and extends into the paracentral lobule on the medial surface of the hemisphere. Inferiorly it reaches just above the sylvian fissure ending in the subcentral gyrus. The precentral sulcus is an interrupted sulcus and is divided by the middle frontal gyrus into superior and inferior portions. The precentral sulcus reaches the interhemispheric fissure on the superomedial surface of the hemisphere and serves as an important surgical landmark. The postcentral sulcus is also interrupted but it does not reach the interhemispheric fissure.

The precentral gyrus corresponds to the primary motor area and the postcentral gyrus corresponds to the primary sensory area. Both the precentral and postcentral gyri are connected to each other at the medial and lateral ends of the central sulcus. The medial connection is the paracentral lobule, which lies on the medial aspect of the hemisphere in the interhemispheric fissure. (Fig. 5) The paracentral lobule is topographically related to the atrium of the lateral ventricle [2, 3]. The lateral connection, which lies above or embedded within the sylvian fissure is the subcentral gyrus. (Fig. 2) The subcentral gyrus forms part of the operculum and covers the posterior part of the insula. It is topographically related to the body of the lateral ventricle. The subcentral gyrus rests over the transverse gyrus of Heschl [2, 3]. The central sulcus has been tradiationally viewed as having two anteriorly convex genus and one middle posteriorly convex genu. Broca described a third middle connection between these two gyri, which lies in the depths of the central suclus as a gyral bridge also known as the "plis de passage" moyen of Broca [2]. This connection corresponds topographically to the middle posterior convex genu of the central sulcus. This middle knee also corresponds to the motor hand area and can be identified at the distal end of the superior frontal sulcus where it reaches the precentral gyrus.

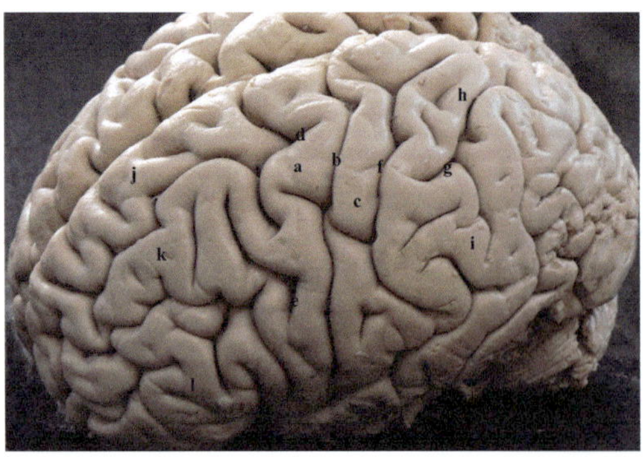

Fig. 11 Superolateral surface of the cerebral hemisphere showing the central lobe. a. Precentral gyrus, b. Central sulcus, c. Postcentral gyrus, d. Superior portion of the precentral sulcus, e. Inferior portion of the precentral sulcus, f. Postcentral sulcus, g. Interparietal sulcus, h. Superior parietal lobule, i. Inferior parietal lobule, J. Superior frontal gyrus, k. Middle frontal gyrus, l. Inferior frontal gyrus

3.2 Subcortical Anatomy

Removal of the superficial gray matter in the region of the precentral gyrus reveals the connections between the superior and middle frontal gyri and the pre-central gyrus. (Fig. 12) Removal

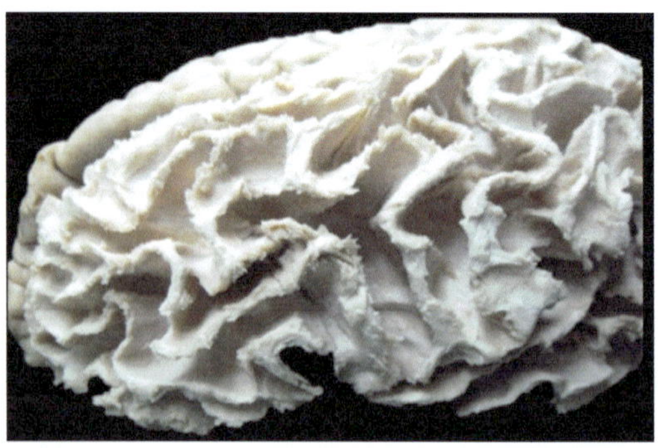

Fig. 12 The superficial gray matter and the gray matter in the depths of the sulci have been removed to reveal the short arcuate fibers. The connections between the middle and inferior gyri and the precentral gyrus are very clearly visible

of the short arcuate fibers reveals from medial to lateral the fibers of the superior longitudinal fasciculus (medial part), the corona radiata, the continuation of the superior longitudinal fasciculus (lateral part) and lastly the arcuate fasciculus. [1] The fibers of the arcuate fasciculus curve downwards just posterior to the post central gyrus in the region of the inferior parietal lobule.

4 Parietal Lobe

The parietal lobe lies between the postcentral gyrus anteriorly, the parieto-occipital sulcus medially and an imaginary line connecting the point of emergence of the parieto-occipital sulcus on the supero-medial surface to the pre-occipital notch [2, 3]. The pre-occipital notch is an apparent cleft on the infero-lateral border of the hemisphere and lies approximately 5 cm anterior to the occipital pole.

The parietal lobe can be divided into a superolateral surface and a medial surface.

4.1 Supero-Lateral Surface

4.1.1 Cortical Anatomy

The supero-lateral surface of the parietal lobe is divided into the superior and inferior parietal lobules by the inter-parietal sulcus. (Fig. 13) The inter parietal sulcus begins from the mid-point of the post central sulcus and traverses vertically almost parallel to the interhemispheric fissure [2]. The inter parietal sulcus gives two vertical offshoots the superior and inferior vertical branches. The superior vertical branch is also known as the transverse parietal sulcus of Brissaud and incompletely divides the superior parietal lobule [2]. The inferior parietal branch is known as the intermediate sulcus of Jensen and divides the inferior parietal lobule into an anterior supramarginal gyrus and a posterior angular gyrus. The supramarginal gyrus is a very distinct curved gyrus that is present around the distal end of the sylvian fissure. Its inferior end is continuous with the superior temporal gyrus. The superior end of the supramarginal gyrus is connected

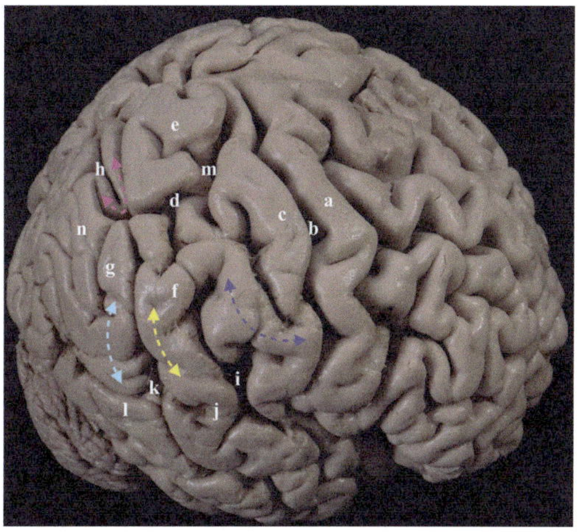

Fig. 13 Superolateral view of the cerebral hemisphere showing the parietal lobe. a. Precentral gyrus, b. Central sulcus, c. Postcentral gyrus, d. Interparietal sulcus, e. Superior parietal lobule, f. Supramarginal gyrus, g. Angular gyrus, h. Parieto-occipital fissure, i. Sylvian fissure, j. Superior temporal gyrus, k. Superior temporal sulcus, l. Middle temporal gyrus, m. Postcentral sulcus, n. Superior occipital gyrus. The colored arrows show the connections between the various gyri. Purple arrow depicts the connection between the supramarginal gyrus and the postcentral gyrus, Yellow arrow depicts the connection between the supramarginal gyrus and the superior temporal gyrus. Blue arrow shows the connection between the angular gyrus and the middle temporal gyrus. Pink arrow shows the parieto-occipital arcus the prominent connection between the superior parietal lobule and the superior occipital gyrus

to the postcentral gyrus by a fold that runs beneath the inferior aspect of the postcentral sulcus. The angular gyrus is not as well defined as the supramarginal gyrus but can be delineated as the gyrus that surrounds the distal limit of the superior temporal sulcus. (Fig. 13) The inferior end of the angular gyrus connects with the middle temporal gyrus. The superior parietal lobule has a quadrangular shape and is bounded laterally by the inter-parietal sulcus and anteriorly by the postcentral sulcus [2]. Medially it is continuous with the precuneus on the medial surface of the hemisphere. Posteriorly it connects with the superior occipital gyrus via a prominent parieto-occipital fold that surrounds the parieto-occipital incisura as it continues on the supero-lateral surface of the hemisphere. The parieto-occipital fold or arcus is a very anatomically constant U shaped convolution that surrounds the parieto-occipital incisura on the supero-lateral surface of the hemisphere. This fold is also known as the first or superior "plis de passage" of Gratiolet [2, 3].

4.1.2 Subcortical Anatomy

Removal of the short U fibers in the region of the superior parietal lobe reveals the superior longitudinal fasciculus and the fibers of the corona radiata from medial to lateral. The inferior parietal lobule consists of the supramarginal and the angular gyrus. The fiber bundles running beneath this lobule are the superior longitudinal fasciculus, the arcuate fasciculus, the middle longitudinal fasciculus, the inferior fronto-occipital fasciculus and the optic radiations [1]. (Fig. 4).

4.2 Medial Surface

4.2.1 Cortical Anatomy

The medial surface of the parietal lobe consists of the precuneus or the quadrangular lobe of Foville [2, 3]. (Fig. 5) It lies posterior to the medial portion of the post-central gyrus and is a continuation of the superior parietal lobule over the supero-medial edge of the hemisphere. The precuneus is limited anteriorly by the marginal branch of the cingulate sulcus, posteriorly by the parieto-occipital sulcus and inferiorly by the subparietal sulcus. Inferiorly the precuneus connects to the isthmus of the cingulate gyrus.

4.2.2 Subcortical Anatomy

Removal of the fibers of the precuneus reveals the superior longitudinal fasciculus, the upward curving callosal fibers and the corona radiata from medial to lateral [1]. (Fig. 6) Beneath the superior longitudinal fasciculus runs the cingulum and beneath the cingulum the transverse fibers of the corpus callosum can be visualized.

5 Occipital Lobe

On the supero-lateral surface of the hemisphere the occipital lobe lies behind an imaginary line that connects the parieto-occipital incisura on the supero-lateral surface to the pre-occipital notch on the infererolateral border of the hemisphere [2]. On the medial surface the occipital lobe lies behind the parieto-occipital sulcus. The occipital lobe has three surfaces: the superolateral, the medial and the basal surfaces.

5.1 Superolateral Surface

5.1.1 Cortical Anatomy

The gyri and sulci on the occipital lobe are somewhat ill defined. But generally there are two or three occipital gyri, which are called as the superior, middle and inferior occipital gyri. (Fig. 14) The superior and inferior occipital gyri are the most consistent. All the three gyri meet posteriorly at the occipital pole. The superior occipital gyrus is oriented vertically parallel to the interhemispheric fissure [2, 3]. On the supero-lateral surface it joins the superior parietal lobule through the parieto-occipital arcus and on the medial surface of the hemisphere it is continuous with the cuneus. It is delineated anteriorly by the parieto-occipital sulcus on the supero-lateral surface of the hemisphere and laterally by the intra-

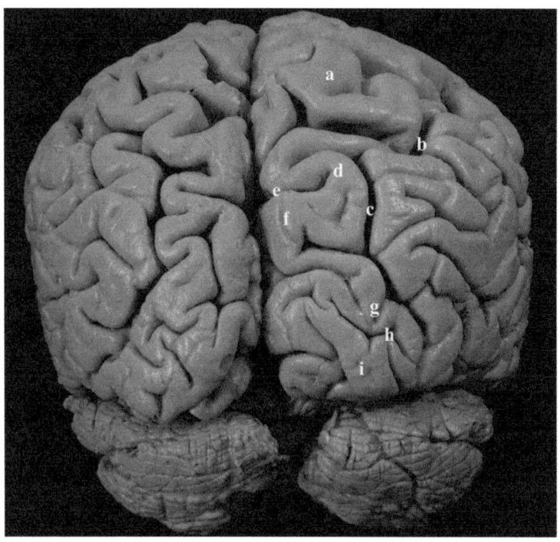

Fig. 14 Posterior view of the cerebral hemisphere. a. Superior parietal lobule, b. Interparietal sulcus, c, Superior occipital sulcus, d. Parieto-occipital arcus, e. Parieto-occipital sulcus, f. Superior occipital gyrus, g. Middle occipital gyrus, h. Inferior occipital sulcus, i. Inferior occipital gyrus

occipital or superior occipital sulcus. This sulcus is an extension of the interparietal sulcus that separates the superior and inferior parietal lobules. The middle occipital gyrus when present lies between an inferior extension of the intra-occipital sulcus and the inferior occipital sulcus. The inferior occipital gyrus is oriented horizontally along the inferior margin of the hemisphere and is continuous anteriorly with the inferior temporal gyrus. The inferior occipital or the lateral occipital sulcus limits it superiorly. Posteriorly the inferior occipital gyrus runs to the occipital pole and then merges with the lingual gyrus on the medial surface of the hemisphere.

5.1.2 Subcortical Anatomy

Beneath the short association fibers of the superior occipital gyrus run the termination of the superior longitudinal fasciculus and the forceps major from medial to lateral [1]. The fiber bundles running beneath the medial and inferior occipital gyri are from medial to lateral: the optic radiations, the terminations of the anterior commissure and inferior fronto-occipital fasciculus and the inferior longitudinal fasciculus.

5.2 Medial Surface

5.2.1 Cortical Anatomy

Unlike the supero-lateral surface of the occipital lobe, the medial surface is rather well defined. (Fig. 15) The medial surface of the occipital lobe is made up of the cuneus and the lingual gyrus separated by the calcarine fissure or sulcus. The occipital lobe is demarcated from the parietal lobe by the parieto-occipital sulcus. These 2 sulci form major landmarks on the medial aspect of the hemisphere. The calcarine sulcus originates beneath the splenium of the corpus callosum and runs posteriorly with a gentle superior curve. At around the apex of the curve the parieto-occipital sulcus originates and extends superiorly [2, 3]. The point of emergence of the parieto-occipital sulcus divides the calcarine sulcus into an anterior and posterior part. The parieto-occipital sulcus separates the pre-cuneus from the cuneus. The cunues has a typical wedge shape and lies

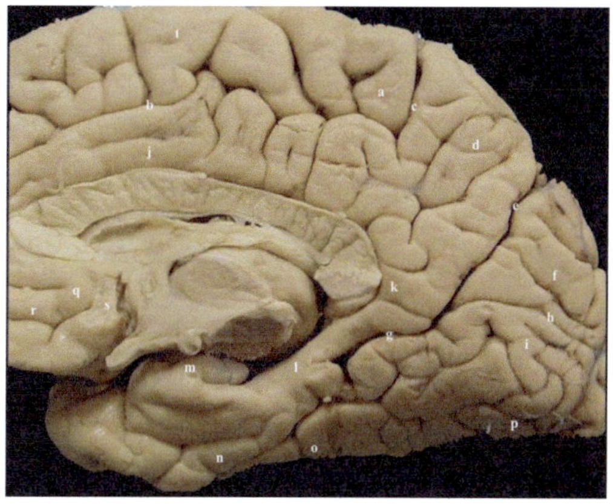

Fig. 15 Medial surface of the cerebral hemisphere. a. Paracentral lobule, b. Cingulate sulcus, c. Marginal ramus of the cingulate sulcus, d. Pre-cuneus, e. Parieto-occipital sulcus, f. Cuneus, g. Anterior part of calcarine sulcus, h. Posterior part of calacarine sulcus, i. Lingual gyrus, j. Cingulate gyrus, k. Isthmus of the cingulate gyrus, l. Parahippocampal gyrus, m. Uncus, n. Fusiform gyrus, o. Inferior temporal gyrus, p. Inferior occipital gyrus, q. Cingulate pole, r. Gyrus rectus, s. Paraolfactory gyri, t. Superior frontal gyrus

between the parieto-occipital sulcus and the posterior part of the calcarine sulcus. The lingual gyrus lies inferior to the calcarine fissure and forms both the basal and medial surfaces of the occipital lobe. Posteriorly in the region of the occipital pole the calcarine fissure ends in a "Y" or "T" shape to form the retrocalcarine sulcus. Along the retrocalcarine sulcus lies the gyrus descendens of Ecker which is made up of striate cortex and forms the real lateral limit of this cortex in the human brain [2].

The primary visual cortex is located on the posterior banks of the calcarine fissure on its superior (cuneal) and inferior (lingual) surfaces. Sometimes there is a cuneo lingual gyrus in the posterior part of the fissure that connects the two gyri [2].

5.2.2 Subcortical Anatomy

Removal of the gray matter and the short association fibers in the region of the cuneus and lingual gyrus revealed a fiber tract running vertically downwards from the precuneus to the lingual gyrus. This fasciculus has been named as the sledge runner fasciculus and has only infre-

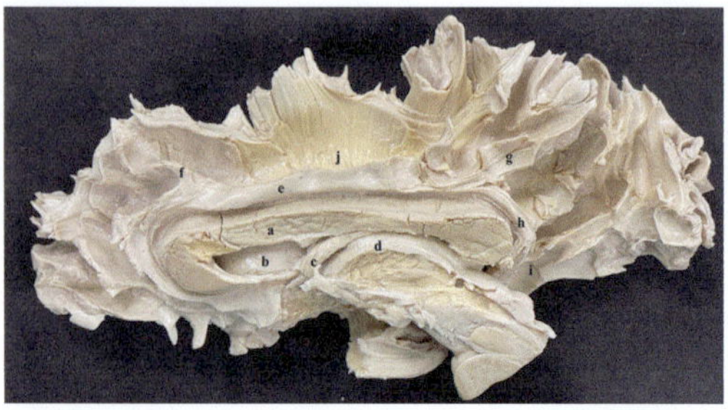

Fig. 16 Fiber dissection of the medial surface of the cerebral hemisphere. a. Corpus callosum, b. Head of caudate nucleus, c. Fornix, d. Thalamus, e. Cingulum, f. Frontal connection of the cingulum, g. Parietal connection of the cingulum, h. Isthmus of the cingulum, i. Forceps major, j. Corpus callosal fibers turning to join the corona radiata

quently been seen in fiber dissection studies [1]. Medial to the sledge runner lies the posterior part of the medial wall of the atrium of the lateral ventricle. Removal of the gray matter in the region of the isthmus of the cingulum and the inferior portion of the precuneus reveals the fibers of the forceps major. (Fig. 16).

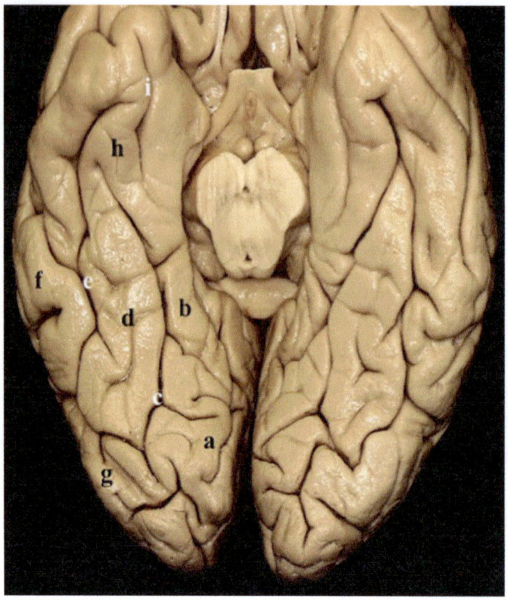

Fig. 17 Basal surface of the occipital and temporal lobes. a. Lingual gyrus, b. Parahippocampal gyrus, c. Collateral sulcus, d. Fusifrom gyrus, e. Occipitotemporal sulcus, f. Inferior temporal gyrus, g. Inferior occipital gyrus, h. Temporo-limbic passage, i. Rhinal sulcus

5.3 Basal Surface

5.3.1 Cortical Anatomy

The basal surface of the occipital lobe is made up of three gyri, the lingual gyrus, the fusiform gyrus and the inferior occipital gyrus from medial to lateral. (Fig. 17) The lingual gyrus turns inferiorly from the medial surface of the hemisphere to continue on its basal surface [1–3]. It is continuous anteriorly with the parahippocampal gyrus and is laterally bordered by the collateral sulcus. The lingual gyrus is also known as the medial temporooccipital gyrus. The fusiform gyrus or the lateral temporooccipital gyrus lies lateral to the lingual gyrus and is separated from it by the collateral sulcus. Laterally the fusiform gyrus is limited by the occipitotemporal sulcus separating it from the inferior occipital gyrus. The inferior occipital gyrus lies lateral to the fusiform gyrus and is continuous anteriorly with the inferior temporal gyrus and medially with the lingual gyrus along the occipital pole.

5.3.2 Subcortical Anatomy

Beneath the short association fibers of the lingual gyrus can be seen the fibers of the forceps major [1]. Running deep to the fusiform gyrus and the inferior occipital gyrus are from medial to lateral the tapetum, the optic radiations, fibers of the anterior commissure and fibers of the inferior fronto-occipital fasciculus (sagittal stratum).

6 Temporal Lobe

The temporal lobe on the superolateral surface lies below the sylvian fissure and in front of an imaginary line that connects the parieto-occipital incisure to the pre-occipital notch. The temporal lobe consists of three surfaces, the superolateral surface, the basal surface and the opercular surface.

6.1 Superolateral Surface

6.1.1 Cortical Anatomy

The lateral surface of the temporal lobe is made up of three main gyri separated by two prominent sulci. The three gyri are the superior, middle and inferior temporal gyri and the sulci are the superior and inferior temporal sulcus. (Fig. 18) The superior and inferior temporal gyri meet at the temporal pole, as the middle temporal gyrus is shorter than these other two gyri. The posterior end of the sylvian fissure ends in the supramarginal gyrus and the superior temporal sulcus runs parallel to the sylvian fissure and terminates in the angular gyrus. The superior temporal sulcus ends as a trifurcation into an ascending sulcal segment, a horizontal segment and an inferior segment. The ascending sulcal segment separates the supramarginal gyrus from the angular gyrus ultimately merging with the interparietal sulcus of Jensen [2]. The superior temporal gyrus is continuous with the supramarginal gyrus and the middle temporal gyrus is continuous with the angular gyrus. The inferior temporal gyrus is continuous with the inferior occipital gyrus. The inferior temporal sulcus separates the middle temporal gyrus from the inferior temporal gyrus.

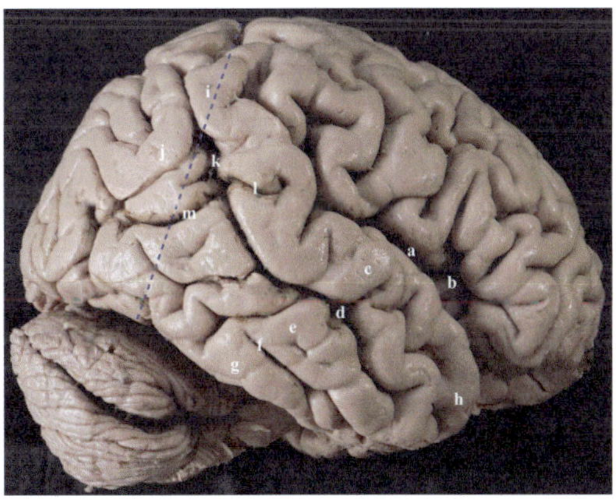

Fig. 18 Supero-lateral surface of the temporal lobe. The purple dashed line indicates an imaginary line connecting the parieto-occipital sulcus to the pre-occipital notch, which separates the temporal from the parietal and occipital lobes. a. Sylvian fissure, b. Anterior sylvian point, c. Superior temporal gyrus, d. Superior temporal sulcus, e. Middle temporal gyrus, f. Inferior temporal sulcus, g. Inferior temporal gyrus, h. Temporal pole, i. Supramarginal gyrus, j. Angular gyrus, k. Ascending sulcul segment of the superior temporal sulcus, l. Horizontal segment of the superior temporal sulcus, m. Inferior segment of the superior temporal sulcus

6.1.2 Subcortical Anatomy

Running beneath the superior temporal gyrus from superficial to deep (and lateral to medial) are the middle longitudinal fasciculus, the temporal fibers of the uncinate fasciculus and the inferior frontooccipital fasciculus, the fibers of the anterior commissure and the sagittal stratum [1]. (Figs. 4, 19, and 20) The fibers of the uncinate fasciculus are in the anterior aspect of the superior temporal gyrus and those of the anterior commissure and the sagittal stratum are in the posterior aspect of the gyrus. The most superomedial portion of the gyrus harbors the temporal portion of the arcuate fasciculus. Removal of the short arcuate fibers in the region of the middle temporal gyrus reveals from superficial to deep the middle longitudinal fasciculus, the fibers of the inferior fronto-occipital fasciculus and the temporal fibers of the anterior commisure as they run posteriorly to form the sagittal stratum. Deeper to these fibers lie the anteriorly curving fibers of the Meyer's loop [1]. All these fibers lie superior to the temporal horn in the anterior aspect of the gyrus and lateral to the temporal

Fig. 19 Fiber dissected specimen of the lateral cerebral hemisphere. a. Lateral part of the superior longitudinal fasciculus, b. Arcuate fasciculus, c. Inferior fronto-occipital fasciculus, d. Uncinate fasciculus, e. Part of the supeiro longitudinal fasciculus has been dissected to show the fibers of the external and internal capsule merging to form the corona radiata. f. Dorsal part of the external capsule, g. Putamen, h. Sagittal stratum

Fig. 20 Fiber dissected specimen of the lateral cerebral hemisphere. The arcuate fasciculus has been lifted to show the sagittal stratum. a. Lateral part of the superior longitudinal fasciculus, b. Arcuate fasciculus, c. The arcuate fasciculus has been lifted. d. Inferior fronto-occipital fasciculus, e. Uncinate fasciculus, f. Dorsal part of the external capsule, g. Putamen, h. Sagittal stratum

horn in the posterior part of the gyrus. The inferior temporal gyrus is relatively devoid of major fiber bundles and hence maybe amenable to a variety of surgical approaches. The fibers that run deep to the inferior temporal gyrus are those of the inferior longitudinal fasciculus.

6.2 Basal Surface

6.2.1 Cortical Anatomy

The basal surface of the temporal lobe consists of the basal portion of the inferior temporal gyrus, the temporal portion of the lateral temporo-occipital gyrus or the fusiform gyrus and the parahippocampal gyrus. (Figs. 7, and 17) The inferior temporal gyrus is separated from the fusiform gyrus by the occipito-temporal sulcus and the fusiform gyrus is separated from the parahippocampal gyrus by the collateral sulcus. The anterior portion of the temporo—occipital sulcus curves medially towards the collateral sulcus thus giving a curved configuration to the anterior aspect of the fusiform gyrus [2]. This anterior turning of the sulcus corresponds medially to the level of the cerebral peduncle. The collateral sulcus is continuous anteriorly with the rhinal sulcus. Sometimes it is separated from it from the so-called temporo-limbic passage. [2] The rhinal sulcus separates the uncus or the entorhinal cortex from the neocortex of the basal surface of the temporal lobe. The parahippocampal gyrus forms part of the limbic lobe and will be discussed below.

6.2.2 Subcortical Anatomy

Removing the fibers of the inferior temporal gyrus will reveal the inferior longitudinal fasciculus as discussed earlier. Deep to the short arcuate fibers of the fusiform gyrus anteriorly lie the inferior frontooccipital fasciculus and the uncinate fasciculus and posteriorly are the fibers of the Meyer's loop and optic radiations medially and the fibers of the inferior fronto-occipital fasciculus, anterior commisure and sagittal stratum laterally [1]. Beneath the region of the parahippocampus lies the radiation of the cingulum, which marks its termination [1, 4]. Deeper to the radiation of the cingulum is the ependyma of the temporal horn. On removing the ependyma the hippocampus with the fimbria and the beginning of the fornix can be visualized.

6.3 Opercular Surface

The opercular surface of the temporal lobe consists of the superior surface of the superior temporal gyrus that is embedded in the sylvian fissure. (Fig. 21) It is made up of a series of trans-

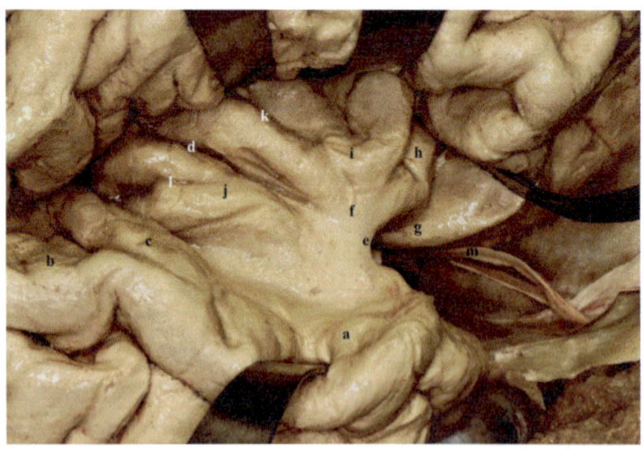

Fig. 21 Image showing the opercular surface of the temporal lobe and the insula. a. Planum polare of the superior temporal gyrus, b. Planum temporale of the superior temporal gyrus, c. Heschl's gyrus, d. Central sulcus of the insula, e. Limen insulae, f. Insular apex, g. Transverse gyrus of the insula, h. Accesory gyrus of insula, i. Anterior, middle and posterior short insular gyri, j. Anterior and posterior long insular gyri, k. Precentral insular sulcus, l. Postcentral insular sulcus, m. Olfactory nerve

verse gyri. The most prominent transverse gyrus is called as the Heschl's gyrus and can be generally identified at the midpoint of the superior temporal gyrus. The Heschl's gyrus divides the opercular surface of the temporal lobe into an anterior polar plane and a posterior temporal plane [2]. The subcentral gyrus usually sits above the sylvian fissure at the level of the Heschl's gyrus.

7 Insular Lobe

7.1 Cortical Anatomy

The insula or the island of Reil lies hidden between the frontal and temporal lobes and is visualized after elevating the frontal and temporal opercula (curtains) [1]. (Fig. 21) The insula can be viewed as an external shield for the central core of the brain. The insula is bounded by the circular sulcus of Reil, which can be divided into three parts: the anterior, superior and inferior limiting or periinsular sulci. (Fig. 2) The insula is shaped like a pyramid with the triangular base lying superiorly and its vertex lying inferiorly constituting the limen insulae. (Fig. 21) It has two surfaces an anterior surface and an inferior surface. The anterior surface of the insula lies beneath the fronto-orbital operculum, which is made up of the pars orbitalis of the inferior frontal gyrus and the posterior portion of the posterior orbital gyrus [2]. The lateral surface of the insula lies underneath the fronto-temporal operculum, which consists of the pars triangularis and pars opercularis of the inferior frontal gyrus, the subcentral gyrus, the anterior portions of the supramarginal gyrus and the temporal plane of the superior temporal gyrus. The limen insula actually consists of a thin layer of olfactory cortex which connects the insula to the anterior perforated substance. The limen insulae extends along he lateral aspect of the lateral olfactory striae [2, 3]. The lateral surface of the insula is divided by the central sulcus of the insula into an anterior and a posterior portion. The central sulcus of the insula runs in the same direction as the central sulcus on the hemisphere.

The peak of the insula is referred to as the insular apex and is present in the anterior part of the insula. The smaller anterior part of the insula is made up of the transverse, accessory and the anterior, middle and posterior short gyri. (Fig. 21) All these gyri arise from the region of the insular apex [2, 3]. The transverse gyrus starts at the insular apex region, traverses along the limen insulae and connects the insula to the posteromedial orbital lobule. The posteromedial orbital lobule is the region where the medial and posterior orbital gyri meet posteriorly on the basal surface of the frontal lobe. The accessory gyrus lies between the anterior short gyrus and the transverse gyrus. The accessory and the transverse gyri form the insular pole. The middle and posterior short gyri are separated by the precentral insular sulcus. The larger posterior part of the insula consists of the anterior and posterior long gyri separated by the postcentral insular sulcus.

7.2 Subcortical Anatomy:

Removal of the insular gyri reveals the extreme capsule. (Fig. 22) The extreme capsule connects the adjacent insular gyri to each other and also consists of vertical fibers, which connect the insular gyri to the frontal parietal and temporal opercula. Beneath the extreme capsule lies the

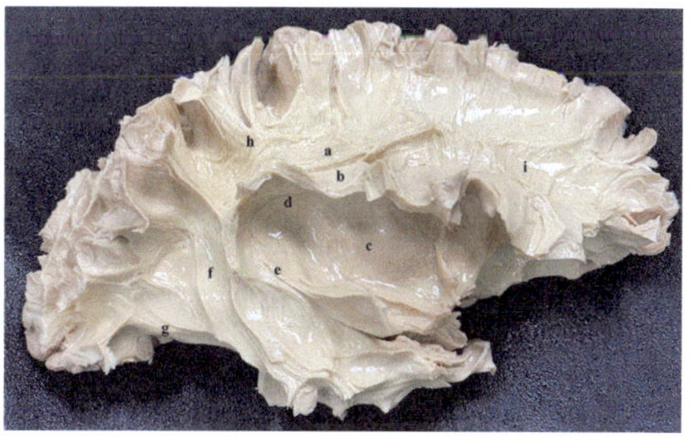

Fig. 22 Fiber dissected specimen of the lateral surface of the cerebral hemisphere. The insular gyri have been removed. a. Lateral part of the superior longitudinal fasciculus, b. Arcuate fasciculus, c, Extreme capsule, d. Superior limiting sulcus of insula, e. Inferior limiting sulcus of insula, f. Middle longitudinal fasciculus, g. Inferior longitudinal fasciculus, h. Corona radiata, i. Frontal aslant tract

claustrum which is a thin layer of gray matter that lies between the extreme capsule laterally and the external capsule medially [1]. Removing the gray matter of the claustrum reveals the external capsule. (Fig. 23) The external capsule consists of two parts the dorsal external capsule and the ventral external capsule. The dorsal external capsule is composed of the claustrocortical fibers. This fiber system connects the clastrum to the frontoparietal cortex between the supplementary motor area and the posterior part of the parietal lobe. The ventral external capsule is composed of the inferior frontooccipital fasciculus superiorly and the uncinate fasciculus inferiorly. (Fig. 23) Removal of the external capsule leads to the gray matter of the lentiform nucleus. (Figs. 24 and 25) The lentiform nucleus is composed of the putamen laterally and the globus pallidus medially. Once the putamen is removed one can see the temporal extension of the anterior commissure and the globus pallidus [1]. The globus pallidus appears paler in comparison to the putamen and hence its name which means "pale globe". The globus pallidus is seen to lie just behind the anterior commissure and faces the genu of the internal capsule medially and the posterior limb of the external capsule posteriorly. Removing the lentiform nucleus exposes the entire extent of the internal capsule, which is discussed in detail later. (Figs. 26 and 27) The external and internal capsules are capsules of the lentiform nucleus [2]. The anterior, superior and inferior limiting

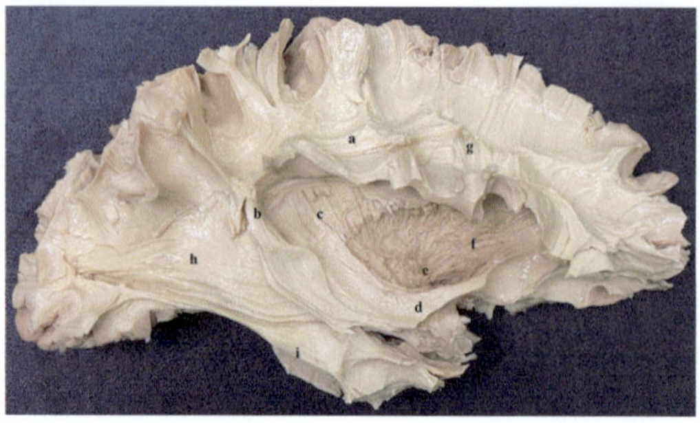

Fig. 24 Further dissection in the region of the insula. A part of the putamen has been removed revealing the internal capsule. a. Lateral part of the superior longitudinal fasciculus, b. Arcuate fasciculus, c, Dorsal external capsule, d. Inferior fronto-occipital fasciculus, e. Part of the putamen has been removed, f. Fibers of the internal capsule, g. Corona radiata, h. Sagittal stratum, i. Inferior longitudinal fasciculus

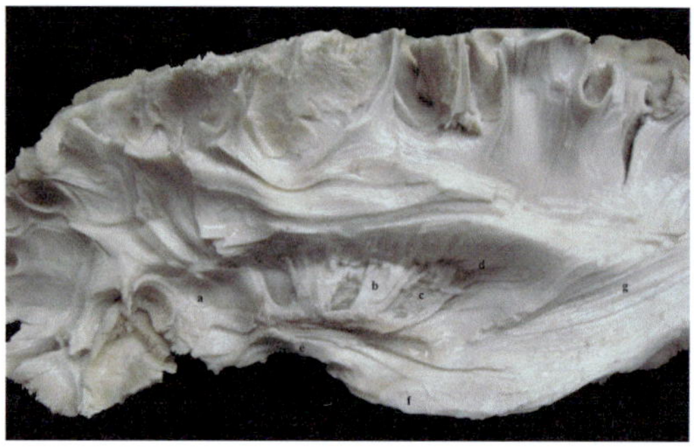

Fig. 25 Insular dissection showing the fibers. a. Extreme capsule. b. Dorsal external capsule, c. Putamen, d. Internal capsule, e. Inferior fronto-occipital fasciculus, f. Meyer's loop of optic radiations, g. Optic radiations

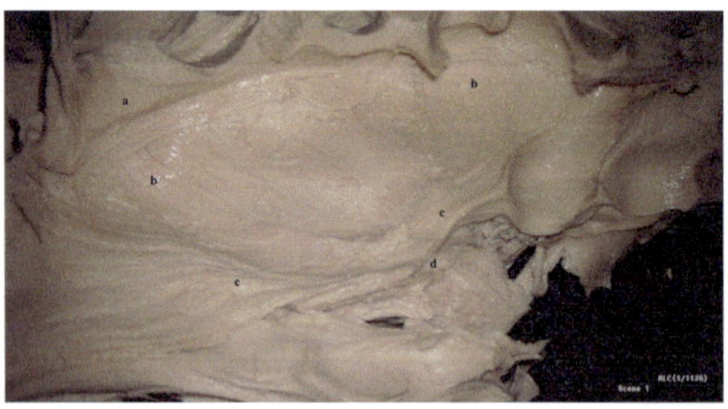

Fig. 23 Magnified image showing the fibers in the region of the insula. The extreme capsule has been removed to show the external capsule. a. Arcuate fasciculus, b. Dorsal external capsule, c. Inferior fronto-occipital fasciculus, d. Uncinate fasciculus

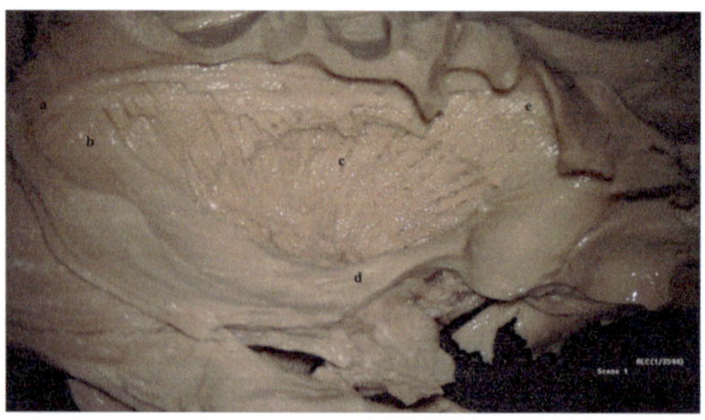

Fig. 26 The putamen has been removed to show the internal capsule. a. Arcuate fasciculus, b. Dorsal external capsule, c. Internal capsule, d. Inferior fronto-occipital fasciculus, e. Fibers of the external and internal capsule merging to form the corona radiata

Fig. 27 Image showing the entire extent of the internal capsule

sulci are very closely related to the internal capsule medially. The anterior insular point (junction of the anterior and superior limiting sulci) is related to the anterior limb of the internal capsule. The posterior insular point (junction of the superior and inferior limiting sulci) is closely related to the posterior limb of the corticospinal tract and superior thalamic radiation and the inferior limiting sulcus is medially related to the optic and auditory radiations.

8 Limbic Lobe

The limbic system is phylogenetically, one of the more primitive parts of the brain and is composed of both gray and white matter. It forms an evolutionary bridge between the primitive olfactory cortex and the neocortex.

The term "limbus" means 'border' or 'margin' in Latin. This region of the brain was first pictured by Thomas Willis and he referred to it as the limbus as early as in 1664 [1]. In 1878 Paul Broca referred to it as '*le grande lobe limbique*' and described it as a cortical ring consisting of the cingulate and parahippocampal gyrus surrounding the hilus of the hemispheres [1]. In his classical paper "*A Proposed Mechanism of Emotion*", James Papez postulated that the anatomical basis of emotions was based in the hypothalamus, anterior thalamic nuclei, cingulate gyrus, hippocampus and in their interconnections [7]. MacLean in 1949, termed these cortical and subcortical systems and their fiber connections as the limbic system [4]. Using silver degeneration techniques, Nauta extended the limits of limbic system to include the midbrain [4].

The various components of the limbic system are arranged in a characteristic 'C' shaped architecture.

As the mammalian CNS evolved, the olfactory lobe, hippocampus and amygdala developed initially over the primitive CNS, which consisted of the spinal cord, brainstem, hypothalamus and the basal ganglia. The olfactory lobes developed to sense the world and the hippocampus and the amygdala along with the hypothalamus developed to interpret and transform this information and modulate behavior [2]. Since these structures surrounded the top of the primitive CNS they were called limbic (limbus = ring) [4]. The limbic lobe contains portions of the archicortex (amygdala and hippocampus), paleocortex (olfactory lobe) and mesocortex (parahippocampal gyrus, cingulate gyrus and the insula).

The gray matter component of the limbic system is composed of the hippocampal formation, amygdaloid nuclei, hypothalamus, nucleus accumbens, cingulate cortex, major areas of the prefrontal cortex, and nuclei of the midbrain [4]. The fiber bundles that are components of the limbic system include the cingulum, longitudinal striae, fornix, anterior commissure, mamillothalamic and mamillotegmental tracts, stria terminalis, striae medullaris thalami, diagonal band of Broca, fasciolus retroflexus, ansa peduncularis, uncinate fasciculus, dorsal longitudinal fasciculus and the medial forebrain bundle [4]. (Fig. 28) The fiber connections of the limbic system will be discussed in a later chapter.

The cingulate gyrus begins in the subcallosal region below the rostrum of the corpus callosum and then follows the curve of the corpus callosum running superiorly over it between the cingulate sulcus superiorly and the callosal sulcus inferiorly. (Figs. 5 and 15) The cingulate gyrus becomes narrow just posterior to the splenium, where it is known as the isthmus of the cingulum and then curves inferiorly to continue as the parahippocampal gyrus. Subcallosally the cingulate gyrus is continuous with the paraolfactory or subcallosal gyrus. The cingulate sulcus curves superiorly as a terminal ascending branch that

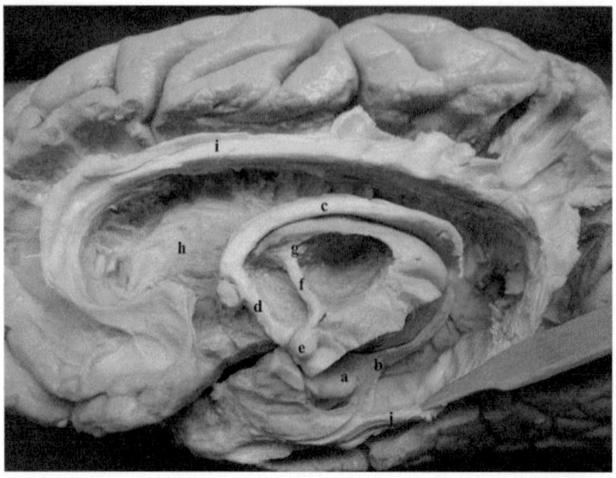

Fig. 28 Dissection of the medial half of the hemisphere to show the various components of the limbic system. a. Hippocampus, b. Fimbria, c. Fornix, d. Columns of the fornix, e. Mammillary body, f. Mammilothalamic tract, g. Anterior nucleus of the thalamus, h. Anterior thalamic radiations, i. Cingulum, j. Radiation of the cingulum

separates the paracentral lobule from the precuneus. (Figs. 5 and 15) The cingulate gyrus is separated from the precuneus by the subparietal sulcus.

The parahippocampal gyrus lies at the junction of the medial and inferior portions of the temporal lobe. Anteriorly and medially it curves upon itself to form the uncus. (Fig. 15) Laterally and anteriorly the parahippocampal gyrus is limited by the rhinal sulcus, which marks the lateral extent of the entorhinal area of the parahippocampal gyrus. The entorhinal area, which, constitutes the anterior one-third portion of the parahippocampal gyrus, is bounded laterally by the rhinal sulcus and posteriorly by the collateral sulcus. The collateral sulcus separates the parahippocampal gyrus from the fusiform gyrus. (Fig. 17) Posteriorly, the parahippocampal gyrus is divided by the anterior calcarine sulcus into the isthmus of the cingulate gyrus superiorly which continues as the cingulate gyrus and the parahippocampal gyrus continuing as the lingual gyrus inferiorly [2, 3]. (Fig. 15).

The uncus is triangular in shape with its tip facing medially. It has an anteromedial surface, which, faces the carotid cistern, and a posteromedial surface, which, faces the cerebral peduncle. (Figs. 7 and 17) The anterior half of the uncus harbors the amygdala and its posterior half harbors the head of the hippocampus. The anterior half of the uncus is made up of two gyri, superiorly the semilunar gyrus and inferiorly the ambient gyrus. The two gyri are separated by the semilunar sulcus in which the anterior choroidal artery runs. The semilunar gyrus is the lateral continuation of the lateral olfactory stria [2]. The posterior half of the uncus is made up of structures that correspond to the extraventricular portion of the head of the hippocampus, the uncinate gyrus anteriorly and the uncal apex posteriorly separated by the band of Giacomini [2, 4]. The uncal apex is also known as the intralimbic gyrus or the hippocampus inversus. The band of Giacomini represents the tail of the dentate gyrus.

The parahippocampal gyrus extends from the inferior surface medially towards the hippocampal fissure where it curves inwards to form the hippocampus, which in turn again curves inwards to form the dentate gyrus. The portion of the parahippocampal gyrus at the edge of the hippocampal fissure is known as the subiculum and the area between the subiculum and the parahippocampal gyrus is known as the presubiculum.

The hippocampus rests on the subiculum on the superior surface of the parahippocampal gyrus [4]. The hippocampus is separated from the parahippocampal gyrus and the subiculum by the hippocampal fissure. Embryologically, the hippocampal formation is laid down along the hippocampal fissure. (Fig. 28) This fissure lies above and parallel to the choroidal fissure. As the temporal lobe develops, both these fissures are carried downwards and forwards to form a C shaped configuration [4]. The hippocampal formation does not develop to the same extent across the whole length of the hippocampal fissure. The dorsal part of the fissure is invaded by fibers of the corpus callosum and the hippocampal formation in this portion becomes a vestigeal band—the indusium griseum [4]. The inferior portion of the hippocampal fissure differentiates into the hippocampal formation, which consists of the hippocampus proper, the dentate gyrus and the subiculum. The hippocampal formation forms the medial part of the floor of the temporal horn. (Fig. 28) From the hippocampus, the fibers form a band known as the alveus (L.Trough or hollow) that is located on its dome. It continues as the fimbria. Lateral to the fimbria and separated from it by the fimbrio-dentate sulcus, lies the dentate gyrus. When viewed from the medial aspect, the dentate gyrus is visualized as a thin band of cortex with a toothed appearance that lies between the fimbria above and the hippocampal fissure below. The fimbria and the dentate gyrus run together posteriory till the splenium. Here, the dentate gyrus loses its denticulations and becomes the fasciolar gyrus which then runs superiorly over the corpus callosum. On the superior surface of the corpus callosum, the fasciolar gyrus becomes a thin lamina of gray matter, namely the indusium griseum, also known as the supracallosal gyrus. Running within the indusium griseum are two bands of myelinated fibers, the medial and lateral longitudinal striae of Lancisii. The indusium griseum and the longitudinal striae course over the entire corpus callosum and terminate in the paraterminal gyrus. The paraterminal gyrus then continues as the diagonal band of Broca [4].

The *amygdala* (Gr. Almond like) lies entirely within the confines of the uncus [4]. The uncus forms the external marker of the amygdala. Superiorly, the amygdala blends with the globus pallidus and there is no clear demarcation between this transition. Inferiorly, the amygdala forms a bulge into the roof of the temporal horn, towards the hippocampal head, anterior to the inferior choroidal point. The amydala and the hippocampal head frequently may be touching or fused with each other. Laterally, the entorhinal cortex is related to the amygdala and the uncus forms its medial relation.

The subcallosal, cingulate, subparietal, anterior calcarine, collateral and rhinal sulci are together also known as the limbic fissure.

9 White Fibers

The white fibers of the brain are of three types: (1) Association fibers that interconnect different cortical regions in the same hemisphere. (2) Commisural fibers connect the cortices of the two hemispheres across the midline and (3) Projection fibers connect the cortex with subcortical regions, the brainstem and spinal cord. The major association fibers are the short association fibers or the U fibers, the superior longitudinal fasciculus, the inferior longitudinal fasciculus, the uncinate fasciculus, the inferior occipitofrontal fasciculus and the cingulum. The corpus callosum forms the largest commisural fiber system. The other commisural fibers consist of the anterior commisure, the posterior commissure, the hippocampal commissure or the psalterium and the habenular commissure. The internal capsule and the thalamic radiations constitute the projection fiber system of the brain.

10 Classification of White Fibers of the Brain

We recently published a novel classification of the structural design of the white fibers of the brain based on the direction and depth of the fiber bundles [1, 8, 9]. We believe that this will help both the novice and the trained neurosurgeon in understanding the intangible anatomy of the white fibers of the brain and in preventing neurological deficits.

Architecturally, the white fibers of the brain can be divided into five groups (four horizontal and one vertical group) [1] (Figs. 29 and 30). The four horizontal layers are the superficial group, the middle group, the deep group, and the central group [1]. The first horizontal layer or the superficial group is that of the short association fibers, the U fibers, or the intergyral fibers. This layer is present through the entirety of the cerebral hemisphere and only after its removal can the next level be approached. The second horizontal layer or the middle group is that of the long association fibers that connect various regions of the cerebral hemispheres. These are the superior longitudinal fasciculus (SLF), the arcuate fasciculus (AF), the middle longitudinal fasciculus, the inferior longitudinal fasciculus (ILF) and the cingulum. Further deep are the UF, the IFOF, and the sagittal stratum [1]. The third horizontal layer or the deep group of fibers is that of the deep association fibers. These fibers lie within the ventricular

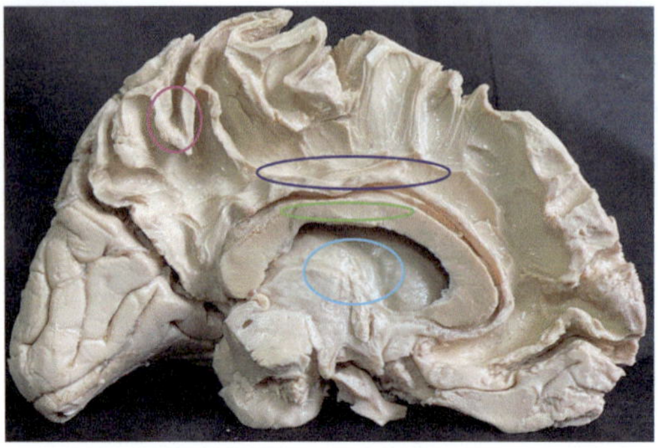

Fig. 29 Medial view of the hemisphere showing the classification of the white fibers into layers. Pink circle—Superficial layer of short arcuate fibers, Purple Circle—Middle layer of long association of fibers, Green circle—Central group of commissural fibers, Blue circle—Deep group of fibers

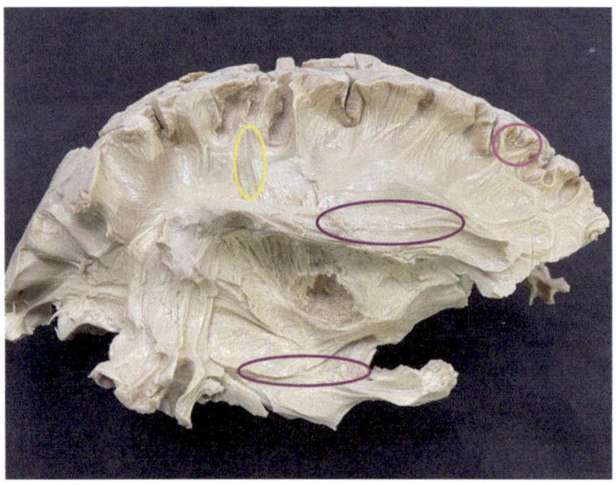

Fig. 30 Lateral view of the hemisphere showing the classification of the white fibers into layers. Pink circle—Superficial layer of short arcuate fibers, Purple Circle—Middle layer of long association of fibers, Yellow circle—Vertical group of projection fibers

system in close relation to the walls of the ventricular cavity. The other characteristic of most of these fibers is that they run in a distinctive "C" formation. These are the fornix, the stria terminalis, striae medullaris thalami, and the medial and longitudinal striae. There are 2 other fiber bundles that lie within the gray matter of the thalamus and the upper midbrain, which are included here. These are the mammillothalamic and the mammillotegmental tracts. The central group of fibers comprises the commissural fibers, which connect the two hemispheres to each other across the midline. These are the corpus callosum, the anterior commissure, the posterior commissure, the habenular commissure, and the hippocampal commissure. The vertical group of fibers are the projection fibers which run rostral to caudal connecting the cortex to the diencephalon and the brainstem. These are the internal capsule and the thalamic radiations. The vertical group of fibers divides the white fibers of the hemisphere into medial and lateral halves. Fibers that are medial to the vertical group do not cross over to the lateral side and those that are lateral to the vertical group do not cross over to the medial side. The only exception is the anterior commissure that traverses below and anterior to the vertical group of fibers as it enters the temporal brain. The deep gray matter of the brain lies straddled on both sides of the vertical group of projection fibers. The central group or the commissural fibers lie between the vertical layer of the projection fibers of both the hemispheres.

10.1 The First Layer/Superficial Group

10.1.1 Short Association or U Fibers

These fibers are visualized over the entire surface of the brain after coring out the gray matter. (Figs. 3, 29 and 30) They interconnect adjoining cortices over the whole hemisphere [1]. These fibers are damaged when a transcortical incision is placed over the hemisphere. Injury to the fibers can be avoided by planning a cortical incision in the direction of these fibers [1]. The turns of the short association fibers can be determined by seeing the adjacent sulcus. The fibers curve in the same direction as the adjacent sulcus.

10.2 The Second Layer/the Middle Group

10.2.1 Superior Longitudinal Fasciculus (SLF)

Course and Connections

The SLF is the most superficial of the long association fibers and runs between the frontal and parietal lobes. (Figs. 4 and 6) Traditionally the superior longitudinal fasciculus and the arcuate fasciculus were considered as part of a single fiber system and described as a single structure with a horizontal frontoparietal segment, a vertical temporo-parietal segment and a horizontal frontotemporal segment or the arcuate fasciculus [10–12]. A few other studies have described the superior longitudinal fasciculus as having three parts: SLF- I, SLF – II and SLF – III from medial to lateral [12]. Our dissections have revealed that the superior longitudinal fasciculus connects the frontal and parietal lobes and consists of two parts viz. the medial SLF and the lateral SLF which lies on either side of the projection fibers [1, 13]. The medial part of the SLF (also known as the SLF – I) begins at the superior frontal gyrus and runs posteriorly to reach the precuneus in the parietal lobe [1]. It runs above the cingulum and its lateral relations are the dorsal radiations of the corpus callosum and the corona radiata. The lateral part of the superior longitudinal fasciculus (SLF-II and SLF – III) connects the middle frontal gyrus to the angular gyrus (SLF – II) and the inferior frontal gyrus to the supramarginal gyrus (SLF – III). It is not very easy to separate these two portions of the SLF as they run between the frontal and parietal lobes as a confluent bundle [1]. The AF lies ventral to this portion of the lateral portion of the SLF.

Function

The SLF I connects the superior frontal gyrus to the superior parietal lobule. The main function of the SLF is to encode location of body parts in a body-centric coordinate system [14, 15]. Thus it seems that the SLF – I is involved in regulating motor behavior, especially conditional associative tasks which select among competing motor tasks based on conditional rules [14, 15]. The SLF II connects the dorsal prefrontal cortex to the caudal inferior parietal lobule. The inferior parietal lobule is known to play a role in spatial attention and visual awareness. Thus the SLF II provides the prefrontal cortex with information regarding perception of visual space [1]. The SLF III connects the supramarginal gyrus to the ventral premotor and prefrontal cortex. This suggests that the SLF III transfers information, such as language articulation, between the ventral premotor cortex, pars opercularis, the supramarginal gyrus and the lateral inferior prefrontal cortex. In the dominant hemisphere this pathway is important for the articulatory aspects of language and in the non-dominant hemisphere it subserves visuospatial attention, prosody and music processing [1].

10.2.2 Arcuate Fasciculus (AF)

Course and Connections

The AF is a reverse C shaped fiber bundle that has been traditionally known to connect the inferior frontal gyrus (Broca's area) and the posterior portion of the superior temporal gyrus (Wernicke's area) [1]. (Figs. 4, 19, and 20) In our dissections the AF was visualized as running beneath the lateral portion of the SLF in the frontal and parietal region then curving downwards in the region of the supramarginal and angular gyrus and then running forwards towards the posterior portion of the superior temporal gyrus [1]. (Figs. 19 and 20) Thus the AF wraps around the posterior edge of the insula in a horseshoe-like configuration beneath the frontal, parietal and temporal opercula and connects the middle and superior temporal gyri to the inferior and middle frontal gyri. It runs beneath the inferior portion of the superior temporal gyrus, the superior portion of the middle temporal gyrus, the angular and supramarginal gyri, the supra-sylvian portions of the pre- and postcentral gyri, and the posterior parts of the middle and inferior frontal gyri [1].

Function

Damage to the AF results in conduction aphasia [1]. This type of aphasia prevents the patient from repeating unfamiliar sounds.

The classical model of language consists of the Broca's area, Wernicke's area and the arcuate fasciculus connecting these two regions. Recently Hickok et al. proposed a dual stream model for language, the dorsal stream and the ventral stream [16]. The dorsal stream is involved with phonological processing and is mediated by the SLF II and III and the AF. The ventral stream of language is involved with semantic processing and consists of the IFOF, the ILF and the middle longitudinal fasciculus.

10.2.3 Ventral Occipital Fasciculus

Course and Connections

This fasciculus runs just posterior to the turn of the arcuate fasciculus. It connects the inferior occipital lobe to the superior occipital lobe and the angular gyrus [1].

Function

This tract is said to have an important role in reading and facial recognition.

10.2.4 Middle Longitudinal Fasciculus

Course and Connections

The existence of the middle longitudinal fasciculus in humans was much debated. Recent fiber dissection studies have clearly demonstrated its presence in the human brain [17]. The middle longitudinal fasciculus connects the superior temporal lobe with the parietal lobe. It is seen to originate in the region of the temporal operculum and is visualized after removing the fibers of the AF [1]. The fibers are seen to run posteriorly and slightly superiorly towards the parietal and occipital lobes and do not have a clear termination (Figs. 4 and 22). They form the uppermost portion of the sagittal stratum.

Function

The middle longitudinal fasciculus is thought to play a role in language processing, however damage to the fasciculus has not been observed to lead to any language impairment. [17]

10.2.5 Inferior Longitudinal Fasciculus (ILF)

Course and Connections

The ILF runs beneath the inferior temporal gyrus to the inferior portion of the occipital lobe. The fibers can be visualized after removing the gray matter of the inferior temporal gyrus and the fusiform gyrus [1]. (Figs. 4 and 22).

Function

The inferior longitudinal fasciculus forms a part of the ventral stream of language processing [1].

10.2.6 Inferior Fronto-occipital fasciculus (IFOF)

Course and Connections

Fibers from the lateral orbital gyri join fibers from the middle and inferior frontal gyri to form the IFOF. (Figs. 5) All the fibers aggregate and pass lateral to the corona radiata fibers and inferior to the lateral portion of the SLF and AF towards the temporal lobe as a constituent of the temporal stem [1]. (Figs. 9, 19, 20, 22, and 24) In the region of the insula IFOF and the UF form the ventral portion of the external capsule with the IFOF lying superiorly and the UF lying inferiorly. After exiting the temporal stem the majority of the IFOF passes through the superior and middle temporal gyri to reach the parietal and occipital gyri. Some fibers of the IFOF turn anteriorly along with the UF to end in the superior and middle temporal gyri. The fibers of the IFOF running posteriorly form a major component of the sagittal stratum and lie lateral to the optic radiations [1].

Function

The IFOF forms the main portion of the ventral semantic pathway. It is believed to be involved in semantic processing of language, which refers to making associations between spoken or written words [1]. It is also believed to have a role in reading, writing, comprehension and production of meaningful speech as well as visual recognition, integration of multimodal sensory input and motor planning. Intra-operative stimulation of the IFOF gives rise to semantic paraphasias (error in the meaning of the target word) [16, 18–20].

10.2.7 Uncinate Fasciculus (UF)

Course and Connections

The UF is a hook or U shaped tract running in the temporal stem connecting the orbitofrontal and temporal regions [1]. The UF originates in the medial and lateral orbital gyri, runs anteroinferior to the IFOF in the temporal stem and ends in the anterior aspect of the superior and middle temporal gyri. (Figs. 9, 19 and 20) The dorsolateral portion of the UF connects the temporal pole to the lateral orbitofrontal gyri and its ventromedial portion connects the temporal pole to the medial orbitofrontal cortex and the septal area. The frontal portion of the UF runs anterior to the anterior perforated substance deep to the posterior orbital gyrus. It covers the infero-medial portion of the nucleus accumbens and ultimately terminates in the region below the genu of the corpus callosum.

Function

The UF is considered to be part of the ventral limbic pathway [1]. The literature refers to three functions of the UF: associative and episodic memory functions, linguistic functions, and social–emotional functions [1, 2]. UF involvement is seen in five disorders namely: anxiety, schizophrenia, psychopathy, epilepsy and frontotemporal dementia [21, 22].

10.2.8 Sagittal Stratum and the Optic Radiations

Course and Connections

The sagittal stratum as the name suggests is a series of layers of fibers that can be seen running from the posterior temporal region to the occipital lobe deep to the AF. (Figs. 19, 20 and 24) It consists of a series of parallel running fibers, which are the MLF, the IFOF, the fibers of the anterior commissure and the optic radiations and the tapetum. The IFOF, fibers of the AF and the optic radiations form the bulk of the sagittal stratum. Though these three fiber systems are difficult to distinguish after they merge with the optic radiations, we have found that the fibers of the anterior commissure and the IFOF lie laterally and the optic radiations lie medially [1, 23]. The point where the three fiber systems converge was clearly identified in our fiber dissection previously [23]. At the point of intersection the fibers of the IFOF, the anterior commissure and the Meyer's loop can be identified from lateral to medial. The optic radiations or the geniculocalcarine tract originates at the lateral geniculate body and ends in the calcarine cortex on the medial aspect of the occipital lobe. [21] The optic radiations are composed of three bundles, anterior, central and posterior. Though by the general direction of the fiber bundles three main bundles can be distinguished it is not usually possible to clearly delineate the three bundles [1, 23]. The anterior bundle, also known as the Meyer's loop, travels anteriorly from the lateral geniculate body to the roof of the temporal horn and then turns posteriorly in the lateral wall of the temporal horn. It continues posteriorly to terminate at the inferior bank of the calcarine sulcus within the lingual gyrus. The central bundle courses laterally over the roof of the temporal horn and then courses posteriorly in the lateral wall of the trigone and occipital horn. It terminates at the occipital pole. The posterior bundle travels directly posteriorly over the trigone as part of the lateral wall and roof to end in the superior bank of the calcarine sulcus within the cuneus [21].

Function

The optic radiations carry visual information from the lateral geniculate body onwards to the visual cortex. They form the sublenticular (anterior bundle) and retrolenticular (central and posterior) portion of the internal capsule. The anterior bundle or Meyer's loop carries visual fibers from the superior quadrant of the contralateral visual field. Damage to these fibers causes a "pie in the sky" (contralateral superior quadrantanopia) type of visual field deficit [1]. The central bundle carries visual inputs from the macula and the posterior bundle is composed of fibers from the inferior quadrant of the contralateral visual field. Damage to the posterior bundle causes a contralateral inferior quadrantanopia or a "pie in the floor" type of visual field deficit.

10.2.9 Cingulum

Course and Connections

The cingulum forms a major component of the limbic system and the Papez circuit. It is a C shaped association bundle which, runs just above and parallel to the corpus callosum [4]. The cingulum originates below the rostrum of the corpus callosum in the region of the subcallosal gyrus [1]. (Figs. 6, 16 and 28) It then runs superiorly bends posteriorly above the genu of the corpus callosum and then continues posteriorly. Above the splenium, the cingulum narrows to form the isthmus of the cingulum, which continues below as the radiation of the cingulum to end in the anterior parahippocampal region adjacent to the hippocampus. The cingulum from its superior surface sends connections to the frontal and parietal cortices.

Function

The cingulum is a major component of the dorsal limbic pathway. Damage to this structure leads to behavioral and emotional disturbances. The cingulum is also a central structure in learning to correct mistakes, and is involved in appraisal of pain and reinforcement of behavior that reduces pain [1].

10.2.10 Medial and Lateral Olfactory Striae

Course and Connections

The medial and lateral olfactory striae represent the terminations of the olfactory tract on the basal surface of the frontal lobe [1]. (Fig. 5) The lateral olfactory striae form the anterolateral border of the anterior perforated substance. The lateral olfactory stria and gyrus pass along the lateral margin of the anterior perforated substance to reach the piriform region [1, 4–6]. These fibers terminate in the piriform cortex and in the corticomedial part of the amygdaloid nuclear complex. The medial olfactory stria becomes continuous with the subcallosal and the paraterminal gyrus. The subcallosal area and the paraterminal gyrus constitute the septal area, beneath which are the septal nuclei [1, 4–6]. The septal region is situated on the medial surface of the cerebral hemisphere, immediately facing the anterior commissure. The medial septal nucleus becomes continuous with the nucleus and tract of the diagonal band of Broca, which then connects with the hippocampal formation and the amygdala. Thus the diagonal band of Broca connects the amygdala to the medial septal region.

Function

Functionally, the septal nuclei are responsible for connecting limbic structures with the hypothalamus and the brainstem, principally via the hippocampal formation.

10.3 The Central Group

10.3.1 Corpus Callosum

Course and Connections

The corpus callosum is the largest commissural fiber system of the brain [1, 9]. When viewed on the medial surface of the brain it appears as a horizontally placed letter “C”. (Fig. 16) It is divided into 4 parts, the rostrum, the genu, the body and the splenium. Superiorly, a thin sheet of gray matter, the indusium griseum covers the surface of the corpus callosum. The indusium griseum is the continuation of the subsplenial gyrus on the surface of the corpus callosum, which ends in the subcallosal region. Within the indusium griseum are two longitudinal striae called as the medial and longitudinal striae of Lancisi. These fibers are considered to be aberrant fibers of the fornix, which leave the fimbria course over the superior surface of the corpus callosum and then join the fornix again anteriorly. Once this layer is removed the transverse running fibers of the corpus callosum are visualized. The transverse running fibers of the corpus callosum curve superiorly and inferiorly and when viewed from a posterior perspective resemble two horizontally placed letters “C” placed one on top of the other. When viewed from above the fibers of the corpus callosum resemble upturned bicycle handles running from one corona radiata to the other corona radiata connecting both the hemispheres [1]. Thus the corpus callosum lies between the vertical layer of fibers in each hemisphere. This whole constellation of fibers of the horizontal spread of the corpus callosum and the two vertical fiber layer of the projection system on each side resembles a hanging rope bridge [1]. Unlike a rope bridge however the transfer of information occurs from side to side rather than from anterior to posterior. The transverse running fibers of the corpus callosum after crossing the cingulum turn anteriorly, superiorly, posteriorly and inferiorly to form the callosal radiations. The fibers running superiorly are known as the dorsal callosal fibers and the fibers running inferiorly are known as the ventral callosal fibers. The fibers arising from the genu of the corpus callosum, curve anteriorly to form the forceps minor. The forceps minor connect the prefrontal and the orbitofrontal areas and form the medial wall of the frontal horn of the lateral ventricle. The callosal fibers emanating from the rostrum form the floor of the frontal horn of the lateral ventricle. These fibers reach the temporal stem on either side of the hemisphere thus connecting the temporal lobes. The horizontal fibers from the body of the corpus callosum course laterally and then curve upwards to merge with the fibers of the corona radiata. They connect the frontal and parietal lobes of each hemisphere to the other. The callosal fibers of the splenium, curve posteriorly to form the forceps major. The forceps major connect the occipital lobes to each other. They form a bulge in the medial wall of the atrium of the lateral ventricle. The ventral fibers emerging from the genu of the corpus callosum form the anteromedial portion of the medial wall of the frontal horn. These fibers also connect the bilateral caudate nuclei to each other [1, 9]. The ventral fibers arising from the body of the corpus callosum curve inferiorly and form the roof of the superior part of the frontal horn and the body of the lateral ventricle. The ventral sheet of fibers that arise from the ventral portion of the splenium run laterally and then curve downwards to form the superior and lateral wall of the atrium and temporal horn. These fibers are known as the tapetum. They separate the temporal and the atrial lateral walls from the optic radiations. Thus the corpus callosum connects similar regions of the hemispheres to each other. It also encompasses the entire ventricular system within its callosal radiations.

Function

The main function of the corpus callosum is to integrate and transfer information from both cerebral hemispheres to process sensory, motor, and high-level cognitive signals [1]. The corpus callosum seems to be topographically organized, involved in the transfer of visual, auditory, and somatosensory information in posterior regions, and higher cognition in anterior regions. The anterior callosal fibers transfer motor information between the frontal lobes, and the posterior fibers are involved in the processing of somatosensory cues (posterior midbody), as well as auditory and visual cues (splenium) by connecting the parietal, temporal, and occipital lobes. The corpus callosum is known to play a vital role in refining motor movements and cognitive functions. Anterior callosal disconnection can cause left unilateral motor apraxia, crossed optic ataxia, agraphia of the left hand, right unilateral constructional apraxia or alien hand syndrome. Posterior callosal disconnection can cause left visual anomia, left hemialexia, left auditory anomia, left tactile anomia and right olfactory anomia [1].

10.3.2 Anterior Commissure

Course and Connections

The anterior commissure resembles a horizontally placed bow and runs anterior to the columns of the fornix [1, 4, 23–25]. (Figs. 31 and 32) The anterior commissure runs in a canal of gray matter (canal of Gratiolet) parallel and deep to the UF and the IFOF [1]. It lies deep to the anterior perforated substance. The fibers of the anterior commissure divide into an anterior and posterior component. Only a small bulk of the fibers run anteriorly while most of the compact bulk of this white fiber tract runs posteriorly. The anterior

Fig. 31 Deep dissection in the region of the insula showing the temporal part of the anterior commissure after removal of the putamen. a. Anterior commissure, b. Inferior fronto-occipital fasciculus, c. Internal capsule

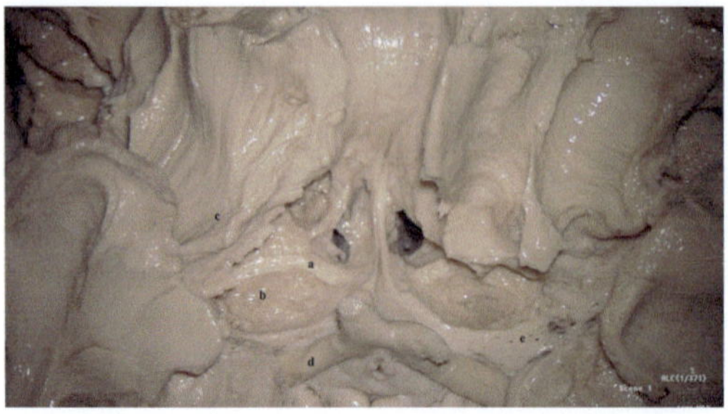

Fig. 32 Basal surface of the frontal lobe showing the anterior commissure. a. Anterior commissure, b. Ventral striatum, c. Uncinate fasciculus, d. Optic tract, e. Anterior perforated substance

fibers connect the anterior olfactory nucleus of one side to the contralateral side. The posterior portion of the anterior commissure runs postero-laterally deep to the lentiform nucleus and middle temporal gyrus and ultimately forms a component of the sagittal stratum. The globus pallidus lies posterior to the temporal component of the anterior commissure. As discussed earlier the anterior commissure is the only fiber bundle that crosses the vertical layer of the projection system.

Function

The role of the anterior commissure is not very well defined. It is a major component of the limbic system and is said to play a role in the inter-hemispheric transfer of visual, auditory, and olfactory information between temporal lobes [1]. It is said to have a key role in pain sensation, sense of smell and chemoreception [25]. Thus the anterior commissure seems to be involved in memory, emotion, speech and hearing, olfaction, instinct, and sexual behavior.

10.3.3 Hippocampal Commissure/Psalterium

Course and Connections

The psalterium (named after an ancient stringed instrument) has been known by various names such as the hippocampal fissure, the commissure of the fornix, forniceal commissure or the inter-ammonic commissure [1]. Anatomically, the psalterium consists of a triangular space between the two crura of the fornix through which run transverse fibers connecting the crura of the two fornices to each other [26]. The psalterium runs beneath the splenium of the corpus callosum and although is in intimate contact with it, it can be separated and distinguished from it by blunt dissection. Beneath the hippocampal commissure lies the velum interpositum and the roof of the third ventricle.

Function

The psalterium normally includes axons from the paleocortex and the subiculum and only a few neurons from the neocortex [1]. It is involved in transmission of information between the two hippocampi [1]. Damage to the psalterium results in memory disturbances.

10.4 The Vertical group

10.4.1 Internal Capsule

Course and Connections

The internal capsule is the largest projection fiber system of the brain and consists of five parts, the anterior limb, the genu, the posterior limb, the retro-lenticular portion and the sub-lenticular portion [1]. (Fig. 27) The internal capsule consists of the corticobulbar fibers, the corticospinal fibers and the thalamic radiations [1]. The corticobulbar and corticospinal tracts lie laterally and the thalamic radiations lie medially. When dissecting from a lateral approach via the insula one encounters the corticospinal and corticobulbar tracts first and when dissecting from a medial perspective, that is, from the lateral ventricle the thalamic radiations are revealed first. (Fig. 33) The anterior limb of the internal capsule lies between the putamen and the head of the caudate nucleus. It consists of the anterior and superior thalamic radiations and the frontopontine fibers [1]. The genu of the internal capsule is the junction of its anterior and posterior limbs and it lies just behind the beginning of the temporal extension of the anterior commissure and in the same coronal plane as the foramen of Monroe [2]. It consists of the superior thalamic radiations, the corticobulbar fibers and the anterior portion of

Fig. 33 Dissection of the medial aspect of the hemisphere showing the thalamic radiations. a. Anterior thalamic peduncle, b Superior thalamic peduncle, c. Posterior thalamic peduncle, d. Inferior thalamic peduncle, e. Thalamus

the corticospinal tract. The posterior limb of the internal capsule lies between the globus pallidus and the thalamus. It consists of the posterior part of the corticospinal tract, the parietal thalamic radiations and the parietopontine fibers. The retrolenticular portion of the internal capsule includes the parietopontine fibers and some portion of the occipital thalamic radiations. The sublenticular portion of the internal capsule consists of the occipitopontine, the temporopontine and the major bulk of the occipital thalamic radiation [1]. The anterior bundle of the optic radiations (occipital thalamic and temporal thalamic radiations) lies in the sublenticular portion of the internal capsule and the central and posterior bundles lies in the retrolenticular portion of the internal capsule [1, 23].

Function

The internal capsule is involved in transmitting motor, sensory and behavioral information from the higher cortical centers to the brainstem and the spinal cord [1]. Damage to the anterior limb of the internal capsule causes behavioral disturbances and damage to the genu and the posterior limb causes contralateral motor and sensory deficits [1].

10.4.2 External Capsule

Course and Connections

The external capsule is a vertical claustro-cortical projection fiber system that connects the dorsal external capsule to the frontal and parietal cortices. (Fig. 5) It lies between the gray matter of the claustrum laterally and the putamen medially.

Function

The external capsule and the dorsal claustrocortical system is involved in the integration of visual, somatosensory and motor information [1]. Bilateral damage to the external capsule causes severe encephalopathy.

10.4.3 Corona Radiata

Course and Connections

The external and internal capsule fibers join together at the upper edge of the putamen to form the corona radiata. The corona radiata run through the majority of the frontal and parietal lobes and bisects the middle group of association fibers into medial and lateral halves [1]. At its medial margin the callosal fibers join the corona radiata to form the centrum semiovale.

Function

Damage to the corona radiata causes varying degrees of contralateral motor and sensory deficits.

10.5 The Third Layer/Deep Group

10.5.1 Fornix

Course and Connections

The fornix is the most prominent fiber system in this group of fibers [1]. The fornix begins on the surface of the hippoacampus as the alveus and the fimbria. The fimbria then curve superolaterally over the pulvinar of the thalamus to form the crus of the fornix (Latin meaning "vault or arch") (Fig. 28) [1, 4]. Further anteriorly, the two crura meet in the midline to form the body of the fornix. The two crura of the fornices are connected to each other by the hippocampal commissure or psalterium. The body of the fornix runs forward to the anterior end of the thalamus, where the fibers of the two sides again separate and arch downwards in front of the intraventricular foramen as the columns of the fornix [1, 4]. The fornix constitutes the main efferent pathway from the hippocampus, though it also carries a few afferent fibers [3]. The fibers of the fornix on approaching the anterior commissure divide into a pre commissural and a post commissural portion. The majority of the fibers constitute the postcommissural portion [1, 4]. These fibers arise mainly from the subiculum and terminate into the mammillary body. These fibers reach the mammillary body and en route also give fibers to the lateral and anterior nucleus of the thalamus and to the lateral septal nuclei. Thus, the anterior nucleus of the thalamus receives fibers from the mamillothalamic tract and also receives direct fibers from the post-commissural fornix. A few fibers pass ventral to the anterior commissure as the pre-commissural fornix [4]. These fibers originate mainly from the pyramidal cells of the hippocampus. Their input is principally to the septal nuclei, the lateral pre-optica rea, the anterior part of the hypothalamus and the nucleus of the diagonal band [1]. The pre-commisural fibers form a small compact bundle that cannot be detected grossly. The post commissural fornix terminates into the medial mammillary nucleus.

Function

The fornix is the major fiber bundle involved in memory processing. Damage to this structure leads to amnesia.

10.5.2 Mammillothalamic and Mammillotegmental Tracts

The mammillothalamic tract of Vicqd'Azyr consists of fibers arising from the medial mammillary nucleus. The fibers climb upwards to the anterior nucleus of the thalamus and from there onwards to the anterior thalamic radiations. Near its origin, the mammillothalamic tract gives rise to another smaller bundle known as the mammillotegmental tract. (Fig. 28) The mammillotegmental tract runs posteriorly and inferiorly towards the brainstem and then terminates into the tegmental nucleus of Gudden [4]. From the anterior nucleus of the thalamus, fibers of the anterior thalamic radiation diverge upward and mingle with the fibers of the anterior limb of the internal capsule to reach the cingulum (Latin meaning "girdle") and the cingulate gyrus (Figs. 28 and 33) [1, 4].

10.5.3 Stria Terminalis

The stria terminalis is a C shaped association fiber that connects the amygdala and the medial forebrain structures. The amygdala gives rise to two major pathways, the stria terminalis and the ventral amygdalofugal pathway [1, 4]. The stria terminalis runs between the caudate nucleus and the thalamus around the lateral ventricle. The fibers terminate in the bed nuclei of the stria terminalis, which lies lateral to the columns of the fornix and superior to the anterior commissure [1]. Some fibers also terminate in the hypothalamus and some join the medial forebrain bundle [3].The bed nucleus of the stria terminalis has widespread connections to the rest of the limbic system and therefore provides an important route by which the amygdala can affect structures beyond the connections of the stria terminalis itself.

10.5.4 Stria Medullaris Thalami

This fiber tract runs along the dorsomedial surface of the thalamus in the floor of the lateral ventricle [1]. It runs lateral to the lower lateral edge of the foramen of Monroe near the anterior commissure. It connects the septal area with the habenula. The impulses from the habenular nuclei are then conveyed via the fasciculus retroflexus to the midbrain. The medial forebrain bundle projects to the midbrain tegmentum, after originating from the septal and lateral preoptic region. It also carries ascending fibers from the midbrain. Ascending fibers to the mamillary body run in the mamillary peduncle. They arise from the dorsal and ventral nuclei of Gudden and traverse to the mamillary body. Some fibers continue rostrally to the lateral hypothalamic regions, medial septal nuclei and preoptic areas via the medial forebrain bundle.

References

1. Shah A, Goel A, Jhawar SS, Patil A, Rangnekar R, Goel A. Neural circuitry: architecture and function-a fiber dissection study. World Neurosurg. 2019;125:e620–38.

2. Ribas G. The cerebral architecture. In: Applied cranial-cerebral anatomy: brain architecture and anatomically oriented microneurosurgery. Cambridge: Cambridge University Press; 2018. p. 15–61. https://doi.org/10.1017/9781316661567.004.
3. Ribas GC. The cerebral sulci and gyri. Neurosurg Focus. 2010;28(2):E2.
4. Shah A, Jhawar SS, Goel A. Analysis of the anatomy of the Papez circuit and adjoining limbic system by fiber dissection techniques. J Clin Neurosci. 2012;19(2):289–98.
5. Goel A, Shah A, Ramdasi R, Patni N. Orbital cortical approach to lesions around the frontal horn of the lateral ventricle: indication and surgical parameters. ActaNeurochir (Wien). 2014;156(4):825–830.
6. Shah A, Dandpat S, Goel A. Anatomic correlates and surgical experience with Goel orbital cortical approach to caudate head tumors. World Neurosurg. 2021;148:e674–9.
7. Papez JW. A proposed mechanism for emotion. Arch Neurol Psychiatr. 1937;38:725–43.
8. Goel A, Shah A, Vutha R, Dandpat S, Hawaldar A. Is "En Masse" tumor resection a safe surgical strategy for low-grade gliomas? Feasibility report on 74 patients treated over four years. Neurol India. 2021;69(2):406–13.
9. Shah A, Jhawar S, Goel A, Goel A. Corpus callosum and its connections: a fiber dissection study. World Neurosurg. 2021;151:e1024–35.
10. Fernández-Miranda JC, Rhoton AL Jr, Alvarez-Linera J, Kakizawa Y, Choi C, deOliveira EP. Three-dimensional microsurgical and tractographic anatomy of the white matter of the human brain. Neurosurgery. 2008;62(6 Suppl 3):989–1026. discussion 1026–8
11. Fernandez-Miranda JC, Pathak S, Engh J, Jarbo K, Verstynen T, Yeh FC, et al. High-definition fiber tractography of the human brain: neuroanatomical validation and neurosurgical applications. Neurosurgery. 2012;71(2):430–53.
12. Yagmurlu K, Vlasak AL, Rhoton AL Jr. Three-dimensional topographic fiber tract anatomy of the cerebrum. Neurosurgery. 2015;11(Suppl 2):274–305. discussion 305
13. Shah A, Goel A, Patil A, Goel A. Letter to the editor. Superior longitudinal fasciculus. J Neurosurg. 2019;132(4):1309–11.
14. Makris N, Kennedy DN, McInerney S, Sorensen AG, Wang R, Caviness VS Jr, et al. Segmentation of subcomponents within the superior longitudinal fascicle in humans: a quantitative, in vivo, DT-MRI study. Cereb Cortex. 2005;15(6):854–69.
15. Kamali A, Flanders AE, Brody J, Hunter JV, Hasan KM. Tracing superior longitudinal fasciculus connectivity in the human brain using high resolution diffusion tensor tractography. Brain Struct Funct. 2014;219(1):269–81.
16. Hickok G, Poeppel D. Towards a functional neuroanatomy of speech perception. Trends Cogn Sci. 2000;4:131–8.
17. Maldonado IL, de Champfleur NM, Velut S, Destrieux C, Zemmoura I, Duffau H. Evidence of a middle longitudinal fasciculus in the human brain from fiber dissection. J Anat. 223(1):38–45.
18. Makris N, Papadimitriou GM, Sorg S, Kennedy DN, Caviness VS, Pandya DN. The occipitofrontal fascicle in humans: a quantitative, in vivo. DT-MRI study Neuroimage. 2007;37(4):1100–11.
19. Schmahmann JD, Pandya DN. The complex history of the fronto-occipital fasciculus. J Hist Neurosci. 2007;16(4):362–77.
20. Almairac F, Herbet G, Moritz-Gasser S, de Champfleur NM, Duffau H. The left inferior fronto-occipital fasciculus subserves language semantics: a multilevel lesion study. Brain Struct Funct. 2015;220(4):1983–95.
21. Kier LE, Staib LH, Davis LM, Bronen RA. MR imaging of the temporal stem: anatomic dissection tractography of the uncinate fasciculus, inferior occipitofrontal fasciculus, and Meyer's loop of the optic radiation. Am J Neuroradiol. 2004;25(5):677–91.
22. Jones DK, Catani M, Pierpaoli C, Reeves SJ, Shergill SS, O'Sullivan M, et al. A diffusion tensor magnetic resonance imaging study of frontal cortex connections in very late onset schizophrenia-like psychosis. Am J Geriatr Psychiatry. 2005;13(12):1092–9.
23. Shah A, Jhawar SS, Goel A. Letter to the editor: optic radiations and anterior commissure. J Neurosurg. 2015;123(3):824–6.
24. Peltier J, Verclytte S, Delmaire C, Pruvo JP, Godefroy O, Le Gars D. Microsurgical anatomy of the temporal stem: clinical relevance and correlations with diffusion tensor imaging fiber tracking. J Neurosurg. 2010;112(5):1033–8.
25. Wilde EA, Bigler ED, Haider JM, Chu Z, Levin HS, Li X, et al. Vulnerability of the anterior commissure in moderate to severe pediatric traumatic brain injury. J Child Neurol. 2006;21(9):769–76.
26. Tubbs RS, Bosmia AN, Gupta T, Chawla K, Loukas M, Sahni D, Cohen-Gadol AA. The enigmatic psalterium: a review and anatomic study with relevance to callosotomy procedures. Neurosurgery. 2015;11(Suppl 2):322–8.

Impact of White Matter Dissection in Microneurosurgical Procedures

Abuzer Güngör, Muhammet Enes Gurses, Oğuz Kağan Demirtaş, Serdar Rahmanov, and Uğur Türe

1 History of White Matter Fiber Dissection

White matter fiber dissection is a technical procedure used in neuroanatomy studies to gain a comprehensive three-dimensional understanding of both gray and white matter anatomy as well as deep nuclei [1]. While difficult and time consuming, this technique is critical for constructing a proper conceptual notion about accurate intrinsic brain anatomy and architecture. This method is the legacy of many early anatomists [2–8].

A. Güngör
Department of Neurosurgery, Yeditepe University School of Medicine, Istanbul, Turkey

Department of Neurosurgery, Istinye University, Istanbul, Turkey

M. E. Gurses
Department of Neurosurgery, Yeditepe University School of Medicine, Istanbul, Turkey

Department of Neurosurgery, Hacettepe University Hospital, Ankara, Turkey

O. K. Demirtaş
Department of Neurosurgery, Yeditepe University School of Medicine, Istanbul, Turkey

Department of Neurosurgery, Sincan Training and Research Hospital, Ankara, Turkey

S. Rahmanov · U. Türe (✉)
Department of Neurosurgery, Yeditepe University School of Medicine, Istanbul, Turkey

Andreas Vesalius was concerned with white matter, and named some parts of it in Book VII of his *Fabrica*, published in 1543 [9–11]. In his 1664 *Cerebri Anatome,* with drawings by Christopher Wren, Thomas Willis (1621–1675) set new standards by properly defining the anatomy of inner white matter components [12]. Raymond Vieussens (1641–1715), a French anatomist, reintroduced the fiber dissection technique [13] and detailed the process, publishing an atlas based on it in 1685 [8] (Fig. 1a). In 1669, Marcello Malpighi (1628–1694) showed that white matter is made up of fibers [17], and Niels Stensen (1638–1686) investigated white matter tracts in 1671 by following "nerve threads" through the brain material to see where they headed [9, 17]. Sir Charles Bell (1774–1842), an Edinburgh anatomist and surgeon, published his brain atlas in 1802 [3]. He illustrated his anatomic works with his engravings, demonstrating his exceptional artistic ability. Johann Christian Reil (1759–1813), a German psychiatrist and neuroanatomist, published a journal in 1809 that showed the internal structures of alcohol-fixed brains as established by fiber dissection (Fig. 1b) [7]. With this technique, Reil gained better gross preservation and dissection. In the history of neuroanatomy, his use of alcohol to preserve and harden the brain was a watershed event. He was also the one who coined the term *corona radiate* [7, 18, 19].

A. Shah et al. (eds.), *Functional Anatomy of the Brain: A View from the Surgeon's Eye*,
https://doi.org/10.1007/978-981-99-3412-6_3

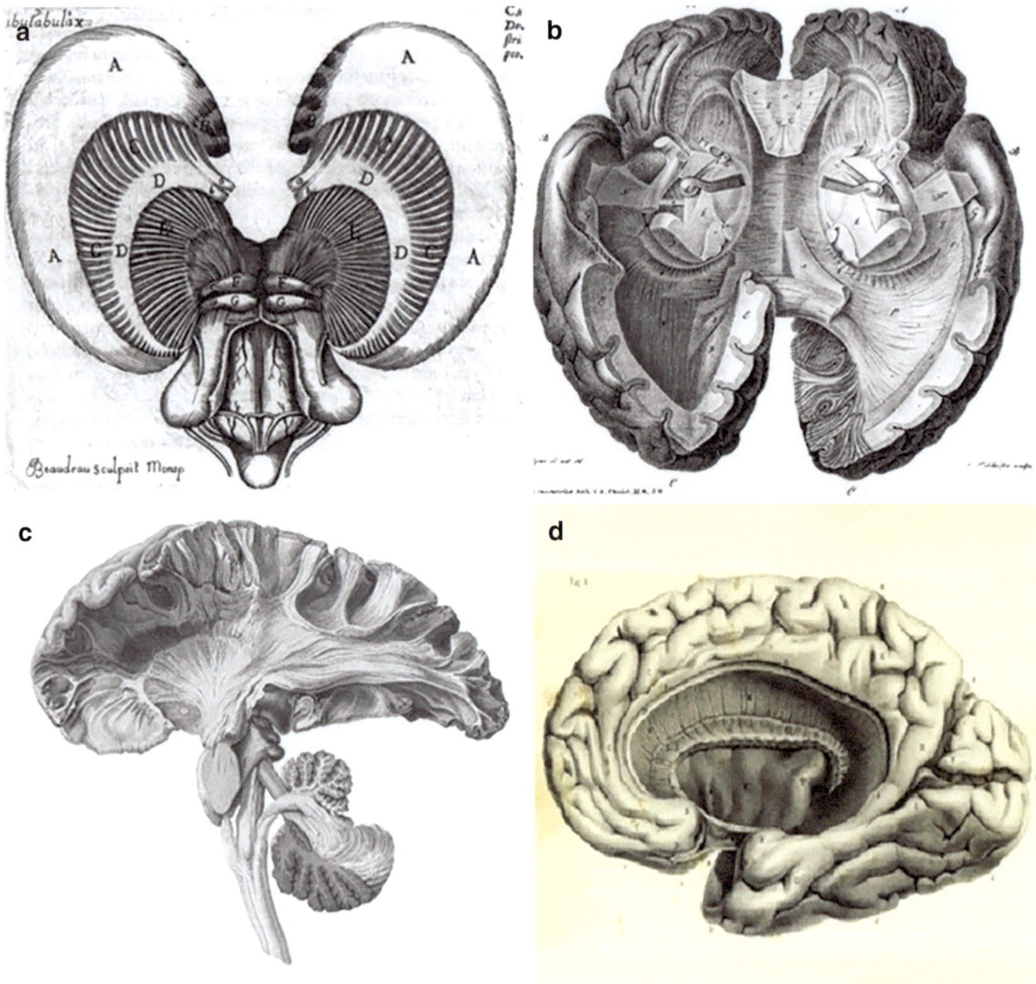

Fig. 1 (**a**) Illustration of the cerebrum internal structures by Raymond Vieussens (ca. 1635–1715). *Neurographia Universalis*. Lyons: Jean Certe, 1684 [8]. (**b**) Drawing the white matter pathways of cerebrum by JC Reil: Fragmente über die Bildung des kleinen Gehirns im Menschen. Arch Physiol Halle 8:1–58, 1807–1808 [7]. (**c**) Drawing by Mayo, demonstrating the internal structures of the brain (1827) *A series of engravings intended to illustrate the structure of the brain and spinal cord in man. Plate IV.* Burgess and Hill, London [6]. (**d**) Illustration of medial side of cerebrum with white matter pathways by ALF Foville: Traité Complet de l'Anatomie, de la Physiologie et de la Pathologie du Système Nerveux Cérébrospinal. Paris, Fortin, Masson et Cie, 1844 [23]

Franz Joseph Gall (1758–1828) and JC Spurzheim (1776–1832) distinguished the important aspects of gray and white matter in 1810 [4], while Karl Friedrich Burdach (1776–1847) focused his three-volume study *Vom Baue und Leben des Gehirns und Rückenmarks* on fiber bundles and identified complicated structures in 1819 [20, 21]. In 1827, Bell's student, English anatomist Herbert Mayo, published some of the best-dissected brain illustrations available at the time (Fig. 1c) [6]. Luigi Rolando (1773–1831), an Italian anatomist, was the first to precisely depict the cerebral sulci and convolutions, including the central sulcus, which carries his name in 1829 [14]. The frontopontine tract (also known as Arnold's tract) extends from the frontal cortex to the anterior limb of the internal capsule, via the medial part of the cerebral peduncle to the pons. This structure was originally demonstrated in 1838 by Friedrich Arnold (1803–1890) [22].

French neurologist Achille L. Foville (1799–1878) published an important treatise on the nervous system, which included an atlas with many magnificent dissections pictured in 1844 [23] (Fig. 1d). Despite his lack of fame, Foville's atlas is likely the most accurate, artistic, and high-quality publication in the neuroscience literature. In 1839–1857, French anatomist Louis Pierre Gratiolet (1815–1865) published an atlas of fiber-dissected brains with his master and friend Francois Leuret (1797–1851) (Fig. 2a) [26]. Gratiolet also explored and named the optic radiation (formerly known as Gratiolet's radiation), which extends from the lateral geniculate body to the visual cortex.

In Vienna in 1872, professor of neurology and psychiatry Theodor H. Meynert (1833–1892) improved Gall's relatively primitive separation of brain fiber systems, for the first time using the phrases "association" and "projection" fibers in their contemporary sense (1872) [27]. He also described the habenulointerpeduncular tract, or fasciculus retroflexus (Meynert's bundle).

Because fiber dissection is a complex and time-consuming method, it was nearly inevitable that it would be forgotten with the invention of the microtome and histological techniques. During the 1920s, white matter tracts began to gain importance in neurosurgery, thanks to Santiago Ramon y Cajal's expressive proposal of

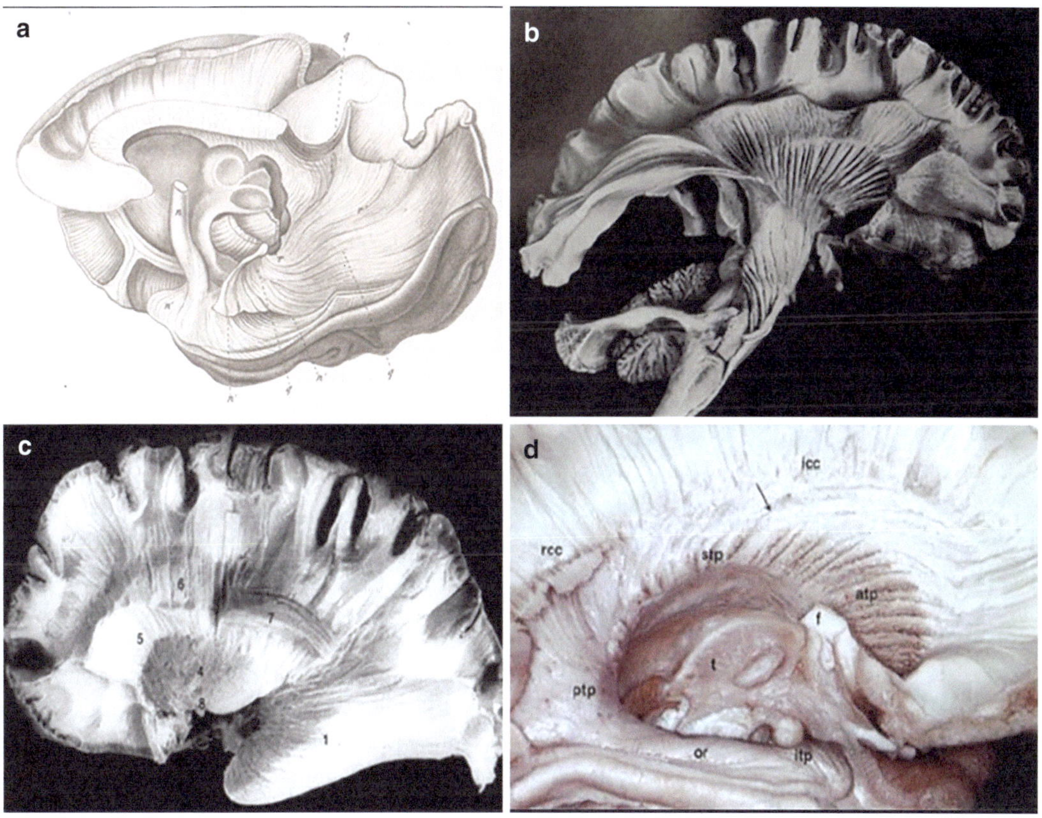

Fig. 2 (**a**) Drawing of white matter tracts by Leuret F, Gratiolet P Anatomie Comparée du Système Nerveux Considéré dans ses Rapports avec l'Intelligence. Paris, Baillière, 1857–1859, vol II [26]. (**b**) Ludwig, E. (Basel), Atlas cerebri humani**,** Der innere Bau des Gehirns, dargestellt auf Grund makroskopischer Präparate/The Inner Structure of the Brain Demonstrated on the Basis of Macroscopical Preparations, 1956 [29]. (**c**) A left-sided presumed optic radiation, which was dissected in 1953 by MG Yaşargil. Reprinted from Hüber A: Eye Symptoms in Brain Tumors, ed. 2, p 1. St. Louis: CV Mosby Co., 1971. With permission from Elsevier. (**d**) White matter dissection photo by Türe U. Türe U, Yaşargil MG, Pait TG. Is there a superior occipitofrontal fasciculus? A microsurgical anatomic study. Neurosurgery. 1997 Jun 1;40 (6):1226–1232 [60]

brain interconnection and circuitry [9]. JW Hultkrantz, a Swedish anatomist, produced an atlas in 1929 that included drawings of fiber-dissected brains and discussed his technique. He demonstrated Vieussens' use of wooden spatulas for dissection. These applied investigations were at the forefront of new tissue preservation techniques, as well as dissection and illustration procedures.

The most significant addition to the fiber dissection technique was provided by Joseph Klingler (1888–1963), an anatomy preparation technician in Basel [5, 28, 29]. In 1935, he developed the process of brain fixation and named the technique for himself [5]. He froze and thawed the brains to separate the fibers. In 1960, he produced a wonderful atlas on fiber dissection, which included thorough anatomic studies of the brain [28] Prof. M. Gazi Yaşargil came under the direct leadership and supervision of Klingler as a medical student in Basel in 1949, through Ludwig [1, 30] (Fig. 2b), and learned Klingler's dissection techniques. Yaşargil was studying three-dimensional brain anatomy using Klingler's models during this time. Despite working with Klingler for only 3 months, Professor Yaşargil was very impressed and decided to pursue a career in neurosurgery using the concepts he had learned. During his neurosurgery training, he brought a few formalin-fixed Klingler brains with him and frequently traveled to Basel to present his anatomic dissections to Klingler. Yaşargil credits his success in neurosurgery in part to his training under Klingler, who impressed him with his comprehensive knowledge, thorough white matter dissection abilities, and excitement with respect to the hippocampus [30]. The senior author of this paper, Dr. Uğur Türe, saw Klingler's specimens while visiting the Anatomical Museum of the University of Basel in December 1990 and acquired a significant interest in this topic. With the guidance and support of Professor Yaşargil, he revitalized the fiber dissection technique even though, in the first years, others had no interest. Later, the introduction of diffusion tensor imaging (DTI) tractography to delineate white matter fibers was widely accepted [1]. In addition, he added the use of the microscope, aspiration system, and step-by-step tutorial to the white matter dissection technique. During this period, Yaşargil and Türe published many articles describing the technique and organized the first hands-on fiber dissection course in St. Louis in 2003 (Fig. 2c, d).

Later, Türe organized many fiber dissection courses all over the world. Today, he and Yaşargil continue to offer these courses in Istanbul (http://istanbul-microneurosurgery.com/). Türe introduced this method to Evandro de Oliviera (1945–2021) and Albert L. Rhoton, Jr. (1932–2016) in the 2000s. They and their fellows have also published many studies of white matter [31–36]. In his commentary on an anatomical study, de Oliveira emphasized that this technique was taught to them by Türe [37]. Later, many groups interested in this subject were established around the world and many studies were conducted [38–42]. Now in the 2020s, Türe and his fellows are publishing many articles about white matter tracts [43–46], and fellows of Yaşargil and Türe continue to work on augmented reality, virtual reality simulations, and three-dimensional (3D) models of white matter [47–54].

2 White Matter Dissection Technique

Currently, the most common anatomical white matter dissection techniques are variations of the original technique published by Klingler in 1935 [28]. According to the initial technical note, the preparation procedure is a three-step process that follows a systematic approach. The three major procedures are usually the same in each laboratory, although the details of each vary. In the first step, the brain is removed from the body and fixed. In the second step, the specimen is frozen. Because the formalin solution did not fully penetrate the myelinated nerve fibers and was discovered at higher concentrations between the fibers, Klingler and colleagues recommended freezing the specimens before dissection [5, 28, 29]. Formalin ice crystals form between the nerve fibers when the specimens are frozen, where they swell and separate the fibers, easing the dissection of fine fiber bundles. The study may be interrupted overnight or longer after dissection has started, as long as the specimen is kept in a 5% formalin solution between sessions. If dissection is postponed for a month or more, the specimen should be frozen for at least 12 h and then thawed as indicated previously before the study can resume. In the third step, the specimen is thawed and white matter tracts are dissected. Klingler used homemade wooden spatulas for most of his dissections but did not use any magnification. The main white matter tracts within the cerebral hemisphere can be dissected with the naked eye, while smaller tracts, such as those within the brain stem, require microscopes and microsurgical equipment [5, 55].

Brains are taken from fresh autopsy specimens (up to 12 h after death) and fixed in a 10% formalin solution for at least 2 months. The basilar artery is ligated and used to suspend each brain in formalin to maintain normal contours. After 2 months, the pia mater, arachnoid membrane, and vessels of the specimens are carefully removed under an operating microscope. Brains are washed under running water for several hours to remove formalin, then filtered and cooled at −10° to −15 °C for 1 week. Before dissection is initiated, the brains are immersed in water and allowed to thaw. Dissection is done under an operating microscope at 6× to 40× magnification.

3 Main Fiber Systems of White Matter

The current classification of white matter fibers as commissural, projection, and association fibers was developed by Theodor Hermann Meynert (1833–1892) in 1885. The fibers interconnecting two distant cortical regions or adjacent cortical structures (short association or U fibers) within the same cerebral hemisphere are defined as *association fibers*. The white matter fibers that connect similar cortical regions in opposite cerebral hemispheres are called *commissural fibers*. Efferent and afferent fibers that provide communication between the cerebral cortex, subcortical structures, cerebellum, brain stem, and spinal cord structures are defined as *projection fibers* [27].

White matter fibers are much more complex than this simple classification, and the actual anatomy goes far beyond this classification. But to better understand the anatomy and connections, we have to simplify it. For this reason, the Meynert classification, which has a history of almost 150 years, continues to be used today.

A limitation of the Meynert classification is that it is based on supratentorial fibers. Thus, brainstem and cerebellar white matter pathways cannot be clearly classified according to this method. Although projection fibers pass especially from the brain stem, other fibers do not show a clear distinction as association or commissural.

3.1 Association Fibers

Short association fibers (U fibers): These fibers originate from the sulcal surface of the gyrus, located just underneath the cerebral cortex, and connect only with the neighboring gyri. Because of the evolutionary development in the lateral surface of the brain, namely the *neocortex*, short association fibers are highly developed. On the other hand, the U fibers on the medial surface of the hemisphere are much thinner and less abundant (Fig. 3a) [1, 24].

Superior longutudinal fasciculus/arcuate complex: The superior longitudinal fasciculis (SLF) is the main bundle system of the lateral hemisphere. It interconnects with the frontal, parietal, occipital, and temporal lobes. It is a C-shaped fiber bundle that rotates around the insular lobe and the Sylvian fissure [1] (Fig. 3b). Although the exact function of all these white matter pathways is still unknown, they have some predicted functions. The SLF/arcuate complex affects articulation, visuospatial attention, and lexical, semantic, and phonological language processing [24, 56]. This complex, the dorsal language stream, has important effects on speech [33, 34].

Extreme capsule: The extreme capsule, located just below the insular cortex, interconnects the adjacent insular and medial opercular gyri with short association fibers [1, 16, 46, 57]. Thus, these fibers provide connections between

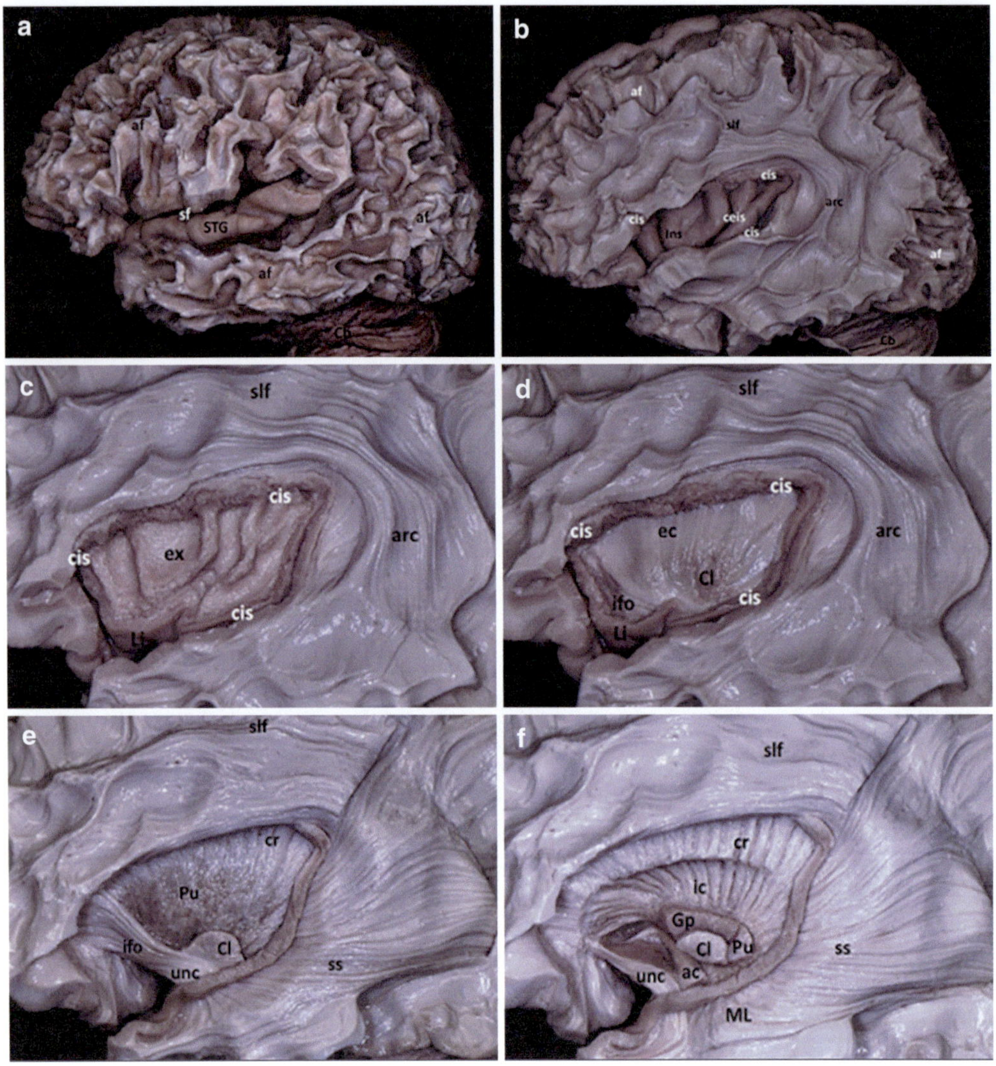

Fig. 3 Stepwise fiber dissection from the lateral to medial direction. (**a**) Lateral view after brain decortication. Removing the cortex reveals the short association fibers (U fibers). (**b**) Removing the short associative fibers exposes the superior longitudinal fasciculus and the arcuate fasciculus. The frontoparietal and temporal opercula are removed to expose the insula. The central insular sulcus divides the insular cortex into the short and long insular gyri and runs deep and almost parallel to the central sulcus on the bulge. (**c**) After decortication of the insular region, the insular U fibers or superficial layer of extreme capsule fibers are exposed. (**d**) Removing the U fibers exposes the deep layer of extreme capsule fibers, such as claustrocortical fibers, inferior fronto-occipital and uncinate fasciculi, and the claustrum. (**e**) The external capsule which consist of deep layer of fronto-occipital fasciculus, uncinate fasciculus and claustrocortical fibers, is between the claustrum and the putamen. Removing the external capsule exposes the putamen and the corona radiata fibers. The posterior part of the superior longitudinal and arcuate fibers are removed to show the sagittal stratum fibers which consists of fronto-occipital fasciculus, anterior commissur fibers, thalamic radiation and optic radiation fibers. (**f**) Removing part of the putamen exposes the globus pallidus and internal capsule. The globus pallidus, medial to the putamen, is differs from the putamen in terms of color and consistency. After the inferior fronto-occipital fascicle is removed, the anterior commissure inferiorly and the globus pallidus are exposed. The anterior commissure, part of the anterior wall of the third ventricle, is located immediately in front of the fornix columns. Removing the posterior part of the inferior fronto-occipital fascicle and the anterior commissure exposes Meyer's loop and the optic radiation. ac, anterior commissure; af, short association fibers; arc, arcuate fasciculus; ceis, central insular sulcus; Cb, cerebellum; cis, circular sulcus; Cl, claustrum; cr, corona radiata; ec, external capsule; ex, extreme capsule; Gp, globus pallidus; ic, internal capsule; ifo, inferior fronto-occipital fascicle; Li, limen insulae; ML, Meyer's loop; Pu, putamen; slf, superior longitudinal fasciculus; ss, sagittal stratum; STG, superior temporal gyrus; unc, uncinate fascicle

the fronto-orbital, frontoparietal, and temporal operculi, and the insular cortex. These short association fibers form the superficial layer of the extreme capsule. The uncinate fasciculus and the fronto-occipital fasciculus are located in the deep layer of the extreme capsule just underneath these short association fibers [58] (Fig. 3c). This structure is thought to be the ventral language stream, as it provides an alternative pathway to the arcute fasciculus/SLF complex between the Broca and Wernicke cortical structures [25, 59].

Uncinate fasciculus: This long association fiber bundle is located in the limen insula just underneath the superficial layer of the extreme capsule. It interconnects the fronto-orbital cortex and the temporal pole (Fig. 3e, f). The uncinate fasciculus influences reward and punishment-based behavioral decisions and is part of the ventral limbic system [24]. It may have a role in episodic memory, language, and social-emotional processing; however, its precise function is unknown [60]. There are also studies showing that this tract is affected in pathological entities such as schizophrenia, anxiety, psychopathy, antisocial personality disorder, epilepsy, and dementia [2, 15, 26, 61–63].

Fronto-occipital fasciculus: This is a long-association pathway that connects the frontal, temporal, and occipital lobes, and is located just dorsomedial to the uncinate fasciculus. Especially in the limen insula region, the distinction between the uncinate fasciculus and the fronto-occipital fasciculus cannot be made clearly as the two fiber bundles intertwine. For this reason, these two fibers are separated according to their originating and terminating regions. As humans do not have a superior-fronto-occipital fasciculus, this fiber was named only as the fronto-occipital fasciculus instead of the inferior fronto-occipital fasciculus. The structure called the superior fronto-occipital fasciculus is misunderstood as the superior thalamic peduncle [46, 60] (Fig. 3d, e). The fronto-occipital fasciculus influences lexical-semantic processes and visual-spatial processing [24, 43, 56].

Inferior longutudinal fasciculus: This fasciculus is deep in the fusiform gyrus and the inferior temporal gyrus, connecting the temporal pole to the occipital pole [33]. The inferior longutudinal fasciculus is believed to have an effect on recognizing and identifying visually perceived objects [64]. This pathway has an impact on brain functions such as reading, visual memory, face recognition, and emotional processes [65].

Cingulum: The cingulum is the main association fiber of the limbic system. It is a long association fiber that starts from the subcallosal area and turns around the corpus callosum deep into the cingulate gyrus where it joins the parahippocampal gyrus under the splenium [33]. The cingulum is the main white matter pathway of the dorsal limbic pathway. It influences the control of emotional processing [24, 26] (Fig. 4b, c). Known to have a great impact on the limbic system and emotional processes, the cingulum has historically been one of the target points of psychosurgery. Thus, cingulotomy has been tried in the treatment of disorders such as obsessive-compulsive disorder and anxiety [66].

Fimbria-fornix: The fimbria-fornix is the main efferent association fiber of the pes hippocampus. The two crus fornices come together under the corpus callosum by arching to form the fornix body and then coursing under the corpus callosum. It then divides into two at the anterior commissure level. The rostral part reaches the septal region as the precommissural fornix while the caudal part reaches the mammillary body and hypothalamus as the postcommissural fornix [44]. The fornix has an effect on memory [64] (Fig. 4a–c). Cognitive disorders presenting with severe episodic memory problems are seen in patients with a fornix injury [67]. Some studies show a structure that can be targeted in the treatment of diseases such as Alzheimer's [68].

3.2 Commissural Fibers

Corpus callosum: The corpus callosum is the main commissural fiber bundle that crosses the midline and connects both cerebral hemispheres (Fig. 4c). Although it is the predominant commissural fiber in humans, both the anterior commissure and the corpus callosum are highly developed in many other mammals [1].

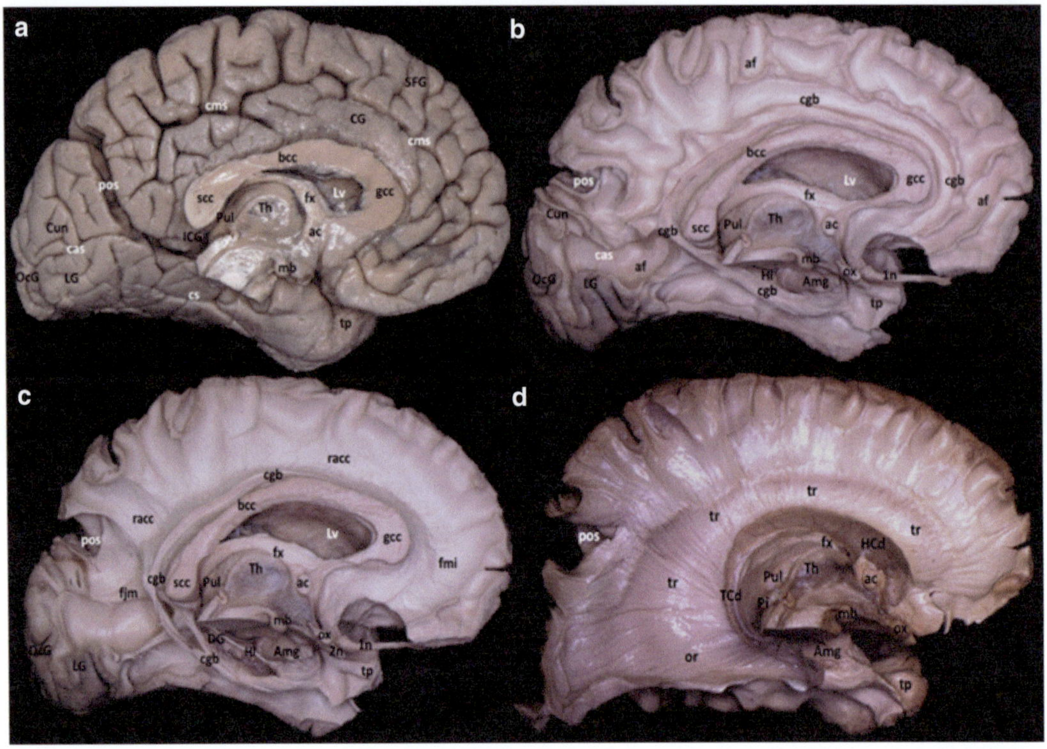

Fig. 4 Stepwise fiber dissection from the medial to lateral direction. (**a**) Identification of anatomical structures and the sulcus and gyrus on the medial surface of the left hemisphere. (**b**) Removing the medial cortex of the medial surface of the cerebrum and U fibers exposes the cingulum. The cingulum fibers start from the subcallosal area and turn around the corpus callosum to join the parahippocampal gyrus. In the parahippocampal gyrus, the cingulum fibers cover the hippocampus laterally and inferiorly. The amygdala is exposed by removing the medial temporal U-fibers. (**c**) Removing the cingulum exposes the corpus callosum radiation. The forceps minor fibers form the superomedial wall and roof of the frontal horn. The forceps major fibers cover the superior half of the medial atrial wall and continue to the occipital lobe. The dentate gyrus is dissected to expose the hippocampus. (**d**) Removing the callosal fibers exposes the caudate nucleus and thalamic radiations, which are divided into five thalamic radiations: anterior frontal, superior frontal, parietal, occipital and inferior. Removing the hippocampus and dissecting the ependyma from the lateral wall of the lateral ventricle and the tapetum exposes the optic radiation fibers. 1n, first cranial nerve; 2n, second cranial nerve; ac, anterior commissure; af, short association fibers; Amg, amygdala; bcc, body of the corpus callosum; cas, calcarine sulcus; CG, cingulum gyrus; cgb, cingulum bundle; cms, callosomarginal sulcus; cs, collateral sulcus; Cun, cuneus; DG, dentate gyrus; fmi, forceps minor of the corpus callosum; fmj, forceps major of the corpus callosum; fx, fornix; gcc, genu of the corpus callosum; HCd, head of the caudate nucleus; Hi, hippocampus; ICG, isthmus of the cingulate gyrus; LG, lingual gyrus; Lv, lateral ventricle; mb, mammillary body; OcG, occipital gyrus; or, optic radiation; ox, optic chiasm; Pi; pineal gland; pos, parieto-occipital sulcus; Pul, pulvinar; racc, radiation of corpus callosum; scc, splenium of the corpus callosum; SFG, superior frontal gyrus; TCd, tail of the caudate nucleus; Th, thalamus; tp, temporal pole; tr, thalamic radiation

Commissural fibers and projection fibers intertwine, especially at the level of the centrum semiovale. Since the fibers are not separated at this level and cannot be followed clearly, it is not possible to have a clear idea about their cortical terminations. For this reason, the centrum semiovale is one of the regions with the highest anatomical limitation in white matter dissection. A similar limitation occurs in magnetic resonance imaging (MRI) tractography studies. With increasing use of 7 T MRI tractography, however, we will have more information about such regions where commissural, projection, and association fibers cannot be separated. Even histological studies have similar limitations in such regions. Since the corpus callosum is a large structure, its functions vary according to the regions it connects. It is believed to have an effect on executive and goal-directed actions [56].

Anterior commissure: The anterior commissure is a major commissural fiber located just frontal to the fornix (Fig. 3f). Although the anterior commissure is usually shown in anatomical drawings and pictures as a small anatomical structure in the midsagittal plane, it is a dense fiber bundle that connects almost all temporal lobes. The fibers of the anterior commissure constitute the majority of the sagittal striata that extend posteriorly [1]. The anterior commissure is thought to have a role in olfactory, auditory, and visual communication [24].

Hippocampal commissure (forniceal commissure): This structure is a bundle of fibers that connects the two fornices just underneath the corpus callosum. It is very small and difficult to observe, especially in humans, but is believed to have an effect on memory.

3.3 Projection Fibers

Corona radiata-internal capsule: The main white matter bundle of the projection system that connects the cerebral cortex to the subcortical areas is the corona radiata (Fig. 3f). When the corona radiata descends to the level of the cerebral core, it is fan-shaped and called the internal capsule.

The sublenticular part of the internal capsule makes a loop as it extends posteriorly. This loop, known as Meyer's loop, is formed during the embryological development of the temporal lobe. Rather than calling this loop the optic radiation, it is more accurate to call it the posterior thalamic peduncle, which also contains the optical radiation. This structure not only consists of optical fibers [69] but is also a two-way pathway for information to and from the cerebral cortex [56].

3.4 Brain Stem

Corticospinal and corticobulbar tracts: The corticospinal and corticobulbar tracts are called pyramidal tracts. These tracts originate in the motor and premotor cortex and successively pass down through the internal capsule and the middle part of the crus cerebri; then, in the ventral part of the pons they are divided into numerous small bundles [70] (Fig. 5c, d). The fibers of the corticobulbar tracts terminate in the motor nuclei of the trigeminal, facial, glossopharyngeal, vagus, accessory, and hypoglossal nerves in the brain stem, which provide voluntary control over the muscles of the face, head, and neck [71]. The corticospinal tract controls primary motor activity for the somatic motor system from the neck to the feet [72]. As they pass through the brain stem, the corticospinal fibers give off numerous collaterals to neighboring structures, particularly the pontine nuclei [70]. In the medulla oblongata, the corticospinal fibers are located anteriorly. In the lower part of the medulla, most corticospinal fibers cross the midline (the motor or pyramidal decussation of Mistichelli) and form a contralateral lateral corticospinal tract, which ends at motor neurons in the anterior horn of the spinal cord. The non-crossing fibers continue ispilaterally in the anterior part of the spinal cord and are called the anterior corticospinal tract, historically known as the bundle of Türck [70, 72].

Corticopontine fibers: Corticopontine fibers arise from all areas of the cortex—the frontal, parietal, temporal, and occipital lobes—and pass through the internal capsule and the crus cerebri to terminate in the deep pontine nuclei. They then pass to the opposite side and are directed to the cerebellum as pontocerebellar fibers within the middle cerebellar peduncle. The corticopontine fibers coordinate planned motor functions (Fig. 5c, d) [73].

Spinothalamic tract: The spinothalamic tract consists of two different tracts conveying different information—the anterior and lateral spinothalamic tracts. The anterior spinothalamic tract transmits information related to rough touch and pressure, while the lateral spinothalamic tract transmits information related to temperature and pain. In the medulla oblongata, the spinothalamic tract lies laterally, far from the medial lemniscus. Further in the pons and the midbrain, the tracts converge and reach the ventral posterolateral thalamic nucleus together. Some fibers of the

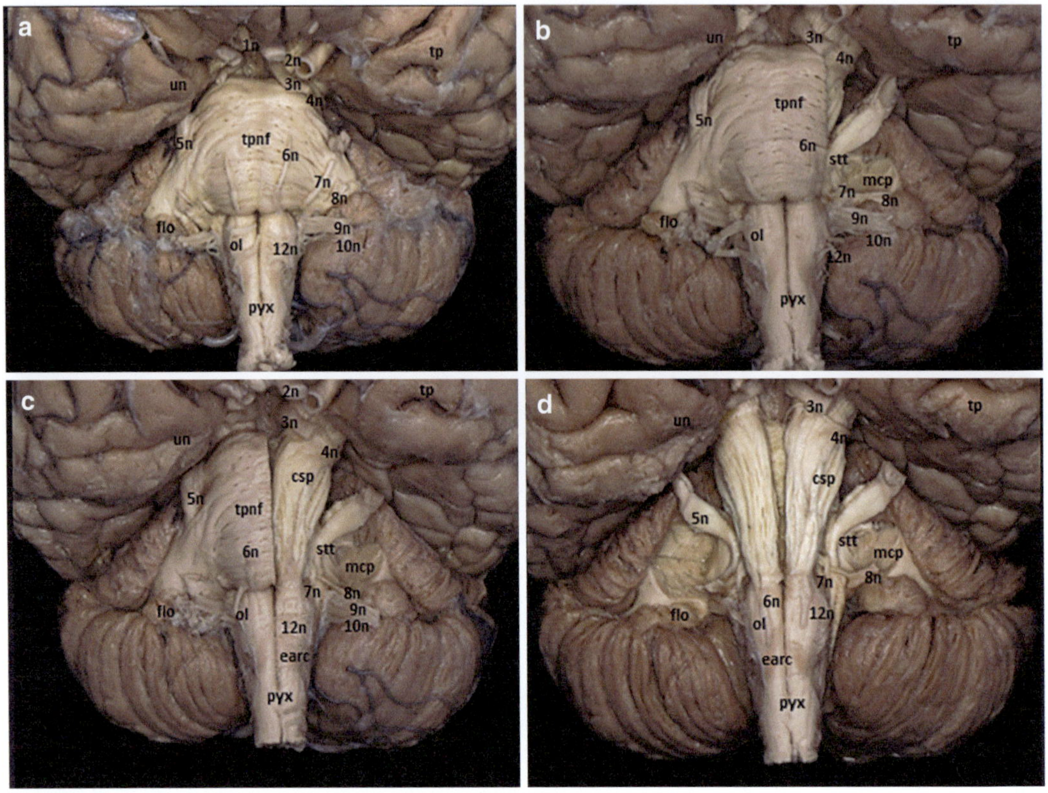

Fig. 5 Stepwise fiber dissection of anterior surface of brain stem. (**a**) Anterior view of the cranial nerves, cerebellum, and brain stem. (**b**) The transverse pontine fibers are partially removed to show the trigeminal nerve root. (**c**) On the left side, the transverse pontine fibers are removed from the midline. The corticospinal tract fibers are exposed till deccusation, the spinal trigeminal tract is dissected, and the facial and vestibulocochlear nerve roots are clearly visible. Ventral external arcuate fibers are seen on the anterior surface of the medulla oblongata, superficial to the pyramidal tract. (**d**) A similar dissection was done in the right side of the brain stem. 2n, second cranial nerve; 3n, third cranial nerve; 4n, fourth cranial nerve; 5n, fifth cranial nerve; 6n, sixth cranial nerve; 7n, seventh cranial nerve; 8n, eighth cranial nerve; 9n, ninth cranial nerve; 10n, tenth cranial nerve; 12n, twelfth cranial nerve; Csp, corticospinal tract; earc, external arcuate fibers; flo, flocculus; mcp, middle cerebellar peduncle; Ol, olive; pyx, pyramidal decussation; stt, spinal trigeminal tract; tp, temporal pole; tpnf, transverse pontine fibers; un, uncus

spinothalamic tract continue to the intralaminar nuclei [70, 72].

Medial lemniscus: The internal and external (ventral and dorsal) arcuate fibers arise from axons of the nucleus gracilis and nucleus cuneatus. The internal arcuate fibers decussate on the contralateral side and form the medial lemniscus. The ventral and dorsal external arcuate fibers cross on the opposite side, circumventing the ventral and dorsal surfaces of the medulla oblangata and then head to the restiform body [70, 72] (Fig. 5c, d). The primary function of the medial lemniscus is to transport the sensory spinothalamic information of conscious proprioception, vibration, fine touch, and two-point discrimination of skin and joints of the body and head [74].

Superior cerebellar peduncle (brachium conjunctivum): The superior cerebellar peduncle is a paired structure that forms the superior lateral wall of the fourth ventricle and connects the cerebellum to the midbrain. It consists mainly of efferent fibers of the dentato-rubro-thalamic tract. It also contains afferent tracts, the most prominent of which is the ventral spinocerebellar tract (Gowers tract) transmitting proprioceptive information from the body to the cerebellum [72, 75] (Fig. 6b–d).

Middle cerebellar peduncle (brachium pontis): The middle cerebellar peduncle consists purely of

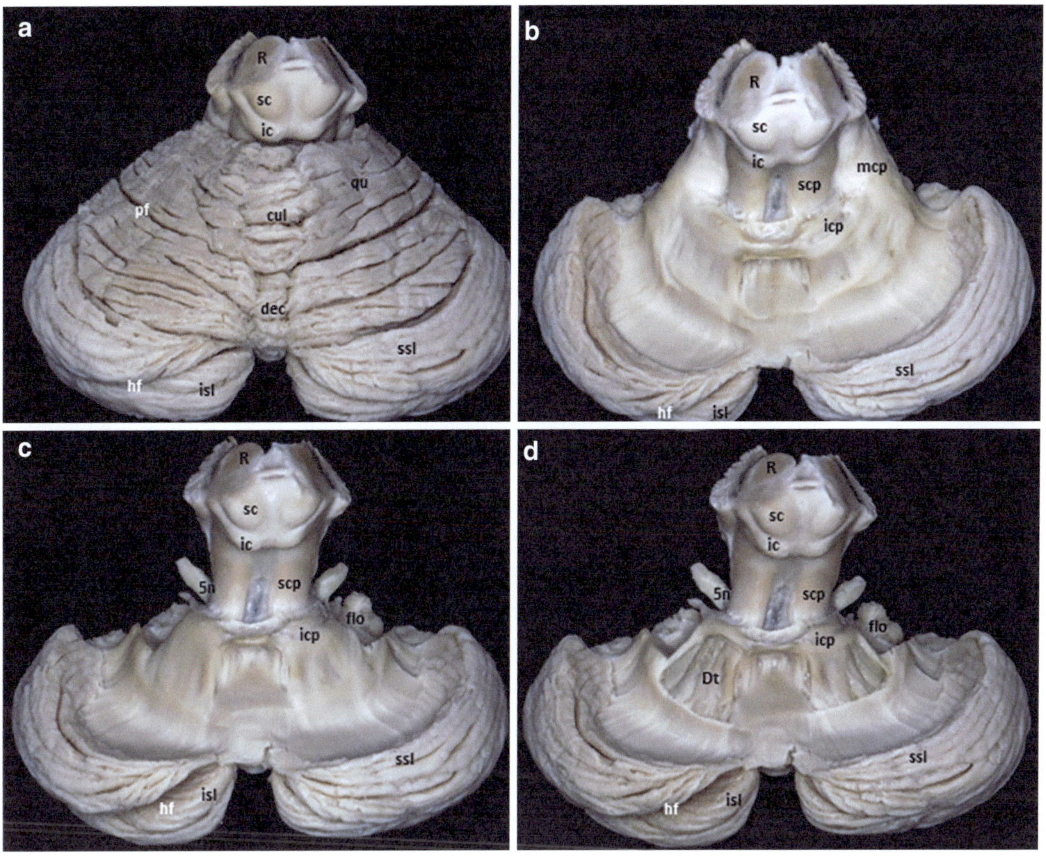

Fig. 6 Stepwise fiber dissection the superior surface of the cerebellum-brainstem complex. (**a**) Posterior superior view of the cerebellum and brain stem. (**b**) The anterior and posterior quadrangular lobes were removed to view the underlying white matter and the cerebellar peduncles. (**c**) The middle cerebellar peduncle fibers are removed and the trigeminal nerve roots and flocculus are seen. (**d**) Dissecting the white matter covering the dentate nucleus posteriorly and the inferior cerebellar peduncle on both sides exposes the upper surface of the dentate nucleus. 5n, fifth cranial nerve; cul, culmen; dec, declive; Dt, dentate nucleus; hf, horizontal fissure; ic, inferior collicul; icp, inferior cerebellar peduncle; isl, inferior semilunar lobule; mcp, middle cerebellar peduncle; pf, primary fissure; qu, quadrangular lobule; R, red nucleus; sc, superior collicul, scp, superior cerebellar peduncle; ssl, superior semilunar lobule

afferent fibers such as transverse pontine fibers or pontocerebellar fibers. Pontocerebellar fibers originate from the contralateral pontine nuclei and travel to the cerebellar cortex (Figs. 5a–c and 6b).

Inferior cerebellar peduncle: The inferior cerebellar peduncle consists of the external and internal divisions: the restiform body and the juxta-estiform body. The restiform body consists of afferent fibers from the brain stem and spinal cord (dorsal spinocerebellar tract, olivocerebellar tract, ventral and dorsal external arcuate fibers) that run lateral to the spinal trigeminal tract of the trigeminal nerve and enter the cerebellum between the middle and superior cerebellar peduncles. The function of the inferior cerebellar tract is to convey proprioceptive sensory input with motor vestibular functions to maintain balance and posture [72, 75] (Fig. 6b–d).

4 Fiber-Dissection Technique and DTI-Based Tractography

Understanding the organization that underpins the white matter is critical in neurosurgery. In dealing with a variety of lesions, such as intraventricular lesions, intracerebral hematomas, cavernomas, arteriovenous malformations, hippocampal sclerosis, and others, the neurosurgeon must advance into the unshaped structure that characterizes the white matter of the brain.

Recently, Klingler's work has received renewed attention as it is highly applicable to relate the results of fiber mapping with diffusion tensor imaging to real anatomical evidence. Klingler created his dissection technique as an interactive tool for easily selecting fiber bundles in the brain and visualizing the naturally curved anatomic structures that run parallel to the imaged fiber bundles. Klingler dissection preparations have been used in a variety of recent neurosurgical anatomic studies, combining diffusion tensor imaging with electrostimulation studies, and combining fiber dissection with diffusion tensor imaging to define the functional implications of the claustrocortical system [5, 28].

The white matter of the brain is composed of myelinated nerve axons. When these axons share a common origin and destination, they form white matter fasciculi or tracts, which are compact structures. White matter is made up of myelinated fibers that are divided into three types of tracts or fasciculi: association fibers, commissural fibers, and projection fibers.

Although contemporary neuroanatomic literature and atlases cover the fiber bundles, there is a lack of anatomic explanations and pictures sufficient to acquire a three-dimensional understanding of the fiber bundles for surgical practice [56, 64, 76]. The recent introduction of DTI-based tractography, a rapid and non-invasive MRI method that can trace white matter tracts in vivo, has facilitated the understanding of white fiber anatomy [77, 78]. The method is based on the fact that the MRI-detectable diffusivity of water molecules is affected by the major orientation of white matter fiber tracts. Diffusion-weighted MRI sequences track such mobility in several directions to determine the distribution inside an image voxel. The principal axis of the associated diffusion ellipsoid coincides with the direction of the largest diffusion coefficient. This coefficient can then be identified through the orientation of the underlying fiber bundle under the premise that this distribution can be mathematically represented by a tensor. The degree to which diffusion is preferred along this path versus other directions is shown through anisotropy measurements [79, 80]. Diffusion MRI can be used to track the orientation of axons throughout the brain noninvasively. This method allows researchers to investigate the anatomy of white matter and the connections between different cortical regions. Based on anatomical knowledge gained from classic anatomical white matter dissections, specific fasciculi can be identified in the entire brain. Anatomical dissection has also confirmed the discovery of potentially novel tracts based on diffusion MRI findings. DTI-based tractography has already been incorporated into the diagnostic neurosurgery armament.

Three techniques (fiber-dissection, DTI-based tractography, and histological anatomy) have the same fundamental goal: to reveal the anatomy of the white fiber tracts. A good anatomic specimen, suitable preparation (including the freezing process), a surgical microscope, and many hours of careful study in the microsurgical laboratory are all required for the fiber-dissection technique [1]. The results may not be perfect at first, but with practice and patience, exquisite dissections can be achieved. The acquisition of unique three-dimensional anatomic expertise on the white matter of the brain is inexorably the result of this manual and intellectual effort.

On the other hand, DTI-based tractography is one of the most exciting radiological breakthroughs in recent decades [32]. It is now possible to not only analyze the intrinsic structure of the brain in vivo in a matter of minutes but also to observe how intracerebral injuries affect the fiber tracts. This research necessitates the use of a diffusion-tensor imaging MRI system as well as three-dimensional tractographic reconstruction software. The ability to "radiologically" investigate and dissect the inside of the normal or disordered brain in vivo has given neurosurgeons a whole new perspective [32]. Because one technique solves the limitations of the other, combining them strengthens both strategies. Because of the complicated relationships between fiber systems, the fiber dissection approach is limited, and demonstrating one fiber system frequently results in the destruction of other fiber systems. The use of DTI-based tractography, which can disclose the complicated interactions among the fiber networks simulataneously, prevents this degradation. As a result, we believe that the three methodologies should be combined not just in neurosurgical training and operative planning, but also as a replacement for and supplement to traditional neuroanatomic research methods. The possible functional relevance of the white matter tracts can be determined using a combination of topographical anatomy of the human cerebral tracts (acquired with the fiber-dissection technique and DTI-based tractography) and data obtained from clinical observation and radiological diagnosis of human white matter lesions, particularly from intraoperative subcortical electric-stimulation studies.

Finally, the three techniques have the same limitation. The extraordinarily complex intracerebral organization, which is integrated through thousands of millions of micrometric axons circulating in three spatial planes, makes it impossible for any one technique to define the structures completely. Thus, fiber dissection and tractography reveal the macroscopic and topographic anatomy of the fiber tract's main stem, but not the specific origin and termination of the fibers that make up each fasciculus. Furthermore, both techniques are open to manipulation. By forcing the dissection in a direction other than the usual axonal planes, artifacts and false bundles might occur. Thus, the techniques should not be done by those who are not experts in white matter routes.

Neurosurgeons must gain a better understanding of the anatomic and functional connection of white matter tracts. Every intracerebral lesion affects the white matter structure of the brain to some extent, and stroke studies have shown that white matter lesions cause more severe and long-term neurological abnormalities than cortical damage [81]. A thorough understanding of the anatomic and functional fasciculi of the white matter and the related disconnection syndromes can help with the interpretation of pre- and post-surgical neurological symptoms and syndromes. The morphology of white matter tracts is very important in the surgical treatment of cerebral gliomas for two reasons: first, gliomas can affect both cortical and subcortical structures, altering connections [81]; and second, most gliomas push the white matter tracts out of place [82, 83]. The spatial relationships of intracerebral lesions with white matter tracts can be studied using DTI-based tractography. Several authors have documented the value of incorporating this technique into the neuronavigation system for locating the pyramidal tract [84–88], primary motor area [85], optic radiations [88, 89], and language pathways [90] during surgery. However, its reliability is questionable in cases of brain shift. Its use in combination with mapping techniques may help preserve eloquent areas during surgery by allowing access to direct connection data between functional regions of the brain [90]. The major purpose of several neurosurgery procedures, such as bilateral cingulectomy or anterior capsulotomy, is to disrupt a specific white matter pathway.

Treatment of a wide range of neurosurgical abnormalities requires a thorough understanding of the microsurgical anatomy of the white matter tracts, and the fiber-dissection technique is the best way for a neurosurgeon to study this three-dimensional architecture [1]. Tractography based on DTI gives an accurate representation of the white matter tracts in the normal human brain, as well as their spatial interactions with intracerebral lesions [32]. Because the anatomic knowledge gained through fiber dissection aids in proper interpretation and analysis of the new tractographic studies, and because the tractographic studies can reveal the complex relationships between the fiber systems at the same time, the combined application of both techniques to study white matter anatomy is reciprocally enriched. An understanding of the possible functional role of the white matter tracts is aided by combining knowledge of the topographical anatomy of the human cerebral tracts with data from neuroanatomic studies in primates, clinical and radiological diagnosis of human white matter lesions, and intraoperative subcortical electric-stimulation studies. Thus, understanding the fiber tracts from anatomic and radiological perspectives can help to explain pre-, intra-, and postoperative clinical "disconnection" syndromes and white matter tumoral invasion patterns, as well as improve presurgical planning and surgical strategy, effectively reducing surgical morbidity.

5 Impact of White Matter Dissection on Microneurosurgical Procedures

In today's conditions, neurosurgical operations can be done more carefully with the support of developing technology, especially tractography. There is no safe area of the brain to freely extract; therefore, the whole brain must be respected. As a result of what we learn from white matter dissection and tractography, we can be more careful and meticulous surgeons.

The location of the white matter tracts is extremely important in deciding the surgical strategy. First, the origin of a tumor is extremely important because, as the tumor volume grows, it pushes the surrounding white matter fibers outward. While planning surgery, we can delineate the new locations of these fibers with the help of tractography.

Postoperative tractography can also confirm the quality of the surgery. With 7 T MR tractography, even the smallest destruction of white matter will be noticed in the postoperative period. The increasing use of this technology means that the next generation of surgeons should be much more careful.

The Human Connectome Project provides a unique compilation of neural data, an interface for navigating this data graphically, and an opportunity to reach never-before-seen conclusions about the living human brain [91]. This sharing of data with the whole world has pioneered the development of this field. The data can be used to manipulate the brain in many ways, navigate the main brain pathways, compare basic circuits, zoom in on a region, and discover the cells that compose brain structures and their functions [92].

5.1 Case Presentations

In the following section, we present case studies to discuss how the white matter dissection technique has an impact on the clinic and planning surgical strategy. All surgeries were performed by senior author UT.

Case 1, globus pallidus tumor: The first case is of a right-sided globus pallidus tumor in a pediatric patient (Figs. 7 and 8). A 5-year-old female presented with slight left-sided hemiparesis. Although globus pallidus tumors of primary origin are rare, they do exist. Because of the important structures around this area, such as the putamen and internal capsule, it is difficult to reach these tumors without damaging normal functional structures. However, the tumors themselves often help by opening a corridor that can be accessed without damaging the surrounding functional parenchyma. If the lateral lenticulostriate arteries are followed, the globus pallidus can be reached. By opening the right sylvian vallecula and following the lateral lenticulostriate arteries, the tumor comes to the surface, making its removal possible. Although this route is more difficult than a lateral intervention, it is a much more physiological approach when performed carefully because it protects the white matter pathways. Thus, when the tumor is removed, the putamen, internal capsule, and even the tumor-free part of the globus pallidus remain intact.

The tumor pathology in this case was a primitive neuroectodermal lesion. Therefore, the patient received adjuvant radiotherapy. The patient has been tumor free and without neurologic deficit for 14 years since surgery. The pathology result was re-evaluated and showed a central nervous sytem embryonal tumor according to the 2016 classification by the World Health Organization.

Case 2, mediobasal temporal region tumor: A 63-year-old female with a right mediobasal temporal glial tumor presented with seizures (Fig. 9). The tumor extended anteriorly to the piriform cortex, posteriorly to the isthmus, superiorly to the temporal horn, medially to the choroidal fissure, and laterally to the collateral sulcus. These tumors usually stay within the borders of the mediobasal region. When these tumors are approached anteriorly, complete removal is not possible without damaging the temporal stem, especially for tumors that extend posteriorly. The lateral approach, on the other hand, is not appopriate for preserving white matter pathways because accessing the mediobasal temporal

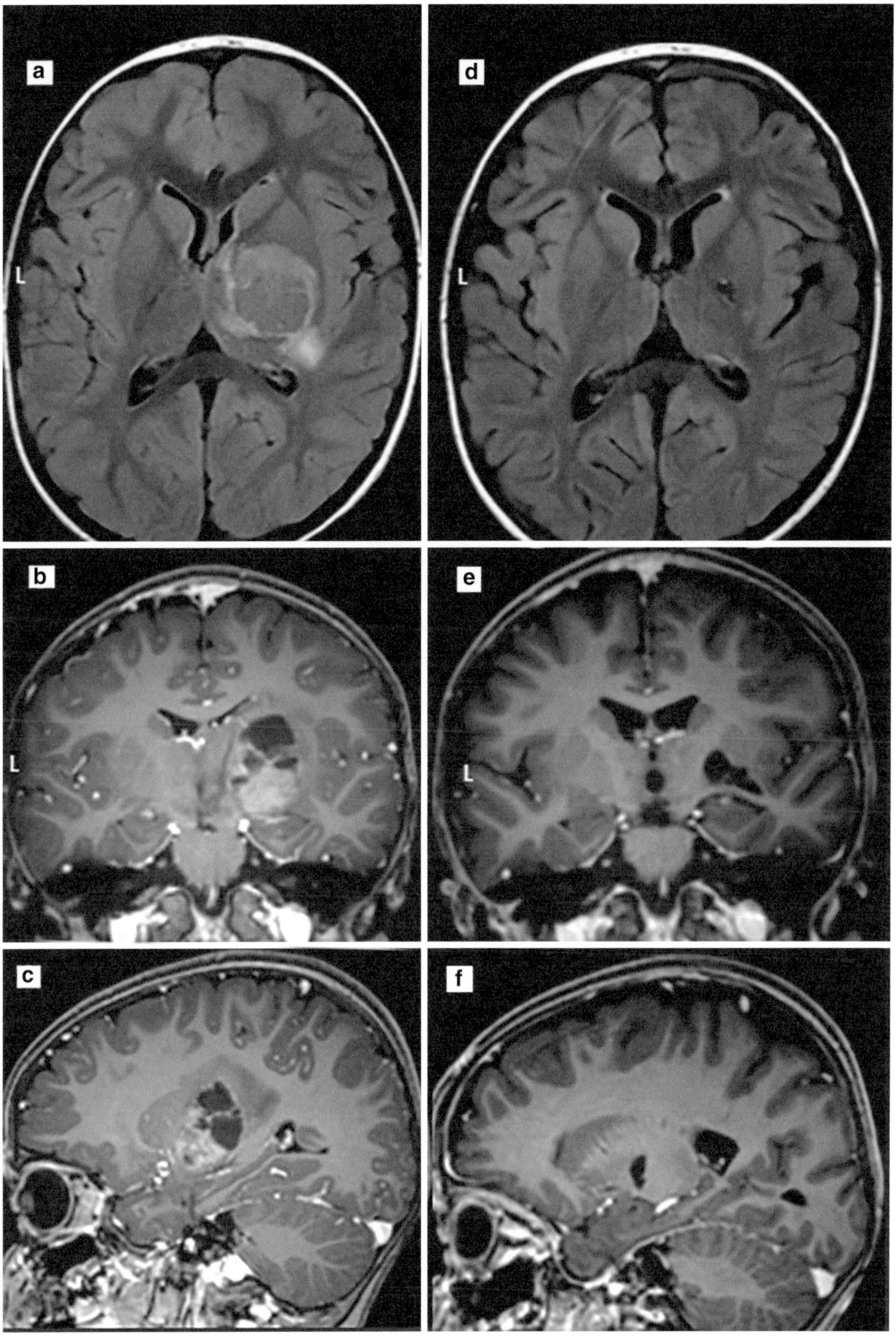

Fig. 7 Contrast-enhanced T1-weighted MRIs of a right-sided globus pallidus tumor. Preoperative axial (**a**), coronal (**b**), and sagittal (**c**) views; postoperative axial (**d**), coronal (**e**), and sagittal (**f**) views

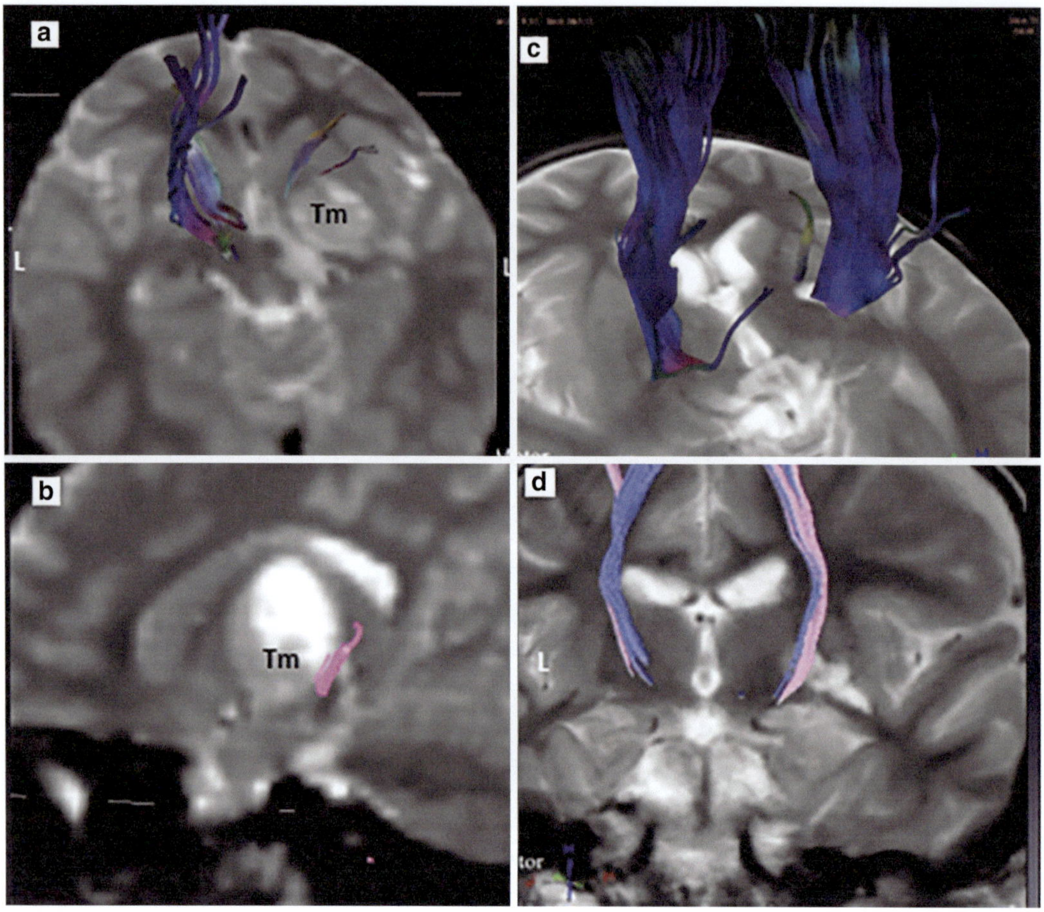

Fig. 8 Right-sided globus pallidus tumor. Preoperative 3D reconstruction of (**a**) axial sensory fibers and (**b**) the sagittal pyramidal tract. Postoperative 3D reconstruction of (**c**) axial sensory fibers and (**d**) the coronal pyramidal tract. L, patient's left side

region laterally requires destroying many white matter pathways, such as the optic radiation, sagital striatum, and middle longitudinal fasciculus. To protect the white matter pathways, the safest approach to this area is from the posterior.

The paramedian supracerebellar transtentorial approach was established by Yaşargil for cavernomas of the mediobasal temporal region. This technique can be used safely along the entire length for mediobasal temporal region pathologies [93]. In this approach, with the patient in a semi-sitting position, access to the entire mediobasal line can be achieved with a single section. As it is difficult in this approach to remove middle fossa tumors with a microscope, endoscopy-assisted surgery is necessary in such cases. The limitation of this approach is access to the isthmus through the tentorium, but it is possible to preserve the uncinate fasciculus, fronto-occipital fasciculus, and posterior thalamic peduncle. After surgery, our patient had no neurologic deficit. The tumor pathology was reported as glioblastoma multiforme. Thus, the patient received adjuvant therapy.

Case 3, mediobasal temporal epilepsy surgery: This 24-year-old female had a 20-year history of epilepsy uncontrolled by four antiepileptic drugs (3–4 times/week) (Fig. 10). As with mediobasal region tumors, a paramedian supracerebellar transtentorial approach is appropriate for

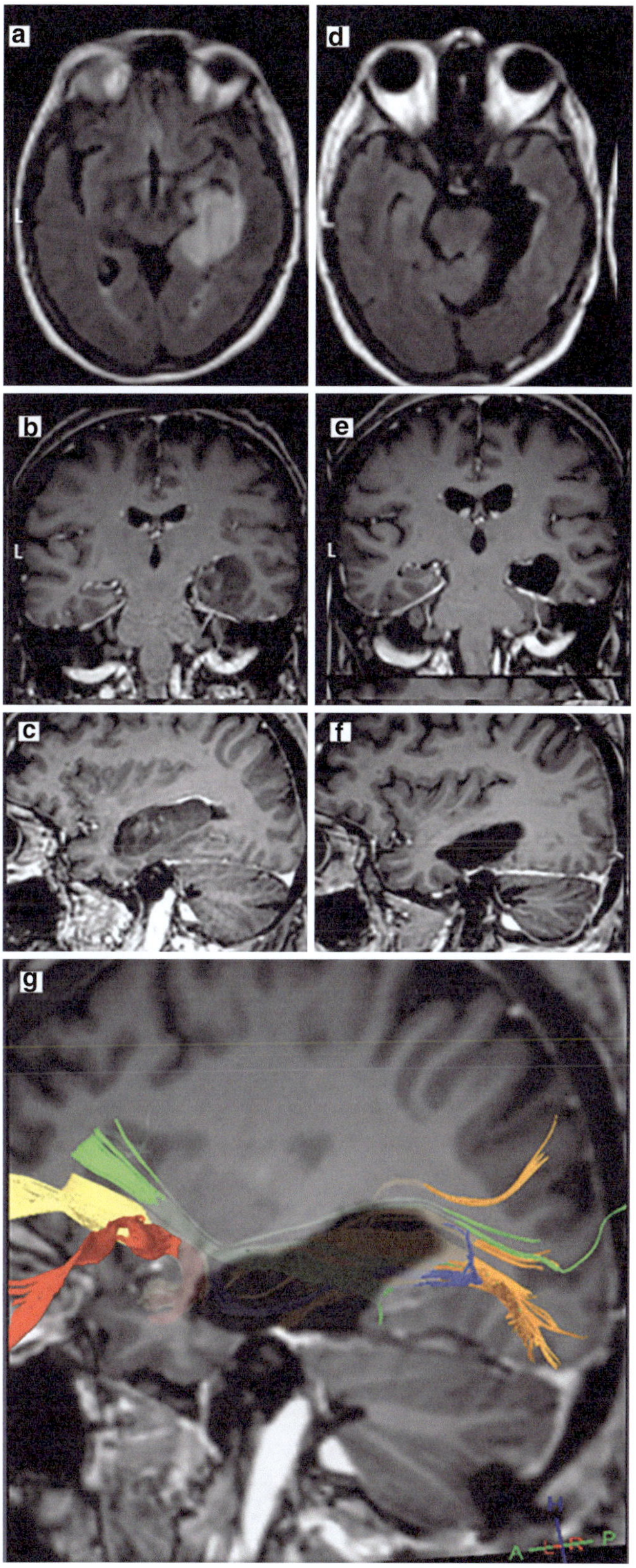

Fig. 9 Right-sided mediobasal temporal glial tumor. (**a**) Preoperative axial FLAIR MRI and (**b**, **c**) coronal and sagittal contrast-enhanced T1-weighted MRI. (**d**) Postoperative axial FLAIR MRI and (**e**, **f**) coronal and sagittal contrast-enhanced T1-weighted MRI. (**g**) 3D reconstruction of the sagittal DTI image. Red and yellow, uncinate fasciculus; green, fronto-occipital fasciculus; orange, posterior thalamic peduncle; blue, Meyer's loop. L, patient's left side

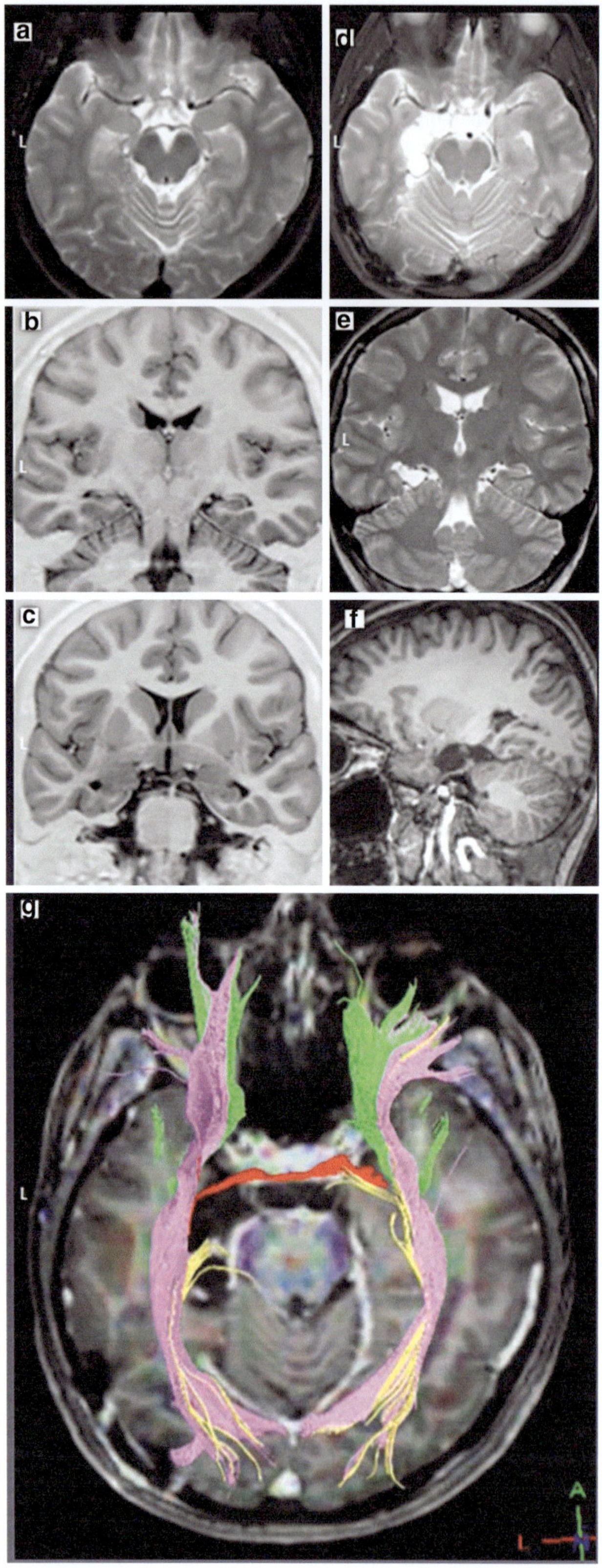

Fig. 10 Images of a patient with left-sided mediobasal temporal lobe epilepsy. (**a**) Preoperative axial T2-weighted MRI and (**b**, **c**) coronal and coronal FLAIR MRI. (**d**) Postoperative axial and (**e**) coronal T2-weighted MRI, and (**f**) sagittal FLAIR MRI. (**g**) 3D reconstruction of the axial DTI image. Red, anterior commissure; green, uncinate fasciculus; purple, fronto-occipital fasciculus; yellow, Meyer's loop; L, patient's left side

selective amygdala-hippocampectomy in this region to preserve white matter pathways. The advantage of this approach is that the entire hippocampal formation, including the tail, can be removed [93], and the surgeon does not have to deal with the neocortex. This approach is also more appropriate as the hippocampus is severely atrophied in epilepsy patients. For this reason, it should be an alternative to the transsylvian approach for patients with epilepsy. The pathology of the patient was reported as hippocampal sclerosis and no postoperative neuromotor deficit was observed. As of this writing, the patient is still free of seizures and antiepileptic drugs.

Even if the transsylvian route is used in such cases, a selective amygdalohippocampectomy can be done through a much more anterior route through an incision in the piriform cortex. Thus, the neocortex and white matter structures such as the optical radiation, uncinate fasciculus and fronto-occipital fasciculus are not damaged [94].

Case 4, cingulate tumor: A 22-year-old female presented with seizures and was found to harbor a glial tumor in the right middle portion of the cingulate gyrus (Fig. 11). We aimed to remove only the tumor and, thus, did not remove all areas of hyperintensity shown on the preoperative MRI FLAIR image. Postoperative imaging at 3 months showed that the tumor was much smaller and that not all hyperintensity seen on the original FLAIR sequence was a tumor. In addition, DTI imaging showed that most of the cingulum fibers on the tumor side were preserved. Pathologic analysis showed the lesion to be a grade 2 subependymal giant cell astrocytoma. The patient had no postoperative neurologic complaints and did not undergo adjuvant treatment. As of this writing, the tumor has not recurreed and all hyperintensity seen on the initial FLAIR disappeared from the 3-month followup images.

Case 5, thalamic tumor: A left thalamic glial tumor was detected in a 14-year-old male who presented with headache, diplopia, and vomiting (Fig. 12). An anterior interhemispheric transventricular approach was used in this patient because hydrocephalus had also developed due to the tumor. The most important feature of thalamic tumors is that they are confined to the thalamus, but they may extensively displace neighboring tissues, such as the internal capsule or hypothalamus, without true infiltration. The vascularization of the internal capsule and thalamus are completely different. The thalamus is supplied by branches of the posterior communicating artery, while the internal capsule is supplied by the anterior choroidal artery and lateral lenticulostriate arteries. These tumors usually do not invade the midbrain; they just push it down, and they always push the white matter pathways laterally. Damage to the internal capsule is the most prevalent cause of morbidity. The risk of such injury can be reduced by avoiding transcortical methods, performing a pure lesionectomy, and sparing the vascular supply to the internal capsule. No neurological deficit developed in the patient after surgery, and he underwent adjuvant therapy for what pathologic analysis defined as a grade 3 astrocytoma.

Case 6, thalamic Tumor: A 13-year-old female presented to another instution with a severe headache, and computed tomography scans showed triventricular hydrocephalus. MRI scans revealed a left pulvinar tumor. Although the paramedian supracerebellar approach was most appropriate for this case because the tumor was on the cisternal surface, we preferred the posterior interhemispheric transtentorial subsplenial approach so we could use intraoperative MRI. Because the tumor was not on the ventricular surface, an anterior interhemispheric transcallosal approach was not appropriate. With the subsplenial approach, the pulvinar could be accessed directly without cutting the splenium. No residual tumor was seen on intraoperative MRI or the 3-month postoperative images (Fig. 13). The tumor pathology was reported as a grade 2 diffuse astrocytoma, and the patient had no neurological deficits after surgery.

Case 7, mid-brain tumor: In this case, a 4-year-old male presented with left-sided hemiparesis stemming from a right tegmental tumor (Figs. 14 and 15). In preoperative planning, the surgical approach to the brain stem should be determined according to its anatomical and tractographic features. The general concept of a "safe entry zone" may cause damage to the white matter pathways because every area of the brain stem

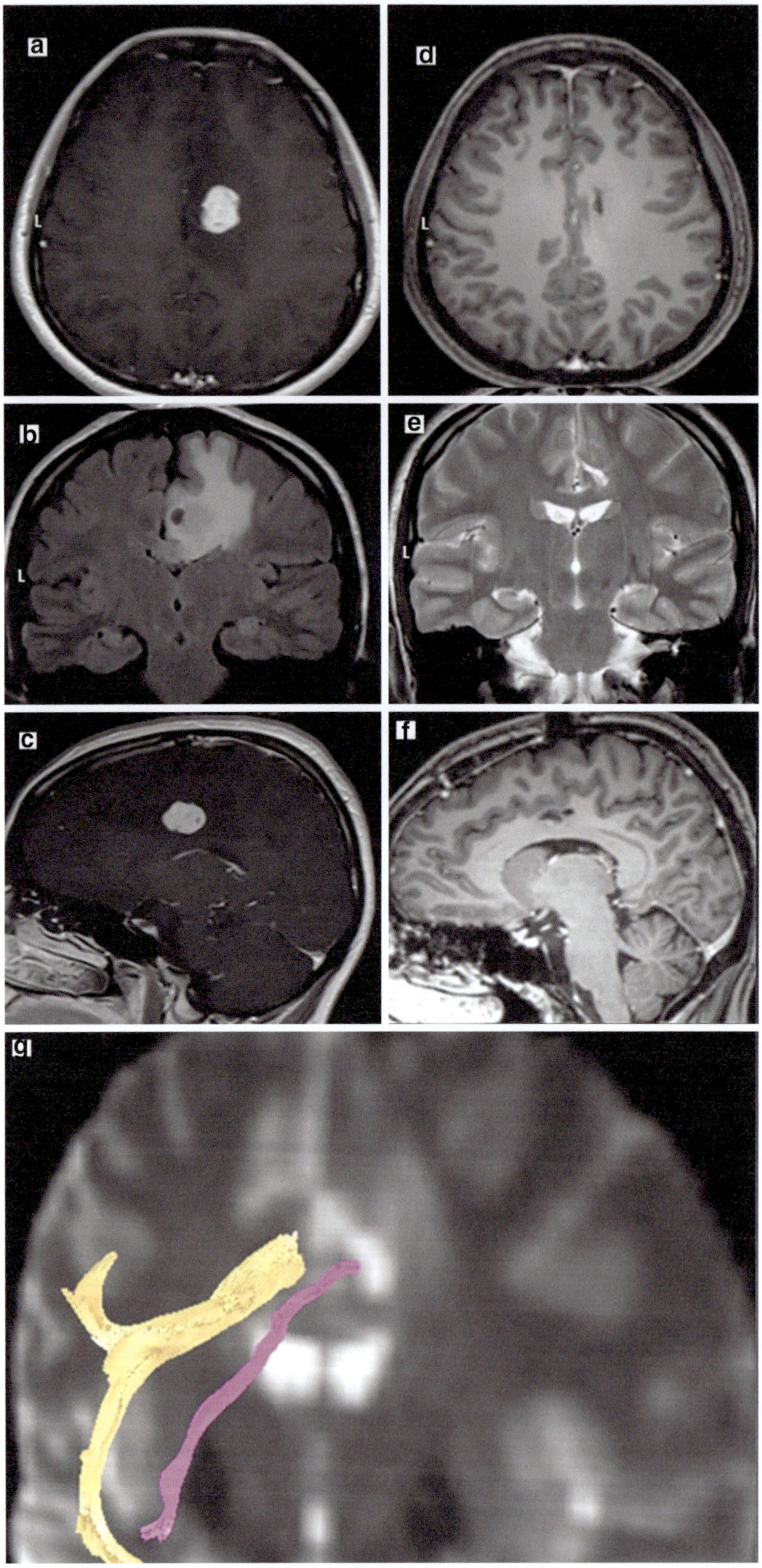

Fig. 11 Right-sided middle cingulate gyrus tumor. (**a**) Preoperative axial contrast-enhanced T1-weighted and (**b**) coronal FLAIR images. (**c**) Sagittal contrast-enhanced T1-weighted MRI. (**d**) Postoperative axial contrast-enhanced T1-weighted MRI. (**e**) Coronal T2-weighted image. (**f**) sagittal contrast-enhanced T1-weighted MRI. (**g**) 3D reconstruction of the coronal DTI image. Purple, left cingulum; yellow, right cingulum; L, patient's left side

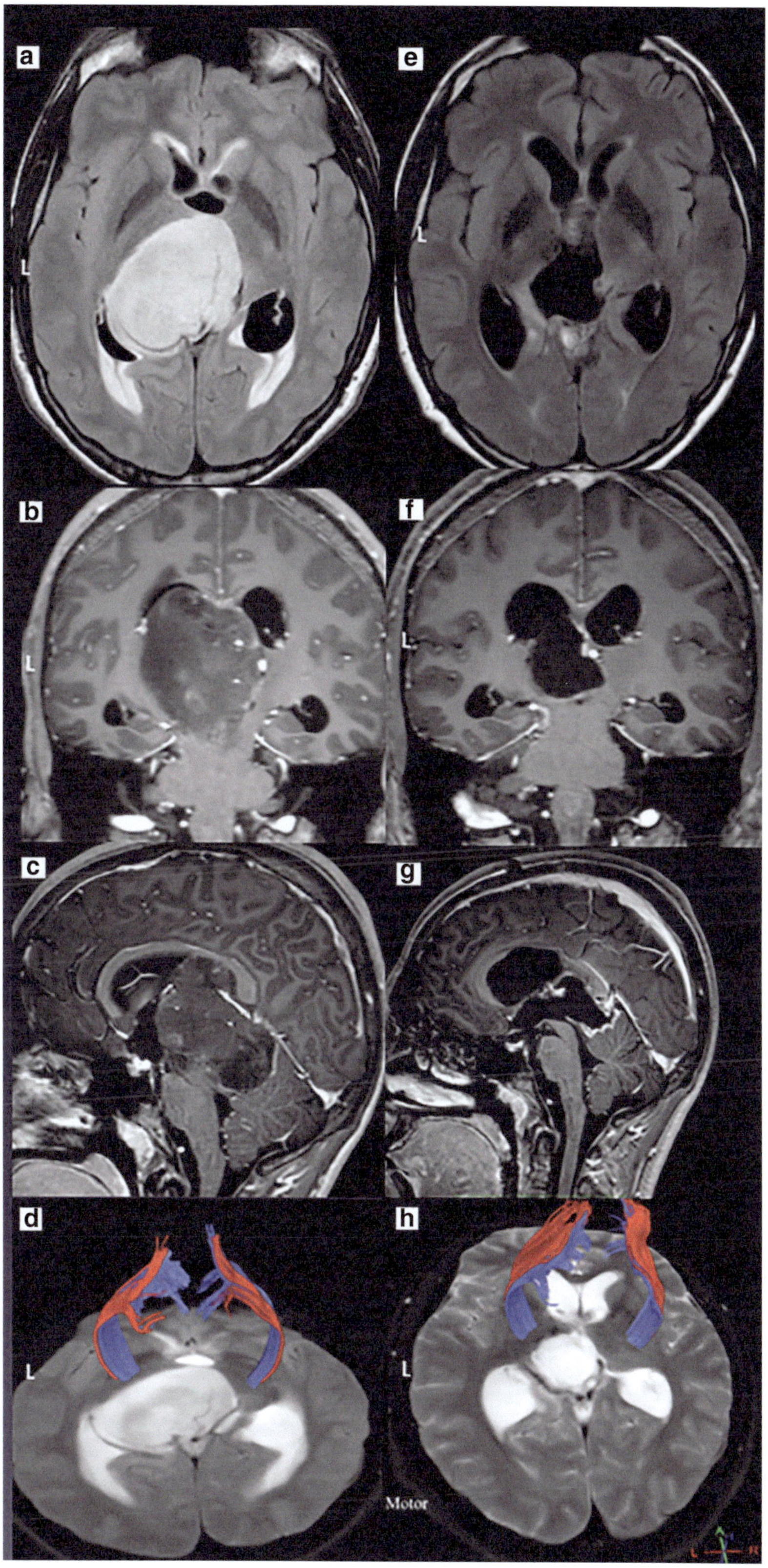

Fig. 12 Left-sided thalamic tumor. (**a**) Preoperative axial FLAIR image, and (**b**) coronal contrast-enhanced T1-weighted MRI. (**c**) Sagittal contrast-enhanced T1-weighted MRI. (**d**) 3D axial reconstruction of the sensory and motor fiber components of the internal capsule. (**e**) Postoperative axial FLAIR image. (**f**) Coronal contrast-enhanced T1-weighted MRI. (**g**) Sagittal contrast-enhanced T1-weighted MRI. (**h**) 3D axial reconstruction of the sensory and motor fiber components of the internal capsule. Red, sensory fibers; blue, motor fibers; L: patient's left side

is extremely functional. For this reason, the surgical approach should be chosen according to the path opened by the tumor itself.

Midbrain tumors usually arise from the tegmentum or tectum. Those originating from the tegmentum are mostly benign. A tractography study showing motor and sensory fibers and the superior cerebellar peduncle before surgical planning is extremely important in determining the approach. In this case, preoperative tractography showed that we could remove the tumor from a lateral direction, so we chose the paramedian supracerebellar approach. No neurological deficit was observed in the postoperative period, and the pathology was reported as a pilocytic astrocytoma.

Case 8, pons tumor: A 35 year-old male patient was treated in another facility and a ventriculo-peritoneal shunt was inserted due to

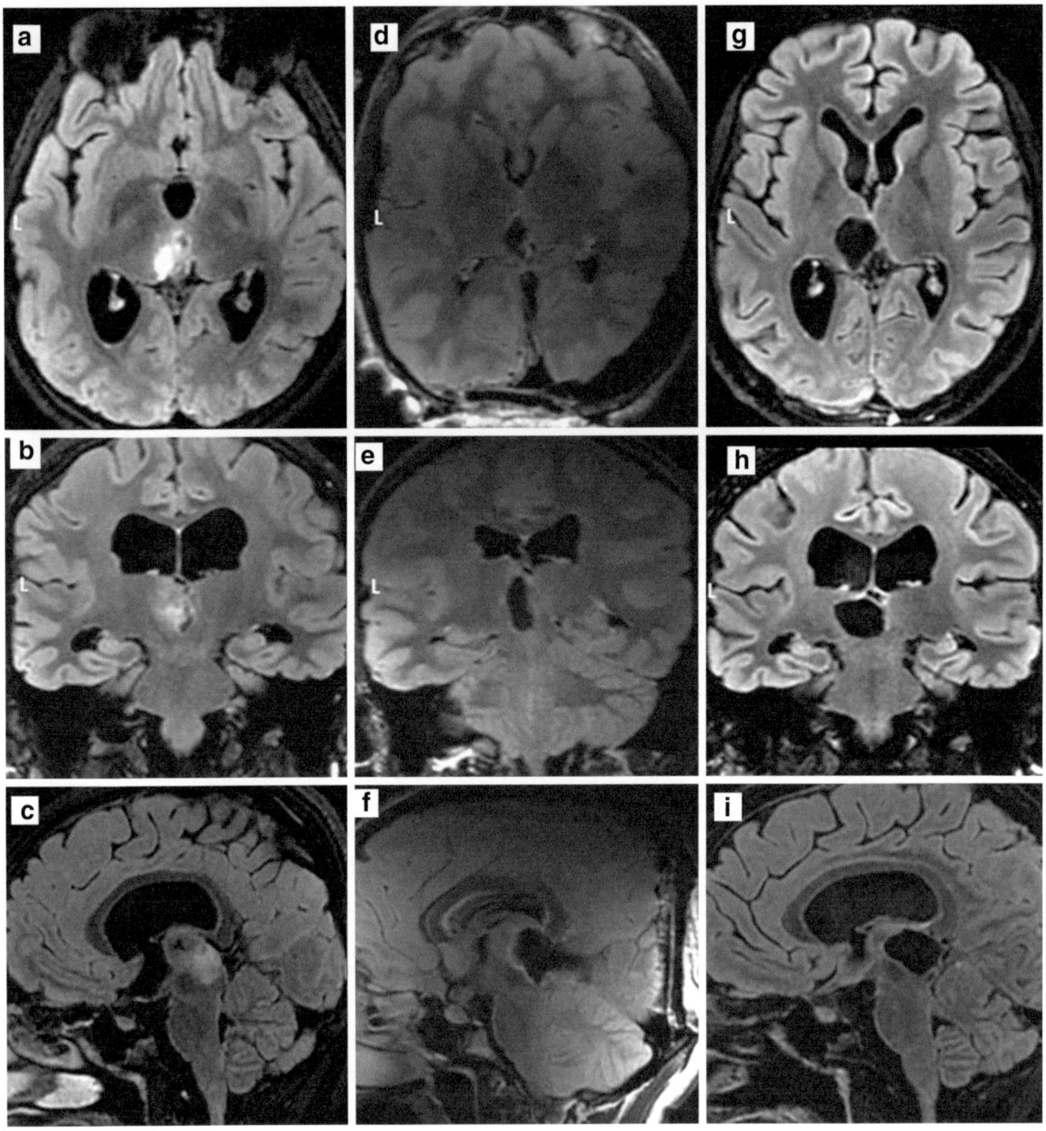

Fig. 13 Left-sided thalamic tumor. Preoperative axial (**a**), coronal (**b**), and sagittal (**c**) contrast-enhanced T1-weighted MRI. Intraoperative axial (**d**), coronal (**e**), and sagittal (**f**) contrast-enhanced T1-weighted MRI. Postoperative axial (**g**), coronal (**h**), and sagittal (**i**) contrast-enhanced T1-weighted MRI. L, patient's left side

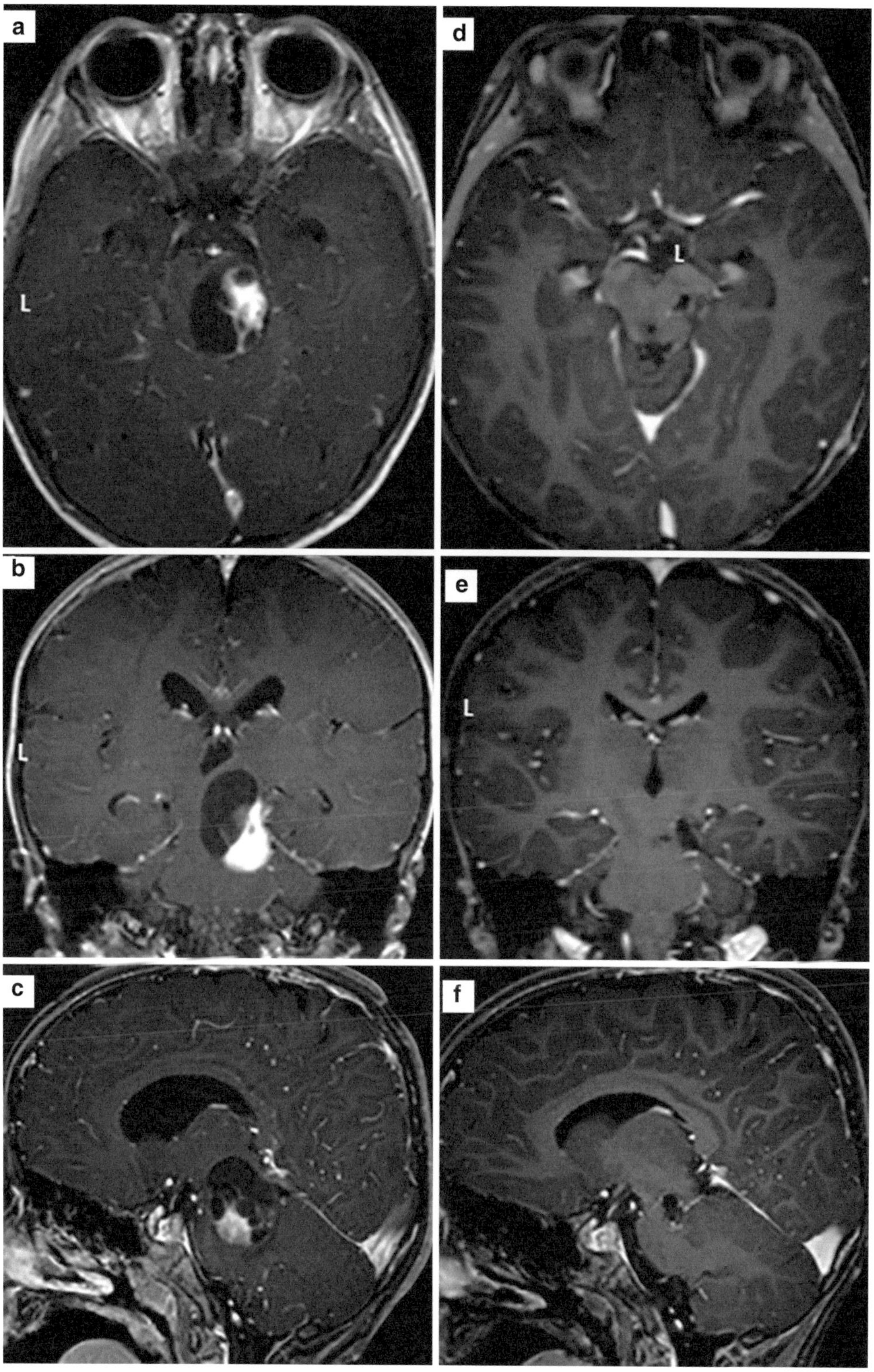

Fig. 14 Right-sided tegmental tumor. Preoperative axial (**a**), coronal (**b**), and sagittal (**c**) contrast-enhanced T1-weighted MRIs. Postoperative axial (**d**), coronal (**e**), and sagittal (**f**) contrast-enhanced T1-weighted MRIs. L, patient's left side

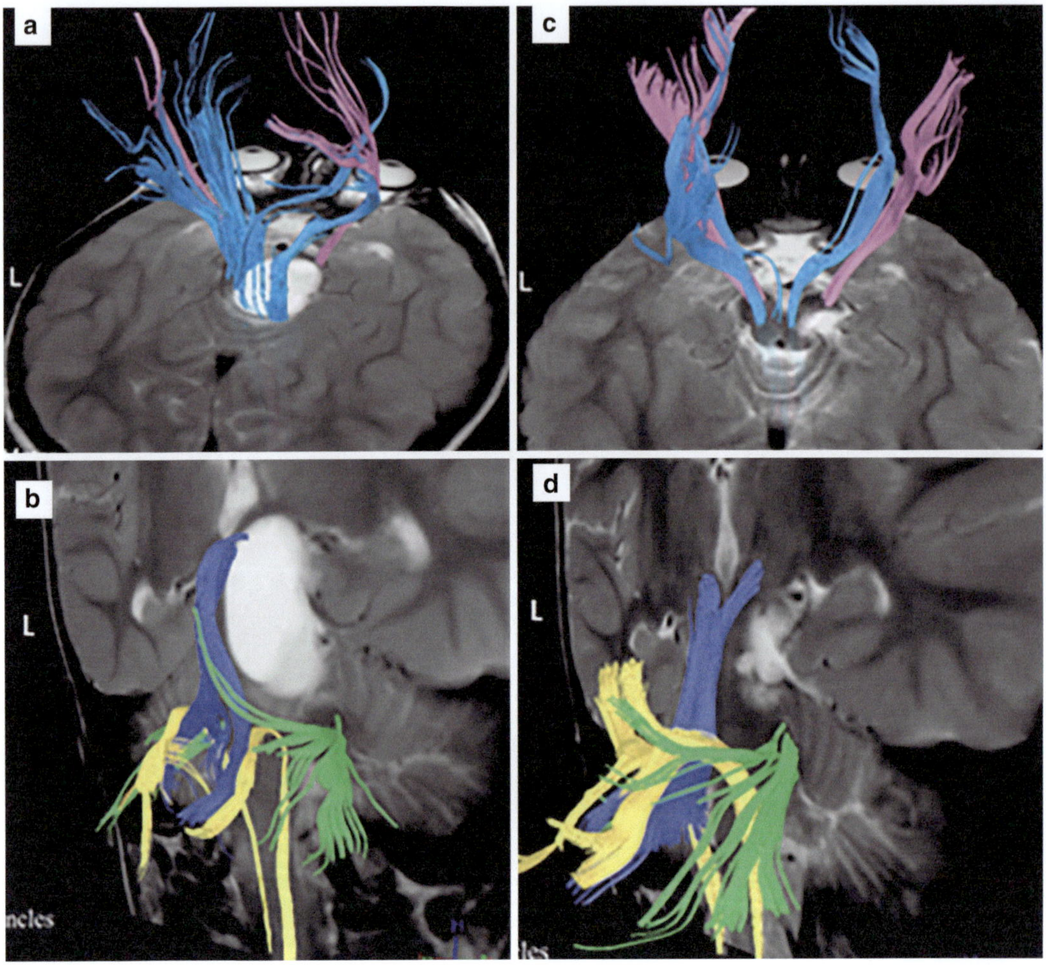

Fig. 15 Right-sided tegmental tumor. (**a**) Preoperative 3D axial reconstruction of the sensory and motor fibers. (**b**), Coronal section of the superior, middle, and inferior cerebellar peduncles. (**c**) Postoperative 3D axial reconstruction of sensory and motor fibers. (**d**) Coronal section of the superior, middle, and inferior cerebellar peduncles. L, patient's left side

the development of hydrocephalus. At our center, the preoperative FLAIR sequence showed the entire brain stem to be hyperintense (Figs. 16 and 17) and the tumor appeared inoperable. With the exception of diffuse pontine gliomas, however, brainstem masses are generally localized. On tractography images, the motor and sensory fibers were pushed anteriorly and the superior cerebellar peduncle was pushed lateraly. This image gave us the courage to try surgical removal of this glial tumor. The transuvulo-tonsillar approach was used with the patient in the semi-sitting position. Radical tumor resection was done, but not all hyperintense areas in the FLAIR sequence were removed. Nonetheless, this hyper-intensity disappeared by the third month after surgery.

Case 9, atrium tumor: A 24-year-old female had an incidentally detected left atrial cavernoma (Fig. 18). Because of its deep location and close proximity to critical neurovascular systems, surgical access to the atrium of the lateral ventricle remains difficult. The atrium is closely associated with the forceps major of the corpus callosum, optic radiation, and cingulum fibers. When we approach the atrium

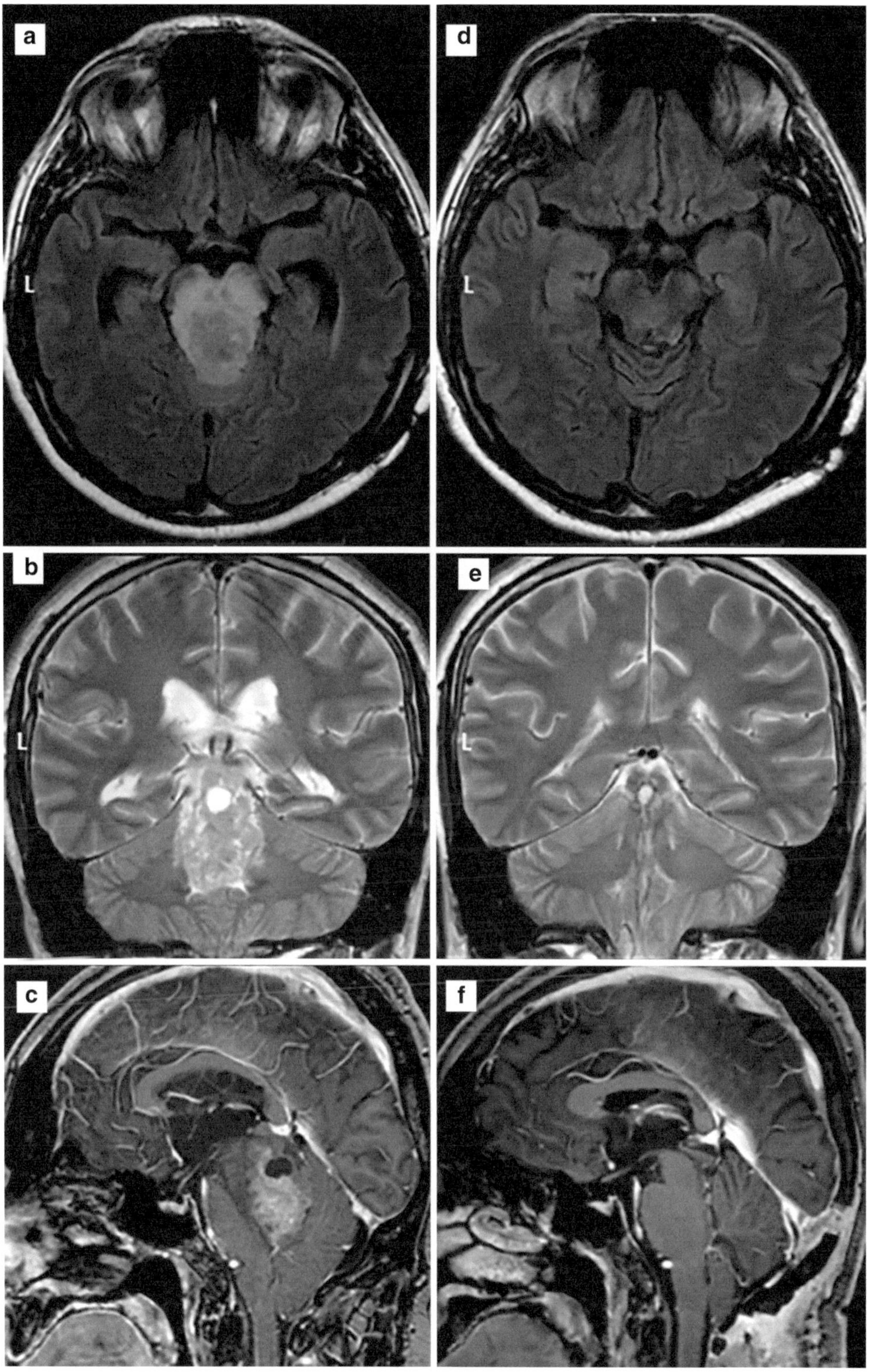

Fig. 16 Pons tumor. (**a**) Preoperative axial FLAIR image. (**b**) Coronal T2-weighted MRI. (**c**) Sagittal contrast-enhanced T1-weighted image. (**d**) Postoperative axial FLAIR. (**e**) Coronal T2-weighted image. (**f**) Sagittal contrast-enhanced T1-weighted MRI. L, patient's left side

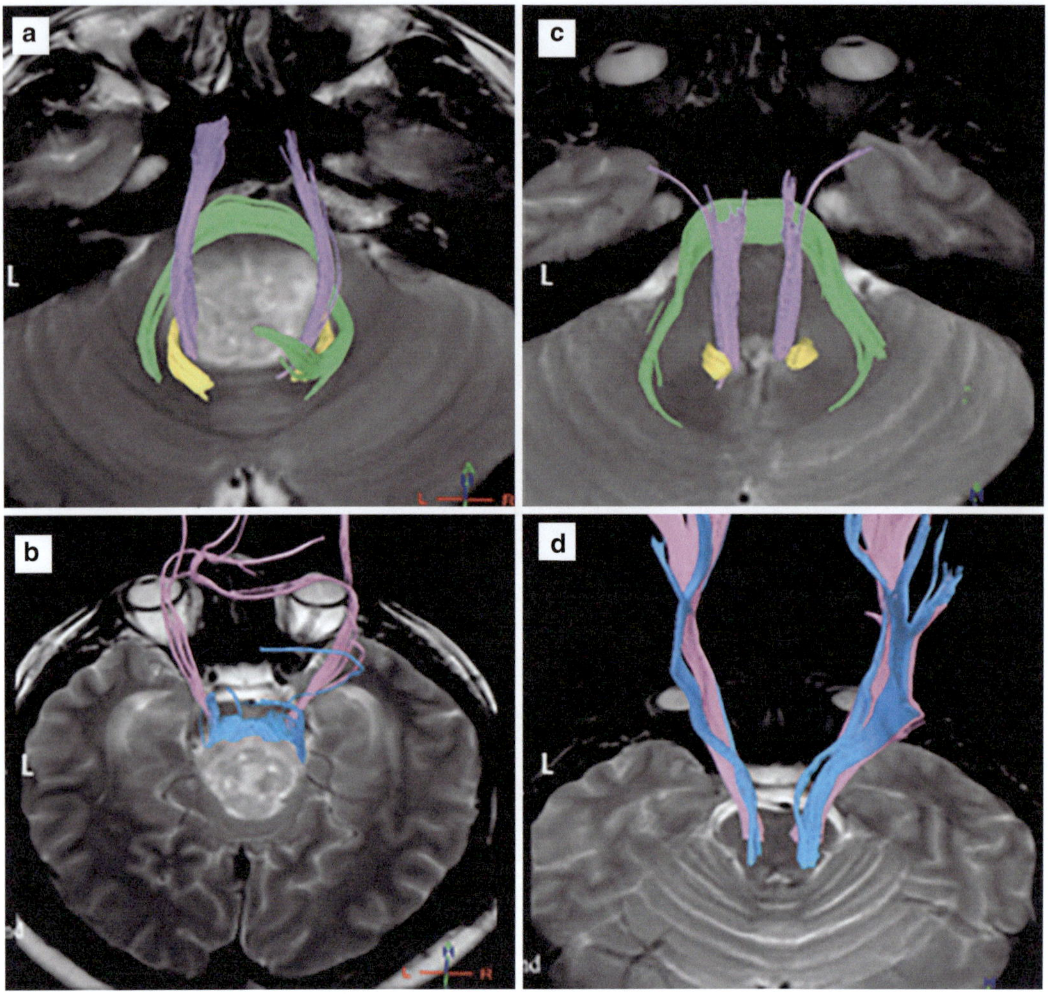

Fig. 17 Pons tumor. (**a**) Preoperative 3D axial reconstruction of the superior, middle, and inferior cerebellar peduncles. (**b**) The sensory and motor fibers. (**c**) Postoperative 3D axial reconstruction of the superior, middle, and inferior cerebellar peduncles. (**d**) The sensory and motor fibers. L, patient's left side

from the parieto-occipital sulcus located between the cuneus and precuneus, no white matter structure is damaged except for simple U fibers [95]. The cavernoma in this patient was removed by entering the atrium through the parieto-occipital sulcus. In this way, no white matter damage occurred except for a small part of the forceps major. No neurological deficit was observed in the patient in the postoperative period.

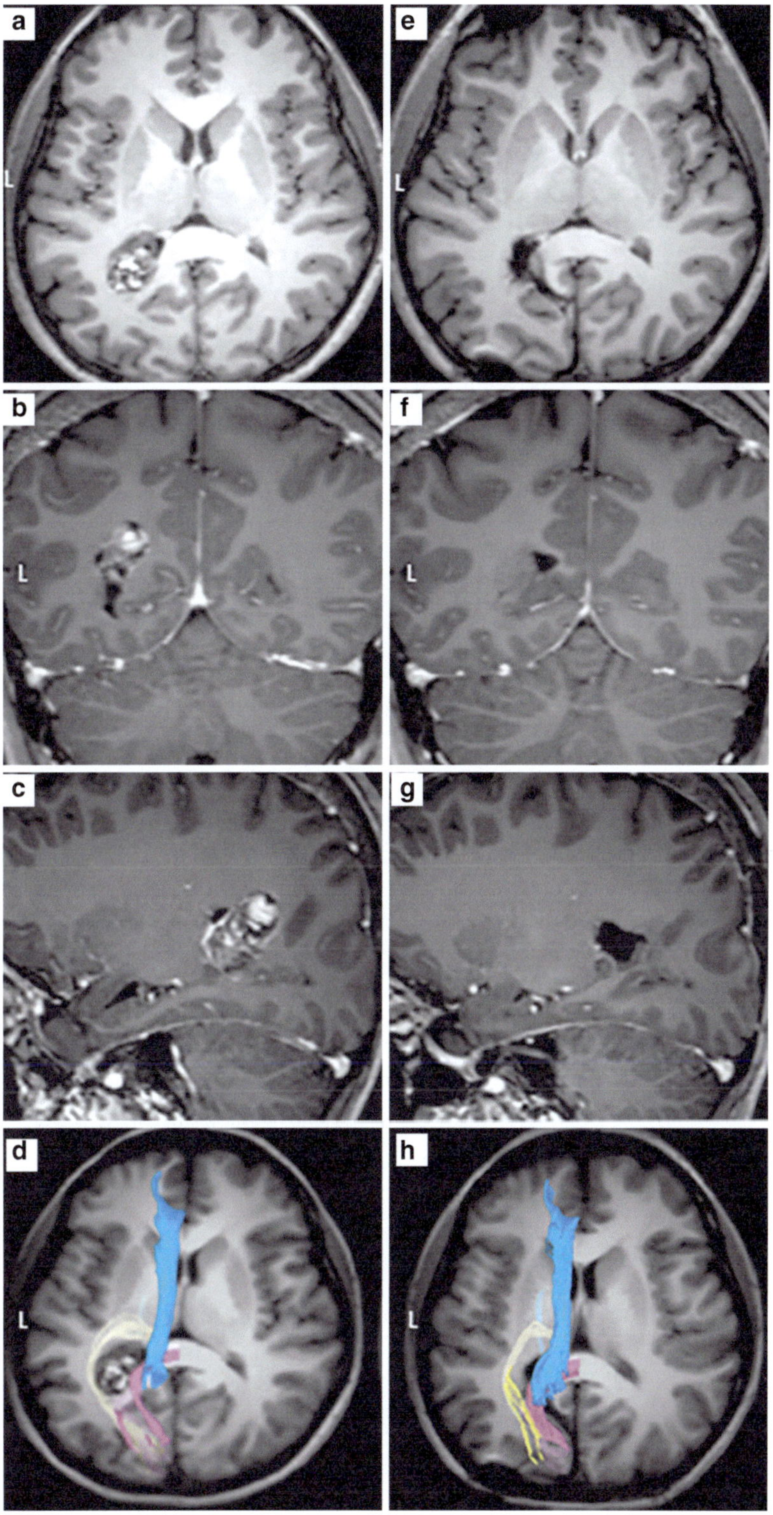

Fig. 18 Cavernoma of the left ventricle atrium. Preoperative axial (**a**), coronal (**b**), and sagittal (**c**) T1-weighted image. (**d**) 3D reconstruction of the preoperative axial DTI image. Postoperative axial (**e**), coronal (**f**), and sagittal (**g**) T1-weighted images. (**h**) 3D reconstruction of postoperative the axial DTI image. Blue, cingulum; purple, forceps major; yellow, Meyer's loop. L, patient's left side

6 Integration of White Matter Dissection, Photogrammetry, 3D Modeling, and Augmented/Virtual Reality Simulations in Neurosurgery and Neuroanatomy Education

With the increasing use of sophisticated imaging and the shift toward minimally invasive surgery, understanding and interpreting 3D anatomical relationships are crucial skills for clinical practice, especially among surgeons and neuroscientists. "Neurophobia" refers to generalized trouble with, and even dread of, specific subjects such as neurology and neuroscience [96]. This condition is exacerbated by the complexity of the neuroanatomical curriculum, which leads to an increase in intimidation and boredom [97]. Not only are there countless discrete structures in the central nervous system, but there are also sophisticated spatial relationships between them that must be recognized and understood. Horizontal and coronal section photos are commonly used in traditional anatomical education about the brain, and atlases and textbook illustrations provide further help.

The ability to observe and manipulate sections is critical for understanding the position of structures, but this does not initiate the cognitive leap required to convert all acquired knowledge into three dimensions. Because of the large number of structures and the intricate relationships between them, three-dimensional vision is extremely challenging to create. The typical educational methods used to teach anatomy include two-dimensional anatomic atlases, lectures, and a brief course in cadaveric dissection. However, increased cost and a lack of cadavers have reduced the practice in dissection classes. As cadaveric-based study declines, surgeons and neuroscientists are being asked to grasp complicated 3D anatomy and spatial relationships from two-dimensional representations seen in standard textbooks. Traditional methods and materials, such as textbooks, coronal and horizontal formalin-fixed brain slices, and photographs, are typically used to teach inner brain structures. However, these various materials are insufficient to help the neurosurgical student comprehend the three-dimensional diversity of white matter architecture. As a result, surgeons and neuroscientists struggle to grasp a clear understanding of white matter morphology.

All physicians should have a working knowledge of neuroanatomy as medical practice will benefit from a better understanding of human brain anatomy. Physicians' knowledge of this anatomy can be improved by adopting the Klingler fiber dissection technique, which focuses on the relationship between white matter tracts and all cortical and related gray matter structures [98, 99]. The extensive removal of surrounding brain tissue could disrupt the correlation of white matter tracts with cortical terminations. The destruction of adjacent brain structures eliminates important cortical landmarks used to establish precise three-dimensional relationships of a specific fiber tract. Furthermore, a dissection can easily be ruined by destroying or cutting through the tract because of the fragility of the fibers, which break readily and are difficult to repair if a split occurs along the surface. Fiber crossings are a concern as these fibers must be cut to continue dissection and isolation of the tract of interest. As a result, this approach cannot be used to dissect many tracts in the same specimen. Only with sufficient knowledge and training would it be possible to obtain well-dissected and informative specimens, and these are also time-consuming [1].

Fiber dissection is a useful technique for dissecting interior brain structures because it provides visualization of white matter fibers, particularly in terms of direction, depth, and relationships to gray-matter structures. This technique has inherent didactic value as it allows for a clear examination of white matter anatomy and delivers information that is now only available from coronal and horizontal slices. As previously noted, this technique takes a long time to master and, more significantly, it is extremely difficult for novice dissectors to achieve good results. Furthermore, with this appropriate educational material, medical students may be able to better understand brain anatomy in the future, reducing "neurophobia."

In the neuroanatomy curriculum, observing brain dissection utilizing the Klingler technique may boost students' confidence for future clinical work, enabling improved diagnosis and treatment of neurological patients. But more research in controlled settings is needed to confirm the applicability and genuine utility of Klingler's technique in teaching and learning human neuroanatomy. When the fiber dissection technique is used with traditional teaching materials, it is possible to appreciate the complex architecture of the white matter. Dissected brains have the advantage of attracting medical students' attention while also supplementing their knowledge through conventional cross-sectional sections, textbook figures, and atlases. These dissected specimens engage students more, making it easier for them to understand the brain's structural relationships, emphasizing the functional foundation of white matter tract systems and, most significantly, preparing them better for the future. The fact that this procedure is carried out in a stratigraphic manner adds value, making it an invaluable tool for gaining a thorough grasp of the complicated anatomical arrangement of the cerebral hemispheres.

In medical education and health care, access to 3D information is becoming increasingly crucial, and an understanding of anatomy is required for the safe and effective use of these technologies. But understanding this application can be difficult, especially in medical neurosciences. The present need is for innovative pedagogical strategies that achieve anatomical learning in less time, yet the ability of students to develop and function inside internal mental models of complicated biological structures remains a barrier. This is especially true in the field of neurosurgery and neuroanatomy. Because the nervous system is one of the most spatially complicated systems in the human body, new techniques that aid in 3D neuroanatomical visualization are extremely useful for improving learning.

Human cadavers are an important part of anatomy training but, in most regions of the world, access to cadaveric specimens and laboratory facilities is limited. As a result, more novel methodologies and widely accessible tools are critical for improving anatomy training. Augmented and virtual reality simulations and 3D models of progressive dissection of white matter tracts have been created and made available worldwide [47]. These models can be freely rotated and moved in multiple planes, as well as viewed from various angles and magnifications. In this way, the relationships of the structures can be easily understood. Better visualization is possible thanks to advanced editing features and the ability to export to a virtual- or augmented-reality simulation. Another advantage is that, once created, these high-quality 3D models can be freely exported and shared electronically. A single institution's rare or unique cadaveric specimen could be shared globally, allowing for wider access to a digital collection of 3D cadaveric specimens [47]. Furthermore, cadaveric specimens can be preserved in digital format, which eliminates the problems of discoloration and degeneration that occur with time in a physical specimen. In the future, emerging technologies such as augmented reality, virtual reality, and mixed reality may provide unique avenues for surgical planning and immersive surgical simulation. Future studies should include 3D modeling in patients before surgery and in practice pathologies. These 3D models can be combined with VR and/or augmented reality. To improve readability, future scientific papers could feature animated 360-degree 3D models, which could be printed in 3D. Human cadavers can be replaced or supplemented with 3D printed cadaveric replicas [47–61].

This technique may also help increase com prehension of the brain's intrinsic three-dimensional structure, which is crucial in both undergraduate and postgraduate neuroanatomy education. Estevez and colleagues [100] found that three-dimensional models outnumber two-dimensional models, implying that 3D physical modeling is an effective tool for teaching spatial relationships of brain anatomy and may better prepare students for visualization of 3D neuroanatomy. In a study by Silva and Andrade, undergraduate neuroanatomy students examined and manipulated specimens created using Klingler's technique [55]. Study participatns were asked to pay special attention to the organization of white matter fibers and their relationships to the ventricular system and grey matter structures. They responded to a poll about how valuable this material was in helping them comprehend the intrinsic three-dimensional anatomy of the white matter of the brain. The students reported that their grasp of the three-dimensional anatomy of the white matter had improved [55]. In this context, fiber dissection can help the student build a spatial perception of relevant anatomical features as well as their capability in dissection, and it could be a useful tool for neurosurgical training and research.

As surgical and neuroscientific studies have shown [58, 101, 102], knowledge of white matter anatomy is also critical for the surgical treatment of eloquent gliomas. During resection of such tumors, it is critical to preserve these structures. As a result, comprehending white matter pathway anatomy is an important aspect of the education of neuroscientists and physicians. This is the driving force for developing an increasing number of white matter dissection courses around the world. These, however, require costly infrastructure and anatomical specimens, which can make such courses prohibitively expensive and scarce. The interface for digital picture displays has improved, as has the use of augmented reality technologies in everyday life. While prototype devices for neuroanatomical data visualization have been on the market for some time, they are now available for purchase. Preprocessed imaging data can be viewed in an office, instructional suite, or conference room using customized goggles and spatial computation. This breakthrough has the potential to improve fiber tract anatomy education in both research and clinical settings. This technique also has the potential to make locating white matter tracts easier, as they can be difficult to spot during cadaver dissection. In the end, systems like these could be useful for teaching neuroscientists and neurosurgeons, as well as for preoperative preparation and planning. The surgical approach might be discussed in greater depth with the entire multidisciplinary team, who can scroll through the scans and view the anatomical objects as a group.

7 Conclusion

The main point surgeons should learn in white matter dissection is to respect every part of the brain parenchyma. Our main philosophy should be to avoid impairing healthy brain parenchyma as much as possible. Thus, we argued in the 1990s that white matter dissection should be a part of basic neurosurgery training [1, 60]. Fortunately, in the 2000s, numerous courses were organized in many countries so that, today, this technique has become a part of neurosurgery training. Although this technique is done under a microscope because of the complexity of the brain, it gives the surgeon a general perspective, but not enough details. For this reason, it is important to supplement this technique with histology and DTI. Although tractography is still a developing technology, it gives us very important information in surgery. The surgeon and neuroradiologist should communicate, and targeted tractography should be done and surgery planned accordingly. The important point for tractography is to define the tracts directly related to the pathology among hundreds of tracts so that only the targeted tracts are selected and confusion is avoided. In addition, surgeons should postoperatively evaluate the surgery they do with tractography and increase the quality of surgery by taking lessons from the experience.

References

1. Türe U, Yaşargil MG, Friedman AH, Al-Mefty O. Fiber dissection technique: lateral aspect of the brain. Neurosurgery. 2000;47:417–27.
2. Baur V, Hänggi J, Jäncke L. Volumetric associations between uncinate fasciculus, amygdala, and trait anxiety. BMC Neurosci. 2012;13:4.
3. Bell C. The anatomy of the brain. London: TN Longman; 1802.
4. Gall FJ, Spurzheim JC. Anatomie et physiologie du systéme nerveux en général, et du cerveau en particulier. Paris: F. Schoell; 1810.
5. Klingler J. Erleichterung der makroskopischen präparation des gehirn durch den gefrierprozess. Schweiz Arch Neurol Psychiatr. 1935;36:247–56.
6. Mayo H. A series of engravings intended to illustrate the structure of the brain and spinal cord in man. London: Burgess and Hill; 1827.

7. Reil JC. Das Balken-System oder die Balken-Organisation im großen Gehirn. Halle: Curtschen Buchhandlung; 1809.
8. Vieussens R. Neurographia universalis: Hoc est, omnium corporis humani nervorum, simul & cerebri, medullaeque spinalis descriptio anatomica. Lyons: Certe; 1684.
9. Clarke E, O'Malley CD. The human brain and spinal cord, a historical study illustrated by writing from antiquity to the twentieth century. Berkeley: University of California Press; 1968.
10. Clarke E, O'Malley CD. The human brain and spinal cord: a historical study illustrated by writings from antiquity to the twentieth century. 2nd ed. San Francisco: Norman Publishing; 1996.
11. Vesalius A. De Humani Corporis Fabrica. Basilae: Johannis Oporini; 1543.
12. Willis T. Cerebri Anatome: Cui Accessit Nervorum Descripto Et Usus. London: Jo Martyn & Ja Allestry; 1664.
13. Marshall LH, Magoun HW. Discoveries in the human brain: neuroscience prehistory, brain structure, and function. Cham: Springer Nature; 1998.
14. Rolando L. Della struttura degli emisferi cerebrali. Torino: Regia Academia delle Scienze di Torino; 1829.
15. Motzkin JC, Newman JP, Kiehl KA, Koenigs M. Reduced prefrontal connectivity in psychopathy. J Neurosci. 2011;31:17348–57.
16. Fernández-Miranda JC, Rhoton AL, Kakizawa Y, Choi C, Álvarez-Linera J. The claustrum and its projection system in the human brain: a microsurgical and tractographic anatomical study. J Neurosurg. 2008;108:764–74.
17. Malpighi M. De Cerebri Cortice. London: Jo Martyn & Ja Allestry; 1669.
18. Reil JC. Die Sylvische Grube oder das Gestreifte Große Hirnganglium, Dessen Kapsel und die Seitentheile des Großen Gehirns. Halle: Curtschen Buchhandlung; 1809.
19. Reil JC. Das Hirnschenkel-System oder die Hirnschenkel-Organisation im Großen Gehirn. Halle: Curtschen Buchhandlung; 1809.
20. Burdach KF. Vom Baue und Leben des Gehirns und Rückenmarks, vol. 3. Leipzig: Dyk'sche Buchhandlung; 1819.
21. Meyer A. Karl Friedrich Burdach and his place in the history of neuroanatomy. J Neurol Neurosurg Psychiatry. 1970;33:553–61.
22. Arnold F. Tabulae anatomicae: quas ad naturam accurate descriptas in lucem edidit (Band 1): Icones cerebri et medullae spinalis: decem tabulae elaboratae et totidem adumbratae. Turici: Impensis Orelli, Fueslini et Sociorum; 1838.
23. Foville AL. Traité complet de l'anatomie, de la physiologie et de la pathologie du système nerveux cérébro-spinal. Paris: Fortin, Masson; 1844.
24. Mai JK, Pazinos G, editors. The human nervous system. 3rd ed. London: Elsevier Academic Press; 2012.
25. Makris N, Pandya DN. The extreme capsule in humans and rethinking of the language circuitry. Brain Struct Funct. 2009;213:343–58.
26. Leuret F, Gratiolet P. Anatomie comparée du système nerveux: considéré dans ses rapports avec l'intelligence. Paris: J-B Baillière et fils; 1839.
27. Meynert TH. Vom gehirn der säugethiere. In: Stricker S, editor. Handbuch der Lehre von den Geweben des Menschen und dier Thiere. Leipzig: Engelmann; 1872.
28. Klingler J, Gloor P. The connections of the amygdala and of the anterior temporal cortex in the human brain. J Comp Neurol. 1960;115:333–69.
29. Ludwig E, Klingler J. Atlas Cerebri Humani: the inner structure of the brain. Little, Brown; 1956.
30. Yasargil MG, Türe U, Yasargil DCH. Impact of temporal lobe surgery. J Neurosurg. 2004;101:725–38.
31. Baydin S, Yagmurlu K, Tanriover N, Gungor A, Rhoton AL Jr. Microsurgical and fiber tract anatomy of the nucleus accumbens. Oper Neurosurg (Hagerstown). 2016;12:269–88.
32. Fernández-Miranda JC, Rhoton AL Jr, Alvarez-Linera J, Kakizawa Y, Choi C, de Oliveira EP. Three-dimensional microsurgical and tractographic anatomy of the white matter of the human brain. Neurosurgery. 2008;62(6 Suppl 3):989–1026. discussion 1026-1028
33. Güngör A, Baydin S, Middlebrooks EH, Tanriover N, Isler C, Rhoton AL Jr. The white matter tracts of the cerebrum in ventricular surgery and hydrocephalus. J Neurosurg. 2017;126:945–71.
34. Yagmurlu K, Middlebrooks EH, Tanriover N, Rhoton AL Jr. Fiber tracts of the dorsal language stream in the human brain. J Neurosurg. 2016;124:1396–405.
35. Yagmurlu K, Rhoton AL Jr, Tanriover N, Bennett JA. Three-dimensional microsurgical anatomy and the safe entry zones of the brainstem. Neurosurgery. 2014;10(Suppl 4):602–19. discussion 619–620
36. Yagmurlu K, Vlasak AL, Rhoton AL Jr. Three-dimensional topographic fiber tract anatomy of the cerebrum. Neurosurgery. 2015;11(Suppl 2):274–305. discussion 305
37. Peuskens D, van Loon J, Van Calenbergh F, van den Bergh R, Goffin J, Plets C. Anatomy of the anterior temporal lobe and the frontotemporal region demonstrated by fiber dissection. Neurosurgery. 2004;55:1174–84.
38. Kamada K, Todo T, Morita A, Masutani Y, Aoki S, Ino K, Kawai K, Kirino T. Functional monitoring for visual pathway using real-time visual evoked potentials and optic-radiation tractography. Neurosurgery. 2005;57(1 Suppl):121–7. discussion 121-127
39. Koutsarnakis C, Liakos F, Kalyvas AV, Sakas DE, Stranjalis G. A laboratory manual for stepwise cerebral white matter fiber dissection. World Neurosurg. 2015;84:483–93.
40. Kucukyuruk B, Richardson RM, Wen HT, Fernandez-Miranda JC, Rhoton AL Jr. Microsurgical anatomy of the temporal lobe and its implications on temporal lobe epilepsy surgery. Epilepsy Res Treat. 2012;2012:769825.

41. Shah A, Jhawar S, Goel A, Goel A. Corpus callosum and its connections: a fiber dissection study. World Neurosurg. 2021;151:e1024–35.
42. Von der Heide RJ, Skipper LM, Klobusicky E, Olson IR. Dissecting the uncinate fasciculus: disorders, controversies and a hypothesis. Brain. 2013;136:1692–707.
43. Demirtaş OK, Güngör A, Çeltikçi P, Çeltikçi E, Munoz-Gualan AP, Doğulu FH, Türe U. Microsurgical anatomy and insular connectivity of the cerebral opercula. J Neurosurg. 2022;137:1509–23. https://doi.org/10.3171/2021.12.JNS212297.
44. Dogan E, Gungor A, Dogulu F, Türe U. The historical evolution of the fornix and its terminology: a review. Neurosurg Rev. 2022;45:979–88.
45. Egemen E, Celtikci P, Dogruel Y, Yakar F, Sahinoglu D, Farouk M, Adiguzel E, Ugur HC, Coskun E, Güngör A. Microsurgical and tractographic anatomical study of transtemporal-transchoroidal fissure approaches to the ambient cistern. Oper Neurosurg (Hagerstown). 2021;20:189–97.
46. Güngör A, Baydın ŞS, Holanda VM, Middlebrooks EH, Isler C, Tugcu B, Foote K, Tanriover N. Microsurgical anatomy of the subthalamic nucleus: correlating fiber dissection results with 3-T magnetic resonance imaging using neuronavigation. J Neurosurg. 2018;130:716–32.
47. Gurses ME, Gungor A, Hanalioglu S, Yaltirik CK, Postuk HC, Berker M, Türe U. Qlone®: a simple method to create 360-degree photogrammetry-based 3-dimensional model of cadaveric specimens. Oper Neurosurg (Hagerstown). 2021;21:E488–93.
48. Gurses ME, Gungor A, Gökalp E, Hanalioglu S, Karatas Okumus SY, Tatar I, Berker M, Cohen-Gadol AA, Türe U. Three-Dimensional Modeling and Augmented and Virtual Reality Simulations of the White Matter Anatomy of the Cerebrum . Operative Neurosurgery. 2022;23(5):355–66.
49. Gurses ME, Gungor A, Rahmanov S, Gökalp E, Hanalioglu S, Berker M, Cohen-Gadol AA, Türe U. Three-Dimensional Modeling and Augmented Reality and Virtual Reality Simulation of Fiber Dissection of the Cerebellum and Brainstem. Operative Neurosurgery. 2022;23(5):345–54.
50. Gurses ME, Hanalioglu S, Mignucci-Jiménez G, Gokalp E, Gonzalez-Romo NI, Gungor A, Cohen-Gadol AA, Türe U, Lawton MT, Preul MC. Three-Dimensional Modeling and Extended Reality Simulations of the Cross-Sectional Anatomy of the Cerebrum, Cerebellum, and Brainstem. Operative Neurosurgery. 2023;25(1):3–10.
51. Şahin MH, Güngör A, Demirtaş OK, Postuk Ç, Fırat Z, Ekinci G, Kadıoğlu HH, Türe U. Microsurgical and fiber tract anatomy of the interthalamic adhesion. Journal of Neurosurgery. 2023;1(aop):1–0.
52. Demirtaş OK, Güngör A, Çeltikçi P, Çeltikçi E, Munoz-Gualan AP, Doğulu FH, Türe U. Microsurgical anatomy and insular connectivity of the cerebral opercula. Journal of Neurosurgery. 2022;137(5):1509–23.
53. Hanalioglu S, Romo NG, Mignucci-Jiménez G, Tunc O, Gurses ME, Abramov I, Xu Y, Sahin B, Isikay I, Tatar I, Berker M. Development and validation of a novel methodological pipeline to integrate neuroimaging and photogrammetry for immersive 3D cadaveric neurosurgical simulation. Frontiers in surgery. 2022;9.
54. Gonzalez-Romo NI, Mignucci-Jiménez G, Hanalioglu S, Gurses ME, Bahadir S, Xu Y, Koskay G, Lawton MT, Preul MC. Virtual neurosurgery anatomy laboratory: A collaborative and remote education experience in the metaverse. Surgical neurology international. 2023;14.
55. Silva SM, Andrade JP. Neuroanatomy: the added value of the Klingler method. Ann Anat. 2016;208:187–93.
56. Carpenter MB. Core text of neuroanatomy. 2nd ed. Williams & Wilkins; 1978.
57. Türe U, Yaşargil DC, Al-Mefty O, Yaşargil MG. Topographic anatomy of the insular region. J Neurosurg. 1999;90(4):720–33.
58. De Witt Hamer PC, Hendriks EJ, Mandonnet E, Barkhof F, Zwinderman AH, Duffau H. Resection probability maps for quality assessment of glioma surgery without brain location bias. PLoS One. 2013;8:e73353.
59. Shekari E, Goudarzi S, Shahriari E, Joghataei MT. Extreme capsule is a bottleneck for ventral pathway. IBRO Neurosci Rep. 2021;10:42–50.
60. Türe U, Yaşargil MG, Pait TG. Is there a superior occipitofrontal fasciculus? A microsurgical anatomic study. Neurosurgery. 1997;40:1226–32.
61. Cooney RE, Atlas LY, Joormann J, Eugène F, Gotlib IH. Amygdala activation in the processing of neutral faces in social anxiety disorder: is neutral really neutral? Psychiatry Res. 2006;148:55–9.
62. Matsuo K, Mizuno T, Yamada K, Akazawa K, Kasai T, Kondo M, Mori S, Nishimura T, Nakagawa M. Cerebral white matter damage in frontotemporal dementia assessed by diffusion tensor tractography. Neuroradiology. 2008;50:605–11.
63. Tröstl J, Sladky R, Hummer A, Kraus C, Moser E, Kasper S, Lanzenberger R, Windischberger C. Reduced connectivity in the uncinate fiber tract between the frontal cortex and limbic subcortical areas in social phobia. Eur Psychiatry. 2011;26:182.
64. Standring S, Ellis H, Healy J, Johnson D, Williams A, Collins P, Wigley C. Gray's anatomy: the anatomical basis of clinical practice. Am J Neuroradiol. 2005;26(10):2703.
65. Herbet G, Zemmoura I, Duffau H. Functional anatomy of the inferior longitudinal fasciculus: from historical reports to current hypotheses. Front Neuroanat. 2018;12:77.
66. Ballantine HT Jr, Cassidy WL, Flanagan NB, Marino R Jr. Stereotaxic anterior cingulotomy for neuropsychiatric illness and intractable pain. J Neurosurg. 1967;26:488–95.
67. Garcia-Bengochea F, Friedman WA. Persistent memory loss following section of the anterior for-

nix in humans. A historical review. Surg Neurol. 1987;27:361–4.
68. Hamani C, McAndrews MP, Cohn M, Oh M, Zumsteg D, Shapiro CM, Wennberg RA, Lozano AM. Memory enhancement induced by hypothalamic/fornix deep brain stimulation. Ann Neurol. 2008;63:119–23.
69. Goga C, Türe U. The anatomy of Meyer's loop revisited: changing the anatomical paradigm of the temporal loop based on evidence from fiber microdissection. J Neurosurg. 2015;122:1253–62.
70. Naidich T, Duvernoy HM, Delman BN, Sorensen AG, Kollias SS, Haacke EM. Duvernoy's Atlas of the human brain stem and cerebellum: high-field MRI: surface anatomy, internal structure, vascularization and 3D sectional anatomy. Vienna: Springer; 2009.
71. Bhardwaj N, Yadala S. Neuroanatomy, corticobulbar Ttact. In: StatPearls. Treasure Island (FL): StatPearls Publishing; 2022.
72. Gray H. Anatomy of the human body. 20th ed. Bartlebycom; 2000.
73. Rea P. Brainstem tracts. In: Rea P, editor. Essential clinical anatomy of the nervous system. Amsterdam: Elsevier; 2015. p. 177–92.
74. Navarro-Orozco D, Bollu PC. Neuroanatomy, medial lemniscus (Reils band, Reils ribbon). Treasure Island (FL): StatPearls Publishing; 2022.
75. Voogd J. Cerebellum and precerebellar nuclei. In: Paxinos G, Mai JK, Mai JK, editors. The human nervous system. 2nd ed. Elsevier Science; 2004. p. 321–92.
76. Nieuwenhuys R, Voogd J, van Huijzen C. The human central nervous system. 3rd ed. Berlin: Springler; 1988.
77. Catani M, Howard RJ, Pajevic S, Jones DK. Virtual in vivo interactive dissection of white matter fasciculi in the human brain. NeuroImage. 2002;17:77–94.
78. Mamata H, Mamata Y, Westin C-F, Shenton ME, Kikinis R, Jolesz FA, Maier SE. High-resolution line scan diffusion tensor MR imaging of white matter fiber tract anatomy. AJNR Am J Neuroradiol. 2002;23:67–75.
79. Basser PJ, Pierpaoli C. Microstructural and physiological features of tissues elucidated by quantitative-diffusion-tensor MRI. J Magn Reson B. 1996;111:209–19.
80. Hofer S, Frahm J. Topography of the human corpus callosum revisited—comprehensive fiber tractography using diffusion tensor magnetic resonance imaging. NeuroImage. 2006;32:989–94.
81. Duffau H. New concepts in surgery of WHO grade II gliomas: functional brain mapping, connectionism and plasticity—a review. J Neuro-Oncol. 2006;79:77–115.
82. Yaşargil M. Microneurosurgery IVA. New York: Thieme; 1996.
83. Yaşargil M. Microneurosurgery IVB. New York: Thieme; 1996.
84. Berman JI, Berger MS, Mukherjee P, Henry RG. Diffusion-tensor imaging-guided tracking of fibers of the pyramidal tract combined with intraoperative cortical stimulation mapping in patients with gliomas. J Neurosurg. 2004;101:66–72.
85. Kamada K, Sawamura Y, Takeuchi F, Kawaguchi H, Kuriki S, Todo T, Morita A, Masutani Y, Aoki S, Kirino T. Functional identification of the primary motor area by corticospinal tractography. Neurosurgery. 2005;56(1 Suppl):98–109. discussion 98-109
86. Nimsky C, Ganslandt O, Fahlbusch R. Implementation of fiber tract navigation. Neurosurgery. 2006;58(4 Suppl):ONS-292-303. discussion ONS-303-304
87. Nimsky C, Ganslandt O, Hastreiter P, Wang R, Benner T, Sorensen AG, Fahlbusch R. Intraoperative diffusion-tensor MR imaging: shifting of white matter tracts during neurosurgical procedures—initial experience. Radiology. 2005;234:218–25.
88. Nimsky C, Ganslandt O, Hastreiter P, Wang R, Benner T, Sorensen AG, Fahlbusch R. Preoperative and intraoperative diffusion tensor imaging-based fiber tracking in glioma surgery. Neurosurgery. 2005;56:130–7. discussion 138
89. Kamada K, Todo T, Masutani Y, Aoki S, Ino K, Takano T, Kirino T, Kawahara N, Morita A. Combined use of tractography-integrated functional neuronavigation and direct fiber stimulation. J Neurosurg. 2005;102:664–72.
90. Henry RG, Berman JI, Nagarajan SS, Mukherjee P, Berger MS. Subcortical pathways serving cortical language sites: initial experience with diffusion tensor imaging fiber tracking combined with intraoperative language mapping. NeuroImage. 2004;21:616–22.
91. Sporns O, Tononi G, Kötter R. The human connectome: a structural description of the human brain. PLoS Comput Biol. 2005;1:e42.
92. Toga AW, Clark KA, Thompson PM, Shattuck DW, Van Horn JD. Mapping the human connectome. Neurosurgery. 2012;71:1–5.
93. Türe U, Harput MV, Kaya AH, Baimedi P, Firat Z, Türe H, Bingöl CA. The paramedian supracerebellar-transtentorial approach to the entire length of the mediobasal temporal region: an anatomical and clinical study. Laboratory investigation. J Neurosurg. 2012;116:773–91.
94. Türe U, Kaya AH, Bingöl CA. Transsylvian selective amygdalohippocampectomy. In: Cataltepe O, Jallo GI, editors. Pediatric epilepsy surgery: preoperative assessment and surgical treatment. 1st ed. Stuttgart: Thieme; 2010. p. 147–55.
95. Panteli A, Güngör A, Fırat Z, Sarıtepe F, Türe H, Türe U. The posterior interhemispheric transparieto-occipital fissure approach to the atrium of the lateral ventricle: a fiber microdissection study with case series. Neurosurg Rev. 2022;45:1663–74.
96. Jozefowicz RF. Neurophobia: the fear of neurology among medical students. Arch Neurol. 1994;51:328–9.
97. Ramos RL, Smith PT. A core neuroanatomy syllabus for diverse student populations. Clin Anat. 2016;29:131.
98. Shah A, Goel A, Jhawar SS, Patil A, Rangnekar R, Goel A. Neural circuitry: architecture and function - a fiber dissection study. World Neurosurg. 2019;125:e620–38.

99. Shah A, Jhawar SS, Nunez M, Goel A, Goel A. Brainstem anatomy: a study on the basis of the pattern of Fiber organisation. World Neurosurg. 2020;134:e826–46.
100. Estevez ME, Lindgren KA, Bergethon PR. A novel three-dimensional tool for teaching human neuroanatomy. Anat Sci Educ. 2010;3:309–17.
101. Ius T, Angelini E, de Schotten MT, Mandonnet E, Duffau H. Evidence for potentials and limitations of brain plasticity using an atlas of functional resectability of WHO grade II gliomas: towards a "minimal common brain". NeuroImage. 2011;56:992–1000.
102. Sanai N, Mirzadeh Z, Berger MS. Functional outcome after language mapping for glioma resection. N Engl J Med. 2008;358:18–27.

An Overview of Connectomics and Glioma Surgery

Nicholas B. Dadario and Michael E. Sughrue

1 Introduction

A principle tenet in the modern neurosurgical care of gliomas is that the extent of resection improves overall survival [1]. While surely justified, this idea has caused the neurosurgical community to aim for maximal, and even supramaximal, resection of tumor tissue whenever surgically feasible. However, large advancements in the field over past decades have continued to offer limited improvements in overall survival without simultaneously causing devasting neurologic deficits [2]. This trade-off in cytoreduction and functional consequences has been termed the "onco-functional" balance, and represents a key concept in modern glioma surgery [3]. In order to optimize patient functioning following resective brain surgery, neurosurgeons have traditionally been taught to avoid damage in the so-called "eloquent" regions, which when removed have readily associated functional deficits [4]. An example of this can be seen with the primary motor strip, which when removed, will likely cause an observable motor deficit. However, this dichotomous model also proposes that non-eloquent regions outside of language and motor cortices are "silent," which hazardously suggests these regions are safe to remove in all patients due to having less overt functional significance [4]. Importantly, use of this simplified, but dominant model is likely a strong reason for the severe deficits glioma patients continue to demonstrate in more subtle, higher-order functions outside of just overt motor or language outcomes [5–8].

Deficits in complex, higher-order functions, like cognition and emotional processing, are increasingly recognized following resective brain surgery and cannot be readily explained by localizationist views of eloquent versus non-eloquent cerebrum. Instead, one concept which has emerged from the general Neuroscientific community is that of the large-scale brain networks. Brain networks provide an organizing principle for the structural and functional connections of the brain in a way which can explain previously less understood, complex functions like cognition [9]. Such higher-order functions rely not on a single piece of brain tissue, but rather an intricate and highly-synchronized group of spatially distant brain regions which dynamically interact to carry out human functioning. By using brain network maps which can define resection boundaries based on functional significance [10], we may further optimize the onco-functional balance in glioma surgery, especially in traditionally less familiar anatomy responsible for higher-order

N. B. Dadario
Robert Wood Johnson Medical School, Rutgers University, New Brunswick, NJ, USA

M. E. Sughrue (✉)
Centre for Minimally Invasive Neurosurgery, Prince of Wales Private Hospital, Randwick, NSW, Australia

Omniscient Neurotechnology, Sydney, NSW, Australia

A. Shah et al. (eds.), *Functional Anatomy of the Brain: A View from the Surgeon's Eye*,
https://doi.org/10.1007/978-981-99-3412-6_4

functions. However, the concept of the large-scale brain networks and connectomics in general has yet to fully enter the regular thinking of neuro-oncological surgeons.

Therefore, the goal of this chapter is to address this gap by introducing connectomics and the concept of brain networks as it relates to glioma surgery. As it is beyond the scope of a single chapter to meticulously describe every single structural and functional connection identified in the human brain, which is still a largely on-going discussion, we aim to provide a broad overview of the large-scale brain networks and focus specifically on their relevance to the more subtle, higher-order cognitive functions.

2 What Is the Brain Connectome?

The brain connectome refers to all of the connections across the human cerebrum. Numerous tools for measuring connectivity, such as functional neuroimaging sequences, have allowed us to characterize the interconnectivity of individual neurons in great detail across a number of organisms, including humans [11]. However, we will focus specifically on the large scale fibers and functional networks which involve the majority of information transfer across the human cerebrum and are of special relevance for neurosurgeons, rather than individual axonal connections which are still under extensive study.

2.1 Functional Connectome

Functional connectivity analyses aim to capture highly synchronized, but spatially distant brain regions which activate or deactivate in statistically meaningful ways during similar time periods. Specifically, they are often measured according to spontaneous changes in blood-oxygenation-level-dependent (BOLD) signals, usually either at rest or during a meaningful task [12]. Importantly, these analyses have continually demonstrated networks of brain regions which are robust and reproducible between studies [13–15].

2.2 Structural Connectome

It has recently become more clear that there are often large-scale white matter bundles which connect functionally synchronous region throughout a structural network—*regions that are wired together fire together* [10]. These connections constitute the structural connectome and are often measured by diffusion tensor imaging (DTI) or diffusion spectrum imaging (DSI) based methods, which estimates the diffusion of water along white matter bundles to reconstruct structural pathways. Importantly, structural connections place important constraints on a functional network as the brain connectome is wired in a way which maximizes information transfer while also minimizing metabolic costs to execute complex functions [16]. Therefore, targeted damage to highly connected regions in a network, such as the core of a network, can carry out disproportionate damage on a networks capacity for information transfer [16], and likely a patient's ability to compensate for functional deficits.

Improved information on the structural and functional connectome can allow for more precise localization of complex symptomology that exceeds previous localizationist views [17], while also prospectively identifying regions which should be spared in surgery [18]. This is important because complex functions, such as executive functioning or attentional processing, do not tend to rely on a single localized brain region. Instead, higher-order, cognitive functions are carried about by dynamic interactions between numerous functionally synchronized and spatially distant regions in a network, which when structurally damaged in surgery, can lead to a loss of that function.

3 Who and What Are the Major Networks We Should Care About?

Significant advancements in the neuroscientific community, such as in the Human Connectome Project (HCP), have led to a fundamental reappraisal of previously outdated neuroanatomical maps, such as Brodmann's areas [11]. Starting from Brodmann's original map of cerebral cortex, newer maps detail over 180 unique cortical parcellations based on functional connectivity, degree of myelination, cortical thickness, and other parcellation schemes. Unsurprisingly, the number of brain networks which these regions associate with can become quite large (i.e., 17) [14], with each network being further subdivided based on highly specialized functions. Instead, we will briefly introduce the seven main large-scale brain networks and provide a framework for neurosurgeons to build on moving forward. These networks are the Default Mode Network (DMN) [13, 19], Central Executive Network (CEN) [20], Salience Network (SN) [15], Dorsal Attention Network (DAN) [21], Sensorimotor Network, Limbic Network, and Visual Network. We provide an illustration of their cores in Fig. 1 and categorize them further below.

3.1 The Cognitive Control Networks

Three canonical resting state networks of which are helpful to first understand are the DMN, CEN, and SN. We have termed these networks the cognitive control networks, as they lie at the top of the network hierarchy and define an axis by which the other networks align to subserve complex human cognition and related functions [9]. Unsurprisingly, disruption of the core these networks is thought to underlie a number of neuro-psychiatric illnesses, like depression [22], due to the inability to allocate cognitive resources between internal and external thoughts or actions [23]. A recent review of the neuroscientific literature conducted by our team identified that damage or dysfunction of the non-traditional large-scale brain networks almost always causes a neurological, cognitive, or emotional deficit, most of which (33%) is specifically the result of

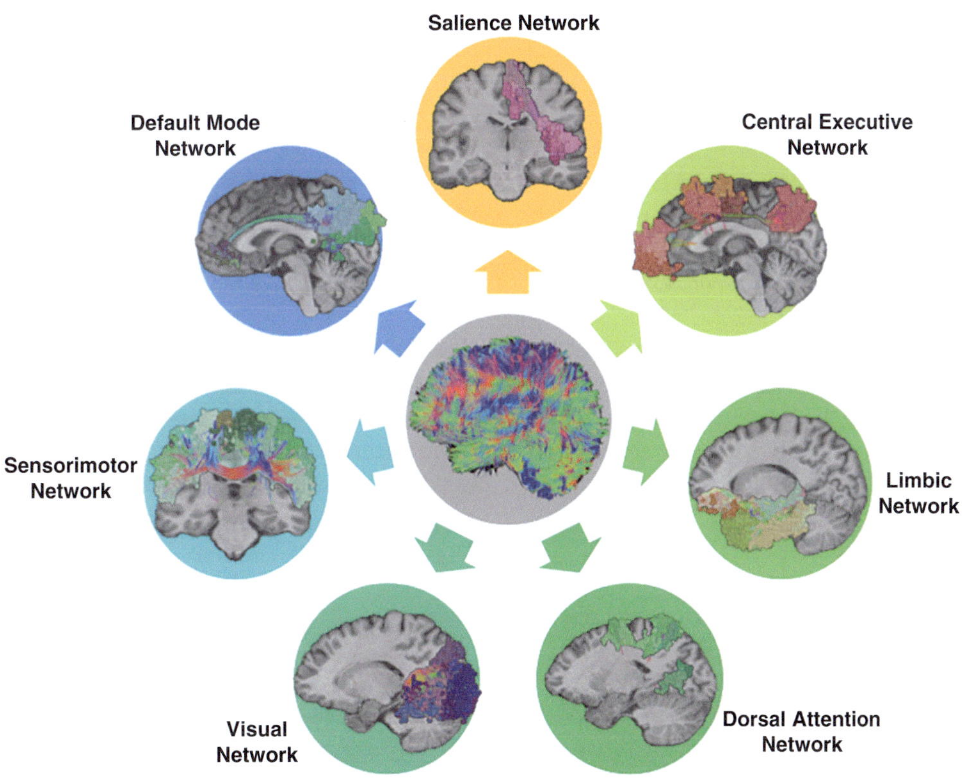

Fig. 1 The seven large-scale brain networks

disruption in a combination of these three networks.

3.1.1 Default Mode Network

The DMN is one of the most recently documented higher-order networks in the literature, and can be thought of as the internal mind responsible for passive thought, theory of mind, imagination, and further roles episodic memory [13, 24]. Similarly, it is often described also as a "task-negative" network because it is active when the subject is at rest. The DMN has three main clusters of parcellations which are structurally connected via the cingulum bundle. Its core includes anterior cingulate, posterior cingulate, and lateral parietal clusters [19].

3.1.2 Central Executive Network

The CEN is also a three-part network with a lateral frontal polar region, posterior aspect of the dorsolateral prefrontal cortex, and part of the inferior parietal lobe in the supramarginal gyrus [20]. The CEN is a superordinate control network and can be thought of as the external mind with key roles in executive functioning and active processes based on integrating information from other networks. It is anti-correlated with the DMN, in which the CEN generally fires during active functioning when the DMN is silent, and generally forces the brain into energetically less favorable states to carry out a complex function. Structurally, its main fiber bundle is the superior longitudinal fascicles (SLF), which is a major fiber bundle that links a number of other networks as well which are involved in attentional processing [25]. Unsurprisingly, the CEN correlates strongly with attentional networks as well [21].

3.1.3 Salience Network

The SN can be thought of as the mediator between our internal (DMN) and external (CEN) minds. It is believed to control the allocation of cognitive resources between the DMN and CEN through task switching, and is highly characterized in a number of mental illnesses, like depression [15, 23]. Anatomically, it includes an anterior insula-operculum cluster, a middle cingulate cluster, and the middle portion of the dorsolateral prefrontal cortex, which are all mostly linked by the frontal aslant tract (FAT) [18].

3.2 Expansion of More Traditional Networks

The rest of the major large-scale brain networks generally align themselves around the actions of the cognitive control networks, specifically the CEN axis or DMN axis. We will not dive into great detail about these more classically recognized networks. However, it is important to note that while they have all been previously recognized, recent connectivity maps suggest they are far more interconnected than we once believed [25].

3.2.1 Sensorimotor Network

The sensorimotor network works closely with the other major networks to integrate information necessary for processing sensory stimuli, and then relays this information elsewhere to execute the appropriate responses [26]. This could include transducing information to the SN for reward processing, CEN for task processing, visual system for sight, and so on. It includes functional areas of the primary motor cortex, primary and sensory cortices in the parietal lobe, part of the cingulate cortex, premotor cortex and most of the supplementary motor area.

3.2.2 Limbic Network

The limbic network is one of the longest known networks and has been classically recognized to control emotional processing. Further study has revealed a number of additional roles, such as in social cognition, emotional learning and memory, and olfaction, among others [27]. It is an extremely widespread network including the amygdala, thalamus, hypothalamus, hippocampus, and many paralimbic structures [28].

3.2.3 Dorsal Attention Network

The DAN is involved in the top-down processing of object and goal-related stimuli [21]. In contrast to the right-hemispheric ventral attention network (VAN) [29], which reorients attention to behaviorally relevant stimuli, the DAN controls more voluntary attentional processing and works closely with the CEN to execute appropriate responses. It is a bilateral network located in the frontal eye fields, intraparietal sulcus, superior parietal lobe, and visual cortex. Similar to the CEN, it is mainly structurally connected by SLF fibers.

4 Brain Network Maps to Navigate the Brain

The HCP project has surely provided the neuroscience community with an extremely valuable and detailed amount of information about the human brain. However, transforming this information into clinical data which may be useful in neurosurgery has required more work [10]. Through machine-learning based approaches, we recently published a connectomic atlas detailing the anatomy, functional connectivity, and structural connectivity of all 180 cortical parcellations from the HCP. This information represents anatomically precise and clinically actionable networks maps that can be leveraged to guide neurosurgical decision making, such as when operating in specific cortical regions [10, 25, 30, 31]. Thus, tumors and surgical plans can be defined against clear boundaries marked by network fibers with known functional significance. With connectome-based neurosurgery, we may begin to make more informed decisions surgery while causing fewer cognitive deficits according to pre-defined surgical and patient goals [9]. At the least, the primary goal of incorporating these maps would include avoiding the core of the networks in Fig. 1 whenever surgically feasible.

It may advantageous if we begin to contextualize glioma surgery as a series of oncological "disconnections" rather than a series of resections in order to limit unnecessary network damage during surgery. In Fig. 2, we demonstrate an

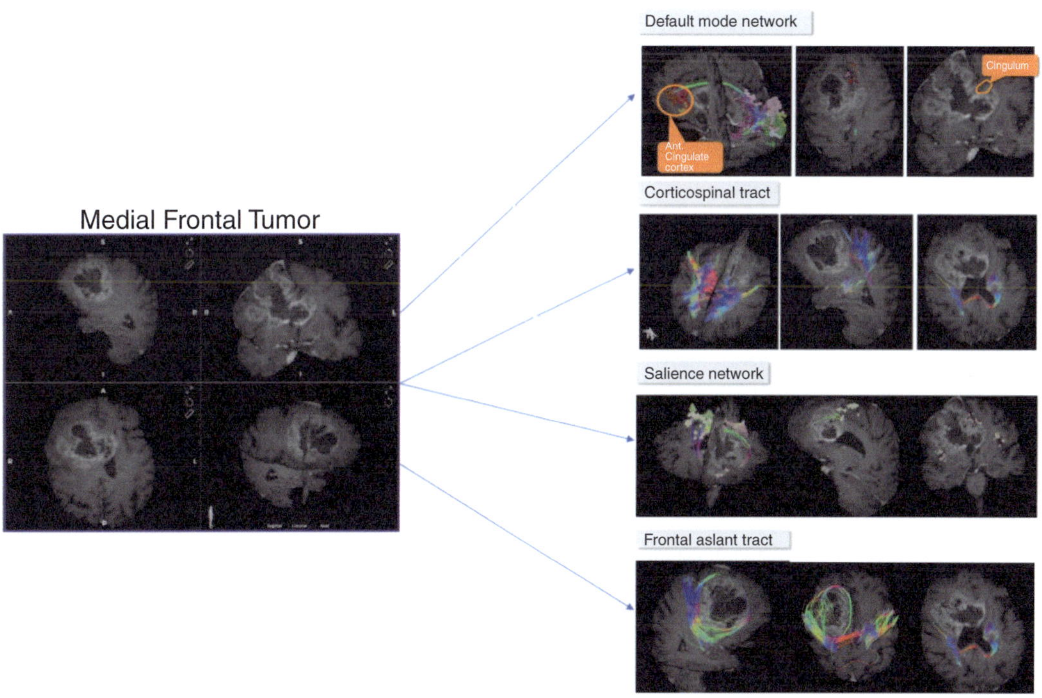

Fig. 2 Defining the risks of the tumor against networks. The current case depicts a right medial frontal glioma. In the right column, important networks which are disrupted by the tumor and at risk during surgery are defined pre-operatively

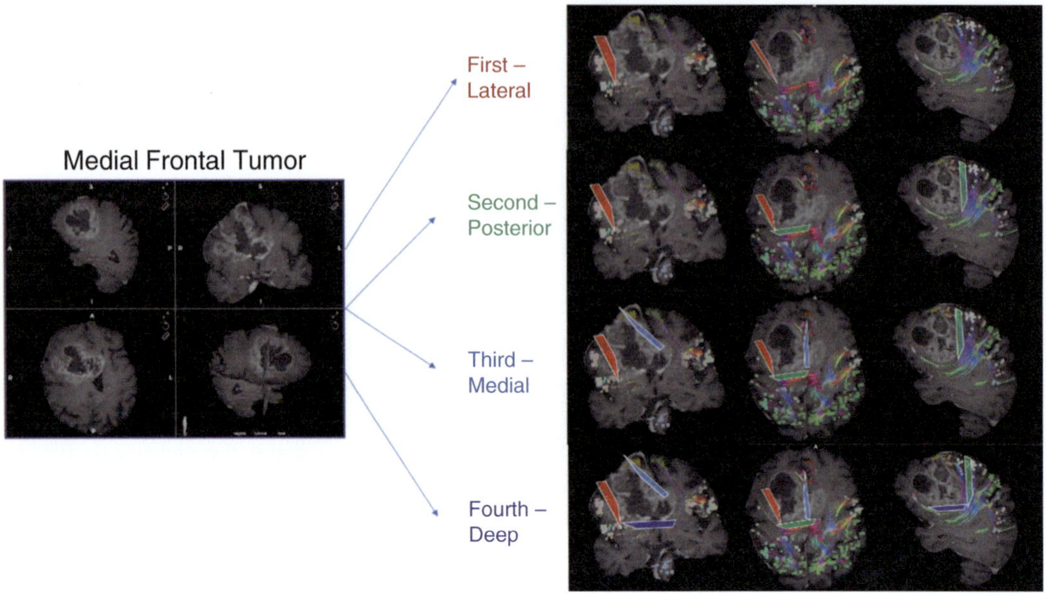

Fig. 3 Planning the disconnection against network boundaries. The current case depicts a right medial frontal glioma. It can be thought of as a series of four disconnections against major network fibers. (1) A lateral disconnection is defined against the tumor while avoiding language fibers. (2) A posterior disconnection is defined which avoids the frontal aslant tract of the salience network and supplementary motor area system. (3) A medial disconnection is defined against the default mode network and cingulum fibers. (4) The final disconnection follows the sulci from superior-inferior without damaging the deep underlying basal ganglia and thalamus structures

example of oncological disconnection surgery for frontal lobe gliomas, which commonly disrupt numerous higher-order brain networks necessary for cognition and emotional processing. Specifically, a medial frontal disconnection is useful for tumors around the superior or middle frontal gyri, and will involve more than just one major disconnection. As one can see, this tumor is an extremely complex tumor which butterflys across midline and has a significant lateral portion. However, the surgical risks of this tumor can be better understood in the context of the surrounding networks, which represent of the most risks of inducing cognitive deficits in this surgical field. In the right column of Fig. 2, we can see that functional regions of concern include the DMN medially, the salience network with its key frontal aslant tract posteriorly, and the corticospinal tract behind the tumor which normally connects the sensorimotor cortex to the spinal cord. We must consider that butterfly portion of the tumor goes under anterior cingulate cluster of the DMN, and if we damage the callosal sulcus for instance, then we can disconnect the DMN from itself and completely stop its function. Of likely the most importance is the SN. We see the tumor has completely disconnected the middle cingulate and lateral insular clusters of the SN and also the tumor completely envelopes the FAT fibers.

After first defining the network boundaries around the tumor, we may then begin to contextualize the surgery as a series of disconnections which aim to cause minimal unnecessary disturbances to a network as surgically feasible. In Fig. 3, we demonstrate first performing a lateral disconnection in order to improve our ability to manipulate the tumor without first damaging the medial DMN system or posterior FAT fibers. In this disconnection, our goal is to separate it from the lateral language SLF and inferior fronto-occipital fasciculus (IFOF) fibers. Then, we must disconnect the tumor from posterior and medial networks, which we can be defined against the salience network and default mode network, respectively. The final disconnections aims to follow the sulci from superior-inferior without damaging these networks or the underlying basal ganglia and thalamus. Ultimately, this information provides clear functional boundaries that can guide the oncological disconnections. If violated in the posterior boundary for instance, damage can cause a Supplementary Motor Area syndrome characterized by problems with the initiation of speech and motor activity [18].

In cases such as the one above, we have previously been held back from our ability to minimize unnecessary cognitive morbidity because neurosurgeons have traditionally had a less thorough familiarity with anatomy responsible for higher-order functions. By using network maps, we can make this anatomy more known, clearly define the risks associated with specific tumors and surgical decisions, and then create surgical plans based on functional boundaries or guide the patient discussion accordingly. This is not to say that we should leave mere gross tumor behind just to save a network. It is likely unwise to preserve a function that may just become disrupted in a few weeks due to the highly malignant growth of the glioma. In many cases, we may choose to completely ignore the networks to focus on the tumor. Nonetheless, brain maps can allow us to better judge the risks associated with extending a 95–98% extent of resection just to achieve a gross total outcome, but possibly damaging a network fiber related to cognitive functions at the same time.

5 Moving Forward

It is clear based on the compelling amount of evidence provided in the literature that damage or dysfunction in a large-scale network will result in severe neurological, cognitive, or emotional deficits in some patients, if not most. Therefore, there is a need to regularly incorporate thinking about preserving the large-scale brain networks whenever performing brain tumor surgery as described above.

However, it is important to note that there remains a number of important questions we have yet to answer. Most obviously:

1. *How much of a specific network can we sacrifice without losing its function?*
2. *How much can the contralateral network compensate following damage to a networks core?*
3. *Can a network always be rehabilitated post-operatively if a network is damaged in surgery?*
4. *How much do peripheral parcellations and fibers really matter?*

These are all important questions moving forward in the field of Connectomics, which are being actively researched by many, including those on our own team. However, it is also important to note that these questions have likely yet to be answered because we have not previously had a common vernacular to discuss these networks or tools to improve our surgical techniques in connectome based neurosurgery to date. It is truly exciting to see the great amount of knowledge and possible benefits which are yet to come when incorporating consideration of the large-scale brain networks into glioma surgery. Moving towards ideas such as disconnection surgery will bring the neurosurgical community closer to optimizing the patient onco-functional balance and cognitive outcomes in brain tumor surgery.

6 Conclusion

Glioma patients continue to present post-operatively with cognitive morbidity following brain tumor surgery. Previously, anatomy for complex functions, outside of just motor or language abilities, has been largely unknown. Large-scale brain networks provide us improved maps which make the relationship between human cerebrum and tumor tissue more understandable so that we may begin to make better surgical decisions while causing fewer deficits in complex functions. Ultimately, it may be most wise if we contextualize glioma surgery as a series of disconnections which can be defined against clear brain network fibers with known functional relevance. By understanding these functional boundaries and connectomic information, we may begin to better understand the risks associated with certain tumors as well as optimize their neurosurgical treatment while meeting patient onco-functional goals.

Acknowledgements We would like to thank Omniscient Neurotechnology for aiding in the creation of the figures utilized in the current work.

References

1. Sanai N, Polley MY, McDermott MW, Parsa AT, Berger MS. An extent of resection threshold for newly diagnosed glioblastomas. J Neurosurg. 2011;115(1):3–8. https://doi.org/10.3171/2011.2.jns10998.
2. Ganau L, Paris M, Ligarotti GK, Ganau M. Management of Gliomas: overview of the latest technological advancements and related behavioral drawbacks. Behav Neurol. 2015;2015:862634. https://doi.org/10.1155/2015/862634.
3. Duffau H, Mandonnet E. The "onco-functional balance" in surgery for diffuse low-grade glioma: integrating the extent of resection with quality of life. Acta Neurochir. 2013;155(6):951–7. https://doi.org/10.1007/s00701-013-1653-9.
4. Kahn E, Lane M, Sagher O. Eloquent: history of a word's adoption into the neurosurgical lexicon. J Neurosurg. 2017;127(6):1461–6. https://doi.org/10.3171/2017.3.JNS17659.
5. Drewes C, Sagberg LM, Jakola AS, Solheim O. Perioperative and postoperative quality of life in patients with glioma–a longitudinal cohort study. World Neurosurg. 2018;117:e465–74. https://doi.org/10.1016/j.wneu.2018.06.052.
6. Dhandapani M, Gupta S, Mohanty M, Gupta SK, Dhandapani S. Trends in cognitive dysfunction following surgery for intracranial tumors. Surg Neurol Int. 2016;7(Suppl 7):S190–5. https://doi.org/10.4103/2152-7806.179229.
7. Herbet G, Moritz-Gasser S. Beyond language: mapping cognition and emotion. Neurosurg Clin N Am. 2019;30(1):75–83. https://doi.org/10.1016/j.nec.2018.08.004.
8. Rijnen SJM, Kaya G, Gehring K, et al. Cognitive functioning in patients with low-grade glioma: effects of hemispheric tumor location and surgical procedure. J Neurosurg. 2019;133:1671–82. https://doi.org/10.3171/2019.8.JNS191667.
9. Dadario NB, Brahimaj B, Yeung J, Sughrue ME. Reducing the cognitive footprint of brain tumor surgery. Front Neurol. 2021;12:711646. https://doi.org/10.3389/fneur.2021.711646.
10. Baker CM, Burks JD, Briggs RG, et al. A connectomic atlas of the human cerebrum-chapter 1: introduction, methods, and significance. Oper Neurosurg (Hagerstown). 2018;15(Suppl 1):S1–9. https://doi.org/10.1093/ons/opy253
11. Glasser MF, Coalson TS, Robinson EC, et al. A multi-modal parcellation of human cerebral cortex. Nature. 2016;536(7615):171–8. https://doi.org/10.1038/nature18933
12. Fornito A, Bullmore ET. What can spontaneous fluctuations of the blood oxygenation-level-dependent signal tell us about psychiatric disorders? Curr Opin Psychiatry. 2010;23(3):239–49.
13. Andrews-Hanna JR, Reidler JS, Sepulcre J, Poulin R, Buckner RL. Functional-anatomic fractionation of the brain's default network. Neuron. 2010;65(4):550–62. https://doi.org/10.1016/j.neuron.2010.02.005.
14. Yeo BTT, Krienen FM, Sepulcre J, et al. The organization of the human cerebral cortex estimated by intrinsic functional connectivity. J Neurophysiol. 2011;106(3):1125–65. https://doi.org/10.1152/jn.00338.2011.
15. Seeley WW, Menon V, Schatzberg AF, et al. Dissociable intrinsic connectivity networks for salience processing and executive control. J Neurosci. 2007;27(9):2349–56. https://doi.org/10.1523/JNEUROSCI.5587-06.2007
16. Bullmore E, Sporns O. The economy of brain network organization. Nat Rev Neurosci. 2012;13(5):336–49. https://doi.org/10.1038/nrn3214.
17. Fox MD. Mapping symptoms to brain networks with the human connectome. N Engl J Med. 2018;379(23):2237–45. https://doi.org/10.1056/NEJMra1706158.
18. Briggs RG, Allan PG, Poologaindran A, et al. The frontal aslant tract and supplementary motor area syndrome: moving towards a connectomic initiation axis. Cancers (Basel). 2021;13(5). https://doi.org/10.3390/cancers13051116.
19. Sandhu Z, Tanglay O, Young IM, et al. Parcellation-based anatomic modeling of the default mode network. Brain Behav. 2021;11(2):e01976. https://doi.org/10.1002/brb3.1976.
20. Niendam TA, Laird AR, Ray KL, Dean YM, Glahn DC, Carter CS. Meta-analytic evidence for a superordinate cognitive control network subserving diverse executive functions. Cogn Affect Behav Neurosci. 2012;12(2):241–68. https://doi.org/10.3758/s13415-011-0083-5.
21. Allan PG, Briggs RG, Conner AK, et al. Parcellation-based tractographic modeling of the dorsal attention network. Brain Behav. 2019;9(10):e01365. https://doi.org/10.1002/brb3.1365.
22. Kaiser RH, Andrews-Hanna JR, Wager TD, Pizzagalli DA. Large-scale network dysfunction in major depressive disorder: a meta-analysis of resting-state functional connectivity. JAMA Psychiat. 2015;72(6):603–11. https://doi.org/10.1001/jamapsychiatry.2015.0071.
23. Menon V. Large-scale brain networks and psychopathology: a unifying triple network model. Trends Cogn Sci. 2011;15(10):483–506. https://doi.org/10.1016/j.tics.2011.08.003.

24. Raichle ME, MacLeod AM, Snyder AZ, Powers WJ, Gusnard DA, Shulman GL. A default mode of brain function. Proc Natl Acad Sci. 2001;98(2):676–82. https://doi.org/10.1073/pnas.98.2.676.
25. Briggs RG, Conner AK, Baker CM, et al. A connectomic atlas of the human cerebrum-chapter 18: the connectional anatomy of human brain networks. Oper Neurosurg (Hagerstown). 2018;15(Suppl 1):S470–80. https://doi.org/10.1093/ons/opy272
26. Chenji S, Jha S, Lee D, et al. Investigating default mode and sensorimotor network connectivity in amyotrophic lateral sclerosis. PLoS One. 2016;11(6):e0157443. https://doi.org/10.1371/journal.pone.0157443.
27. Raj Mohan V, Mohandas E. The limbic system. CME. Indian J Psychiatry. 2007;49(2):132–9. https://doi.org/10.4103/0019-5545.33264
28. Koikegami H, Hirata Y, Oguma J. Studies on the paralimbic brain structures. Psychiatry Clin Neurosci. 1967;21(3):151–80. https://doi.org/10.1111/j.1440-1819.1967.tb01290.x.
29. Allan PG, Briggs RG, Conner AK, et al. Parcellation-based tractographic modeling of the ventral attention network. J Neurol Sci. 2020;408:116548. https://doi.org/10.1016/j.jns.2019.116548.
30. Baker CM, Burks JD, Briggs RG, et al. A connectomic atlas of the human cerebrum-chapter 4: the medial frontal lobe, anterior cingulate gyrus, and orbitofrontal cortex. Oper Neurosurg (Hagerstown). 2018;15(Suppl 1):S122–74. https://doi.org/10.1093/ons/opy257
31. Baker CM, Burks JD, Briggs RG, et al. A connectomic atlas of the human cerebrum-chapter 9: the occipital lobe. Oper Neurosurg (Hagerstown). 2018;15(Suppl 1):S372–406. https://doi.org/10.1093/ons/opy263.

Intraoperative Mapping and Monitoring Techniques for Intra-Axial Brain Tumors

Aliasgar Moiyadi, Prakash Shetty, Parthiban Velayutham, Vikas Kumar Singh, and Kanchi Jain

1 Introduction

Maximal safe resection of gliomas whenever feasible is currently the treatment of choice and has a positive influence on the natural history of the disease [1–3]. However, due to the frequent occurrence of supratentorial gliomas near, or within the eloquent regions, and their infiltrative pattern of spread, maximal resection can often be associated with higher risk of neurologic sequelae. In order to optimize the risk-benefit ratio of surgery, intraoperative functional mapping and functional neuroimaging (FNI) techniques have been extensively used over the past few decades. Considerable inter-individual anatomico-functional variability may exist even in healthy volunteers and this variability is further accentuated in cases of gliomas especially in identification of the language areas [4–6]. It has also been observed that due to neurovascular decoupling, bold response near gliomas does not reflect the neuronal signal as accurately as it does in healthy tissue [7]. FNI is also unable to differentiate areas that are essential for the function, which should be surgically preserved, from those "bystander" areas, which can be functionally compensated and therefore, potentially resected without permanent deficit [5]. Due to these limitations, it is necessary to use additional invasive electrophysiological adjuncts like direct electrical stimulation (DES). Techniques of brain mapping can broadly be classified into intraoperative neurophysiological guided techniques used under anaesthesia and awake brain mapping techniques. This chapter describes the two techniques including the technical details and applications with respect to intra-axial brain and brainstem tumors.

2 Historical Overview

Humans have been fascinated by the brain since pre-historic times [8].Archaeological records indicate that awake craniotomies in the form of trephination were successfully practiced using cocoa leaves long before the advent of general anesthesia. The earliest human attempts to understand brain function can be traced back as far as the ancient Greeks. Many theories were propounded although direct evidence for functional localization eluded scientists until very recently. Anatomico-functional correlation depended heavily on indirect methods as used by Broca, Djerine and Wernicke's to identify putative func-

A. Moiyadi (✉) · P. Shetty · P. Velayutham
V. K. Singh · K. Jain
Neurosurgical Oncology Services, Department of Surgical Oncology, Tata Memorial Centre, Mumbai, India

Division of Health Sciences, Homi Bhabha National Institute, Mumbai, India

A. Shah et al. (eds.), *Functional Anatomy of the Brain: A View from the Surgeon's Eye*,
https://doi.org/10.1007/978-981-99-3412-6_5

tional areas of the brain. Broadmann then described cytoarchitectural maps of the cortex, though no functionality was proposed [9].

The first direct attempts to localize brain function using stimulation mapping were animal studies using galvanic current [10–13]. Robert Bartholow first used direct cortical electrical stimulation in humans using electrodes over the central region through the dura in a patient with an eroding tumor of the scalp and skull [14]. Subsequent work by Cushing and Foerster using faradic stimulation in humans during epilepsy surgery paved the way for systematic application of the bipolar awake technique as pioneered by Wilder Penfield and popularized by Ojemann [15, 16]. Introduction of regional anaesthesia in neurosurgery (scalp block) by Harvey Cushing provided an opportunity to keep patients awake during surgery and monitor their functions [17]. In many ways, the modern era of awake craniotomies began more than 60 years ago when Wilder Penfield and André Pasquet published their landmark paper on the surgical and anesthetic aspects of surgery after administration of local and intermittent sedation and analgesia [18, 19]. At about the same time, the introduction of electromyography, electroencephalography and evoked potentials ushered in a new era of intraoperative neuromonitoring. The major ground-breaking study by Merton and Morton in 1980, using high intensity stimulation through a pair of electrodes transcranially positioned over the primary motor cortex in awake patients, showed that compound action potentials could be elicited in the contralateral limbs [20]. In 1993, Taniguchi et al. [21] for the first time, were able to record MEPs in an anesthesized patient using a high frequency(250-500HZ), multi-pulse (train of 5) stimulation technique leading to popularization of transcranial electrical stimulation elicited motor evoked potentials (TES-MEPs) as a method for monitoring corticospinal tract activity. These techniques have evolved significantly over the last decade. Simultaneously, greater understanding of brain function and techniques for testing cognitive functions have expanded the scope of awake mapping and monitoring.

3 Intraoperative Neurophysiological Techniques for Brain Tumors Operated under Anesthesia

IONM under general anesthesia (GA) consist of techniques which measure recordable electrical responses, either due to spontaneous activity or those evoked by stimulation (evoked potentials). There are a gamut of such techniques available for various neurosurgical (and non-neurosurgical) procedures.

They can be broadly classified into mapping and monitoring techniques. Mapping refers to identifying the specific functional area of the brain before proceeding with the resection and can be repeated as and when required, in an intermittent fashion, alternating with the surgical manipulation. Monitoring refers to repeated (and sometimes near-continuous) testing of the entire pathway during surgery to establish its functional integrity. These are complimentary to each other and used in combination during surgery. The present section will focus on techniques used during supratentorial and intra-axial brainstem tumor surgeries.

3.1 General Considerations

IONM is performed using specific stimulation and recording electrodes. The motor evoked potentials are recorded peripherally in the muscles after cranial stimulation. The sensory evoked potentials (somatosensory, visual, auditory) are recorded cranially after stimulation from peripheral sites. The recording in the peripheral muscles is done by needle electrodes, while stimulation of peripheral nerves is done by a surface electrode. The central (skull/scalp) electrode placement is according to the standard 10–20 system. This system is based on the relationship between the location of an electrode and the underlying area of the brain, specifically the cerebral cortex. It was developed internationally to maintain uniformity among various testing methods like electroencephalography (EEG), electrooculography

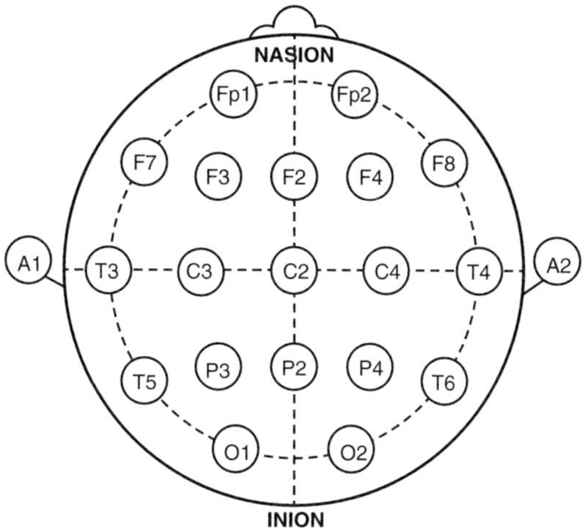

Fig. 1 Illustrative diagram showing the electrode locations of international the standard 10–20 system. A letter is used to identify the lobe, or area of the brain, where the electrodes are placed e.g. pre – frontal (Fp), frontal (F), temporal (T), parietal (P), occipital (O) and central (C) i.e., vertex of the brain. A "Z"(zero) refers to an electrode placed on the midline sagittal plane of the skull, (FpZ, Fz, Cz, Oz) and is used as a reference/measurement point. Similarly, A1 and A2 denote the auricular points. In this measurement system, even-numbered electrodes [2, 4, 6, 8] refer to electrode placement on the right side of the head, whereas odd numbers [1, 3, 5, 7] refer to those on the left side of the head

(EOG) and intraoperative neuromonitoring (IONM). The "10" and "20" refers to the actual distances between adjacent electrodes that are either 10% or 20% of the total antero- posterior or right–left distance of the skull with reference to the Z (zero) sites (Fig. 1).

3.2 Various IONM Modalities Used: Neuro-Physiological Basis and Technical Aspects

3.2.1 Motor System

The motor pathways consist of two major subsystems- the central and the lateral pathways. The central system comprises reticulospinal, tectospinal, and vestibulospinal pathways, which control the axial muscles for posture maintenance. Most of the activity in this system is involuntary (except for initiation and stopping). The lateral system is primarily the cortico-spinal tract (pyramidal) with a small contribution from the rubrospinal tract. The pyramidal system primarily controls the distal musculature and is the voluntary system. It originates from M1(primary motor cortex) but has a varying contribution from M2 (supplementary motor area), M3 (premotor area), and S1(primary somatosensory area). The fibers originate from the giant pyramidal cells of M1 (Betz cells), and the majority of them cross over in the medulla (pyramid) and connect to alpha motor neurons (lower motor neuron) of the ventral horn of spinal cord either directly or through interneurons at the level of exit. The alpha motor neuron finally connects with the muscle via the motor endplate. The motor unit consists of muscle fibers attached to a single motor neuron. Large proximal muscles usually have fewer motor neurons than distal muscles, the latter requiring finer control. Modulatory interactions with the basal ganglia, thalamus, cerebellum, and sensory system contribute to the overall motor control system.

Motor Evoked Potentials (MEPs) (transcranial, cortical, and subcortical)

Motor evoked potentials (MEPs) are generated by electrical stimulation of the brain either transcranially or directly along the corticospinal pathway (cortical and subcortical level). Electrical stimulation depolarizes the corticospinal tract proximal to the level of surgery (in supratentorial tumors) and induces evoked potentials distally which are recorded from the upper and lower limb muscles (Fig. 2). The exact point at which the CST is activated depends on the nature of the stimulation. Transcranial stimulation is thought to stimulate the CST axons at the axon hillock, at a depth which varies depending on the stimulation intensity used, whereas direct cortical/ subcortical stimulation activates them at the point of application of the stimulus. It is important to include all muscles at risk in the recording setup for a particular surgery. This allows the assessment of real-time functional integrity of the motor pathways (monitoring), as well as identification of the motor pathways within the surgical field by intermittent stimulation of a region of interest (mapping). Various

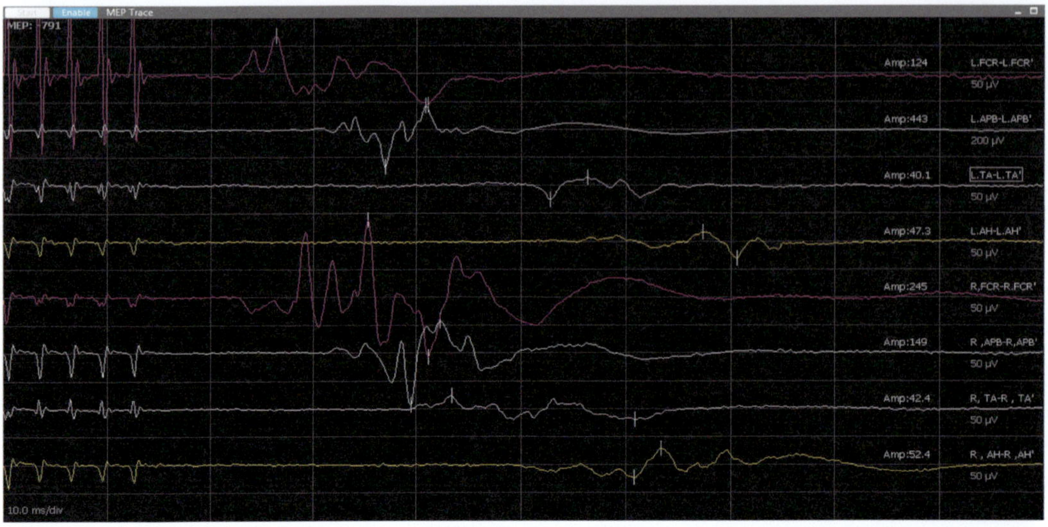

Fig. 2 Trace of Transcranial electrical stimulation induced bilateral MEPs responses obtained from both upper limb and lower limbs muscles

IONM modalities using the principle of MEP include Transcranial MEP, Cortical/Strip MEP, and subcortical MEP. Triggered EMG is the recording of motor unit potentials obtained by direct nerve stimulation, and spontaneous free run EMG as the name suggests, records spontaneous muscle activity. The stimulation and recording parameters for MEPs are shown in Table 1. Cortico-bulbar MEPs (coMEPs) are a special type of transcranial MEPs with double train technique, used to monitor the UMN connections of the motor cranial nerves, during surgery in and around the brainstem in addition to direct nerve stimulation (Fig. 3). This technique tests the entire pathway of the motor cranial nerves including the central pathways, which is in contrast to direct nerve stimulation which may give false negative results in case more proximal/central nerve injuries exist.

As compared to MEPs which are recorded from the muscles, the Direct wave (D-wave) originates from the spinal cord. D wave is recorded from fast conducting corticospinal motor fibers [22] of the spinal cord by the electrical stimulation of motor cortex transcranially. D wave monitoring is usually used in conjunction with TcMEPs. The D wave recording electrode is linear with 2 or 3 contact points and is usually placed in the epidural space or subdural space of the spinal cord, caudal to the surgical site. The D wave is recorded from the axons of first order neurons of the corticospinal pathways before they synapse with the interneurons and alpha motor neurons of the anterior horn cells and is the summation of fast conducting volley signals with no synapse. As a result, they are unaffected by anesthetic drugs which usually work on synapse blockade. As it records over the surface of the spinal cord, it is difficult to predict the laterality. Usually, D wave electrodes are placed rostral and caudal to the site of the surgery in the spinal cord or lower brainstem. The rostral electrode serves as an internal control. D wave recording is difficult from the rostral side when the lesion is in the lower brainstem and upper cervical region. In such cases, only the caudal response is recorded. D wave monitoring has been reported for motor monitoring in supratentorial tumors [23]. Yet, the invasive nature of the recording makes its use during supratentorial surgeries uncommon.

Technical Aspects of MEPs

Monopolar versus bipolar stimulation: Monopolar stimulation delivers current through a monopolar electrode and exits through a ground electrode placed distally, resulting in a wide field of current activation. Bipolar stimulation is a more focal stimulation between the tips of bipo-

Table 1 Stimulation and recording parameters for various IONM modalities

Modalities	Stimulation site	Recording site (10–20 EEG International system)	Stimulus intensity / parameters	Stimulus Repetition rate	Stimulus duration	Filter set up (Hz)
Transcranial motor evoked potentials (TES-MEPs)	C1/C2 or C3/C4 (biphasic) Corkscrew scalp electrodes	Contralateral deltoid, FCR, APB muscles for upper limbs and quadriceps, TA, and AH muscles for lower limbs using EMG needle electrodes	100 V – 600 V; train of 5–9 pulses; inter stimulus interval – 3 - 4 msec	–	500 μs	30–1500 Hz
Direct cortical stimulation (strip - motor evoked potentials)	1 × 4 contact strip electrode placed on the hand area of the exposed cortex. Also using handheld monopolar probe		5–20 mA; train of 4–5 pulses; inter stimulus interval – 4 msec	0.5–1 Hz	500 μs	30–1500 Hz
Subcortical motor evoked potentials	Subcortical using monopolar probe or suction stimulator		Train of 4–5 pulses; inter stimulus interval – 3 -4 msec. Starting with 15 mA and step down of 3 mA to reach 3 mA	2 Hz	500 μs	30–1500 Hz
Corticobulbar motor evoked potentials (CoMEPs)	C3/C4 (biphasic) Corkscrew scalp electrodes	Corresponding lower cranial nerve innervated muscles using EMG needle electrodes	100 V – 400 V; double train of 5:1 pulses; inter stimulus interval – 3 followed by 90 msec for single stimulation	–	500 μs	30–1500 Hz
Triggered electromyography (triggered EMG for cranial nerve)	Direct nerve structure/ cranial nerve nucleus in brainstem using handheld probe (mono- or bipolar). * in free run EMGs, there is no electrical stimulus. Spontaneous discharges are recorded during tissue handling	Corresponding cranial nerve innervated muscles using EMG needle electrodes	0.1–3.0 mA	2–4 Hz	100 -200 μs	30–500 Hz
Somatosensory evoked potentials (SSEPs) for central sulcus mapping (phase reversal)	Bipolar stimulation in contralateral median (or ulnar) nerve	1 × 4 contact strip electrode on the exposed cortex	5-25 mA	4.1 or 4.7 Hz	100-200 μs	10–500 Hz

(continued)

Table 1 (continued)

Modalities	Stimulation site	Recording site (10–20 EEG International system)	Stimulus intensity / parameters	Stimulus Repetition rate	Stimulus duration	Filter set up (Hz)
Somatosensory evoked potentials (SSEPs) for upper limb	Bipolar stimulation in contralateral median (or ulnar) nerve	Active at C3' or C4' Reference at Cz	5-25 mA	4.1 or 4.7 Hz	100-200 μs	10–500 Hz
Somatosensory evoked potentials (SSEPs) for lower limb	Bipolar stimulation in posterior tibial (or peroneal) nerve bilateral	Active at Cz' Reference at FpZ	15–40 mA	4.1 or 4.7 Hz	100-200 μs	10–500 Hz
Brainstem auditory evoked potentials (BAEPs)	Auditory click stimuli on the ear	Active at Cz and reference at mastoid bone; ground at FpZ	85–110 dB; white noise 65 dB	10 – 12 Hz	100 μs	100 – 1000 Hz
Visual evoked potentials (VEPs)	Photic stimulation via goggles on the eye	Active at O1 / O2 / Oz Reference – A1 /A2/ Cz	10,000–20,000 lx	1 – 2 Hz	10–20 msec	5 – 500 Hz
Electroencephalography (EEG) / Electrocorticography (EcoG)	–	Scalp – Frontal, central, temporal and occipital bilaterally (for EEG); strip electrode on the exposed cortex (for EcoG); reference - A1 & A2 (ear lobe); ground – FpZ	–	–	–	1–35 Hz

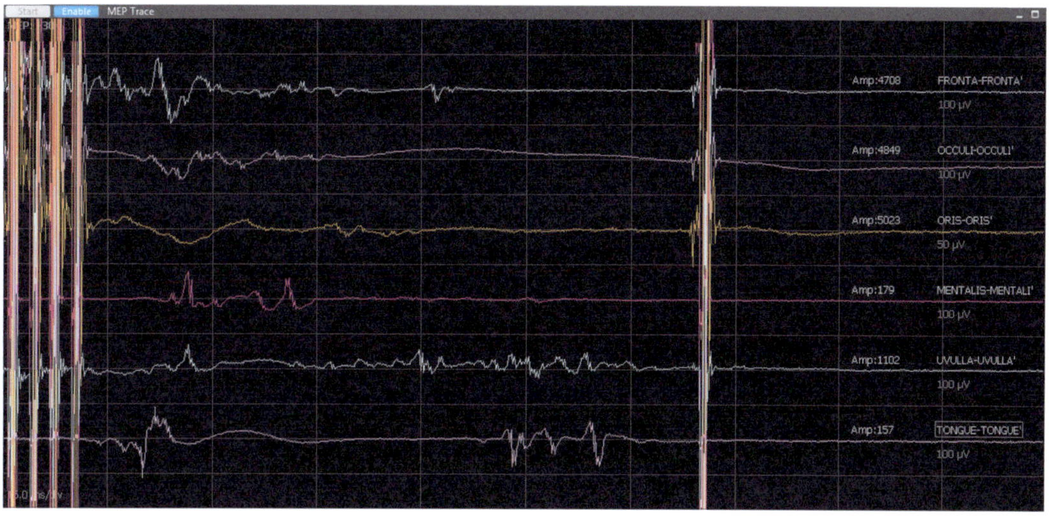

Fig. 3 Trace of corticobulbar MEPs from seventh, ninth and twelfth cranial nerves

lar electrode and is described in more detail in the subsequent section of awake mapping. Both are used for cortical as well as subcortical stimulation, either with a hand held probe or using strip electrodes.

Low-frequency versus high-frequency stimulation: Cortical or subcortical stimulation can be performed using either a low- or high-frequency paradigm. Low-frequency (50-60 Hz) stimulation technique is generally used with bipolar stimulation and is preferentially used during awake surgery. It can be used to elicit motor responses (even muscle twitches) under anaesthesia; however, the stimulation thresholds are generally high (with a significant risk of seizures) and the recorded electrical responses are not time-locked MEPs.

High-frequency stimulation consists of a monophasic pulse wave, 0.5 ms in duration, delivered as a train (usually train of 5 pulses with inter stimulus interval of 2–4 ms – multipulse technique) most often using a monopolar electrode. This results in summation of excitatory post synaptic potentials leading to neuromuscular depolarisation and MEP generation under general anaesthesia. High-frequency stimulation (250-500 Hz) has low ictogenicity due to very short trains delivery (10.5–18.5 msec) whereas low frequency has longer trains (1–4 sec) of 60 Hz stimuli. This reduces the charge concentrated over the tissue stimulated, with wider area of excitation and shorter duration of current application as compared to low frequency stimulation [24].

Slow and Fast charge delivery: During transcranial electrical stimulation (high frequency, multipulse technique), the delivery of electrical charge is one of the most important parameters to achieve reliable muscle MEPs. This can be delivered as a fast charge (1.0 Coulomb/s) or slow charge (0.1 Coulomb/s). The fast charge technique delivers more charge in lesser time; hence the mean duration of stimulation is lesser compared to slow charge (50 micro seconds vs 500 micro seconds) as shown in the Fig. 4a, b. Both techniques produce reliable MEPs, but the fast charge technique is more efficient, requiring 35% less charge delivery [25].

Monophasic and Biphasic Stimulation: Transcranial MEPs can be elicited with either monophasic or biphasic stimulation. A monophasic stimulation with anode on the left and cathode on the right side of cranium (head) would typically stimulate the anodal side (left) resulting in MEP recording from the right-side limbs (Fig. 5a). To record from the left side limbs, the anodal and cathodal polarity has to be reversed. In a typical biphasic stimulation, there is a second reversed phase stimulation immediately following a first stimulation (Fig. 5b), hence both sides of the brain are stimulated sequentially in quick alternating cycles. Hence it helps to moni-

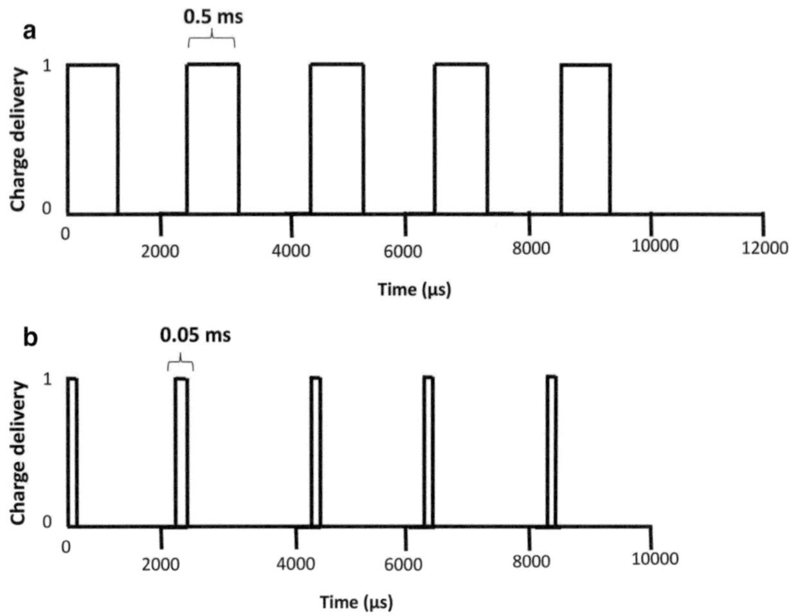

Fig. 4 Schematic drawing representation of (**a**) slow charge and (**b**) fast charge delivery during transcranial MEP stimulation

tor both sides simultaneously. However, it is less accurate compared to the monophasic stimulation particularly in supratentorial tumors [26, 27]. Generally, for supratentorial tumors, the region of interest is on one side and hence a monophasic paradigm for direct strip MEPs, cortical and subcortical stimulation is sufficient. Biphasic stimulation can be used in this situation to monitor the muscles ipsilateral to the side of surgery and with the contralateral recordings as negative control to rule out surgical, technical and anaesthetic causes for changes in MEPs.

Free-Run electromyography and Triggered electromyography

Free-run EMG (fEMG) recording comprises of real-time evaluation of the spontaneous muscle activity from the muscles of interest. This depends on the integrity and function of the muscle fibers, neuromuscular junction, peripheral (or central) nerve, ventral root, alpha motor neuron, and its spinal (or brainstem) interneuronal synapses. In intraoperative neurophysiology, this monitoring tool is used for detecting surgically driven mechanical irritation of the peripheral nervous system and of the cranial nerves, often before irreversible damage occurs to these structures (Fig. 6a). Triggered EMG consist of electrical stimulus applied directly on the peripheral (or central) motor nerves or roots, for eliciting compound muscle action potentials (CMAPs) which are recorded from the corresponding muscles (Fig. 6b). This mapping tool can be used to detect the location of peripheral or cranial nerves which may sometimes be difficult to localize in the tumor or surrounding tissues during resection. EMG responses are not very useful for supratentorial tumors but are important in surgery around the brainstem and cranial nerves. The stimulation and recording parameters for free run EMG and triggered EMG are shown in Table 1.

3.2.2 Sensory System

Sensory System and Somatosensory Evoked Potentials

The sensory pathways consist of the dorsal column system (fine touch, vibration and proprioception) and the spinothalamic pathway (protopathic sensations like pain). The dorsal column pathways can be monitored by way of somatosensory evoked potentials (SSEPs). The sensations are conveyed through the dorsal root ganglia, axons of which ascend uncrossed in the gracile and cuneate fascicle to reach the nuclei in the medulla. The second order neu-

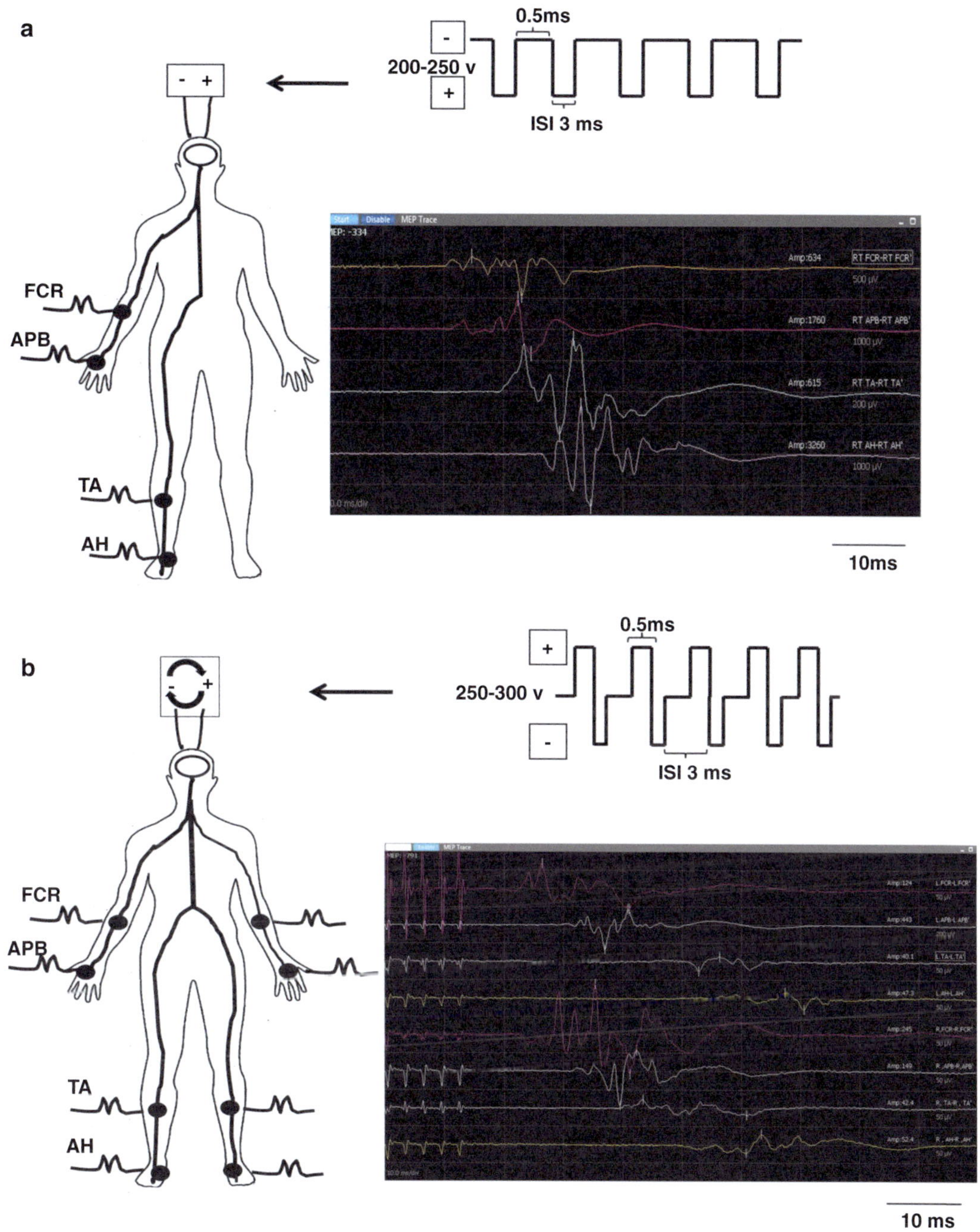

Fig. 5 Various paradigms for transcranial electrical stimulation of motor-evoked potentials.(**a**) Monophasic Pattern with left side anode and right-side recordings from the muscles. (**b**) Biphasic Pattern: Reverse polarity stimulation and bilateral recordings from the muscles

rons from the nuclei cross and ascend as medial lemniscus to the ventrobasal nucleus of thalamus. From the thalamus the third order neurons reach the primary somatosensory areas. SSEPs used to monitor the integrity of the dorsal column at different levels along this pathway. During SSEP monitoring, a peripheral nerve is stimulated (median or ulnar nerve for upper

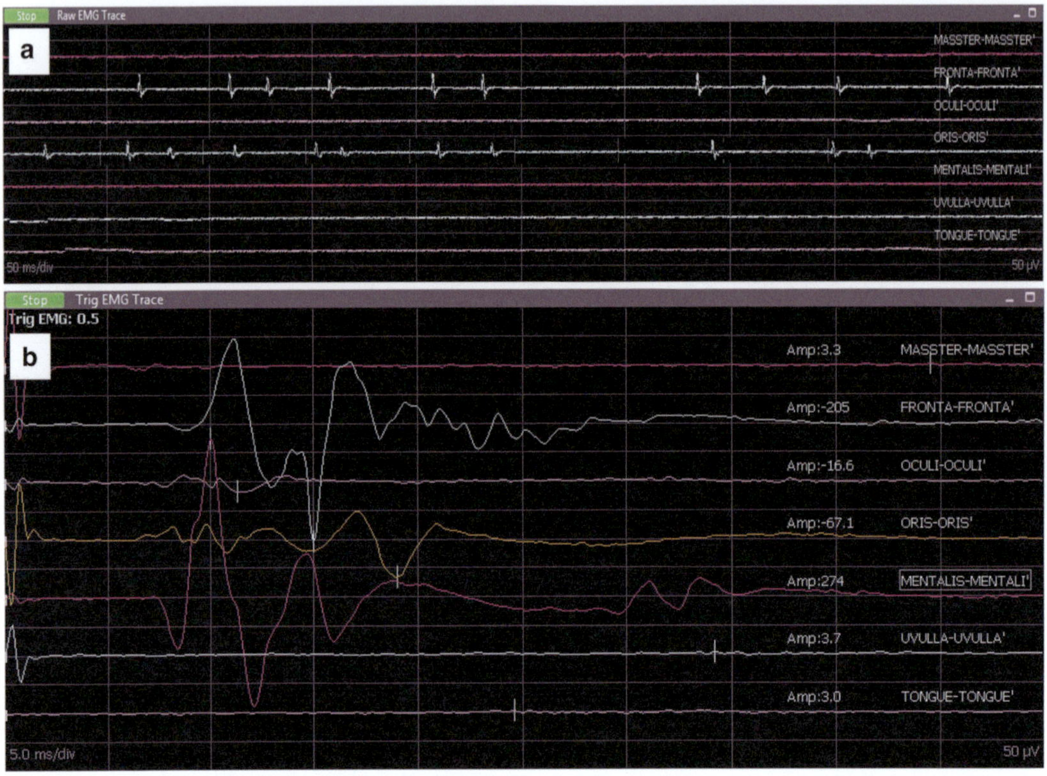

Fig. 6 EMG recordings of facial muscles during vestibular schwannoma surgery. Screen recording of (**a**) mechanical irritation responses on free run EMG, and (**b**) triggered EMG responses with 0.5 mA current stimulation

limbs and posterior tibial or peroneal nerve for lower limbs) and the electrical response evoked is measured along the ascending pathway at various points (spinal cord/nerve - near field responses; or distally on the scalp over the somato-sensory area -far field responses). The responses are in the form of positive or negative waves which are time-locked events (related to the distance from the site of stimulation). The stimulation and recording parameters for SSEPs is shown in Table 1. The negative potential measured on the scalp corresponding to the upper extremities occurs at 20 milliseconds (N20), and a positive potential corresponding to the lower extremities occurs at 37 milliseconds (P37) as shown in Fig. 7a, b.

Phase reversal: This is a special application of SSEP used to identify the central sulcus using cortically recorded SSEPs. Central sulcus (CS) is a prominent sulcus that runs down the middle of the lateral surface of the brain (perirolandic region), separating the frontal lobe (motor) from the parietal lobe (sensory). Surgical procedures in the perirolandic region require accurate identification of the CS to minimize the risk of inadvertent injury to the sensorimotor system. This can be pre-operatively achieved using fMRI or Magnetoencephalogram. However, these maybe inadequate to precisely identify the CS intraoperatively due to distortion of the anatomical landmarks [28]. SSEPs can be used to identify the CS by reversal of the polarity (Phase reversal) at N20 and P20 latencies in the sensorimotor cortex with contralateral median nerve stimulation [29]. The stimulation and recording parameters are shown in Table 1. A strip electrode is passed along the surface of the cortex across the presumed central sulcus. Detection of the phase reversal across any two contacts indicates the precise position of the central sulcus.

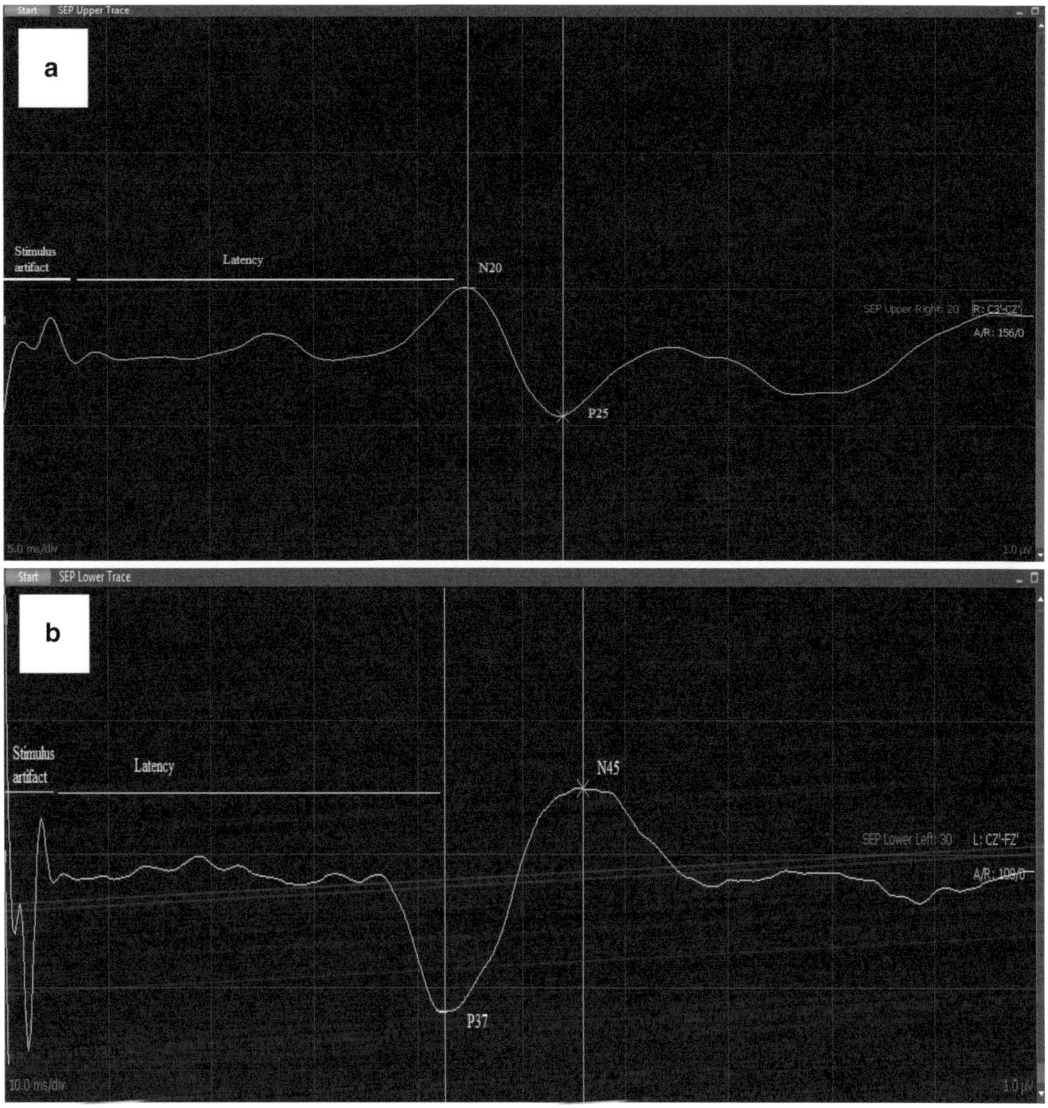

Fig. 7 Screenshot of SSEP responses from (**a**) Upper limb obtained from Median nerve stimulation at 15 mA and response recorded from the scalp with peak-to-peak latency obtained from N20 and P25 milliseconds; and (**b**) lower limb obtained from posterior tibial nerve stimulation at 25 mA and response recorded from the scalp with peak-to-peak latency obtained from P37 and N45 milliseconds

Visual Pathways and VEPs

VEPs (visual evoked potentials) are used to assess the functional integrity of the afferent visual pathways from the retina to the brain's visual cortex in response to a photic stimulus delivered via special light-emitting diodes or glasses during surgery. This photic stimulus evokes neural responses in the retina which are then transmitted via the optic nerve, optic chiasma, optic tract, lateral geniculate body of the thalamaus, optic radiations, to finally reach to visual cortex of the occipital lobe [30] from where they are recorded via surface electrodes placed on the scalp. Intraoperative VEP responses are recorded from O_1, O_2 and O_z with reference to A_1 and A_2 (as per EEG 10–20 international electrode placement system) is shown in Fig. 8. The stimulation and recording parameters for VEPs is

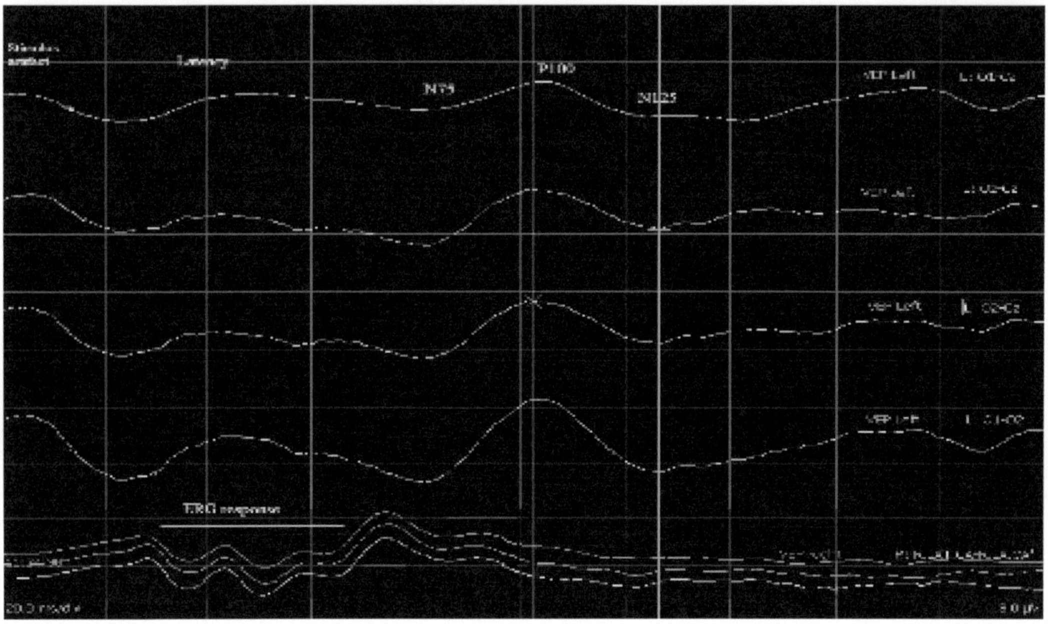

Fig. 8 Visual evoked potential responses (N75-P100-N125ms) with electroretinogram (ERG) activity during pituitary adenoma surgery

shown in Table 1. The waveforms, latencies and amplitudes of the various peaks are important parameters during interpretation of VEPs.

Brainstem Auditory Evoked Potentials (BAEPs)

BAEPs monitor the function of the auditory nerve and auditory pathways in the brainstem. The auditory signal travels from the cochlear hair cell to the primary auditory cortex via the auditory pathway (vestibulocochlear nerve, superior olivary complex, lateral lemniscus, inferior colliculus, and medial geniculate body) [31]. The response is evoked using a click stimulus via earphones. These evoked responses are far-field potentials and are recorded from the scalp (vertex). The stimulation and recording parameters for BAEPs are shown in Table 1. Wave I and II represent the conduction in distal and proximal part of cochlear nerve. Wave III represents transmission at level of cochlear nucleus. Wave IV and V represent intra-brainstem transmission till the midbrain. Wave V is recognized as the most prominent negative potentials after 5.5 msec, with latency ranges between 5.3–6.5 msec.

3.2.3 Spontaneous and triggered brain activity recording:

(a) **Electroencephalography (EEG):** The spontaneous electrical activity measured by EEG is generated by groups of pyramidal neurons, which have cell bodies in the third and fifth layer of the cerebral cortex [32]. These are recorded from the scalp. It is used exclusively to monitor cerebral function. In conditions, which affect the cerebral circulation, change in the EEG recordings provide an early warning of impending ischemia.

(b) **Electrocorticography (ECoG):** is used to monitor stimulation-triggered epileptiform discharges and after discharges (AD) before and during electrophysiologic functional mapping and is recorded directly from the surface of the brain. ADs can sometimes mimic stimulation induced deficits (in awake surgery) and monitoring ADs can rule out such confounding causes. It can also pick up early seizure activity before it becomes overt. The illustrative picture of spike like discharges (AD) is shown in Fig. 9.

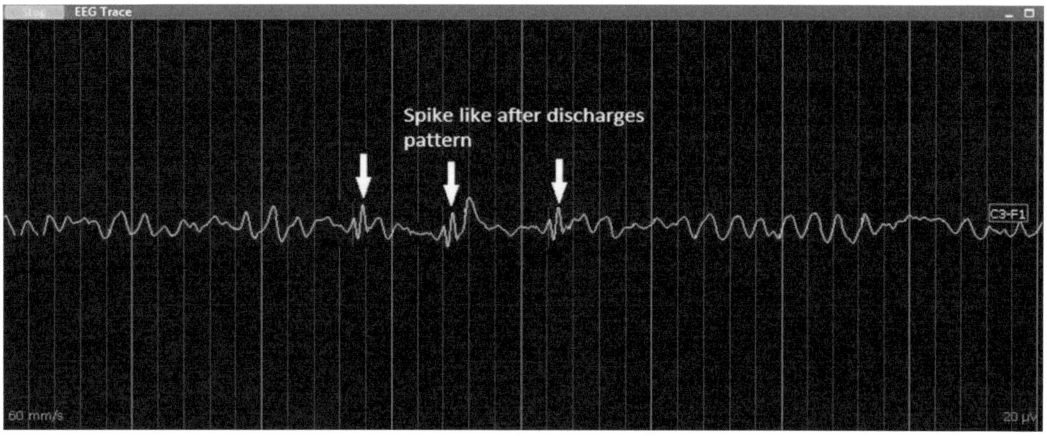

Fig. 9 Screenshot showing spike-like after discharges (AD) in awake surgery

3.2.4 Monitoring of Brainstem Reflexes

The existing IONM modalities sometimes may not be able to monitor all the brainstem functions, especially afferent components of cranial nerves. Brainstem reflexes which involve the entire reflex arc and are multisegmental, can be monitored to assess integrity of brainstem pathways. These include blink reflex, masseter reflex and laryngeal adductor reflex.

Blink reflex: This measures the component waveform in patients under general anaesthesia [33]. Application of 1–7 rectangular constant current stimuli with an interstimulus interval of 2 ms, an intensity of 20–40 mA with repetition rate of 0.4 Hz on the supraorbital nerve is used to produce a consistent R1 component of BR. The R1 oligosynaptic reflex arc includes the trigeminal afferents, brainstem connections between the trigeminal and facial nuclei, and ends with the facial nerve.

Masseter reflex: Masseter reflex can be elicited by stimulating the masseteric nerve (branch of the fifth nerve) through a pair of monopolar EMG needle electrodes under the zygomatic arch (0.5 cm lateral to the temporomandibular joint). A single stimulus with a duration of 0.2–0.5 ms followed with a steady increase in intensity is applied with a repetition rate of 0.7–1.0 Hz. The H reflex and M responses can be recorded with twisted pair subdermal needle electrode in the masseter and temporalis muscles [34]. This is an important technique for the monitoring of brainstem functional integrity, particularly in the midbrain and mid-pons, in addition to the trigeminal nerve path.

Laryngeal adductor reflex (LAR): Laryngeal adductor reflex is a newly developed IONM technique to assess the brain-stem modulated involuntary airway protective reflex arc in response to stimuli on the laryngeal mucosa. The superior laryngeal nerve serves as the afferent limb and the recurrent laryngeal nerve as the efferent reflex limb. The stimulus results in vocal fold closure. The LAR is evoked by single stimulation with duration of 0.5–0.7 ms and the current intensity range of 3 – 5 mA to the right and left laryngeal mucosa using embedded EMG tracheal tube electrodes. Contralateral LAR responses can be recorded with same EMG tracheal tube [35]. However, the application of LAR technique in brainstem neurosurgical procedures is still evolving.

3.2.5 Newer IONM Modalities

Long Latency responses (LLRs): Monitoring speech traditionally requires the patient to be awake. However recently it has been demonstrated that stimulation of M1 area for laryngeal muscles produces SLR (short latency responses) which can be recorded from laryngeal muscles [36–39]. Stimulation of caudal opercular areas (Inferior frontal gyrus -pars opercularis) results in recording of LLR (long latency responses) from the laryngeal muscles [36, 40]. These LLR could be surrogate markers of activity in the speech modulatory areas. However, these studies are at an experimental level and require further elucidation before becoming routinely used.

Cortico-cortical Evoked Potentials (CCEPs): Cortico-cortical evoked potentials "CCEPs", is a relatively novel in vivo brain mapping technique, utilizing low frequency stimulation to track and trace the cortico-cortical networks intraoperatively. A single electrical stimulus is applied directly to the cortex, and CCEPs are recorded from the remote cortex. In contrast to 50–60 Hz stimulation, a single pulse (1 Hz) electrical stimulation provides a unique opportunity to track the functional connectivity between different cortical regions. This CCEPs method has been successfully used to define the intercortical network subserving higher-order language, motor and parietofrontal networks [41]. Further development and understanding of CCEPs can potentially help map largescale brain networks, hitherto unmappable under anesthesia.

3.3 Application of IONM Mapping and Monitoring Modalities for Intra-Axial Brain/Brainstem Tumor Resection

3.3.1 Anesthesia Considerations for IONM

Evoked potentials (EPs) are the electrophysiologic responses of the nervous system to sensory (SSEP, VEP, BAER) or motor pathway stimulation (MEP). Since various anesthetic agents directly act on these neural substrates, it is important to understand their effects on these modalities. In general, evoked potentials are more affected by volatile inhalational agents (halogenated gases & nitrous oxide) compared to intravenous agents (e.g., propofol, barbiturates, benzodiazepines). All the anesthetic agents affect the intraoperative evoked potentials by either increasing the latency and or decreasing the amplitude in a dose dependent manner and progressing to burst suppression (1.5 MAC for inhalation gases) and electrical silence [42]. The influences of various anaesthetic agents on IONM modalities are shown in Table 2. However, BAER is highly resistant to most of the anesthetic agents. Neuromuscular blocking agents (NMBA) can block the neuromuscular junction abolishing the myogenic MEPs. For cases requiring only SSEP monitoring, a low dose IV agent (e.g., propofol) along with low dose inhalation agent (< 0.5 MAC) and relatively high dose opioid can be used. If SSEPs cannot be recorded then we need to avoid inhalational anaesthesia and convert to total intravenous anesthesia (TIVA). For MEP monitoring, a similar approach can be followed in addition to restricting the use of neuromuscular blockade during the monitoring period. Anaesthetists should also exercise appropriate precautions to avoid tongue bite injury, and injuries occurring due to patient movement especially when MEPs are being monitored.

Intraoperative EP changes may result either from surgical injury or ischemia of the specific neuronal pathway and it is important to ensure

Table 2 Effect of various anaesthetic agents on latency and amplitude during intraoperative neuromonitoring

	SSEP		MEPs(Transcranial/Strip/Cortical/subcortical)		Triggered EMG		VEPs		BAEPs	
Anaesthetic drugs	Lat	Amp	Lat	Amp	Lat	Amp	Lat	Amp	Lat	Amp
Inhalational agents	Yes	Yes	Yes	Yes	Yes	Yes	Yes	Yes	No	No
Propofol	Yes	Yes	Yes	Yes	Yes	Yes	Yes	Yes	No	No
Barbiturates	Yes	Yes	Yes	Yes	Yes	Yes	Yes	Yes	No	No
Opioids	No	No	No	No	No	No	No	No	No	No
Benzodiazepines	Yes	Yes	Yes	Yes	Yes	Yes	Yes	Yes	No	No
Ketamine	Enhanced									

Lat latency, *Amp* amplitude

that the changes recorded reflect true surgical events. Anaesthetic drugs and sedatives are the most common pharmacologic causes of nonspecific intraoperative EP changes during the surgery. However, these changes may also be caused by physiologic factors like temperature, blood pressure, hematocrit, acid-base balance, and oxygen and carbon dioxide tensions. Other factors causing intra-operative EP changes include technical factors like equipment malfunction or failure. Thus, a neuroanesthetist with understanding of intra-operative neurophysiology is important not only for anesthesia management, but also for troubleshooting when there are changes in the EPs. Usually, the surgical causes are focal or ipsilateral, whereas those resulting from physiological or anesthesia-related causes are global. Neuroanesthesists not only maintain intraoperative cerebral metabolism and blood flow, but also aid in optimal performance of electrophysiological monitoring during the surgery. Hence, preoperative planning and a good understanding between the surgeon, anaesthetist and neurophysiologist are imperative to establish an optimal environment for intraoperative neuromonitoring procedures.

3.3.2 IONM for Supratentorial Tumors

The IONM modalities used for supratentorial tumor surgery depend on the location of the tumor and surrounding structures at risk, which need to be monitored and mapped. The modalities for motor monitoring include TcMEP, and Strip MEPs (cortical MEPs) and uncommonly D-waves. For TcMEP, either a monophasic or a biphasic montage can be used for stimulation. The cork screws can be arranged to stimulate bilateral lower limbs (C1-C2) or upper limbs along with lower limbs (C3-C4), depending on the electrode placement. A base line reading is established where optimal recording from all monitorable muscles is obtained. Usually, stable MEPs during the surgery are associated with preserved motor outcomes. TcMEP is ideal for monitoring motor system during brainstem and spinal surgeries, but its use during supratentorial surgery has some fallacies which should be borne in mind. During supratentorial surgery, there may a false positive alarm (MEP loss- but post-operative power preserved), which typically occurs with brain sagging after debulking of a large tumor. Also, there may be false negative results (MEPs preserved, but post-operative power decreased). This can occur in two situations – when the TcMEP stimulates CST axons distal to the location of the tumor (and therefore distal to the potential site of iatrogenic injury during surgery) as is common during supratentorial lobar tumors in and around the Rolandic region; and also when there is vascular injury to cortex/internal capsule, but TcMEPs may still be obtained by increasing the voltage, due to stimulation of the motor fibres lower down [43] rather than in the cortex leading to a false sense of preserved MEPs though in reality the most proximal pathway is damaged. Various methods like threshold stimulation, suprathreshold and supramaximal stimulation have been used to try and optimize the utility of TcMEPs in supratentorial procedures [44–47]. However, interpretation of MEP changes (and hence the warning criteria) vary for these different techniques and generalizations cannot be made. As described earlier, D-waves may also be recorded by transcranial stimulation, however their use in supratentorial surgery is still unclear. Direct cortical stimulation induced MEPs overcomes many of the shortcomings of the TcMEPs in the context of supratentorial surgery. Stimulation is more focal and usually proximal to the tumor location, allowing accurate monitoring of the entirety of the motor pathways. For DCS-MEPs the motor cortex needs to be exposed or in the close vicinity of the craniotomy which is generally not a major problem. A shortcoming of the strip electrodes is that it monitors only one of the anatomical regions (arm/face or the leg) and hence the region to be monitored has to be judiciously selected based on the location of the tumor. The patient may develop a deficit in the unmonitored region/muscles though the DCS-MEPs may be preserved. Figure 10 shows a typical recording of DCS-MEP from a strip electrode. Brain deformation and shift can sometimes result

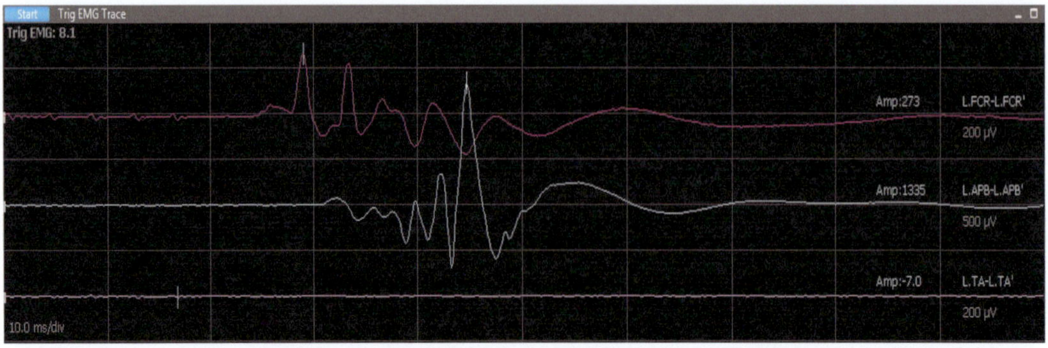

Fig. 10 Direct cortical stimulation induced MEP responses from upper limb FCR and APB muscles using 1 × 4 strip electrode in supratentorial surgery. The active contact was placed on the hand knob area. Note, that unlike TcMEPs, only hand muscles (and no other muscle) groups are activated on stimulation

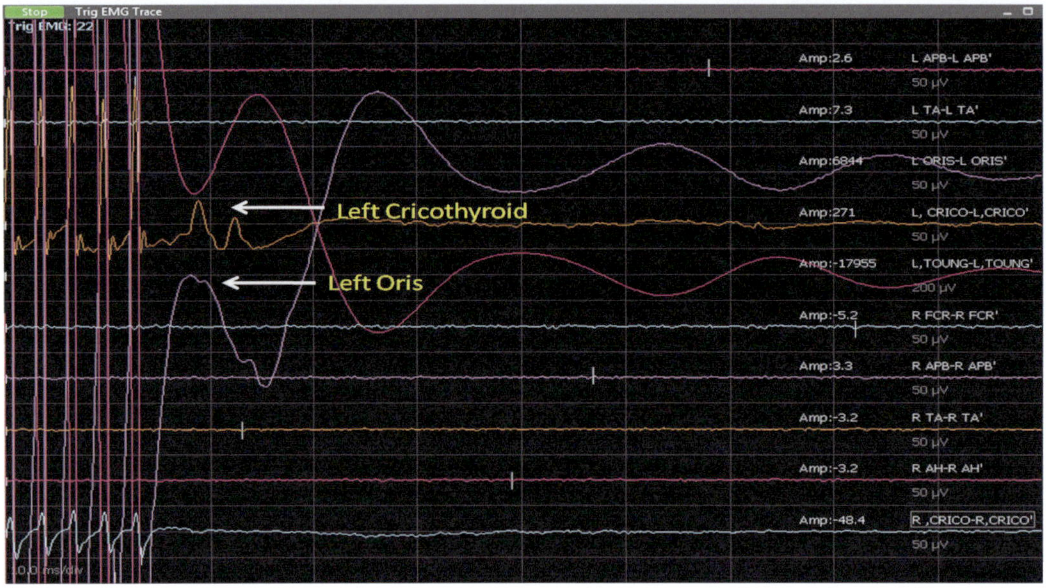

Fig. 11 Direct anodal monopolar cortical stimulation induced localization of functional areas reflected in the form of MEPs from orbicularis oris and cricothyroid muscles

in movement of the strip and changes in the DCS-MEPs, but can be easily overcome by careful technique and readjustment of the strip.

Besides monitoring of MEPs, motor mapping techniques are commonly used during supratentorial surgery. Mapping helps define the eloquent boundaries. Both, cortical and subcortical mapping (monopolar or dynamic suction mapping) can be used. Cortical mapping is generally performed using a hand held monopolar stimulator using anodal current polarity (Table 1). Starting with a low stimulation current, the intensity is increased in a step wise manner to establish a threshold at which stable MEPs are elicited. Figure 11 shows a typical cortical stimulation recording of face and cricothyroid muscles. As resection progresses into the subcortical white matter, mapping of white matter pathways is very important to prevent inadvertent damage. Intermittent subcortical electrical stimulation with a handheld monopolar probe is the conventional method to map the corticospinal tract. Current polarity is cathodal and the stimulation threshold for subcortical fibres is different from cortical grey matter areas. It has been shown that using monopolar subcortical cathodal stimulation, there is a relatively uniform relationship between the current strength and the distance from the white matter tract (1 mA = 1 mm), providing a reliable estimate of proximity to the cortico-spinal tracts depending on the strength of current used for stimulation [48, 49]. The current is kept high initially (15–20 mA to stimulate the CST at distance of 1.5–2 cm). Once the tract direction is localized, and tumor debulked, the current is gradually reduced to a predefined limit of 3–5 mA. Radical resections beyond this require judicious decision making as the current-distance relationship may be violated at very low levels of current intensity [50–52]. To overcome the intermittency of the mapping technique a dynamic subcortical mapping technique has been described [52]. The monopolar stimulator is built in with a suction tip, as result of which the subcortical area (in the resection cavity) can be continuously stimulated during tumor resection with an added feature of audio feedback when MEPs are elicited. Nevertheless, it is best to use this in conjunction with monitoring techniques like TcMEP or strip MEPs as the subcortical mapping directly stimulates the pathway distally and may not pick up proximal injury to cortex which the MEP monitoring will identify [53]. Though motor mapping and monitoring are the most common techniques used, other functions can also be evaluated during surgery. Phase reversal can be used to localise the motor strip using the SSEP technique previously described. In addition, SSEP are used in tumors close to the sensory cortex, while VEPs are used in surgeries where visual fibres are in proximity to the tumor. In addition, ECoG is used to pick up after discharges and regular cortical activity to monitor the seizure activity and level of anaesthesia. The modalities are summarised in Table 3. Fig. 12 shows a typical case of supratentorial tumor with motor mapping and MEP monitoring.

3.3.3 IONM for Intra-Axial Brainstem Surgeries

Tumors in (and around) the brainstem put the long tracts (especially motor) and cranial nerve nuclei at risk. The long tracts passing through the brainstem are monitored using TcMEP, D-wave(motor), SSEPs and BAEPs (sensory pathways). TcMEPs using a biphasic paradigm are preferred as they monitor bilateral tracts. D wave monitoring is additionally useful in the lower brainstem and cervicomedullary region. CoMEPs help monitor integrity (including central pathways) of motor cranial nerves and just like TcMEPs, preservation of CoMEPs has correlated with good cranial nerve outcomes [37, 38, 54]. BAEP is used to monitor the auditory path-

Table 3 Selection of IONM modalities for various anatomical sites

Modalities	Supratentorial	Supratentorial –Insular	Infratentorial –intra-axial
Motor (mapping/monitoring)			
Transcranial motor evoked potentials (MEP)	✓	✓	✓
Strip MEP	✓	✓	–
Dynamic subcortical mapping	✓	✓	–
D wave	✓/–	–	✓
Direct nerve stimulation	–	–	✓
Monopolar/ bipolar mapping	✓	✓	✓
Spontaneous EMG	–	–	✓
Corticobulbar MEP	–	–	✓
Sensory			
Somatosensory evoked potentials (EP)	✓	✓	✓
Visual EP	✓	–	–
Brainstem auditory EP	–	–	✓
Electrocorticography	✓	✓	–

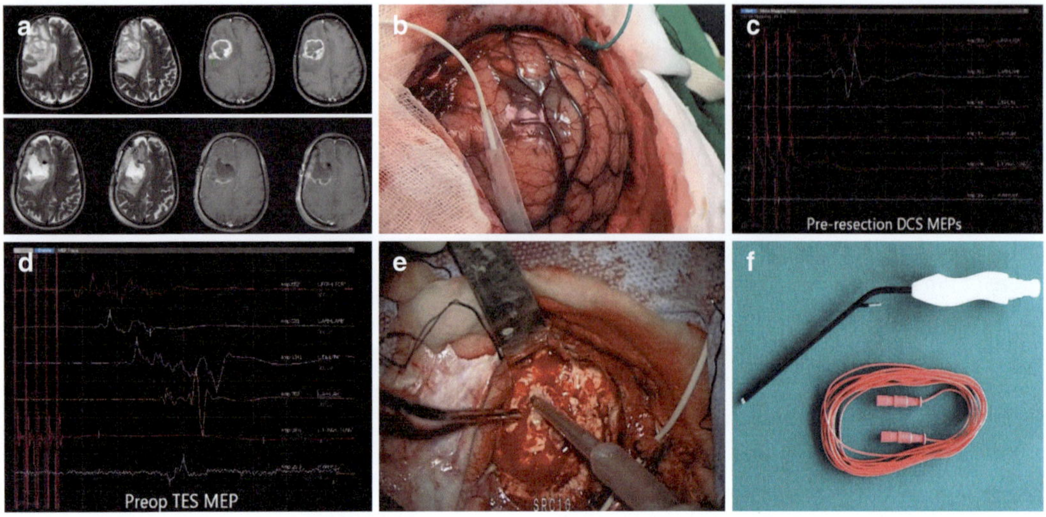

Fig. 12 A case of right frontal high grade glioma in a 65-year-old male patient. (**a**) shows pre-operative (upper row) and post-operative (lower row) T2 and T1 contrast MRI images. The patient was operated using IONM modalities for motor monitoring which included (**b**) strip electrode with its recording (**c**) of opposite side limbs and face. (**d**)-shows the transcranial MEP recording. (**e**) and (**f**) show the use of dynamic suction monopolar for continuous sub-cortical mapping during tumor debulking. Intra-operative neuro-monitoring parameters remained stable and there was no worsening of neurological status post-operatively. Histopathology was IDH negative Glioblastoma

way, which help to monitor the integrity of brainstem during surgery for CP angle tumors and brainstem lesions. SSEP is used in intramedullary spinal cord surgeries and brainstem surgeries to monitor the dorsal column pathway. Besides evoked potentials, free run (spontaneous) EMG is used to monitor any surgical traction induced neurotonic/myotonic discharges from manipulation or inadvertent nerve handling during brainstem surgeries. Monitoring of brainstem reflexes like blink reflex, masseter reflex and laryngeal adductor reflex can provide valuable information regarding integrity of the central pathways of the cranial nerves during brainstem surgeries. Besides monitoring, the motor cranial nerve nuclei are mapped on the brainstem floor using spontaneous and triggered EMG. The cranial motor nerve nuclei are stimulated in the floor of the brainstem using hand-held probes (monopolar or bipolar) to identify safe entry points to approach the tumor. These nuclei may be distorted and displaced by the tumor and stimulation mapping helps in accurate localisation [55, 56]. In addition, direct nerve stimulation is done by using monopolar or bipolar stimulation (triggered EMG technique) to identify cranial nerves in proximity to tumors in and around the brainstem. Figure 13 illustrates a case of cervicomedullary tumor operated with multimodal IONM.

3.3.4 Interpretation of IONM Changes and Warning Criteria

The purpose of using IONM is twofold. Whereas mapping techniques help identify eloquent structures, monitoring assesses the integrity of the various functional pathways. Therefore, changes in the monitoring recordings reflect potential disturbances in these pathways. Technical, anesthesia-related and other physiological causes should always be considered and ruled out before attributing any IONM changes to a surgical cause (Table 4). Whereas some variability in recordings is expected due to the complex intraoperative environment where ideal conditions for IONM recordings are not always possible, a reasonable threshold of changes can reliably point towards a problem, prompting corrective action. Therein lies the benefit of IONM. Though a complete loss of previously recordable potentials points to an almost certain neurological worsening, using an

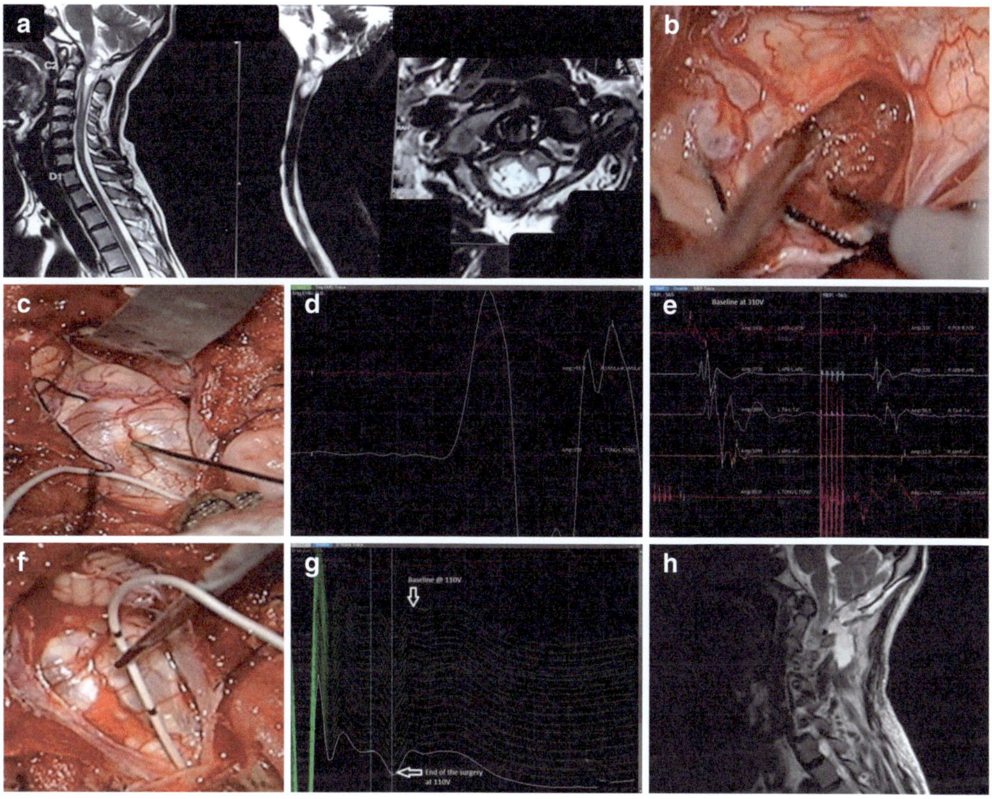

Fig. 13 An exophytic brainstem glioma in a 34-year-old patient who presented with long standing weakness of right upper limb and numbness (**a**) -shows a pre-operative T2 axial and sagittal image showing an exophytic brainstem lesion from the right medulla. He underwent a sub-occipital craniotomy and tumor excision under IONM guidance (**b**). A monopolar probe (**c**) was used to map the brainstem surface and recording from ipsilateral tongue is shown in image (**d**). Transcranial MEPs (**e**) were measured during the surgery which showed 30% drop at the end of surgery. A D- wave electrode was placed (**f**) and the D wave (**g**) was monitored which remained stable. Near total excision was achieved(**h**) on the post –operative images and final histopathology was ependymoma grade 2. The patients neurological condition remained stable post-operatively

Table 4 Causes of intraoperative evoked potential changes, troubleshooting and corrective manoeuvres

Type of change in Evoked potential	Common Causes	Prevention/ Remedies
Focal /ipsilateral	**Surgical handling or focal ischemia**	1. Stop /reverse the surgical maneuver 2. Wait till the changes recover 3. Avoid coagulation 4. **TIPP** procedure – **T**emperature increment, **i**rrigation with saline, mean arterial **p**ressure increment, **P**apaverine application to alleviate local ischemia
Global /bilateral	**Anesthesia**	Check for any changes in anesthetic plan, any boluses of drug given.
	Physiological causes	
	Blood pressure	Check for blood loss, hypotension and correct it. Blood pressure can be raised if there is fall in EP to improve micro perfusion.
	Temperature	Check for hypothermia and correct if present.
	Oxygenation	Check the ventilation and blood gases to optimize the oxygenation in tissues
	Hematocrit	Correct the oxygen carrying capacity by transfusion if required.
Focal or global	**Technical**	1. Check the connections- leads, channel box, machine. 2. Check for electrical interference from cautery, electrical circuits.

all-or-none approach may not be appropriate if early and subtle events are to be picked up. A variety of so-called "warning criteria" have therefore been evolved. An ideal warning criterion is one which predicts and pre-empts the appearance of a permanent deficit accurately. A very strict criterion (one which takes cognizance of minor changes in IONM recordings) may lead to over-diagnosis of a potential deficit (often leading to frequent and unnecessary stoppages during surgery, and sometimes even premature termination of surgery and false positives), whereas a very lenient criterion (which provides large leeway in the IONM changes) could miss potential deficits and lead to false negatives. A judicious balance must be struck in order to optimize the resection with function preservation. For each of the IONM modalities, different criteria have been proposed.

The intra-operative changes in the MEP characteristic, denote changes occurring in the vicinity or in the cortico-spinal tract. These criteria are based on changes in the MEP amplitude, threshold of stimulation, combination of both or a change in the wave form. They also depend on the technique of MEP (Transcranial or cortical, suprathreshold versus supramaximal) as well as the location of the tumor (supratentotial versus brainstem). Coupled with the inherent baseline variability of MEPs and thir susceptibility to anesthetic agents, interpretation of MEP changes requires assessment of various factors and is often multifactorial. Whereas the negative predictive value of MEP monitoring is excellent (preserved MEPs almost always indicate intact neurological function postoperatively), the positive predictive value can be variable. Generally, a complete loss or a 50% reduction in amplitude of MEPs is always to be regarded as a major warning sign prompting immediate corrective action [57]. A significant increase in threshold (much higher stimulation needed to regain the MEPs as compared to baseline) is also an important warning sign. For sensory modalities like SSEP, VEP, and BAER, the warning criteria frequently used are a 10% increase in the latency or 50% reduction in the amplitude from the baseline. More subtle changes need to be interpreted with caution and after assessing their temporal relationship to any potentially causative surgical manoeuvre. These warning criteria are not sacrosanct but can be tailored according to the unit's protocol and a priori understanding between the members of the team. There is always a delicate interplay during surgery, whereby dynamic surgical decisions need to be made with changes occurring in the evoked potentials. There may be reduction in EPs which may reverse with application of methods listed in the Table 4. Sometimes however, the recovery may not occur at all or may not reach the baseline, at which point a difficult decision to continue the surgery needs to be made. Also, many a times the fall in MEPs may not be gradual and occur suddenly with complete disappearance without providing any preceding warning, probably due to ischemia or sub cortical direct injury. It is important to remember that IONM techniques cannot replace but provide a valuable adjunct, to meticulous microsurgical principles and judicious decision making by a trained neurosurgeon.

3.3.5 Challenges During IONM and Complication Avoidance

IONM is generally a safe technique, However, due care and precautions are essential to minimize and avoid potential problems. It is important to screen and rule our contra-indications for IONM (such as implantable pacemakers and deep brains stimulation devices). Meticulous care while setting up the electrodes and connections is crucial to avoid electrical injuries and burns. Mechanical injuries can occur due to needle electrode placement (bleeding and hematoma) especially with patients on anticoagulants, as can infections if asepsis is not followed. During MEP monitoring (without muscle relaxants) there is always a possibility of muscle jerks which can sometimes interfere with the surgical procedure, and if severe, can even cause mechanical injuries to the patient (including fractures). Soft bite blocks should always be used to prevent tongue bites due to forceful jaw closures during MEP monitoring. Finally, seizures due to direct cortical stimulation can occur and cold irrigation should be readily available at hand. As compared to IONM in adults for which there is robust experience, IONM for paediatric cases comes with its own set of challenges due to a developing nervous system and different physiology compared to adults. This requires titration of anesthetic agents as well as different IONM stimulation parameters (usually higher due to immature myelination and slower conduction speeds). The literature for paediatric IONM is still evolving careful individualization of stimulation and recording parameters is preferable to ensure reliable information and avoid complications.

4 Awake Mapping and Monitoring

4.1 Background: Relevant Neuroanatomical Concepts and Basis of Use of Direct Electrical Stimulation (DES)

Our understanding of brain function has evolved exponentially in the recent past. This has led to the elucidation of various networks which subserve different functions. Whereas some cortical regions can have specific functions (such as the primary sensorimotor or visual modalities), most are believed to serve as relay or coordinating centres within interconnected networks (the so-called plurimodal or multimodal areas). Both, the cortical regions, as well as the connecting pathways (tracts), together constitute functional units. Perturbation of any part of this functional unit can lead to a loss of function of the entire network. Often, functional units can subserve more than one function, each with different inter-network connections and levels of compensation, adding complexity to their roles. Neuronal plasticity also influences effects of disruption of these networks and can modulate and alter existing ones in patients with brain tumors (especially slow growing low grade gliomas). In the awake patient, these functions can be interrogated using DES and assessing responses of a conscious and cooperative patient.

Basis of use of DES: The use of DES during AC is based on the understanding that DES produces a transient perturbation in function of the neuronal circuitry manifested as an observable response (spontaneous or on the performance of a particular task in an awake patient). This perturbation indicates the functionality of the stimulated area and the type of dysfunction induced reflects the potential role of the stimulated area in the functional pathway. The exact physiological basis of this action of DES remains incompletely understood [58]. DES induces a focal transient depolarization of the neurons/ axons mediated mostly via gamma aminobutyric acid (GABA). However, more than the direct effect of the DES, it is the indirect effect resulting from its non-focal (and often far-reaching) propagation of the impulse which then affects more diffuse brain networks and manifests as the perturbation of function noted [59]. This also highlights the importance of the subcortical connectivity in driving function and reinforces the need to combine cortical and subcortical DES during awake surgery. This has resulted in a paradigm shift from the localizationist model of brain function to the more network oriented hodotopological model that we now know [60, 61]. Absence of function perturbation can generally be taken as absence of function (though there are exceptions) and this forms the basis for using DES during awake mapping. DES is used to identify eloquent areas before tumor resection due to its effectiveness and reliability. DES allows individualized mapping of critical pathways (which may differ across patients due to plasticity). It is worthwhile to remember that the deficit seen intraoperatively may not always be the same as the one that can result from surgical lesioning (for example DES of the primary motor areas elicits involuntary muscle twitches, whereas a surgical lesion would cause weakness). Yet, DES remains the most reliable method to "simulate" a possible surgical deficit. A testament to the importance of DES as a clinical tool is the fact that it has remained the gold standard for mapping functions in the human brain in the neurosurgical context nearly 50 years [62]. Not only is DES a very reliable mapping tool, but it has also emerged as a very useful technique to enhance cognitive neuroscience by way of understanding brain circuitries and their interactions.

False positive and false negative results with DES are possible. Optimization of technical parameters and technique are imperative to reduce these. When DES positive sites (especially subcortical) are elicited intraoperatively, it is usual to encounter transient postoperative worsening (true positives with DES) which has been observed to eventually recover [63, 64]. Often, in such cases, one also encounters intraoperative deficits in the corresponding function [65]. Sometimes however, intraoperative deficits (which may lead to termination of surgery) are seen to recover in the postoperative period. One of the reasons for this could be related to patient performance and fatigue especially towards the end of the awake phase and reflects a false positive mapping result; on the other hand, it is also possible that immediate short-term network plasticity results in redistribution of function which may not have been evident intraoperatively (due to the extremely transient nature of the DES and lack of time to observe this physiological reorganization). More often, the plasticity is slow and delayed recovery is seen (as happens when one encounters a supplementary motor syndrome where contralateral homologue areas compensate for the functional loss over time). Rarely, antidromic (and non-physiological) propagation of the DES may result in an observed intraoperative deficit which a surgical lesion would not have induced. This can theoretically lead to premature termination of resection which could have otherwise been continued (false positive). This remains a potential limitation of DES. In case of absence of DES responses (negative mapping), one would expect no postoperative deficits (provided the correct functions were mapped using the correct technique). This is either because the stimulated substrate is truly non eloquent or is part of a network where immediate reorganization (and plasticity) has occurred allowing one to resect it without functional deficit. These are the true negatives with DES. Occasionally, the transient DES does not produce any immediately evident deficit (and network appears well compensated or non-eloquent), but a permanent surgical lesion can induce a delayed reorganization of the network unearthing deficits which manifest later (often hours to days after the surgery) producing a false negative DES result. Of course, delayed edema and venous congestion can also produce delayed postoperative worsening though these should not be classified as false negatives of DES. Understanding the mechanism of DES and the brain's capacity of plasticity is an essential paradigm for the application and interpretation of DES results.

4.2 Technical Aspects of DES during Awake Mapping

The current concept of application of DES for mapping brain function involves local stimulation of brain tissue with direct current [60]. DES can be applied via a handheld bipolar or monopolar stimulator, or through implanted grids or strips. The handheld bipolar stimulation technique was developed by Wilder Penfield and George Ojemann [66, 67] and is preferred during awake craniotomies, while the monopolar technique is used more often under anesthesia. The bipolar stimulator probe with 2 tips separated by 6–10 mm has become standard and routine in use for brain mapping procedure in awake surgery (Fig. 14a, b). An interelectrode distance >10 mm seems to favour the elicitation of large pyramidal tract neurons [68]. Stimulation with a bipolar probe creates an electric field in which the current density is more uniform and the electric field lines between both poles are close to delivery points limiting spread of current and improving focality of stimulation and thereby the accuracy of DES. Though the DES bipolar technique is used universally for the awake surgical procedure, the application of mapping at different centers involves a diversity of approaches. This includes the variability in the technical aspects like stimulator device; stimulation parameters consisting of the individual pulse type, pulse duration, frequency of stimulation, current intensity and the stimulating probe being used to deliver the current [69].

A constant current stimulator device is considered safer and more reliable for brain mapping procedures since it delivers the preset current

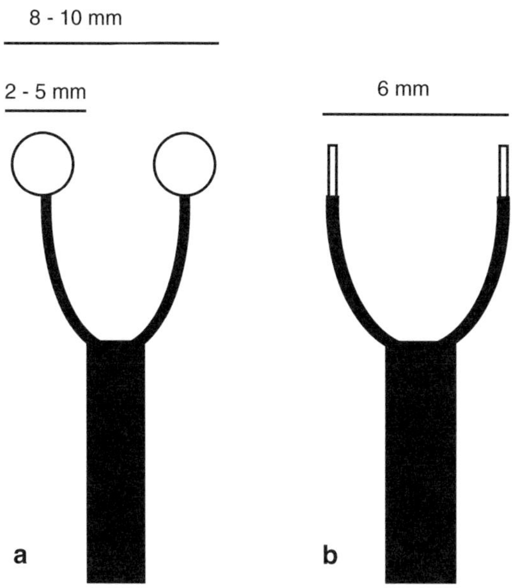

Fig. 14 Schematic pictorial representation of bipolar stimulator types with ball tip (a) and straight tip (b) showing the inter electrode distance

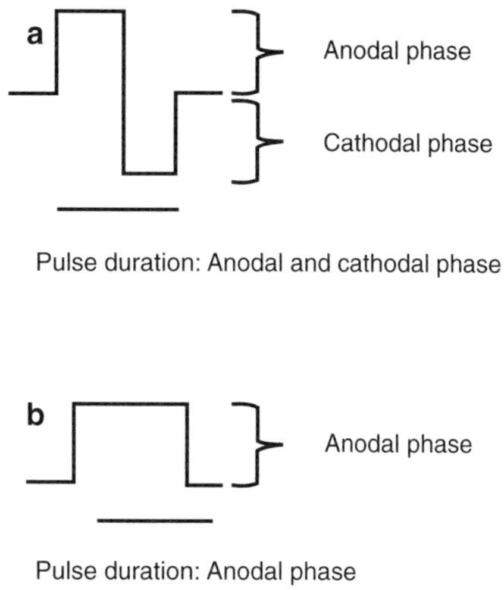

Fig. 15 Schematic representation biphasic (a) and mono phasic (b) pulse

independent of the impedance of the cortical and subcortical surface. Conversely, in constant-voltage stimulators, the delivered current depends on the tissue impedance. If the impedance decreases during stimulation (as it does when the operative site is irrigated), the amount of current delivery will dramatically increase and compromise the safety of the surgical procedure. Vice versa, if the impedance increases (layer of clot or a very dry field), there may be inadequate current delivery leading to potentially false negative results.

For awake DES brain mapping, a low frequency stimulation paradigm is used consisting of the application of short pulse trains with frequencies of 25–60 Hz. The most used frequencies are 50 Hz (Europe) and 60 Hz (North America and Asia). The individual pulse is a rectangular pulse of duration 1 millisecond (ms), either biphasic or monophasic as shown in Fig. 15a, b. Historically, a biphasic pulse has been found to be less neurotoxic as the reversed pulse could neutralize the charge delivery and minimize the deleterious effects of the stimulus. However, considering the extremely transient nature of intraoperative DES, this may not be clinically significant. Some stimulators may only display the positive phase of the pulse (and therefore the actual current intensity delivered may be twice the displayed value). The stimulus strength of DES generally ranges from 0.5–6 mA (and double that for biphasic pulse). Duration of stimulation should be around 1 second for sensorimotor areas and about 4 seconds for language and other cognitive functions. Bipolar stimulation during awake surgery usually elicits responses between 1–4 mA (whereas during asleep stimulation intensities increase upto 12-15 mA with only motor responses elicitable). The risk of seizures with bipolar mapping increases with increase in stimulation intensities and duration (especially during surgeries under anesthesia) but can be readily controlled by profuse irrigation with ice-cold Ringer or saline.

4.3 Technique of Awake Brain Mapping – Setting up and Awake Surgery

4.3.1 Preoperative Planning

The preparation for the intraoperative mapping testing paradigm begins much before, in the preoperative period. Careful case selection and individualized assessment is important. A checklist should be instituted to meticulously address all concerns and plan the procedure effectively. Patients with anticipated difficult airway such as those having obstructive sleep apnoea or morbid obesity; lesions with anticipated massive blood loss with hemodynamic instability; and potentially uncooperative patients (extremes of age, patients with drug abuse, chronic pain syndrome and anxiety disorders and claustrophobia) may not be ideal for an awake procedure. The mapping protocol is always customized to suit individual patient's requirements. The patient's expectations, preoperative functional (and occupational) status, handedness and lifestyle influences the type of functions essential to be mapped and preserved. For example, the ability to calculate is crucial for a mathematics teacher or an engineer but may not be very important for someone else; the ability to differentiate various musical notes is imperative for a musician. This must be factored in while designing the mapping strategy. The patient's preoperative evaluation is very important in designing the optimal mapping protocol. A thorough neurological evaluation and neuropsychological assessment provides clues about the customization of intraoperative testing [70, 71]. Besides basic sensori-motor and speech mapping, complex cognitive functions (visuospatial, memory, visual fields, attentional processing and emotion) may occasionally need to be mapped. In such cases, presence of a neuropsychologist and/or a speech pathologist or a neuro-linguist during the surgery is preferable. Thorough preoperative counselling of the patient should be performed to ensure good cooperation during the surgery. The patient should be aware and familiar with the personnel as well as the procedure that will be used (surgical as well as testing paradigm) so that he/she remains relaxed and comfortable throughout. The entire team (surgical, nursing, anesthesiological, neuropsychologist, neurophysiologist) should be coordinated and in sync all the time. A mock rehearsal of the entire procedure (simulation) may be very useful in acquainting the patient with the eventual surgery. The tests being planned intraoperatively should be administered preoperatively to check for baseline performance. In case of extremely poor performance, the test should be excluded from the intraoperative protocol for that patient. Occasionally, in case of severe neurocognitive deficits preoperatively, awake surgery may not be a good option (reliability of mapping is reduced significantly).

4.3.2 Perioperative Management

Excessive preoperative fasting leads to perioperative discomfort. 6 hours of fasting is required for solids. However clear liquids are allowed till 2 hours preoperatively to prevent fasting induced headache, gastritis. Anxiolytics on the night before can alleviate much of the patient's anxiety. Routine antibiotic and antiepileptic prophylaxis should be implemented. Intraoperatively, the patient is administered dexamethasone and ondansetron as prophylaxis for perioperative nausea and vomiting. The OT layout should accommodate all necessary equipment leaving adequate space for the monitoring team with unhindered access to the patient's face and airway. The position of the patient is generally lateral, though supine with head rotated minimally is also used. It is important to ensure complete patient comfort and adequate care should be taken with positioning. The entire setting up and anesthesia time should be kept to a minimum to ensure maximum cooperation of the patient and avoid unduly long times in the operating room. Silence and proper etiquette in the room is imperative.

4.3.3 Anesthesia Techniques and Management

The great challenge for the anesthetist during awake craniotomy is to choose the technique which provides not only an adequate sedation, analgesia, and respiratory and hemodynamic control, but also an awake and cooperative patient for neurological testing. The cornerstone of the anaesthesia during awake craniotomy is the scalp block. An effective block minimizes haemodynamic responses associated with applications of Mayfield clamps and surgical stimuli to the skull. This also decreases requirement of opioid analgesics and sedatives. A complete block requires bilaterally blocking all six nerves innervating the scalp (supraorbital, supratrochlear, auriculotemporal, zygomaticotemporal and lesser and greater occipital nerves). Since this block involves multiple pricks bilaterally, at the very beginning of the procedure, it can be performed under sevoflurane anaesthesia with patient breathing spontaneously or under conscious sedation using dexmedetomidine infusion to make the experience less traumatic. 0.25% Levobupivacaine (maximum dose - 2.5 mg/kg) provides an effective block and 0.5–1% Lignocaine can be infiltrated locoregionally as and when required. Whereas local anaesthesia is crucial and uniformly applied in all cases, the management of the phase prior to and following the awake brain mapping can vary [72]. Two commonly utilized strategies (Table 5) are sleep-awake-sleep (SAS) and awake-awake-awake (AAA). In SAS, general anesthesia (TIVA – total intravenous anesthesia) is given and the airway is controlled using a laryngeal mask airway (LMA) until the tumor is exposed. The anesthetic drugs are gradually reduced, and the airway device is removed when the patient regains upper airway reflexes. At the time of performing neurocognitive tests and intra-operative mapping all the anesthetic agents are stopped. After the resection is completed, the anesthetic drugs (TIVA) are restarted till closure. Reinsertion of airway device can be challenging especially when the head is fixed on clamps. Alternately the patient can be managed with conscious sedation for the remaining period. AAA on the other hand utilizes the principle of conscious sedation and monitored anesthesia care throughout all phases of the surgery. The patient maintains spontaneously breathing with careful titration of the TIVA. After the exposure of the tumour is complete, the sedation is gradually tapered so that patient is alert at the time of mapping. The patient breaths spontaneously under supplemental oxygen. As the sedation requirement and pain thresholds may vary, some patients may experience pain during certain steps (especially while performing the craniotomy and lifting the bone flap from the dura). Patients may also be subject to early fatigue in this technique, which may interfere with mapping towards the latter part of the surgery. Depending on the duration of surgery planned and the comfort of the anesthesiological team, either technique can be used. Most reports suggest that the two techniques (and variations of the same) are comparable in terms of outcomes [73–75].

Table 5 Comparison of the common techniques of anesthesia management during awake surgery

Awake – Awake – Awake {Monitored anesthesia care (MAC)}	Sleep-Awake-Sleep {SAS}
Patient is awake when positioned. So more likely to be in a comfortable position throughout the surgery and provide feedback.	Patient is positioned under anesthesia. They may experience discomfort due to uncomfortable positioning. This can be circumvented by intubating the patient after positioning the patient appropriately.
As patient is under conscious sedation, they can experience pain during painful steps of surgery especially craniotomy and elevating the bone flap	As patient is under anesthesia, they do not experience pain during this stage and remain comfortable.
Patient is alert and amenable to mapping within few minutes of stopping the sedation	Patient may take slightly longer to be fully awake for mapping after extubation
Patient may get exhausted early and may not be cooperative for mapping during latter stages of the surgery	Patient are likely to stay cooperative through the surgery
More suitable for short duration, small-moderately sized tumours	More suitable for longer duration, large sized tumours
Total operative time less	Total operative time may be more

Many TIVA techniques have been described in literature for awake craniotomy [76–78]. Neuroleptic analgesia has been shown to be associated with excessive sedation and a higher incidence of pain and convulsions [79]. Use of propofol has led to decreased incidence of convulsions but with a higher incidence of respiratory depression [76]. However, when used in conjunction with laryngeal airway mask (LMA) it is well tolerated and provides an efficient method of airway control in case of respiratory depression [77, 78]. The combination of remifentanil and propofol has also been used successfully for awake craniotomy in spontaneously breathing patients [80]. Remifentanil provides rapid control of depth of anesthesia and greater hemodynamic stability. However, these patients may also develop respiratory depression, which can be circumvented by use of LMA for controlled ventilation. Use of Dexmedetomidine which is a highly selective alpha-2 agonist can cause some sedation without cognitive impairment, which can be particularly helpful when awake –awake –awake surgery technique is planned; however, it has a relatively slow offset and can cause oversedation [81]. Our preferred regimen for monitored anesthesia care comprises of - Propofol @ 25-150mic/kg/min infusion; Fentanyl @ 0.5-1mic/kg/hr. and Dexmedetomidine loading dose of 1 mic/kg over 10 min followed by infusion @ 0.2–1 mic/ kg/hr. Occasionally, in very cooperative patients, sedation can be minimized or completely eliminated [82]. Alternately techniques like hypnotic suggestion have been described to effectively substitute the need for sedation [83].

4.3.4 Selection of Tasks

By far, the most crucial element in the awake mapping procedure is the selection of appropriate tasks to be tested and then the optimal execution of the plan intraoperatively. The tasks selected should be appropriate for the anatomical substrate at risk, easy and quick to administer and provide a distinct, reliable, and interpretable response. Task selection relies primarily on the lobe and subcortical substrates involved or likely to be encountered during the resection [84–86]. For this, a detailed preoperative assessment of the imaging (anatomical sequences as well as tractography and fMRI) is essential. A thorough preoperative neurocognitive assessment using a neuropsychological battery can unearth latent deficits in domains which may need to be evaluated intraoperatively. Our current understanding of brain function mandates consideration of mapping functions beyond the traditional language modalities to include functions subserved by the so called "non-language dominant" hemisphere as well as functions dominant bilaterally. This has expanded the scope and application of awake mapping to both right and left hemispheric tumors [87]. Further, depending on the patient's socio-economic profile and occupation/professional needs, the tests can be customized [84, 85]. There exists no standard "test battery" for intraoperative testing. Each service should have a customized array of intraoperative tests to be administered for assessing various functions. Given the constraints of time and patient cooperation, a smartly planned set of tests (both for mapping and monitoring) along with an adept monitoring expert (preferably a neuropsychologist and/or a neuro-linguist) provide for the best results. These functions can be broadly categorized as primary input/output modalities (inviolable and non-compensable) such as primary sensorimotor, visual, visuospatial perception and speech output [61, 88]. Most of these are subserved by anatomically well-defined cortical regions and their primary connections (projection fibres). Intermediate intra-network functions are a result of specific known connected networks subserving a common domain (such as the dorsal phonological stream and ventral semantic streams for language subserved by the arcuate fasciculus and the inferior fronto-occipital fasciculus respectively; or the motor modulatory network by the bilateral fronto-striatal tracts or the speech modulatory network in the dominant front-aslant tract). Damage to these lead to deficits which may be missed by cursory neurological examination but can lead to potentially long standing and debilitating deficits. Finally inter-network connections that drive higher executive control, emotion and metacognition, require more nuanced testing (choice and decision making, reading the mind of the eyes and so on). Though this may sound complex, it is often possible to simplify the protocol by selecting tests which can evaluate many of the various functions needed. For example, the simple picture naming test can assess anarthria (ventral premotor cortex), dysarthria (laryngeal motor areas), anomia (lexical retrieval), phonological paraphasias (dorsal stream) and semantic paraphasias (ventral stream) when administered as a verbal response test. The same test when administered with a leading phrase cue "This is a….." can evaluate reading. And when combined with a semantic association task (pyramids and palm tree test) can evaluate verbal as well as non-verbal (when asked to indicate the correct response without verbally speaking it out) semantic functions. Here we outline our unit's protocol (Table 6). These may be customized to suit individual situations and needs. It is however important that the same set of tests be repeated serially during a par-

Table 6 Outline of the intraoperative testing paradigm for awake surgery followed at our centre

	Assessment Tool	Response	Substrate
Spontaneous (no task)			
1	–	Muscle twitch/ movement	Primary motor cortex and projection fibres (corticospinal)
2	–	Paresthesia	Primary sensory cortex and thalamocortical projections
3	–	Phosphenes, photopsia	Primary visual cortex and optic radiations
4	–	Vertigo	Parietal lobe, especially subcortical
5	–	Saccadic eye movements	Frontal eye fields (bilateral middle frontal gyri)
Task based			
1	Number counting	Anarthria	Ventral precentral motor cortex (VPMC), terminations of horizontal segment of arcuate fasciculus
		Dysarthria	Primary oro-laryngeal motor area and subcortical projections
2	Picture naming	Anomia	Lexical retrieval areas (posterior temporal), perisylvian region and underlying white matter
		Phonological paraphasia	Dorsal language network and associated regions
		Semantic (verbal) Paraphasia	Ventral language network and associated regions
		Slowing/ delay/ apraxias	Dominant premotor/ dorsolateral prefrontal/ SMA and negative motor networks of speech (frontal aslant) - transcortical motor aphasia
		Perseverative paraphasia	Caudate and fronto-striatal connections
	+reading task (leading phrase like "this is a ….")	Alexia	Visual word recognition areas – Posterior basal temporal; dorsal premotor/frontal regions
	+ picture recall	Inability to recall	Verbal memory circuits (fronto-temporo parietal)
	+ 2 pictures in opposite quadrants (visual filed assessment)	Inability to see one picture	Optic radiations
3	Repetition task	Inability to repeat spoken sentence	Conduction aphasia (dominant arcuate F), Wernicke's (dominant SMG/AG), Inf frontal gyrus- pars Opercularis (dominant)
4	Semantic association task – Visual (pyramid and palm tree test - PPT)	Inability to verbally name the correct picture associated with the cue picture	Dominant ventral language stream (especially anterior IFOF)
	PPT – Non-verbal (identify the correct word without speaking it out)	Asemantism	Bilateral IFOF
5	Complex motor task/ bimanual tasks	Slowness/delay/clumsiness	Negative motor modulatory networks, bilateral frontostriatal connections, callosal fibres
		Apraxias	Bilateral fronto-parietal connections

(continued)

Table 6 (continued)

	Assessment Tool	Response	Substrate
6	Line bisection	Deviation to (right)	Non dominant parietal lobe and underlying white mater (frontoparietal and perisylvian)
7	Dual task – Simultaneous picture naming and motor task	Inability to perform both, though individually each possible	Attentional and executive function – Dorsolateral prefrontal regions, bilateral perisylvian networks (arcuate F)
8	Calculation – Simple arithmetic (single digit numbers with single operand)	Acalculia	Dominant angular gyrys area and underlying perisylvian connections
9	Reading the mind in the eyes (mentalization task)	Emotional disconnect	Bilateral cingulum (medial hemispheric regions) and arcuate F and right IFOF

ticular procedure. Moreover, to compare across surgeries it is advisable to have a fixed protocol and testing paradigms. This eventually builds the expertise of the team leading to more reliable execution of the procedure. Figures 16 and 17 illustrate two cases of awake mapping (dominant and nondominant hemispheres).

4.3.5 Performance of Tasks and Assessment: Practical Tips and Tricks

1. It is important to have the patient comfortably positioned and draped with a clear line of sight for both the patient (to see the examiner and the tests administered) as well as for the examiner (to see the patient response and record it correctly)
2. The tasks should be simple, reproducible and easy to administer. The dictum is to keep the testing protocol as simple as possible.
3. The tasks can be administered manually or via a computerized setup. It is essential to coordinate and synchronize the DES (this should ideally be 1–2 s before) and the administration of the intraoperative test stimulus (picture naming and related tasks) in most cases while performing mapping (except when the patient is performing a spontaneous task like number counting, reading or movement). A timed audio beep or a computerized synchronized system would help facilitate this. It also allows temporally synchronized visualization and recording of patient responses for offline evaluation and documentation.
4. Careful, systematic, and thorough mapping of the cortex is the first step. This generally involves establishing a positive mapping threshold using motor response (twitches) over M1 or complete anarthria when the ventral premotor cortex (VPMC) is stimulated. The lowest threshold eliciting this response should be looked for. Following this the exposed cortex can then me mapped systematically using the picture naming test (most often), or any other specific tests depending on the location and as per the decided plan of mapping.
5. Consistent response must be obtained on 3 non-consecutive trials in the same area before labelling it positive. Consecutively stimulating the same area can lead to seizures. Moreover, if a DES is applied during the electrically refractory period of the site, a false negative may be encountered.
6. It is preferable to use sterile marker strips to identify and mark the areas which are positive. One must be careful to avoid dislodging the markers while irrigating and manipulating instruments during the surgery.
7. During subcortical stimulation a slightly higher stimulation threshold may be applied (though most prefer using the same threshold as for cortical mapping), and tests administered as per the location and anticipated fibre tracts in the vicinity. DTI and fibre tracking can be particularly useful to anticipate the location of the tracts which may then be easily confirmed and mapped in that focused area. DTI acts as a useful

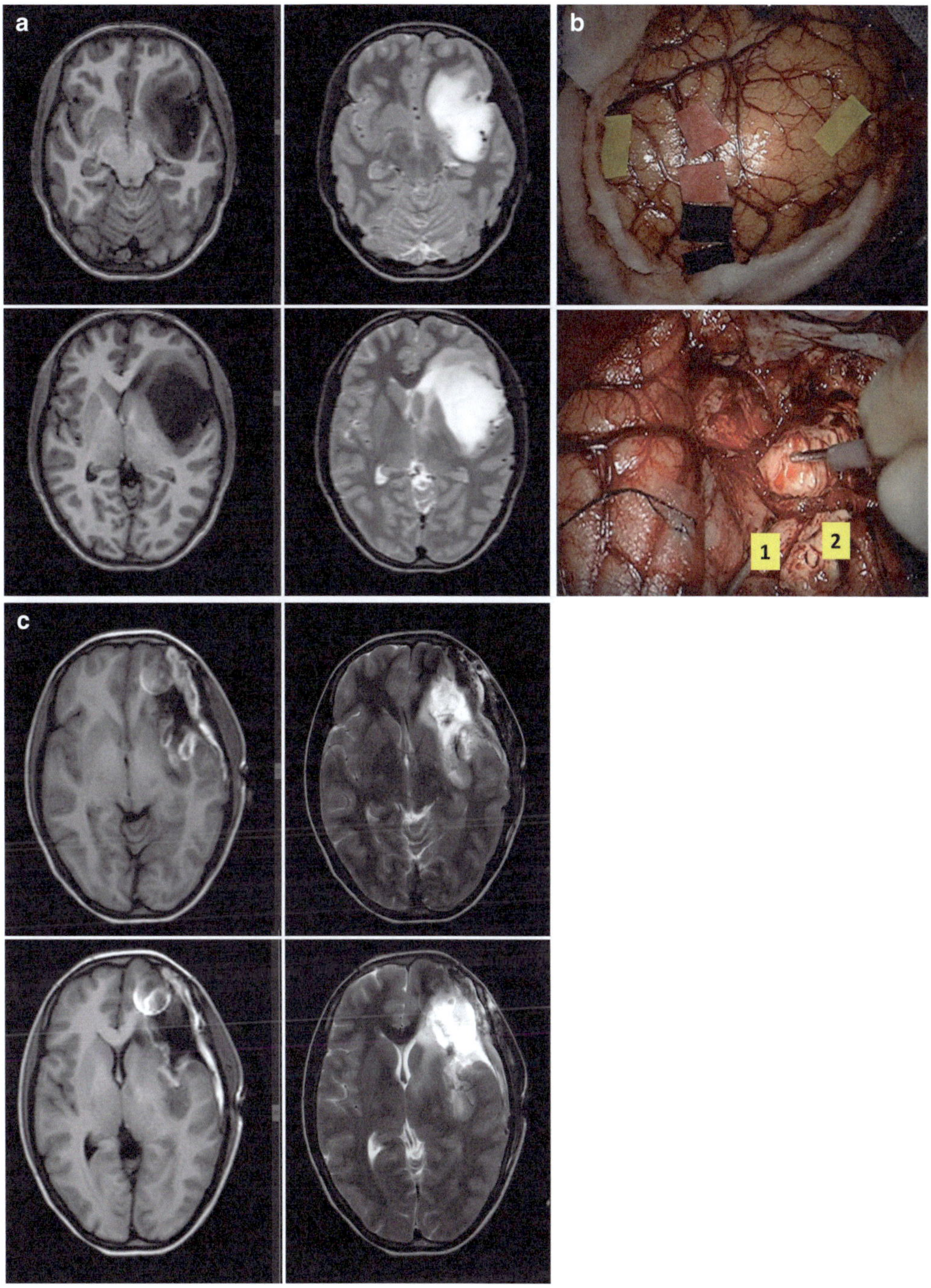

Fig. 16 Case of a 35 year old female with a large left (dominant) fronto-insular glioma who underwent awake mapping and trans-opercular resection. (**a**) Preoperative T1 and T2 weighted MR images showing a giant left fronto-insular non-enhancing glioma. (**b**) intraoperative microscope snapshots pre-resection (top) and post-resection (bottom). Pre-resection markers show the eloquent cortical surface areas revealed by awake bipolar mapping at a threshold of 1.5 mA (red indicates anarthria obtained over the ventral premotor region, black denotes motor twitches in upper limb and face and yellow indicate language eloquent areas). Post resection, multiple cortical entry windows are seen. Yellow markers indicate subcortical eloquent sites (1- anarthria along the rostral bank of the ventral precentral gyrus and 2 – semantic paraphasia along the ventral bank of the middle frontal gyrus) forming the limits of resection. (**c**) Postoperative MR shows near-complete resection with infiltrating residue along eloquent deep white matter and caudate nucleus

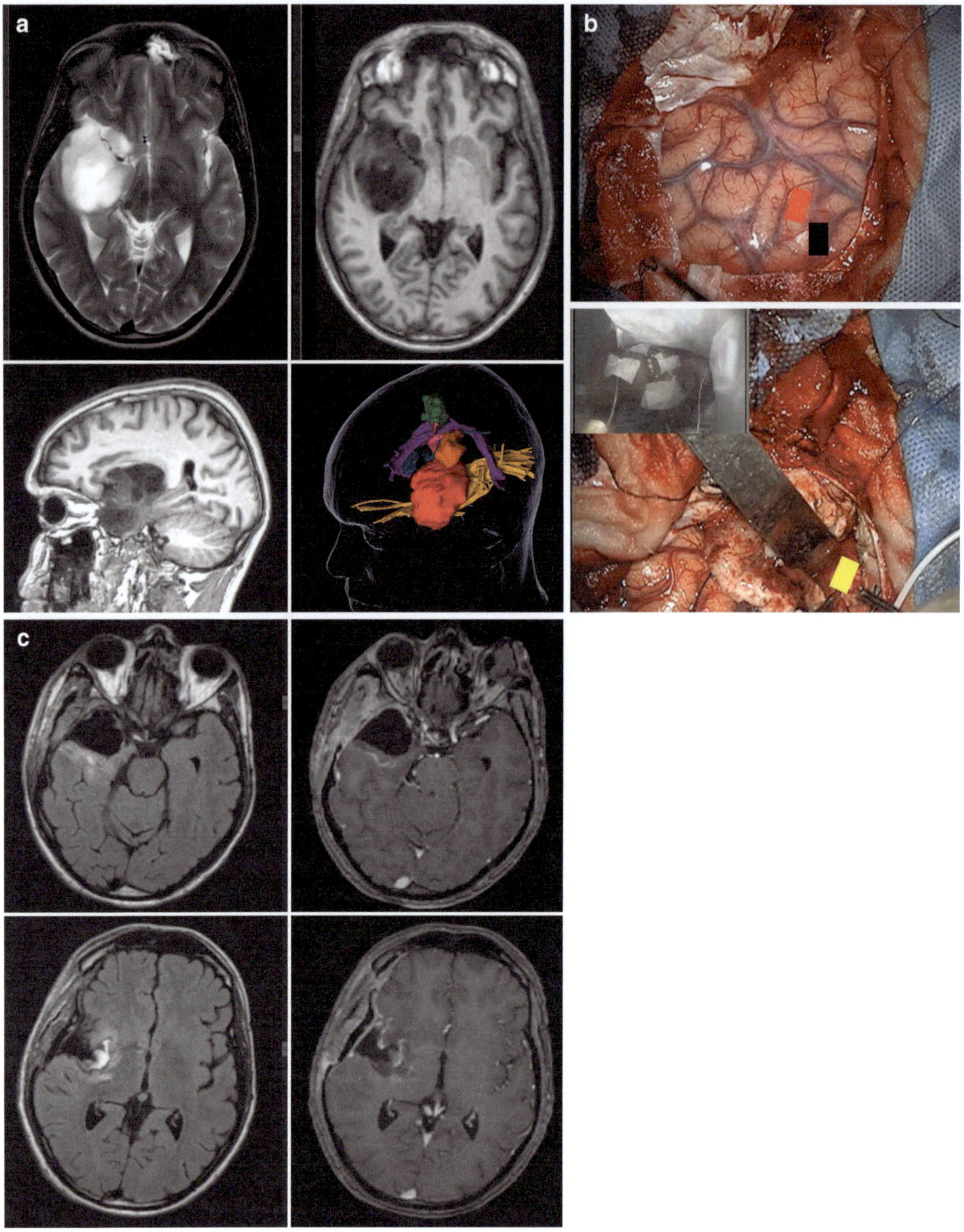

Fig. 17 Case of a 30 year old male with a right temporo-insular glioma who underwent awake mapping and trans-opercular resection. (**a**) Preoperative T1 and T2 weighted MR images showing a non-enhancing glioma. Bottom panel of A shows the relationship of the tumor to the subcortical tracts. (**b**) intraoperative microscope snapshots pre-resection (top) and post-resection (bottom). Pre-resection markers show the eloquent cortical surface areas revealed by awake bipolar mapping at a threshold of 1.8 mA (red indicates anarthria obtained over the ventral premotor region, black denotes motor twitches in upper limb and face). Post resection cavity shows a yellow marker indicating the region of the inferior fronto-occipital fasciculus where delays during pyramid and palm tree test (semantic association) were encountered forming the limits of resection. Patient's response is captured in the inset. (**c**) Postoperative MR shows gross total resection

guide but may not always be accurate and DES confirmation is mandatory.

8. It is important to use the lowest threshold stimulus intensity which consistently elicits the response. Higher intensities may produce false positive responses and elicit seizures. Electrocorticography (ECoG) if available, is very useful in detecting electrical seizures, much before they become clinically evident. ECoG also detects after-discharges (ADs). Sometimes AD-induced functional deficits may be falsely mistaken for positive stimulation findings and can be misinterpreted if ECoGs are not used.
9. One must be aware that the same test administered may yield variable responses depending on which part of the cortex/white matter is being stimulated. For example, when picture naming is administered and cortical stimulation is done it elicits a dysarthria if it is over the M1, complete anarthria (speech arrest with loss of all voluntary speech related oro-pharyngeal movements) when it is over the VPMC (ventral premotor cortex), and anomia when over the dominant supramarginal gyrus. On the other hand, during subcortical stimulation, the same test can yield speech apraxia (lateral dorsal dominant perisylvian white matter—SLF), phonological aphasia (medial dorsal dominant perisylvian white matter—AF) or anomia (posterior vertical component of AF). Though categorization of the deficit is less important (a deficit, regardless of the type indicates perturbation of function and means a positive response necessitating decision making as described below), it is desirable in order to understand the complex functions subserved by the various networks.
10. Some groups prefer a strategy of negative mapping [89]. This is usually when a smaller, tailored craniotomy is performed and absence of DES responsive sites within the exposed region is taken as a reliable indicator of absence of eloquence. Though this technique is effective when used by experienced teams, there are caveats which one must remember. Performance of negative mapping requires meticulous recording of after-discharges (AD) at each stimulated site. Absence of a response when DES threshold is just below the threshold which elicits an AD is a reliable indicator of absence of function. Nevertheless, thresholds of ADs may vary across the cortex and subcortical regions and there always remains a danger of false negatives. More importantly, in low grade gliomas, where the aim is to resect tumor and tumor-infiltrated brain parenchyma often beyond the imaging-defined abnormality, a strategy of negative mapping could yield suboptimal resection rates with potentially inferior oncological outcomes [90].

4.3.6 Monitoring

Besides DES mapping, continuous monitoring of neuro-cognitive functions during the surgery is very important. Just like the mapping protocol, the monitoring too is tailored to the individual surgery keeping in mind the patient's profile and the location of the tumor. The presence of a trained neuropsychologist and/or neuro-linguist in the theatre is desirable to ensure optimal monitoring. The intensity of monitoring may vary from the performance of a series of tasks (such as spontaneous speech, fluency and reading tasks, complex movements, or dual tasks) in a repetitive and intermittent manner to a more intense composite neuropsychological battery [91]. However, it must be borne in mind that patients may fatigue over time and adequate rest periods during "non-critical" phases of the surgery is advisable. For continuous motor monitoring, MEP recordings can be combined with awake mapping as described in the previous section.

4.3.7 Intraoperative Decision Making

Awake brain mapping relies on the identification of eloquent substrates based on the responses elicited by DES. When a response is obtained, a "positive" mapping is deemed and this reliably confirms eloquence. If these responses are seen within tumor tissue, a subtotal resection will result. At the subcortical level, even after complete tumor excision, no positive responses may be obtained in many cases. It is argued that in such cases, especially in the context of low-grade gliomas, one can extend the resection into "normal-appearing" brain tissue, till a positive response is obtained indicating close proximity to eloquent white matter tracts [90]). Whereas this supramarginal resection may be appropriate in low grade gliomas (where there is time to institute rehabilitative measures postoperatively and wait for reversal of deficits which do occur frequently), it may not be a wise approach in high grade gliomas where timely adjuvant therapy is a crucial prognostic factor. In such cases, a more conservative approach and a "safe margin" approach may be more advisable. It must also be remembered, that with a radical, supramarginal strategy, one is more likely to encounter postoperative neurological deficits, most of which are transient and recover over time. Further, it is important to know that certain positive cortical and subcortical sites are essential and non-compensable. These cannot be violated at any cost. These include primary sensorimotor, visual, speech output cortices and their white matter tracts as well as the perisylvian white matter fibre bundles (arcuate complex and the inferior fronto-occipital fasciculus) [88]. The corollary to this is that certain other so called eloquent sites may be "sacrificed" with no/little long-term sequelae due to the phenomenon of brain plasticity. The classical Broca's area and supplementary motor area (non-dominant hemisphere) are such examples. DES mapping unveils this plasticity and helps decide truly critical and eloquent regions. In addition to positive responses elicited by DES, one often encounters intraoperative deficits. It is hypothesized that intraoperative deficits occur most often because of proximity to eloquent sites and the incidence is therefore higher in series with positive mapping (supramarginal resections) [65]. However, unlike DES-induced deficits, spontaneous intraoperative deficits may or may not be reversible. Being able to identify reversibility of such deficits is crucial during awake surgery. Chang et al. proposed a method to predict reversibility of new onset linguistic deficits using a battery of intraoperative testing [92]. Edema and retraction/handling injury contribute to most such deficits [93], majority of which would be likely to reverse. Sometimes, these deficits are not evident intraoperatively, but progress and manifest in the immediate postoperative period leading to the phenomenon of immediate postoperative worsening. Most of these are known to improve over a longer follow up.

4.4 Complication Avoidance

Though generally safe, as with any tumor surgery, complications can occur during awake surgery. Careful planning and anticipation can avoid and minimize these. Nausea and vomiting can be prevented by prophylaxis. Expeditious pain control is vital to ensure patient comfort and cooperation and reduce all the potential complications. Intraoperative seizures can be prevented by adequate preoperative antiepileptic prophylaxis and appropriate cortical stimulation strategies that minimize the cortical stimulation current, duration of stimulation, and avoid repetitive stimulation of the same cortical sites. Intraoperative brain swelling during can occur (especially as the patient is awake or during the transition from asleep to awake phase). Repositioning of the head, hyperventilation, and use of mannitol is helpful. Brain swelling can also be minimized by opening dura only when patient is completely awake and cooperative. This helps patients calm down and prevents straining. Rapid tumor debulking (or cyst tapping) can alleviate brain tension in some cases. Rarely if brain swelling is refractory, conversion to anesthesia or deferment of surgery may be needed. Postoperative neurological worsening is often seen, especially after radical "supra-marginal" resections with positively mapped subcortical regions. Most of these are expected and known to recover over time. Understanding the rationale of DES and potential false positives and negatives as elaborated in the earlier section (excluding technical related confounders) is very important to avoid unexpected postoperative complications.

5 Clinical Benefits and Relevance of Functional Mapping and Monitoring

Surgical resection of gliomas in close vicinity of eloquent regions remains challenging. Ascertaining resectability based purely on preoperatively defined imaging criteria can be fallacious [94] and deprive patients of what could have potentially been a prognostically better radical resection. This reflects the inherent variability of functional eloquence in the human brain across individuals, as also the tumor-induced plasticity (particularly in low grade gliomas) that could alter the functional map of a particular patient's brain [3]. Intraoperative mapping is therefore important to confirm (and sometimes to rule out) eloquence. Awake mapping and subsequently mapping under anesthesia, have evolved into very reliable techniques. A large meta-analysis analysed the effect of intra-operative stimulation mapping on supratentorial infiltrative glioma resection [64]. This found that the early severe neurological deficits were more with the use of mapping (36% vs 11.3%). However, these deficits were significantly lower on long term follow-up (3.4%vs 8.3%) indicating that most of such deficits are transient. Further, despite having more eloquently located tumors in the mapping group, gross total excision rates were higher (75% vs 58%). Awake craniotomy specifically has also shown to improve resections in gliomas close to eloquent areas, and reduce postoperative deficits, when compared with surgery under anesthesia [5, 95, 96]. Transient postoperative deficits may be seen more often, especially if positive sites are identified at surgery [5, 65]; however, long term deficits are significantly lower. Depending on the functions that need to be mapped, awake or asleep mapping can be used. For pure motor function mapping, though both monopolar mapping (high frequency) and bipolar mapping (awake, low frequency) have been seen to be equally effective [97], the monopolar technique may be more reliable and effective in certain situations (depending on location, age, seizure history and prior treatment) and therefore is probably more versatile [98, 99]. Of course, advanced motor mapping for tumors in the premotor or parietal areas may require mapping under awake condition to assess complex motor functions (initiation, coordination and execution of complex movements) [100]. Non-motor functions require awake mapping with a tailored intraoperative assessment protocol. Awake surgery has also been shown to economize resource utilization, especially duration of hospital stay and may be a cost-effective solution [101, 102]. Finally, awake mapping using DES has the [potential to unravel the complex neurocognitive functions of the brain and further out the field of cognitive neuroscience. Besides mapping, monitoring is essential to ensure functional integrity of the tracts. In the awake patient, continuous clinical and neuropsychological assessment is performed and under anaesthesia, long tract monitoring (MEP, SSEP, VEP, BAER) correlate reliably with postoperative neurological outcomes, both for supratentorial as well as infratentorial tumors [53, 103, 104].

References

1. Berger MS, Deliganis AV, Dobbins J, Keles GE. The effect of extent of resection on recurrence in patients with low grade cerebral hemisphere gliomas. Cancer. 1994;74(6):1784–91.
2. Keles GE, Lamborn KR, Berger MS. Low grade hemispheric gliomas in adults: a critical review of extent of resection as a factor influencing outcome. J Neurosurg. 2001 Nov;95(5):735–45.
3. Duffau H. New concepts in surgery of WHO grade II gliomas: functional brain mapping, connectionism and plasticity – a review. J Neuro-Oncol. 2006 Aug;79(1):77–115.
4. Brett M, Johnsrude IS, Owen AM. The problem of functional localization in the human brain. Nat Rev Neurosci. 2002;3(3):243–9.
5. Duffau H. Functional recovery after surgical resection of low grade gliomas in eloquent brain: hypothesis of brain compensation. J Neurol Neurosurg Psychiatry. 2003;74(7):901–7.
6. Roux F-E, Boulanouar K, Lotterie J-A, Mejdoubi M, LeSage JP, Berry I. Language functional magnetic resonance imaging in preoperative assessment of language areas: correlation with direct cortical stimulation. Neurosurgery. 2003;52(6):1335–47.
7. Aubert A, Costalat R, Duffau H, Benali H. Modeling of pathophysiological coupling between brain electrical activation, energy metabolism and hemody-

namics: insights for the interpretation of intracerebral tumor imaging. Acta Biotheor. 2002;50(4):281–95.
8. Folzenlogen Z, Ormond DR. A brief history of cortical functional localization and its relevance to neurosurgery. Neurosurg Focus. 2019;47(3):E2.
9. Loukas M, Pennell C, Groat C, Tubbs RS, Cohen-Gadol AA. Korbinian Brodmann (1868-1918) and his contributions to mapping the cerebral cortex. Neurosurgery. 2011;68(1):6–11. discussion 11
10. Fritsch G, Hitzig E. Electric excitability of the cerebrum (Über die elektrische Erregbarkeit des Grosshirns). Epilepsy Behav. 2009;15(2):123–30.
11. Ferrier D. Experimental researches in cerebral physiology and pathology. J Anat Physiol. 1873;8(Pt 1):152–5.
12. Sherrington CS, Grunbaum ASF. An ADDRESS on LOCALISATION in the "MOTOR" CEREBRAL CORTEX: delivered to the pathological Society of London, December 17th, 1901. BMJ. 1901;2(2139):1857–9.
13. Molnár C, Nemes C, Szabó S, Fülesdi B. Harvey Cushing, a pioneer of neuroanesthesia. J Anesth. 2008;22(4):483–6.
14. Bartholow R. Experimental investigations into the functions ofthe. Human Erain Am J Med Sci. 1874;67:305–13.
15. Ladino LD, Rizvi S, Téllez-Zenteno JF. The Montreal procedure: the legacy of the great Wilder Penfield. Epilepsy Behav EB. 2018;83:151–61.
16. Ojemann GA, Whitaker HA. Language localization and variability. Brain Lang. 1978;6(2):239–60.
17. Cushing H. A NOTE upon the faradic stimulation of the postcentral gyrus in conscious patients. Brain. 1909;32(1):44–53.
18. Penfield W. Combined regional and general anesthesia for craniotomy and cortical exploration. I. Neurosurgical considerations. Curr Res Anesth Analg. 1954;33(3):145–55.
19. Pasquet A. Combined regional and general anesthesia for craniotomy and cortical exploration: part II. Anesthetic considerations. Int Anesthesiol Clin. 1986;24(3):12–20.
20. Merton PA, Morton HB. Stimulation of the cerebral cortex in the intact human subject. Nature. 1980;285(5762):227.
21. Taniguchi M, Cedzich C, Taniguchi M, Cedzich C, Schramm J. Modification of cortical stimulation for motor evoked potentials under general anesthesia. Neurosurgery. 1993;32(2):219–26.
22. Deletis V, Sala F. Intraoperative neurophysiological monitoring of the spinal cord during spinal cord and spine surgery: a review focus on the corticospinal tracts. Clin Neurophysiol Off J Int Fed Clin Neurophysiol. 2008;119(2):248–64.
23. Fujiki M, Furukawa Y, Kamida T, Anan M, Inoue R, Abe T, et al. Intraoperative corticomuscular motor evoked potentials for evaluation of motor function: a comparison with corticospinal D and I waves. J Neurosurg. 2006;104(1):85–92.
24. Riva M, Fava E, Gallucci M, Comi A, Casarotti A, Alfiero T, et al. Monopolar high-frequency language mapping: can it help in the surgical management of gliomas? A comparative clinical study. J Neurosurg. 2016;124(5):1479–89.
25. Hausmann ON, Min K, Boos N, Ruetsch YA, Erni T, Curt A. Transcranial electrical stimulation: significance of fast versus slow charge delivery for intraoperative monitoring. Clin Neurophysiol Off J Int Fed Clin Neurophysiol. 2002;113(10):1532–5.
26. Mortimer JT, Shealy CN, Wheeler C. Experimental nondestructive electrical stimulation of the brain and spinal cord. J Neurosurg. 1970;32(5):553–9.
27. Pudenz RH, Bullara LA, Dru D, Talalla A. Electrical stimulation of the brain. II. Effects on the blood-brain barrier. Surg Neurol. 1975;4(2):265–70.
28. Pujol J, Deus J, Acebes JJ, Villanueva A, Aparicio A, Soriano-Mas C, et al. Identification of the sensorimotor cortex with functional MRI: frequency and actual contribution in a neurosurgical context. J Neuroimag Off J Am Soc Neuroimag. 2008;18(1):28–33.
29. Gregorie EM, Goldring S. Localization of function in the excision of lesions from the sensorimotor region. J Neurosurg. 1984;61(6):1047–54.
30. Hayashi H, Kawaguchi M. Intraoperative monitoring of flash visual evoked potential under general anesthesia. Korean J Anesthesiol. 2017;70(2):127–35.
31. Petrova LD. Brainstem auditory evoked potentials. Am J Electroneurodiagnostic Technol. 2009;49(4):317–32.
32. Britton JW, Frey LC, Hopp JL, Korb P, Koubeissi MZ, Lievens WE, et al. Electroencephalography (EEG). In: St. Louis EK, Frey LC, editors. An introductory text and atlas of Normal and abnormal findings in adults, children, and infants [internet]. Chicago: Am Epilepsy Society; 2016. [cited 2022 Feb 13]. Available from: http://www.ncbi.nlm.nih.gov/books/NBK390354/.
33. Deletis V, Urriza J, Ulkatan S, Fernandez-Conejero I, Lesser J, Misita D. The feasibility of recording blink reflexes under general anesthesia. Muscle Nerve. 2009;39(5):642–6.
34. Deletis V, Fernández-Conejero I. Intraoperative monitoring and mapping of the functional integrity of the brainstem. J Clin Neurol Seoul Korea. 2016;12(3):262–73.
35. Sinclair CF, Téllez MJ, Tapia OR, Ulkatan S, Deletis V. A novel methodology for assessing laryngeal and vagus nerve integrity in patients under general anesthesia. Clin Neurophysiol Off J Int Fed Clin Neurophysiol. 2017;128(7):1399–405.
36. Deletis V, Rogić M, Fernández-Conejero I, Gabarrós A, Jerončić A. Neurophysiologic markers in laryngeal muscles indicate functional anatomy of laryngeal primary motor cortex and premotor cortex in the caudal opercular part of inferior frontal gyrus. Clin Neurophysiol Off J Int Fed Clin Neurophysiol. 2014;125(9):1912–22.

37. Deletis V, Fernandez-Conejero I, Ulkatan S, Costantino P. Methodology for intraoperatively eliciting motor evoked potentials in the vocal muscles by electrical stimulation of the corticobulbar tract. Clin Neurophysiol Off J Int Fed Clin Neurophysiol. 2009;120(2):336–41.
38. Deletis V, Fernández-Conejero I, Ulkatan S, Rogić M, Carbó EL, Hiltzik D. Methodology for intraoperative recording of the corticobulbar motor evoked potentials from cricothyroid muscles. Clin Neurophysiol Off J Int Fed Clin Neurophysiol. 2011;122(9):1883–9.
39. Espadaler J, Rogić M, Deletis V, Leon A, Quijada C, Conesa G. Representation of cricothyroid muscles at the primary motor cortex (M1) in healthy subjects, mapped by navigated transcranial magnetic stimulation (nTMS). Clin Neurophysiol Off J Int Fed Clin Neurophysiol. 2012;123(11):2205–11.
40. Rogić Vidaković M, Jerković A, Jurić T, Vujović I, Šoda J, Erceg N, et al. Neurophysiologic markers of primary motor cortex for laryngeal muscles and premotor cortex in caudal opercular part of inferior frontal gyrus investigated in motor speech disorder: a navigated transcranial magnetic stimulation (TMS) study. Cogn Process. 2016;17(4):429–42.
41. Enatsu R, Matsumoto R, Piao Z, O'Connor T, Horning K, Burgess RC, et al. Cortical negative motor network in comparison with sensorimotor network: a cortico-cortical evoked potential study. Cortex J Devoted Study Nerv Syst Behav. 2013;49(8):2080–96.
42. Velayutham P, Cherian VT, Rajshekhar V, Babu KS. The effects of propofol and isoflurane on intraoperative motor evoked potentials during spinal cord tumour removal surgery - a prospective randomised trial. Indian J Anaesth. 2019;63(2):92–9.
43. Rothwell J, Burke D, Hicks R, Stephen J, Woodforth I, Crawford M. Transcranial electrical stimulation of the motor cortex in man: further evidence for the site of activation. J Physiol. 1994;481(Pt 1):243–50.
44. Abboud T, Huckhagel T, Stork J H, Hamel W, Schwarz C, Vettorazzi E, et al. Why does threshold level change in transcranial motor-evoked potentials during surgery for Supratentorial lesions? J Neurosurg Anesthesiol. 2017;29(4):393–9.
45. Okamoto E, Ishikawa E, Yamamoto T, Matsuda M, Nakai K, Matsushita A, et al. Variability in amplitude and stimulation threshold values in motor evoked potential (MEP) monitoring during the resection of brain lesions. Clin Neurophysiol Off J Int Fed Clin Neurophysiol. 2015;126(6):1271–8.
46. Kombos T, Süss O, Vajkoczy P. Subcortical mapping and monitoring during insular tumor surgery. Neurosurg Focus. 2009;27(4):E5.
47. Tanaka S, Tashiro T, Gomi A, Takanashi J, Ujiie H. Sensitivity and specificity in transcranial motor-evoked potential monitoring during neurosurgical operations. Surg Neurol Int. 2011;2:111.
48. Nossek E, Korn A, Shahar T, Kanner AA, Yaffe H, Marcovici D, et al. Intraoperative mapping and monitoring of the corticospinal tracts with neurophysiological assessment and 3-dimensional ultrasonography-based navigation. Clinical article. J Neurosurg. 2011;114(3):738–46.
49. Prabhu SS, Gasco J, Tummala S, Weinberg JS, Rao G. Intraoperative magnetic resonance imaging-guided tractography with integrated monopolar subcortical functional mapping for resection of brain tumors. Clinical article. J Neurosurg. 2011;114(3):719–26.
50. Schucht P, Seidel K, Jilch A, Beck J, Raabe A. A review of monopolar motor mapping and a comprehensive guide to continuous dynamic motor mapping for resection of motor eloquent brain tumors. Neurochirurgie. 2017;63(3):175–80.
51. Han SJ, Morshed RA, Troncon I, Jordan KM, Henry RG, Hervey-Jumper SL, et al. Subcortical stimulation mapping of descending motor pathways for perirolandic gliomas: assessment of morbidity and functional outcome in 702 cases. J Neurosurg. 2018;131(1):201–8.
52. Raabe A, Beck J, Schucht P, Seidel K. Continuous dynamic mapping of the corticospinal tract during surgery of motor eloquent brain tumors: evaluation of a new method. J Neurosurg. 2014;120(5):1015–24.
53. Seidel K, Beck J, Stieglitz L, Schucht P, Raabe A. The warning-sign hierarchy between quantitative subcortical motor mapping and continuous motor evoked potential monitoring during resection of supratentorial brain tumors. J Neurosurg. 2013;118(2):287–96.
54. Morota N, Ihara S, Deletis V. Intraoperative neurophysiology for surgery in and around the brainstem: role of brainstem mapping and corticobulbar tract motor-evoked potential monitoring. Childs Nerv Syst ChNS Off J Int Soc Pediatr Neurosurg. 2010;26(4):513–21.
55. Fahlbusch R, Strauss C. Surgical significance of cavernous hemangioma of the brain stem. Zentralbl Neurochir. 1991;52(1):25–32.
56. Morota N, Deletis V. The importance of brainstem mapping in brainstem surgical anatomy before the fourth ventricle and implication for intraoperative neurophysiological mapping. Acta Neurochir. 2006;148(5):499–509. discussion 509
57. Macdonald DB, Skinner S, Shils J, Yingling C, American Society of Neurophysiological Monitoring. Intraoperative motor evoked potential monitoring - a position statement by the American Society of Neurophysiological Monitoring Clin Neurophysiol Off J Int Fed 2013 124(12):2291–2316.
58. Vincent M, Rossel O, Hayashibe M, Herbet G, Duffau H, Guiraud D, et al. The difference between electrical microstimulation and direct electrical stimulation – towards new opportunities for innovative functional brain mapping? Rev Neurosci. 2016;27(3):231–58.

59. Mandonnet E, Winkler PA, Duffau H. Direct electrical stimulation as an input gate into brain functional networks: principles, advantages and limitations. Acta Neurochir. 2010;152(2):185–93.
60. Duffau H. Contribution of cortical and subcortical electrostimulation in brain glioma surgery: methodological and functional considerations. Neurophysiol Clin Neurophysiol. 2007;37(6):373–82.
61. Duffau H. What direct electrostimulation of the brain taught us about the human connectome: a three-level model of neural disruption. Front Hum Neurosci. 2020;14:315.
62. Borchers S, Himmelbach M, Logothetis N, Karnath H-O. Direct electrical stimulation of human cortex - the gold standard for mapping brain functions? Nat Rev Neurosci. 2011;13(1):63–70.
63. Duffau H. Contribution of intraoperative electrical stimulations in surgery of low grade gliomas: a comparative study between two series without (1985-96) and with (1996-2003) functional mapping in the same institution. J Neurol Neurosurg Psychiatry. 2005;76(6):845–51.
64. De Witt Hamer PC, Robles SG, Zwinderman AH, Duffau H, Berger MS. Impact of intraoperative stimulation brain mapping on glioma surgery outcome: a meta-analysis. J Clin Oncol Off J Am Soc Clin Oncol. 2012;30(20):2559–65.
65. Kim SS, McCutcheon IE, Suki D, Weinberg JS, Sawaya R, Lang FF, et al. Awake craniotomy for brain tumors near eloquent cortex: correlation of intraoperative cortical mapping with neurological outcomes in 309 consecutive patients. Neurosurgery. 2009;64(5):836–45. discussion 345–346
66. Penfield W, Boldrey E. Somatic motor and sensory representation in the cerebral cortex of man as studied by electrical stimulation. Brain. 1937;60(4):389–443.
67. Ojemann GA. Individual variability in cortical localization of language. J Neurosurg. 1979;50(2):164–9.
68. Katayama Y, Tsubokawa T, Maejima S, Hirayama T, Yamamoto T. Corticospinal direct response in humans: identification of the motor cortex during intracranial surgery under general anaesthesia. J Neurol Neurosurg Psychiatry. 1988;51(1):50–9.
69. Szelényi A, Bello L, Duffau H, Fava E, Feigl GC, Galanda M, et al. Intraoperative electrical stimulation in awake craniotomy: methodological aspects of current practice. Neurosurg Focus. 2010;28(2):E7.
70. Herbet G, Rigaux-Viodé O, Moritz-Gasser S. Peri- and intraoperative cognitive and language assessment for surgical resection in brain eloquent structures. Neurochirurgie. 2017;63(3):135–41.
71. Klein M, Duffau H, De Witt Hamer PC. Cognition and resective surgery for diffuse infiltrative glioma: an overview. J Neuro-Oncol. 2012;108(2):309–18.
72. Lobo FA, Wagemakers M, Absalom AR. Anaesthesia for awake craniotomy. Br J Anaesth. 2016;116(6):740–4.
73. Frost EAM, Booij LHDJ. Anesthesia in the patient for awake craniotomy. Curr Opin Anaesthesiol. 2007;20(4):331–5.
74. Peruzzi P, Bergese SD, Viloria A, Puente EG, Abdel-Rasoul M, Chiocca EA. A retrospective cohort-matched comparison of conscious sedation versus general anesthesia for supratentorial glioma resection. Clinical article. J Neurosurg. 2011;114(3):633–9.
75. Eseonu CI, ReFaey K, Garcia O, John A, Quiñones-Hinojosa A, Tripathi P. Awake craniotomy anesthesia: a comparison of the monitored anesthesia care and asleep-awake-asleep techniques. World Neurosurg. 2017;104:679–86.
76. Herrick IA, Craen RA, Gelb AW, Miller LA, Kubu CS, Girvin JP, et al. Propofol sedation during awake craniotomy for seizures: patient-controlled administration versus neurolept analgesia. Anesth Analg. 1997;84(6):1285–91.
77. Tongier WK, Joshi GP, Landers DF, Mickey B. Use of the laryngeal mask airway during awake craniotomy for tumor resection. J Clin Anesth. 2000;12(8):592–4.
78. Fukaya C, Katayama Y, Yoshino A, Kobayashi K, Kasai M, Yamamoto T. Intraoperative wake-up procedure with propofol and laryngeal mask for optimal excision of brain tumour in eloquent areas. J Clin Neurosci Off J Neurosurg Soc Australas. 2001;8(3):253–5.
79. Danks RA, Rogers M, Aglio LS, Gugino LD, Black PM. Patient tolerance of craniotomy performed with the patient under local anesthesia and monitored conscious sedation. Neurosurgery. 1998;42(1):28–34. discussion 34–36
80. Johnson KB, Egan TD. Remifentanil and propofol combination for awake craniotomy: case report with pharmacokinetic simulations. J Neurosurg Anesthesiol. 1998;10(1):25–9.
81. Bekker AY, Kaufman B, Samir H, Doyle W. The use of Dexmedetomidine infusion for awake craniotomy. Anesth Analg. 2001;92(5):1251–3.
82. Hansen E, Seemann M, Zech N, Doenitz C, Luerding R, Brawanski A. Awake craniotomies without any sedation: the awake-awake-awake technique. Acta Neurochir. 2013;155(8):1417–24.
83. Zemmoura I, Fournier E, El-Hage W, Jolly V, Destrieux C, Velut S. Hypnosis for awake surgery of low-grade gliomas. Neurosurgery. 2016;78(1):53–61.
84. Bu L, Lu J, Zhang J, Wu J. Intraoperative cognitive mapping tasks for direct electrical stimulation in clinical and neuroscientific contexts. Front Hum Neurosci. 2021;2(15):612891.
85. Wager M, Rigoard P, Bouyer C, Baudiffier V, Stal V, Bataille B, et al. Operating environment for awake brain surgery – choice of tests. Neurochirurgie. 2017;63(3):150–7.
86. Fernández Coello A, Moritz-Gasser S, Martino J, Martinoni M, Matsuda R, Duffau H. Selection of

intraoperative tasks for awake mapping based on relationships between tumor location and functional networks. J Neurosurg. 2013 Dec;119(6):1380–94.
87. Duffau H. New philosophy, clinical pearls, and methods for intraoperative cognition mapping and monitoring "à la carte" in brain tumor patients. Neurosurgery. 2021;88(5):919–30.
88. Ius T, Angelini E, Thiebaut de Schotten M, Mandonnet E, Duffau H. Evidence for potentials and limitations of brain plasticity using an atlas of functional resectability of WHO grade II gliomas: Towards a "minimal common brain". NeuroImage. 2011;56(3):992–1000.
89. Sanai N, Mirzadeh Z, Berger MS. Functional outcome after language mapping for glioma resection. N Engl J Med. 2008;358(1):18–27.
90. Gil-Robles S, Duffau H. Surgical management of World Health Organization grade II gliomas in eloquent areas: the necessity of preserving a margin around functional structures. Neurosurg Focus. 2010;28(2):E8.
91. Skrap M, Marin D, Ius T, Fabbro F, Tomasino B. Brain mapping: a novel intraoperative neuropsychological approach. J Neurosurg. 2016;125(4):877–87.
92. Chang W-H, Pei Y-C, Wei K-C, Chao Y-P, Chen M-H, Yeh H-A, et al. Intraoperative linguistic performance during awake brain surgery predicts postoperative linguistic deficits. J Neuro-Oncol. 2018;139(1):215–23.
93. Bertani G, Fava E, Casaceli G, Carrabba G, Casarotti A, Papagno C, et al. Intraoperative mapping and monitoring of brain functions for the resection of low-grade gliomas: technical considerations. Neurosurg Focus. 2009;27(4):E4.
94. Chang EF, Clark A, Smith JS, Polley M-Y, Chang SM, Barbaro NM, et al. Functional mapping–guided resection of low-grade gliomas in eloquent areas of the brain: improvement of long-term survival: clinical article. J Neurosurg. 2011;114(3):566–73.
95. Sacko O, Lauwers-Cances V, Brauge D, Sesay M, Brenner A, Roux F-E. Awake craniotomy vs surgery under general anesthesia for resection of supratentorial lesions. Neurosurgery 2011 68(5):1192–8; discussion 1198–1199.
96. Pinsker M, Nabavi A, Mehdorn H. Neuronavigation and resection of lesions located in eloquent brain areas under local anesthesia and neuropsychological-neurophysiological monitoring. Min - Minim Invasive Neurosurg. 2007;50(5):281–4.
97. Suarez-Meade P, Marenco-Hillembrand L, Prevatt C, Murguia-Fuentes R, Mohamed A, Alsaeed T, et al. Awake vs. asleep motor mapping for glioma resection: a systematic review and meta-analysis. Acta Neurochir. 2020;162(7):1709–20.
98. Nossek E, Matot I, Shahar T, Barzilai O, Rapoport Y, Gonen T, et al. Intraoperative seizures during awake craniotomy: incidence and consequences: analysis of 477 patients. Neurosurgery. 2013;73(1):135–40. discussion 140
99. Gonen T, Grossman R, Sitt R, Nossek E, Yanaki R, Cagnano E, et al. Tumor location and IDH1 mutation may predict intraoperative seizures during awake craniotomy. J Neurosurg. 2014;121(5):1133–8.
100. Rossi M, Fornia L, Puglisi G, Leonetti A, Zuccon G, Fava E, et al. Assessment of the praxis circuit in glioma surgery to reduce the incidence of postoperative and long-term apraxia: a new intraoperative test. J Neurosurg. 2018;130(1):17–27.
101. Taylor MD, Bernstein M. Awake craniotomy with brain mapping as the routine surgical approach to treating patients with supratentorial intraaxial tumors: a prospective trial of 200 cases. J Neurosurg. 1999 Jan;90(1):35–41.
102. Blanshard HJ, Chung F, Manninen PH, Taylor MD, Bernstein M. Awake craniotomy for removal of intracranial tumor: considerations for early discharge. Anesth Analg. 2001 Jan;92(1):89–94.
103. Kodama K, Javadi M, Seifert V, Szelényi A. Conjunct SEP and MEP monitoring in resection of infratentorial lesions: lessons learned in a cohort of 210 patients. J Neurosurg. 2014 Dec;121(6):1453–61.
104. Slotty PJ, Abdulazim A, Kodama K, Javadi M, Hänggi D, Seifert V, et al. Intraoperative neurophysiological monitoring during resection of infratentorial lesions: the surgeon's view. J Neurosurg. 2017,126(1):281–8.

Neural Basis of Language, a Comprehensive Update for Neurosurgeons

Masazumi Fujii and Mudathir Bakhit

1 The Mental Lexicon (Semantics)

How does the brain process speak language input to produce meaning? And how to produce speech? To answer these questions, several essential elements must be considered. Among these is the **storage** element: The brain must store representations of words and their associated concepts. A spoken word in language has two major properties: a **meaning** and a **phonological** (sound-base) form [1].

The mental store is commonly known among psycholinguistics and neuropsychologist as the **mental lexicon** or the mental dictionary (lexicon is the Greek word for a book). The Mental Lexicon essentially contains all necessary items a person knows and needs to communicate. These items can be the **semantic** information (i.e., the word's meaning), the **syntactic** information (how the words combine to form sentences), the **word form** (spelling and sound elements) [1]. Whether language production and comprehension share the same mental lexicon or not is still a matter of debate.

The mental lexicon has mainly three functions [1]:

Lexical access. The American Psychologist Association (APA) defines it as the process by which individuals produce a specific word from their mental lexicon or recognize it when others use it. This shall include these words' semantic and syntactic elements.

Lexical selection. As the name says, a stage in which the best match representation to the input is selected.

Lexical integration. In the final stage, words are integrated into the complete sentence, discourse, or larger context.

However, the mental lexicon must be well organized to work efficiently and smoothly. The organization of the mental lexicon is based on particular principals [1].

- The smallest unit of representation in the mental lexicon is called a **morpheme**. For example, consider the words luck, luck(y), and (un) luck(y). The root of these words, luck, forms one morpheme; the word lucky consists of two morphemes, and the word unlucky consists of three morphemes.
- More frequently used words are accessed quickly.
- The lexicon is organized in neighborhoods consisting of words that differ from one another by a single letter or a **phoneme**. A phoneme is the smallest unit of sound that makes a difference in meaning (e.g., hat, fat, cat, etc.). As we will see later, in a particular

M. Fujii · M. Bakhit (✉)
Department of Neurosurgery, Fukushima Medical University, Fukushima city, Japan
e-mail: fujiim@fmu.ac.jp; m-bakhit@fmu.ac.jp

A. Shah et al. (eds.), *Functional Anatomy of the Brain: A View from the Surgeon's Eye*,
https://doi.org/10.1007/978-981-99-3412-6_6

condition called phonemic paraphasia, the patient makes errors by replacing phonemes.

- Representation in the mental lexicon is organized based on the semantic relationship between words. This was based on a result that came out in the seventies from a semantic priming study [2]. In these, the volunteers were presented with a pair of words, the first was a prime, and the next was a target. The volunteers have to confirm whether the target word was a real word or not. The authors provided a prime word associated with the target, e.g., BREAD-BUTTER and NURSE-DOCTOR, or unassociated words, e.g., BREAD-DOCTOR and NURSE-BUTTER. They found that the response reaction was faster when semantically associated words were presented.

Several mental lexicon models have been proposed, and the most famous model was proposed by Collins and Loftus [3]. In their model, words meaning are represented in semantic network nodes connected. Fig. 1 shows an adaptation of the original schema made by Collins and Loftus. The edges' length represents the strength of semantic association between each pair of words. The shorter the edge, the stronger the association.

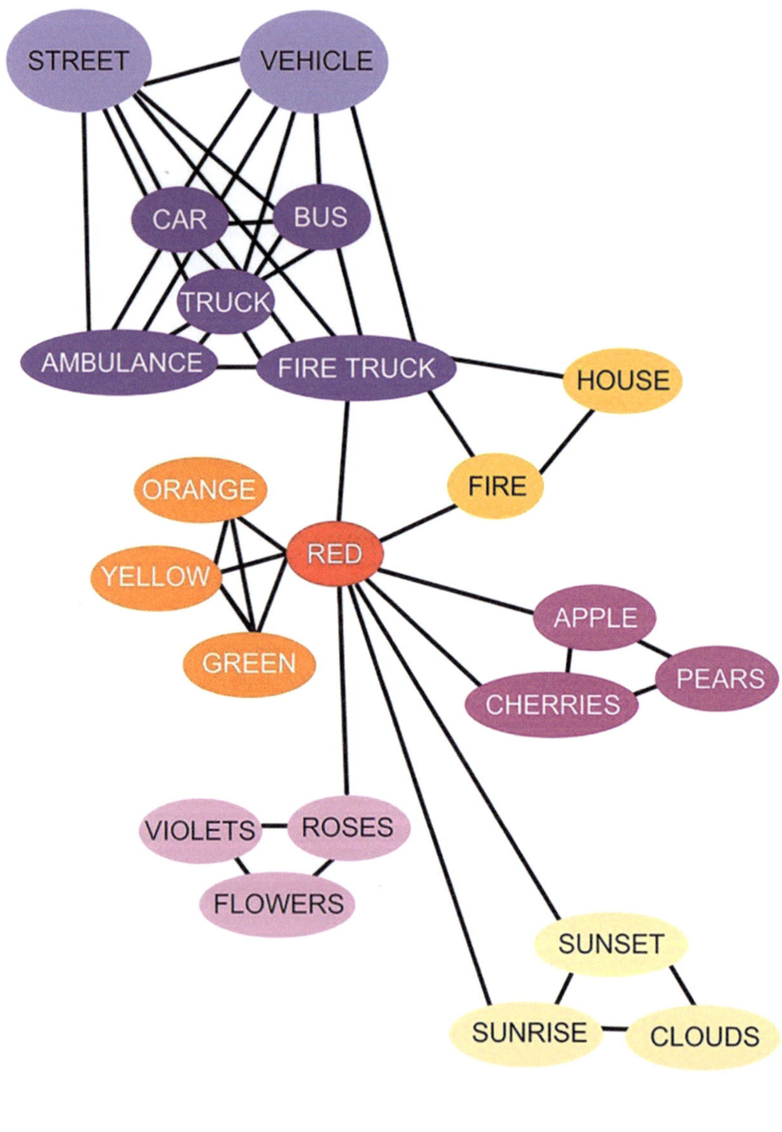

Fig. 1 A schematic representation of the mental lexicon semantic network. Adapted from [3]

For example, the word *apple* will have a strong association with *pears*. However, this model was updated in the following reports, and many other models have been proposed, and the model structure is still a matter of investigation.

Another question to be asked is how the storage of the mental lexicon semantics is organized in the human brain. To answer this question, Huth et al., in a breakthrough study, studied how the brain represents the semantic data [4]. An fMRI data were collected from seven healthy adults while listening to two hours stories taken from the Moth Radio Hour. The stories' words were coded with a transcript to follow the activity in the brain and correlate it to the common 985 words in the stories. They systematically mapped semantic selectivity across the human cerebral cortex using voxel-wise modeling. The semantic system seems to be organized into complicated patterns consistent across individuals. These details were used to create a semantic map atlas (Fig. 2). A significant finding is that the semantic map was not restricted to certain regions as was previously thought. A wide surface area beyond the commonly thought semantic areas of the brain is involved in the process. For an interactive version of this map, visit https://gallantlab.org/huth2016.

To summarize, language function can be divided into (1) phonetics/phonology process, related mainly to speech sounds and production; (2) semantic process, the processing of word meaning; (3) orthography process, the representation of written characters; and (4) syntax process, the methods for combining words into sentences [5]. Also, the semantics organization is not restricted to classical language areas in the left hemisphere but can involve a wide area of the whole cerebral cortex.

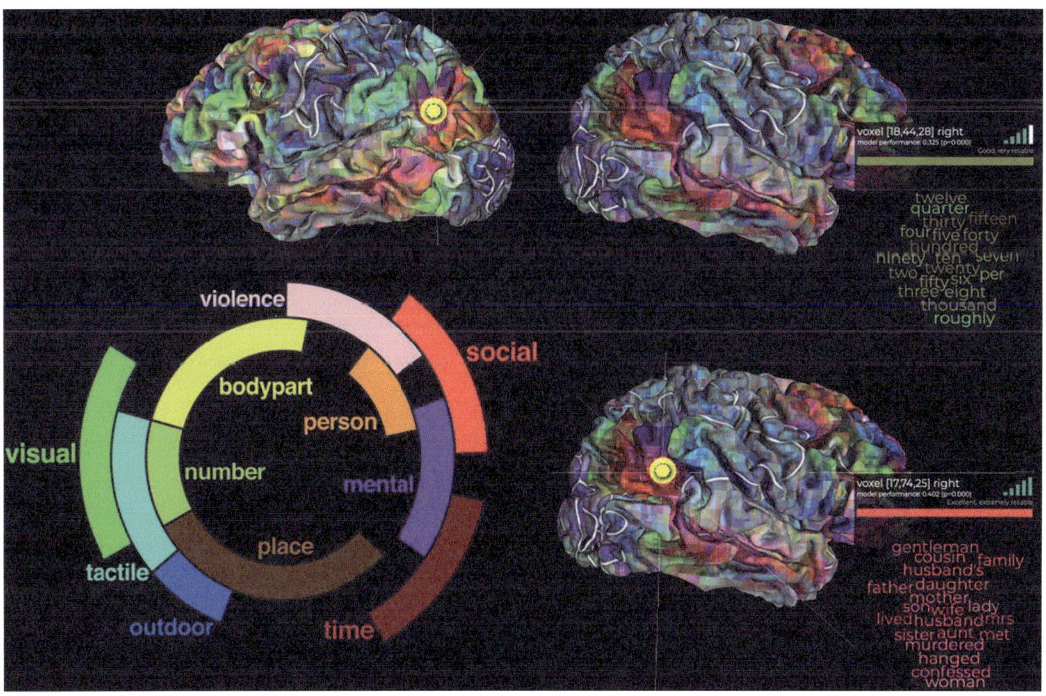

Fig. 2 The Semantic map of Subject 2, from [4]. A voxel selection (yellow circle) at the left SMG showed selectivity mainly for numbers semantic category (Upper). A voxel selection (yellow circle) at the right temporoparietal junction showed selectivity mainly for the social semantic category (Upper). The level of reliability of the word selection is shown as signal bars at the right of the image. The different colors represent different word categories (lower left). For an interactive version of this map, visit https://gallantlab.org/huth2016

2 The Language-Relevant Cortical Regions

This left-dominant perisylvian language network comprises parts of the inferior frontal gyrus (IFG), premotor cortex (PMC) including ventral part of the precentral gyrus, superior temporal lobe, and a temporoparietal area, including the supramarginal gyrus. However, other functionally connected brain regions are involved in the language process, including the middle frontal gyrus (MFG), the supplementary motor area (SMA), fusiform gyrus, middle and inferior temporal gyrus, angular gyrus, and others.

2.1 Language-relevant frontal regions

Inferior Frontal Gyrus (IFG) and Ventral Premotor Cortex (vPMC). Traditionally, language production was thought to be executed and controlled in the frontal lobe by Broca's area alone, corresponding to Brodmann's area (BA) 44 and 45. However, this concept has dramatically changed, and several other frontal regions have been found to play a significant role in speech [6]. These areas of interest in the ventral part of the frontal lobe span along a caudal-rostral gradient from the vPMC and BA6 in the precentral gyrus to the pars opercularis (BA44) and the pars triangularis (BA45). The latter two are customarily known to form the Broca's area in the IFG. Then they continue to the more anterior located pars orbitalis (BA47) (Fig. 3). Furthermore, ventromedial to the BA 44/45 lies the frontal operculum (FOP). A recent receptor architectonic analysis has divided the BA44 and BA45 further into BA44 dorsal/ventral (BA44d/BA44v) and BA45 anterior/posterior (BA45a/BA45p), respectively [7] (Fig. 4).

Based on fMRI studies, these areas support different aspects of language processing [8–10]. BA6 is involved in phonology and articulation (see below); BA44 and BA45p are engaged in syntactic structure building; and area BA45a, together with BA47, supports semantic processes [10]. Uddén et al. reviewed the literature on left IFG activation across different cognitive tasks, including language, music, and the action domain. They proposed a rostro-caudal abstraction gradient in the left IFG. In their model (Fig. 4), BA44d and rostro-ventral BA6 (BA6v) are engaged in phonological processing; BA44v and BA45p are involved in processing syntactic rules at the word, and sentence processing level and perhaps also semantic rules at the sentence level; and BA45a and BA47 are engaged in the integration of these processes as well as information from other domains [8].

It is assumed that aphasia due to damage to the "classical" Broca's area will produce apraxia of

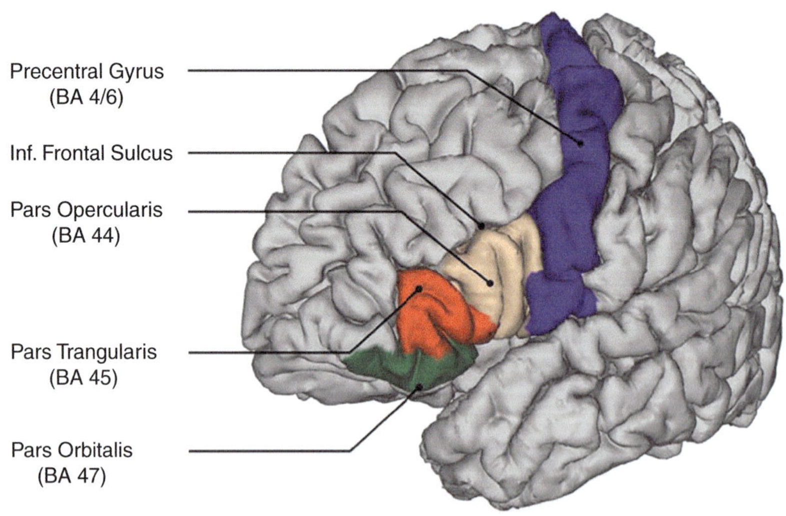

Fig. 3 The language function relevant inferior frontal areas. Image constructed using data from the Human Connectome Project, including Desikan-Killiany Atlas and Human Connectome Project workbench software

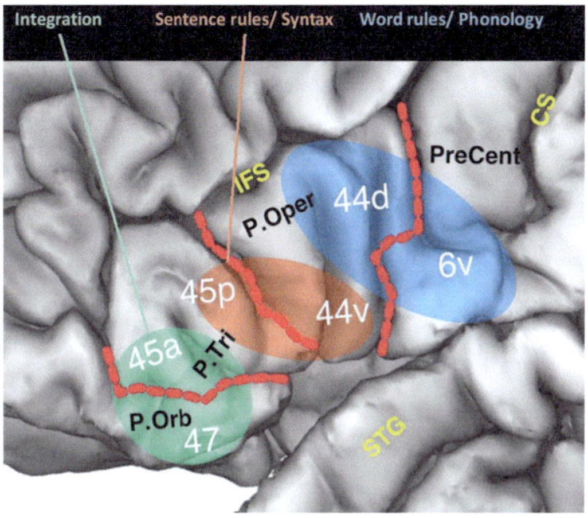

Fig. 4 The inferior frontal cortex rostro-caudal abstraction functional gradient model by [8]. *CS* Central sulcus, *IFS* Inferior frontal sulcus, *P.Oper* Pars opercularis, *P.Orb* Pars orbitalis, *P.Tri* Parstriangularis, *STG* Superior temporal gyrus

speech and agrammatism. After introducing the CT scan followed by MRI, we now understand that this assumption is not valid [11]. Lesion extension to the insula, lower motor cortex, and subjacent subcortical and periventricular white matter are required to develop such a language deficit [12]. Isolated lesions to the traditional Broca's area (BA44/45) are known to produce only transient mutism that resolves within 3 to 6 weeks [11], or mild forms of aphasia, such as anomic aphasia. Functional neuroimaging studies have demonstrated that language production activates several regions, not only BA44/45, but also a wide circuit including BA46, BA47, the anterior insula, the supplementary motor areas (SMA), and subcortical structures [12]. In brief, neither lesions to Broca's area alone result in a classical Broca's aphasia nor Broca's aphasic patients must have lesions in "classical" Broca's area.

The caudal portion of vPMC (BA 6, may include 4) lies directly anterior to the orofacial movement PMC and plays a critical role in phonologic processing, bilaterally [13]. Duffau and his colleagues reported that stimulation of the left and right vPMC produced dysarthria/anarthria in up to 83% and 55% of patients, respectively [14]. Other direct electrical stimulation (DES) studies have consistently shown the vPMC to have the highest rate of speech arrest of any cortical site [13]. Bizzi et al. found a five-fold increase in aphasia in patients with gliomas of the vPMC compared to IFG (classic Broca's area) [15]. Damage in the anterior part of the dominant vPrCG (BAs 6 and 4) tends to enhance loss of prosody or mal-connection of phonemes more. In contrast, damage in the posterior part (BA 4) can distort phonemes (Fig. 5) [16]. Also, the left vPMC, especially the ventral precentral gyrus (BA6), plays an essential role in the programming of the articulation process that is upstream and is a higher function to the process of motor movement of the articulatory organs generated by the motor cortex (BA4).

Although there is bilateral representation, with a left side dominance, the vPMC has a limited capacity for plasticity compared to other language regions. Studies have shown that reorganization of vPMC typically occurs only within the remaining portions of the ipsilateral vPMC. As such, lasting post-operative speech deficits correlate directly with the percentage of the vPMC resected. This limitation is probably due to an anatomical constraint, being the necessity for the left vPMC to remain connected to the superior longitudinal system (SLF) [17].

Middle Frontal Gyrus (MFG). The MFG, especially the left, is involved in the language function and even comparable with the Broca's area in determining hemispheric dominance for language and can be used instead of the IFG as a noninvasive modality in patients with gliomas to evaluate hemispheric dominance of language before tumor resection [18]. Also, the left MFG seems to have a role in word production as it may serve as a temporal perceptual information storage space, supporting the hierarchical state feedback control model of the word production [19]. The caudal part of the MFG (cMFG) seems to play a major role in language switching in bilinguals [20]. Direct stimulation of the cMFG in brain tumor patients caused involuntary switching from Catalan to Spanish in one patient while altering the switching process in another [21].

The cMFG possesses several cytoarchitectonic areas, including part of the BA6 (dPMC), BA 8, 9,

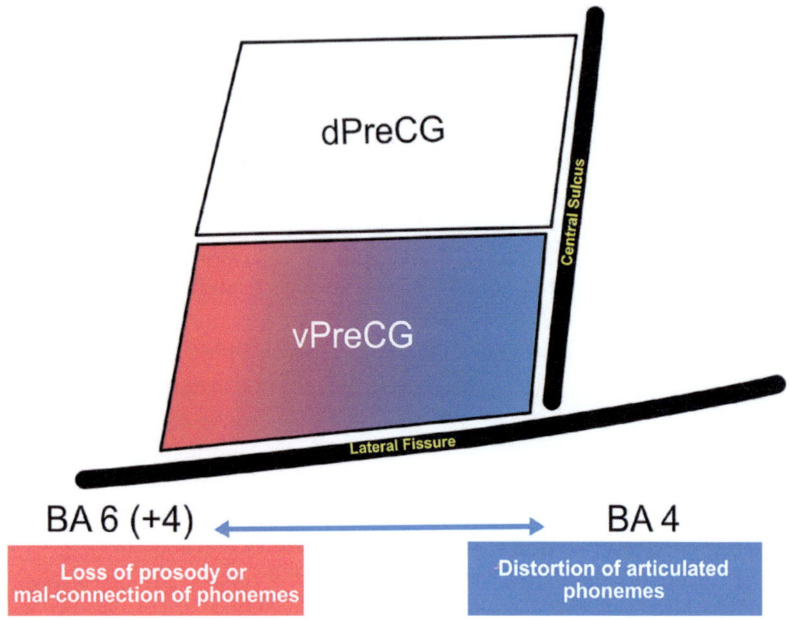

Fig. 5 Articulation in the ventral precentral gyrus (vPrCG, or vPMC). The lesion in the anterior part can cause loss of prosody, while posterior lesions can cause distortion of articulated phonemes

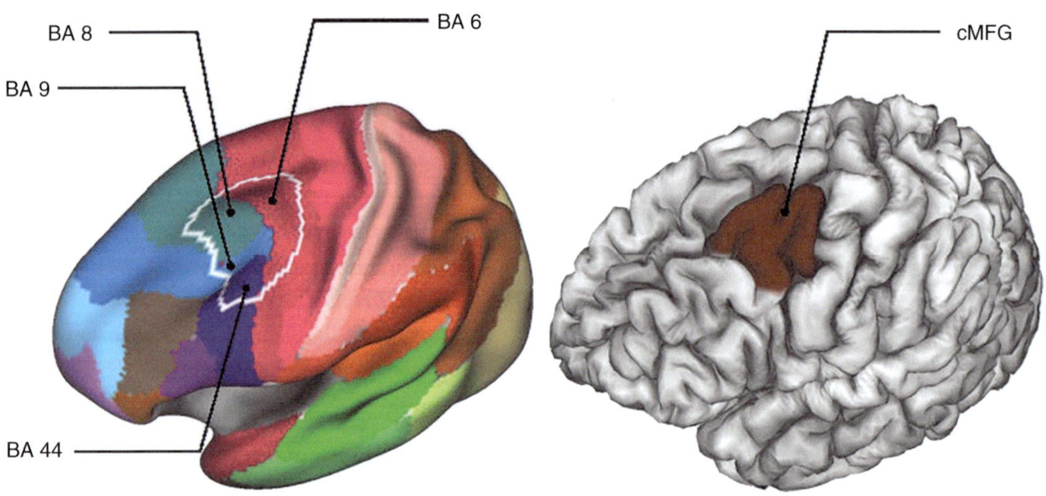

Fig. 6 The caudal middle frontal gyrus (cMFG). The approximate location of the cMFG (brown) overlaid over a left brain surface template (right). The borders of the cMFG (white) and the Brodmann areas within it (left). Image constructed using data from the Human Connectome Project, including Desikan-Killiany and Brodmann Atlases and Human Connectome Project workbench software

and 44 (Fig. 6). Also, through a different cytoarchitecture, the cMFG composes a region known as area 55b found by the neuroanatomist Adolf Hof in 1956 to have unique cytoarchitecture [22]. Area 55b is delimited by the frontal eye field (FEF) superiorly, the premotor eye field (PEF) inferiorly, the primary motor cortex posteriorly, and the remaining part of the MFG anteriorly (Fig. 7) [23]. Damage to this area has been linked to the occurrence of pure apraxia of speech (AoS) in several cases, despite being labeled as a premotor cortex lesion in some; however, the images in these reports point towards the cMFG [25, 26]. Trans magnetic stimulation (TMS) and cortical stimulation of the 55b area have caused speech hesitations, phonemic and semantic errors, and even speech arrest (Fig. 8) [27].

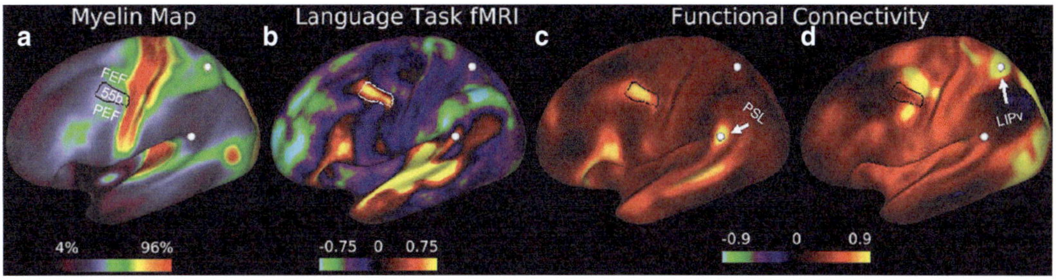

Fig. 7 Area 55b. The border of 55b is indicated by a white or black outline. (**a**) Cortical myelin map showing area 55b with low myelination in contrast to the moderately myelinated FEF and heavily myelinated motor cortex. (**b**) Group average beta map from the language "Story versus Baseline" task contrast shows activation of the 55b in association with the language areas in the IFG and STG. (**c**) Functional connectivity correlation maps from seed in the perisylvian language area (PSL) (white sphere, arrow) showed correlated activations in the common language areas. (**d**) As a contrast to the language network, another seed in the lateral intraparietal- ventral area (LIPv) (white sphere, arrow) showed a correlated activation of the fronto-parietal network with apparent inactivity of the 55b. The image was made with the human connectome project workbench software. Image constructed using open-source data from the Human Connectome Project [23, 24]

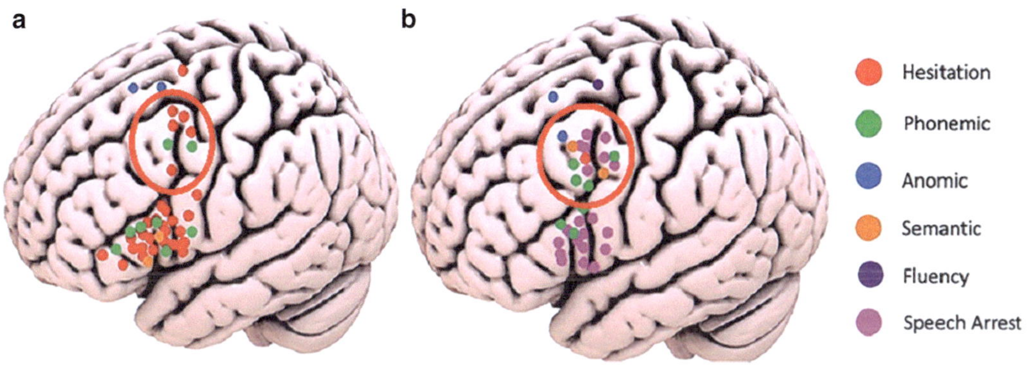

Fig. 8 Schematic representation of the positive navigated transcranial magnetic stimulation responses and intraoperative cortical responses by [27]. (**a**) Positive TMS responses. (**b**) Positive cortical intraoperative responses. Red circle: posterior MFG. Adapted from [27] with permission

cMFG possesses some of the frontal terminations of the arcuate fasciculus and the superior longitudinal fasciculus II (SLF II) [28, 29]. Several subcortical language white matter tracts are running beneath the MFG and are at risk of damage during MFG tumor excision. We will discuss these tracts in detail later in this chapter.

Other functions of the MFG related to language are sentence processing, management of discourse coherence, top-down predictive mechanisms during speech processing, speaker-listener interaction, integration of prosody (bilateral MFG), advanced lexical processing, and processing of non-literal language meaning [30]. These functions are processed in combination with other brain regions such as the supplementary motor area (SMA), parietal cortex, the limbic system, the cerebellum, and the basal ganglia.

Regarding tumor excision in the area of the MFG, Morshed et al. reported the clinical outcome for patients undergoing resection of MFG diffuse glioma in 39 patients [31]. No patients had positive cortical mapping sites overlying the tumor, and 33.3% of the cases had positive subcortical stimulation sites. Eight patients developed variable language deficits after surgery; all were cases with lesions on the left side; however, deficits recovered by 90 days. The temporary language deficit was the drop in naming accuracy rate, non-fluent spontaneous speech, verbal paraphasia, inability to repeat, and speech hesitancy.

In Fig. 6, we can see the cMFG compose of some part of the dorsal premotor cortex (dPMC, BA6). Other areas of the dPMC beyond the cMFG have roles in the language function. Studies of brain-damaged patients, neuroimaging studies, and meta-analyses all support the existence of specific regions of the left dPMC, whose localization is remarkably similar across studies and is mandatory for the production of graphemes through handwriting, known as Graphemic/Motor Frontal Area (GMFA) or Exner's area [32, 33].

The supplementary motor area (SMA). The SMA is a superordinate control region for speech and language processing. Anatomically, the SMA is roughly located in the medial part of BA 6 and 8 and comprises at least two subareas, the rostral pre-SMA and the caudal SMA-proper [34, 35]. Probabilistic cytoarchitectonic maps of areas SMA-proper and pre-SMA showed that they are located dorsal to the cingulate sulcus in the medial view. However, there is a relatively significant inter-individual variability in the size of the areas (see Fig. 9, voxels coding for a low overlap in blue and green colors) [36].

The SMA-proper serves mainly motor-related functions, whereas the pre-SMA is involved in higher-order cognitive control mechanisms. Regarding speech, SMA proper is primarily involved in speech motor control as an initiation and timing interface. By contrast, pre-SMA can be considered an area of cognitive control managing complex sequencing, ambiguity resolution, and task switching [34].

Damage to the SMA results in the SMA syndrome characterized by contralateral hemiparesis and/or hemiplegia with various degrees of speech deficits in the language-dominant hemisphere, ranging from speech hesitancy to complete mutism [37–40]. SMA syndrome usually recovers within a few weeks to months.

Recently, the evolution of fMRI activations during recovery of the language component of the SMA syndrome has been comprehensively described [40]. FMRI images in a patient with left SFG oligodendroglioma were performed four times, including preoperative imaging and three follow-up studies. On day 12 after surgery, there was a complete shift of blood oxygen level-dependent (BOLD) activation to the contralateral right SMA as well as coequal activation and an increased volume of activation in the left and right IFG. The activity in the right IFG was thought to be a compensation mechanism, which may temporarily interfere with language SMA recovery. After 3 months, fMRI showed the return of activation to the left pre-SMA with a coinciding decrease in activation in the right pre-SMA. After 5 months, dominant BOLD activation was returned in the left IFG. The authors, in their discussion, attributed part of these activities to the structural connectivity that exists between each SMA and the bilateral IFG through the frontal aslant tract (FAT, see below) and a subgroup of commissural fibers that intermingle with the association fibers of the FAT to connect the SMA to the contralateral Broca's area. These interhemispheric connections are thought to play a crucial role in the recovery of the SMA syndrome [28, 29].

2.2 The Insula

The insula, especially the anterior part (BA13), is known to have significant functional connectivities with the primary and secondary language areas. Anterior insula connections included not only areas involved in language production (such as the Broca's area) and language comprehension (such as the Wernicke's area) but also areas involved in language repetition (such as the supramarginal gyrus, SMG) and other linguistic functions such as BA9 in the left prefrontal lobe (part of the prefrontal cortex probably involved in complex language processes) and BA37 (involved in lexico-semantic associations) [6]. The anterior insular cortex of the dominant language hemisphere shows activation after language task in fMRI in the prep-operative fMRI [41].

However, whether the insula has a major role in language processing remains unknown [42]. Wernicke predicted conduction aphasia was linked to damage to the insula, a cortical area that, in his

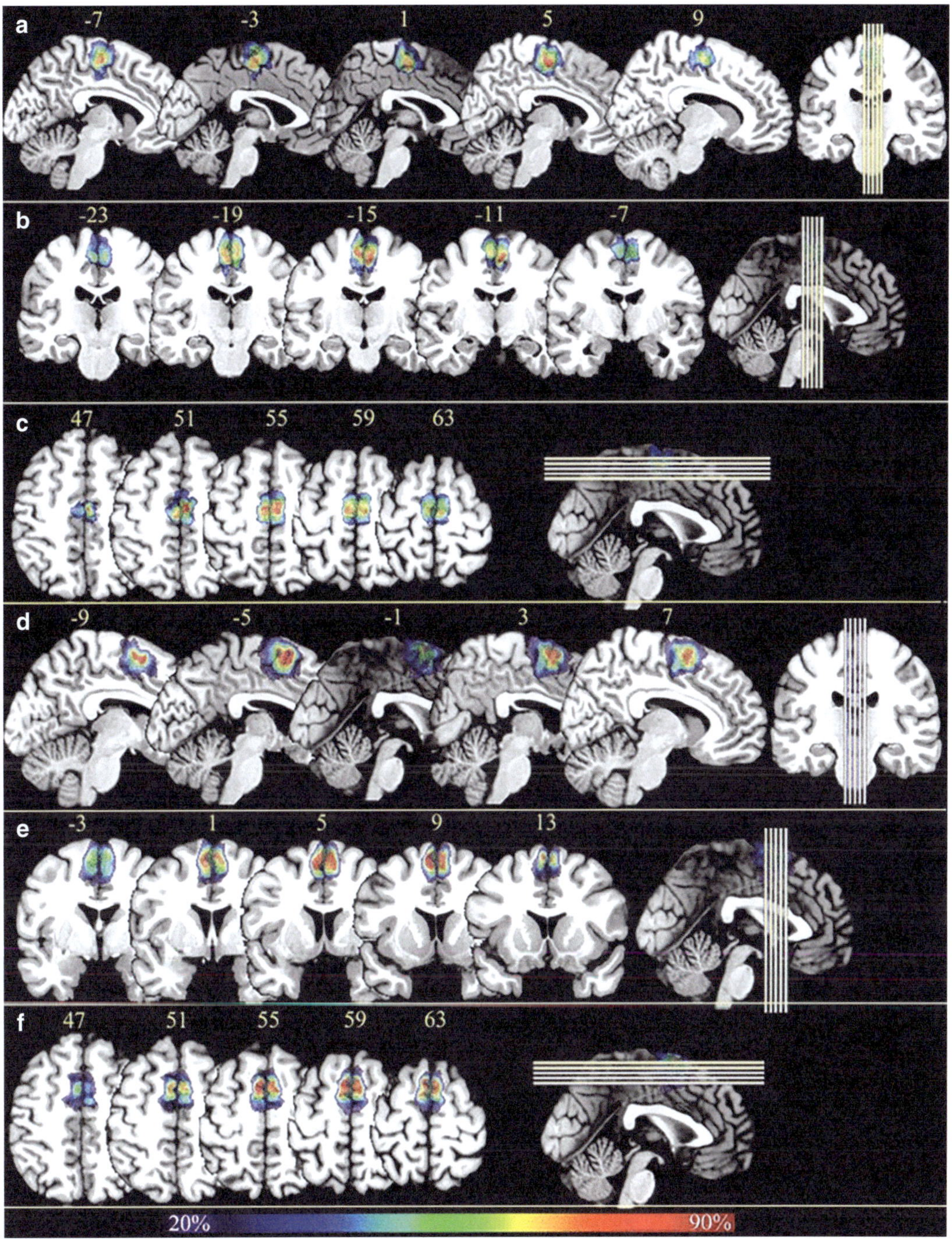

Fig. 9 Continuous probability maps of areas SMA and pre-SMA in anatomical MNI reference brain. Cytoarchitectonic probability map of area SMA-proper in the sagittal (**a**), coronal (**b**), horizontal (**c**), and sectioning planes, as well as of pre-SMA, also in the sagittal (**d**), coronal (**e**), horizontal (**f**), respectively. The number of overlapping brains for each voxel is color coded. The left and right sides of each section in b, c, e, and f indicate the left and right hemispheres, respectively. The yellow numbers in a–f indicate the x-, y-, and z-coordinates, respectively. The images at the extreme right of panels show the locations of corresponding sections. From [37] with permission

opinion, interconnected rostral and caudal perisylvian regions. He abandoned this concept several years later [43]. However, the role of the insula in language process was brought again to focus after Dronkers presented a lesion-overlap study in stroke patients that proposed damage of the short insular gyrus-posterior (SIG-P, Fig. 10), also known as the precentral gyrus of the insula, can result in apraxia of speech (AoS) [44]. However, this study's methodology was criticized because of its methods since lesion-overlap investigations cannot rule out "that all patients with apraxia of speech have insular damage, but that few patients with insular damage have apraxia of speech" [43, 45].

Ackermann et al., in their review of the contributions of the insula to speech production, have mentioned that isolated ischemic damage restricted, by and large, to the intrasylvian cortex represents a highly rare event, and also, it is bewildering that motor planning of speech might depend upon the rostral, i.e., "para- limbic" component of intra-Sylvian cortex [43]. Cereda et al. studied the clinical presentation of four patients with first-ever acute

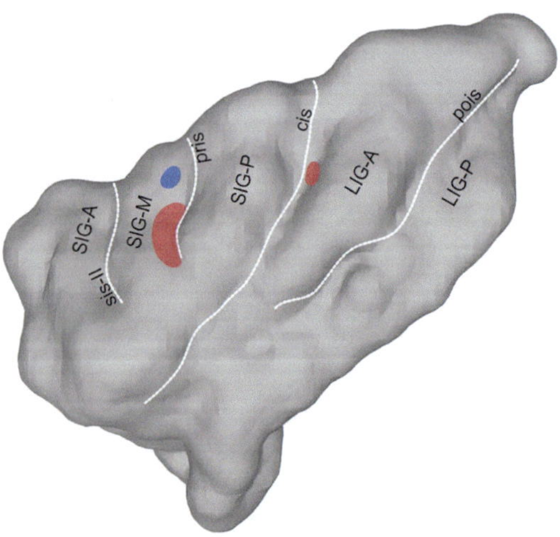

Fig. 10 A volumetric template of the left insula with an overlay of the Afif et al. study's results. The red and blue regions represent the areas that, after sub-gyral electrode stimulation, elicited speech arrest and lower speech volume, respectively. LIG-A: long insular gyrus-anterior, LIG-P: long insular gyrus-posterior, SIG-A: short insular gyrus-anterior, SIG-M: short insular gyrus-middle, SIG-P: short insular gyrus-posterior, cis: central insular sulcus, pris: precentral insular sulcus, pois: postcentral insular sulcus, sis-II: short insular sulcus-II

stroke restricted to the insula on MRI, and none of their cases presented with AoS; however, all speech disturbances reported were related to the posterior insular cortex (aphasia (left), dysarthria, and transient somatoparaphrenia (right))[46].

Besides cerebrovascular accidents (CVS), in one case series of brain tumors, speech articulatory disturbances were elicited by DES of the left insular cortex in 20% of cases [47, 48]. However, patients with insular low grade glioma who undergo awake brain surgery usually do not develop permanent language deficits [47–49]. Electrical stimulation in drug resistance epilepsy patients through stereotactically implanted electrodes within the gyral substructures of the short insular gyrus-middle (SIG-M, Fig. 10) in the anterior insula elicited speech disturbances [50]. These speech disturbances were in the form of either speech arrest or lowering of speech volume. The remaining insular regions did not show any speech disturbance after stimulation.

Damage to the insular cortex seems not to cause permanent language deficit; however, at the sub-cortical level, several crucial white matter structures run beneath the insula that have major roles in the language process, including the extreme capsule, and attention and care will be needed when dealing with tumors in this area (see below).

In literature, there are a few reported cases of resection of glioma of the insulo-opercular structures induced transient Foix-Chavany-Marie syndrome, a rare type of suprabulbar palsy characterized by an automatic voluntary dissociation of the orofacial musculature causing a bilateral facio-linguo-pharyngo-laryngal palsy with a reversible inability of the patient to speak and swallow [51, 52]. The most frequent causes of this syndrome are bilateral strokes affecting both opercula, but the syndrome has also been reported after infections, developmental abnormalities, status epilepticus, neurodegenerative disorders, and head injury [53].

In conclusion, significant language deficits because of isolated damage to the insula are rare, and the evidence is unreliable. However, its strategic location between the anterior and posterior language areas can be crucial to playing a role in the language coordinating function [6].

2.3 Language-relevant temporal regions

In contrast to what we did in the frontal language-relevant regions, we will approach the temporal regions from a functional perspective rather than an anatomical perspective. This is because Wernicke's area boundaries are poorly understood.

The Auditory Cortices. From the medial geniculate nucleus (MGN), the acoustic radiation exits to terminate in the primary auditory cortex (PAC, BA 41), also known as A1 or the core region, which is in Heschl's gyrus (HG) (also known as transverse temporal gyrus) in the dorsal surface of the superior temporal gyrus (STG) (Fig. 11). HG is entirely hidden in the lateral fissure, with a caudal-medial to a rostro-lateral axis. The planum polare (PP) and planum temporale (PT) of the STG are delimiting the HG anteriorly and posteriorly, respectively. The PAC is surrounded by a secondary auditory cortex (SAC) made of two regions known as the belt (BA 42) and parabelt (BA 22) [54]. The parabelt is part of the larger Wernicke's area [10]. To know more about the mapping of the human's auditory cortex, see [54].

The classical hierarchical model of speech perception assumes that sound information is first received in the HG and then transformed into surrounding more complex regions in the STG. However, a recent breakthrough study added new insight to this concept. The study recorded the intracranial signals in the human primary and non-primary auditory cortex through implanting high-density electrode grids in 9 patients (a total of 636 electrode sites) undergoing left hemisphere insular or opercular tumor resection or phase II monitoring for intractable epilepsy. This study has shown the presence of an "onset zone" in the lateral STG that possesses a parallel activation to the PAC (Fig. 12a, black zone) [55, 56]. These parallel strong transient responses at the onset were seen in the lateral STG (pSTG-onset) and not on the superior temporal plane where the PAC is located. In the same study, they did a latency analysis on trial-averaged high-gamma data within each region of interest (ROI, Fig. 12a) and applied a cross-correlation analysis to determine lead-lag relationships. They found fast-latency responses to sentences in posteromedial HG (pmHG), pSTG-onset, and PT regions and more extended latency responses in anterolateral HG (alHG), pSTG non-onset, mSTG, and PP regions (Fig. 12b).

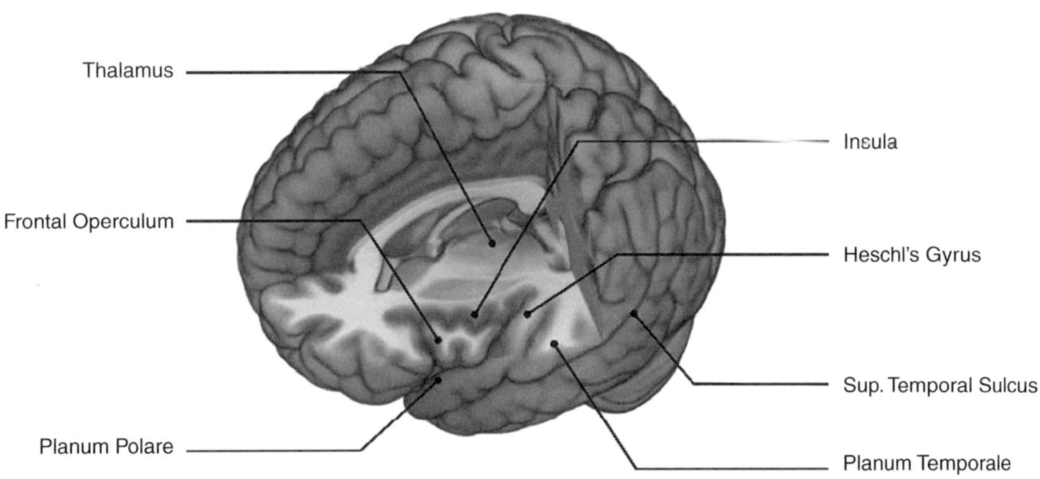

Fig. 11 Overview of the primary auditory cortex and its surrounding anatomical structures Beyond the primary and secondary auditory cortices, the STG has other language-relevant regions, the anterior and posterior STG (aSTG and pSTG) and anterior and posterior superior temporal sulcus (aSTS and pSTS)

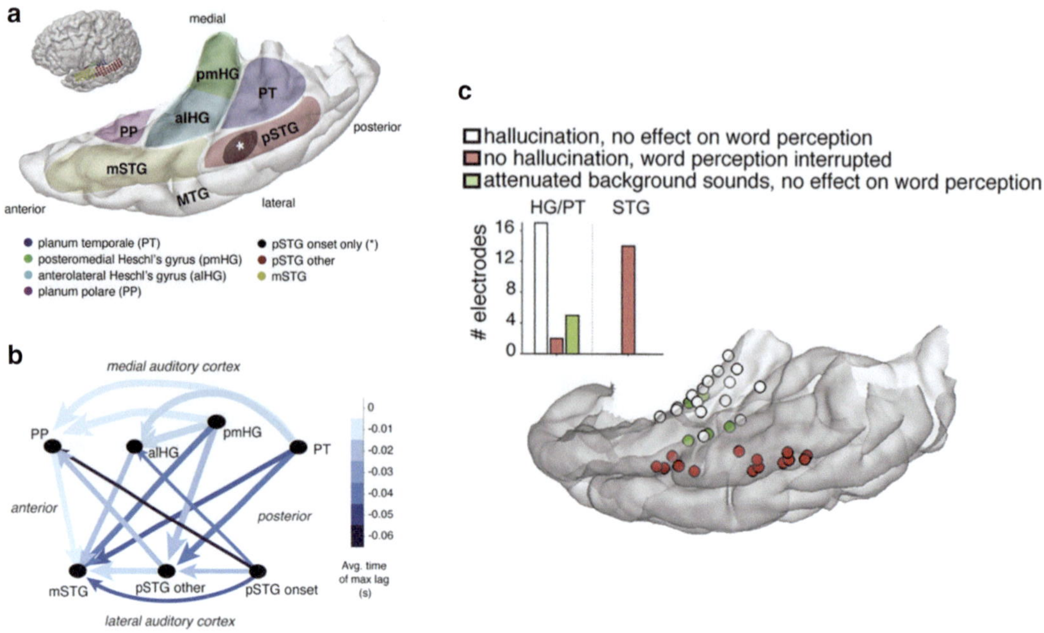

Fig. 12 The auditory cortex. (**a**) An example of participant's anatomical regions of interest on the left hemisphere temporal lobe. *STG* superior temporal gyrus, *MTG* middle temporal gyrus. (**b**) The onset of fast-latency responses in the pSTG is indistinguishable from the onset of responses in primary auditory areas. The fast responses in PT, pmHG and pSTG onset preceded non-onset regions of the pSTG and mSTG as well as PP. (**c**) Focal electrocortical stimulation shows double dissociation between effects of stimulation on HG and lateral STG. Stimulation in HG evoked auditory hallucinations but did not interfere with word perception and repetition. Participants could not perceive words during stimulation on lateral STG, but no other sound hallucinations were evoked. From [55] with permission

Furthermore, in the same study, direct focal electrocortical stimulation suggested that HG is neither necessary nor sufficient for speech perception (Fig. 12c). Stimulation of HG evoked clear contralateral sound percepts without interruption or distortion of word perception. In contrast, stimulation of the lateral STG did not evoke any sounds but interfered significantly with speech processing. These results suggest that medial and lateral auditory areas may be part of parallel sound analysis pathways rather than part of a commonly assumed single serial, hierarchical pathway. HG does not appear to be required for speech perception, whereas STG does. The authors suggested that these connections play an essential role in speech perception and, therefore, may not require PAC processing; thus, they concluded in their work an alternative model to the mainstream model of a core-belt-parabelt cortical pathway that is thought to underlie hierarchical processing of progressively complex, abstract features.

Part of these new and interesting findings can be explained by the fact that the SAC receives direct and distinct thalamic projections (non-lemniscal) from the medial geniculate nucleus MGN, and pulvinar, an anatomical finding that has been proven in animal studies [57–60].

Damage to the auditory cortex and its vicinity has been related to pure word deafness (PWD), a very rare condition denotes individuals with selective damage to speech input processing and intact reading comprehension, speech, and writing [61]. Even though bilateral temporal damage is by far the most frequent cause of PWD, cortical acoustic disorders have also been described following a single left temporal lesion (usually a stroke) [61, 62]. In the latter situation, it was considered by Geschwind as a disconnection syndrome due to the cortico-subcortical or entirely subcortical damage would sever intrahemispheric connections between the PAC and SAC, as well as transcallosal connections between primary acoustic areas and between the Wernicke's area and its right- hemisphere homologue [63]. Lesion analysis of PWD patients supports the importance of the posterior STG in the auditory input processing [61]. Sometimes PWD is the result of a Wernicke's aphasia after recovery.

Wernicke's area. The origins of Wernicke's aphasia can be traced to Wernicke's publication titled *Der Aphasische Symptomencomplex* in 1987 [64], where he presented two cases' (Case 2: Susanne Rother and Case 8: Louise Funke) symptoms of what is known now as Wernicke's aphasia. Wernicke, in his work, did not provide an illustration to show where exactly this area was; until 1881, in his book *Lehrbuch der Gehirnkrankheiten* he drew a brain surface with a shaded area overlaying the middle two-thirds (not posterior!) of the STG (see Fig. 13-right, label ***x***) [66] [67]. For unclear reasons, the location of this region was pushed back into the posterior third of the STG by Dejerine, a relocation endorsed by Wernicke in his last major work on the aphasia [67]. Moreover, other scientists kept expanding Wernicke's area to include other regions from the MTG and ITG [68, 69].

Bogen and Bogen [70] defined Wernicke's area as "the area where a lesion will cause language comprehension deficit." Not an anatomical definition, but rather a functional one. This was based on their review of the multiple pieces of literature at that time (until the mid-seventies), which showed the possibility of the Wernicke's area in various anatomical locations such as posterior STG (BA 22), the supramarginal gyrus (SMG) (BA 40) and angular gyrus (AG) (BA 39) of the inferior parietal lobe (IPL), the middle temporal gyrus (MTG) (BAs 21 and 37), and the inferior temporal gyrus (ITG) (BAs 20 and 37). In other words, language comprehension is not localized but can involve a wide area of the left temporal and inferior parietal lobe [65, 71].

Despite what has been mentioned earlier, the anatomical description of Wernicke's area is still widely seen in the literature as the region surrounding the posterior ascending ramus of the Sylvian fissure, i.e., including the posterior third of the STG extending caudally to the SMG. However, there is significant evidence that the region currently labeled the Wernicke's area might play little or no important role in language comprehension [65, 71]. Studying the pre-CT era literature on this topic shows that lesions associated with Wernicke aphasia (including Wernicke's two original cases) are relatively large, typically including portions of the MTG and AG in addition to STG and SMG [65, 71]. In Fig. 13-left panel, based on Wernicke's descriptions, Binder had drawn an estimation of the lesion site in Wernicke's two original cases [65]. We can see that Wernicke's description of the lesions in his own words is surprisingly extensive! How

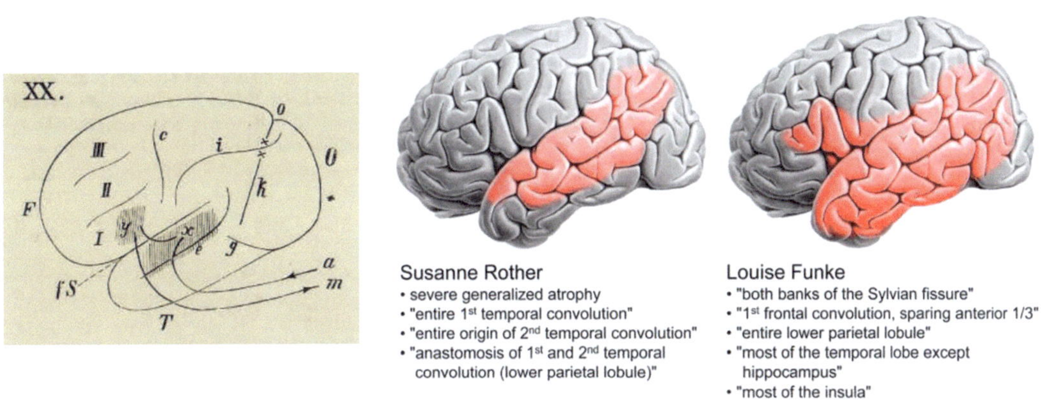

Fig. 13 Wernicke's area. The first ever known description of the sensory speech area (labeled with an x) was made by Wernicke in 1881 (Right) (Figure is public domain from wellcomecollection.org). A depiction of the lesions in Wernicke's two original cases, inferred from his text description, with the principal areas cited as damaged listed below each drawing from [65] (Left), with permission from Springer Nature

Wernicke concluded his finding of a sensory speech area in STG remains unclear to us.

As we saw in Wernicke's two cases, MTG and ITG may have a major role in speech comprehension. Other studies, including meta-analyses, have suggested that two posterior language areas can be distinguished; one is a core Wernicke's area (language auditory processing area in the left hemisphere), including BAs 22 and BA21 (as usually accepted), but also BA41 and BA42; a "language associations area" roughly corresponding to BAs 20, 37, 38, 39, and 40 ("extended Wernicke's area" or "Wernicke's system") [72]. Binder et al. refer to the existence of left hemisphere temporoparietal language areas outside the traditional "Wernicke area," namely, in the middle temporal, inferior temporal, fusiform, and angular gyri; it means a kind of "extended Wernicke's area (Fig. 14).

Many quantitative lesion-deficit mapping studies have proved an association between comprehension deficit and damage to MTG, AG, anterior STG, and several areas in the left prefrontal cortex, but no association with damage to posterior STG or SMG; however, in contrast, reports describing lesions affecting the left MTG, ITG, FG, and AG, sparing the STG and SMG were associated with transcortical sensory aphasia, a syndrome in which language comprehension is impaired but phonological processing is intact (See [11, 65, 73]). Pillay et al. used voxel-based lesion-symptom mapping (VLSM) to localize impairments specific to multiword (phrase and sentence) spoken language comprehension in 51 right-handed patients with chronic left hemisphere stroke. They found that damage to the mid-to-posterior MTG is specifically associated with spoken language comprehension impairment [74].

The posterior half of the MTG probably serves a major role in comprehension. Lesions in this area are likely responsible for most language comprehension impairments observed in patients with "Wernicke aphasia" [65]. Turken and Dronkers compared the resting state functional connectivity of the major region of interest relative to the language [73]. In their result, the posterior MFG has the most connectivity pattern consistent with a characterization of the functional anatomy of the verbal semantic system (Fig. 15).

Due to its proximity to the semantic processing pathways (dorsal AF, IFOF, and ILF, see below) and also to the phonologic areas (STG), the MTG (BA 21 and 37) is most likely a hub of the semantic pathway linking phonologic and semantic functions [13]. However, isolated damages to the MTG are rare [13], but DES of both posterior and middle parts of the MTG can illicit semantic paraphasia [75]. If damage occurs, it can cause transcortical sensory aphasia (relatively intact repetition and fluent speech with frequent semantic paraphasias) [13]. All this evidence suggests that language comprehension depends on an extensive, distributed network beyond the posterior STG and SMG boundaries.

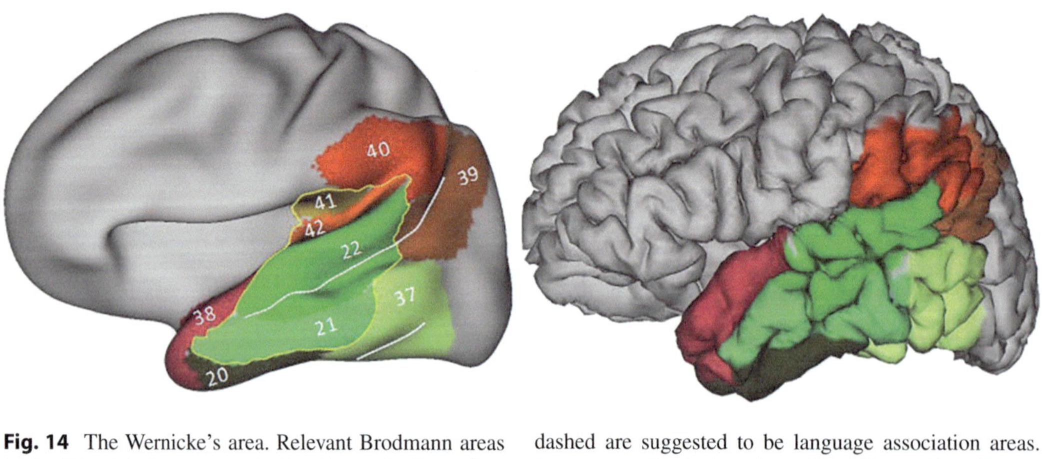

Fig. 14 The Wernicke's area. Relevant Brodmann areas to the Wernicke's area are displayed on a brain surface (pial and inflated). The white dashed lines represent the STS and ITS. The yellow dashed lines represent the Core Wernicke's area (see [12, 72]). Regions outside the yellow dashed are suggested to be language association areas. Image constructed using data from the Human Connectome Project, including Brodmann Areas Atlas and Human Connectome Project workbench software [23, 24]

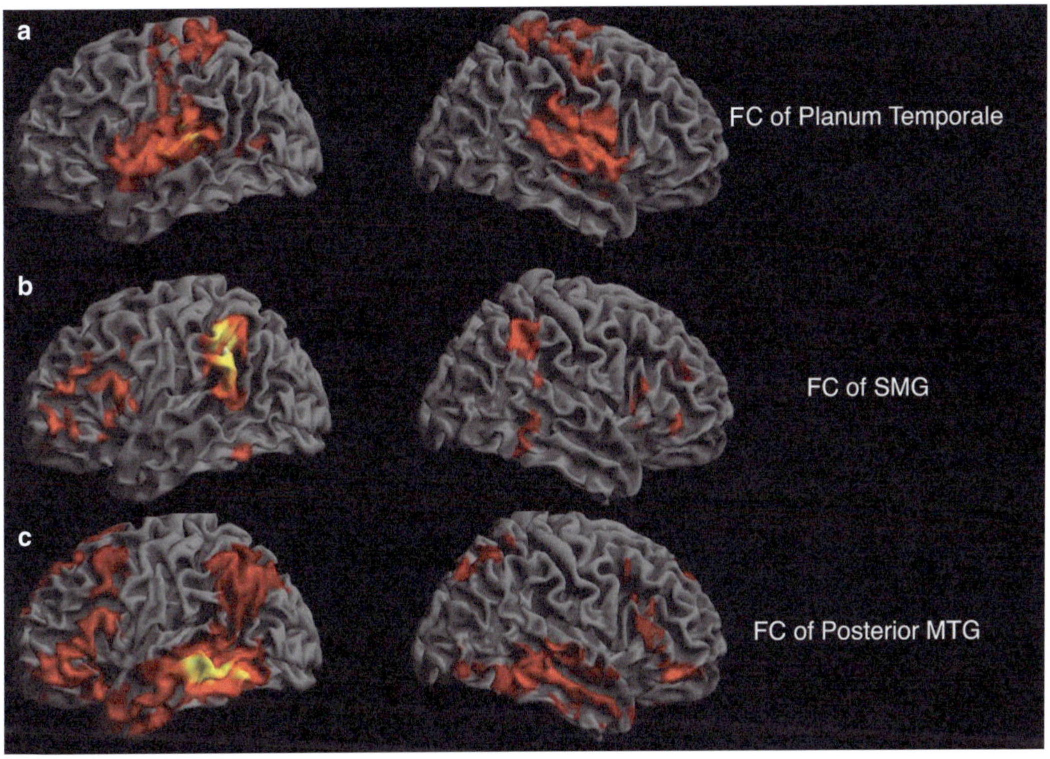

Fig. 15 Resting-State Functional connectivity profile of the (**a**) left planum temporale, (**b**) left posterior supramarginal gyrus, and (**c**) left posterior middle temporal gyrus. Colors indicate t-values (dark red = lowest, yellow-white = highest, with the voxels within the ROI showing the highest correlation). Adapted from [73]

The posterior STG and SMG (Classical Wernicke's area) are involved in phonologic retrieval, also known as phonological representation, a necessary stage prior to all speech output tasks, including spontaneous speech, naming, reading aloud, and repetition [65]. Phonological paraphasia occurs with disruption of the phonological representation, leading to several types of errors in the phoneme sequence like deletion, duplication, and reordering. Damage in this area can cause phonemic paraphasia, also referred to as phonological paraphasia or literal paraphasia, which can be seen in both Wernicke aphasia and conduction aphasia. This occurs when the patient has an inappropriate substitution of part of the word for one with a similar sound. For example, the patient says "pish" instead of "fish" or "dat" instead of "hat." Thus Wernicke aphasia, though often thought of as a syndrome affecting comprehension, also includes a prominent speech production impairment [71].

CT and MRI studies have confirmed that conduction aphasia is a typical result of focal lesions of the posterior STG and SMG, sometimes including the posterior insula. This type of deficit is characterized by severe impairment of repetition and fluent expressive language functions with frequent phonemic paraphasias, repetitive self-corrections, word-finding difficulties, and paraphrasing [76]. A similar deficit can be elicited by DES of these regions and/or the white fiber running beneath them, the arcuate fasciculus (AF) [75, 77–79].

The Anterior Temporal lobe (ATL). The ATL is considered a core component of the semantic and syntactic cognitive system [80, 81]. Surgical resection of the ATL is currently the treatment of choice for intractable temporal lobe epilepsy (TLE), achieving long-term seizure freedom in up to 50–60% of patients [82]. However, after a unilateral ATL resection, a mild decline in naming or verbal memory ability is seen in roughly 30–50% of patients [83]. Left TLE patients showed weaker performance on tasks that required naming or accessing semantic information from a written word. In comparison, right

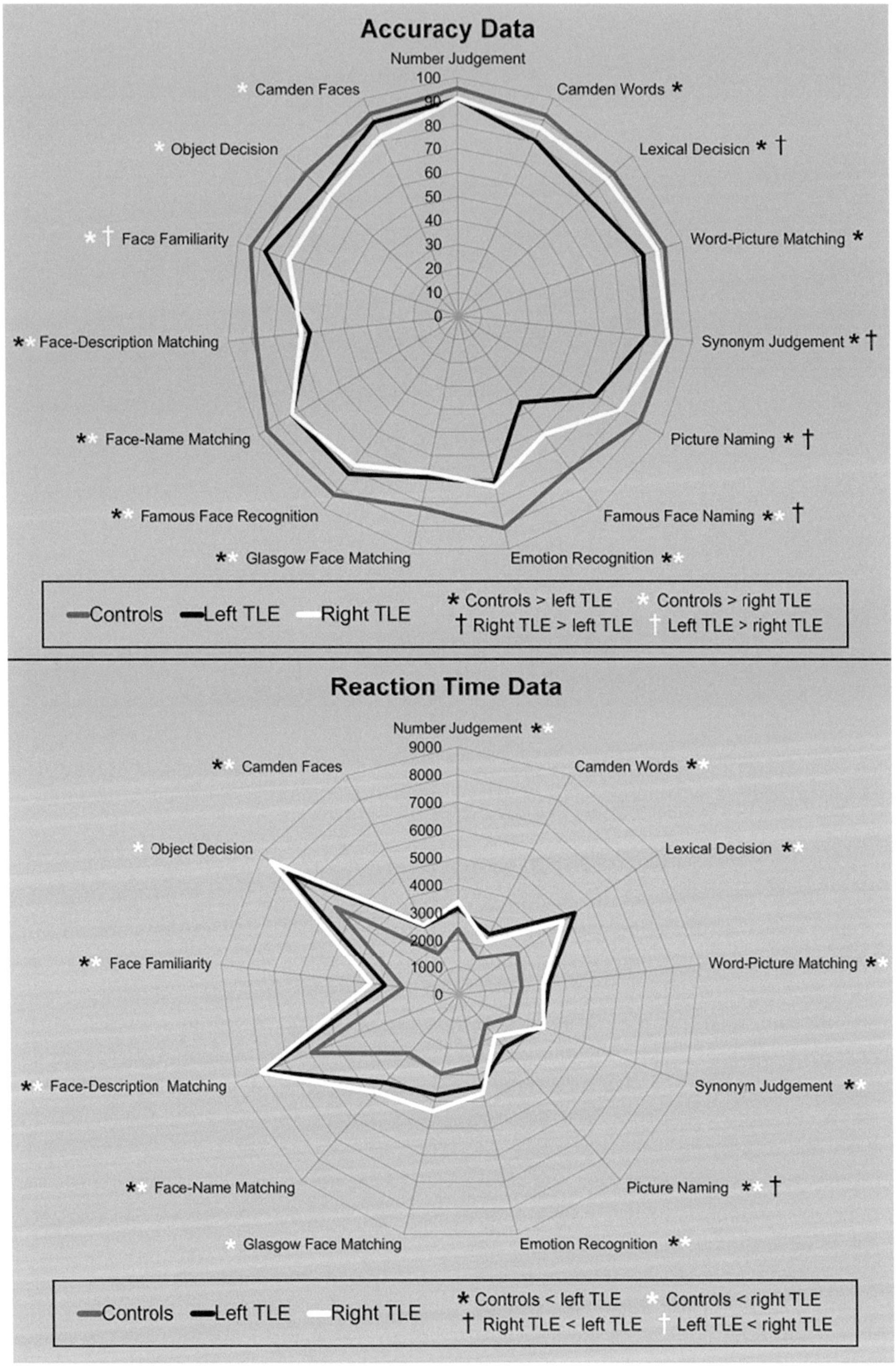

Fig. 16 Summary of accuracy (**a**) and reaction time (**b**) data from the neuropsychological battery conducted by [84]. They assessed multiple aspects of semantic cognition in a large group of postsurgical temporal lobe epilepsy (TLE) patients with left versus right anterior temporal lobectomy ($n = 40$). Data for the control participants are shown in grey, and data from the left and right TLE patients are shown in white and black, respectively. Accuracy data are reported as percentages, and correct response times are reported in milliseconds. Significant differences between the groups based on one-way between group ANOVAs are noted with asterisks; the color of the asterisk denotes the direction of the effect. *The image is under Creative Commons Attribution License (CC BY 4.0)*

TLE patients showed weaker performance in face recognition, and bilateral ATL damage causes severe semantic impairments [84]. Figure 16 gives an idea about the possible semantic cognition impairment in cases with ATL resection.

2.4 Language-Relevant Parietal Regions

The inferior Parietal lobule (IPL). The posterior parietal cortex (PPC), particularly the IPL, is a principal convergence zone of diverse mental capacities, potentially most developed in the human species, including semantic memory and modes of social exchange [85]. The IPL consists of three prominent functional and anatomical subdivisions, the angular gyrus (AG), the supramarginal gyrus (SMG), and the banks of the adjacent sulci, including the interparietal sulcus (IPS) (Fig. 17). We discussed the SMG in the Wernicke's area section (see above), and here will focus on the AG (BA 39).

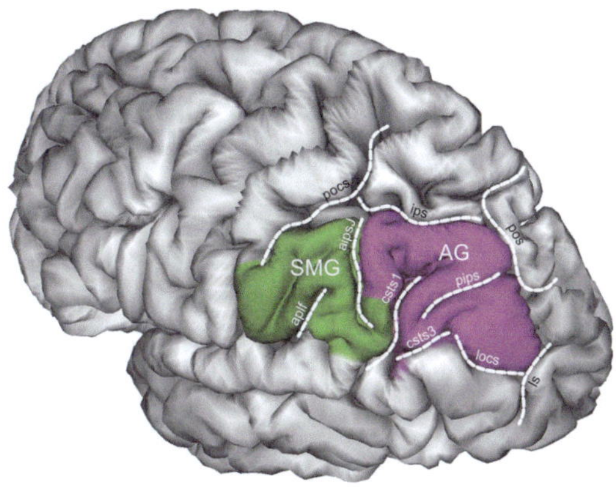

Fig. 17 The inferior parietal lobe comprises the supramarginal gyrus (SMG), angular gyrus (AG), and the surface of the adjacent sulci. aipsJ: anterior intermediate parietal sulcus of Jenssen, aplf: ascending posterior ramus of the lateral fissure, csts1–3: caudal superior temporal sulcus- first and third segment (in this sample, the second segment is missing), ips: interparietal sulcus, locs: lateral occipital sulcus, ls: lunate sulcus, pips: posterior intermediate sulcus, pocs: postcentral sulcus, pos: paro-occipital sulcus. Image constructed using data from the Human Connectome Project, including the Desikan-Killiany Atlas and Human Connectome Project workbench software [23, 24]

The AG shares a direct connection with the primary language areas. It is connected with the IFG (BA 44/45) via the superior longitudinal fasciculus (SLF-III). Furthermore, AG is connected to the MFG (BA 46/9) via the SLF-II. Also, it is connected to the STG (BA 22) via the middle longitudinal fasciculus (MdLF).

Regarding its role in language functions, AG is involved in semantic processing, specifically, the left AG [85]. Left AG is activated during semantic tasks on auditory and visual stimuli, and these findings have been replicated with high consistency and reliability across many other studies. In a meta-analysis of 120 fMRI studies with semantic contrasts, the left AG was the area with the densest concentration of activation foci in the whole brain with a less robust effect in the right AG [86].

Another meta-analysis of fMRI data found that AG and SMG were involved more than any other language area in phoneme discrimination in the sublexical speech perception [87]. The activation of the IPL in phoneme discrimination can be because it plays a role in phonological working memory, which is required for these tasks [87].

Furthermore, the left AG is also involved in reading and comprehension. Dejerine distinguished two alexic syndromes, alexia with agraphia and alexia without agraphia. Regarding the alexia with agraphia, his patients had acquired a deficit in reading and writing with damage to the left AG [88]. The left AG was therefore linked to memories of visual word forms, and Geschwind defined it as the visual memory center for words that turns written language into spoken language and vice versa [89]. However, fMRI studies showed that the AG is not sensitive to visual word forms unless the words make up sentences [90]. Another study suggested that the AG is likely to support semantic feature knowledge rather than mapping from semantics to phonology [91].

This activity of the AG is absent for words relative to fixation, which can be explained by the strong task-independent deactivation typically seen in bilateral AG. This deactivation relative to fixation and rest has been attributed to more significant semantic associations during free thought than during reading, i.e., as part of the default mode network (DMN) of the resting state [92]. In other words, the AG manipulates conceptual knowledge and mental representations when the mind wanders during rest [90].

The DMN was initially described by Shulman et al. [93] and subsequently by Raichle et al. [94], who observed that specific brain regions experienced increased metabolic activity during rest and decreased activity when engaged in goal-directed behavior [93, 94]. The DMN includes mainly the bilateral IPL, medial frontal, and posterior cingulate cortex (Fig. 18, for more about the spatial designated areas of the resting state networks, see [95]). An interaction between the DMN and the Language network has been reported in the literature [96, 97]. Such an interaction is hypothesized to represent linguistic input into the internally oriented functions of the DMN [97]. The AG may represent an anatomical hub for this DMN-Language interaction. Another significance of the DMN for preoperative neurosurgical planning is that most DMN areas overlap with semantic brain regions.

2.5 Language-relevant right hemispheric (RH) regions

Recent scientific evidence came out to confirm the RH has a role in the language, specifically in prosody, the semantic processing of words and discourse, the processing of context, and pragmatic abilities, in addition to its involvement in executive functions, attention, and working memory [98].

In a large meta-analysis of 125 fMRI studies, RH participation in ortholinguistic tasks was identified in only 33% of the studies. It mainly consisted of co-activations with LH homotopic areas (66%), demonstrating that RH involvement is dominated by the inter-hemispheric interactions [99]. Many studies have proven that language comprehension is associated with an extensive brain activation network involving not just the

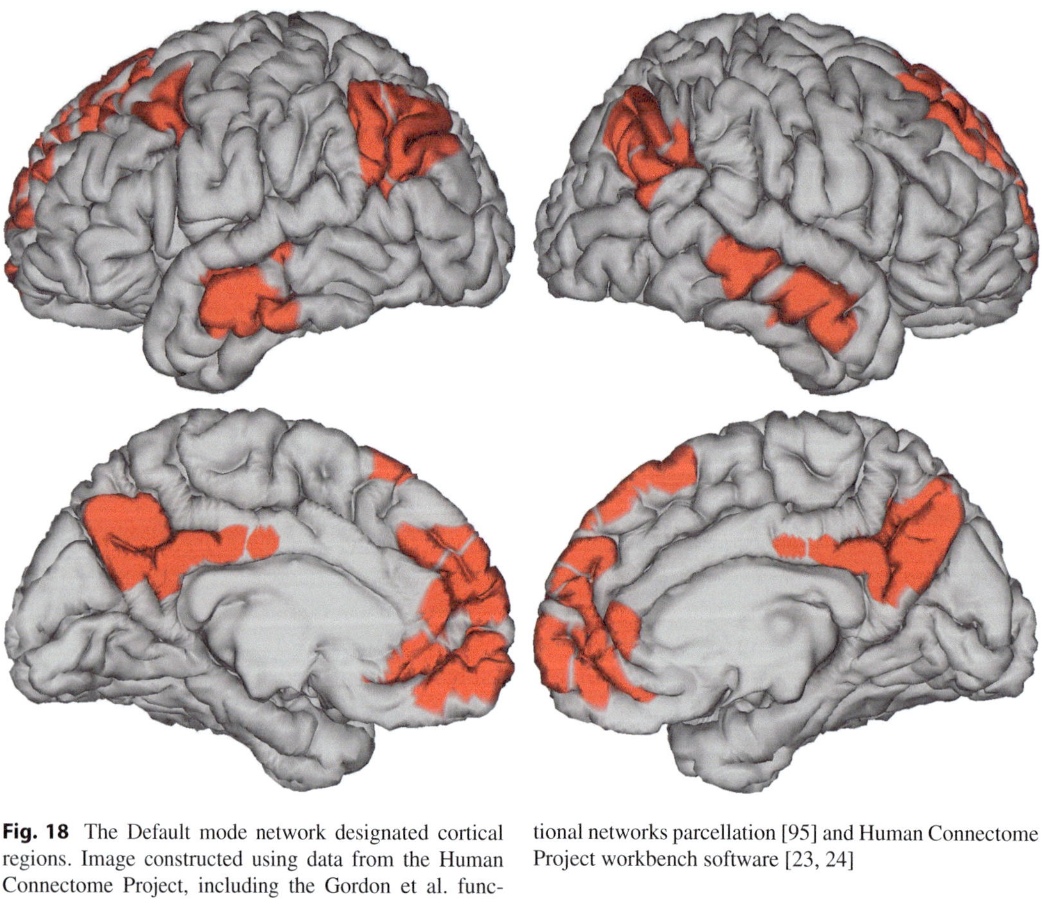

Fig. 18 The Default mode network designated cortical regions. Image constructed using data from the Human Connectome Project, including the Gordon et al. functional networks parcellation [95] and Human Connectome Project workbench software [23, 24]

left but also the right hemisphere. In a meta-analysis of 23 neuroimaging studies on text comprehension, it was found that the bilateral temporal poles were consistently activated [100].

Split visual field experiments have reported consistent right visual field-left hemisphere advantages on a test of written word recognition, the lexical decision paradigm (i.e., deciding whether a letter string represents a real word or not). However, left visual field-right hemisphere performances are typically better with words that elicit vivid images and nouns with high emotional valence, pictographic stimuli such as logos or kanji, and distantly related stimuli all boost left visual field-right hemisphere reading performance compared with their respective alternatives [101].

Another issue related to hemispheric dominance is the relation between language laterality and handedness. Rasmussen and Milner found that in the non-right-handers, the speech was represented in the left cerebral hemisphere in nearly a third of the group, in the right hemisphere in half the group, and bilaterally in the remainder [102]. Accordingly, it might be advisable to do awake brain surgery in non-right-handed patients with lesions in the eloquent areas [98].

Duffau reported his experience with language cortical and subcortical mapping in nine patients with right low grade glioma [103]. In frontal regions, cortical stimulation elicited anarthria in the vPMC (9 out of 9), anomia in the dPMC (1 out of 5), speech arrest in the pars opercularis (1 out of 5), and semantic paraphasia in the dorsolateral prefrontal cortex (2 out of 5). Insular stimulation generated dysarthria (2 out of 2), IPL stimulation caused phonemic paraphasias (2 out of 2), and temporal stimulation had semantic paraphasias (2 out of 2).

3 The Language-Relevant White Fiber Tracts

In this section, we will be discussing, in brief, the fibers that are relevant to language cognitive function. However, we will discuss one projection fiber only, the acoustic radiation. The tracts will be demonstrated using tractography construction.

Acoustic radiation (AR). The AR represents a highly-myelinated group of axonal projections, carrying auditory information from the MGN, the thalamic relay for auditory information, ascending from the inferior colliculus (IC) [60, 104]. Neural tracing studies in animals' brains found the AR is made of two types of projections, a primary (or 'lemniscal') pathway from the central IC passes through the ventral division of MGN, which in turn projects to the PAC. The belt and parabelt regions (SAC) receive input from the dorsal division of the MGN but little or no input from its ventral part (non-lemniscal) [60, 105].

For technical reasons, the AR remained missing from most tractography studies investigating auditory processing and the integration of language and non-language auditory inputs, as well as from diffusion-based atlases [59, 104]. The AR lies in a nearly axial plane crossing some of the major fiber systems of the human brain: internal capsule, external capsule, and posterior thalamic radiation (See [104] for details).

A typical course of AR is made out of three segments [104] (Fig. 19). First, the 'Genu,' the segment projecting from the posterior thalamus and curves towards the lateral side to enter the most caudal part of the posterior limb of the internal capsule. The genu represents the most sophisticated segment since it densely intermingles with the vertical fibers projecting to the somatosensory cortex (anterior part of the posterior thalamic radiation (PTR)), with the IC fibers and the posterior part of the PTR. Next is the 'Stem' segment that runs rostro-lateral and enters the substances of the HG. The stem is delimited posteriorly by the fibers of the arcuate fasciculus (AF) and anteriorly by the posterosuperior margin of the circular sulcus of the insula. The fibers within the stem of the AR expand at the horizontal plane and create a 'Fan' segment that carries the terminal fibers within the auditory cortex.

Regarding the upcoming white fibers in the section, the images will be demonstrated using an average template from 1065 healthy adults' data obtained from the human connectome project, an open-source database [24, 107]. Images were

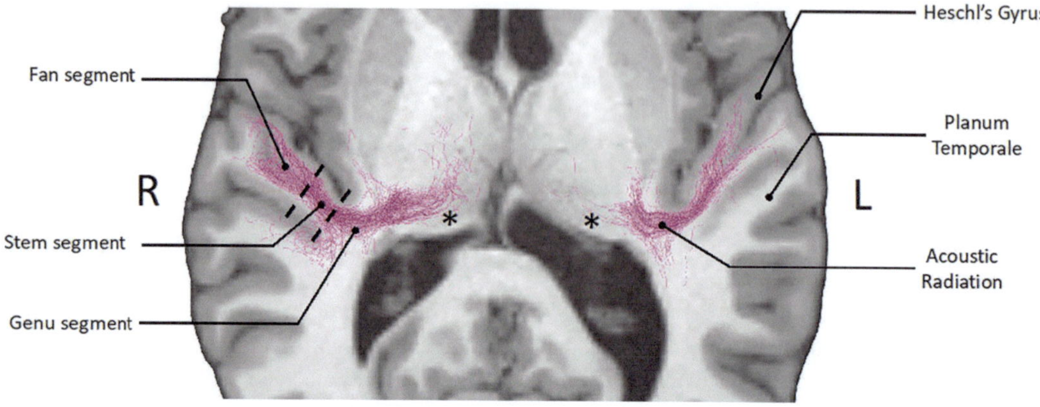

Fig. 19 The acoustic radiation and its segments. The image was reconstructed using MRITrix software [106] and data of a healthy subject from the Human Connectome Project [23, 24]

constructed using the open source DSI software, an open-source software [108, 109]. Also, we will discuss the possible and known functions of these white matter tracts from a lesional point of view.

The arcuate fasciculus (AF). AF is sometimes labeled as a synonym or a part of the superior longitudinal fasciculus (SLF, see below), and there is confusion about its nomenclature. In this chapter, we consider the AF an independent bundle from the SLF system since it connects the temporal and frontal lobes, while the SLF system represents a frontoparietal network.

The left AF is significantly larger than the right (Figs. 20 and 21), with the left AF volume twice as large as the right [110]. There are four frontal lobe termination points: pars opercularis, pars triangularis, ventral precentral gyrus, and MFG. The bulk of the temporal terminations is in the MTG. As can be seen in Figs. 20d and 21, the STG segment is left-lateralized, a fact been outlined in many other studies [110, 111]. Novel tractography studies have concluded that the AF has no connections with the middle or anterior parts of the STG [110]. For more details about the connectivity and segmentation of the AF, read [110].

Functional segmentation of the left AF has been proposed in the literature. Glasser et al. proposed a two-segment model of the left AF, with an MTG-IFG 'lexicosemantic' segment and an STG-IFG 'phonological' segment [111]. Other studies suggested two AF streams; one projects from the posterior temporal cortex to the PMC and is involved in sensorimotor function, and the other projects from the posterior temporal cortex to the posterior IFG and is involved in the complex syntactic processing [112]. Despite all of this, the whole functions of AF are still unknown. Furthermore, the right AF was suggested to carry prosodic information to the inferior frontal lobe, which is active in the same prosodic tasks as the right temporal lobe [111].

Direct cortical stimulation of the AF can induce transient symptoms observed in conduction aphasia, i.e., phonemic paraphasia and repetition disorders [98]. Also, grammatical errors were elicited through AF stimulation to support its role in the syntactic processing [98].

The posterior segment of the AF (the temporoparietal tract, TP). This bundle nomenclature is one of the most controversial. In several articles, it is considered as one of the SLF systems (named SLF-V), and in other reports was regarded as an indirect posterior segment of the AF. However, from a connectional point of view, this bundle does not fit in either nomenclature system because the TP bundle connects the temporal and parietal lobes (Fig. 22); neither the SLF nor the AF does. Panesar et al. have proposed this bundle as a total distinction from the AF system and re-name it the 'temporoparietal aslant tract'[113]. The TP mainly connects the IPL (especially the SMG) with the MTG and ITG.

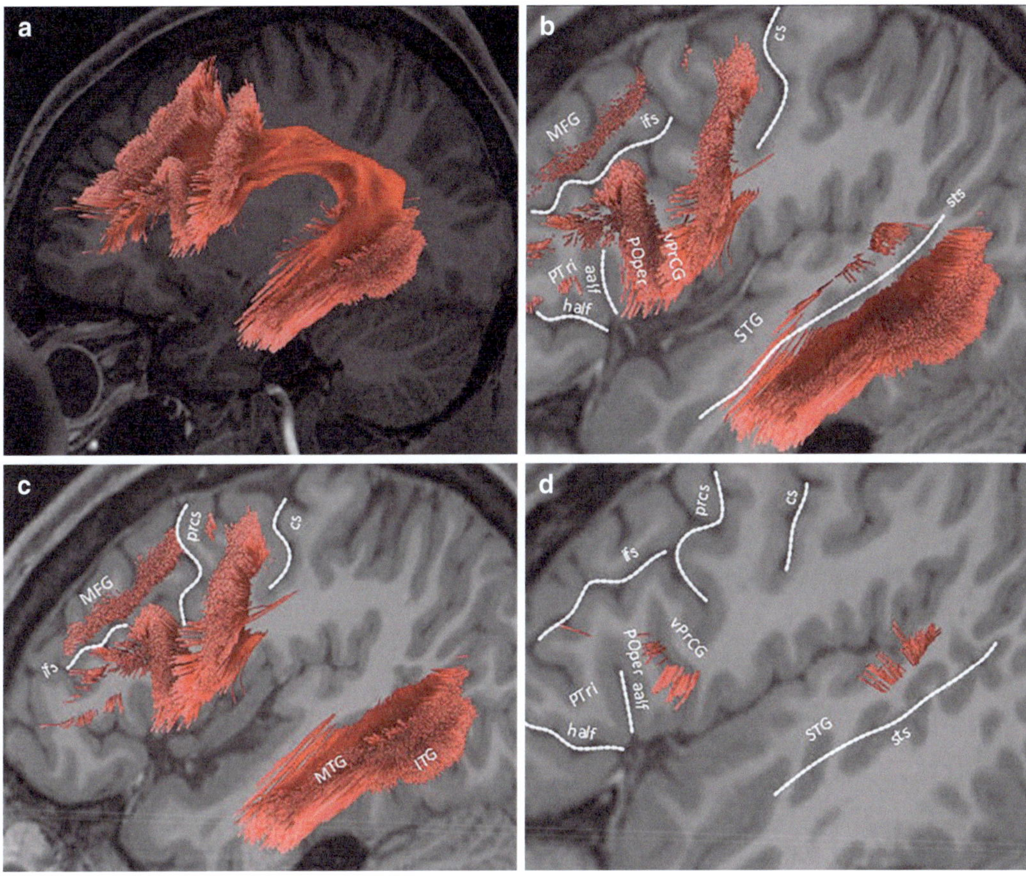

Fig. 20 The Left Arcuate Fasciculus (AF). The whole structure of the AF (**a**). In the temporal lobe, the AF terminates within the three temporal lobes, while in the frontal lobe, it terminates in the inferior and middle lobes (**b**). The AF segment from the MTG and ITG primarily terminates in frontal designated areas (**c**). In contrast, the STG segment terminates exclusively in the IFG, especially in the Pars Opercularis (BA 44) (**d**). *ITG* Inferior temporal gyrus, *MFG* Middle frontal gyrus, *MTG* Middle temporal gyrus, *POper* Pars opercularis, *PTri* Pars triangularis, *STG* Superior temporal gyrus, *vPrCG* Ventral precentral gyrus, *aalf* anterior ascending ramus of the lateral fissure, *cs* central sulcus, *half* horizontal ascending ramus of the lateral fissure, *ifs* inferior frontal sulcus, *prcs* precentral sulcus, *sts* superior temporal sulcus

3% of samples will have terminations in the superior parietal lobule (SPL) [113].

Catani suggested that a lesion of the TP would alter connectivity between Wernicke's area and the parietal region, failing auditory semantic comprehension, resembling transcortical sensory aphasia [114].

The Frontal Aslant tract (FAT). The FAT was first described in 2012 by Catani [115, 116]. It connects the SMA (both proper and pre-SMA) with the Broca's area (ventral BA 6 and BA 44), and a lesser number of fibers can reach the pars triangularis (BA 45) (Fig. 23). Catani found the FAT to be left-lateralized in right-handed subjects. As discussed in the SMA section, the commissural fibers intermingle with FAT, connecting the SMA with the contralateral IFG (Fig. 23e). These fibers are hypothesized to play an essential role in the recovery of the SMA syndrome after surgery.

The FAT plays a role in language through verbal fluency, initiation, and inhibition of speech, sentence production, and lexical decision [117]. DES of the left FAT can induce speech arrest or speech initiation delay [118, 119]. Stimulation of the FAT usually does not cause semantic errors

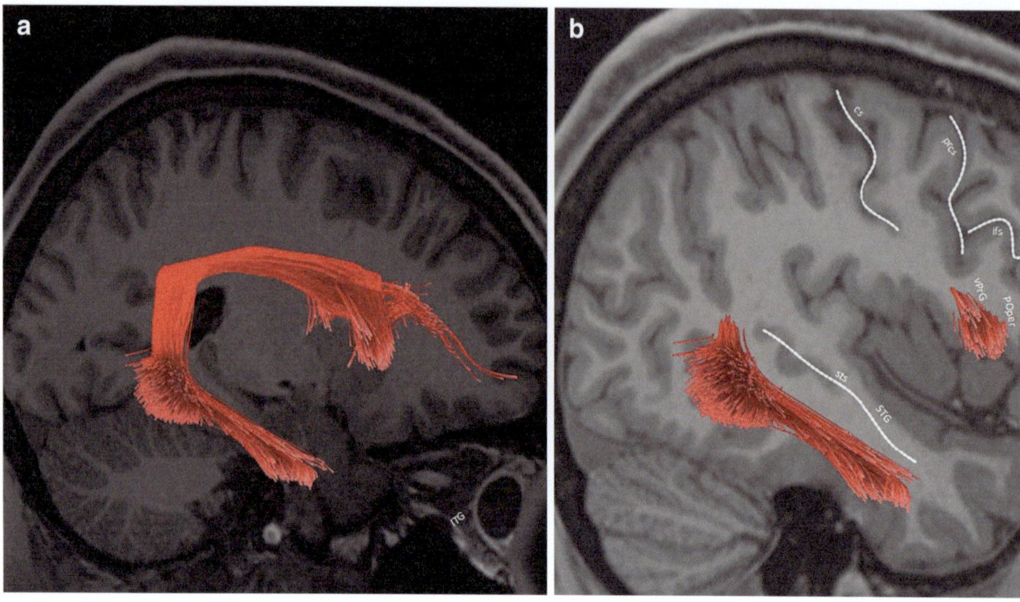

Fig. 21 The Right Arcuate Fasciculus. The whole structure is significantly smaller than the left AF (**a**). The right AF, here, lacks termination in the STG, and its frontal terminations are in the vPrG and POper (**b**). *ITG* Inferior frontal gyrus, *POper* Pars opercularis, *STG* Superior temporal gyrus, *vPrG* Ventral precentral gyrus, *cs* central sulcus, *ifs* inferior frontal sulcus, *prcs* precentral sulcus, *sts* superior temporal sulcus

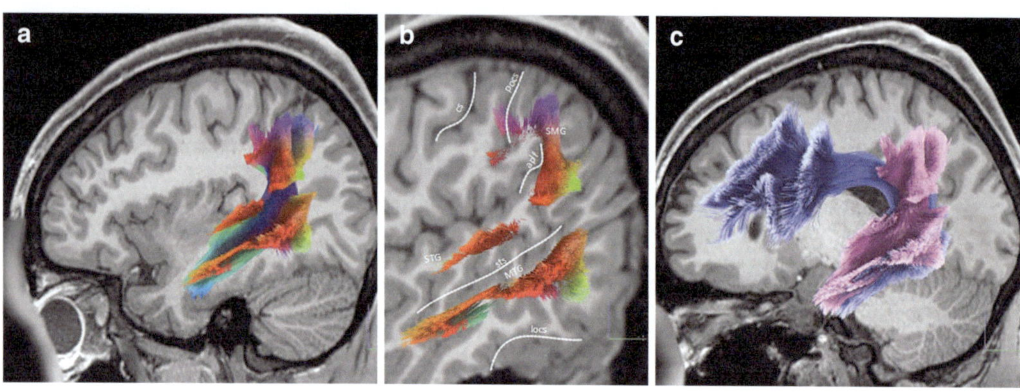

Fig. 22 The Temporoparietal tract (TP, also known as the AF posterior segment and SLF-V). The TP connects the posterior and middle temporal lobe with the inferior parietal lobule (IPL) (**a**). Most parietal terminations are in the SMG, while the temporal terminations are mainly in the middle and posterior thirds of the MTG and, for less extent, in the posterior STG (**b**). The TP (pink) lies immediately lateral to the posterior part of the AF (blue) (**c**). This average template shows no terminations in the SPL. *MTG* Middle temporal gyrus, *SMG* Supramarginal gyrus, *STG* Superior temporal gyrus, *aplf* ascending posterior ramus of the lateral fissure, *cs* central sulcus, *locs* lateral occipital sulcus, *pocs* postcentral sulcus, *sts* superior temporal sulcus

during picture naming; however, stimulation of the adjacent frontostriatal tract might do [120]. For a systemic review regarding the role of FAT in language, see [117].

The Superior Longitudinal Fasciculus II and III (SLF II and SLF III). The most common pattern of connectivity for the SLF II is between the AG (and occasional contributions from the superior parietal lobule) to the cMFG and the dorsal PrCG (Fig. 24); for the SLF III from the SMG to the vPrCG and pars opercularis (Fig. 25) [121]. Both the SLF II & III are right lateralized

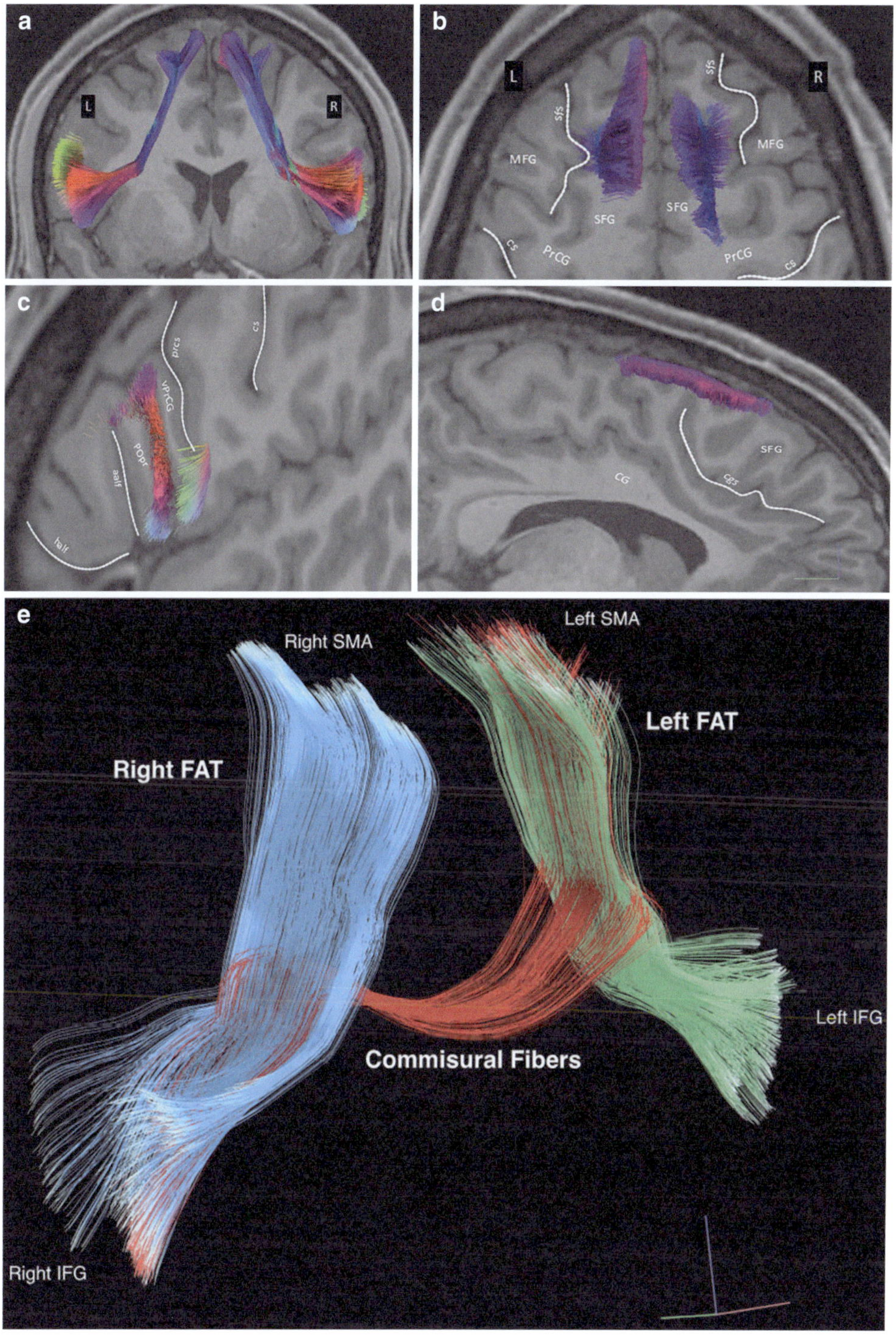

Fig. 23 The frontal Aslant Tract (FAT). The FAT (**a**) runs vertically and connects the SFG (**b**) with the ventral precentral cortex and pars opercularis (**c**). The frontal termination extends medially towards the supplementary motor area in the SFG (**d**). Some commissural fibers (red) intermingle with the FAT to connect the SMA with the contralateral IFG (**e**). *CG* Cingulate gyrus, *MFG* Middle frontal gyrus, *POpr* Pars opercularis, *PrCG* Precentral gyrus, *SFG* Superior Frontal Gyrus, *vPrCG* Ventral precentral gyrus. *aalf* anterior ascending ramus of the lateral fissure, *cgs* cingulate sulcus, *cs* central sulcus, *half* horizontal ascending ramus of lateral fissure, *prcs* precentral sulcus, *sfs* superior frontal sulcus

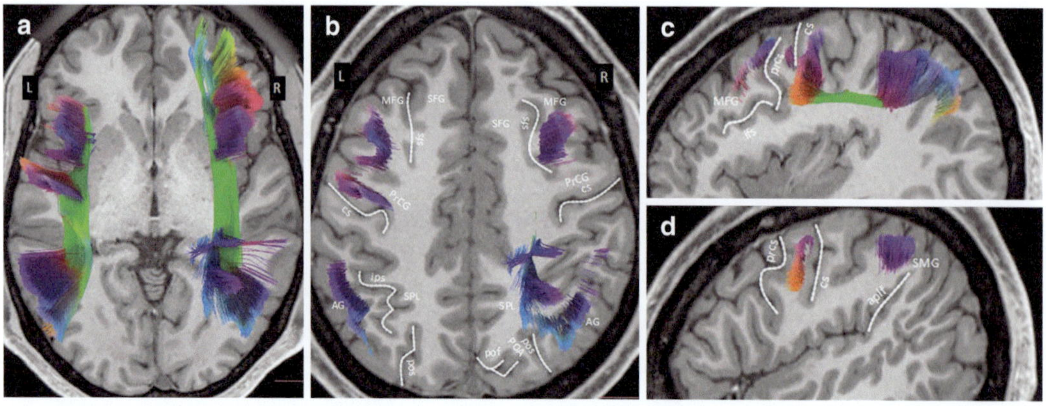

Fig. 24 The superior frontal longitudinal fasciculus II (SLFII). The SLFII runs between the parietal and frontal lobes, the right larger than the left (**a**). The frontal terminations are mainly in the PrCG and MFG, while the parietal terminations are in the SMG, AG, and SPL (**b**–**d**). The Right side, since it is larger, has the termination even in the SPL. *AG* angular gyrus, *MFG* middle frontal gyrus, *POA* paroccipital arcus, *PrCG* precentral gyrus, *SFG* superior Frontal Gyrus, *SMG* supramarginal gyrus, *SPL* superior parietal lobule, *aplf* ascending posterior ramus of the lateral fissure, *cs* central sulcus, *ifs* inferior frontal sulcus, *ips* interparietal sulcus, *pof* parieto-occipital fissure, *pos* paroccipital sulcus, *prcs* precentral sulcus, *sfs* superior frontal sulcus

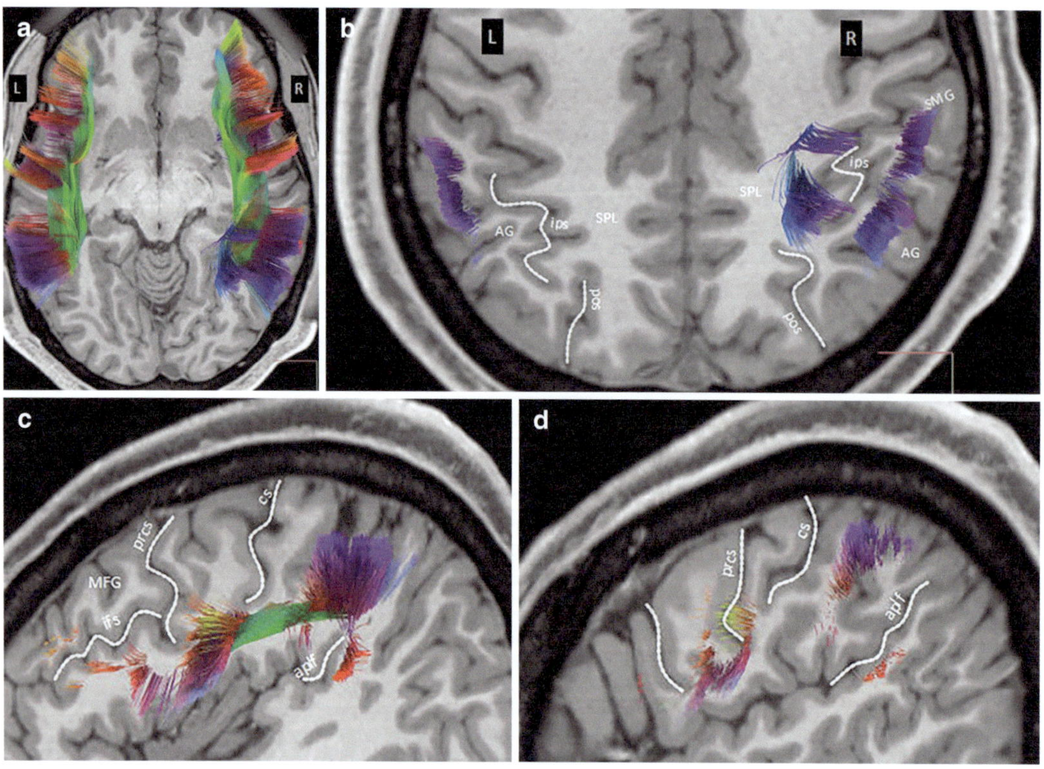

Fig. 25 The superior frontal longitudinal fasciculus III (SLFIII). The SLFIII runs between the parietal and frontal lobes, the right larger than the left (**a**). The frontal terminations are mainly in the subcentral area, vPrCG, and to a lesser extent, the pars opercularis and MFG, while the parietal terminations are in the SMG, AG, and SPL (**b**–**d**). The Right side, since it is larger, has the termination even in the SPL. *AG* angular gyrus, *MFG* middle frontal gyrus, *vPrCG* ventral precentral gyrus, *SMG* supramarginal gyrus, *SPL* superior parietal lobule, *aplf* ascending posterior ramus of the lateral fissure, *cs* central sulcus, *ifs* inferior frontal sulcus, *ips* interparietal sulcus, *prcs* precentral sulcus

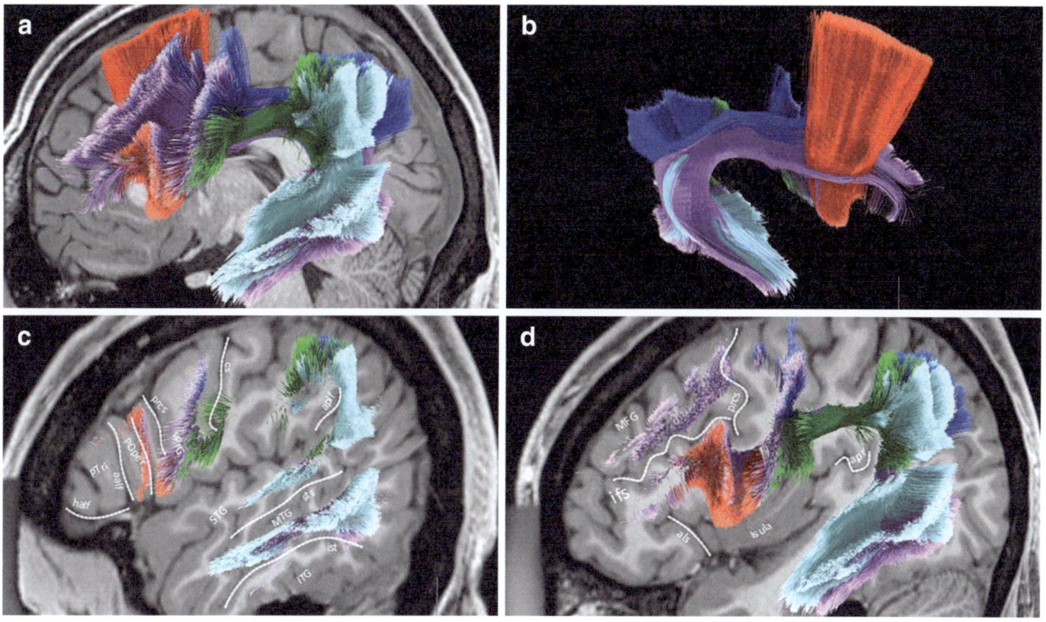

Fig. 26 The Peri-Sylvian language white matters and FAT. The TP and SLFIII are lateral to the AF, while the SLFII is dorsal (**a**). The AF is the most medially located in the peri-Sylvian fibers (**b**). The AF and TP share almost the same temporal terminations; in the inferior frontal, the AF and FAT share their terminations, and in the MFG, the AF with SLFII (**c** and **d**)

[122]. For more details about the SLFs' anatomy and connectivity, see [121, 123].

The SLF, specifically the connection of the SLF III (IPL to Broca's area), participates in articulation and phonological processing during speech perception and production. Duffau induced speech arrest during stimulation of both pars opercularis (BA 44), the left SMG (BA 40), and the subcortical white matter running between them [51, 52]. Catani suggested that a lesion of the SLF III would alter connectivity between Broca's area and the parietal region, resulting in a failure to vocalize semantic content, resembling transcortical motor aphasia [114] (Fig. 26).

The Inferior Fronto-occipital Fasciculus (IFOF). The anterior ends of the IFOF are found in the frontal pole, orbitofrontal cortex, IFG, MFG, and even the SFG. The posterior terminations are widely distributed through parietal, temporal, and occipital lobes (Fig. 27). Sarubbo et al. claimed the IFOF has termination in Wernicke's area [124]. For more about the connectivity and subcomponents of the IFOF, see [123–125]. The IFOF volume has no asymmetry towards any hemisphere [125].

Semantic paraphasias are induced either due to stimulation of the frontal, insular, parietal, or occipitotemporal portion of the IFOF [120]. Stimulation of the temporal part of left IFOF has been reported to induce associative (e.g., "key" for "padlock") or coordinate semantic paraphasia (e.g., "tiger" for "lion") [126]. Also, stimulation of the frontal, insular or temporal portion of the left IFOF can elicit coordinate semantic paraphasia or the use of a hyperonym (e.g., "bird" for "eagle") [127].

In a series of right low-grade glioma patients with lesions located within or close to the right ventral stream, the IFOF was stimulated and induced semantic association impairments during a visual non-verbal association task (the Pyramids and Palm Trees Test (PPTT)) in 12 out of 13 patients [128]. Duffau has suggested that the IFOF is a 'direct' pathway of the ventral semantic system for language (see below in the ILF section).

The Uncinate fasciculus (UF). The UF (Fig. 28) connects the orbital and lateral frontal cortex with the ATL, and its volume shows a right laterality [129]. To know more about the UF's subcomponents and connectivity, see [123, 129].

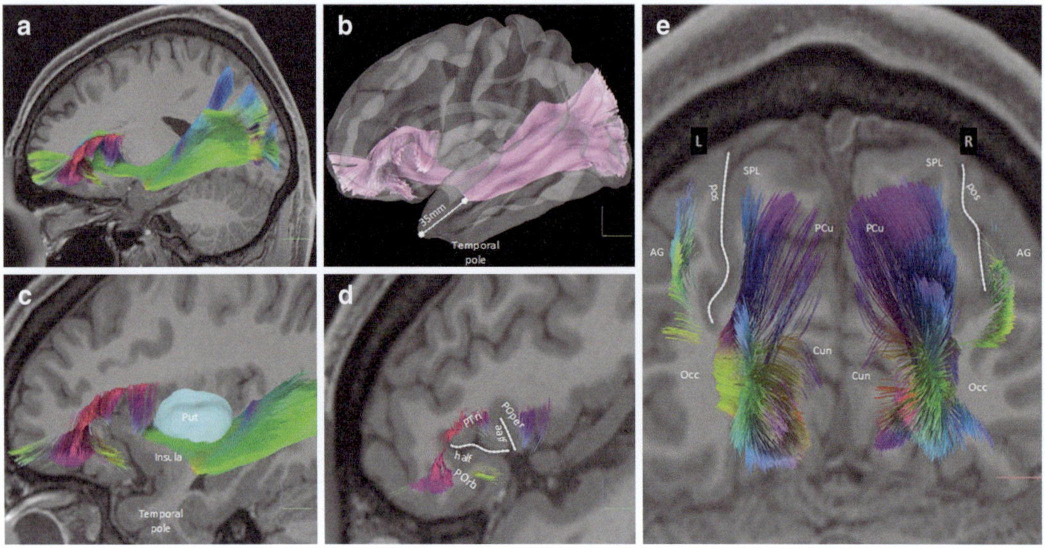

Fig. 27 The inferior fronto-occipital fasciculus (IFOF). The IFOF, as the name describes, runs between the frontal and the occipital lobe; however, it also has some terminations in the parietal lobe (**a**). The IFOF temporal segment starts approximately 35 mm caudal to the temporal pole (**b**). The IFOF develops a narrow neck when passing in the insular zone; this part lies lateral and ventral to the putamen (**c**). The IFOF frontal terminations are mainly the three gyri of the inferior frontal cortex (**d**), while the posterior terminations are in the occipital lobe, the SPL, and the AG (**e**). These posterior terminations spread around the paroccipital sulcus; however, most end in the occipital lobe, and some reach the cuneus. *AG* angular gyrus, *Cun* cuneus, *Occ* Occipital lobe, *Pcu* precuneus, *POper* pars opercularis, *POrb* pars orbitalis, *Ptri* pars triangularis, *Put* putamen, *SPL* superior parietal lobule, *aalf* ascending anterior ramus of the lateral fissure, *half* horizontal ascending ramus of the lateral fissure, *pos* paroccipital sulcus

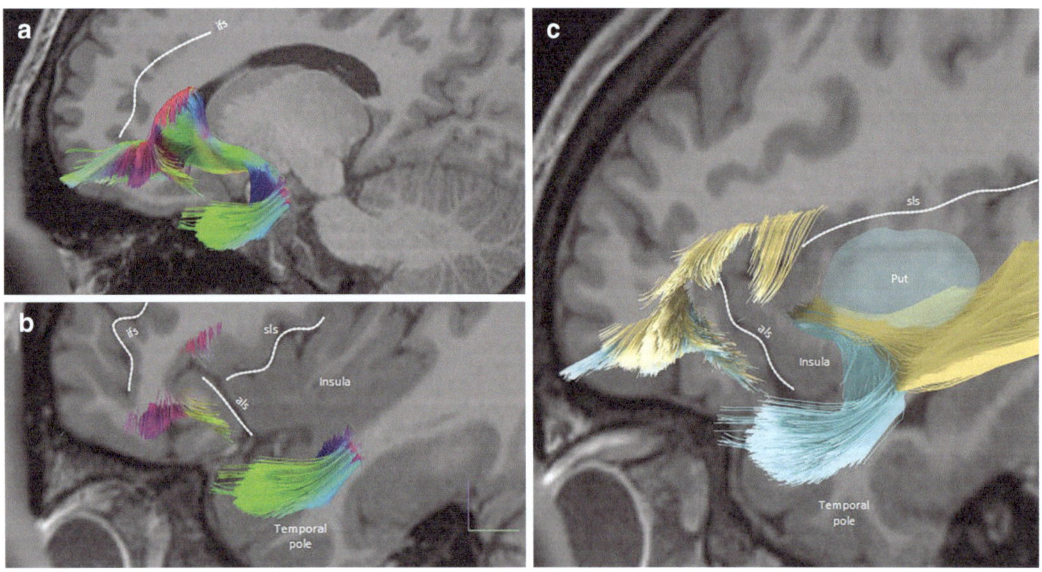

Fig. 28 The Uncinate Fasciculus (UF). The UF connects the inferior frontal and orbitofrontal cortex with the anterior temporal lobe (**a** and **b**). The UF and IFOF share the same anatomical narrow neck region in the insular zone, both lateral and ventral to the putamen (**c**), *als* anterior limiting sulcus of the insula, *ifs* inferior frontal sulcus, *sls* superior limiting sulcus of the insula

The UF seems to be associated with semantic processing during language, given its strong connectivity with the ATL, a proposed 'hub' for semantic processing (see above in the ATL section). The loss of semantic function supported this role (e.g., picture naming deficits) following the resection of this bundle [130, 131]. However, other reports have challenged this concept by documenting that no semantic deficit occurs after ATL excision [126]. Also, semantic functions of the UF after excision usually is compensated by other pathways (see below in the ILF section).

The Middle Longitudinal Fasciculus (MdLF). The human MdLF (Fig. 29) mainly connects the STG with the SPL and parietooccipital region [132, 133]. The MdLF has no hemispheric asymmetry [132, 133]. The termination of the MdLF in the IPL remained a debatable topic in literature. The controversy is based on results from old tractography model studies that presented MdLF with fibers terminating in the AG (e.g., tensor models). However, these results were not replicable by modern tractography studies or recent cadaveric studies. Wang et al. studied the anatomy of the MdLF using high definition tractography and white fiber dissection and identified a few MdLF fibers that reach the AG [133]. Maldonado et al., in their white fiber dissection study, concluded that the AG does not receive any fibers from the MdLF [134]. However, this author claimed the MdLF is purely a temporo-occipital tract without SPL terminations. Kalyvas et al. did an extensive and comprehensive DTI and fiber dissection study of the MdLF connectivity, found no terminations in the IPL, and confirmed the existence of the SPL fibers [132]. In our personal experience, we confirmed the MdLF to have terminations in the SPL, occipital lobe, and inconsistently among samples have terminations in the AG.

The involvement of MdLF in language is also controversial. In a series of eight patients with

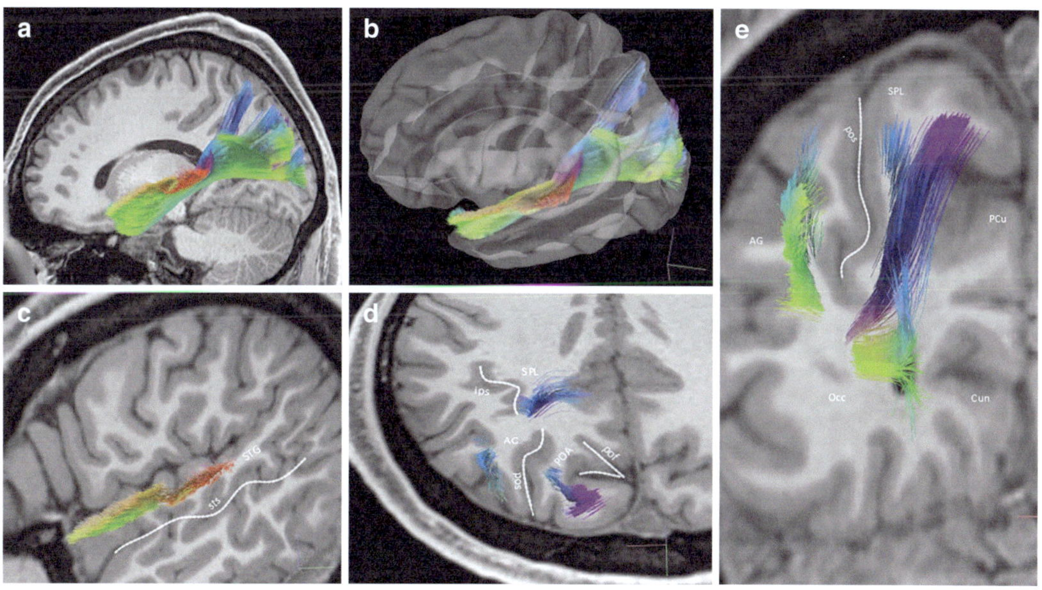

Fig. 29 The Middle Longitudinal Fasciculus (MdLF). The MdLF connects the superior temporal gyrus with the parieto-occipital area (**a** and **b**). The anterior termination is mainly in the middle and anterior parts of the superior temporal gyrus (**c**). The posterior terminations are mainly in the AG, POA, SPL, and to a lesser extent, the occipital lobe (**d** and **e**). Same as the IFOF, the MdLF posterior termination split around the paroccipital sulcus (**e**). *AG* angular gyrus, *Cun* cuneus, *Occ* occipital lobe, *Pcu* precuneus, *POA* paroccipital arcus, *SPL* superior parietal lobule, *STG* superior temporal gyrus, *ips* interparietal sulcus, *pof* parieto-occipital fissure, *pos* paroccipital sulcus, *sts* superior temporal sulcus

glioma resection involving the STG of the left dominant hemisphere, the resection cavity was extensive enough to include a large part of the MdLF in all patients. Intra-operative stimulation of the MdLF failed to show any naming errors. Postoperatively, no new permanent language deficits are detected by systematic language examination. Accordingly, Duffau and his colleagues suggested that the MdLF may participate but is not essential for language processing [135]. Also, in a recent systemic review of the white matter architecture underlying semantic processing, the functional significance of the MdLF in semantic processing was found to be unclear [120].

Several factors tend to shift the MdLF functional hypothesis away from language function. These include the MdLF lack or has scarce fibers terminating in the IPL, has no leftward asymmetry, and the lack of language deficit in patients after excision of part of the MdLF. However, based on its anatomical connectivity, a proposal can be made that the MdLF might serve a significant role in the "what" and "where" auditory pathway, parallel cortical processing streams, similar to those identified in the visual cortex existing in the auditory system [132, 133].

The Inferior Longitudinal Fasciculus (ILF). The ILF (Figs. 30 and 31) connects the ATL mainly with the posterior part of ITG and the Occipital lobe. Some posterior termination exists in the AG too. In association with the IFOF, MdLF, optic radiation, posterior thalamic radiation, and the tapetum of the corpus callosum, they form the sagittal stratum, the lateral wall of the ventricular atrium and a large part of the temporal horn and occipital horn (Fig. 32).

Linguistic deficits due to DES of the left ILF presented either as alexia during reading tasks or as semantic paraphasias during picture naming [120]. In contrast, Duffau and his colleagues, in a series of 12 patients, found that subcortical stimulation of ILF did not elicit any language disturbances. After a transient language disturbance after surgery, all patients recovered [136]. Based on this finding,

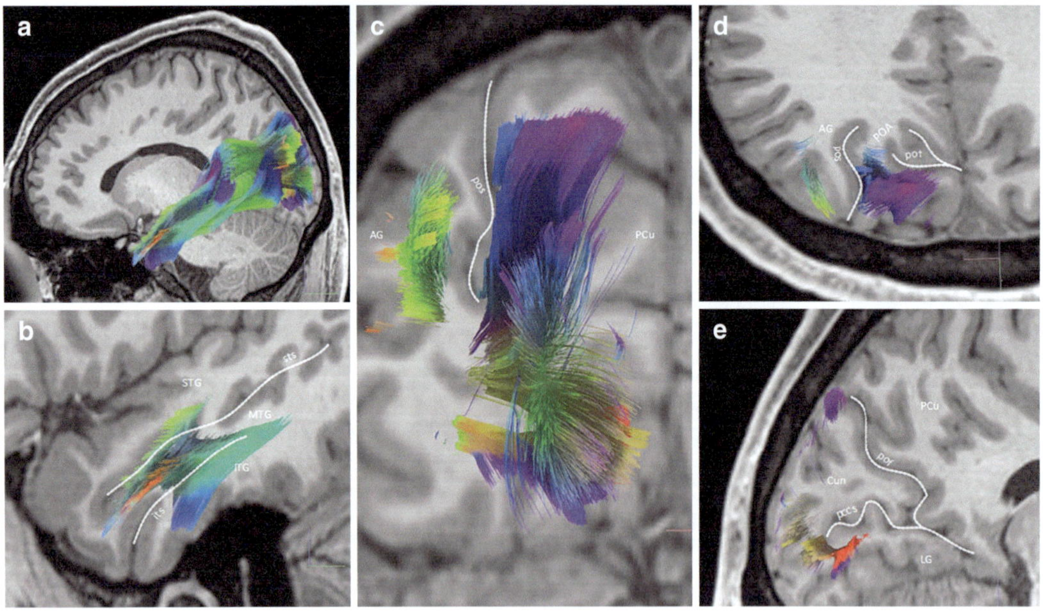

Fig. 30 The inferior longitudinal fasciculus (ILF). The ILF runs in the ventral surface of the temporo-occipital cortex connecting the temporal lobe with the occipital and inferior parietal lobule (**a**). The temporal terminations are found in the three gyri (**b**). Same as the IFOF and MdLF, the ILF posterior termination splits around the paroccipital sulcus to reach the AG, POA, and medially the cuneus (**c-e**), *AG*: angular gyrus, *Cun*: cuneus, *LG*: lingual gyrus, *Pcu*: precuneus, *POA*: paroccipital arcus, *STG*: superior temporal gyrus, ips: interparietal sulcus, *its*: inferior temporal sulcus, *pccs*: posterior calcarine sulcus, *pof*: parieto-occipital fissure, *pos*: paroccipital sulcus, *sts*: superior temporal sulcus

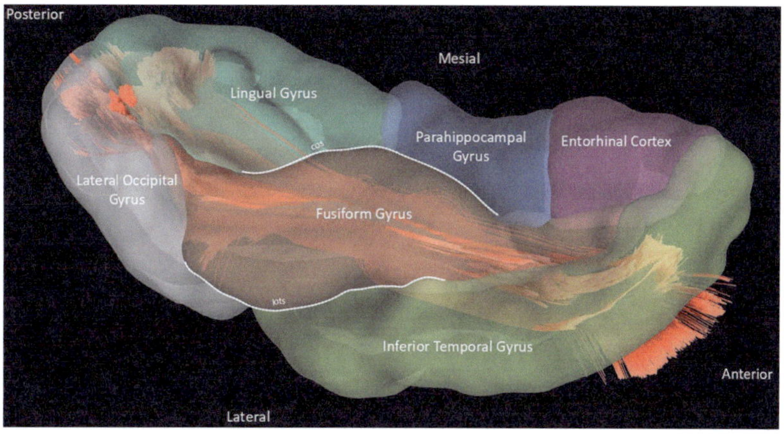

Fig. 31 The ventral occipito-temporal and ILF. The ILF is delimited ventrally by the lateral occipital and lingual gyri in the posterior third and by the fusiform and inferior temporal gyrus in the anterior two-thirds. *Lots* lateral occipital sulcus, *cos* collateral sulcus

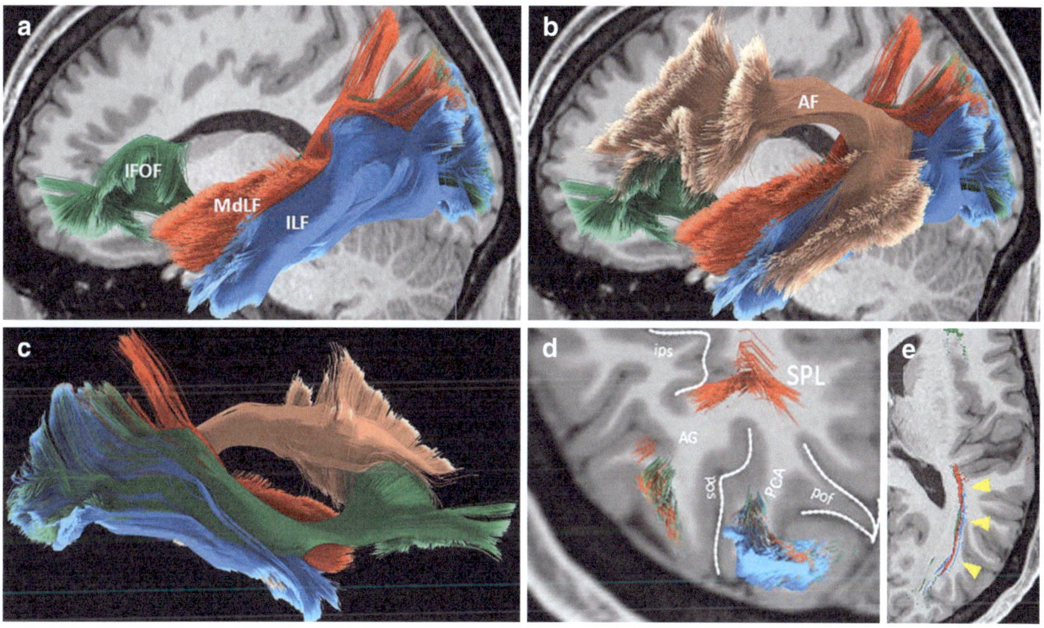

Fig. 32 The IFOF, MdLF, and ILF contribute to the Sagittal Stratum (SS, Yellow arrowheads). The SS lies medial to the posterior segment of the AF (**b** and **c**). *AG*: angular gyrus, *POA*: paroccipital arcus, *SPL*: superior parietal lobule, *ips*: interparietal sulcus, *pof*: parieto-occipital fissure, *pos*: paroccipital sulcus

they proposed that the ventral semantic stream is composed of at least two pathways—first, a direct pathway operated by the IFOF since eliciting semantic paraphasia when stimulated. Second, an indirect pathway, operated by the ILF, which connects the posterior occipitotemporal areas and the temporal pole, is then relayed by the UF that connects the ATL to the frontal lobe. The indirect pathway is not systematically indispensable since it can be compensated during stimulation and after surgical resection [47, 48].

4 Neural Primary Network of Language Processing (a Simplified Model)

A Historical note. In the 1860s and 1870s, when Pierre Paul Broca [137] and Karl Wernicke [64] [66] reported their findings in patients with certain types of language deficits caused by focal lesions in the brain, their reports were taken as evidence that localized regions of the brain are specialized in specific cognitive functions (Localizationist model). This concept was challenged and turned out to be unfavorable against more sophisticated models such as the holistic model of brain function proposed by Sigmund Freud [138] and the associationist model elaborated by Meynert [139] and Wernicke [64].

Hodotopic framework. However, recent methodological advances in subcortical characterization have highlighted the crucial role played by white matter. Consequently, the functional contribution of a given area depends on its integration with the activity of anatomically interconnected regions. This complementary approach between topography (cortical function epicenters) and hodology (connectivity between areas) was recently summarized by Marco Catani and Dominic H. ffytche in the so-called hodotopic framework [140]. In this framework, the brain, inside which the gray and white matters are part of the same interactive process, is organized as complex, multimodal, and integrated parallel distributed networks, with a cognitive process consisting of a continuous and often redundant stream of information dynamically modulated by experience and external environment. When considering the neural basis of a cognitive function such as language, it is important to understand not only the cortical areas but also the subcortical white matter network.

Here, we attempt to describe the neural basis of language in an updated model [141] based on recent advancements in neuroscience such as patients' lesion analysis, neuroimaging, cadaveric fiber dissection studies, and electrophysiological studies with cortical and subcortical stimulation during awake surgery. However, this model is still not perfect. Regardless of the great advancement in neuroscience, many aspects of the language process remain unknown to us.

For the sake of simplicity, the current model structure considers only the association white fibers involved in language and their relevant cortical areas (Hodotopical). The reader shall keep in mind the other structures involved in language processes, such as the frontostriatal tract, which is necessary for speech production, and many other structures not mentioned here. Also, the model can be related to the dual stream model concept [142]. The dorsal stream (DS) composes mainly of the AF/SLF complex and is responsible for the phonological process [143]. The ventral stream (VS) mainly by the IFOF as a direct pathway, and the ILF and UF as the indirect pathway and is responsible for the semantic processing (Fig. 33). However, as we discussed earlier, the dorsal part of the AF and its TP might have a role in semantic processing. Other works include the FAT and its speech production function in the DS [13]. Also, the MdLF has been included in the semantic stream [141]. It is worth mentioning that the ventral stream is represented bilaterally; however, the dorsal stream is mainly left-lateralized.

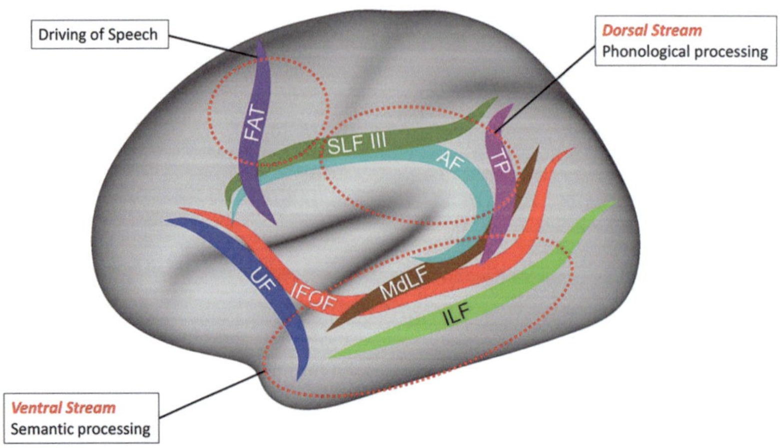

Fig. 33. A simplified schematic illustration of the white matter tracts of the Dual Stream Model of Language

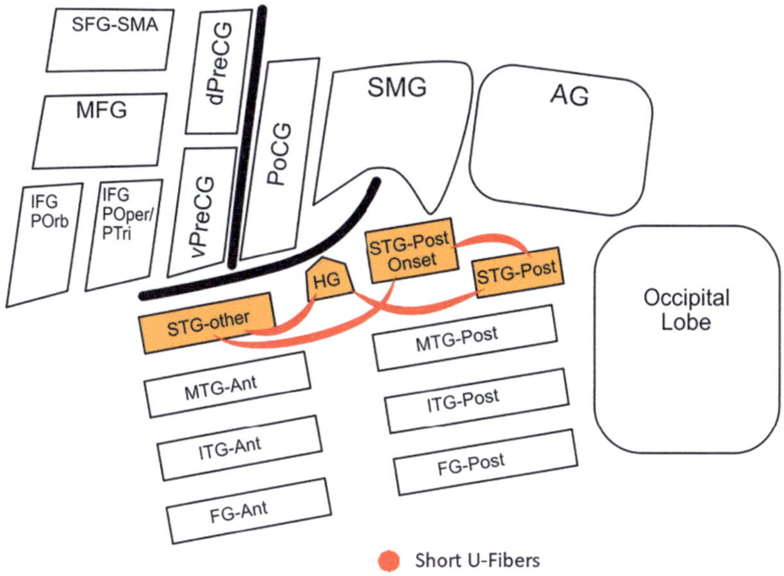

Fig. 34 From sound to phoneme

A. **From sound to phoneme** (Fig. 34**).** The first step of auditory comprehension is converting sound data into phonemes. This process was thought to be hierarchical and processed only in the PAC, including the HG and the adjacent STG. However, as discussed earlier in this chapter, there is evidence that alternative parallel sound analysis pathways exist. In this model, the posterior medial part of the HG is an onset area for words and another onset area for sentences in the SAC can be found in the posterior STG (Fig. 12) [55, 56]. These regions are further connected through short association fibers with sub-networks in the STG (e.g., the middle STG, the planum polare, etc.).
B. **Access to the lexicon and semantic processing** (Fig. 35). The series of phonemes perceived are then transferred to the MTG. As discussed earlier, the semantic cortical areas mainly include the MTG (especially the posterior part), the ITG, AG, the IFG (especially the pars triangularis and pars orbitalis), and the MFG. However, recent studies showed a much broader semantic map.

   A white matter network serving this process will include the dorsal part of the AF [111], the TP [114], the IFOF [120, 126–128], the ILF [120], the UF [130, 131], the extreme capsule [144], and other temporal short U-fibers [144]. The AF and TP are not classically part of the ventral (semantic) stream; however, they seem to have roles in this process. The MdLF's role in the semantic process is controversial. However, short association fibers in its axis might have a role in the semantic process.
C. **Phonological processing** (Fig. 36). Phonological process involves the STG, SMG, ventral sensor-motor cortex, pars opercularis, and pars triangularis. The white matter network is mainly conducted through the dorsal stream (Fig. 33), in which the frontal, parietal, and temporal lobes are all connected by the SLF/AF complex.
D. **Drive of Speech** (Fig. 37**).** The speech motor control as an initiation and timing interface depends on the integrity of white matter fiber connecting the SMA with the vPreCG and IFG, the FAT.

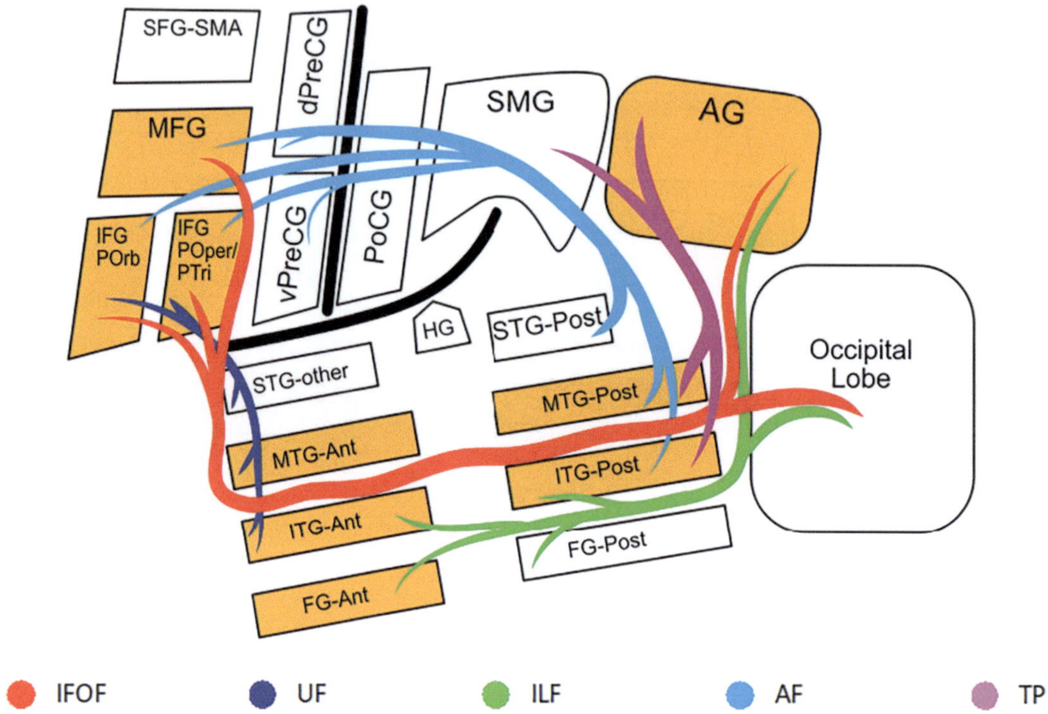

Fig. 35 Access to the lexicon and semantic processing

Fig. 36 Phonological processing

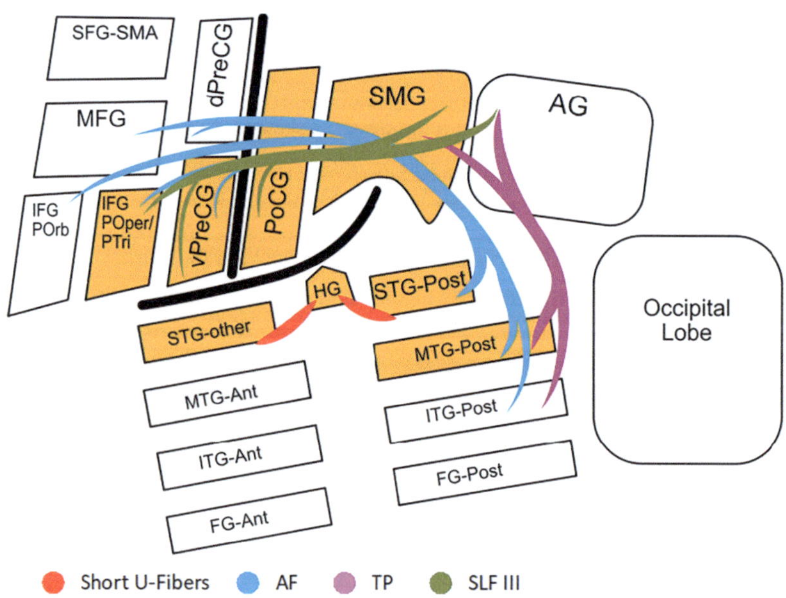

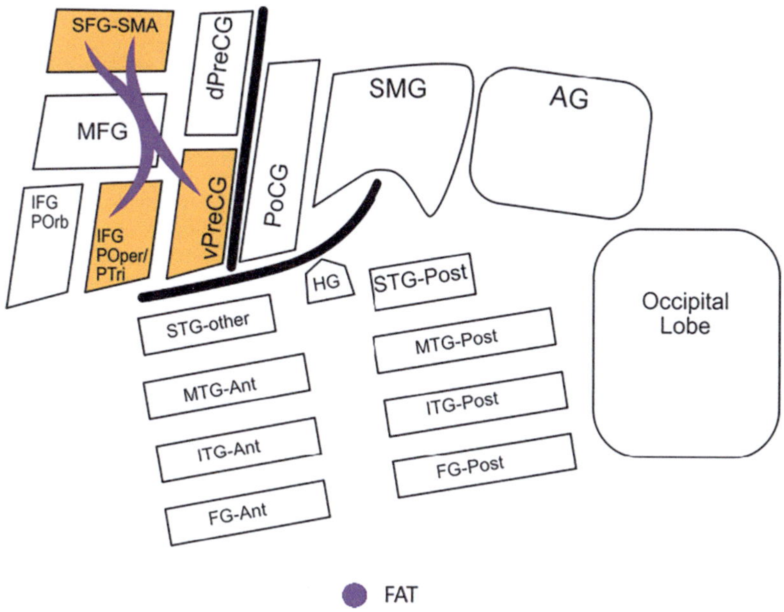

Fig. 37 Drive of Speech. The FAT conducts the function, connecting the SMA with the vPreCG and IFG

References

1. Gazzaniga MS, et al. Cognitive Neuroscience. The biology of the mind. Norton: New York. 2019.
2. Meyer DE, Schvaneveldt RW. Facilitation in recognizing pairs of words: evidence of a dependence between retrieval operations. J Exp Psychol 1971;90(2):227–34.
3. Collins AM, Loftus EF. A spreading-activation theory of semantic processing. Psychol Rev. 1975;82(6):407–28.
4. Huth AG, de Heer WA, Griffiths TL, Theunissen FE, Gallant JL. Natural speech reveals the semantic maps that tile human cerebral cortex. Nature. 2016;532(7600):453–8. https://doi.org/10.1038/nature17637.
5. Mark LP. Imaging of Cortical and White Matter Language Processing. Semin Ultrasound CT MR. 2015;36(3):249–59.
6. Ardila A, Bernal B, Rosselli M. How localized are language brain areas? A review of Brodmann areas involvement in Oral language. Arch Clin Neuropsychol. 2015;31(1):112–22. https://doi.org/10.1093/arclin/acv081.
7. Amunts K, Lenzen M, Friederici AD, Schleicher A, Morosan P, Palomero-Gallagher N, Zilles K. Broca's region: novel organizational principles and multiple receptor mapping. PLoS Biol. 2010;8(9):e1000489. https://www.ncbi.nlm.nih.gov/pmc/articles/PMC2943440/pdf/pbio.1000489.pdf
8. Uddén J, Bahlmann J. A rostro-caudal gradient of structured sequence processing in the left inferior frontal gyrus. Philos Trans R Soc Lond B Biol Sci. 2012;367(1598):2023–32. https://www.ncbi.nlm.nih.gov/pmc/articles/PMC3367683/pdf/rstb20120009.pdf
9. Friederici AD. The brain basis of language processing: from structure to function. Physiol Rev. 2011;91(4):1357–92. https://doi.org/10.1152/physrev.00006.2011.
10. Friederici AD. White-matter pathways for speech and language processing. Handb Clin Neurol. 2015;129:177–86.
11. Dronkers NF, Wilkins DP, Van Valin Jr RD, Redfern BB, Jaeger JJ. Lesion analysis of the brain areas involved in language comprehension. Cognition. 2004;92(1–2):145–77.
12. Ardila A, Bernal B, Rosselli M. Why Broca's area damage does not result in classical Broca's aphasia [opinion]. Front Hum Neurosci. 2016b;10 https://doi.org/10.3389/fnhum.2016.00249.
13. Middlebrooks EH, Yagmurlu K, Szaflarski JP, Rahman M, Bozkurt B. A contemporary framework of language processing in the human brain in the context of preoperative and intraoperative language mapping. Neuroradiology. 2017;59(1):69–87. https://link.springer.com/content/pdf/10.1007/s00234-016-1772-0.pdf
14. Tate MC, Herbet G, Moritz-Gasser S, Tate JE, Duffau H. Probabilistic map of critical functional regions of the human cerebral cortex: Broca's area revisited. Brain. 2014;137(Pt 10):2773–82. https://doi.org/10.1093/brain/awu168.
15. Bizzi A, Nava S, Ferrè F, Castelli G, Aquino D, Ciaraffa F, Broggi G, DiMeco F, Piacentini S. Aphasia induced by gliomas growing in the ventrolateral frontal region: assessment with diffusion MR tractography, functional MR imaging and neu-

ropsychology. Cortex. 2012;48(2):255–72. https://doi.org/10.1016/j.cortex.2011.11.015.
16. Otsuki M. [anarthrieとapraxia of speech]anarthrieの症候学 [symptomatology of anarthria]. Japanese J Neuropsychol. 2005;21(3):172–82.
17. van Geemen K, Herbet G, Moritz-Gasser S, Duffau H. Limited plastic potential of the left ventral premotor cortex in speech articulation: evidence from intraoperative awake mapping in glioma patients. Hum Brain Mapp. 2014;35(4):1587–96. https://www.ncbi.nlm.nih.gov/pmc/articles/PMC6869841/pdf/HBM-35-1587.pdf
18. Gohel S, Laino ME, Rajeev-Kumar G, Jenabi M, Peck K, Hatzoglou V, Tabar V, Holodny AI, Vachha B. Resting-state functional connectivity of the middle frontal gyrus can predict language lateralization in patients with brain tumors. Am J Neuroradiol. 2019;40(2):319–25. https://www.ncbi.nlm.nih.gov/pmc/articles/PMC6375738/pdf/zj4319.pdf
19. Wen J, Yu T, Liu L, Hu Z, Yan J, Li Y, Li X. Evaluating the roles of left middle frontal gyrus in word production using electrocorticography. Neurocase. 2017;23(5–6):263–9. https://doi.org/10.1080/13554794.2017.1387275.
20. Sierpowska J, Fernandez-Coello A, Gomez-Andres A, Camins À, Castañer S, Juncadella M, Gabarrós A, Rodríguez-Fornells A. Involvement of the middle frontal gyrus in language switching as revealed by electrical stimulation mapping and functional magnetic resonance imaging in bilingual brain tumor patients. Cortex. 2018;99:78–92. https://doi.org/10.1016/j.cortex.2017.10.017.
21. Sierpowska J, Gabarrós A, Ripollés P, Juncadella M, Castañer S, Camins Á, Plans G, Rodríguez-Fornells A. Intraoperative electrical stimulation of language switching in two bilingual patients. Neuropsychologia. 2013;51(13):2882–92. https://doi.org/10.1016/j.neuropsychologia.2013.09.003.
22. Hopf A. Distribution of myeloarchitectonic marks in the frontal cerebral cortex in man. J Hirnforsch. 1956;2(4):311–33. (Uber die Verteilung myeloarchitektonischer Merkmale in der Stirnhirnrinde beim Menschen.)
23. Glasser MF, Coalson TS, Robinson EC, Hacker CD, Harwell J, Yacoub E, Ugurbil K, Andersson J, Beckmann CF, Jenkinson M, Smith SM, Van Essen DC. A multi-modal parcellation of human cerebral cortex. Nature. 2016;536(7615):171–8. https://doi.org/10.1038/nature18933.
24. Van Essen DC, Smith SM, Barch DM, Behrens TE, Yacoub E, Ugurbil K. The WU-Minn human connectome project: an overview. NeuroImage. 2013;80:62–79. https://doi.org/10.1016/j.neuroimage.2013.05.041.
25. Altschuler EL. Commentary: pure apraxia of speech after resection based in the posterior middle frontal gyrus. Neurosurgery. 2020, Sep 1;87(3):E390–e391. https://doi.org/10.1093/neuros/nyaa036.
26. Chang EF, Kurteff G, Andrews JP, Briggs RG, Conner AK, Battiste JD, Sughrue ME. Pure apraxia of speech after resection based in the posterior middle frontal gyrus. Neurosurgery. 2020, Sep 1;87(3):E383–e389. https://doi.org/10.1093/neuros/nyaa002.
27. Hazem SR, Awan M, Lavrador JP, Patel S, Wren HM, Lucena O, Semedo C, Irzan H, Melbourne A, Ourselin S, Shapey J, Kailaya-Vasan A, Gullan R, Ashkan K, Bhangoo R, Vergani F. Middle frontal gyrus and area 55b: perioperative mapping and language outcomes [original research]. Front Neurol. 2021;12. https://www.frontiersin.org/article/10.3389/fneur.2021.646075
28. Baker CM, Burks JD, Briggs RG, Sheets JR, Conner AK, Glenn CA, Sali G, McCoy TM, Battiste JD, O'Donoghue DL, Sughrue ME. A Connectomic atlas of the human cerebrum—chapter 3: the motor, premotor, and sensory cortices. Operative Neurosurgery. 2018a;15(Suppl_1):S75–S121. https://doi.org/10.1093/ons/opy256.
29. Baker CM, Burks JD, Briggs RG, Smitherman AD, Glenn CA, Conner AK, Wu DH, Sughrue ME. The crossed frontal aslant tract: a possible pathway involved in the recovery of supplementary motor area syndrome. Brain Behav. 2018b;8(3):e00926. https://doi.org/10.1002/brb3.926.
30. Hertrich I, Dietrich S, Blum C, Ackermann H. The role of the dorsolateral prefrontal cortex for speech and language processing. Front Hum Neurosci. 2021;15:217.
31. Morshed RA, Lee AT, Wang EJ, Young JS, Cha S, Hervey-Jumper SL, Berger MS. Functional outcomes after resection of middle frontal gyrus diffuse gliomas. J Neurosurg. 2021;137:1–8. https://doi.org/10.3171/2021.8.JNS211624.
32. Purcell J, Turkeltaub P, Eden G, Rapp B. Examining the central and peripheral processes of written word production through meta-analysis [review]. Front Psychol. 2011;2 https://doi.org/10.3389/fpsyg.2011.00239.
33. Roux FE, Dufor O, Giussani C, Wamain Y, Draper L, Longcamp M, Démonet JF. The graphemic/motor frontal area Exner's area revisited. Ann Neurol. 2009;66(4):537–45. https://doi.org/10.1002/ana.21804.
34. Hertrich I, Dietrich S, Ackermann H. The role of the supplementary motor area for speech and language processing. Neurosci Biobehav Rev. 2016;68:602–10.
35. Schwartze M, Rothermich K, Kotz SA. Functional dissociation of pre-SMA and SMA-proper in temporal processing. NeuroImage. 2012;60(1):290–8. https://doi.org/10.1016/j.neuroimage.2011.11.089.
36. Ruan J, Bludau S, Palomero-Gallagher N, Caspers S, Mohlberg H, Eickhoff SB, Seitz RJ, Amunts K. Cytoarchitecture, probability maps, and functions of the human supplementary and pre-supplementary motor areas. Brain Struct Funct. 2018;223(9):4169–86. https://doi.org/10.1007/s00429-018-1738-6.
37. Abel TJ, Buckley RT, Morton RP, Gabikian P, Silbergeld DL. Recurrent supplementary motor

area syndrome following repeat brain tumor resection involving supplementary motor cortex. Neurosurgery. 2015;11(03):447–56. https://doi.org/10.1227/NEU.0000000000000847.

38. Quirarte JA, Kumar VA, Liu H-L, Noll KR, Wefel JS, Lang FF. Language supplementary motor area syndrome correlated with dynamic changes in perioperative task-based functional MRI activations: case report. J Neurosurg JNS. 2021;134(6):1738–42. https://doi.org/10.3171/2020.4.JNS193250.
39. Rostomily RC, Berger MS, Ojemann GA, Lettich E. Postoperative deficits and functional recovery following removal of tumors involving the dominant hemisphere supplementary motor area. J Neurosurg. 1991;75(1):62–8. https://doi.org/10.3171/jns.1991.75.1.0062.
40. Shamov T, Al-Hashel J, Rousseff RT. Postoperative supplementary motor area syndrome: clinical evolution and prognosis in nine patients after left hemispheric tumor resection. Hippokratia. 2020;24(1):38–42. https://pubmed.ncbi.nlm.nih.gov/33364738 https://www.ncbi.nlm.nih.gov/pmc/articles/PMC7733362/ https://www.ncbi.nlm.nih.gov/pmc/articles/PMC7733362/pdf/hippokratia-24-38.pdf
41. Duffau H. Brain hodotopy: new insights provided by intrasurgical mapping. In: Duffau H, editor. Brain mapping: from neural basis of cognition to surgical applications. Springer Vienna; 2011a. p. 335–47. https://doi.org/10.1007/978-3-7091-0723-2_26.
42. Gasquoine PG. Contributions of the insula to cognition and emotion. Neuropsychol Rev. 2014;24(2):77–87. https://link.springer.com/content/pdf/10.1007/s11065-014-9246-9.pdf. [Record #979 is using a reference type undefined in this output style.]
43. Ackermann H, Riecker A. The contribution (s) of the insula to speech production: a review of the clinical and functional imaging literature. Brain Struct Funct. 2010;214(5):419–33.
44. Dronkers NF. A new brain region for coordinating speech articulation. Nature. 1996;384(6605):159–61. https://doi.org/10.1038/384159a0.
45. Hillis AE, Work M, Barker PB, Jacobs MA, Breese EL, Maurer K. Re-examining the brain regions crucial for orchestrating speech articulation. Brain. 2004;127(Pt 7):1479–87. https://doi.org/10.1093/brain/awh172.
46. Cereda C, Ghika J, Maeder P, Bogousslavsky J. Strokes restricted to the insular cortex. Neurology. 2002;59(12):1950–5. https://doi.org/10.1212/01.wnl.0000038905.75660.bd.
47. Duffau H, Gatignol P, Moritz-Gasser S, Mandonnet E. Is the left uncinate fasciculus essential for language? A cerebral stimulation study. J Neurol. 2009a;256(3):382–9. https://doi.org/10.1007/s00415-009-0053-9.
48. Duffau H, Moritz-Gasser S, Gatignol P. Functional outcome after language mapping for insular World Health Organization grade II gliomas in the dominant hemisphere: experience with 24 patients. Neurosurg Focus FOC. 2009b;27(2):E7. https://doi.org/10.3171/2009.5.Focus0938.
49. Lang FF, Olansen NE, DeMonte F, Gokaslan ZL, Holland EC, Kalhorn C, Sawaya R. Surgical resection of intrinsic insular tumors: complication avoidance. J Neurosurg. 2001;95(4):638–50. https://doi.org/10.3171/jns.2001.95.4.0638.
50. Afif A, Minotti L, Kahane P, Hoffmann D. Anatomofunctional organization of the insular cortex: a study using intracerebral electrical stimulation in epileptic patients. Epilepsia. 2010;51(11):2305–15. https://doi.org/10.1111/j.1528-1167.2010.02755.x.
51. Duffau H, Gatignol P, Denvil D, Lopes M, Capelle L. The articulatory loop: study of the subcortical connectivity by electrostimulation. Neuroreport. 2003a;14(15) https://journals.lww.com/neuroreport/Fulltext/2003/10270/The_articulatory_loop__study_of_the_subcortical.26.aspx:2005–8.
52. Duffau H, Karachi C, Gatignol P, Capelle L. Transient Foix-Chavany-Marie syndrome after surgical resection of a right insulo-opercular low-grade glioma. Neurosurgery. 2003b;53(2):426–31. https://academic.oup.com/neurosurgery/article-abstract/53/2/426/2735869?redirectedFrom=fulltext
53. Martino J, de Lucas EM, Ibáñez-Plágaro FJ, Valle-Folgueral JM, Vázquez-Barquero A. Foix-Chavany-Marie syndrome caused by a disconnection between the right pars opercularis of the inferior frontal gyrus and the supplementary motor area: Case report. J Neurosurg JNS. 2012;117(5):844–50. https://doi.org/10.3171/2012.7.Jns12404.
54. Sweet RA, Dorph-Petersen KA, Lewis DA. Mapping auditory core, lateral belt, and parabelt cortices in the human superior temporal gyrus. J Comp Neurol. 2005;491(3):270–89. https://onlinelibrary.wiley.com/doi/pdfdirect/10.1002/cne.20702?download=true
55. Hamilton LS, Oganian Y, Hall J, Chang EF. Parallel and distributed encoding of speech across human auditory cortex. Cell. 2021;184(18):4626–4639.e4613. https://doi.org/10.1016/j.cell.2021.07.019.
56. Hamilton LS, Edwards E, Chang EF. A spatial map of onset and sustained responses to speech in the human superior temporal gyrus. Curr Biol. 2018;28(12):1860–1871.e1864. https://doi.org/10.1016/j.cub.2018.04.033.
57. Hackett TA, Stepniewska I, Kaas JH. Thalamocortical connections of the parabelt auditory cortex in macaque monkeys. J Comp Neurol. 1998;400(2):271–86. https://doi.org/10.1002/(sici)1096-9861(19981019)400:2<271::aid-cne8>3.0.co;2-6.
58. Lee CC. Exploring functions for the non-lemniscal auditory thalamus [perspective]. Front Neural Circuits. 2015;9 https://doi.org/10.3389/fncir.2015.00069.
59. Maffei C, Sarubbo S, Jovicich J. A missing connection: a review of the macrostructural anatomy and Tractography of the acoustic radiation [review].

Front Neuroanat. 2019;13 https://doi.org/10.3389/fnana.2019.00027.
60. Scott BH, Saleem KS, Kikuchi Y, Fukushima M, Mishkin M, Saunders RC. Thalamic connections of the core auditory cortex and rostral supratemporal plane in the macaque monkey. J Comp Neurol. 2017;525(16):3488–513. https://doi.org/10.1002/cne.24283.
61. Maffei C, Capasso R, Cazzolli G, Colosimo C, Dell'Acqua F, Piludu F, Catani M, Miceli G. Pure word deafness following left temporal damage: behavioral and neuroanatomical evidence from a new case. Cortex. 2017;97:240–54. https://doi.org/10.1016/j.cortex.2017.10.006.
62. Takahashi N, Kawamura M, Shinotou H, Hirayama K, Kaga K, Shindo M. Pure word deafness due to left hemisphere damage. Cortex. 1992;28(2):295–303.
63. Geschwind N. Disconnexion syndromes in animals and man. I. Brain. 1965;88(2):237–94. https://doi.org/10.1093/brain/88.2.237.
64. Wernicke, C. (1874). Der aphasische symptomencomplex.
65. Binder JR. Current controversies on Wernicke's area and its role in language. Curr Neurol Neurosci Rep. 2017;17(8):1–10.
66. Wernicke C. Lehrbuch der gehirnkrankheiten für aerzte und studirende, vol. 1. Fischer; 1881.
67. Mesulam M-M, Thompson CK, Weintraub S, Rogalski EJ. The Wernicke conundrum and the anatomy of language comprehension in primary progressive aphasia. Brain. 2015;138(8):2423–37. https://doi.org/10.1093/brain/awv154.
68. Marie P. The third left frontal convolution plays no special role in the function of language. Sem Méd. 1906;26:241–7.
69. Penfield W, Roberts L. Speech and brain mechanisms. Princeton University Press; 2014.
70. Bogen JE, Bogen GM. Wernicke's region–Where is it? Ann N Y Acad Sci. 1976;280:834–43. https://doi.org/10.1111/j.1749-6632.1976.tb25546.x.
71. Binder JR. The Wernicke area: modern evidence and a reinterpretation. Neurology. 2015;85(24):2170–5. https://doi.org/10.1212/WNL.0000000000002219.
72. Ardila A, Bernal B, Rosselli M. How extended is Wernicke's area? Meta-analytic connectivity study of BA20 and integrative proposal. Neurosci J. 2016a;2016:4962562. https://doi.org/10.1155/2016/4962562.
73. Turken A, Dronkers N. The neural architecture of the language comprehension network: converging evidence from lesion and connectivity analyses [original research]. Front Syst Neurosci. 2011;5. https://www.frontiersin.org/article/10.3389/fnsys.2011.00001
74. Pillay SB, Binder JR, Humphries C, Gross WL, Book DS. Lesion localization of speech comprehension deficits in chronic aphasia. Neurology. 2017;88(10):970–5. https://doi.org/10.1212/WNL.0000000000003683.
75. Corina DP, Loudermilk BC, Detwiler L, Martin RF, Brinkley JF, Ojemann G. Analysis of naming errors during cortical stimulation mapping: implications for models of language representation. Brain Lang. 2010;115(2):101–12. https://doi.org/10.1016/j.bandl.2010.04.001.
76. Bartha L, Benke T. Acute conduction aphasia: an analysis of 20 cases. Brain Lang. 2003;85(1):93–108. https://doi.org/10.1016/S0093-934X(02)00502-3.
77. Anderson J, Gilmore R, Roper S, Crosson B, Bauer RM, Nadeau S, Beversdorf DQ, Cibula J, Rogish M III, Kortencamp S. Conduction aphasia and the arcuate fasciculus: a reexamination of the Wernicke–Geschwind model. Brain Lang. 1999;70(1):1–12.
78. Berthier ML, Lambon Ralph MA, Pujol J, Green C. Arcuate fasciculus variability and repetition: the left sometimes can be right. Cortex. 2012;48(2):133–43. https://doi.org/10.1016/j.cortex.2011.06.014.
79. Quigg M, Fountain N. Conduction aphasia elicited by stimulation of the left posterior superior temporal gyrus. J Neurol Neurosurg Psychiatry. 1999;66(3):393–6.
80. Hertrich I, Dietrich S, Ackermann H. The margins of the language network in the brain [review]. Front Commun. 2020;5. https://www.frontiersin.org/article/10.3389/fcomm.2020.519955
81. Patterson K, Nestor PJ, Rogers TT. Where do you know what you know? The representation of semantic knowledge in the human brain. Nat Rev Neurosci. 2007;8(12):976–87. https://doi.org/10.1038/nrn2277.
82. Fong JS, Jehi L, Najm I, Prayson RA, Busch R, Bingaman W. Seizure outcome and its predictors after temporal lobe epilepsy surgery in patients with normal MRI. Epilepsia. 2011;52(8):1393–401. https://doi.org/10.1111/j.1528-1167.2011.03091.x.
83. Binder JR, Gross WL, Allendorfer JB, Bonilha L, Chapin J, Edwards JC, Grabowski TJ, Langfitt JT, Loring DW, Lowe MJ. Mapping anterior temporal lobe language areas with fMRI: a multicenter normative study. NeuroImage. 2011;54(2):1465–75. https://www.ncbi.nlm.nih.gov/pmc/articles/PMC2997157/pdf/nihms240637.pdf
84. Rice GE, Caswell H, Moore P, Hoffman P, Lambon Ralph MA. The roles of left versus right anterior temporal lobes in semantic memory: a neuropsychological comparison of postsurgical temporal lobe epilepsy patients. Cereb Cortex. 2018;28:1487–501. https://doi.org/10.1093/cercor/bhx362.
85. Numssen O, Bzdok D, Hartwigsen G. Functional specialization within the inferior parietal lobes across cognitive domains. eLife. 2021;10:e63591. https://doi.org/10.7554/eLife.63591.
86. Binder JR, Desai RH, Graves WW, Conant LL. Where is the semantic system? A critical review and meta-analysis of 120 functional neuroimaging studies. Cereb Cortex. 2009;19(12):2767–96. https://doi.org/10.1093/cercor/bhp055.
87. Turkeltaub PE, Coslett HB. Localization of sublexical speech perception components. Brain Lang. 2010;114(1):1–15. https://doi.org/10.1016/j.bandl.2010.03.008.

88. Dejerine J. Contribution à l'étude anatomo-pathologique et clinique des différents variétés de cécité verbale. Mémoires de la Société de Biologie. 1892;4:61–90.
89. Geschwind N. The organization of language and the brain. Science. 1970;170(3961):940–4.
90. Price CJ. The anatomy of language: contributions from functional neuroimaging. J Anat. 2000;197(Pt 3):335–59. https://doi.org/10.1046/j.1469-7580.2000.19730335.x.
91. Graves WW, Desai R, Humphries C, Seidenberg MS, Binder JR. Neural systems for reading aloud: a multiparametric approach. Cereb Cortex. 2010;20(8):1799–815. https://doi.org/10.1093/cercor/bhp245.
92. Binder JR, Frost JA, Hammeke TA, Bellgowan PS, Rao SM, Cox RW. Conceptual processing during the conscious resting state. A functional MRI study. J Cogn Neurosci. 1999;11(1):80–95. https://doi.org/10.1162/089892999563265.
93. Shulman GL, Fiez JA, Corbetta M, Buckner RL, Miezin FM, Raichle ME, Petersen SE. Common blood flow changes across visual tasks: II. Decreases in cerebral cortex. J Cogn Neurosci. 1997;9(5):648–63. https://doi.org/10.1162/jocn.1997.9.5.648.
94. Raichle ME, MacLeod AM, Snyder AZ, Powers WJ, Gusnard DA, Shulman GL. A default mode of brain function. Proc Natl Acad Sci. 2001;98(2):676–82. https://doi.org/10.1073/pnas.98.2.676.
95. Gordon EM, Laumann TO, Adeyemo B, Huckins JF, Kelley WM, Petersen SE. Generation and evaluation of a cortical area Parcellation from resting-state correlations. Cereb Cortex. 2016;26(1):288–303. https://doi.org/10.1093/cercor/bhu239.
96. Gordon EM, Laumann TO, Gilmore AW, Newbold DJ, Greene DJ, Berg JJ, Ortega M, Hoyt-Drazen C, Gratton C, Sun H. Precision functional mapping of individual human brains. Neuron. 2017;95(4):791–807. e797. https://www.ncbi.nlm.nih.gov/pmc/articles/PMC5576360/pdf/nihms892846.pdf
97. Gordon EM, Laumann TO, Marek S, Raut RV, Gratton C, Newbold DJ, Greene DJ, Coalson RS, Snyder AZ, Schlaggar BL, Petersen SE, Dosenbach NUF, Nelson SM. Default-mode network streams for coupling to language and control systems. Proc Natl Acad Sci. 2020;117(29):17308–19. https://doi.org/10.1073/pnas.2005238117.
98. Duffau H. Brain mapping: from neural basis of cognition to surgical applications. Springer Science & Business Media; 2011b.
99. Vogel AC, Miezin FM, Petersen SE, Schlaggar BL. The putative visual word form area is functionally connected to the dorsal attention network. Cereb Cortex. 2012;22:537–49. https://doi.org/10.1093/cercor/bhr100.
100. Ferstl EC, Neumann J, Bogler C, von Cramon DY. The extended language network: a meta-analysis of neuroimaging studies on text comprehension. Hum Brain Mapp. 2008;29(5):581–93. https://doi.org/10.1002/hbm.20422.
101. Taylor KI, Regard M. Language in the right cerebral hemisphere: contributions from Reading studies. Physiology. 2003;18(6):257–61. https://doi.org/10.1152/nips.01454.2003.
102. Rasmussen T, Milner B. The role of early left-brain injury in determining lateralization of cerebral speech functions. Ann N Y Acad Sci. 1977;299:355–69. https://doi.org/10.1111/j.1749-6632.1977.tb41921.x.
103. Duffau H, Leroy M, Gatignol P. Cortico-subcortical organization of language networks in the right hemisphere: an electrostimulation study in left-handers. Neuropsychologia. 2008;46(14):3197–209. https://doi.org/10.1016/j.neuropsychologia.2008.07.017.
104. Maffei C, Jovicich J, De Benedictis A, Corsini F, Barbareschi M, Chioffi F, Sarubbo S. Topography of the human acoustic radiation as revealed by ex vivo fibers micro-dissection and in vivo diffusion-based tractography. Brain Struct Funct. 2018;223(1):449–59. https://doi.org/10.1007/s00429-017-1471-6.
105. Carbajal GV, Malmierca MS. The neuronal basis of predictive coding along the auditory pathway: from the subcortical roots to cortical deviance detection. Trends Hearing. 2018;22:2331216518784822. https://doi.org/10.1177/2331216518784822.
106. Tournier JD, Calamante F, Connelly A. MRtrix: diffusion tractography in crossing fiber regions. Int J Imaging Syst Technol. 2012;22(1):53–66.
107. Yeh F-C, Panesar S, Fernandes D, Meola A, Yoshino M, Fernandez-Miranda JC, Vettel JM, Verstynen T. Population-averaged atlas of the macroscale human structural connectome and its network topology. NeuroImage. 2018;178:57–68. https://doi.org/10.1016/j.neuroimage.2018.05.027.
108. Yeh F-C. DSI studio. 2016. http://dsi-studio.labsolver.org.
109. Yeh FC, Wedeen VJ, Tseng WY. Generalized q-sampling imaging. IEEE Trans Med Imaging. 2010;29(9):1626–35. https://doi.org/10.1109/tmi.2010.2045126.
110. Fernández-Miranda JC, Wang Y, Pathak S, Stefaneau L, Verstynen T, Yeh F-C. Asymmetry, connectivity, and segmentation of the arcuate fascicle in the human brain. Brain Struct Funct. 2015;220(3):1665–80. https://link.springer.com/article/10.1007/s00429-014-0751-7
111. Glasser MF, Rilling JK. DTI tractography of the human brain's language pathways. Cereb Cortex. 2008;18(11):2471–82. https://doi.org/10.1093/cercor/bhn011.
112. Brauer J, Anwander A, Perani D, Friederici AD. Dorsal and ventral pathways in language development. Brain Lang. 2013, Nov;127(2):289–95. https://doi.org/10.1016/j.bandl.2013.03.001.
113. Panesar SS, Belo JTA, Yeh F-C, Fernandez-Miranda JC. Structure, asymmetry, and connectivity of the human temporo-parietal aslant and vertical occipital fasciculi. Brain Struct Funct. 2019;224(2):907–23. https://doi.org/10.1007/s00429-018-1812-0.

114. Catani M, Jones DK, Ffytche DH. Perisylvian language networks of the human brain. Ann Neurol. 2005;57(1):8–16. https://doi.org/10.1002/ana.20319.
115. Catani M, Dell'acqua F, Vergani F, Malik F, Hodge H, Roy P, Valabregue R, Thiebaut de Schotten M. Short frontal lobe connections of the human brain. Cortex. 2012, Feb;48(2):273–91. https://doi.org/10.1016/j.cortex.2011.12.001.
116. Catani M, Mesulam MM, Jakobsen E, Malik F, Martersteck A, Wieneke C, Thompson CK, Thiebaut de Schotten M, Dell'Acqua F, Weintraub S, Rogalski E. A novel frontal pathway underlies verbal fluency in primary progressive aphasia. Brain. 2013, Aug;136(Pt 8):2619–28. https://doi.org/10.1093/brain/awt163.
117. La Corte E, Eldahaby D, Greco E, Aquino D, Bertolini G, Levi V, Ottenhausen M, Demichelis G, Romito LM, Acerbi F, Broggi M, Schiariti MP, Ferroli P, Bruzzone MG, Serrao G. The frontal aslant tract: a systematic review for neurosurgical applications [systematic review]. Front Neurol. 2021;12 https://doi.org/10.3389/fneur.2021.641586.
118. Fujii M, Maesawa S, Motomura K, Futamura M, Hayashi Y, Koba I, Wakabayashi T. Intraoperative subcortical mapping of a language-associated deep frontal tract connecting the superior frontal gyrus to Broca's area in the dominant hemisphere of patients with glioma. J Neurosurg. 2015;122(6):1390–6.
119. Kinoshita M, de Champfleur NM, Deverdun J, Moritz-Gasser S, Herbet G, Duffau H. Role of fronto-striatal tract and frontal aslant tract in movement and speech: an axonal mapping study. Brain Struct Funct. 2015;220(6):3399–412. https://doi.org/10.1007/s00429-014-0863-0.
120. Cocquyt E-M, Lanckmans E, van Mierlo P, Duyck W, Szmalec A, Santens P, De Letter M. The white matter architecture underlying semantic processing: a systematic review. Neuropsychologia. 2020;136:107182.
121. Wang X, Pathak S, Stefaneanu L, Yeh FC, Li S, Fernandez-Miranda JC. Subcomponents and connectivity of the superior longitudinal fasciculus in the human brain. Brain Struct Funct. 2016;221(4):2075–92. https://doi.org/10.1007/s00429-015-1028-5.
122. Thiebaut de Schotten M, Dell'Acqua F, Forkel SJ, Simmons A, Vergani F, Murphy DG, Catani M. A lateralized brain network for visuospatial attention. Nat Neurosci. 2011;14(10):1245–6. https://doi.org/10.1038/nn.2905.
123. Shah A et al. World Neurosurgery 2019;125: e620–638.
124. Sarubbo S, De Benedictis A, Maldonado IL, Basso G, Duffau H. Frontal terminations for the inferior fronto-occipital fascicle: anatomical dissection, DTI study and functional considerations on a multi-component bundle. Brain Struct Funct. 2013;218(1):21–37. https://doi.org/10.1007/s00429-011-0372-3.
125. Wu Y, Sun D, Wang Y, Wang Y. Subcomponents and connectivity of the inferior fronto-occipital fasciculus revealed by diffusion spectrum imaging fiber tracking. Front Neuroanat. 2016;10:88. https://www.ncbi.nlm.nih.gov/pmc/articles/PMC5033953/pdf/fnana-10-00088.pdf
126. Moritz-Gasser S, Herbet G, Duffau H. Mapping the connectivity underlying multimodal (verbal and non-verbal) semantic processing: a brain electrostimulation study. Neuropsychologia. 2013;51(10):1814–22. https://doi.org/10.1016/j.neuropsychologia.2013.06.007.
127. Duffau H, Gatignol P, Mandonnet E, Peruzzi P, Tzourio-Mazoyer N, Capelle L. New insights into the anatomo-functional connectivity of the semantic system: a study using cortico-subcortical electrostimulations. Brain. 2005;128(4):797–810. https://doi.org/10.1093/brain/awh423.
128. Herbet G, Moritz-Gasser S, Duffau H. Direct evidence for the contributive role of the right inferior fronto-occipital fasciculus in non-verbal semantic cognition. Brain Struct Funct. 2017;222(4):1597–610. https://doi.org/10.1007/s00429-016-1294-x.
129. Hau J, Sarubbo S, Houde JC, Corsini F, Girard G, Deledalle C, Crivello F, Zago L, Mellet E, Jobard G. Revisiting the human uncinate fasciculus, its subcomponents and asymmetries with stem-based tractography and microdissection validation. Brain Struct Funct. 2017;222(4):1645–62. https://link.springer.com/content/pdf/10.1007/s00429-016-1298-6.pdf
130. Holland R, Lambon Ralph MA. The anterior temporal lobe semantic hub is a part of the language neural network: selective disruption of irregular past tense verbs by rTMS. Cereb Cortex. 2010;20(12):2771–5. https://doi.org/10.1093/cercor/bhq020.
131. Papagno C, Miracapillo C, Casarotti A, Romero Lauro LJ, Castellano A, Falini A, Casaceli G, Fava E, Bello L. What is the role of the uncinate fasciculus? Surgical removal and proper name retrieval. Brain. 2011;134(2):405–14. https://doi.org/10.1093/brain/awq283.
132. Kalyvas A, Koutsarnakis C, Komaitis S, Karavasilis E, Christidi F, Skandalakis GP, Liouta E, Papakonstantinou O, Kelekis N, Duffau H. Mapping the human middle longitudinal fasciculus through a focused anatomo-imaging study: shifting the paradigm of its segmentation and connectivity pattern. Brain Struct Funct. 2020;225(1):85–119. https://link.springer.com/content/pdf/10.1007/s00429-019-01987-6.pdf
133. Wang Y, Fernández-Miranda JC, Verstynen T, Pathak S, Schneider W, Yeh F-C. Rethinking the role of the middle longitudinal fascicle in language and auditory pathways. Cereb Cortex. 2012;23(10):2347–56. https://doi.org/10.1093/cercor/bhs225.
134. Maldonado IL, de Champfleur NM, Velut S, Destrieux C, Zemmoura I, Duffau H. Evidence of a middle longitudinal fasciculus in the human brain

from fiber dissection. J Anat. 2013;223(1):38–45. https://www.ncbi.nlm.nih.gov/pmc/articles/PMC3798102/pdf/joa0223-0038.pdf
135. De Witt Hamer PC, Moritz-Gasser S, Gatignol P, Duffau H. Is the human left middle longitudinal fascicle essential for language? A brain electrostimulation study. Hum Brain Mapp. 2011;32(6):962–73. https://www.ncbi.nlm.nih.gov/pmc/articles/PMC6870476/pdf/HBM-32-962.pdf
136. Mandonnet E, Nouet A, Gatignol P, Capelle L, Duffau H. Does the left inferior longitudinal fasciculus play a role in language? A brain stimulation study. Brain. 2007;130(Pt 3):623–9. https://doi.org/10.1093/brain/awl361.
137. Broca P. Sur les mots aphemie, aphasie et aphrasie; Lettre a M. le Professeur Trousseau. Gazette des hopitaux, 1864;23.
138. Freud, S. (1891). Contribution à la conception des aphasies: une étude critique Paris: PUF, 1983. *Edição original.*
139. Meynert T. Psychiatry; a clinical treatise on diseases of the fore-brain based upon a study of its structure, functions, and nutrition. GP Putnam's Sons; 1885.
140. Catani M, Ffytche DH. The rises and falls of disconnection syndromes. Brain. 2005;128(10):2224–39. https://doi.org/10.1093/brain/awh622.
141. Fujii M, Maesawa S, Ishiai S, Iwami K, Futamura M, Saito K. Neural basis of language: an overview of an evolving model. Neurol Med Chir. 2016;56(7):379–86. https://doi.org/10.2176/nmc.ra.2016-0014.
142. Hickok G, Poeppel D. Dorsal and ventral streams: a framework for understanding aspects of the functional anatomy of language. Cognition. 2004;92(1–2):67–99.
143. Coello AF, Moritz-Gasser S, Martino J, Martinoni M, Matsuda R, Duffau H. Selection of intraoperative tasks for awake mapping based on relationships between tumor location and functional networks: a review. J Neurosurg JNS. 2013;119(6):1380–94. https://doi.org/10.3171/2013.6.JNS122470.
144. Saur D, Schelter B, Schnell S, Kratochvil D, Küpper H, Kellmeyer P, Kümmerer D, Klöppel S, Glauche V, Lange R, Mader W, Feess D, Timmer J, Weiller C. Combining functional and anatomical connectivity reveals brain networks for auditory language comprehension. NeuroImage. 2010;49(4):3187–97. https://doi.org/10.1016/j.neuroimage.2009.11.009.

Optic Radiations Monitoring in Awake Glioma Surgery: Intraoperative Campimetry

Pablo Gonzalez-Lopez,
Carlos Martorell-Llobregat,
Maria Dolores Coves Piqueres,
and Eduardo Fernández

1 Introduction

<<Vision is the dominant sense of human beings, as it allows humans to connect with our surroundings, stay safe, and helps on keeping our minds in a sharp state. Vision provides the most relevant source of information about our environment. As a reflection of this important role, vision brain regions account for more than a quarter of our cortex, not only as a primary visual area, but mainly including the areas involved on visual process, being the widest of all senses>>.

Vision, as other highly functional, in brain process terms, can be affected by a multitude of diseases, including brain tumors. A contemporary multimodal treatment of lesions in highly functional areas requires different but related subspecialities. Therefore, understanding the proper localization, stablishing a diagnosis, designing a treatment, and defining prognosis, requires following some relevant protocols.

Most of these glial-origin tumors, show an infiltrative pattern through the axons and perivascular spaces aided by neuropil. This seems to be the main reason why surgical resection is not curative at all, with most patients suffering a local progression sooner or later.

However, surgery still maintains a key role at treatment, as the neoplastic cells volume reduction provided by a large resection alleviates neurological symptoms and intracranial hypertension, provides tissue for diagnosis, and contributes to increase the efficacy of complementary treatments. Recently, many contemporary series supported the concept of wider tumor resections, with some of them having proved that a bigger resection provides a better control of the disease. Nevertheless, larger resections might increase the rate of postoperative neurological complications, fact that is really challenging in the case of the a forecited eloquent regions of the brain.

New techniques have been incorporated to the diagnostic and therapeutic arsenal with the aim of minimizing the potential associated morbidity while extending the resection to the so-called supramarginal region. The objective of this philosophy consists on removing the tumor core, as well as the surrounding tissue, that might be macroscopically healthy during surgery, but with a high suspicion of being invaded by neoplastic cells, or by cells carrying molecular disorders

P. Gonzalez-Lopez (✉) · C. Martorell-Llobregat
Department of Neurosurgery, Alicante General University Hospital, Alicante Institute of Health and Biomedical Research (ISABIAL), Alicante, Spain
e-mail: gonzalez_pab@gva.es

M. D. C. Piqueres
Department of Clinical Neurophysiology, Alicante General University Hospital, Alicante Institute of Health and Biomedical Research (ISABIAL), Alicante, Spain

E. Fernández
Biomedical Neuro Engineering Department, University Miguel Hernández and CIBER-BBN, University Miguel Hernández, Elche, Spain

A. Shah et al. (eds.), *Functional Anatomy of the Brain: A View from the Surgeon's Eye*,
https://doi.org/10.1007/978-981-99-3412-6_7

that could transform them into tumor ones. Thus, the tumor removal should be moved to the so-called functional limit, where there might be a neurological deterioration [1].

This modern strategy has been named as the "onco-functional balance" of surgery. Neurosurgeons must implement this philosophy and modify the classic approach, which means carrying out the removal based on cortical and subcortical functionality, and not only to anatomical and oncological limits (functional resection limit) [1].

Anatomy by itself cannot predict the functional eloquence. This is the main reason why certain tumors must not be classified as unresectable, just based on anatomical appearance. We, as surgeons, must squeeze awake mapping techniques and modern technology to develop individual maps and precisely design management strategies. Including the tumor surrounding brain in the resection plan is nowadays a must, with the condition that the underlying neural networks removed might compensate due to the so-called brain plasticity, diminishing potential complications and finally quality of life [2–6].

Although awake brain tumor surgery is a standard procedure nowadays, it was not until the late nineteenth century when it was carried out as a surgeon decision. In 1874 a case of electrical stimulation of a human brain was reported, causing a substantial ethical discussion, insomuch as the method was invasive and without clear therapeutic benefit at that time [7, 8]. In 1937, Penfield and Boldrey described, for the first time, a standardized procedure combining electrical stimulation with awake surgery for epilepsy [9]. Later on, George Ojemann presented in the 70's a systemized testing to detect and avoid damaging functional tissues [10]. In the next years, Ojemann and Berger adapted this method in cases of central nervous system lesions and, over this time frame, the importance of subcortical mapping was better evaluated [11].

Since the first attempts to save human functions during brain surgery, the focus has been moved to preserve language and motor eloquence. Surprisingly, human vision studies have been mainly focused on restoring vision in blind patients, instead of saving it during brain tumor surgery.

This chapter is a review of the methodology followed by the authors to make the diagnosis, plan the surgical procedure, and intraoperatively identify and respect visual functions on awake patients during brain tumor surgery. Specifically, we will focus on lesions appearing into the temporo-parieto-occipital 'hot-spot' area. This is an area completely covered by highly eloquent cortex, with relevant underlying crossing fibers with anatomically different directions and functions.

2 Anatomy of the Visual Pathway

The Primary Visual Area (PVA) lies in the mediobasal region of the occipital lobe, around the calcarine fissure. Visual inputs are transferred from the retina through the optic nerve and tract, reaches the lateral geniculate body (LGB), where it synapses. Then, the electrical information travels to the occipital cortex through the geniculocalcarine fibers, which consists of three bundles (Fig. 1).

The first group of fibers leaves the LGB backwards over the roof of the atrium and occipital horn; they travel through a direct path to the upper lip of the calcarine fissure and are described as the posterior bundle [12–15], which carries impulses from the inferior visual quadrant [16].

The next group suffer an anterior and mediolateral curve, but do not extend to the anterior aspect of the temporal horn. This group of fibers (central bundle) runs slightly laterally and posteriorly over the temporal horn, lateral wall and roof of the trigone, and occipital horn to finally end up at the PVA in the occipital pole. This fascicle is mainly made of the macular fibers (central areas of the retina) and is the largest of the three of them [12–16].

The third one (anterior bundle), initially moves forward for 8–16 mm, above the temporal horn and tail of the caudate and below the lentiform nuclei, before turning posteriorly along the temporal horn lateral aspect. It continues below

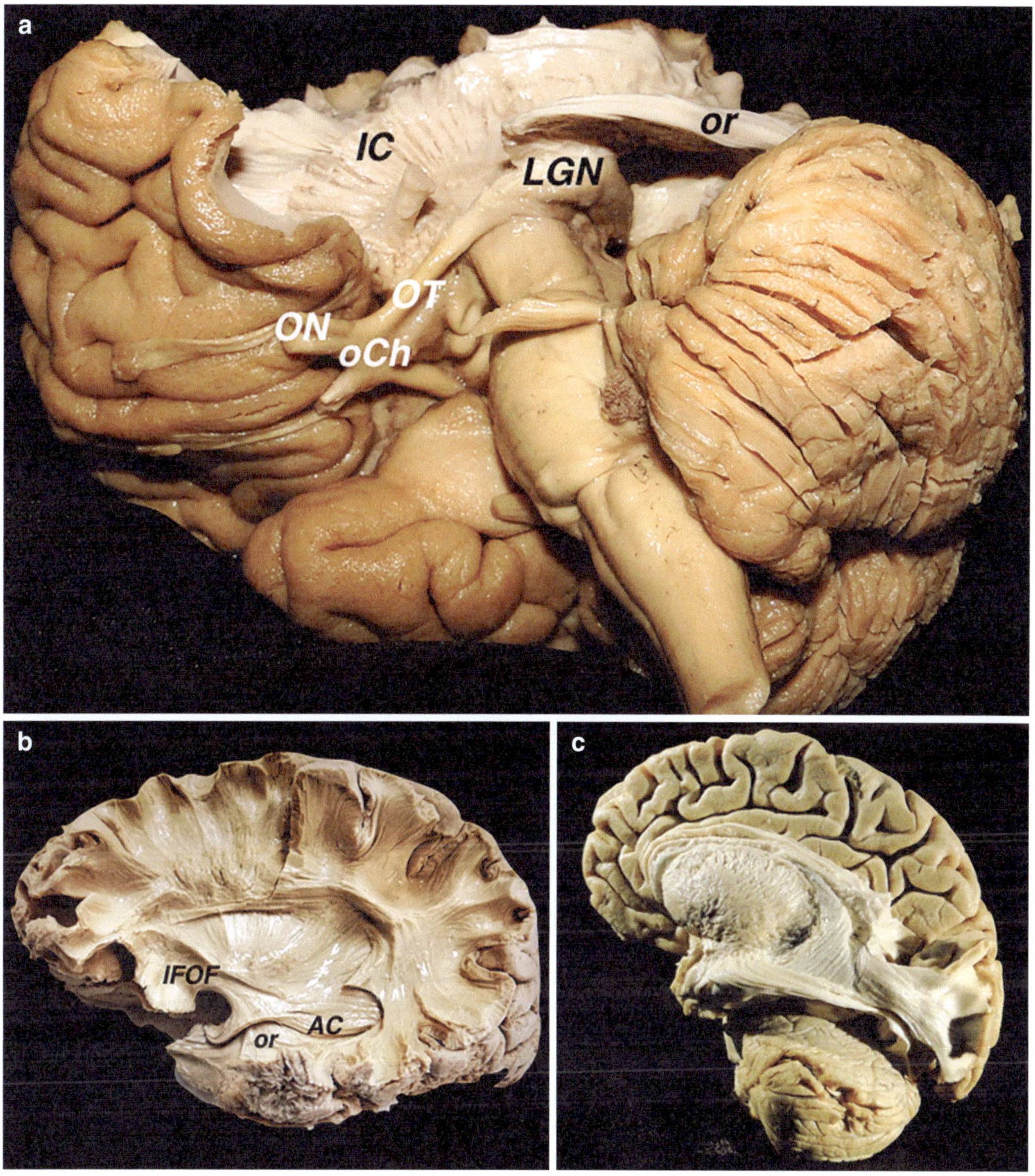

Fig. 1 Left hemisphere human brain dissection different steps. (**a**) Basal view showing the whole intracranial extension of the visual pathways and its close relationship with the temporal horn, atrium and occipital horn. (**b**) Close relationship between the main components of the sagittal stratum: inferior frontooccipital fascicle, posterior limb of the anterior commissure and optic radiations. (**c**) Lateral view of the Meyer's loop and sublenticular aspect of the internal capsule to the putamen and the tapetum fibers running through the splenium and covering the lateral wall of the atrium and temporal horns. *AC*, anterior commissure; *IC*, internal capsule; *IFOF*, inferior frontooccipital fascicle; *LGN*, lateral geniculate nucleus; *oCh*, optic chiasm; *ON*, optic nerve; *or*, optic radiations; *OT*, optic tract

the atrium and occipital horn, curving medially to reach the calcarine fissure inferior lip [17]. These fibers are commonly referred as the Meyer's loop and use to carry activity from the superior visual quadrant [12–16] (Fig. 2).

An accurate understanding of the internal capsule (IC) anatomy is crucial to describe the optic radiations (OR) [14, 15]. The different parts of the IC are the anterior limb, between the head of the caudate and the lenticular nucleus; the poste-

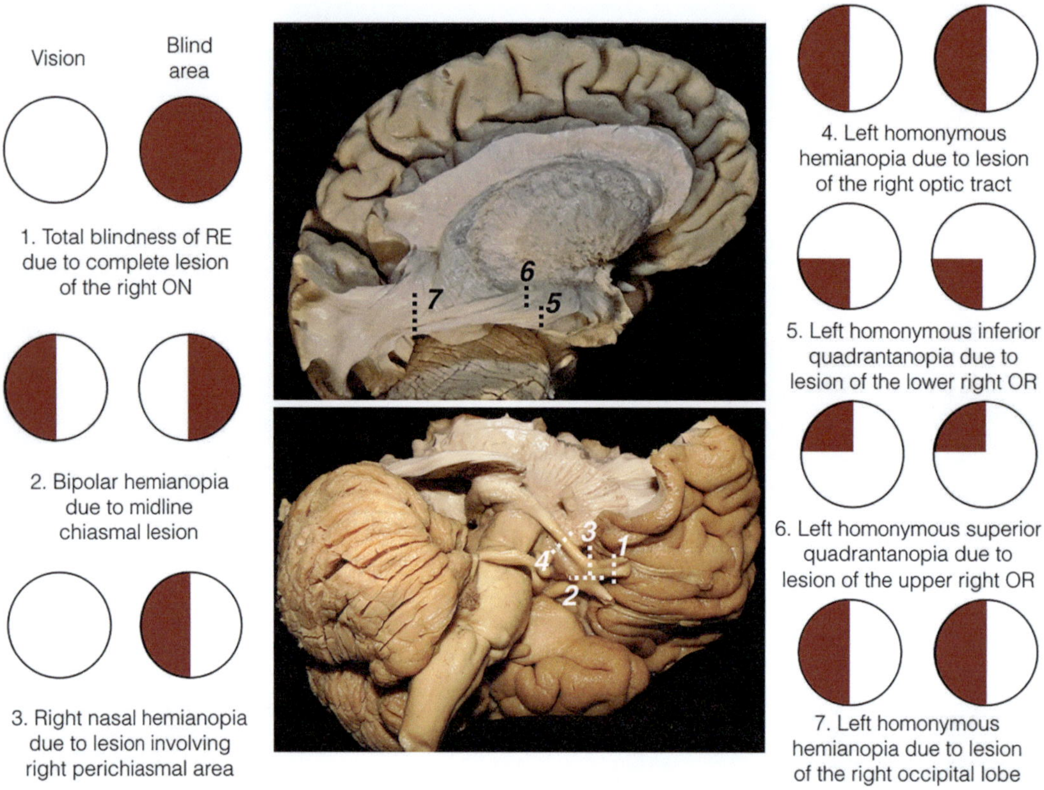

Fig. 2 Different potential visual field deficits depending on the area of damage and/or area of stimulation under awake surgical conditions. (1), optic nerve; (2), central optic chiasm; (3), lateral optic chiasm; (4), optic tract; (5) lower optic radiations; (6) upper optic radiations; (7), pericalcarine region

rior limb between the thalamus and the lenticular nucleus; genu between the aforementioned limbs; a sublenticular component below the lenticular nucleus; and a retrolenticular one, just posterior to the lenticular nucleus. The sublenticular and the retrolenticular parts are directly related with the OR.

The retrolenticular aspect is made of the occipito- and parietopontine fibers, while the posterior thalamic peduncle, including part of the optic radiations, together with the connections between the pulvinar and parieto-occipital cortex. This group of fibers are in close relationship and, even for some authors, belong to the sagittal stratum, which is shaped by two layers: inner (occipito- and parietopontine fibers), and outer (OR) (Fig. 3). The inferior frontooccipital fasciculus (IFOF), and the temporo-occipital component of the anterior commissure are located laterally and are generally described as the-most-lateral component of the sagittal stratum in several studies. From an anatomical and radiological perspective, it is tedious and somehow impossible to separate these components into single ones, due to the intrinsic limitation of these methods [12–21].

On the other hand, the sublenticular component of the IC is made of the Meyer's loop, temporo-pontine fibers, as well as the auditory radiations [18].

Another important consideration regarding this anatomical area known as the temporo-parieto-occipital region, is the fact that this deep layer of longitudinally oriented fibers is nicely covered by the arcuate fascicle and vertical component of the SLF, which are tracts mainly related with verbal fluency, nomination and comprehension.

Periatrial intrinsic brain tumors may grow through or in between these fibers. This is the

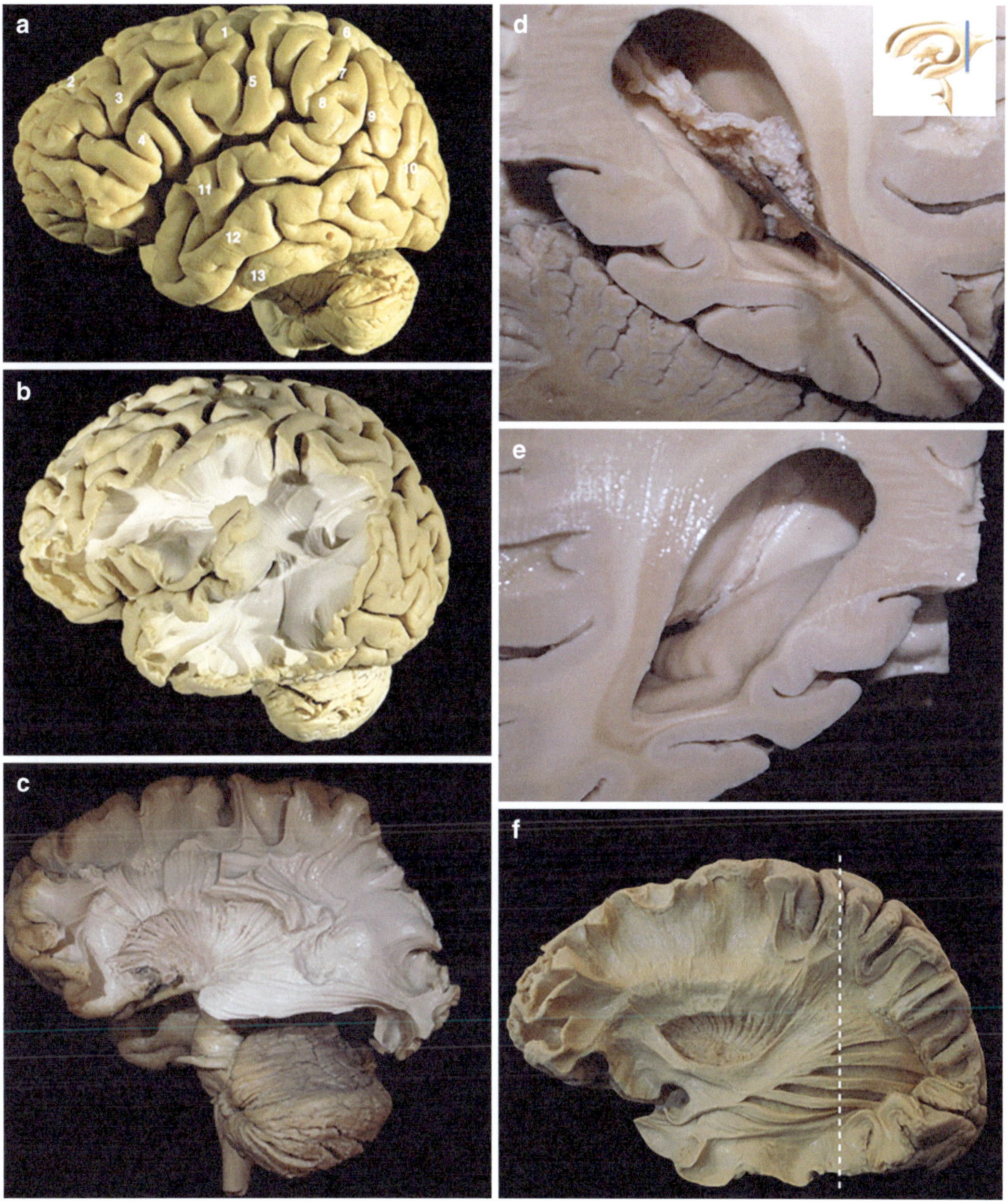

Fig. 3 Left hemisphere human brain dissection different steps. (**a**) Lateral view of a human brain left hemisphere where the main cortical landmarks are numbered. Note the location of the temporo-parieto-occipital junction (6, 7, 8, 9, 10). (**b**) The periSylvian gyri have been resected exposing the SLF-III, SLF-II as well as the arcuate fascicles. (**c**) The IFOF and anterior commissure posterior limb have been cut and peeled away, as well as the lateral basal ganglia, exposing the whole extension of the internal capsule including the optic radiations. (**d & e**) Coronal cut of the left hemisphere at the level of the atrium in the temporo-parieto-occipital junction showing the intraventricular anatomy as well as its lateral wall composed by the tapetum and sagittal stratum. (**f**) Left hemisphere where following the fiber dissection technique, the IFOF, anterior commissure and optic radiations are exposed in different stratum. (1), precentral gyrus; (2), superior frontal gyrus; (3), middle frontal gyrus; (4), inferior frontal gyrus; (5), post central gyrus; (6), superior parietal lobe; (7), intermediate sulcus of Jensen; (8), supramarginal gyrus; (9), angular gyrus; (10), occipital lobe; (11), superior temporal gyrus; (12), middle temporal gyrus; (13), inferior temporal gyrus

reason why brain surgeons must know the close anatomical relationship of the OR to the atrium, temporal and occipital horns. Firstly, the lateral wall of the ventricle from the temporal horn to the trigone and occipital horn is covered by OR. The temporal horn roof is also upholstered by OR. The medial wall of the trigone is devoid of these fibers. Posteriorly, the occipital horn shows a close relationship with medial structures as the calcarine fissure, which is functionally relevant. Finally, the temporal horn floor, just anterior to the lateral geniculate body, is another area free of optic pathways [12, 17]. All these features must be balanced when deciding the optimal route to deal with periventricular tumors.

3 First Contact. Visual Tests & Imaging

3.1 Clinical Records and Neurological Exam

The initial neurological exam is key in patients suffering from neurological diseases. Nevertheless, this is of great importance when facing a brain tumor in or around the visual areas, as our goal as surgeons will remain on removing as much tissue as possible, preserving or even improving our patient's initial neurological status.

Patients affected by brain tumors are commonly admitted after presenting irritative symptoms and/or neurological deficits. Occipital lobe and periatrial tumors generally cause visual disturbances as contralateral hemianopia in most of the cases (Fig. 2). Other potential deficits in these cases are visual agnosia, or inability to recognize objects, faces (prosopagnosia), motion (akinetopsia), or alexia, or an inability to read [1].

3.2 Visual Field Testing

Visual field tests, although subjective, represent a useful technique to detect diseases affecting the extent or the intactness of the area that a patient can perceive [22]. Thus, this simple test is of great help to understand the kind of visual deficit the patient is suffering. The term "visual field" refers to the total sum of visual perceptions for a given eye, which remains fixed on a stationary object. The binocular visual field is the combined visual perception of both eyes under the same restrictions of movements. Indeed, its intrinsic subjectivity makes this test indispensable to start creating the intraoperative tests to be carried out, as well as the potential postoperative scenario.

It can be simply performed in the patient's bed with a confrontational visual field test by assessing the detection of colored objects showed to the patient, or by the patient counting the tester fingers.

Automated perimetry is used for a more precise analysis, in which the patient receives an output through a kind of visual stimulus, responding when it is detected (with a button-click or verbally).

The Goldmann and Humphrey tests are used to for a perimetry assessment. In the Goldmann test, the subject is asked to focus on a screen target located one meter away in the center of the visual field. Then, the tester moves an object toward the target, and the subject is asked to respond when he starts perceiving it 'at the periphery'. The Humphrey test employs a computer screen where the patient must again focus on its center, and respond whenever a flash is perceived into the periphery [23] (Fig. 4).

3.3 Imaging Studies

Recent advances in neuroimaging have become a must to diagnose, localize and evaluate brain abnormalities, as well as to characterize the disease, perform the surgical planning, predict potential postoperative neurological deficits, provide information about the tumor metabolism, understand its relative location regarding the cortical and subcortical eloquent regions, and even to evaluate the extent of resection intra and postoperatively. Therefore, the amount of information we can extract from a simple MRI sequence is enormous [1].

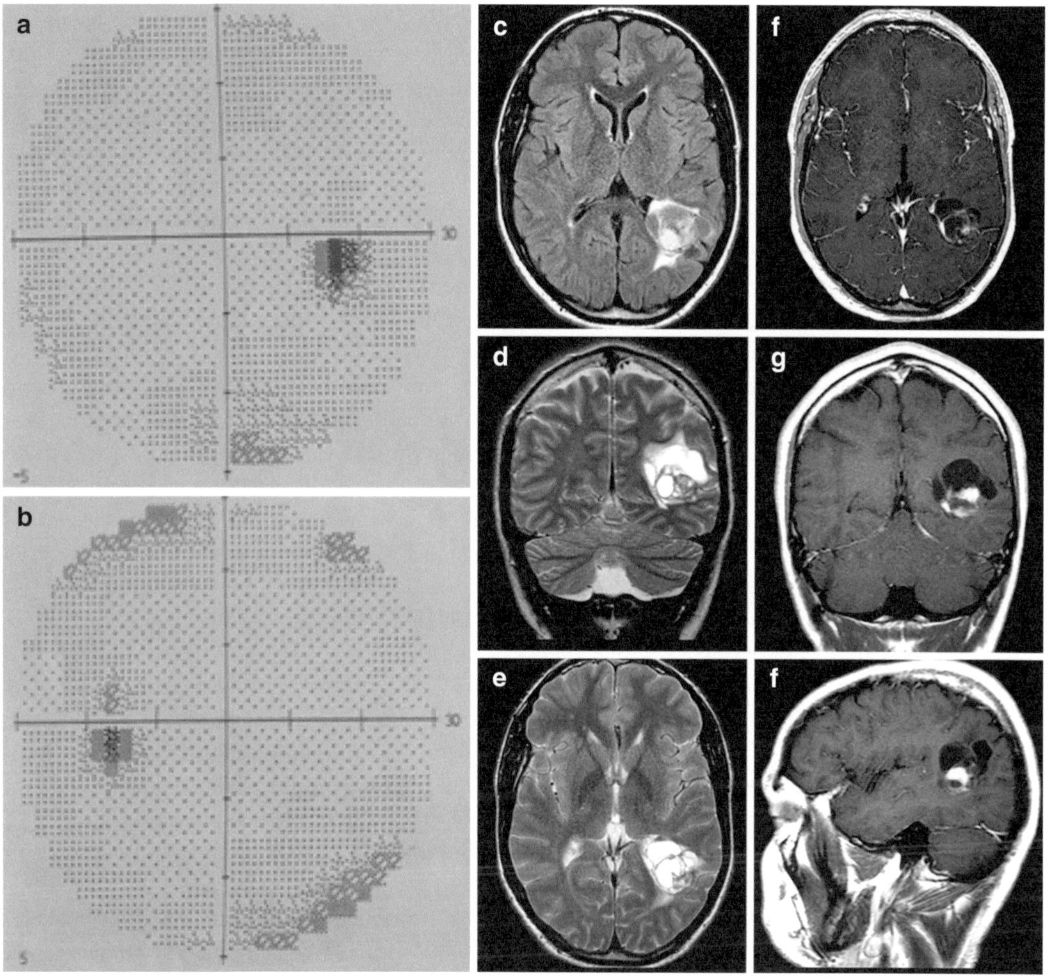

Fig. 4 (**a & b**) Close to normal right and left visual fields tested by a preoperative campimetry. (**c**) MRI FLAIR sequence axial slice at the level of the internal capsule and thalamus, where an heterogenous signal intensity is shown lateral to the atrium in the left lateral ventricle in close relationship with the sagittal stratum. (**d & e**) T2-weighted coronal and axial slices where the tumor is highlighted showing a cystic aspect and mild perilesional edema. (**f-h**) T1-weighted GD enhanced axial, coronal and sagittal slices showing the heterogenous enhancing behavior of the aforementioned tumor

3.3.1 Anatomical MRI Sequences

The classic FLAIR, T1-, T2-weighted and FIESTA MRI provide a detailed anatomical information. They are used to identify, localize, characterize the lesion, as well as to determine its cortical and subcortical extension.

Our protocol to achieve an accurate tumor anatomical localization and definition, includes at least the FLAIR, T2-weighted, and T1- with and without Gadolinium sequences. The primary visual area is identified through the calcarine and parieto–occipital fissures, which are better delineated on a midsagittal cut of the hemisphere. The lateral occipital surface sulcogyral architecture is difficult to delineate due to its heterogeneity. However, it may be defined by an imaginary line between the midline parieto-occipital sulcus projection and the tentorial notch in to the latero-basal limit of the hemisphere (Fig. 4).

Some other MRI modalities as the functional MRI and DTI—based tractography, have the capability to enrich our understanding of the previously delineated anatomy, by providing priceless functional and subcortical anatomical information.

3.3.2 Functional MRI (fMRI)

fMRI has been thought to provide direct functional information of different cortical areas of the brain. It detects changes in brain blood flow by somehow measuring the balance of deoxy- and oxy- hemoglobin at the capillary level. These changes, after some computer processing, are seen in s dynamic way as the result of 'cortical activation' after carrying out specific tasks [1].

However, we must still consider intraoperative electrophysiology the gold-standard to detect brain function as, when comparing it with fMRI, there was only a 71% and 33% positive correlation for motor and language function, respectively [6, 24, 25]. In a different study, Kuchsinski et al., compared preoperative fMRI with direct cortical stimulation for language, defining a 37.1% and 83.4% sensitivity and specificity of fMRI, respectively [26]. Such discrepancies can be explained by wrong tasks (mistaken design regarding the glioma anatomical location), by a neurovascular disengage in glioma patients (capillary changes in the vicinity of gliomas might not reflect the signal as accurately as it does in healthy tissue) or even due to problems as a wrong threshold definition [27].

Additionally, fMRI is not able to distinguish functionally essential areas (which should be preserved in the surgery) from areas of the brain which are not functionally crucial (which may be surgically removed with relative safety). Therefore, there is a double risk of taking decisions only based on the fMRI findings: (1) good candidates for surgical excision might be discarded because of a false positive of the fMRI; (2) tumors might be removed from a functionally essential region because of a false negative result of the fMRI [6, 27]. Moreover, patients with primary focal deficits or cognitive processing problems, fMRI might be very laborious, or even impossible to be done.

fMRI of the visual system can provide a specific image of the cortical visual organization with a large amount of functional organizations, depending on the tested perceptual qualities: color, motion, face recognition, places bodies, visual words, objects, … It also gives us the possibility of measuring visual field maps and demonstrate visual perceptions. The complex arrangement and large number of connections into the striate cortex is responsible for some of the encountered gaps when performing a fMRI. In this sense, specific phenomena as attention and conscious perceptions, might affect its signals and analysis [28, 29].

We routinely include fMRI visual studies into our protocol, specifically testing motion, face recognition and visual words to create a preoperative map orienting us about the relative location of the functional active areas in selected patients. These cortical areas will be later on used to assess the subcortical connective map process through DTI (Fig. 5).

3.3.3 DTI: Based Tractography

After delineating the anatomy of the brain tumor and having an orientation about its relationship with the eloquent regions based on fMRI, the study of the connectivity is crucial to decide the surgical approach and technique. In this sense, diffusion tensor imaging (DTI) has the ability to detect the water diffusion in tissues and 'draw' its particular Brownian movement. Water diffusion is aimed by an ordered distribution of the membranes covering the neural axons. Therefore, this modality provides an accurate display of brain axons, by choosing the areas with the same values of Fractional Anisotropy as "seed" to start highlighting the tracts in a 3D space.

Thus, DTI provides valuable images about the normal course, damaged tracts due to vasogenic edema or tumor infiltration, and the disruption or displacement of bundles around the tumor. Its use in Neurosurgery, does not only remain on understanding the relative location of the relevant fiber tracts and the tumor, but also to localize selected association fibers. In this sense, tractography is of great help when fMRI cannot be easily carried out, and also in vision analysis, in which the results of fMRI are not as precise as in other functions [1].

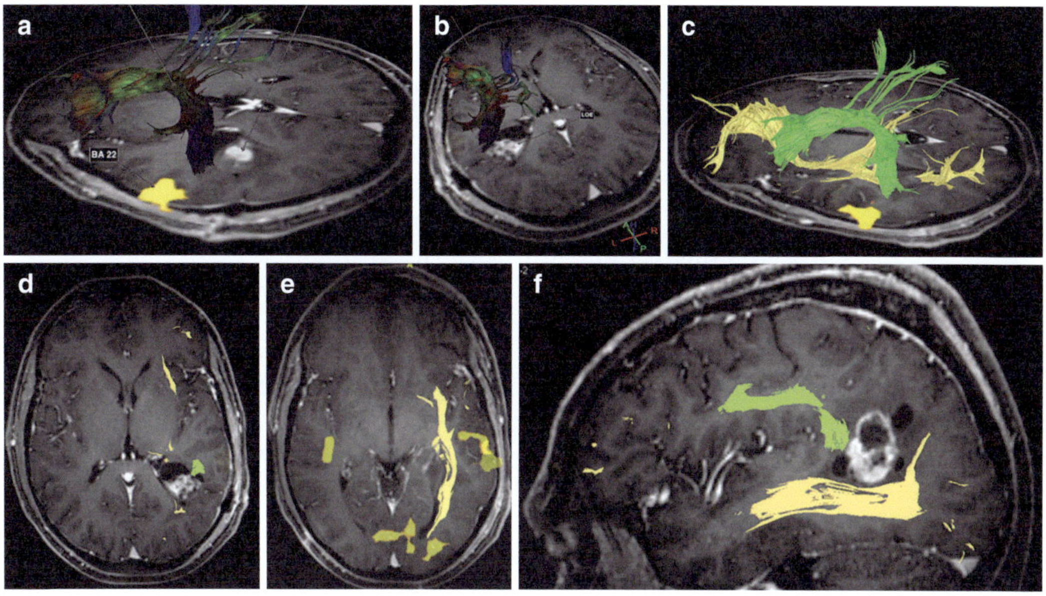

Fig. 5 (**a-b**) T1-weigthed GD enhanced axial cut showing the close relationship of the tumor with the arcuate fascicle posterior curve by DTI-tractography as well as the representation of language comprehension area by fMRI to the tumor lateral cortical projection. (**c**) T1-weigthed GD enhanced axial cut showing the relative position of language comprehension area (orange), arcuate fascicle (green) and IFOF (yellow). (**d-e**) T1-weigthed GD enhanced axial cuts showing the periatrial tumor itself as well as the arcuate fascicle (green), optic radiations (yellow) and reading task. (**f**) T1-weigthed GD enhanced sagittal cut showing the periatrial tumor itself as well as the arcuate fascicle (green) and inferior longitudinal fascicle (yellow)

DTI tractography of the visual pathway represents a key tool of our preoperative protocol, as even being aware of its major limitations in accuracy, it gives us a primary orientation about the relative course of the OR and its relationship to the lesion itself. The aggressiveness of a given tumor and/or the perilesional edema may make the process difficult in some cases. However, even in those cases, the color maps are helpful to visualize the relative location of the stratum sagittale and the embedded OR (Fig. 5).

In most of the cases our fiber reconstruction procedure is based on the choice of a couple of seeds. The first one includes the optic chiasm. The second seed is located into the intraorbital portion of both optic nerves. This protocol will theoretically only show up the fibers related to the visual system. In some cases, the result may be weak in fiber tracking terms. These are the cases in which a third or even a fourth seed can be placed into the lateral geniculate nucleus or even in the sagittal stratum area in a coronal cut [30].

Some intrinsic limitations must be considered: (1) DTI analysis is subjected to a complex mathematical calculation, the acquisition of data and the employed software [27, 31]. Indeed, differences in 3D reconstructions have being found after having used distinct software and models by different teams, but using the same data. Thus, we must be cautious about applying tractography results intraoperatively [6, 27]. In this sense, our strong recommendation is that the neurosurgeon should be involved on the DTI processing. (2) When dealing with distorted or abnormal tracts, particularly in gliomas, the risk a wrong tractography result is augmented, as diffusion features may be altered by the tumor itself. When comparing DTI and direct subcortical stimulation, some studies have shown a positive concordance in 82% of cases. As a matter of repetition, DTI allows the analysis of the white matter tracts anatomy and location, but not necessarily their function [6, 27, 31–36].

Thereupon, despite the progress in fMRI an DTI, intraoperative cortical and subcortical stimulation protocols, still remain as the gold-standard methodology when performing surgery in eloquent structures.

4 Second Contact. Functional Preoperative Tests

The identification of functional areas pre- and intraoperatively through some neurophysiological tests improves their definition to maximize the tumor removal with higher safety by avoiding eloquent areas damage.

Different functional preoperative tests are divided into preoperative and intraoperative techniques, depending on when they are performed [1].

4.1 Preoperative Mapping: Visual Evoked Potentials (VEP)

Regarding the visual areas, the main preoperative mapping test we trust in, are the VEP. After performing some specific visual stimuli (pattern reversal and/or flash stimuli), the electrical potentials generated are transmitted to the visual cortex, which are recorded from the scalp. VEP waveforms are registered on an electroencephalogram (EEG), with the recording electrode usually located on the occipital midline scalp (Oz referenced to Fz according to the International 10/20 System).

VEPs are mainly used to evaluate in an objective way, the anatomo-fuctional integrity of the visual system between the retina and the visual cortical areas into the occipital lobe, eliminating the factor of subjectivity and patient cooperation needed for perimetry. Any disturbance into the visual pathways and visual cortex, will alter the VEP.

There are a series of negative and positive waves in Flash VEP (with leds inserted in googles). The N2 and P2 peaks are the most robust components and appear around 120 ms and 90 msec. P2 amplitude is considered from N2 to P2. Flash VEPs are useful when patients who are unable or unwilling to cooperate (or surgery under general anesthesia).

The utility of VEPs has been largely discussed and agreed that this method may measure the anatomo-functional integrity of the whole optic system more precisely than any imaging technique such as the MRI [37–39].

5 Third Contact. Functional Intraoperative Tests

5.1 Intraoperative Mapping

Contemporary oncological neurosurgery has helped developing and refining some intraoperative mapping techniques, through which neurosurgeons are able to test brain functions with the goal of maximizing the extent of resection and minimizing the postoperative deficits, especially in and around eloquent areas. Some of the tested functions, as sensory and motor can be even monitored in sleep anesthetic conditions. However, others as language and the complex process of the visual function requires its test under awake conditions for a better objective response measurement [1].

5.1.1 Visual Evoked Potentials (VEP)

Despite the already mentioned need for awake conditions when intraoperatively monitoring the visual pathways, VEP can be useful in cases when an awake surgery is difficult to perform for any reason.

Flash VEP are performed by some goggles that release specific stimuli to the patient's retina. Those luminous stimuli are transformed into electrical activity which is finally transferred to the occipital lobe and registered by scalp electrodes. The retinal macular area, especially its central region, is that one with a bigger cortical representation into the occipital lobe. This is the reason why VEP provokes the activation of most of the fibers into this area. Nevertheless, it has not been widely proved the amount of axonal injury that might affect the VEP results and consequently be detected by this test. A consistent and repeated reduction bigger than 50% in the VEP waveform has been advocated as an intraoperative alarm sign.

Despite the improvement in anesthesia techniques and technical advances in the field of VEPs, there are several important limitations to be considered.

First of all, changes in VEP may represent alterations in visual acuity, but not necessarily visual field defects. Therefore, a damage in small areas of the OR might be hidden by normal conduction of the stimulus thought undamaged white matter fibers, keeping the potentials without changes notwithstanding the possible visual field lost.

The intraoperative use of VEP to assess the integrity of the visual pathways, has represented a matter of debate in the last years. Different studies have tried to address an answer to this question, being most of them quite reluctant to the use of scalp electrodes to record the VEP. According to Ota et al., VEPs recorded with subdural grids are much more reliable to monitor the functional integrity of the visual pathways during surgery [40–44].

These strong limitations regarding the integrity of the OR are influencing our standards, moving our protocol to trust in direct cortical and subcortical stimulation when dealing with lesions in their proximity.

5.1.2 Direct Stimulation: Towards the Maximal Oncological Resection

Subcortical stimulation is of great importance when approaching tumors that grew subcortically in the surroundings or in really close contact with important white matter fiber bundles.

After debulking the tumor core, an oncological concept involves extending the resection to the farthest points even including 'non-functional' healthy tissue, towards a supramarginal limit. This goal can be better accomplished in relatively non-functional subcortical areas, but when dealing with tumors in close contact with highly functional tracts, the supramarginal resection might cause unacceptable neurological deficits.

The gold-standard intraoperative tool to reach the goal of a supramarginal resection is the subcortical stimulation. This technique allows neurosurgeons to check the tumor periphery areas during the resection. Therefore, the stimulation of the cavity walls every 3–4 mm through a probe during 2–4 sec, with intensities between 2 and 10 mA has been proved as the ideal method [1].

To achieve this goal, the visual tracts must be checked when in close contact with the tumor itself. Tumors in the 'hot-spot' area defined by the temporo-parieto-occipital junction, require a combination of cortical stimulation, to detect the speech-related areas, and a continuous subcortical mapping to detect the arcuate fascicle and deeply, the OR once approaching the lateral ventricle.

Duffau et al. in 2004, reported for the first time this technique in a case of a low-grade glioma, in which the OR were found in the posteromedial limit by a subcortical stimulation positive response, helping to determine that functional limit. The intraoperative anatomo-functional identification of the OR helped in avoiding a postoperative hemianopia, just causing an 'acceptable quadrantanopia'. Since this first case, very few reports have mentioned this technique [41, 42, 44, 45, 47].

Lately, Gras-Combe et al. Reported a series of cases in whom the OR were detected intraoperatively by subcortical stimulation under awake conditions. All cases presented with a glioma near the visual pathways. The authors concluded that subcortical stimulation of the OR is efficient in avoiding postoperative hemianopsia, considering the quadrantanopia as an acceptable functional result [32, 44].

Recently, Chan- Seng et al. [46], reported their experience based on five cases in which a superior quadrantanopia was achieved, but avoiding hemianopia by intraoperative subcortical stimulation.

In most of these cases, the visual mapping was performed by presenting two objects in a diagonal way over a slide which was divided in four parts or quadrants. This is a kind of modified picture-naming task. Both objects were situated in diagonal as one was placed in the quadrant to be studied and potentially respected, while the other image was located on the opposite quadrant to be used as the control.

The subcortical stimulation casued a transient visual disturbance (shadows, phosphenes, blurred vision or even visual illusions), and thus a difficulty to perceive the objetc located in the chosen quadrant. This fact allowed the surgeon to define a safe margin for resection [23].

The temporo-parieto-occipital junction is considered a hot spot, it is covered by relevant cortical areas as the supramarginal and angular gyri, as well as highly functional white matter tracts, in addition to the OR.

Therefore, our protocol consists of performing a cortical stimulation at the angular and supramarginal gyri. A neuropsychologist or a neurophysiologist usually work on the operating theatre interpreting the results of the stimulation on language mapping in this first stage. A preoperative training is mandatory to be carried out, and this represents a key part of the procedure, as during this training the patient gets familiar to the tests to be performed intraoperatively. Moreover, the protocol is cancelled in the case that the patient fails more than 25% of the presented tasks. Several language tasks are performed during cortical stimulation, such as naming tests (measuring the expressive language), counting, asking the patient read sentences and/or name objects presented on a screen. Some variables as familiarity, age, and education might influence the choice of the tasks to be performed [1]. In our protocol, the most common ones are the Boston and Token's Tests, as they help on choosing pictures of items that the patient can name quickly and easily during the test. We employ a monopolar electrode, applying electrostimulation with a monophasic current intensity between 3–6 mA (250-Hz pulse frequency, 3–8 seconds of tissue contact). A cortical point is considered as eloquent when a 'positive response' happens (speech arrest, paraphasia or speech slowness) at least in three different trains of stimuli on the same point (Fig. 6).

At the subcortical level, the arcuate fascicle is not the only one located in this area. As previously described, the IFOF and the inferior longitudinal fasciculi (ILF) are intimately related with the OR. Indeed, OR run posteriorly in a space just medial and superior to the ILF. The ILF has important functions as reading and visual recognition. Just covering laterally both fascicles, the IFOF is more involved in semantic processing [32, 46]. Therefore, we also perform a subcortical stimulation to control these three fascicles related with language. We normally use the same language tests as for cortical stimulation and employ a monopolar electrode, applying the same parameters that for cortical mapping. Equally, when we detect a transient deficit, either speech arrest, paraphasia or speech slowness, at least three times in the same area, we consider this zone as a functional boundary (Fig. 6).

Once the resection is reaching the OR relative location based on our preoperative DTI tractography study, the subcortical stimulation to detect the optic fibers starts. In order to prevent any potential visual field deficit (hemianopia but even also a quadrantanopia) a new method is proposed. Our protocol uses a modified picture-naming task, with a tactile screen divided into four parts, but with an image or video in each quadrant. The images and videos displayed have been previously shown to the patient, and some of them represent past records of himself. In order to accurately define the perimetry, the patient must focus sight during the procedure at the accurate center of the displayed images, where a small red cross is placed. This fact is of extreme importance, and that is why the patient must be precisely trained preoperatively on this task. Direct stimulation is repeated without the patient being aware of, by a bipolar electrode, with a 5-mm space tips, applying electrostimulation with a biphasic current, 1-msec single-pulse phase duration, 50/60-Hz pulse frequency, intensity between 2 and 4 mA in 1 mA increments (4–10 seconds of tissue contact). Every point is stimulated three times during 10 seconds. The patient was learned to '*draw* shadows and/or to '*point*' phosphenes in the tactile screen. Any subcortical region, where the stimuli provoked a transient visual deficit (shadows, phosphenes, blurred vision or even visual illusions) in any quadrant, was considered as a functional boundary for the tumor resection [32, 45] (Fig. 6).

A relevant aspect during this 'intraoperative campimetry' remains on the fact that the patients should always focus on the center of the image.

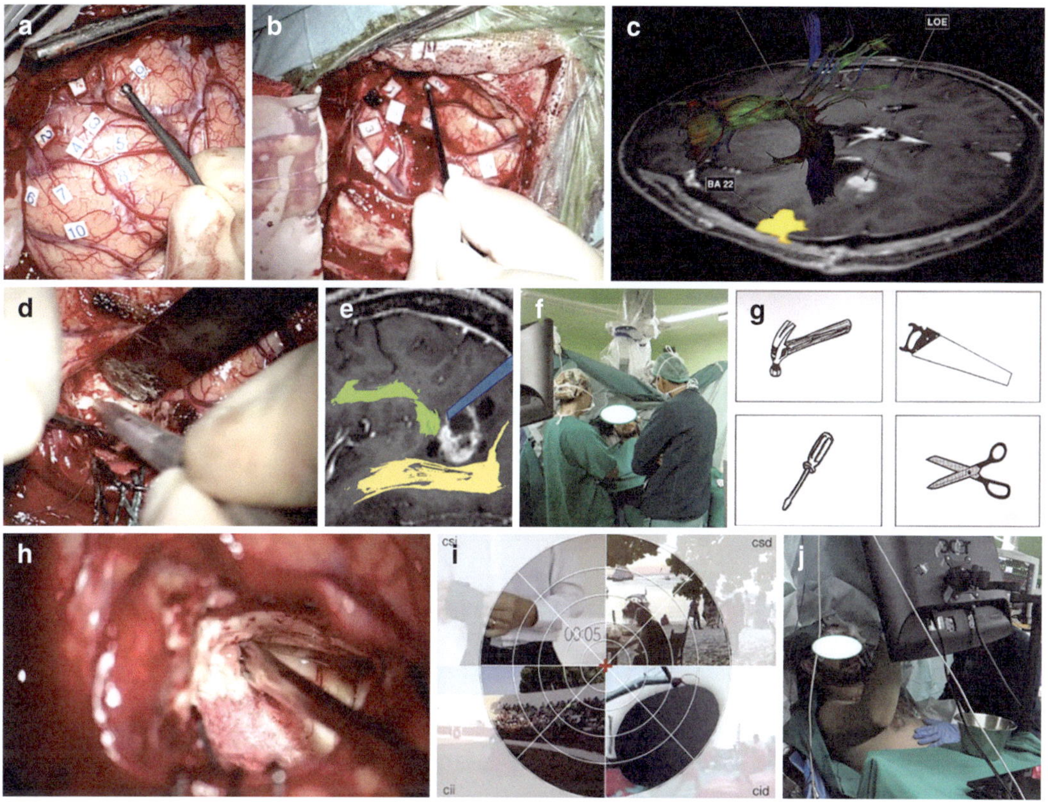

Fig. 6 Progressive steps during a left periatrial glioma surgery. (**a**) Speech mapping after dural opening. (**b**) Supramarginal gyrus stimulation (area labelled as 2) causing anomia to the awake patient, and matching quite accurately with the orange area into the fMRI (**c**). (**d**) Ultrasonic aspirator direct subcortical stimulation causing a speech arrest during the anterior pole of the tumor resection. (**e**) T1-weigthed GD enhanced sagittal cut showing the periatrial tumor itself as well as the arcuate fascicle (green) and inferior longitudinal fascicle (yellow). The blue arrow is pointing the area where the arrest happened. (**f**) Intraoperative view of the operating room setup. Note the semisitting position in an effort to gain patient's comfort. (**g**) Boston Naming test slide used to hold the subcortical stimulation speech tasks. (**h**) Intraoperative subcortical stimulation into the periatrial region. Optic radiations stimulation causing a right homonymous inferior quadrantanopia (**i**), and how the patient is perceiving the deficit, drawing it by himself into the tactile campimetry screen (**j**)

Therefore, any response found when the patient is not focusing on this center, is considered as invalid and should be repeated. In previous reports, this eye movements were controlled by a neuropsychologist, who is just reminding the patient to focus on the red cross. To check the eye movement in every moment and to avoid mistakes during the visual field evaluation, we have implemented an eye tracker system (Tobii Pro Glasses 2, Tobii) designed to capture what a person is looking at in real time, providing a "true view" of what users see wirelessly. The system integrates four eye cameras for gaze control (2 cameras per eye), a Gyroscope, an accelerometer and a wide-angle HD scene camera. Furthermore, it allows for synchronization with a broad spectrum of physiological data. In the case of eye movement away from the center of the image, the point should be reevaluated again (Figs. 7 and 8).

Once the OR have been functionally identified into the periventricular area, the resection is extended towards this functional limit, helping to extend the tumor removal while diminishing the chances to have a visual postoperative problem (Fig. 9).

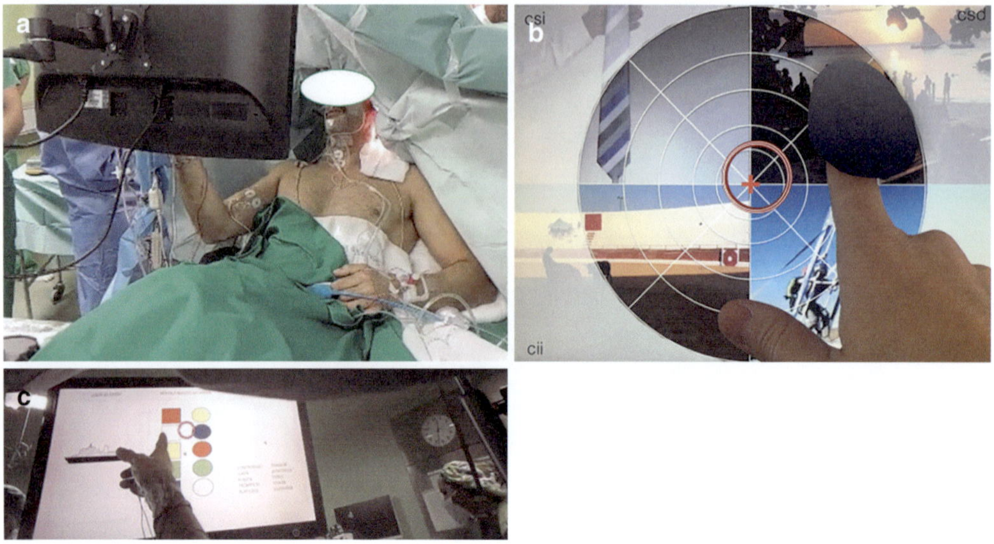

Fig. 7 (**a**) Intraoperative pictures during optic radiations stimulation. (**b**) An apparent deficit into the right superior quadrant is drawn by the patient himself, while the optic radiations are being stimulated, thus causing a 'shadow' as described by the patient, who is surely looking at the center of the image as the red circle is showing. (**c**) Snapshot taken from the patient's googles

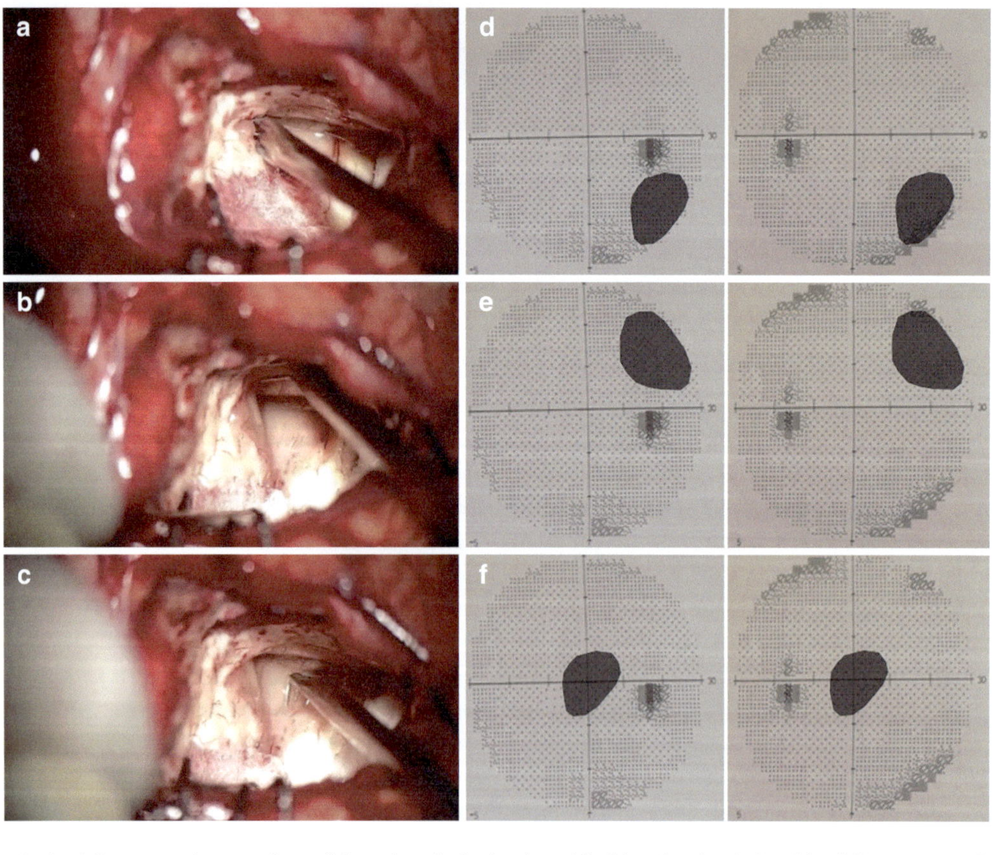

Fig. 8 (**a-c**) Intraoperative snapshots of the subcortical stimulator blocking the visual signal in different parts of the visual pathway, and (**d-f**) their correspondent 'shadows' described and drawn by the patient into the perimetry

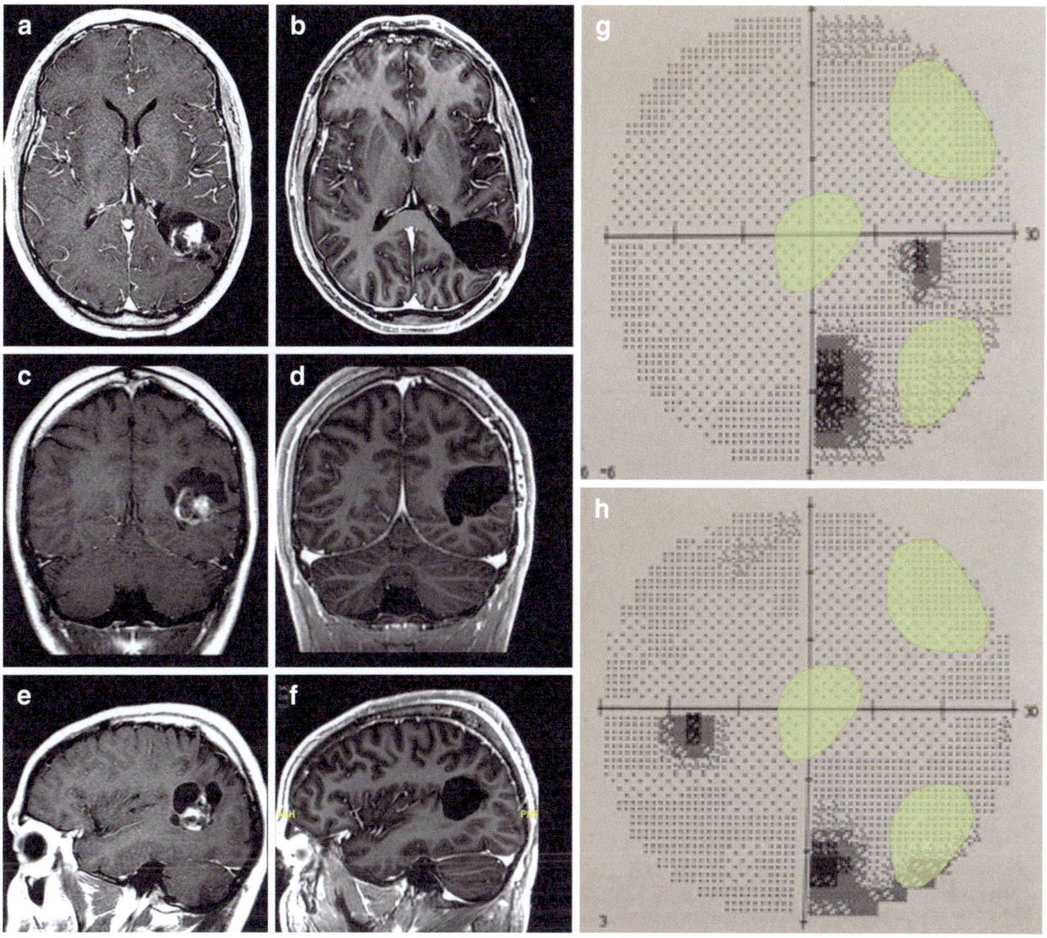

Fig. 9 (**a-f**) T1-weigthed GD enhanced axial, coronal and sagittal cuts showing the preoperative (first column) and postoperative (second column) images showing a complete resection of the tumor. (**g-h**) Postoperative campimetry showing a minor deficit into the right inferior quadrant, as well as the areas where the patient detected 'shadow' during the stimulation

6 Conclusions

Vision is one of the most important and highly developed human functions. Indeed, it strongly helps to defines us as human beings. It helps us to interact with our environment, understand and integrate what is surrounding us, and probably and most importantly it lets us creating social relationships and enjoying our human life. Therefore, identifying and preserving the visual pathways during brain tumor surgery, should be a must, and any technology helping to reach this goal should be tested and implemented.

References

1. González-Darder JM, González-López P. Management of Brain Tumors in eloquent areas. In: Abujamra AL, editor. Diagnostic techniques and surgical management of brain tumors. InTech; 2011. https://doi.org/10.5772/21796.
2. Ferracci FX, Duffau H. Improving surgical outcome for gliomas with intraoperative mapping. Expert Rev Neurother. 2018;18(4):333–41. https://doi.org/10.1080/14737175.2018.1451329. Epub 2018 Mar 20
3. Duffau H. Preserving quality of life is not incompatible with increasing overall survival in diffuse low-grade glioma patients. Acta Neurochir. 2015;157:165–7.
4. Soffietti R, Baumert BG, Bello L, Von Deimling A, Duffau H, Frénay M, Grisold W, Grant R, Graus F,

Hoang-Xuan K, Klein M, Melin B, Rees J, Siegal T, Smits A, Stupp R, Wick W. Guidelines on management of low-grade gliomas: report of an EFNS-EANO task force. Eur J Neurol. 2010 Sep;17(9):1124–33. https://doi.org/10.1111/j.1468-1331.2010.03151.x.
5. Duffau H, Mandonnet E. The "onco-functional balance" in surgery for diffuse low-grade glioma: integrating the extent of resection with quality of life. Acta Neurochir. 2013;155(6):951–7. https://doi.org/10.1007/s00701-013-1653-9. Epub 2013 Feb 28
6. Duffau H. Surgery for diffuse low-grade gliomas (DLGG) functional Ouctomes. In: Duffau H, editor. Diffuse low-grade gliomas in adults. Springer; 2013. p. 497–533.
7. Gogos AJ, Young JS, Morshed RA, Hervey-Jumper SL, Berger MS. Awake glioma surgery: technical evolution and nuances. J Neuro-Oncol. 2020;147(3):515–24. https://doi.org/10.1007/s11060-020-03482-z. Epub 2020 Apr 8
8. Bartholow R. Experimental investigations into the functions of the human brain. Am J Med Sci. 1874;7:305–13.
9. Penfeld W, Boldrey E. Somatic motor and sensory representation in the cerebral cortex of man as studied by electrical stimulation. Brain. 1937;60(4):389–443. https://doi.org/10.1093/brain/60.4.38919.
10. Whitaker HA, Ojemann GA. Graded localisation of naming from electrical stimulation mapping of left cerebral cortex. Nature. 1977;270(5632):50–1. https://doi.org/10.1038/270050a0.
11. Berger MS, Kincaid J, Ojemann GA, Lettich E. Brain mapping techniques to maximize resection, safety, and seizure control in children with brain tumors. Neurosurgery. 1989;25(5):786–92. https://doi.org/10.1097/00006123-198911000-00015.
12. Sincoff EH, Tan Y, Abdulrauf SI. White matter fiber dissection of the optic radiations of the temporal lobe and implications for surgical approaches to the temporal horn. J Neurosurg. 2004;101(5):739–46. https://doi.org/10.3171/jns.2004.101.5.0739.
13. Peltier J, Travers N, Destrieux C, Velut S. Optic radiations: a microsurgical anatomical study. J Neurosurg. 2006;105(2):294–300. https://doi.org/10.3171/jns.2006.105.2.294.
14. Shah A, Goel A, Jhawar SS, Patil A, Rangnekar R, Goel A. Neural circuitry: architecture and function - a Fiber dissection study. World Neursurg. 2019;125:eg20–e638.
15. Shah A, Jhawar SS, Goel A. Letter to the editor: optic radiations and anterior commissure. J Neurosurg. 2015;123(3):824–6.
16. Peuskens D, van Loon J, Van Calenbergh F, van den Bergh R, Goffin J, Plets C. Anatomy of the anterior temporal lobe and the frontotemporal region demonstrated by fiber dissection. Neurosurgery. 2004;55(5):1174–84. https://doi.org/10.1227/01.neu.0000140843.62311.24.
17. Mahaney KB, Abdulrauf SI. Anatomic relationship of the optic radiations to the atrium of the lateral ventricle: description of a novel entry point to the trigone. Neurosurgery. 2008;63(4 Suppl 2):195–202; discussion 202–3. https://doi.org/10.1227/01.NEU.0000313121.58694.4A.
18. Fernández-Miranda JC, Rhoton AL Jr, Alvarez-Linera J, Kakizawa Y, Choi C, de Oliveira EP. Three-dimensional microsurgical and tractographic anatomy of the white matter of the human brain. Neurosurgery. 2008;62(6 Suppl 3):989–1026; discussion 1026-8. https://doi.org/10.1227/01.neu.0000333767.05328.49.
19. Türe U, Yaşargil MG, Friedman AH, Al-Mefty O. Fiber dissection technique: lateral aspect of the brain. Neurosurgery. 2000;47(2):417–26; discussion 426–7. https://doi.org/10.1097/00006123-200008000-00028.
20. Yasargil MG, Türe U, Yasargil DC. Impact of temporal lobe surgery. J Neurosurg. 2004;101(5):725–38. https://doi.org/10.3171/jns.2004.101.5.0725.
21. Maldonado IL, Destrieux C, Ribas EC, de Abreu S, Brito Guimarães B, Cruz PP, Duffau H. Composition and organization of the sagittal stratum in the human brain: a fiber dissection study. J Neurosurg. 2021;8:1–9. https://doi.org/10.3171/2020.7.JNS192846. Epub ahead of print
22. Kiviranta P. Mapping the visual field: an empirical study on the user experience benefits of gaze-based interaction in visual field testing. [MDP in human-technology interaction, University of Tampere] 2017. http://urn.fi/URN:NBN:fi:uta-201710092561
23. Wolfson R, Soni N, Shah AH, Hosein K, Sastry A, Bregy A, Komotar RJ. The role of awake craniotomy in reducing intraoperative visual field deficits during tumor surgery. Asian. J Neurosurg. 2015;10(3):139–44. https://doi.org/10.4103/1793-5482.161189.
24. Bartos R, Jech R, Vymazal J, Petrovický P, Vachata P, Hejcl A, et al. Validity of primary motor area localization with fMRI versus electric cortical stimulation: a comparative study. Acta Neurochir. 2009;151:1071–80.
25. Petrovich N, Holodny AI, Tabar V, Correa DD, Hirsch J, Gutin PH, et al. Discordance between functional magnetic resonance imaging during silent speech tasks and intraoperative speech arrest. J Neurosurg. 2005;103:267–74.
26. Kuchcinski G, Mellerio C, Pallud J, Dezamis E, Turc G, Rigaux-Viodé O, et al. Three-tesla functional MR language mapping: comparison with direct cortical stimulation in gliomas. Neurology. 2015;84:560–8.
27. Duffau H. Awake mapping and tumor surgery. In: Duffau H, editor. Brain mapping. Springer; 2011. https://doi.org/10.1007/978-3-7091-0723-2_24.
28. Kollias SS. Functional magnetic resonance imaging of the human visual system. In: Baert AL, Sartor K, Müller-Forell WS, editors. Imaging of orbital and visual pathway pathology. Medical radiology (diagnostic imaging). Berlin, Heidelberg: Springer; 2006. https://doi.org/10.1007/3-540-27989-X_4.
29. Dumoulin S. Functional MRI of the visual system. In: Uludag K, Ugurbil K, Berliner L, editors. fMRI: from nuclear spins to brain functions. Biological mag-

netic resonance, vol 30. Boston, MA: Springer; 2015. https://doi.org/10.1007/978-1-4899-7591-1_15.
30. Hana A, Husch A, Gunness VR, Berthold C, Hana A, Dooms G, Boecher Schwarz H, Hertel F. DTI of the visual pathway–white matter tracts and cerebral lesions. J Vis Exp. 2014;90:51946. https://doi.org/10.3791/51946.
31. Feigl GC, Hiergeist W, Fellner C, Schebesch KM, Doenitz C, Finkenzeller T, et al. Magnetic resonance imaging diffusion tensor tractography: evaluation of anatomic accuracy of different fiber tracking software packages. World Neurosurg. 2014;81:144–50.
32. Gras-Combe G, Moritz-Gasser S, Herbet G, Duffau H. Intraoperative subcortical electrical mapping of optic radiations in awake surgery for glioma involving visual pathways. J Neurosurg. 2012;117(3):466–73. https://doi.org/10.3171/2012.6.JNS111981. Epub 2012 Jul 13
33. Pujol S, Wells W, Pierpaoli C, Brun C, Gee J, Cheng G, et al. The DTI challenge: toward standardized evaluation of diffusion tensor imaging tractography for neurosurgery. J Neuroimaging. 2015;25:875–82.
34. Kinoshita M, Yamada K, Hashimoto N, Kato A, Izumoto S, Baba T, et al. Fiber-tracking does not accurately estimate size of fiber bundle in pathological condition: initial neurosurgical experience using neuronavigation and subcortical white matter stimulation. NeuroImage. 2005;25:424–9.
35. Leclercq D, Duffau H, Delmaire C, Capelle L, Gatignol P, Ducros M, Chiras J, Lehéricy S. Comparison of diffusion tensor imaging tractography of language tracts and intraoperative subcortical stimulations. J Neurosurg. 2010;112:503–11.
36. Duffau H. The dangers of magnetic resonance imaging diffusion tensor tractography in brain surgery. World Neurosurg. 2014;81:56–8.
37. Creel DJ. Visually evoked potentials. Handb Clin Neurol. 2019;160:501–22. https://doi.org/10.1016/B978-0-444-64032-1.00034-5.
38. Visually CD, Potentials E. In: Kolb H, Fernandez E, Nelson R, editors. Webvision: the organization of the retina and visual system [internet]. Salt Lake City (UT): University of Utah Health Sciences Center; 2012. p. 1995.
39. Sharma R, Joshi S, Singh KD, Kumar A. Visual evoked Potentials: normative values and gender differences. J Clin Diagn Res. 2015;9(7):CC12–5. https://doi.org/10.7860/JCDR/2015/12764.6181. Epub 2015 Jul 1
40. Fountas K, Kapsalaki EZ. Surgical management of occipital gliomas: a practical atlas. In: Fountas K, Kapsalaki EZ, editors. Epilepsy surgery and intrinsic brain tumor surgery. Springer; 2019.
41. Verst SM, et al. Awake surgery versus VEP in tumors of visual pathway: case report. Interdisciplinary Neurosurg. 2020;20:100675.
42. Shahar T, Korn A, Barkay G, Biron T, Hadanny A, Gazit T, Nossek E, Ekstein M, Kesler A, Ram Z. Elaborate mapping of the posterior visual pathway in awake craniotomy. J Neurosurg. 2018;128(5):1503–11. https://doi.org/10.3171/2017.2.JNS162757. Epub 2017 Aug 25
43. Ota T, Kawai K, Kamada K, Kin T, Saito N. Intraoperative monitoring of cortically recorded visual response for posterior visual pathway. J Neurosurg. 2010;112:285–94.
44. Steňo A, Hollý V, Fabian M, Kuniak M, Timárová G, Steňo J. Direct electrical stimulation of the optic radiation in patients with covered eyes. Neurosurg Rev. 2014;37(3):527–33. https://doi.org/10.1007/s10143-014-0535-9. discussion 533. Epub 2014 Feb 28
45. Duffau H, Velut S, Mitchell MC, Gatignol P, Capelle L. Intra-operative mapping of the subcortical visual pathways using direct electrical stimulations. Acta Neurochir. 2004;146(3):265–9. https://doi.org/10.1007/s00701-003-0199-7. discussion 269–70. Epub 2004 Jan 8
46. Chan-Seng E, Moritz-Gasser S, Duffau H. Awake mapping for low-grade gliomas involving the left sagittal stratum: anatomofunctional and surgical considerations. J Neurosurg. 2014;120(5):1069–77. https://doi.org/10.3171/2014.1.JNS132015. Epub 2014 Jan 31
47. Nguyen HS, Sundaram SV, Mosier KM, Cohen-Gadol AA. A method to map the visual cortex during an awake craniotomy. J Neurosurg. 2011;114(4):922–6. https://doi.org/10.3171/2010.11.JNS101293. Epub 2011 Jan 14

Visualization and Characterization of Glioma and Brain Activity to Maximize Safe Tumor Resection in Awake Surgery

Hirokazu Takami

1 Introduction

Studies have shown that the extent of resection (EOR) is positively correlated with prognosis; therefore, maximal safe resection is the key to favorable treatment outcomes in patients with diffuse glioma, regardless of tumor grades [1–5]. Maximal EOR translates into improved prognosis, reduces the risk of malignant transformation of low-grade gliomas (LGG) and also achieves better seizure control [2, 6, 7]. However, difficult-to-access tumors, anatomically important structures surrounding the tumor, and diffuse infiltrative gliomas that cannot easily be distinguished from surrounding healthy tissues limit the EOR. Usually, surgeons overestimate the resectability of gliomas preoperatively [8]. Postoperative adverse neurological effects negatively affect patient survival [9]. Therefore, maximal safe resection is paramount for favorable outcomes.

Differentiation of normal brain parenchyma from tumor tissue using white light is often challenging even for experienced neurosurgeons, based on subtle differences in appearance including color and bleeding characteristics and tactile features. Therefore, glioma resection primarily involves image-guided surgery, either based on preoperative or a combination of pre- and intraoperative imaging, with the goal of resecting the tumor depicted as imaging changes, including fluid-attenuated inversion recovery (FLAIR)-high areas for non-enhancing gliomas or enhancing areas in T1-weighted gadolinium-enhanced magnetic resonance imaging (MRI). Preoperative imaging modalities include MRI, magnetic resonance spectroscopy, and positron emission tomography (PET). Neuronavigation using pre- and intraoperative MRI enables accurate tumor localization. Intraoperative imaging modalities include ultrasonography, intraoperative MRI, fluorescence including 5-aminolevulinic acid (5-ALA), and cutting-edge technological adjuncts for intraoperative histological diagnosis. Combinations of existing and advanced technology are useful for accurate tumor identification.

In addition to optimal visualization of tumors, clear imaging of surrounding brain functions is important. Preoperative techniques include diffusion tensor imaging (DTI), functional MRI (fMRI), magnetoencephalography (MEG), and transcranial magnetic stimulation (TMS). Furthermore, intraoperative cortical and subcortical electrical stimulation to localize brain functions during awake craniotomy enables identification of functional areas and facilitates real-time monitoring of the patient's neurological status to avoid injury to critical structures.

A balance between tumor resection and function preservation is the cornerstone of glioma

H. Takami (✉)
Department of Neurosurgery, Faculty of Medicine, The University of Tokyo Hospital, Tokyo, Japan
e-mail: takami-tky@umin.ac.jp

A. Shah et al. (eds.), *Functional Anatomy of the Brain: A View from the Surgeon's Eye*,
https://doi.org/10.1007/978-981-99-3412-6_8

surgery. Recent studies have focused on resection beyond the tumor, including the tissue surrounding the tumor; this approach is referred to as "supra-total" resection. A FLAIR-documented abnormality surrounding an enhancing lesion is shown to contain tumor cells in high-grade gliomas (HGG) [10, 11]. Supra-total resection is shown to have a positive effect on progression-free survival (PFS) in LGG [12]. Reportedly, resection of areas that show FLAIR-documented abnormalities (referred to as "FLAIRectomy") prolongs overall survival (OS) in patients with HGG, and extensive resection of FLAIR abnormalities (>53%) is known to result in better OS (median 20.7 months) compared with outcomes in patients who do not undergo this treatment (15.5 months, $p < .01$) [13]. Another study, which showed a similar outcome, has corroborated the finding of OS prolongation following FLAIRectomy [14]. Supra-total resection includes extension of surgical resection until eloquent areas are encountered to ensure function preservation.

Technological advances and availability of innovative equipment have enabled better visualization of tumors as well as accurate identification and preservation of critical anatomical structures to ensure maximal safe resection of brain tumors [15]. Application of the techniques is necessary for safe maximization of the EOR, regardless of surgeries performed in primary or recurrent cases [16]. This chapter outlines the available and prospective technologies and modalities for glioma surgery and illustrates their usefulness in preoperative planning and intraoperative guidance/decision-making for maximization of safe tumor resection. We have additionally presented a literature review, as well as the limitations of these technologies.

1.1 White Matter Fiber Visualization

Imaging technologies that aid with visualization of white matter fibers are an indispensable component of surgical planning for accurate interpretation of the anatomical and spatial relationship between tumors and the surrounding functional fibers. DTI is the standard method used for this purpose, based on a deterministic algorithm that utilizes diffusion-weighted MRI. This technology tracks eigenvectors based on the Brownian movement of the water molecule, which parallels the white matter fiber orientation [17]. With regard to accuracy of this method, studies have verified that the distance between DTI-documented corticospinal tracts and those observed on direct subcortical electrical stimulation was 8.7 ± 3.1 mm [18] or a maximum distance of 8 mm [19], which suggests that a 9–10 mm margin may be considered safe for imaging-guided tumor resection. A randomized controlled study that delineated the corticospinal tracts with and without the use of DTI in 238 patients with glioma showed that DTI application improved the gross total resection (GTR) rate (74 vs. 33%, $p < .001$), decreased the postoperative motor deficit (15 vs. 33%, $p < .001$), improved the 6-month Karnofsky performance scores (86 vs. 74, $p < .001$), and prolonged OS (21 vs. 14 months, $p = .048$) in the HGG group [20].

Following are the limitations of this technology: (a) The method is not useful for delineation of multiple crossing, kissing, or branching fibers that course in different directions in any one region, as well as for detection of curved fibers. (b) Placement of the region of interest (ROI) is subjective and the tracking algorithm is variable, which may lead to heterogeneity in image reconstruction depending on the technical skill of observers. A study in which 8 different institutions constructed the corticospinal tracts on the same images showed significant inter-algorithm variability, which indicates limited utility of DTI in clinical practice [21].

High angular resolution diffusion-weighted imaging with a q-ball algorithm represents an alternative method for visualization of white matter fibers. This technique is useful to resolve intravoxel white matter fiber architecture and can therefore accurately delineate multiple fibers coursing in different directions in the same region [22]. However, the reliability of this technology remains currently unverified, and image reconstruction requires high-level expertise (Fig. 1).

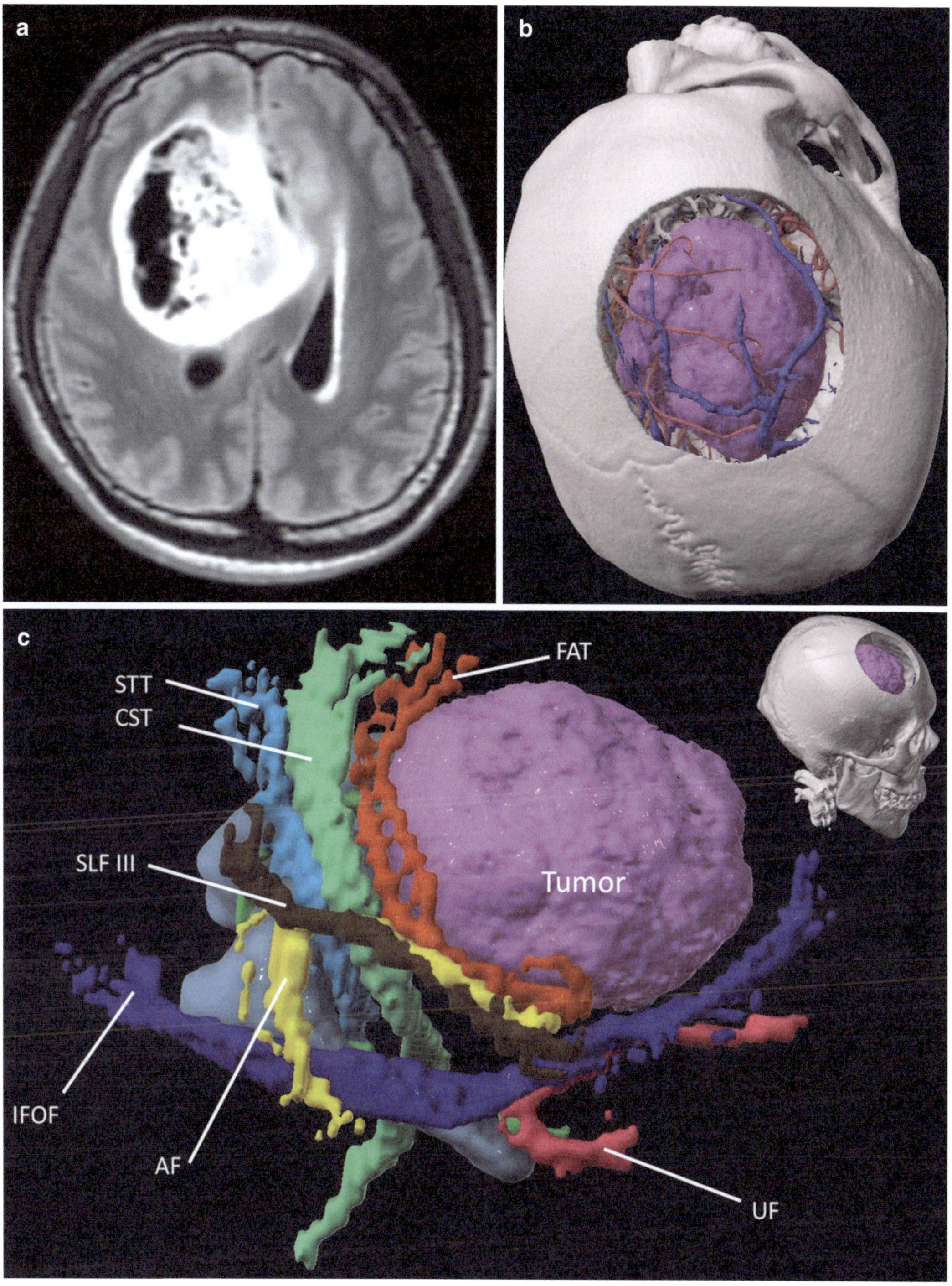

Fig. 1 A case of right frontal diffuse glioma (grade 3). FLAIR images (**a**), three-dimensional reconstructed images combining MRI and CT images with planned craniotomy site information (**b**), and fiber visualization by q-ball algorithm enabled the clear interpretation of the relationship between tumor and surrounding white matter fibers (**c**)

1.2 Functional Magnetic Resonance Imaging (Task-Based and Resting-State)

Notably, fMRI detects changes in the oxyhemoglobin-deoxyhemoglobin level, referred to as the blood oxygen level-dependent (BOLD) signal, which serves as a surrogate marker of neuronal activity [23]. This noninvasive technique identifies the cortex associated with task-related activities, such as motor function and language.

Multiple tasks which can be used in the MRI room, include motor tasks like finger-tapping and language tasks including sentence comprehension [24] and verb generation [25], among others. Changes in the MRI signal are usually subtle; therefore, multiple steps are necessary to improve sensitivity and specificity. Data processing is necessary for complete image reconstruction following image acquisition.

Based on direct cortical stimulation (DCS) mapping results, fMRI showed sensitivity and specificity of 71–100% and 68–87%, respectively, for motor tasks [26–28] and 59–80% and 68–97%, respectively, for language tasks [26, 27, 29]. The wide range observed in sensitivity and specificity is attributable to the heterogeneity of the tasks and imaging protocols adopted by the aforementioned studies [30].

A fMRI uses the BOLD signal, which is affected by blood flow changes. Therefore, tissues in close proximity to a tumor may be affected by neurovascular uncoupling; eloquent areas in the vicinity of the tumor may not increase the BOLD signal and lead to false-negative results [26].

The resting-state fMRI (rsfMRI) is similar to a task-based fMRI with regard to detection of a cortical BOLD signal; however, it does not require the patient's cooperation during imaging acquisition. The method involves high temporal correlation of the BOLD signal change in multiple associated areas, which are identified as functional connectivity [31]. Data can be acquired within a short period ranging between 6 and 12 min while patients are awake or asleep. The optimal manner of data processing remains debatable.

A study that compared rsfMRI and DCS in 98 patients who underwent both sensorimotor and language function evaluation showed that DCS elicited a positive response in 96 ± 11% within 10 mm, and 84 ± 24% within 5 mm of motor area identified by rsfMRI. The accuracy was inferior for language, and the positive response from electrical stimulation was 92 ± 21% within 10 mm of positive area identified rsfMRI and 70 ± 41% within 5 mm [32]. Intraoperative rsfMRI to localize the motor area showed high accuracy verified by DCS, with sensitivity and specificity of 62 and 94%, respectively [33].

Task-based and rsfMRI are complementary methods to compute the probable eloquent areas preoperatively. Their efficacy is limited owing to low statistical power, variability in data analysis, software errors, and lack of direct replication, among other such factors [34]. DCS is the gold standard for intraoperative mapping of the brain functions.

1.3 Magnetoencephalography

MEG detects cortical functional areas based on determination of the electroneurophysiological activity of the neuronal synapses in association with brain activity, which causes changes in the magnetic field [35]. Visualization requires registration of the data using a three-dimensional (3D) MRI sequence. This represents task-based imaging performed preoperatively and detects cortical areas associated with motor, language, visual, and auditory functions.

A study that used intraoperative DCS to localize the central sulcus showed that MEG was superior to fMRI [36], and two small studies have also shown its usefulness to localize motor function areas [37, 38].

Furthermore, MEG-based images combined with high-resolution MRI can be utilized for surgical planning to determine the optimal approach [39]. A study performed to locate functionally active areas in 106 patients with gliomas reported that tumor involved these areas in 17–18% of patients with grade 2–3 lesions and in only 8% of patients with grade 4 lesions [40]. MEG can predict tumor resectability in patients in whom the surrounding brain shows decreased functional connectivity with other areas [41].

Although MEG is a useful tool in preoperative prediction of functional areas, availability of this modality, technical expertise for accurate interpretation of results, and high costs limit its application in clinical practice.

1.4 Transcranial Magnetic Stimulation

TMS is based on stimulation/inhibition of the brain cortex through creation of a magnetic field that passes through the scalp, skull, and dura, which generates a cortical current, followed by neuronal activation. TMS combined with neuronavigation (navigated TMS [nTMS]) enables visualization of the cortical functional areas [42].

A study that compared nTMS with MEG and intraoperative DCS for detection of motor areas reported that the distance between the nTMS- and DCS-documented areas was 2.13 ± 0.29 mm and that between the nTMS and MEG-documented areas was 4.71 ± 1.08 mm, which suggests superiority of nTMS over MEG for identification of the motor cortex [43]. Another study also showed a comparable accuracy margin between the nTMS-defined and DCS-defined motor areas (5.7 ± 4.6 mm in newly diagnosed patients and 6.0 ± 6.0 mm in patients with recurrent glioma) [44]. A study that compared MEG in 250 patients vs. matched historical controls demonstrated that the GTR rate increased from 42 to 59% ($p < .05$), and the median PFS improved from 15.4 to 22.4 months in the LGG group ($p < .05$). The study also showed that preoperative nTMS expands the surgical indications by reducing the probability of inaccurate identification of the motor eloquent areas [45].

The advantage of this method is that it is robust against tumor neurovascular uncoupling and does not require the patient's cooperation. TMS data can be combined with DTI tractography; the ROI is selected as the nTMS region in contrast to manually selected ROI to structure DTI for greater reliability [46]. In a study for visualization of language-related fiber tracts, nTMS-based DTI was observed to be superior to a ROI-based protocol with 76 vs. 40%, respectively [47]. Combination of TMS with fMRI for language mapping showed stronger correlation with DCS compared with use of these techniques alone, with sensitivity of 98% and specificity of 83% to overcome the low specificity of TMS [48].

The limitation of this technology is the need for experience and expertise and lack of easy availability of the device.

1.5 Three-Dimensional Imaging

Notably, 3D imaging facilitates integration of multiple imaging modalities including MRI, PET, DTI, and functional MRI and is therefore useful for accurate preoperative analysis of anatomical relationships between a tumor and the surrounding normal structures. This information provides clarity regarding anatomical structures and aids with surgical decision-making to adopt the optimal strategy, plan intraoperative electrical stimulation and avoid inadvertent injury to peritumoral functional structures (Fig. 1B-C). Tablet application allows manual rotation of images, and surgeons can add or remove parts of structures such as the cerebral cortex, vasculature, and the tumor to clearly visualize deep-seated white matter fibers, pre- and intraoperatively (personal communication), which provides a practical update for interpretation of structural localization of the tumor and the brain and can achieve greater surgical efficiency.

1.6 Electrophysiological Monitoring

Electrophysiological mapping (brain mapping) allows real-time continuous intraoperative feedback of functional status depending on the proximity of the tumor to eloquent areas. Electrical stimulation is performed using monopolar or bipolar stimulators with variable electrical currents placed over cortical and subcortical regions.

Electrical stimulation causes local excitation or inhibition, which elicits functional responses [49]. Brain mapping enables localization of functional anatomical structures at both the cortical and white matter levels. This technique also highlights less functional areas that may safely be sacrificed to approach the tumor and delineates the EOR at the tumor-normal tissue margin. Tumor removal as close as possible to the normal functional structures is preferred to ensure a balance between function preservation and increased EOR [50]. The use of mapping for safe but aggressive tumor resection is shown to increase the EOR in cases of LGG [51]. Mapping efficacy is impaired in patients with preoperative neurological deficits. Functional brain mapping during awake surgery is affected by the patient's cooperation, which is associated with age, and preoperative performance status, among other factors [52].

1.7 Neuronavigation

The neuronavigation system is a ubiquitous tool available in neurosurgery operating theaters. This device can incorporate multiple imaging scans including computed tomography images, various MRI sequences (such as DTI and fMRI) and PET images, among others and is therefore useful to preoperatively plan the surgical approach, as well as for intraoperative confirmation of the neuroanatomy and tumor localization [53]. A study that included 104 patients with glioblastoma (GBM), who underwent resection with and without neuronavigation support reported a higher GTR rate (31 vs. 19%, nonsignificant), lower residual volume ($p = .01$), and longer survival time (18.3 vs. 10.3 months, $p < .0001$) in the navigation group [54].

Precision achieved at the time of registration is usually <3 mm, depending on the patient's position. Baseline MRIs performed after fixation of the patient's head improves accuracy in patients in whom intraoperative MRI is available [55]. Opening the dura is associated with loss of accuracy, secondary to a brain shift caused by cerebrospinal fluid (CSF) drainage, mannitol administration, and tumor resection, among other factors, which serves as a major drawback of neuronavigation. This limitation can be overcome using intraoperative MRI (described later in this chapter).

1.8 Ultrasonography

Intermittent ultrasonography reflects real-time visualization of tumors for intraoperative identification of a residual tumor and is unaffected by navigation inaccuracy caused by brain shift [56]. Furthermore, this is a cost-effective and widely-available tool. Calcified lesions, such as oligodendrogliomas are more easily visualized ultrasonographically. Ultrasonography can detect possible intraoperative complications such as intraparenchymal hematomas (Fig. 2).

Ultrasound-guided surgery was shown to be associated with better prognosis, with 1- and 2-year OS of 59 and 33%, respectively, compared with 43 and 13%, respectively, in cases of HGG in the historical control group in which ultrasonography was not performed ($p < .05$) [57]. Other studies have corroborated the beneficial effect of intraoperative ultrasonography in prolonging PFS and OS in patients with GBM [58, 59].

Color Doppler ultrasonography provides information regarding the peritumoral vasculature, including tumor-feeding arteries, which enables prompt initiation of measures to decrease blood flow to the tumor. Contrast-enhanced ultrasonography involves intravenous administration of a contrast agent to visualize the vasculature [60]. Compared with a conventional curved or sector array ultrasound transducer, high-frequency linear probes provide higher tissue resolution, with a higher rate of residual tumor detection [61]. A study has reported higher sensitivity of high-frequency linear probes for detection of GBM after attempted resection compared with intraoperative MRI and conventional ultrasonography (76, 55, and 24%, respectively). Notably, no significant difference was observed in specificity compared with intraoperative MRI (58 and 74%, respectively), although specificity was lower than that of conventional ultrasonography (96%) [62].

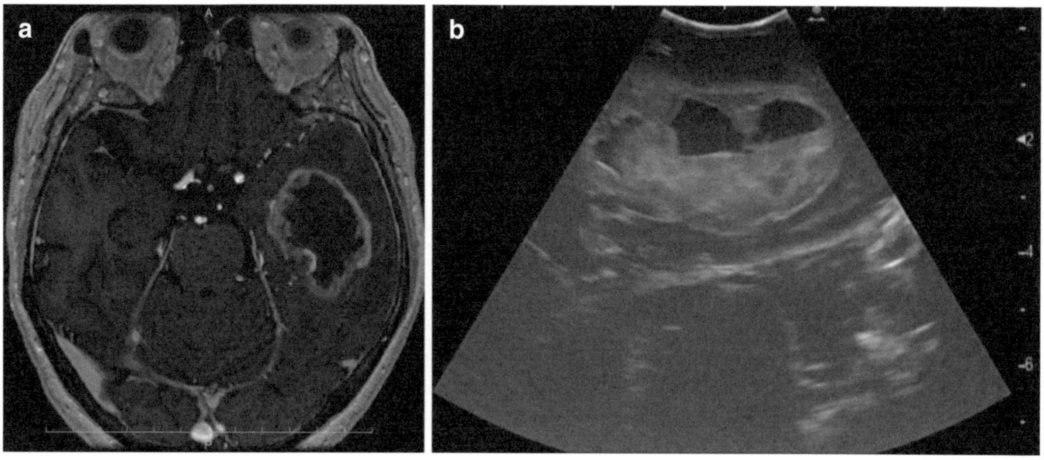

Fig. 2 Left temporal glioblastoma on T1-weighted gadolinium-enhanced MRI image (**a**) and the intraoperative ultrasonography image before resection (**b**). Ultrasonography distinguishes necrotic tissue and intratumoral fluid collection as in this case

Intraoperative images can be combined with preoperative MRI to update the neuronavigation system to compensate for brain shifts [63, 64].

Compared with neuronavigation, ultrasonography requires greater experience for image interpretation, and distinguishing between the residual tumor and injured brain tissue, edema, or hemostatic materials attached to the resection cavity is often challenging [65]. Ultrasonography should be cautiously performed and interpreted in cases of recurrent HGG, because irradiation obscures the ultrasonographic appearance of tissues [66]. Tumor resection based solely on ultrasonographically documented high-echoic signals can lead to inadvertent neurological sequelae [67].

1.9 Illuminating Probes and Fluorescence-Guided Surgery

Fluorescence aids with intraoperative glioma visualization, and fluorescence-navigated surgery is referred to as fluorescence-guided surgery (FGS). Indocyanine green (ICG), fluorescein sodium (FS), and 5-aminolevulinic acid (5-ALA) are among the several fluorophores used in routine clinical practice or preclinical studies [68]. Compared with other surgical adjuncts such as intraoperative MRI or neuronavigation, fluorescence is a more feasible tool that can be used in routine practice as a relatively low-cost option, provided the drugs are approved by national regulations [69]. Fluorescence labeling is completely unaffected by brain shift, which is another advantage it offers over image-guided surgery. Additionally, fluorescence is useful to identify the tumorous portions in samples prepared for histopathological evaluation and diagnosis or for preservation of samples for later clinical trials or research.

1.9.1 5-Aminolevulinic Acid

5-ALA is a natural biochemical precursor of hemoglobin that promotes synthesis and accumulation of intracellular fluorescence of protoporphyrin IX (PPIX) in tumor cells in cases of HGG, secondary to reduced ferrochelatase and/or coproporphyrinogen III oxidase activity [70, 71]. This agent is administered orally before anesthesia induction followed by tumor resection 4–5 h after administration. In response to excitation with ultraviolet-A to visible blue radiation (375–445 nm), 5-ALA emits red fluorescence (600–740 nm) (Fig. 3) [72]. Peak fluorescence is observed approximately 6–8 hours after administration and the effect lasts for 12 hours [73, 74]. Use of this technique for surgical time > 12 hours increases the risk of false-negative results. A

Fig. 3 5-ALA emits red fluorescence at the GBM tissue under a violet-blue light, which enables identification of the boundary between tumor and normal tissue and residual tumor

meta-analysis that investigated fluorescence positivity in 800 histopathological samples of GBM reported 82.6% sensitivity and 88.9% specificity [75]. Fluorescence intensity indicates tumor cell density in tissues; a study has reported that strong fluorescence showed 100% positive predictive value (PPV) and indicated a proliferative tumor, and weak fluorescence showed 95% PPV and indicated an infiltrating tumor [76]. Other studies that have investigated the intensity of 5-ALA fluorescence have also shown a strong correlation with cellularity and the Ki-67 index [77, 78]. PPV for strong fluorescence was 92% and that for weak fluorescence was 82% in cases of recurrent HGG, which suggests that 5-ALA is useful for recurrent HGG [79], although another study has reported false-positive results in the area of treatment effect in cases of recurrent HGG [80, 81]. A randomized controlled study in which patients underwent either 5-ALA-guided resection or conventional microsurgery for HGG (n = 161, each) showed a higher percentage of complete resection (65 vs. 36%, $p < 0.0001$) and better 6-month PFS (41 vs. 21%, $p = .0003$), without any differences in the adverse event rate in the 5-ALA group [73]. A study that included 33 patients with GBM used a combination of 5-ALA-FGS, and intraoperative MRI vs. intraoperative MRI alone and observed increased GTR rates in the combined group (100 vs. 82%, $p < .01$) [82]. With regard to detection of tumor infiltration at the tumor-normal tissue boundary, compared with intraoperative MRI, 5-ALA showed superior sensitivity (91 vs. 66%) and specificity (80 vs. 60%) ($p < .001$) [83].

Disadvantages and practical difficulties associated with 5-ALA use include false-negative results in patients with LGG [84] and failure of fluorescence labeling owing to presence of blood products in the resection cavity. Notably, periventricular tissues, particularly in the subependymal area are likely to fluoresce, as PPIX accumulates outside the blood-brain barrier (BBB), which can lead to false-positive results [85–87]. Furthermore, reactive astrocytes, scattered atypical cells, and inflammation can lead to false-positive results [88].

Attempts have been made to quantify PPIX using a spectroscopic method to overcome the subjectivity of fluorescence evaluation, and this approach is shown to be effective even in cases of LGG [89].

1.9.2 Fluorescein Sodium

FS is injected intravenously intraoperatively and accumulates in hypervascular areas, particularly in the extracellular space secondary to BBB disruption. FS improves tumor visualization using yellow-green fluorescence, either with white-light illumination or more precisely with the use of a specific filter attached to the surgical microscope [90]. HGG is commonly well visualized with this agent [91]. A multicenter prospective phase II trial (FLUO-GLIO) that investigated 46 patients who underwent FGS for HGG resection showed that the sensitivity and specificity of FS for tumor identification were 81% and 79%, respectively, and 83% of the patients achieved GTR [92]. Usefulness of FS is limited by its tendency of extracellular accumulation, which leads to fluorescence in edematous tissue and extravasated FS in response to tumor resection [93].

A few studies that compared 5-ALA and FS have reported superiority of 5-ALA with regard to both sensitivity and specificity; however, these studies were of limited power owing to a small sample size [94, 95]. Of note, the cost per quality-adjusted life years (QALY) was significantly higher in the 5-ALA group [94]. Combined use of 5-ALA and FS to augment a healthy brain background to highlight tumor visualization was shown to be a feasible alternative [96].

1.9.3 Indocyanine Green

ICG has been widely used in surgeries for vascular lesions such as hemangioblastomas, dural arteriovenous fistulas, and arteriovenous malformations. ICG is intravenously administered intraoperatively and illuminates blood vessels and the extracellular space in areas of BBB disruption through emission of light in a near-infrared spectrum. Administration of a high dose of ICG 24 hours preoperatively improves visualization of contrast-enhanced lesions (such as gliomas) on MRI owing to increased permeability of intratumoral blood vessels, lack of lymphatic drainage, and a hypoxic microenvironment. This method was shown to successfully identify gadolinium-enhanced gliomas with sensitivity of 98% and specificity of 45% [97]. Limitations of ICG administration include false-positive results associated with endothelial injury observed in a setting of inflammation and necrosis, which leads to ICG accumulation within these non-neoplastic tissues [98]. Additionally,it is necessary to switch between a bright-light and near-infrared field, because proceeding with surgery in a dark field is technically challenging.

1.9.4 New Fluorescence

Recent research highlights a novel attempt to use a hydroxymethyl rhodamine green-based fluorescence labeling system administered as a topical spray intraoperatively for detection of GBM [99]. Topical application of fluorescence can be used repeatedly intraoperatively, it is a nearly real-time approach, and it shows lesser false-positive results, all of which serve as advantages of this approach. However, the study was performed in mouse models and in vitro, and further investigations are warranted to confirm its safety and clinical utility in humans.

A phase I clinical trial has investigated the safety profile of tozuleristide (BLZ-100), a near-infrared imaging agent composed of the peptide chlorotoxin and ICG [100]. In phase I/II trials, chlorotoxin, a 36-amino-acid disulfide-containing peptide derived from the venom of a scorpion was shown to cross the BBB and preferentially bind to glioma tissue [101]. Epidermal growth factor receptor (EGFR)-targeted fluorescent affibody (ABY-029) was investigated in a preclinical study [102]. Another EGFR-targeted fluorescence, cetuximab-IRDye800 was effective in contrasting the tumor-background in a human study [103]. Cancer-selective alkylphosphocholine analogs using green or near-infrared fluorescence have been investigated in animal models and showed a similar or superior tumor-to-brain fluorescence ratio compared with 5-ALA [104]. Further comparative studies between conventional probes and novel fluorescence labelling agents are warranted.

Limitations of fluorescence agents include the lack of fluorescence observed in most LGGs. Excessive dependence on FGS should be avoided in HGG, and balancing the fluorescence findings and other information obtained from anatomical interpretation, imaging, and neurophysiological monitoring is essential to avoid false-positive results, which may lead to inadvertent injury to functional structures. Tumor resection in the vicinity of functional structures warrants extreme caution.

1.10 Intraoperative Magnetic Resonance Imaging

Intraoperative MRI is useful to confirm the EOR and to identify a potentially resectable residual tumor. Neuronavigation is useful to plan the craniotomy site for necessary and sufficient tumor exposure and assists in tumor resection; however, "brain shift" invariably occurs after the dura is opened with drainage of CSF and edema, which significantly impair accuracy of navigation during tumor resection [105, 106]. Intraoperative MRI and re-registration of neuronavigation can improve accuracy [107]. Additionally, intraoperative MRI enables detection of intraoperative complications such as ischemia and hematoma.

The early experience was reported in 1999, and intraoperative MRI use was shown to increase the EOR of HGG [108]. Another study showed that intraoperative MRI led to an increased EOR of both enhancing and non-enhancing gliomas from 76 to 96% ($p < .001$) [109]. Several studies have reported increased GTR rates in both GBM (47 to 56%) [110, 111] and non-enhancing gliomas [111, 112]. The initial evidence was produced in 2011 in a study that included 58 patients with contrast-enhanced gliomas, who were randomized into the following groups: surgery using intraoperative MRI and conventional surgery using a microscope. The GTR rate was higher in the intraoperative MRI group than in the conventional surgery group (96 vs. 68%, $p = .02$) [113]. A systematic review that reported the use of intraoperative MRI showed improved quality of life, EOR, and OS in patients who underwent this procedure [114]. A recent retrospective study in a large cohort of 640 patients with GBM, who underwent surgery with or without intraoperative MRI reported that the EOR increased by 6.3% and the GTR rate from 31 to 72%, which suggests that intraoperative MRI potentially prolongs OS in patients with GBM [115]. Moreover, intraoperative MRI is shown to be a cost-effective option for calculation of QALY [116].

A major drawback of intraoperative MRI is the fact that it is time-consuming and requires an additional 0.5–2 hours for surgery and anesthesia [117]. Furthermore, intraoperative MRI is expensive with regard to the equipment and potential modifications required in the operating theater infrastructure.

DTI based on intraoperative MRI such as corticospinal tract reconstruction, which was shown to have high sensitivity (100%) and moderate specificity (72%) verified by direct electrical subcortical stimulation in 20 patients is an evolving technique in clinical practice [118]. A study has reported application of DTI using intraoperative MRI to identify the arcuate fasciculus in an attempt to resect a dominant-side temporoparietal HGG during nonawake surgery [119]. More evidence is necessary to conclusively establish whether intraoperative MRI-based DTI can replace direct subcortical stimulation in non-awake surgery settings.

1.11 Intraoperative Diagnosis (Stimulated Raman Scattering Microscopy and Confocal Microscopy)

Frozen tissue samples are useful for rapid intraoperative histopathological diagnosis; this method also facilitates decision-making to establish the surgical strategy. However, neuropathologists usually require 30–60 min to confirm diagnosis after the specimen is submitted, and results may be non-diagnostic. Therefore, intraoperative histopathological diagnosis is not useful to determine the boundary between normal and tumorous tissue across multiple sites.

Stimulated Raman scattering spectroscopy is a useful imaging technique that is recently introduced in the operating theater. This spectrographic method utilizes the Raman effect based on inelastic scattering of photons by matter, which tends to differ depending on the molecular properties of tissues and objects. Raman scattering spectroscopy provides information of the chemical composition of tissues, which is useful to intraoperatively differentiate between normal and tumorous tissue. This technique has been adopted for various cancers, including breast, gastrointestinal, and cervical malignancies [120–123]. It is also known to be effective in differentiating between normal gray matter and necrotic and tumorous tissues in patients with GBM [124]. Stimulated Raman scattering microscopy processes images of tissues acquired intraoperatively and provides images of simulated hematoxylin and eosin staining [125]. A hand-held Raman-based probe can detect abnormal tissue around the resection cavity during surgery, with both sensitivity and specificity >90% [126–128]. This label-free method does not require administration of dyes or injections. Furthermore, a combination of this method and artificial intelligence provides real-time histopathological diagnosis of tissues, with accuracy that is non-inferior to diagnosis by pathologists [129]. A major drawback of this technique includes a relatively small area of surveillance, with a typical depth of 100 μm [126]. Currently, the availability of this modality is limited. Standardization of this technology may be necessary to improve its usefulness as an adjunct in the operating theater to maximize safe EOR.

Confocal microscopy or hand-held microendoscopy is another intraoperative technique that is recently introduced for tumor diagnosis [130]. Confocal microscopy utilizes a confocal pinhole, which allows the passage of light of a specific frequency among the photons emitted by fluorophores of tissues in reaction to a laser beam. This technique selectively obtains images at a certain depth by rejecting out-of-focus light using a confocal pinhole. The optical sectioning capability enables high-resolution imaging of thick biological samples, such as tissues in vivo, without tissue excision, fixation, and sectioning as required for conventional histological processing. Following adjustment in the confocal pinhole, the software combines multiple images at different depths to produce 3D images [131, 132]. A fluorescent agent is administered to enhance tissue contrast [133, 134]. Reportedly, intraoperative confocal microscopy can visualize 5-ALA-induced fluorescence at the cellular level in LGG; 5-ALA does not usually fluoresce under a conventional microscope [135]. Confocal microscopy enables real-time interpretation of the boundary between normal and tumorous tissue on the surface of the resection cavity. Reportedly, the sensitivity of this technique in accurate diagnosis of HGG was 81% and that for LGG was 90% [136]. Notably, tumor tissue in direct proximity to the device is maximally visualized as opposed to that beneath the normal cortex. Interpretation of the acquired images requires expertise in histopathological diagnosis [68].

1.12 Exoscope

An exoscope has recently been introduced in the operating theater. Its body is smaller than that of a microscope, and surgeons and operating theater staff can view the same images on the heads-up displays. An exoscope offers better magnification and illumination of the surgical area with light-emitting diode light at lower light intensity with lesser glare than a microscope, which uses xenon lighting. It is ergonomically friendly because the exoscope head can change position in line with the surgical pathway, and the surgeon's position is more comfortable compared with the position required for the use of a microscope, which requires positioning of the surgeon and the head of the microscope based on the surgical pathways [137]. A study that used a robotically-controlled exoscope for resection of GBM showed an excellent outcome with high EOR (95%) and good prognosis (6-month PFS of 86%) [138]. A disadvantage of an exoscope is that all systems are not equipped with a filter for FGS, which should be addressed to incorporate an exoscope among the standard instruments used for glioma surgery.

2 Conclusion

Anatomical interpretation of a tumor in relation to the eloquent areas is the key to successful glioma resection, which is aided by preoperative imaging. The anatomical relationship alters spatially and chronologically during surgery, and intraoperative modalities are necessary to gain a deeper understanding. Excessive reliance on a single technology is associated with a risk, because no modality shows 100% sensitivity and specificity, and utilizing only one modality may prevent knowledge of potentially valuable information from other sources.

Recent studies propose a surgical strategy that uses the functional boundary as opposed to imaging boundary to guide tumor resection, which is enabled by dedicated preoperative imaging and intraoperative mapping during awake surgery. This strategy is shown to postpone malignant transformation of LGG [12]. However, caution is essential because careful selection of tasks that are used intraoperatively needs significant expertise, brain function defined today does not necessarily mean the entire brain function and the function sacrificed due to the limitation of the tasks may be observed. Neurosurgeons have to be humble to the yet-to-be-known universe of the brain function.

Currently available technologies work synergistically with surgeons' deep understanding and knowledge of the normal baseline anatomy. Journal articles, books and online resources, as well as white matter dissection courses for off-the job training contribute to the vast knowledge database that provides information regarding cortical and subcortical brain functions and their anatomical locations.

Evolution of these technologies has significantly altered our approach to glioma resection. The modalities outlined in this chapter constitute the essential armamentarium of neurosurgeons useful for glioma surgeries.

References

1. Sanai N, Berger MS. Glioma extent of resection and its impact on patient outcome. Neurosurgery. 2008;62(4):753–66.
2. Smith JS, Chang EF, Lamborn KR, et al. Role of extent of resection in the long-term outcome of low-grade hemispheric gliomas. J Clin Oncol. 2008;26(8):1338–45.
3. Keles GE, Lamborn KR, Berger MS. Low-grade hemispheric gliomas in adults: a critical review of extent of resection as a factor influencing outcome. J Neurosurg. 2001;95(5):735–45.
4. Kavouridis VK, Boaro A, Dorr J, et al. Contemporary assessment of extent of resection in molecularly defined categories of diffuse low-grade glioma: a volumetric analysis. J Neurosurg. 2019;133(5):1291–301.
5. Pichlmeier U, Bink A, Schackert G, Stummer W. Resection and survival in glioblastoma multiforme: an RTOG recursive partitioning analysis of ALA study patients. Neuro-Oncology. 2008;10(6):1025–34.
6. Ius T, Isola M, Budai R, et al. Low-grade glioma surgery in eloquent areas: volumetric analysis of extent of resection and its impact on overall survival. A single-institution experience in 190 patients. J Neurosurg. 2012;117(6):1039–52.
7. Wang Y-C, Lee C-C, Takami H, et al. Awake craniotomies for epileptic gliomas: intraoperative and postoperative seizure control and prognostic factors. J Neuro-Oncol. 2019;142(3):577–86.
8. Orringer D, Lau D, Khatri S, et al. Extent of resection in patients with glioblastoma: limiting factors, perception of resectability, and effect on survival. J Neurosurg. 2012;117(5):851–9.
9. McGirt MJ, Mukherjee D, Chaichana KL, Than KD, Weingart JD, Quinones-Hinojosa A. Association of surgically acquired motor and language deficits on overall survival after resection of glioblastoma multiforme. Neurosurgery. 2009;65(3):463–70.
10. Gill BJ, Pisapia DJ, Malone HR, et al. MRI-localized biopsies reveal subtype-specific differences in molecular and cellular composition at the margins of glioblastoma. Proc Natl Acad Sci. 2014;111(34):12550–5.
11. Kelly PJ, Daumas-Duport C, Kispert DB, Kall BA, Scheithauer BW, Illig JJ. Imaging-based stereotaxic serial biopsies in untreated intracranial glial neoplasms. J Neurosurg. 1987;66(6):865–74.
12. Duffau H. Long-term outcomes after supratotal resection of diffuse low-grade gliomas: a consecutive series with 11-year follow-up. Acta Neurochir. 2016;158(1):51–8.
13. Li YM, Suki D, Hess K, Sawaya R. The influence of maximum safe resection of glioblastoma on survival in 1229 patients: can we do better than gross-total resection? J Neurosurg. 2016;124(4):977–88.
14. Pessina F, Navarria P, Cozzi L, et al. Maximize surgical resection beyond contrast-enhancing boundaries in newly diagnosed glioblastoma multiforme: is it useful and safe? A single institution retrospective experience. J Neuro-Oncol. 2017;135(1):129–39.
15. Ius T, Mazzucchi E, Tomasino B, et al. Multimodal integrated approaches in low grade glioma surgery. Sci Rep. 2021;11(1):1–12.
16. Barbagallo GMV, Certo F, Di Gregorio S, et al. Recurrent high-grade glioma surgery: a multimodal intraoperative protocol to safely increase extent of tumor resection and analysis of its impact on patient outcome. Neurosurg Focus. 2021;50(1):E20.
17. Nucifora PG, Verma R, Lee S-K, Melhem ER. Diffusion-tensor MR imaging and tractography: exploring brain microstructure and connectivity. Radiology. 2007;245(2):367–84.
18. Berman JI, Berger MS, Chung S, Nagarajan SS, Henry RG. Accuracy of diffusion tensor magnetic resonance imaging tractography assessed using intraoperative subcortical stimulation mapping and magnetic source imaging. J Neurosurg. 2007;107(3):488–94.
19. Zolal A, Hejčl A, Vachata P, et al. The use of diffusion tensor images of the corticospinal tract in intrinsic brain tumor surgery: a comparison with direct subcortical stimulation. Neurosurgery. 2012;71(2):331–40.
20. Wu J-S, Zhou L-F, Tang W-J, et al. Clinical evaluation and follow-up outcome of diffusion tensor imaging-based functional neuronavigation: a prospective, controlled study in patients with gliomas involving pyramidal tracts. Neurosurgery. 2007;61(5):935–49.
21. Pujol S, Wells W, Pierpaoli C, et al. The DTI challenge: toward standardized evaluation of diffusion tensor imaging tractography for neurosurgery. J Neuroimaging. 2015;25(6):875–82.
22. Tuch DS. Q-ball imaging. Magnetic Resonance Med. 2004;52(6):1358–72.
23. Khanna N, Altmeyer W, Zhuo J, Steven A. Functional neuroimaging: fundamental principles and clinical applications. Neuroradiol J. 2015;28(2):87–96.

24. Ashtari M, Perrine K, Elbaz R, et al. Mapping the functional anatomy of sentence comprehension and application to presurgical evaluation of patients with brain tumor. Am J Neuroradiol. 2005;26(6):1461–8.
25. Vannest J, Rasmussen J, Eaton K, et al. FMRI activation in language areas correlates with verb generation performance in children. Neuropediatrics. 2010;41(05):235–9.
26. Bizzi A, Blasi V, Falini A, et al. Presurgical functional MR imaging of language and motor functions: validation with intraoperative electrocortical mapping. Radiology. 2008;248(2):579–89.
27. Meier M, Ilmberger J, Fesl G, Ruge M. Validation of functional motor and language MRI with direct cortical stimulation. Acta Neurochir. 2013;155(4):675–83.
28. Bartoš R, Jech R, Vymazal J, et al. Validity of primary motor area localization with fMRI versus electric cortical stimulation: a comparative study. Acta Neurochir. 2009;151(9):1071–80.
29. Roux F-E, Boulanouar K, Lotterie J-A, Mejdoubi M, LeSage JP, Berry I. Language functional magnetic resonance imaging in preoperative assessment of language areas: correlation with direct cortical stimulation. Neurosurgery. 2003;52(6):1335–47.
30. Giussani C, Roux F-E, Ojemann J, Sganzerla EP, Pirillo D, Papagno C. Is preoperative functional magnetic resonance imaging reliable for language areas mapping in brain tumor surgery? Review of language functional magnetic resonance imaging and direct cortical stimulation correlation studies. Neurosurgery. 2010;66(1):113–20.
31. Hart MG, Price SJ, Suckling J. Functional connectivity networks for preoperative brain mapping in neurosurgery. J Neurosurg. 2016;126(6):1941–50.
32. Cochereau J, Deverdun J, Herbet G, et al. Comparison between resting state fMRI networks and responsive cortical stimulations in glioma patients. Hum Brain Mapp. 2016;37(11):3721–32.
33. Qiu T-m, Gong F-y, Gong X, et al. Real-time motor cortex mapping for the safe resection of glioma: an intraoperative resting-state fMRI study. Am J Neuroradiol. 2017;38(11):2146–52.
34. Poldrack RA, Baker CI, Durnez J, et al. Scanning the horizon: towards transparent and reproducible neuroimaging research. Nat Rev Neurosci. 2017;18(2):115–26.
35. Sutherling W, Crandall P, Darcey T, Becker D, Levesque M, Barth D. The magnetic and electric fields agree with intracranial localizations of somatosensory cortex. Neurology. 1988;38(11):1705–14.
36. Korvenoja A, Kirveskari E, Aronen HJ, et al. Sensorimotor cortex localization: comparison of magnetoencephalography, functional MR imaging, and intraoperative cortical mapping. Radiology. 2006;241(1):213–22.
37. Kirsch HE, Zhu Z, Honma S, Findlay A, Berger MS, Nagarajan SS. Predicting the location of mouth motor cortex in patients with brain tumors by using somatosensory evoked field measurements. J Neurosurg. 2007;107(3):481–7.
38. Nagarajan S, Kirsch H, Lin P, Findlay A, Honma S, Berger MS. Preoperative localization of hand motor cortex by adaptive spatial filtering of magnetoencephalography data. J Neurosurg. 2008;109(2):228–37.
39. Firsching R, Bondar I, Heinze H-J, et al. Practicability of magnetoencephalography-guided neuronavigation. Neurosurg Rev. 2002;25(1):73–8.
40. Schiffbauer H, Ferrari P, Rowley HA, Berger MS, Roberts TP. Functional activity within brain tumors: a magnetic source imaging study. Neurosurgery. 2001;49(6):1313–21.
41. Guggisberg AG, Honma SM, Findlay AM, et al. Mapping functional connectivity in patients with brain lesions. Ann Neurol. 2008;63(2):193–203.
42. Curra A, Modugno N, Inghilleri M, Manfredi M, Hallett M, Berardelli A. Transcranial magnetic stimulation techniques in clinical investigation. Neurology. 2002;59(12):1851–9.
43. Tarapore PE, Tate MC, Findlay AM, et al. Preoperative multimodal motor mapping: a comparison of magnetoencephalography imaging, navigated transcranial magnetic stimulation, and direct cortical stimulation. J Neurosurg. 2012;117(2):354–62.
44. Krieg SM, Shiban E, Buchmann N, Meyer B, Ringel F. Presurgical navigated transcranial magnetic brain stimulation for recurrent gliomas in motor eloquent areas. Clin Neurophysiol. 2013;124(3):522–7.
45. Frey D, Schilt S, Strack V, et al. Navigated transcranial magnetic stimulation improves the treatment outcome in patients with brain tumors in motor eloquent locations. Neuro-Oncology. 2014;16(10):1365–72.
46. Raffa G, Conti A, Scibilia A, et al. The impact of diffusion tensor imaging fiber tracking of the corticospinal tract based on navigated transcranial magnetic stimulation on surgery of motor-eloquent brain lesions. Neurosurgery. 2018;83(4):768–82.
47. Negwer C, Sollmann N, Ille S, et al. Language pathway tracking: comparing nTMS-based DTI fiber tracking with a cubic ROIs-based protocol. J Neurosurg. 2017;126(3):1006–14.
48. Ille S, Sollmann N, Hauck T, et al. Combined noninvasive language mapping by navigated transcranial magnetic stimulation and functional MRI and its comparison with direct cortical stimulation. J Neurosurg. 2015;123(1):212–25.
49. Hamer PDW, Robles SG, Zwinderman AH, Duffau H, Berger MS. Impact of intraoperative stimulation brain mapping on glioma surgery outcome: a meta-analysis. J Clin Oncol. 2012;30(20):2559–65.
50. Morshed RA, Young JS, Lee AT, Hervey-Jumper SL. Functional mapping for glioma surgery, part 2: intraoperative mapping tools. Neurosurg Clin. 2021;32(1):75–81.
51. De Benedictis A, Moritz-Gasser S, Duffau H. Awake mapping optimizes the extent of resection for low-grade gliomas in eloquent areas. Neurosurgery. 2010;66(6):1074–84.

52. Takami H, Khoshnood N, Bernstein M. Preoperative factors associated with adverse events during awake craniotomy: analysis of 609 consecutive cases. J Neurosurg. 2020;134(5):1631–9.
53. Schulz C, Waldeck S, Mauer UM. Intraoperative image guidance in neurosurgery: development, current indications, and future trends. Radiol Res Pract. 2012;2012:1–9.
54. Rainer Wirtz WS, Albert FK, Schwaderer M, et al. The benefit of neuronavigation for neurosurgery analyzed by its impact on glioblastoma surgery. Neurol Res. 2000;22(4):354–60.
55. Tanaka S, Puffer RC, Hoover JM, et al. Increased frameless stereotactic accuracy with high-field intraoperative magnetic resonance imaging. Oper Neurosurgery. 2012;71(suppl_2):ons321–8.
56. Morin F, Courtecuisse H, Reinertsen I, et al. Brain-shift compensation using intraoperative ultrasound and constraint-based biomechanical simulation. Med Image Anal. 2017;40:133–53.
57. Wang J, Liu X, Ba YM, et al. Effect of sonographically guided cerebral glioma surgery on survival time. J Ultrasound Med. 2012;31(5):757–62.
58. Neidert MC, Hostettler IC, Burkhardt J-K, et al. The influence of intraoperative resection control modalities on survival following gross total resection of glioblastoma. Neurosurg Rev. 2016;39(3):401–9.
59. Saether C, Torsteinsen M, Torp S, Sundstrøm S, Unsgård G, Solheim O. Did survival improve after the implementation of intraoperative neuronavigation and 3D ultrasound in glioblastoma surgery? A retrospective analysis of 192 primary operations. J Neurol Surg Part A Central Eur Neurosurg. 2012;73(02):073–8.
60. Prada F, Perin A, Martegani A, et al. Intraoperative contrast-enhanced ultrasound for brain tumor surgery. Neurosurgery. 2014;74(5):542–52.
61. Coburger J, König RW, Scheuerle A, et al. Navigated high frequency ultrasound: description of technique and clinical comparison with conventional intracranial ultrasound. World Neurosurg. 2014;82(3–4):366–75.
62. Coburger J, Scheuerle A, Kapapa T, et al. Sensitivity and specificity of linear array intraoperative ultrasound in glioblastoma surgery: a comparative study with high field intraoperative MRI and conventional sector array ultrasound. Neurosurg Rev. 2015;38(3):499–509.
63. Mercier L, Del Maestro RF, Petrecca K, et al. New prototype neuronavigation system based on preoperative imaging and intraoperative freehand ultrasound: system description and validation. Int J Comput Assist Radiol Surg. 2011;6(4):507–22.
64. Prada F, Del Bene M, Mattei L, et al. Fusion imaging for intra-operative ultrasound-based navigation in neurosurgery. J Ultrasound. 2014;17(3):243–51.
65. Sastry R, Bi WL, Pieper S, et al. Applications of ultrasound in the resection of brain tumors. J Neuroimaging. 2017;27(1):5–15.
66. Mursch K, Scholz M, Brück W, Behnke-Mursch J. The value of intraoperative ultrasonography during the resection of relapsed irradiated malignant gliomas in the brain. Ultrasonography. 2017;36(1):60–5.
67. Rygh OM, Selbekk T, Torp SH, Lydersen S, Hernes TAN, Unsgaard G. Comparison of navigated 3D ultrasound findings with histopathology in subsequent phases of glioblastoma resection. Acta Neurochir. 2008;150(10):1033–42.
68. Belykh E, Martirosyan NL, Yagmurlu K, et al. Intraoperative fluorescence imaging for personalized brain tumor resection: current state and future directions. Front Surg. 2016;3:55.
69. Weissleder R, Pittet MJ. Imaging in the era of molecular oncology. Nature. 2008;452(7187):580–9.
70. Krieg RC, Fickweiler S, Wolfbeis OS, Knuechel R. Cell-type specific Protoporphyrin IX metabolism in human bladder cancer in vitro¶. Photochem Photobiol. 2000;72(2):226–33.
71. Ferreira GC, Franco R, Lloyd SG, Moura I, Moura JJ, Huynh BH. Structure and function of ferrochelatase. J Bioenerg Biomembr. 1995;27(2):221–9.
72. Colditz MJ, Jeffree RL. Aminolevulinic acid (ALA)–protoporphyrin IX fluorescence guided tumour resection. Part 1: clinical, radiological and pathological studies. J Clin Neurosci. 2012;19(11):1471–4.
73. Stummer W, Pichlmeier U, Meinel T, et al. Fluorescence-guided surgery with 5-aminolevulinic acid for resection of malignant glioma: a randomised controlled multicentre phase III trial. Lancet Oncol. 2006;7(5):392–401.
74. Kaneko S, Suero Molina E, Ewelt C, Warneke N, Stummer W. Fluorescence-based measurement of real-time kinetics of protoporphyrin IX after 5-aminolevulinic acid administration in human in situ malignant gliomas. Neurosurgery. 2019;85(4):E739–46.
75. Eljamel S. 5-ALA fluorescence image guided resection of glioblastoma multiforme: a meta-analysis of the literature. Int J Mol Sci. 2015;16(5):10443–56.
76. Stummer W, Tonn J-C, Goetz C, et al. 5-Aminolevulinic acid-derived tumor fluorescence: the diagnostic accuracy of visible fluorescence qualities as corroborated by spectrometry and histology and postoperative imaging. Neurosurgery. 2014;74(3):310–20.
77. Idoate MA, Díez Valle R, Echeveste J, Tejada S. Pathological characterization of the glioblastoma border as shown during surgery using 5-aminolevulinic acid-induced fluorescence. Neuropathology. 2011;31(6):575–82.
78. Roberts DW, Valdés PA, Harris BT, et al. Coregistered fluorescence-enhanced tumor resection of malignant glioma: relationships between δ-aminolevulinic acid–induced protoporphyrin IX fluorescence, magnetic resonance imaging enhancement, and neuropathological parameters. J Neurosurg. 2011;114(3):595–603.

79. Five-aminolevulinic acid for fluorescence-guided resection of recurrent malignant gliomas: a phase II study. Neurosurgery. 2009;65(6):1070–7.
80. Valdés PA, Roberts DW, Lu F-K, Golby A. Optical technologies for intraoperative neurosurgical guidance. Neurosurg Focus. 2016;40(3):E8.
81. La Rocca G, Sabatino G, Menna G, et al. 5-Aminolevulinic acid false positives in cerebral neuro-oncology: not all that is fluorescent is tumor. A case-based update and literature review. World Neurosurg. 2020;137:187–93.
82. Coburger J, Hagel V, Wirtz CR, König R. Surgery for glioblastoma: impact of the combined use of 5-aminolevulinic acid and intraoperative MRI on extent of resection and survival. PLoS One. 2015;10(6):e0131872.
83. Coburger J, Engelke J, Scheuerle A, et al. Tumor detection with 5-aminolevulinic acid fluorescence and Gd-DTPA–enhanced intraoperative MRI at the border of contrast-enhancing lesions: a prospective study based on histopathological assessment. Neurosurg Focus. 2014;36(2):E3.
84. Jaber M, Wölfer J, Ewelt C, et al. The value of 5-aminolevulinic acid in low-grade gliomas and high-grade gliomas lacking glioblastoma imaging features: an analysis based on fluorescence, magnetic resonance imaging, 18F-fluoroethyl tyrosine positron emission tomography, and tumor molecular factors. Neurosurgery. 2016;78(3):401–11.
85. Roberts DW, Valdés PA, Harris BT, et al. Glioblastoma multiforme treatment with clinical trials for surgical resection (aminolevulinic acid). Neurosurg Clin N Am. 2012;23(3):371–7.
86. Olivo M, Wilson BC. Mapping ALA-induced PPIX fluorescence in normal brain and brain tumour using confocal fluorescence microscopy. Int J Oncol. 2004;25(1):37–45.
87. Müther M, Stummer W. Ependymal fluorescence in fluorescence-guided resection of malignant glioma: a systematic review. Acta Neurochir. 2020;162(2):365–72.
88. Lau D, Hervey-Jumper SL, Chang S, et al. A prospective phase II clinical trial of 5-aminolevulinic acid to assess the correlation of intraoperative fluorescence intensity and degree of histologic cellularity during resection of high-grade gliomas. J Neurosurg. 2016;124(5):1300–9.
89. Valdés PA, Jacobs V, Harris BT, et al. Quantitative fluorescence using 5-aminolevulinic acid-induced protoporphyrin IX biomarker as a surgical adjunct in low-grade glioma surgery. J Neurosurg. 2015;123(3):771–80.
90. Diaz RJ, Dios RR, Hattab EM, et al. Study of the biodistribution of fluorescein in glioma-infiltrated mouse brain and histopathological correlation of intraoperative findings in high-grade gliomas resected under fluorescein fluorescence guidance. J Neurosurg. 2015;122(6):1360–9.
91. Acerbi F, Broggi M, Eoli M, et al. Is fluorescein-guided technique able to help in resection of high-grade gliomas? Neurosurg Focus. 2014;36(2):E5.
92. Acerbi F, Broggi M, Schebesch K-M, et al. Fluorescein-guided surgery for resection of high-grade gliomas: a multicentric prospective phase II study (FLUOGLIO). Clin Cancer Res. 2018;24(1):52–61.
93. Stummer W. Poor man's fluorescence? Acta Neurochir. 2015;157(8):1379–81.
94. Kaneko S, Eljamel MS. Fluorescence image-guided neurosurgery. Future Oncol. 2017;13(26):2341–8.
95. Della Puppa A, Munari M, Gardiman MP, Volpin F. Combined fluorescence using 5-aminolevulinic acid and fluorescein sodium at glioblastoma border: intraoperative findings and histopathologic data about 3 newly diagnosed consecutive cases. World Neurosurg. 2019;122:e856–63.
96. Molina ES, Wölfer J, Ewelt C, Ehrhardt A, Brokinkel B, Stummer W. Dual-labeling with 5–aminolevulinic acid and fluorescein for fluorescence-guided resection of high-grade gliomas. J Neurosurg. 2017;128(2):399–405.
97. Lee JY, Thawani JP, Pierce J, et al. Intraoperative near-infrared optical imaging can localize gadolinium-enhancing gliomas during surgery. Neurosurgery. 2016;79(6):856–71.
98. Cho SS, Salinas R, Lee JY. Indocyanine-green for fluorescence-guided surgery of brain tumors: evidence, techniques, and practical experience. Front Surg. 2019;6:11.
99. Kitagawa Y, Tanaka S, Kamiya M, et al. A novel topical fluorescent probe for detection of glioblastoma. Clin Cancer Res. 2021;27:3936–47.
100. Patil CG, Walker DG, Miller DM, et al. Phase 1 safety, pharmacokinetics, and fluorescence imaging study of tozuleristide (BLZ-100) in adults with newly diagnosed or recurrent gliomas. Neurosurgery. 2019;85(4):E641–9.
101. Cohen G, Burks SR, Frank JA. Chlorotoxin—a multimodal imaging platform for targeting glioma tumors. Toxins. 2018;10(12):496.
102. Ribeiro de Souza AL, Marra K, Gunn J, et al. Optimizing glioma detection using an EGFR-targeted fluorescent Affibody. Photochem Photobiol. 2018;94(6):1167–71.
103. Miller SE, Tummers WS, Teraphongphom N, et al. First-in-human intraoperative near-infrared fluorescence imaging of glioblastoma using cetuximab-IRDye800. J Neuro-Oncol. 2018;139(1):135–43.
104. Swanson KI, Clark PA, Zhang RR, et al. Fluorescent cancer-selective alkylphosphocholine analogs for intraoperative glioma detection. Neurosurgery. 2015;76(2):115–24.
105. Nimsky C, Ganslandt O, Hastreiter P, et al. Preoperative and intraoperative diffusion tensor imaging-based fiber tracking in glioma surgery. Neurosurgery. 2005;56(1):130–8.

106. Romano A, D'Andrea G, Calabria LF, et al. Pre-and intraoperative tractographic evaluation of corticospinal tract shift. Neurosurgery. 2011;69(3):696–705.
107. Martin C, Alexander E, Wong T, Schwartz R, Jolesz F, Black PM. Surgical treatment of low-grade gliomas in the intraoperative magnetic resonance imager. Neurosurg Focus. 1998;4(4):E10.
108. Knauth M, Wirtz CR, Tronnier VM, Aras N, Kunze S, Sartor K. Intraoperative MR imaging increases the extent of tumor resection in patients with high-grade gliomas. Am J Neuroradiol. 1999;20(9):1642–6.
109. Hatiboglu MA, Weinberg JS, Suki D, et al. Impact of intraoperative high-field magnetic resonance imaging guidance on glioma surgery: a prospective volumetric analysis. Neurosurgery. 2009;64(6):1073–81.
110. Kuhnt D, Becker A, Ganslandt O, Bauer M, Buchfelder M, Nimsky C. Correlation of the extent of tumor volume resection and patient survival in surgery of glioblastoma multiforme with high-field intraoperative MRI guidance. Neuro-Oncology. 2011;13(12):1339–48.
111. Mohammadi AM, Sullivan TB, Barnett GH, et al. Use of high-field intraoperative magnetic resonance imaging to enhance the extent of resection of enhancing and nonenhancing gliomas. Neurosurgery. 2014;74(4):339–50.
112. Pamir MN, Özduman K, Dinçer A, Yildiz E, Peker S, Özek MM. First intraoperative, shared-resource, ultrahigh-field 3-tesla magnetic resonance imaging system and its application in low-grade glioma resection. J Neurosurg. 2010;112(1):57–69.
113. Senft C, Bink A, Franz K, Vatter H, Gasser T, Seifert V. Intraoperative MRI guidance and extent of resection in glioma surgery: a randomised, controlled trial. Lancet Oncol. 2011;12(11):997–1003.
114. Kubben PL, ter Meulen KJ, Schijns OE, ter Laak-Poort MP, van Overbeeke JJ, van Santbrink H. Intraoperative MRI-guided resection of glioblastoma multiforme: a systematic review. Lancet Oncol. 2011;12(11):1062–70.
115. Shah AS, Sylvester PT, Yahanda AT, et al. Intraoperative MRI for newly diagnosed supratentorial glioblastoma: a multicenter-registry comparative study to conventional surgery. J Neurosurg. 2020;1(aop):1-10:1–10.
116. Abraham P, Sarkar R, Brandel MG, et al. Cost-effectiveness of intraoperative MRI for treatment of high-grade gliomas. Radiology. 2019;291(3):689–97.
117. Makary M, Chiocca EA, Erminy N, et al. Clinical and economic outcomes of low-field intraoperative MRI-guided tumor resection neurosurgery. J Magn Reson Imaging. 2011;34(5):1022–30.
118. Javadi SA, Nabavi A, Giordano M, Faghihzadeh E, Samii A. Evaluation of diffusion tensor imaging–based tractography of the corticospinal tract: a correlative study with intraoperative magnetic resonance imaging and direct electrical subcortical stimulation. Neurosurgery. 2017;80(2):287–99.
119. D'Andrea G, Familiari P, Di Lauro A, Angelini A, Sessa G. Safe resection of gliomas of the dominant angular gyrus availing of preoperative FMRI and intraoperative DTI: preliminary series and surgical technique. World Neurosurg. 2016;87:627–39.
120. Haka AS, Shafer-Peltier KE, Fitzmaurice M, Crowe J, Dasari RR, Feld MS. Diagnosing breast cancer by using Raman spectroscopy. Proc Natl Acad Sci. 2005;102(35):12371–6.
121. Shim MG, Wong Kee Song LM, Marcon NE, Wilson BC. In vivo near-infrared Raman spectroscopy: demonstration of feasibility during clinical gastrointestinal endoscopy¶. Photochem Photobiol. 2000;72(1):146–50.
122. Mahadevan-Jansen A, Mitchell MF, Ramanujam N, Utzinger U, Richards-Kortum R. Development of a fiber optic probe to measure NIR Raman spectra of cervical tissue in vivo. Photochem Photobiol. 1998;68(3):427–31.
123. Cui S, Zhang S, Yue S. Raman spectroscopy and imaging for cancer diagnosis. J Healthc Eng. 2018;2018:1–11.
124. Kalkanis SN, Kast RE, Rosenblum ML, et al. Raman spectroscopy to distinguish grey matter, necrosis, and glioblastoma multiforme in frozen tissue sections. J Neuro-Oncol. 2014;116(3):477–85.
125. Orringer DA, Pandian B, Niknafs YS, et al. Rapid intraoperative histology of unprocessed surgical specimens via fibre-laser-based stimulated Raman scattering microscopy. Nat Biomed Eng. 2017;1(2):1–13.
126. Ji M, Lewis S, Camelo-Piragua S, et al. Detection of human brain tumor infiltration with quantitative stimulated Raman scattering microscopy. Sci Transl Med. 2015;7(309):309ra163.
127. Ji M, Orringer DA, Freudiger CW, et al. Rapid, label-free detection of brain tumors with stimulated Raman scattering microscopy. Sci Transl Med. 2013;5(201):201ra119.
128. Jermyn M, Mok K, Mercier J, et al. Intraoperative brain cancer detection with Raman spectroscopy in humans. Sci Transl Med. 2015;7(274):274ra219.
129. Hollon TC, Pandian B, Adapa AR, et al. Near real-time intraoperative brain tumor diagnosis using stimulated Raman histology and deep neural networks. Nat Med. 2020;26(1):52–8.
130. Sankar T, Delaney PM, Ryan RW, et al. Miniaturized handheld confocal microscopy for neurosurgery: results in an experimental glioblastoma model. Neurosurgery. 2010;66(2):410–8.
131. Mooney MA, Zehri AH, Georges JF, Nakaji P. Laser scanning confocal endomicroscopy in the neurosurgical operating room: a review and discussion of future applications. Neurosurg Focus. 2014;36(2):E9.

132. Ragazzi M, Longo C, Piana S. Ex vivo (fluorescence) confocal microscopy in surgical pathology: state of the art. Adv Anat Pathol. 2016;23(3):159–69.
133. Martirosyan NL, Cavalcanti DD, Eschbacher JM, et al. Use of in vivo near-infrared laser confocal endomicroscopy with indocyanine green to detect the boundary of infiltrative tumor. J Neurosurg. 2011;115(6):1131–8.
134. Foersch S, Heimann A, Ayyad A, et al. Confocal laser endomicroscopy for diagnosis and histomorphologic imaging of brain tumors in vivo. PLoS One. 2012;7(7):e41760.
135. Sanai N, Snyder LA, Honea NJ, et al. Intraoperative confocal microscopy in the visualization of 5-aminolevulinic acid fluorescence in low-grade gliomas. J Neurosurg. 2011;115(4):740–8.
136. Breuskin D, Szczygielski J, Urbschat S, Kim Y-J, Oertel J. Confocal laser endomicroscopy in neurosurgery—an alternative to instantaneous sections? World Neurosurg. 2017;100:180–5.
137. Langer DJ, White TG, Schulder M, Boockvar JA, Labib M, Lawton MT. Advances in intraoperative optics: a brief review of current exoscope platforms. Oper Neurosurg. 2020;19(1):84–93.
138. Baron RB, Lakomkin N, Schupper AJ, et al. Postoperative outcomes following glioblastoma resection using a robot-assisted digital surgical exoscope: a case series. J Neuro-Oncol. 2020;148:519–27.

Insular Gliomas: A Review of Surgical Anatomy and Operative Considerations

Ramin A. Morshed, Jacob S. Young, Alexander F. Haddad, and Mitchel S. Berger

1 Introduction

The insula is region of cerebral cortex folded deep within the sylvian fissure and is a common area of involvement for low and high-grade gliomas [1]. Tumors may emanate from the insula directly or may extend into the insula from the frontal, temporal, or parietal lobes. It is well established that greater extent of resection improves survival in patients with low- and high-grade gliomas [2–5]. Yet, surgical management of insular gliomas is complex due to proximity to critical vascular structures as well as eloquent cortical and subcortical regions. In this chapter, we review general anatomy of the insula, relevant surgical anatomy relevant for insular gliomas, and clinical outcomes after resecting tumors in this region.

R. A. Morshed · J. S. Young · A. F. Haddad · M. S. Berger (✉)
Department of Neurological Surgery, University of California, San Francisco, CA, USA
e-mail: Mitchel.Berger@ucsf.edu

2 Anatomy of the Insula

2.1 Gryi of the Insula and Relevant Sulcal Boundaries

The insula is enveloped by overlying eloquent frontal and temporal cortical regions which are referred to as the frontal and temporal operculum, respectively (Fig. 1). Superiorly, the suprasylvian insula is covered by the inferior frontal gyrus, which can be segmented into the pars orbitalis, pars triangularis, and pars opercularis from anterior to posterior (i.e. the frontal operculum). In the dominant hemisphere, the frontal operculum is frequently the site of language function, although permanent deficits are rarely, if ever, encountered with pars orbitalis resections [6]. Continuing to move posteriorly along the frontal operculum, the inferior aspect of the precentral and postcentral gyrus cover the superior and posterior portion of the insula. The inferior (or infrasylvian) portion of the insula is covered by the superior temporal gyrus.

Removal of the opercula or dissection of the sylvian fissure with retraction of overlying cortex reveals the insula. The insular cortex itself is comprised of five gyri, three anterior short gyri and two posterior long gyri (Fig. 1). These short gyri form an inverted pyramid with the most prominent point called the insular apex. The limen insula is the inferior convergence of all

A. Shah et al. (eds.), *Functional Anatomy of the Brain: A View from the Surgeon's Eye*,
https://doi.org/10.1007/978-981-99-3412-6_9

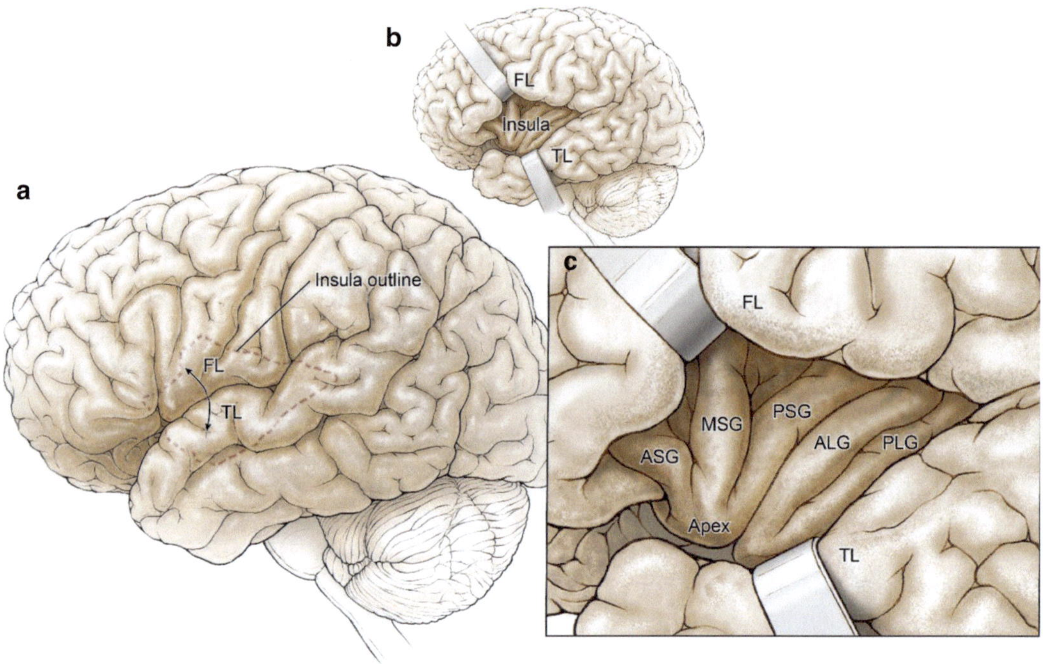

Fig. 1 Anatomy of the insula. (**a**) The insula cortex lies beneath the frontal and temporal operculum. (**b**) Retraction of the operculum reveals the insula which is bound by the limiting sulci. (**c**) The insular cortex is comprised of five gyri, three anterior short gyri and two posterior long gyri. The short gyri form an inverted pyramid with the most prominent point called the insular apex. The limen insula is the inferior convergence of all these gyri and marks the anterior limit of the temporal stem. *FL* frontal lobe, *TL* temporal lobe, *ASG* anterior short gyrus, *MSG* middle short gyrus, *PSG* posterior short gyrus, *ALG* anterior long gyrus, *PLG* posterior long gyrus

these gyri and marks the anterior limit of the temporal stem. The long and short gyri are separated by the central sulcus of the insula which parallels the central sulcus dividing frontal and parietal cortical lobes. The insula is also bound superiorly, anteriorly, and posteriorly by the limiting sulcus. Typically, the medial border of insular glioma resections stops just short of the basal ganglia structure with avoidance of injury to the lateral lenticulostriate arteries.

2.2 Subcortical Tracts Adjacent to the Insula

There are several relevant subcortical white matter tracts that run deep or adjacent to the insular cortex (Fig. 2). Knowledge of these tracts is essential to insular glioma surgery given the risks involved to language and motor function. Prior categorization of subcortical white matter pathways has included (1) fronto-opercular fasciculi including the superior longitudinal fasciculus (SLF), the arcuate fasciculus (AF), and medial longitudinal fasciculus, (2) fascicli crossing through the insula including the uncinate (UF) and inferior fronto-occipital fasciculi (IFOF), and (3) temporo-opercular fasciculi including the inferior longitudinal fasciculus, arcuate fasciculus, and optic radiations.

These large white matter association fibers are especially relevant in the dominant hemisphere as many are implicated in either the dorsal or ventral processing streams of language [7]. The SLF and AF are the major fiber tracts involved in the dorsal stream of language processing. The SLF III is particularly relevant for insular glioma surgery as it connects the premotor frontal region with the supramarginal gyrus. Direct electrical stimulation of the SLF III can cause impairments in repetition and phonological processing. The AF classically connects the inferior frontal gyrus (classical

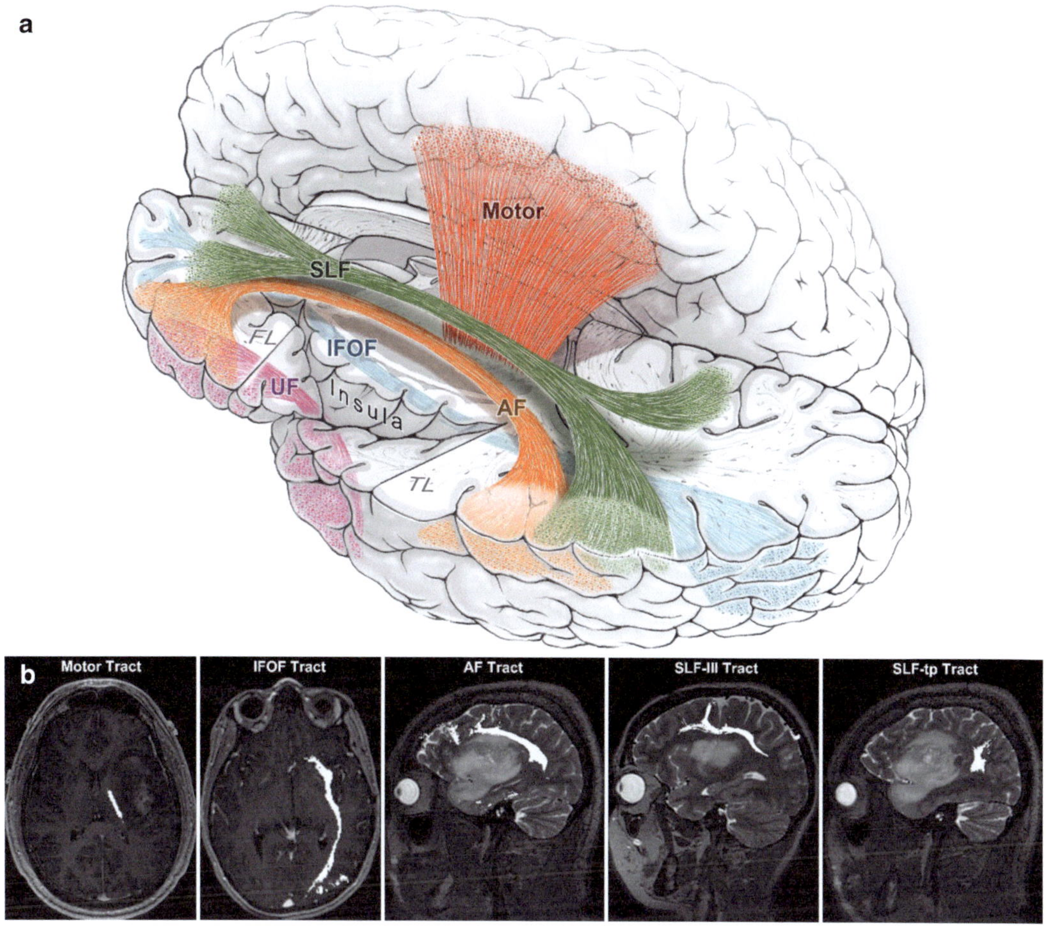

Fig. 2 Key subcortical tracts adjacent to insular gliomas. (**a**) In this illustration, several critical subcortical tracts are displayed in relationship to the insula. The corticospinal tract is medial and posterior to the insula. The AF and SLF run superior and then posterior to the body of the insula in an arc fashion. The IFOF runs medial to the insula along its inferior half, and the UF runs inferior to the inferior and anterior portion of the insula. (**b**) MRI images with tractography demonstrate real-world correlates in a patient with a giant insula involving all Berger-Sanai zones. *AF* arcuate fasciculus, *SLF* superior longitudinal fasciculus, *IFOF* inferior fronto-occipital fasciculus, *UF* uncinate fasciculus, *FL* frontal lobe, *TL* temporal lobe

Broca's area) to the posterior superior temporal gyrus (classical Wernicke's area), and stimulation of this tract has traditionally resulted in phonological paraphasias and syntactical errors.

The IFOF and UF are components of the ventral stream of language processing. The IFOF connects the inferior frontal lobe with the occipital lobe, extending through the subcortical insula in the external capsule. and fanning through the temporal lobe to send radiating fibers to the middle and inferior temporal gyri on its way to the occipital lobe. Intraoperative stimulation of the IFOF frequently causes a semantic paraphasia or perseveration, and injury to this tract can result in a permanent deficit. The UF extends from the inferior frontal gyrus to the anterior superior temporal gyrus in a C-shape under the Sylvian fissure. The importance of the UF in language processing has been debated, with direct electrical stimulation failing to elicit language disturbances and only transient deficits reported after resection of part of the UF [8].

It is important to have an understanding of the spatial relationship of these fibers to the insula and the overlay of these fibers on top of each other (Fig. 2). After peeling away the frontotemporal opercular coverings, the most superficial of the dorsal association fiber tracts are the AF and SLF systems, which curve in a C-shape around the superior and posterior aspect of the insular cortex [9]. Meanwhile, the white matter medial to the insular cortex is organized into three capsules (the extreme, external, and internal capsules) that are separated by two gray matter structures (claustrum and lenticular nucleus). Ventrally, the IFOF and the UF extend through the subcortical region of the insula and comprise the ventral part of the extreme capsule (Destrieux 2018) and contribute fibers to the external capsule. The internal capsule contains the corticospinal tract in its posterior limb with injury to this leading to hemiplegia.

The temporo-opercular fascicules are encountered less frequently during the resection of insular gliomas but provide important anatomical boundaries/landmarks. The ILF is a group of fibers originating in the fusiform gyrus of the temporal pole and runs in the posterolateral direction below the temporal and occipital ventricular horns to the occipital pole [10]. Meanwhile, the optic radiations are deep to the ILF, IFOF, and UF and the ventro-anterior component of the optic radiations, called Meyer's loop, curves around the roof of the temporal horn before running posteriorly toward primary visual centers in the calcarine cortex.

2.3 Connectivity of the Insula

The insula has been described as an anatomical hub with connectivity to a number of cortical and subcortical brain regions; it serves to integrate information from various functional systems. The subcortical and cortical regions that connect to the insula contribute to a broad range of processes, including olfaction, language, sensory, motivation, craving, addiction, emotional, awareness and other cognitive functions [11–15]. A meta-analyses of 1768 functional neuroimaging experiments by Kurth et al. demonstrated four functionally unique regions of the insula which can broadly be categorized into sensorimotor (mid-posterior), social-emotional (anterior-ventral), olfacto-gustatory (central), and cognitive (anterior-dorsal) [16]. However, there is also significant overlap between these functional regions of the insula, hinting at its role as an integration center in the brain [16, 17].

Connectivity to the insula also emanates from many different brain regions. Sensory inputs into the insula originate from direct thalamic afferents which carry somatosensory, auditory, olfactory, gustatory, and visual information [13]. Information regarding emotion stems from links between the insula, the middle and inferior temporal cortex, and the anterior cingulate cortex [17]. In addition, the ventral anterior and posterior portions of the insula also receive input from the amygdala [18]. Cognitive and executive functions are represented in the insula by connections with frontal brain regions such as the anterior cingulate [13]. Meanwhile, connections between the insula and nucleus accumbens and caudate are thought to contribute to motivation and reward functions [13]. Details of the insular connectome continue to be investigated, and the cognitive impact of resection of gliomas in this region requires further investigation.

3 Management of Insular Gliomas

3.1 Berger-Sanai Zone Classification of Insular Gliomas

When approaching insular gliomas, consideration should be given to the specific region of the insula that is involved in addition to adjacent cortical regions that these tumors may extend. Insular gliomas can be classified according to a quadrant-based Berger-Sanai classification system which has a high inter-user reliability [2]. The insular zone should be defined to determine both head positioning and overlying opercular cortex that needs to be removed to access the tumor. In this system, tumors are anatomically defined based on a perpendicular axis with the foramen of Monro and an axis in-line with the Sylvian fissure. Zone 1 and 2 tumors are anterior and posterior to the foramen of Monro axis, respectively, and both superior to the Sylvian fissure axis thus requiring removal of the frontal operculum to be accessed. Zone 4 and 3 tumors are anterior and posterior to the foramen of Monro axis, respectively, and both inferior to the Sylvian fissure axis and thus requiring removal of the temporal operculum to access. Zone 1 and 2 tumors require removal of the frontal operculum for access while Zone 3 and 4 tumors require removal of the temporal operculum. The most common single-zone site for insular tumors is Zone 1 (35% of insular gliomas), yet nearly 40% of tumors are equally represented within 2 separate zones. Furthermore, 13% of insular gliomas are "giant", extending into all 4 quadrants [4]. Extension into neighboring cortical lobes should also be noted to guide the window of access.

The Berger-Sanai classification system for insular gliomas helps with positioning considerations intraoperatively and correlates with extent of resection and morbidity outcomes [2, 4]. Zone 3 and 4 tumors (below the Sylvian fissure) require the vertex of the head to be tipped about 15 degrees towards the floor whereas Zone 1 and 2 tumors require the vertex to be tipped about 15 degrees towards the ceiling. Zone 2 and 3 tumors, the head is positioned 15 degrees upward to maximize visualization underneath functional cortical areas.

3.2 Surgical Approaches to Insular Gliomas

There are two main approaches to insular gliomas: a transcortical approach through the operculum (i.e. transopercular) and a transsylvian approach. Some prior reports have utilized a combination of these two approaches [19]. Yasargil et al. first described the transslyvian approach for limbic and paralimbic tumors in 1992 of which a subset included insular tumors [20]. The transsylvian approach involves a wide arachnoid dissection of the sylvian fissure with retraction of the frontal and temporal opercular cortex superiorly and inferiorly for visualization of the insula. The approach requires meticulous microsurgical dissection of the M3 and M2 middle cerebral arteries which are followed to expose "windows" between vessels into the insular cortex [21]. The posterior end of the Sylvian fissure dissection tends to be more difficult as the pial margins of the frontal and temporal opercula become adherent or interlocked. This approach spares resection of the frontal and temporal opercula, minimizing injury to critical motor and language tracts. Downsides to the approach include potential sacrifice of Sylvian veins which may place the patient at risk of venous infarct. For giant insular tumors, opercular retraction may be needed placing the opercula cortex at risk of retraction-related ischemia/contusion [22]. Furthermore, gliomas, especially when high-grade, may distort normal fissure anatomy posing risk to veins and arteries during dissection.

The transcortical approach has gained popularity with improvements in intraoperative mapping techniques. This approach involves removing stimulation-negative opercular cortex in order to gain access to the insula. Overlying cortex is first mapped with direct electrical stimulation to rule-out functional motor or language sites. Negative opercular cortex overlying the insular glioma can then be removed, exposing the insular windows between the M2 branches. The insula is then entered through these windows, and resection cavities between windows are connected in order to achieve a complete resection. Prior work has demonstrated that the risk of neurologic deficits with this approach using modern mapping techniques is low [2, 23]. Once the insula has been entered via either approach, care must be taken not to enter the deep nuclei of the basal ganglia or injure the lateral lenticulostriate arteries which constitute the medial boundary of resection.

Selection of the approach has typically been based on surgeon preference. A side-by-side comparison of pros and cons for each approach can be found in Table 1. Based on prior reports, it appears that surgeons are more likely to choose the transcortical approach over the transsylvian approach for larger gliomas and those located in

Table 1 Pros and cons for transcortical versus transsylvian approaches for insular gliomas

Transcortical approach		Transsylvian approach	
Pros	Cons	Pros	Cons
Wide exposure of insula	Potential injury to critical motor and language tracts	Sparing of opercular cortex	Potential need for retraction of opercula
Negative map guides transcortical corridor with minimal deficit risk	Mapping techniques required adding to operative time		Distorted sylvian anatomy due to tumor
Appropriate approach for giant and zone II tumors			Potential sacrifice of sylvian veins for adequate exposure
			Potential for arterial injury and infarct during dissection

Zone III. Zone II tumors that underwent transcortical approach were associated with a significantly lower incidence of tissue ischemia on postoperative MRI, potentially related to excessive retraction of the frontal opercula or sacrifice of Sylvian vessels in order to gain access [24]. Anatomic studies in cadaveric subjects have demonstrated that the transcortical approach provides for more surgical freedom to posterior zones (Zones II and III) of the insula when compared to the transsylvian approach [22].

3.3 Insular Glioma Surgical Outcomes

Prior studies have examined clinical outcomes including extent of resection and morbidity after surgical resection of insular glioma via either a transcortical or transsylvian approach. In a study involving 218 patients undergoing a transcortical approach, Hervey-Jumper et al. reported a median EOR of 85% and long-term language and motor deficit rates of 0.8% and 1.6%, respectively, without any surgical mortality. Zone I and IV gliomas had the highest extent of resection rates as did small tumors [2]. Another study by Rossi et al. examined outcomes specifically in 95 patients with giant insular gliomas undergoing a transcortical approach and identified a mean EOR of 92.3% with a permanent language deficit rate of 3.2% and no mortality events [25]. Thus, despite cortical transgression, negative mapping provides safe corridors to the insula that pose minimal morbidity to patients.

Modern cohorts utilizing a transsylvian approach are more limited. A recent report utilizing the transsylvian approach demonstrated ≥90% EOR achieved in 38% of cases and < 90% in 62% of cases and a persistent neurologic deficit rate of 6.3%. Zone II tumors have lower EOR via the transsylvian approach likely related to their posterior position within the insula [26].

Comparison of outcomes between these two approaches has demonstrated similar outcomes except for posterior insular gliomas which may be approached more safely transcortically. Pryzbylowski et al. compared insular glioma outcomes between patients undergoing a transsylvian versus a transcortical approach. Mean EOR was not significantly different (transsylvian vs transcortical: 91.6% vs 88.6%) despite patients in the transcortical group having larger tumors overall. There were not significant differences in rates of persistent neurological deficits by 1 year between approaches [24].

4 Conclusions

Insular gliomas are daunting surgical lesions that are associated with critical subcortical pathways and traversing arteries that pose significant risks to patients when injured. However, novel mapping techniques and meticulous microsurgical dissection can allow for extensive resection of these tumors with minimal patient morbidity. Care must be taken to note the specific Berger-Sanai Zones of involvement, identify at-risk subcortical tracts, and tailor surgical resection techniques including intraoperative cortical and subcortical mapping to achieve a maximal extent of resection to improve patient prognosis.

References

1. Duffau H, Capelle L. Preferential brain locations of low-grade gliomas. Cancer. 2004;100(12):2622–6. https://doi.org/10.1002/cncr.20297.
2. Hervey-Jumper SL, Li J, Osorio JA, Lau D, Molinaro AM, Benet A, Berger MS. Surgical assessment of the insula. Part 2: validation of the Berger-Sanai zone classification system for predicting extent of glioma resection. J Neurosurg. 2016; https://doi.org/10.3171/2015.4.JNS1521.
3. Molinaro AM, Hervey-Jumper S, Morshed RA, Young J, Han SJ, Chunduru P, Zhang Y, Phillips JJ, Shai A, Lafontaine M, Crane J, Chandra A, Flanigan P, Jahangiri A, Cioffi G, Ostrom Q, Anderson JE, Badve C, Barnholtz-Sloan J, et al. Association of Maximal Extent of resection of contrast-enhanced and non-contrast-enhanced tumor with survival within molecular subgroups of patients with newly diagnosed glioblastoma. JAMA Oncol. 2020;6(4):495–503. https://doi.org/10.1001/jamaoncol.2019.6143.
4. Sanai N, Polley M-Y, Berger MS. Insular glioma resection: assessment of patient morbidity, survival, and tumor progression. J Neurosurg. 2010; https://doi.org/10.3171/2009.6.JNS0952.
5. Smith J, Chang E, Lamborn K, Chang S, Prados M, Cha S, Tihan T, VandenBerg S, McDermott M, Berger M. Role of extent of resection in the long-term outcome of low-grade hemisperic gliomas. J Clin Oncol. 2008;26(8):1338–45. https://doi.org/10.1200/JCO.2007.13.9337.
6. Rolston JD, Englot DJ, Benet A, Li J, Cha S, Berger MS. Frontal operculum gliomas: language outcome following resection. J Neurosurg. 2015;122(4):725–34. https://doi.org/10.3171/2014.11.JNS132172.
7. Chang EF, Raygor KP, Berger MS. Contemporary model of language organization: an overview for neurosurgeons. J Neurosurg. 2015;122(2):250–61. https://doi.org/10.3171/2014.10.JNS132647.
8. Zhang J, Zhong S, Zhou L, Yu Y, Tan X, Wu M, Sun P, Zhang W, Li J, Cheng R, Wu Y, Yu Y, Ye X, Luo B. Correlations between dual-pathway white matter alterations and language impairment in patients with aphasia: a systematic review and meta-analysis. Neuropsychol Rev. 2021;31(3):402–18. https://doi.org/10.1007/s11065-021-09482-8.
9. Dziedzic TA, Bala A, Marchel A. Anatomical aspects of the insula, opercula and peri-insular white matter for a transcortical approach to insular glioma resection. Neurosurg Rev. 2022;45(1):793–806. https://doi.org/10.1007/s10143-021-01602-5.
10. Pastor-Escartín F, García-Catalán G, Holanda VM, Muftah Lahirish IA, Quintero RB, Neto MR, Quilis-Quesada V, Ibaoc KB, González Darder JM, de Oliveira E. Microsurgical anatomy of the insular region and Operculoinsular association Fibers and its neurosurgical application. World Neurosurg. 2019;129:407–20. https://doi.org/10.1016/j.wneu.2019.05.071.
11. Craig ADB. How do you feel--now? The anterior insula and human awareness. Nat Rev Neurosci. 2009;10(1):59–70. https://doi.org/10.1038/nrn2555.
12. Ghaziri J, Nguyen DK. Structural connectivity of the insula. In: *Island of Reil (insula) in the human brain: anatomical, functional, clinical and surgical aspects*; 2018. p. 77–83. https://doi.org/10.1007/978-3-319-75468-0_7.
13. Gogolla N. The insular cortex. Curr Biol. 2017;27(12):R580–6. https://doi.org/10.1016/J.CUB.2017.05.010.
14. Gu X, Hof PR, Friston KJ, Fan J. Anterior insular cortex and emotional awareness. J Comp Neurol. 2013;521(15):3371. https://doi.org/10.1002/CNE.23368.
15. Hervey-Jumper SL, Berger MS. Insular glioma surgery: an evolution of thought and practice: JNSPG 75th anniversary invited review article. J Neurosurg. 2019;130(1):9–16. https://doi.org/10.3171/2018.10.JNS181519.
16. Kurth F, Zilles K, Fox PT, Laird AR, Eickhoff SB. A link between the systems: functional differentiation and integration within the human insula revealed by meta-analysis. Brain Struct Funct. 2010;214(5):519–34. https://doi.org/10.1007/S00429-010-0255-Z.
17. Cauda F, D'Agata F, Sacco K, Duca S, Geminiani G, Vercelli A. Functional connectivity of the insula in the resting brain. NeuroImage. 2011;55(1):8–23. https://doi.org/10.1016/J.NEUROIMAGE.2010.11.049.
18. Ghaziri J, Tucholka A, Girard G, Boucher O, Houde JC, Descoteaux M, Obaid S, Gilbert G, Rouleau I, Nguyen DK. Subcortical structural connectivity of insular subregions. Sci Rep. 2018;8(1):1–12. https://doi.org/10.1038/s41598-018-26995-0.
19. Simon M, Neuloh G, von Lehe M, Meyer B, Schramm J. Insular gliomas: the case for surgical management. J Neurosurg. 2009;110(4):685–95. https://doi.org/10.3171/2008.7.JNS17639.
20. Yaşargil MG, von Ammon K, Cavazos E, Doczi T, Reeves JD, Roth P. Tumours of the limbic and paralimbic systems. Acta Neurochir. 1992;118(1–2):40–52. https://doi.org/10.1007/BF01400725.
21. Safaee MM, Englot DJ, Han SJ, Lawton MT, Berger MS. The transsylvian approach for resection of insular gliomas: technical nuances of splitting the sylvian fissure. J Neuro-Oncol. 2016;130(2):283–7. https://doi.org/10.1007/s11060-016-2154-5.
22. Benet A, Hervey-Jumper SL, Sánchez JJG, Lawton MT, Berger MS. Surgical assessment of the insula. Part 1: surgical anatomy and morphometric analysis of the transsylvian and transcortical approaches to the insula. J Neurosurg. 2016;124(2):469–81. https://doi.org/10.3171/2014.12.JNS142182.
23. Hervey-Jumper SL, Berger MS. Insular glioma surgery: an evolution of thought and practice. J Neurosurg.

2019;130(1):9–16. https://doi.org/10.3171/2018.10.JNS181519.
24. Przybylowski CJ, Baranoski JF, So VM, Wilson J, Sanai N. Surgical morbidity of transsylvian versus transcortical approaches to insular gliomas. J Neurosurg. 2019;132(6):1731–8. https://doi.org/10.3171/2018.12.JNS183075.
25. Rossi M, Gay L, Conti Nibali M, Sciortino T, Ambrogi F, Leonetti A, Puglisi G, Howells H, Zito P, Villa F, Ciroi G, Riva M, Bello L. Challenging Giant insular gliomas with brain mapping: evaluation of neurosurgical, neurological, neuropsychological, and quality of life results in a large mono-institutional series. Front Oncol. 2021;11:629166. https://doi.org/10.3389/fonc.2021.629166.
26. Pitskhelauri D, Bykanov A, Konovalov A, Danilov G, Buklina S, Sanikidze A, Sufianov R. Transsylvian insular glioma surgery: new classification system, clinical outcome in a consecutive series of 79 cases. Operat Neurosurg (Hagerstown, Md). 2021;20(6):541–8. https://doi.org/10.1093/ons/opab051.

Interpreting Imaging for Pre-Operative Planning and Execution of Glioma Surgery

Jun Muto, Zeya Aung Moe Thein, and Yuichi Hirose

1 Introduction

Glioblastomas are infiltrative brain tumors with poor prognoses despite multimodal treatment, and maximum safe resection is recommended whenever feasible. The extent of resection is positively correlated with survival [1].

The current standard for resection guidance is T2-weighted (T2W)-fluid attenuation inversion recovery magnetic resonance imaging (MRI) for non-enhancing gliomas and T1-weighted (T1W) MRI with a gadolinium-based contrast agent for enhancing gliomas.

The neurosurgeon's goal is to find the balance between maximizing the extent of resection and preserving brain function since surgery-induced neurological deficits result in lower quality of life and shorter duration of survival.

We have discussed the white fiber network, especially the superior longitudinal fasciculus (SLF) I, II, and III; arcuate fasciculus (AF); inferior frontooccipital fasciculus (IFOF); inferior longitudinal fasciculus (ILF); uncinate fasciculus (UF); cingulum; and corpus callosum (CC). Tractography requires that the region of interest (ROI)s are drawn over a specific fiber. And we also highlight how to set and draw the ROI for each essential white fiber.

This chapter reviews the understanding of the white matter network for each function and the course and radiographic findings for glioma surgery based on postmortem human brain dissection and tractography on MRI to aid pre-operative planning and surgical decision-making.

2 Case Presentation

2.1 Case 1

A 41-year-old male was referred to our hospital for headaches and vomiting that had lasted for 1 month. MRI of the brain with gadolinium administration showed a ring enhanced lesion in the right frontal lobe (Fig. 1a). The neuropsychological evaluation on admission revealed a severe headache. The rest of the neurologic examination findings were normal. The surgery involved a right frontotemporal approach to resect the tumor. Postoperative MRI demonstrated the total removal of the enhanced lesion (Fig. 1b). The essential related white fibers, in this case, were SLF I, II, and III; AF; and IFOF.

2.2 Case 2

An 81-year-old male presenting with headache and vomiting was brought to the emergency room.

J. Muto (✉) · Z. A. M. Thein · Y. Hirose
Department of Neurosurgery, Fujita Health University, Toyoake, Aichi, Japan
e-mail: muto@2002.jukuin.keio.ac.jp; muto@fujita-hu.ac.jp

A. Shah et al. (eds.), *Functional Anatomy of the Brain: A View from the Surgeon's Eye*,
https://doi.org/10.1007/978-981-99-3412-6_10

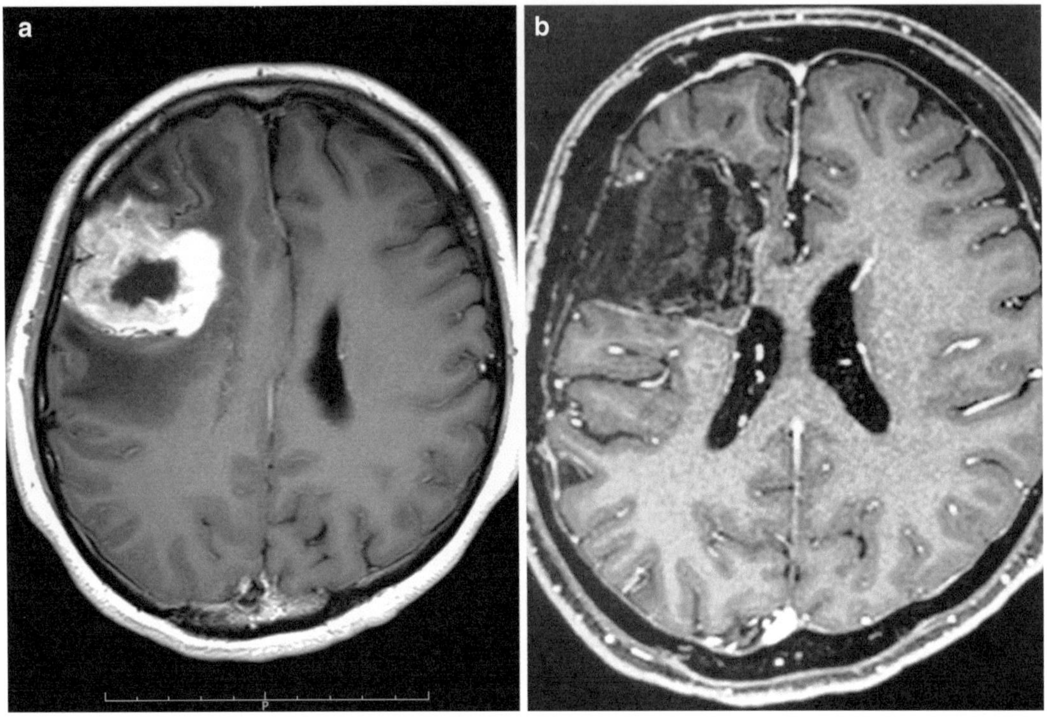

Fig. 1 Case 1 (**a**) A 41-year-old female with a ring-like enhanced lesion in the right frontal lobe on Gd-enhanced MRI before surgery. (**b**) Postoperative MRI showed complete removal of the enhanced lesion

MRI showed a ring enhanced lesion in the temporofrontal lobe with an extension to the parietal lobe with a high density on fluid attenuation inversion recovery MRI (Fig. 2a). The neuropsychological evaluation on admission revealed disorientation of time and place and difficulties with visual naming. The rest of the neurologic examination findings were normal. Surgery was performed, and a right frontotemporal approach was used to remove the enhanced lesion (Fig. 2b). The related essential white fibers were ILF, IFOF and UF.

2.3 Case 3

An 84-year-old female presenting with headache and vomiting was brought to the emergency room. MRI showed a ring-like enhanced lesion in the right parietal lobe (Fig. 3a). Surgery was performed, and a parietal transcortical approach was used to remove the tumor (Fig. 3b). The related white fibers were the cingulum, CC, corona radiata, and AF.

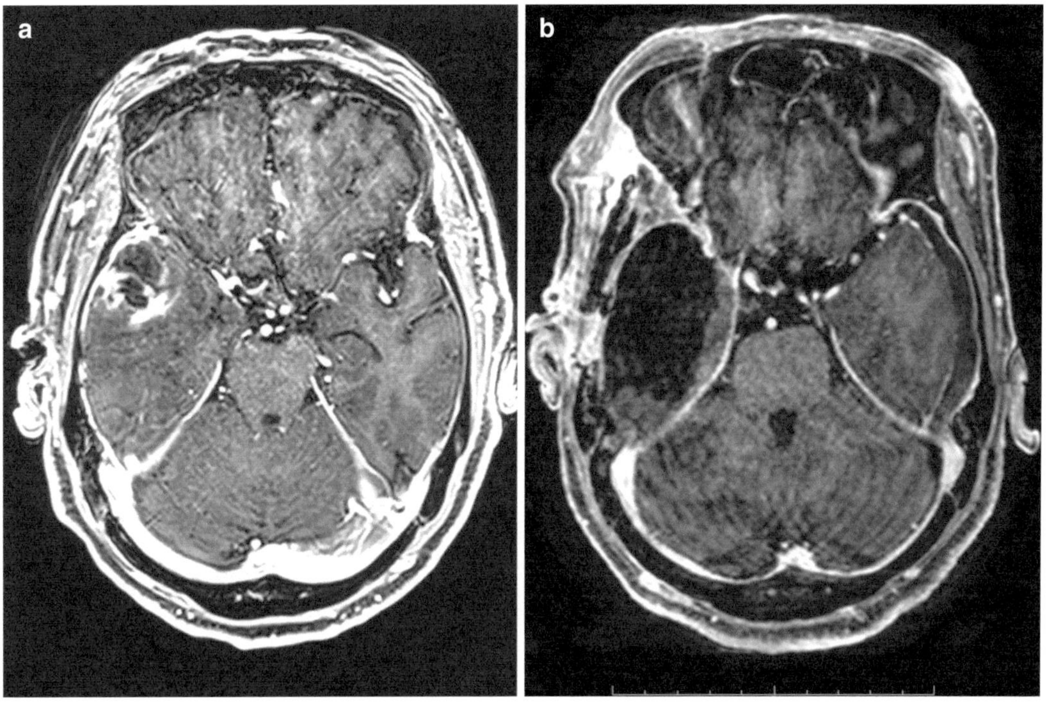

Fig. 2 Case 2. (**a**) An 81-year-old female with a ring-like enhanced lesion in the right temporal lobe on preoperative Gd-enhanced MRI. (**b**) Postoperative MRI showed complete removal of the enhanced lesion

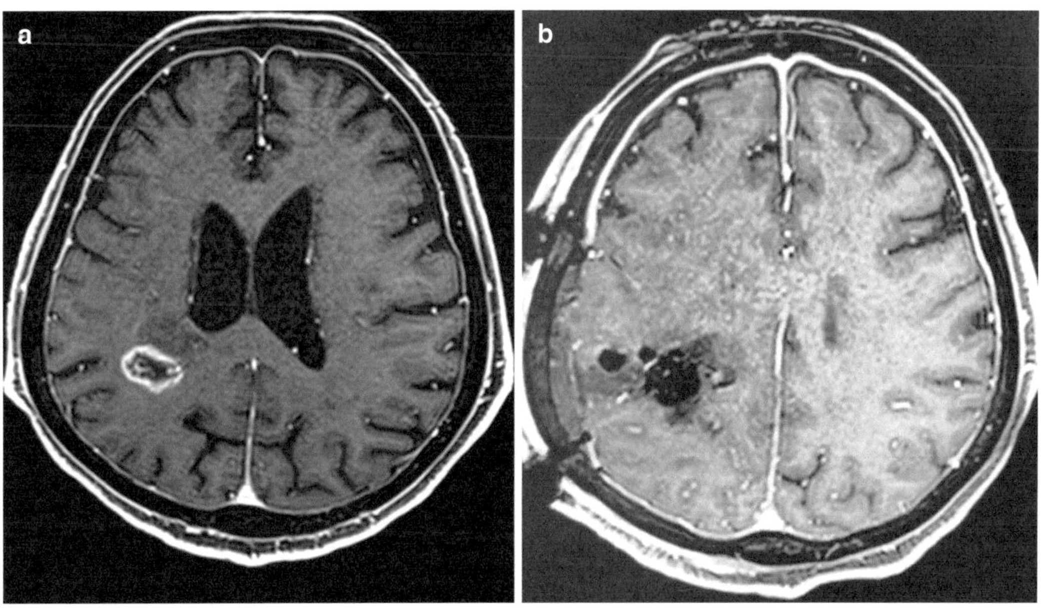

Fig. 3 Case 3. (**a**) An 84-year-old female with a ring-like enhanced lesion in the right parietal lobe on preoperative Gd-enhanced MRI. (**b**) Postoperative MRI showed complete removal of the enhanced lesion

3 White Matter Fibers

3.1 SLF I

The SLF I provides connections between the superior parietal lobe, which is important for recognizing limb and trunk locations in the body-centered coordinate space, and the premotor areas engaged in the higher aspects of motor behavior and the supplementary motor area, which is essential for intention and the initiation of motor activity. The SLF I runs from the precuneus (medial Brodmann area (BA) 7) and from the superior parietal lobule (BA 7) to the superior frontal gyrus (BA 8 and 9), supplementary motor area (BA 6), and anterior cingulate cortex, passing deep into the postcentral (BA 5) and precentral gyrus (BA 4). Therefore, the SLF I may be associated with higher-order control of body-centered action and the initiation of motor activity [2, 3].

On white-matter dissection, the SLF I is located within the medial part of the hemisphere and courses within the upper bank of the cingulate sulcus above the cingulum and medial to the callosal fibers. The cortex is above the cingulate sulcus, where the SLF I can be found.

During tractography, we reconstructed a system comprising three parallel longitudinal branches passing through the dorsolateral portion of the white matter situated above the ventricles and the lateral sulcus and connecting the parietal and the frontal lobes. The superior branch runs from the superior parietal lobule and the precuneus (BA7 and BA5) to the superior frontal gyrus (BA6, BA8, BA9 up to BA10). ROIs for the SLF I can be set within the suprafrontal gyrus (SLF1 I in green in Fig. 4a) and parietal lob (SLF I in green in Fig. 4b) in the coronal section and delete Tf in blue in Fig. 4c to allow the selection of the fiber running anteroposteriorly [4, 5] (Fig.4a–d).

3.2 SLF II

The SLF II is located within the middle part of the occipitoparietofrontal region, and it shows right dominance in 65% of the cerebral [6]. The SLF II extends from the occipital lobe (BA 19) and/or the upper part of the angular gyrus (BA 39) to the middle part of the middle frontal gyrus (dorsolateral prefrontal cortex; BA 46 and 9 or the lateral frontopolar cortex (BA10)) [7]. Functionally, the SLF II can regulate the focusing of attention on different parts of space because of its course, which connects the prefrontal cortex to the posterior parietal lobule. Likewise, damage to the SLF II is postulated to result in spatial working memory disorders such as neglect syndrome [2, 6].

Tractography shows that the middle branch connects the angular gyrus (BA39) to the posterior regions of the middle frontal gyrus (BA6) ventrally. Few projections continue further into the middle frontal gyrus up to BA46. ROIs for SLF II can be set within the middle frontal gyrus (SLF II in green in Fig. 4a) and parietal lobe (SLF II in green in Fig. 4b) in the coronal section and delete Tf in blue in Fig. 4c to allow the selection of the fiber running anteroposteriorly [5, 8] (Fig. 4a–c, e).

3.3 SLF III

The SLF III courses within the frontoparietal operculum and extends from the supramarginal gyrus (BA 40) to the pars opercularis (BA44) [6]. Functionally, the left SLF III is involved in phonological working memory, which is divided into a phonological store (in the left supramarginal gyrus) and an articulatory rehearsal module (in the left inferior frontal gyrus). Stimulation of the left SLF III produces articulatory disorders such as dysarthria or anarthria [9]. However, the right SLF III is involved in visuospatial attention, prosody, and music processing [6].

During white matter dissection, beginning from the level of the middle frontal gyrus, the cortex is removed to reveal the SLF. At first, the temporoparietal SLF (SLF-tp) can be exposed. It is recommended to preserve the white matter of the operculum, as the fibers of the SLF can easily be disrupted. SLF II can be located within the middle frontal gyrus with horizontal fibers and SLF III within the inferior frontal gyrus connect-

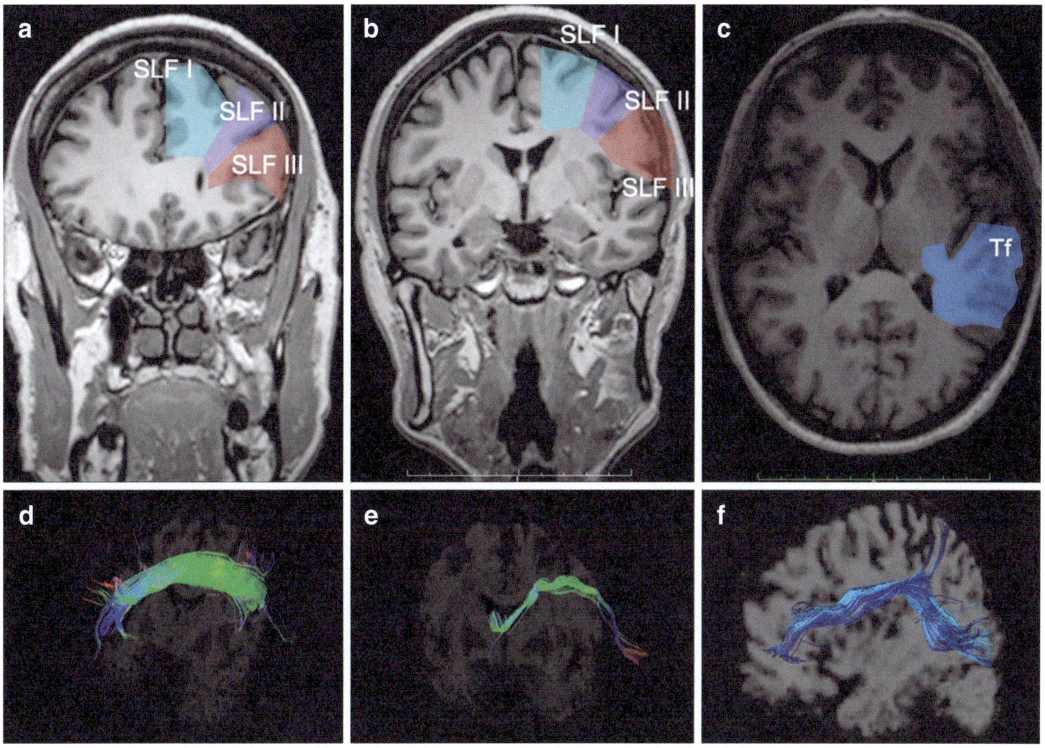

Fig. 4 Superior longitudinal fasciculis (SLF) I, II, and III. Delineation of the regions of interest (ROIs) used for tractography of the three subcomponents of parietofrontal connections. (**a**) Three ROIs in the left frontal lobe (SLF I, II, and III). (**b**) Parietal ROIs in the left hemispheres (SLF I in the superior frontal gyrus, SLF II in the middle frontal gyrus, SLF III in the inferior frontal gyrus). (**c**) Temporal ROI used to exclude the connections belonging to the temporofrontal arcuate fasciculus in the left (Tf) hemispheres. SLF I is the blue ROI in A and B, NOT Tf in C. SLF II is the purple ROI in (**a**) and (**b**), NOT Tf in C. SLF III is the red ROI in A and B, NOT Tf in (**c**). The multiple ROI approach was used to isolate the three components of the SLF (i.e. SLF I, SLF II, and SLF III). Three coronal ROIs were delineated around the white matter of the superior, middle, and inferior/precentral frontal gyri, and three other 'AND' ROIs were delineated posteriorly in the parietal region. Streamlines of the arcuate fasciculus projecting to the temporal lobe were excluded using an axial 'NOT' ROI in the temporal white matter (the arcuate is not part of the longitudinal system as it forms an arc to reach the temporal lobe). Tractography of SLF I (**d**), SLF II (**e**), and SLF III (**f**)

ing the supramarginal gyrus (BA 40) with the pars opercularis (BA44) (Figs. 6 and 7).

Tractography shows that the inferior branch originates in the supramarginal gyrus (BA40) and terminates in the pars opercularis (BA44), triangularis (BA45), and the inferior frontal gyrus (BA47) [8]. ROIs for the SLF III can be set within the inferior frontal gyrus (SLF III in green in Fig. 4a) and parietal lobe(SLF III in green in Fig. 4b) in the coronal section and delete Tf in blue in Fig. 4c to allow the selection of the fiber running anteroposteriorly [5, 8] (Fig.4a–c, f).

3.4 AF

The SLF and AF are the most superficial long association fibers visible after removing the cortex. The AF connects Broca's area, which is located within the frontal lobe, and Wernicke's area, which is located within the temporal lobe. The AF is divided into the ventral and dorsal segments based on its relationship with the supramarginal and angular parts of the inferior parietal lobe.

The AF ventral segment extends from the posterior (BA 22, BA 41, and BA 42) and the middle parts (BA 21) of the middle temporal gyrus, passes deep into the lower part of the supramarginal gyrus (BA 40), and continues ventrally with the SLF III (ventral pathway) in the frontoparietal operculum to reach the pars opercularis (BA 44), triangularis (BA 45), and ventral premotor cortex (BA 6).

The AF dorsal segment extends from the posterior part of the middle (BA 37) and inferior (BA 20) temporal gyri, passes deep to the lower part of the angular gyrus (BA 39), and continues ventrally with the SLF II (middle pathway) to reach the pars opercularis (BA 44), ventral premotor cortex (BA 6), and posterior part of the middle frontal gyrus (BA 6) [7].

Functionally, the AF ventral segment is associated with phonological language processing in the dominant hemisphere, while the AF dorsal segment is associated with lexical and semantic language processing. In the inferior frontal cortex, cortical stimulation has been suggested to underlie separate language function subsystems for phonological (basic sounds that make up words), syntactic (principles and processes by which sentences are constructed), and semantic (concepts and meanings of words) processing. The phonological region is located posteriorly in the inferior frontal gyrus at the border of the pars triangularis (BA 45) and pars opercularis (BA 44). The semantic region is located anterosuperiorly in the inferior frontal gyrus and includes pars orbitalis (BA 47), triangularis (BA 45), and the dorsolateral prefrontal cortex (middle part of the middle frontal gyrus) (BA 46/9) [10]. In the non-dominant hemisphere, the AF is involved in visuospatial processing [11] and some aspects of language such as prosody and semantics [12]. In the superior temporal cortex, functional studies suggest that the posterior part of the superior temporal gyrus (BA 22) is involved in phonological processing bilaterally, and the middle temporal gyrus (BA 21 and 37) is involved in lexical and semantic processing bilaterally, but prosody is exclusively processed in the right hemisphere [13].

During white matter dissection, the AF is a broad white matter tract that encircles the insula in the frontal, parietal, and temporal opercula. It connects the frontal, parietal, occipital, and temporal cortices and can readily be dissected [8] (Figs. 5, 6, and 7).

During tractography, the frontoparietal portion of the AF encompasses a group of fibers with anteroposterior directions running laterally to the projection fiber of the corona radiata. A single ROI (A) is used for the dissection of the AF. (A in

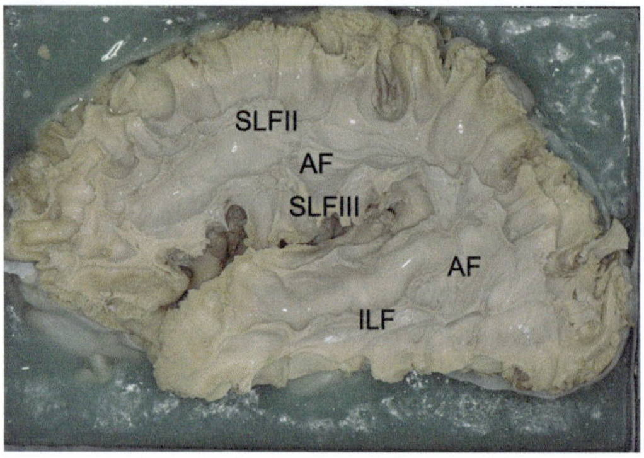

Fig. 5 Stepwise white matter dissection using the lateral approach. Dissecting the cortex from the frontal and temporal lobes. First, SLF-tp was identified in the temporoparietal lesion. SLF-tp originates posterior to the superior temporal gyrus and ascends to reach the superior parietal lobe. After removing SLF-tp, SLF II in the middle frontal gyrus, SLF III in the inferior frontal gyrus, AF around the perisylvian area, and ILF in the middle temporal gyrus can be observed. *AF* arcuate fasciculus, *ILF* Inferior longitudinal fasciculus, *SLF* Superior longitudinal fasciculus

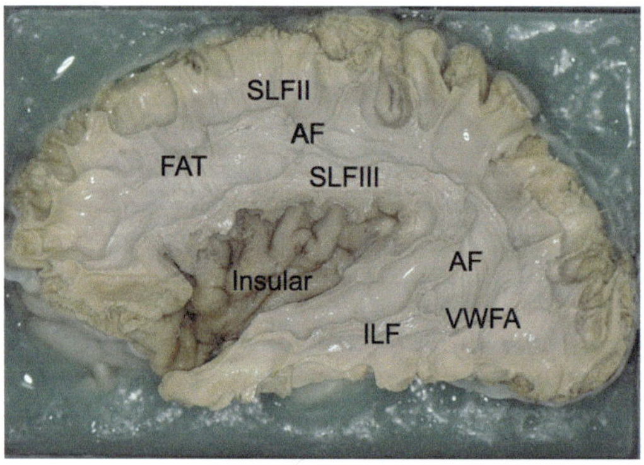

Fig. 6 Stepwise white matter dissection using the lateral approach. Removing the white matter of the operculum of the frontal and temporal lobes exposes the insula. Gradual dissection discloses the SLF more precisely. *AF* arcuate fasciculus, *FAT* Frontal aslant tract, *ILF* Inferior longitudinal fasciculus, *SLF* Superior longitudinal fasciculus, *VWFA* visual word forming area

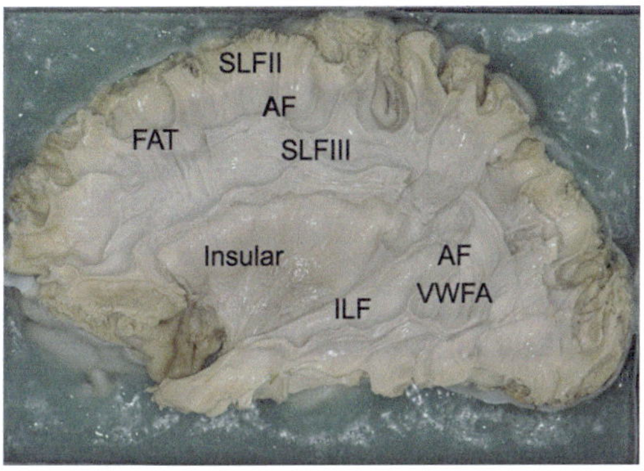

Fig. 7 Stepwise white matter dissection using the lateral approach. After acknowledging the insular surface anatomy and removing the gray matter of the insular, the fibers of the extreme capsule are exposed, and the gray matter of the claustrum can be observed. Removing the gray matter of the limen insular exposes the superficial layer of the uncinate fasciculus that is a part of the extreme capsule at this level. *AF* arcuate fasciculus, *FAT* Frontal aslant tract, *ILF* Inferior longitudinal fasciculus, *SLF* Superior longitudinal fasciculus, *VWFA* visual word forming area

Fig. 8a, b) A large half-moon-shaped region is delineated on the most dorsal part of the arcuate, usually one or two slices above the body of the CC. The lowest region is delineated around the posterior temporal stem. The medial border of the region is easy to identify from the FA maps as a black line between the arcuate and the corona radiata. The lateral border of the ROI passes through the bottom part of the frontal, parietal, and temporal sulci. The precentral sulcus demarcates the anterior border of the ROI, and the intraparietal sulcus is its posterior border [8] (Fig. 8a–c).

3.5 IFOF

Schmahmann and Pandya et al. [14] stated that the inferior occipitofrontal fasciculus corresponds to the extreme capsule, and the IFOF is one of the longest association fiber tracts of the

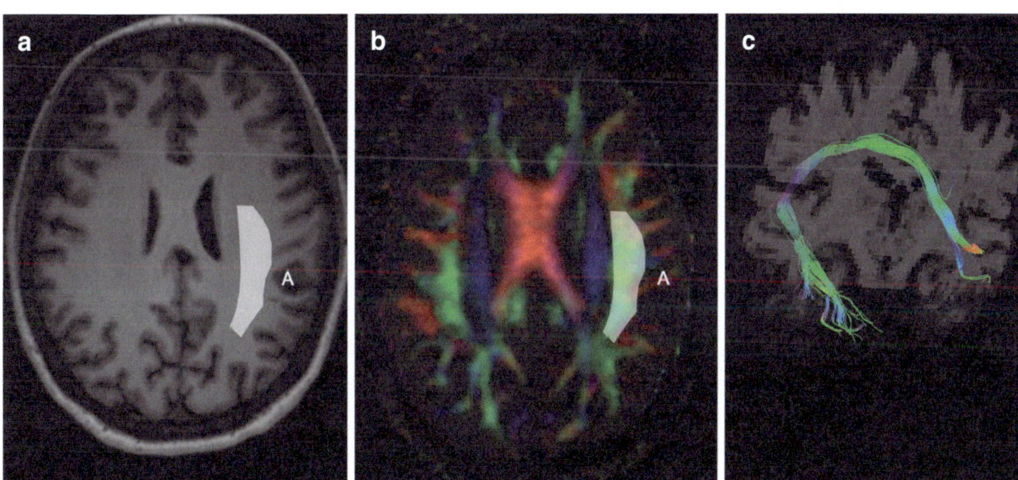

Fig. 8 AF. Delineation of the ROIs used for the tractography of the anterior and the long segments of the arcuate fasciculi. The frontoparietal portion of the arcuate fasciculus encompasses a group of fibers with anteroposterior directions (green) running laterally to the projection fiber of the corona radiata. A single ROI (**a**) is used for the dissection of the arcuate fasciculus in the T1-weighted image (**a**) and FA map (**b**). A large half-moon-shaped region is defined on the most dorsal part of the arcuate, usually one or two slices above the body of the corpus callosum. The lowest region is defined around the posterior temporal stem. The medial border of the region is easy to identify in the FA maps as a black line between the arcuate and the corona radiata. The lateral border of the ROI passes through the bottom part of the frontal, parietal, and temporal sulci. The precentral and intraparietal sulci demarcate the anterior and posterior borders of the ROI [4]. (**c**) tractography of right AF

brain. The IFOF may be split into superficial and deeper layers [15]. The superficial layer connects the frontal lobe to the superior parietal lobule and the posterosuperior part of the occipital lobe, whereas the deep layer connects the frontal lobe with the posteroinferior part of the occipital lobe and the posterobasal temporal areas.

Recent observations related to intraoperative stimulation during glioma surgery implicated the dominant ventral (anteroinferior) insular cortex in planning for motor speech [16, 17]. At the subcortical level, direct intraoperative stimulation of the ventral external (and extreme) capsule elicited paraphasic errors and repetition disorders characteristic of conductive aphasia [16, 18]. The inferior occipitofrontal fasciculus forms part of the ventral extreme and the external capsule and travels under the ventral insular cortex from the prefrontal to the posterotemporal regions. This fasciculus is the anatomic substrate of the ventral semantic pathway described by Duffau et al. [18] in contraposition to the dorsal phonological pathway. The IFOF is involved in reading [19], attention [11], and visual processing [20]. The IFOF may only exist in the human brain, and it is involved in verbal and visual semantic cognition [18, 21] and mentalizing [22].

During white matter dissection, the IFOF and UF can be observed beneath the limen insula after dissecting SLF I, II, II, AF, and the insular. The IFOF connects the frontal lobe to different territories of the posterior cortex. The extreme capsule is exposed when the anteroinferior portion of the insular cortex is removed. At the anteroinferior portion of the extreme and external capsules, the UF and the IFOF are exposed. In this region, both fasciculi narrow in section, and the IFOF is located dorsal and posterior to the UF. (Fig. 9).

During tractography, a two-ROI approach is used to dissect the fibers of the IFOF. The first region is delineated around the occipital lobe (O)) (Fig. 10a, b). The criteria used for the delineation of this occipital region have already been described in the paragraph of the ILF. The second region is delineated around the external/extreme capsule (E) (Fig. 10c, d), and the criteria for this delineation have already been explained in the paragraph of the UF [8]. (Fig. 10e, f).

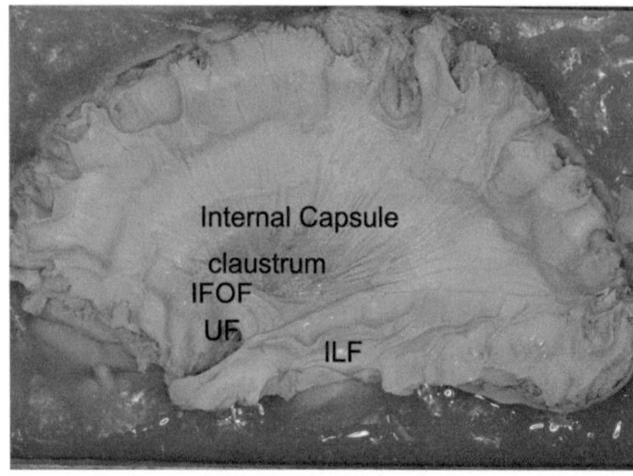

Fig. 9 Stepwise white matter dissection using the lateral approach. Dissecting the extreme capsule reveals the gray matter of the claustrum, whereas the IFOF and UF are revealed at the level of the limen insula. These white matter pathways are parts of the external capsule at the level of the limen insula. *AF* arcuate fasciculus, *IFOF* Inferior frontooccipital fasciculus, *ILF* Inferior longitudinal fasciculus, *UF* Uncinate fasciculus

3.6 ILF

The ILF connects the temporal pole with the dorsolateral occipital cortex and passes deep into the inferior temporal and inferior occipital gyri at the inferolateral edge of the temporal lobe. The ILF runs inferolateral to the temporal horn and atrium. The ILF in the right hemisphere forms a ventral occipitotemporal processing pathway with the IFOF, which is involved in visual identification and pattern discrimination, connecting regions of the core system of facial recognition; the fusiform gyrus seems to be especially important for processing facial identity [23, 24].The ILF in the left hemisphere forms the visual word form area (VWFA) at the crossing point of the posterior part of AF. VWFA is associated with alexia. Subcortical stimulation studies have demonstrated that (1) resection of the left ILFa does not induce alexia; (2) resection of the left ILFp leads to "pure alexia," namely alexia without aphasia

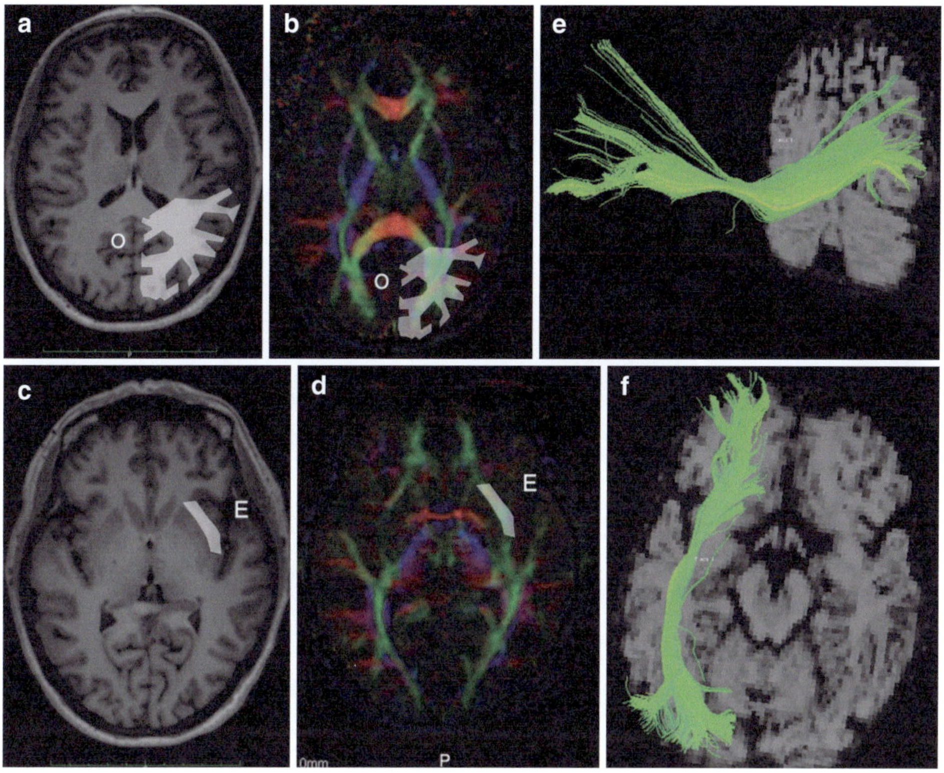

Fig. 10 IFOF. Delineation of the ROIs used for the tractography of the uncinate and the inferior frontooccipital fasciculus (IFOF). (**a**, **b**) Occipital lobe (O) in the hemispheres on the T1-weighted image (**a**) and FA map (**b**). (**c**, **d**) External/Extreme capsule ROIs (e) in the hemisphere on the T1-weighted image (**c**) and FA map (**d**). A two-ROI approach (O) AND (**e**) was used to isolate the inferior frontooccipital fasciculus. (**e**, **f**) tractography of IFOF

and agraphia, by disconnecting the VWFA from its main visual inputs; (3) the most anterior part of the VWFA is required to allow "addressed phonology" through the lexical-semantic pathway, i.e., to read irregular words; and (4) inferior termination of the AFp in the ITGp is a crucial part of the network involved in assembled phonology and in reading pseudowords and irregular words aloud [25]. The ILF is involved in facial recognition [26], visual perception [27], reading [19], visual memory [28], and other functions related to language [8].

During white matter dissection, removing the vertical part of the SLF exposes a group of horizontal fibers (such as the sagittal stratum) situated within the posterior temporal and occipital areas. This compact bundle comprises the fibers of the inferior occipitofrontal fasciculus, anterior commissure, ILF. The gray matter of the insula can be peeled away until its connective fibers, which constitute the superficial layer of the extreme capsule, are evident. Removing the gray matter of the limen insula, in particular, exposes the superficial layer of the UF, which is essentially a part of the extreme capsule at this level. Dissecting the extreme capsule reveals the ventral part of the external capsule. The claustrum becomes apparent at the level of the insular apex, whereas the occipitofrontal and uncinate fasciculi are depicted at the level of the limen insula [29]. (Figs. 5, 6, 7, and 9).

During tractography, a two-ROI approach is used to dissect the ILF. The first ROI (T in Fig. 11a, b)is delineated around the white matter

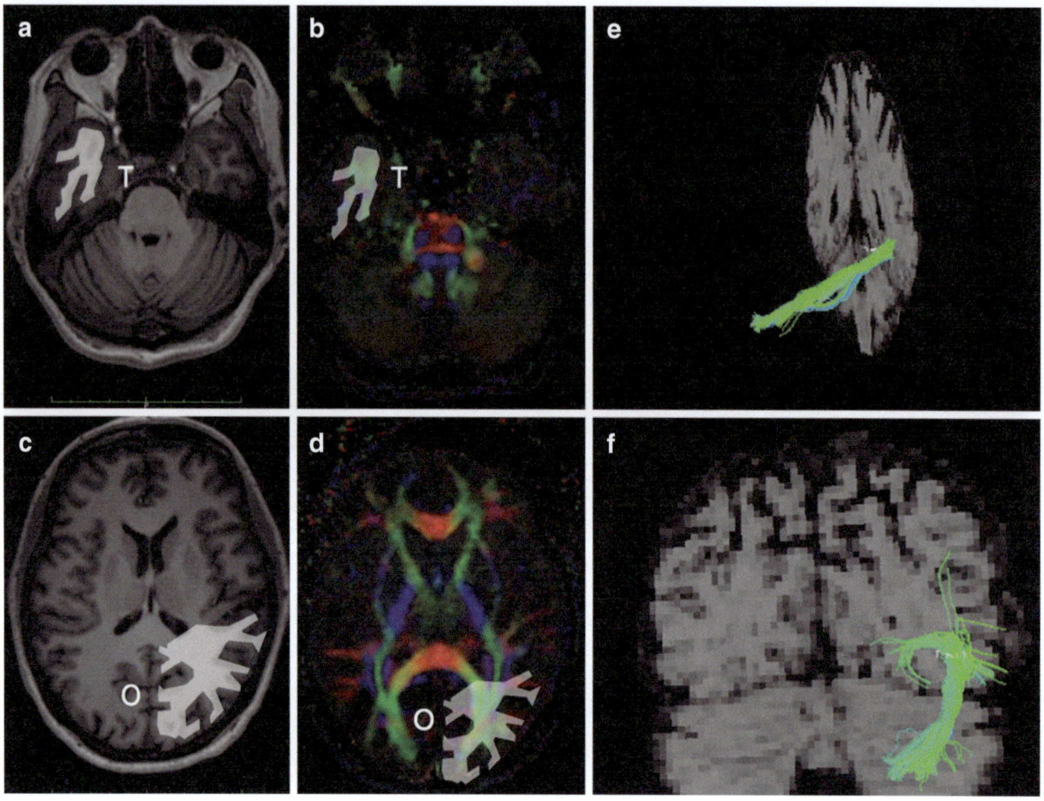

Fig. 11 ILF. Delineation of the ROIs used for the tractography of the inferofrontal longitudinal fasciculus. (**a**, **b**) Temporal lobe ROI (T) in the hemispheres on T1-weighted image (**a**) and FA map (**b**). (**c**, **d**) Occipital lobe ROI (O) in the hemisphere on a T1-weighted image (**c**) and FA map (**d**). A two-ROI approach was employed to isolate the inferior frontal longitudinal fasciculus. For each hemisphere, we delineated an axial ROI (T) AND (O). (**e**, **f**) tractography of ILF

of the anterior temporal lobe. The first region of the temporal ROI is delineated on a slice located three slices below the slice containing the junction between the frontal and temporal branches of the anterior floor of the external capsule. It is important to exclude the most posterior white matter of the temporal lobe to prevent contamination from fibers belonging to other tracts. The second ROI (O in Fig. 11c, d) is delineated around the white matter of the occipital lobe. The anterior border should always be posterior to the tip of the occipital horn or cingulum [4]. (Fig. 11e, f).

3.7 UF

The UF is a bidirectional long-range white matter tract that connects the lateral orbitofrontal cortex and BA 10 with the anterior temporal lobes.

Abnormalities in the UF have been associated with several psychiatric disorders, and previous studies have suggested that it plays an important role in episodic memory, language, and social-emotional processing; however, its exact function is not well-understood.

Von Der Heide RJ et al. proposed that the role of the UF is to allow temporal lobe-based mnemonic associations, such as an individual's name + face + voice, to modify behavior through interactions with the lateral orbitofrontal cortex, which provides valence-based biasing of decisions. The bidirectionality of the UF information flow allows orbital and frontal cortex-based reward and punishment history to rapidly modulate temporal lobe-based mnemonic representations. UF is involved in emotion processing, memory [28, 30]. and language functions [8]. The disruption of the uncinate may cause problems in the expression of memory to guide decisions and the acquisition of certain types of learning and memory. Moreover, uncinate perturbation should cause problems that extend beyond memory to include social and emotional problems due to people and objects being stripped of personal value and emotional history and lacking in higher-level motivational value [31]. During white matter dissection, the UF can be found after removing the gray matter of the insular and the limen insular, which is essentially a part of the extreme capsule at this level. Superficially, the extreme capsule, claustrum in gray matter, external capsule, putamen in gray matter, and the internal capsule can be found individually. After dissecting the putamen, which has a spongy consistency, the firm density of the globus pallidus is identified. The occipital frontal and UF are left in place. (Fig. 9).

During tractography, a two-ROI approach is used to dissect the UF. The first ROI (temporal, T) (Fig. 12a, b) is placed within the anterior temporal lobe, as described for the ILF. A second ROI (external/extreme capsule, E) (Fig. 12c, d) is delineated around the white matter of the anterior floor of the external/extreme capsule. The insula delineates the lateral border of the ROI, with the lenticular nucleus as its medial border. We suggest beginning the delineation of this second ROI from the slice where the temporal and frontal branches of the external/extreme capsule join (Fig. 12e, f).

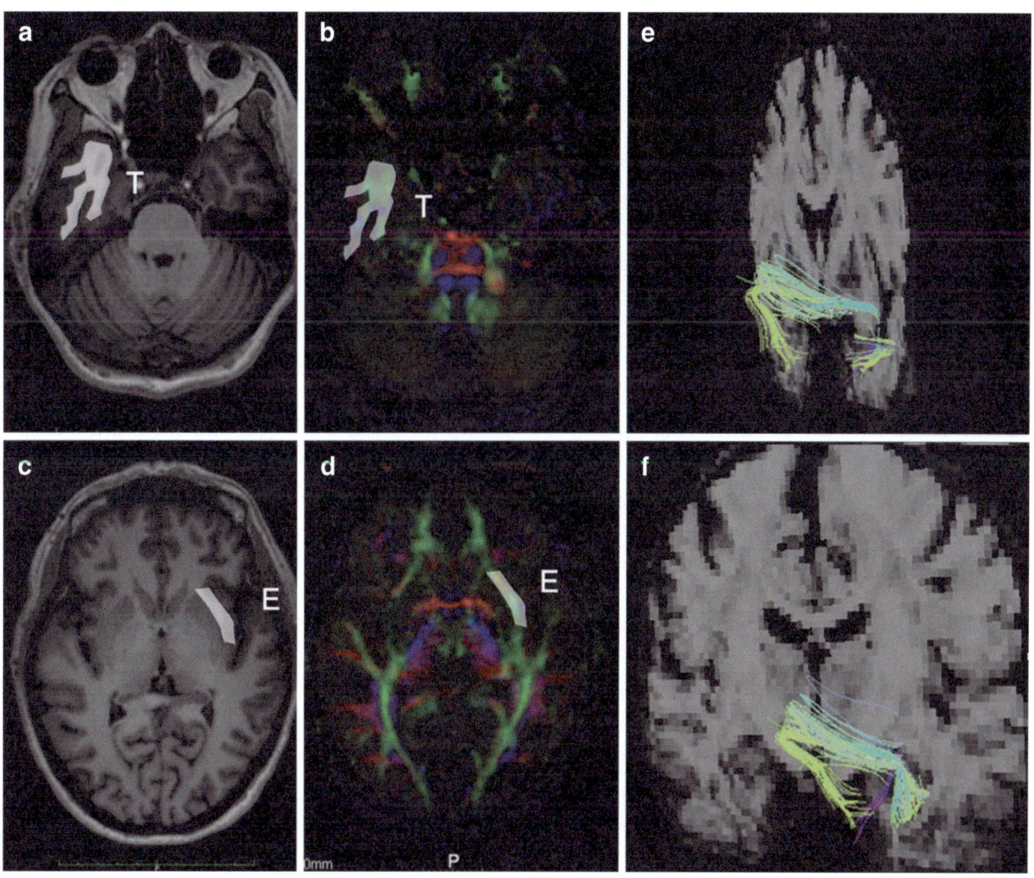

Fig. 12 UF. Delineation of the ROIs used for the tractography of the uncinate fasciculus. (**a**, **b**) Temporal lobe ROI (T) in the hemispheres on T1-weighted image (**a**) and FA map (**b**). (**c**, **d**) External/Extreme capsule ROIs (**e**) in the hemisphere on a T1-weighted image (**c**) and FA map (**d**). A two-ROI approach was employed to isolate the inferior frontal longitudinal fasciculus. For each hemisphere, we delineated an axial ROI (T) AND (E). (**e**, **f**) tractography of UF

3.8 CC

The CC is the largest white matter structure in the brain, consisting of 200–250 million contralateral axonal projections and the major commissural pathway connecting the hemispheres of the human brain. Four named regions are described by anatomists for the CC. These are, from the anterior to the posterior, the rostrum, the genu, the body, and the splenium [32]. While this classification is still widely used today, it does not define the boundaries between the different segments. A second tentatively better parcellation was based on histological measures of the midsagittal axonal diameter but it could not be systematically generalized across individuals [33]. The enforcement of a geometrical grid of seven parcels has been used to somewhat circumvent the limitations of generalization [34]. All these anatomical, histological, and geometrical parcellations of the midsection of the CC do, however, not consider the functional organization of the brain areas it connects.

Since the functions of the CC are defined by the areas that it connects to, tract-tracing in non-human primates or diffusion-weighted imaging tractography in humans is likely the most promising approach to investigate the functional organization of the CC. Friedrich et al [35] provided the first comprehensive map of the callosal cross-section that displays the distribution of callosal connections according to a hierarchical organization of cognitive functions without enforced boundaries [35].

During white matter dissection, the CC is on the medial side. The gray matter of the medial cerebral surface should be peeled away starting from the depth of the anterior part of the cingulate sulcus [36]. Subsequently, the CC can be resected, leaving a thin strip attached to the cingulum. The "U" fibers and the white matter of the superior arm of the cingulum should be dissected to reveal the radiating fibers of the CC.

During tractography, a single ROI (CC) (Fig. 13a, b) is delineated around the body, the genu, and the splenium of the CC. The shape of the regions follows the anatomy of the different parts of the CC: butterfly shape for the body and horseshoe shape for the genu and splenium. Alternatively, a single ROI delineated around the CC on a midsagittal slice can be used (Fig. 13c, d).

3.9 Cingulum

The cingulum is part of the limbic system and is involved in attention, memory, and emotions [20].

The cingulum bundle (CB) forms connections between the frontal lobe, parietal lobe, and temporal lobe, which are associated with cognitive and emotional processing implicated in the pathophysiology of numerous psychiatric disorders. CB-I ran from the subrostral areas to the precuneus and splenium, encircling the CC. CB-II arched around the splenium and extended anteriorly above the CC to the medial aspect of the superior frontal gyrus (SFG). CB-III connected the superior parietal lobule (SPL) and precuneus with the medial aspect of the SFG. CB-IV was a relatively minor subcomponent from the SPL and precuneus to the frontal region. CB-V, the parahippocampal cingulum, stemmed from the medial temporal lobe and fanned out to the occipital lobes [37]. CB-V supposedly participates as a major subcortical stream within the default mode network, possibly subserving the transfer of multimodal cues relevant to visuospatial, facial, and mnemonic information to the precuneal hub [38].

During white matter dissection, the cingulum is observed on the medial side [39, 40]. The gray matter of the medial cerebral surface should be peeled away, starting from the depth of the anterior part of the cingulate sulcus and gradually extending posteriorly to the precuneus. Attention should be paid to the dissection of the cingulum, as it is easily disrupted, especially in its inferior arm such as the parahippocampal gyrus.

During tractography, a single ROI (Ci) (Fig. 14a, b) is used to dissect the cingulum. A single cigar-shaped region is defined on the top three slices. When the cingulum separates into two branches, an anterior and posterior region is

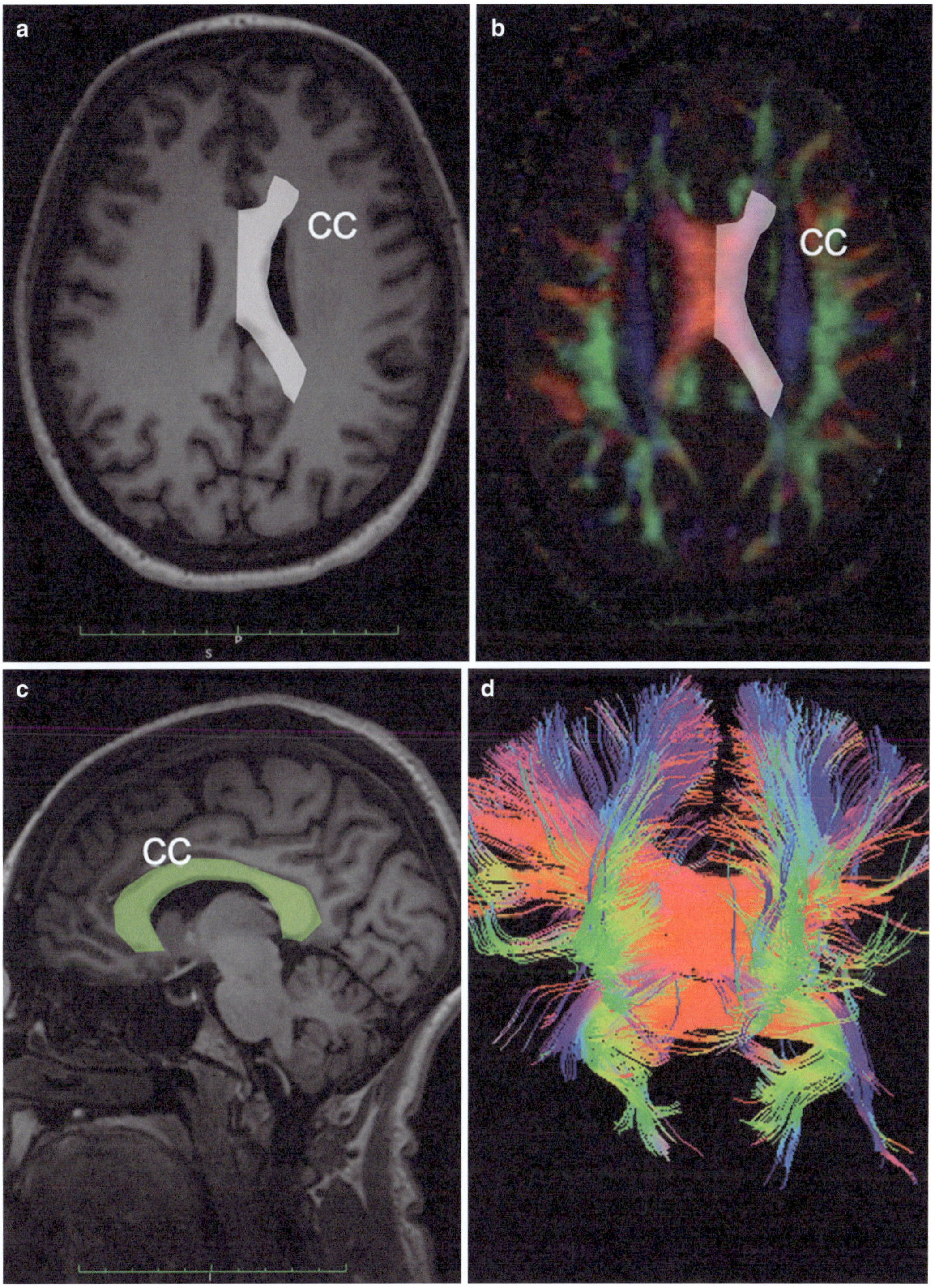

Fig. 13 Corpus callosum. Delineation of the ROI: A single ROI (CC) is delineated around the body, the genu, and the splenium of the corpus callosum on the T1-weighted image (**a**) and FA map (**b**). The shapes of the regions follow the anatomy of the different parts of the corpus callosum: butterfly shape for the body and horseshoe shape for the genu and splenium. (**c**) Alternatively, a single ROI around the corpus callosum on a midsagittal slice can be used. (**d**) tractography of the Corpus callosum

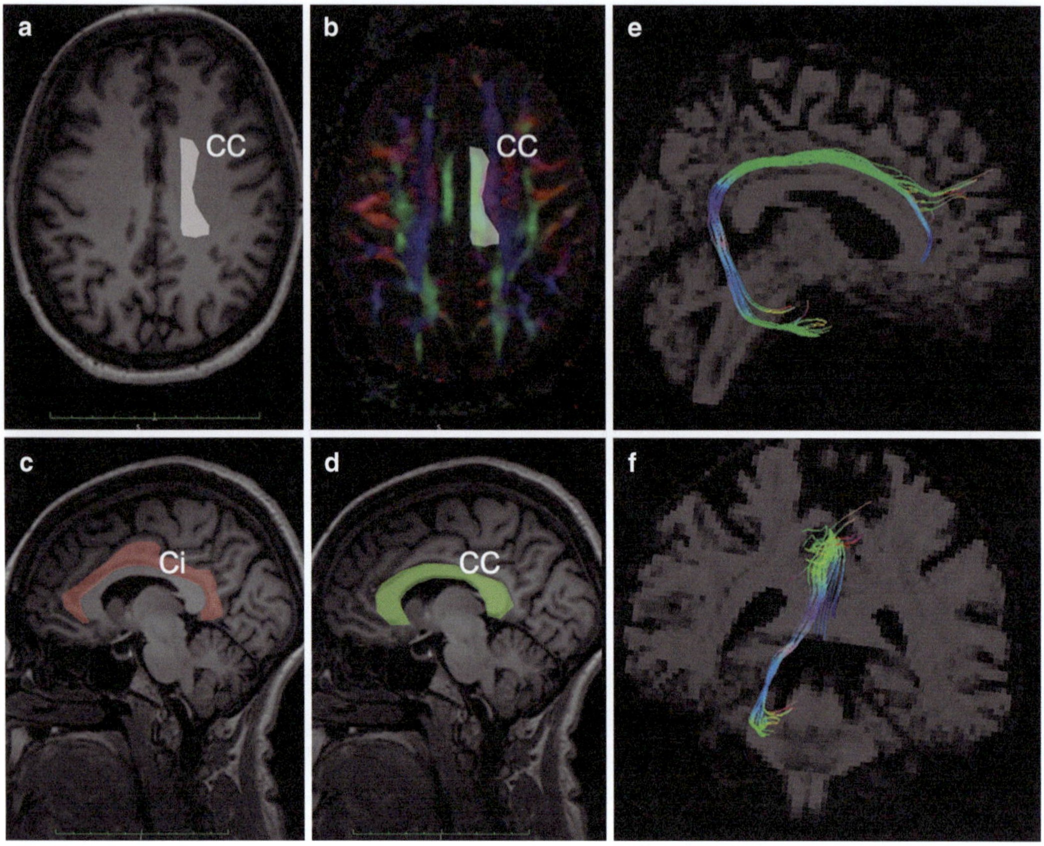

Fig. 14 Cingulum. Delineation of the ROIs used for tractography of the cingulum. A single ROI (Ci) is used to dissect the cingulum. A single cigar-shaped region is delineated in an axial slice on the T1 weighted image (**a**) and FA map (**b**). It is important to remember that the majority of the fibers of the cingulum are short U-shaped fibers connecting adjacent gyri. The use of a two-ROI approach excludes the majority of these short fibers from the analysis. (**c**) Cingulum ROI (Ci) in the hemisphere. (**d**) Corpus callosum ROI (CC) in the medial section to exclude the connections belonging to the corpus callosum. A two-ROI approach was used to isolate the cingulum fasciculus (**c**) NOT (**d**). A sagittal ROI was set around the white matter of the cingulum for the hemispheres. Streamlines of the corpus callosum projecting to the hemisphere opposite to the ROI were excluded using a sagittal 'NOT' ROI on the corpus callosum. (**e**, **f**) tractography of the Cingulum

delineated on each slice. It is important to remember that the majority of the fibers of the cingulum are short U-shaped fibers connecting the adjacent gyrus. The use of a two-ROI approach excludes the majority of these short fibers from the analysis. For this reason, the use of the one-ROI approach ROI (Ci) (Fig. 14a, b) or ROI Ci (Fig. 14c) without fibers of ROI CC (Fig. 14d), which includes all fibers of the cingulum is recommended (Figs. 14e, f, 15, and 16).

Ethical Approval All procedures performed in this study involving human participants were in accordance with the ethical standards, of the institutional research committee of the hospital of Fujita Health University with the 1964 Helsinki declaration and its later amendments or comparable ethical standards.

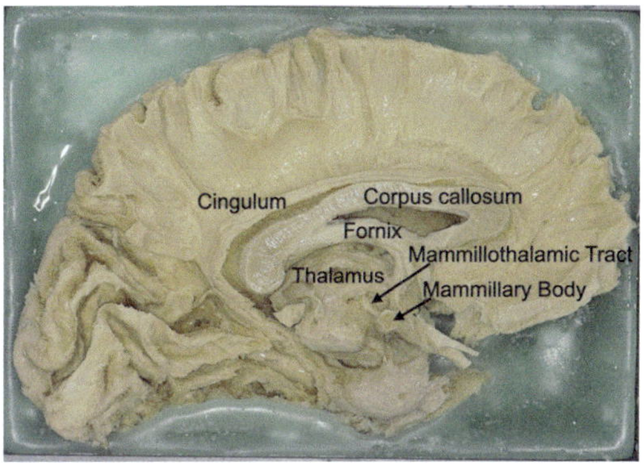

Fig. 15 Stepwise white matter dissection using the medial approach. The gray matter of the medial cerebral surface should be peeled away. Attention must be given to the dissection of the cingulum, as it is easily disrupted, especially in the inferior arm of the parahippocampal gyri. Subsequently, the cingulum fibers will be exposed

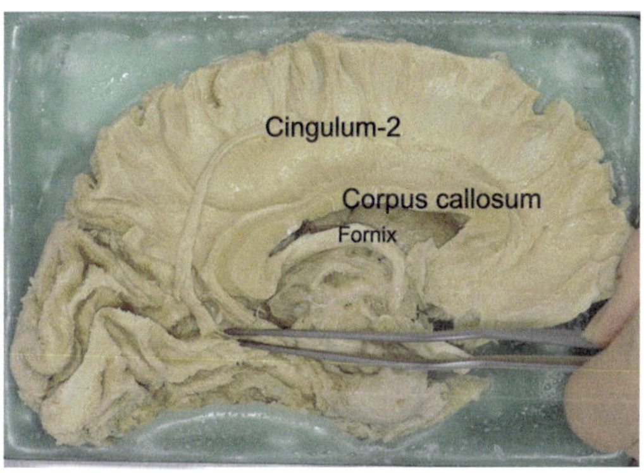

Fig. 16 Stepwise white matter dissection using the medial approach. The cingulum fibers consist of five layers. The CB2 (the second part of the cingulum) was held by the forceps

References

1. Sanai N, Polley MY, McDermott MW, Parsa AT, Berger MS. An extent of resection threshold for newly diagnosed glioblastomas. J Neurosurg. 2011;115(1):3–8. https://doi.org/10.3171/2011.2.JNS10998.
2. Makris N, Kennedy DN, McInerney S, et al. Segmentation of subcomponents within the superior longitudinal fascicle in humans: a quantitative, in vivo, DT-MRI study. Cereb Cortex. 2005;15(6):854–69. https://doi.org/10.1093/cercor/bhh186.
3. Petrides M, Pandya DN. Efferent association pathways originating in the caudal prefrontal cortex in the macaque monkey. J Comp Neurol. 2006;498(2):227–51. https://doi.org/10.1002/cne.21048.
4. Catani M, Thiebaut de Schotten M. A diffusion tensor imaging tractography atlas for virtual in vivo dissections. Cortex. 2008;44(8):1105–32. https://doi.org/10.1016/j.cortex.2008.05.004.
5. Rojkova K, Volle E, Urbanski M, Humbert F, Dell'Acqua F. Thiebaut de Schotten M. Atlasing the frontal lobe connections and their variability due to age and education: a spherical deconvolution tractography study. Brain Struct Funct. 2016;221(3):1751–66. https://doi.org/10.1007/s00429-015-1001-3.
6. Thiebaut de Schotten M, Dell'Acqua F, Forkel SJ, et al. A lateralized brain network for visuospatial attention. Nat Neurosci. 2011;14(10):1245–6. https://doi.org/10.1038/nn.2905.
7. Yagmurlu K, Middlebrooks EH, Tanriover N, Rhoton AL Jr. Fiber tracts of the dorsal language stream in the human brain. J Neurosurg. 2016;124(5):1396–405. https://doi.org/10.3171/2015.5.Jns15455.
8. Catani M, Mesulam M. The arcuate fasciculus and the disconnection theme in language and aphasia: history and current state. Cortex. 2008;44(8):953–61. https://doi.org/10.1016/j.cortex.2008.04.002.
9. Mandonnet E, Duffau H. Understanding entangled cerebral networks: a prerequisite for restoring brain function with brain-computer interfaces. Front Syst Neurosci. 2014;8:82. https://doi.org/10.3389/fnsys.2014.00082.
10. Kaplan E, Naeser MA, Martin PI, et al. Horizontal portion of arcuate fasciculus fibers track to pars opercularis, not pars triangularis, in right and left hemispheres: a DTI study. NeuroImage. 2010;52(2):436–44. https://doi.org/10.1016/j.neuroimage.2010.04.247.
11. Doricchi F, Thiebaut de Schotten M, Tomaiuolo F, Bartolomeo P. White matter (dis)connections and gray matter (dys)functions in visual neglect: gaining insights into the brain networks of spatial awareness. Cortex. 2008;44(8):983–95. https://doi.org/10.1016/j.cortex.2008.03.006.
12. Catani M, Mesulam MM, Jakobsen E, et al. A novel frontal pathway underlies verbal fluency in primary progressive aphasia. Brain. 2013;136(Pt 8):2619–28. https://doi.org/10.1093/brain/awt163.
13. Poeppel D, Guillemin A, Thompson J, Fritz J, Bavelier D, Braun AR. Auditory lexical decision, categorical perception, and FM direction discrimination

differentially engage left and right auditory cortex. Neuropsychologia. 2004;42(2):183–200. https://doi.org/10.1016/j.neuropsychologia.2003.07.010.
14. Schmahmann JDPD. Fiber pathways of the brain. Oxford University Press; 2006.
15. Martino J, Brogna C, Robles SG, Vergani F, Duffau H. Anatomic dissection of the inferior fronto-occipital fasciculus revisited in the lights of brain stimulation data. Cortex. 46(5):691–9. https://doi.org/10.1016/j.cortex.2009.07.015.
16. Duffau H. New concepts in surgery of WHO grade II gliomas: functional brain mapping, connectionism and plasticity--a review. J Neuro-Oncol. 2006;79(1):77–115. https://doi.org/10.1007/s11060-005-9109-6.
17. Duffau H, Capelle L, Lopes M, Faillot T, Sichez JP, Fohanno D. The insular lobe: physiopathological and surgical considerations. Neurosurgery. 2000;47(4):801–10.; discussion 810–1. https://doi.org/10.1097/00006123-200010000-00001.
18. Duffau H, Gatignol P, Mandonnet E, Peruzzi P, Tzourio-Mazoyer N, Capelle L. New insights into the anatomo-functional connectivity of the semantic system: a study using cortico-subcortical electrostimulations. Brain. 2005;128(Pt 4):797–810. https://doi.org/10.1093/brain/awh423.
19. Epelbaum S, Pinel P, Gaillard R, et al. Pure alexia as a disconnection syndrome: new diffusion imaging evidence for an old concept. Cortex. 2008;44(8):962–74. https://doi.org/10.1016/j.cortex.2008.05.003.
20. Rudrauf D, Mehta S, Grabowski TJ. Disconnection's renaissance takes shape: formal incorporation in group-level lesion studies. Cortex. 2008;44(8):1084–96. https://doi.org/10.1016/j.cortex.2008.05.005.
21. Moritz-Gasser S, Herbet G, Duffau H. Mapping the connectivity underlying multimodal (verbal and non-verbal) semantic processing: a brain electrostimulation study. Neuropsychologia. 2013;51(10):1814–22. https://doi.org/10.1016/j.neuropsychologia.2013.06.007.
22. Buchsbaum MS, Schoenknecht P, Torosjan Y, et al. Diffusion tensor imaging of frontal lobe white matter tracts in schizophrenia. Ann General Psychiatry. 2006;5:19.
23. Grossi D, Soricelli A, Ponari M, et al. Structural connectivity in a single case of progressive prosopagnosia: the role of the right inferior longitudinal fasciculus. Cortex. 2014;56:111–20. https://doi.org/10.1016/j.cortex.2012.09.010.
24. Beis JM, Keller C, Morin N, et al. Right spatial neglect after left hemisphere stroke: qualitative and quantitative study. Neurology. 2004;63(9):1600–5.
25. Zemmoura I, Herbet G, Moritz-Gasser S, Duffau H. New insights into the neural network mediating reading processes provided by cortico-subcortical electrical mapping. Hum Brain Mapp. 2015;36(6):2215–30. https://doi.org/10.1002/hbm.22766.
26. Fox CJ, Iaria G, Barton JJ. Disconnection in prosopagnosia and face processing. Cortex. 2008;44(8):996–1009. https://doi.org/10.1016/j.cortex.2008.04.003.
27. Ffytche DH. The hodology of hallucinations. Cortex. 2008;44(8):1067–83. https://doi.org/10.1016/j.cortex.2008.04.005.
28. Ross ED. Sensory-specific amnesia and hypoemotionality in humans and monkeys: gateway for developing a hodology of memory. Cortex. 2008;44(8):1010–22. https://doi.org/10.1016/j.cortex.2008.02.002.
29. Koutsarnakis C, Liakos F, Kalyvas AV, Sakas DE, Stranjalis G. A laboratory manual for stepwise cerebral white matter fiber dissection. World Neurosurg. 2015;84(2):483–93. https://doi.org/10.1016/j.wneu.2015.04.018.
30. Gaffan D, Wilson CR. Medial temporal and prefrontal function: recent behavioural disconnection studies in the macaque monkey. Cortex. 2008;44(8):928–35. https://doi.org/10.1016/j.cortex.2008.03.005.
31. Yordanova YN, Duffau H, Herbet G. Neural pathways subserving face-based mentalizing. Brain Struct Funct. 2017;222(7):3087–105. https://doi.org/10.1007/s00429-017-1388-0.
32. Crosby E, Humphrey T, Lauer EW. *White matter of the hemisphere*. Pt 3Correlative anatomy of the nervous system. Macmillan; 1962.
33. Aboitiz F, Montiel J. One hundred million years of interhemispheric communication: the history of the corpus callosum. Braz J Med Biol Res. 2003;36(4):409–20. https://doi.org/10.1590/s0100-879x2003000400002.
34. Witelson SF. Hand and sex differences in the isthmus and genu of the human corpus callosum. A postmortem morphological study. Brain. 1989;112(Pt 3):799–835. https://doi.org/10.1093/brain/112.3.799.
35. Friedrich P, Forkel SJ, Thiebaut de Schotten M. Mapping the principal gradient onto the corpus callosum. NeuroImage. 2020;223:117317.
36. Shah A, Jhawar S, Goel A, Goel A. Corpus callosum and its connections: a finer dissection study. World Neurosurg. 2021;151:e1024–35.
37. Wu Y, Sun D, Wang Y, Wang Y. Subcomponents and connectivity of the inferior Fronto-occipital fasciculus revealed by diffusion Spectrum imaging fiber tracking. Front Neuroanat. 2016;10:88. https://doi.org/10.3389/fnana.2016.00088.
38. Skandalakis GP, Komaitis S, Kalyvas A, et al. Dissecting the default mode network: direct structural evidence on the morphology and axonal connectivity of the fifth component of the cingulum bundle. J Neurosurg. 2020;134(3):1334–45. https://doi.org/10.3171/2020.2.JNS193177.
39. Shah A, Goel A, Jhawar SS, Patil A, Rangnekar R, Goel A. Neural circuitry: architecture and function - a finer dissection study. World Neurosurg. 2019;125:e620–38.
40. Shah A, Jhawar SS, Goel A. Analysis of the anatomy of the Papez circuit and adjoining limbic system by fibre dissection techniques. J Clin Neurosci. 2012;19(2):289–98.

Origin and Spread of Gliomas and Surgical Strategy of Gliomas Based on a Novel Surgical Classification

Abhidha Shah and Atul Goel

1 Introduction

The treatment of low grade gliomas has evolved over the last two decades from clinical observation to radical excision. These advances have been possible due to recent strides in imaging and in molecular analysis and the use of histological and molecular markers for disease prognostication [1]. The current standard of care includes radical, total or supra-maximal resection of the gliomas to decrease tumor burden and to achieve improved life-quality and a longer disease free survival [2–5]. Despite the advances in surgery, imaging, neurophysiology and molecular biology, we are still vexed by the lack of understanding in the origin and spread of these tumors. We speculate that gliomas arise from a named white matter tract and spreads along it and displaces the normal adjacent white matter bundles.

A. Shah (✉)
Department of Neurosurgery, K.E.M. Hospital and Seth G.S. Medical College, Parel, Mumbai, India

A. Goel
Department of Neurosurgery, Lilavati Hospital and Research Center, Bandra, Mumbai, India

Department of Neurosurgery, R.N. Cooper Hospital, Mumbai, India

Bombay Hospital Institute of Medical Sciences and Research Center, Mumbai, India

2 Origin of Gliomas

Yasargil believed that the initial sites of glioma formation in the cerebrum were subcortical, subgyral, lobar, callosal and limbic or paralimbic [6]. According to him the tumors grew on the basis of their segmental vascular supply and embryological glial pathways towards the ventricle. He believed that the tumors rarely if ever spread along the long association fibers such as the superior longitudinal fasciculus, inferior fronto-occipital fasciculus or the cingulum.

Much of the information of spread of gliomas is based on post-mortem studies from older neuropathological and neurosurgical literature when there was access to a large number of untreated gliomas. The spread of gliomas was believed to be along the path of blood vessels and also along myelinated axons. It was speculated that due to the high affinity of the gliomas for myelinated tracts distant spread may occur to the optic radiations through the corpus callosum or the anterior commissure thus resulting in bilateral spread of tumors. Giese et al. in their paper noted that dissemination of gliomas seemed to be confined to the white matter and only limited by the cortex and deep grey matter nuclei [7]. However, all this information was derived from the study of tumors that were in their terminal stages of growth. In this era of modern imaging with early detection of gliomas, we are able to decipher and speculate on

A. Shah et al. (eds.), *Functional Anatomy of the Brain: A View from the Surgeon's Eye*,
https://doi.org/10.1007/978-981-99-3412-6_11

the origin and spread of gliomas. We believe that gliomas arise and grow along white matter and do so in a well-organized manner following the trajectory of the fiber tracts.

Recently bio-mathematical *in-silico* models have been proposed to elucidate the growth pattern of gliomas [8–11]. These models identify that the growth of gliomas occurs through two mechanisms: proliferation and diffusion. During proliferation there is growth of tumor cells by cellular division, leading to expansion. Diffusion causes the cells to infiltrate into the surrounding parenchyma especially around the white matter tracts. Proliferation gives the glioma a spherical or a "localized" geometrical shape whereas diffusion gives the tumor a convoluted or a "diffuse" shape mimicking the white matter pathways. Duffau et al. proposed an anatomical classification for paralimbic or insular low grade gliomas based on white matter invasion patterns along the involved fiber tract viz., the uncinate fasciculus, the arcuate fasciculus and the fronto-occipital fasciculus and the optic radiations [8].

Gliomas have been variously classified based on cell type, histological grade, molecular markers and their location in the brain. Sawaya et al. proposed a classification system for pre-operative planning based on eloquence, where Sawaya I included tumors in functionally silent areas of the brain, Sawaya II included tumors adjacent to functionally eloquent areas and Sawaya III included tumors infiltrating functionally eloquent areas [12]. Similarly Friedlein et al. published a classification to assess tumor operability based on two gradations. Friedlein grading A included lesions at a safe distance from eloquent regions and could be completely resected and Friedlein grading B referred to tumors that could only be partially resected or biopsied due to their location [13]. None of these classifications take into consideration the subcortical networks of the brain, which have their own functional eloquence as has come to light in recent years.

We proposed a novel anatomical classification of white fibers of the brain based on their course and depth [14–17]. According to this classification as we have seen earlier, the white fibers were divided into four horizontal groups and one vertical group. The first horizontal group is that of the short arcuate or short association fibers. The second horizontal group is that of the middle group which consists of the long association fibers. The fibers in this group are the superior longitudinal fasciculus, the arcuate fasciculus, the inferior fronto-occipital fasciculus, the middle and inferior longitudinal fasciculus, the uncinate fasciculus and the cingulum. The third horizontal layer is that of the deep group which mainly consists of association fibers in proximity of the ventricle such as the fornix, the stria terminalis and the striae medullaris thalami. The fourth horizontal group is the central group, which essentially comprises of the commissural fibers. The vertical group of fibers consists of the internal capsule and the thalamic radiations. It was observed that the long association fibers on either side of the vertical group do not cross over to the other side of the vertical group of fibers. The anterior commissure is the only fiber tract that crosses the vertical group of fibers inferiorly as it goes to the temporal lobe. The corpus callosum envelops the entire ventricular system within its dorsal and ventral radiations.

Based on this fiber tract model we speculate that gliomas arise from a named white fiber tract i.e. from the white matter of the short arcuate fibers, the commissural fibers and long association fibers [18–20]. Gliomas arising primarily from the projection fibers that are the internal capsule, the optic radiations and the thalamic radiations are rare. Gliomas both low and high grade can be seen to have a defined pattern of origin and extension. The extension is disciplined and is along a named white fiber tract and its haphazard extension is restricted or limited by the adjoining traversing tracts. The fibers around the tumor are edematous and the edema is seen to extend along the course of the involved fiber [18–20].

3 Morphological Classification of Gliomas

Gliomas (both low and high grade) can be morphologically classified into two broad categories: localized and diffuse [18–20] (Fig. 1). The

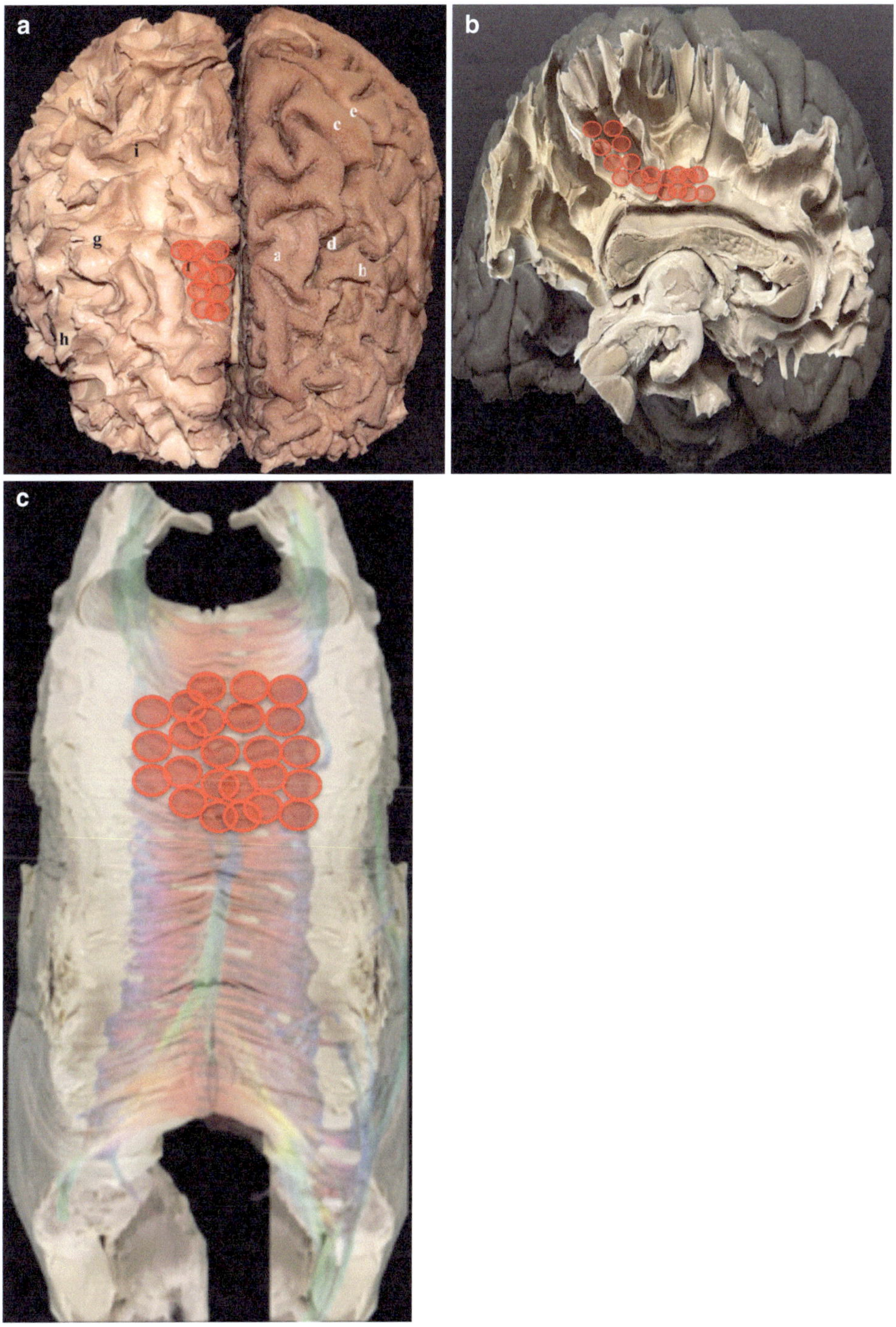

Fig. 1 Images showing the classification of the gliomas. (**a**) Image of the superior view of the cadaveric human brain. On the right half of the hemisphere superficial dissection has been performed to reveal the short arcuate fibers. The red circles denote a localized glioma arising from the short arcuate fibers in the region of the superior frontal gyrus. (**b**) Medial view of the fiber dissected cerebral hemisphere. The red circles depict a diffuse glioma arising from the cingulum. (**c**) Superior view of the dissected corpus callosum. The red circles denote a diffuse glioma arising from the commissural fibers and growing bilaterally

localized gliomas arise from the short arcuate fibers of the involved gyrus and are named according to the gyrus they are associated with. The diffuse gliomas arise from the long association fiber bundles or the commissural fibers.

Gliomas that arise from the short arcuate fibers are named as medial, lateral and basal frontal, precentral and post central, medial and lateral parietal, medial and lateral occipital and medial, basal temporal and lateral temporal depending on the location of the short arcuate fibers. The insular tumors are considered as a separate entity because insula harbors tumors arising both from the short association fibers and the long association fibers [18–20].

As a general rule, localized gliomas remain confined to the particular gyrus that contains the involved short arcuate fibers. When the tumor involves the adjacent gyrus, it is through the short arcuate connections between the two neighboring gyri. Gliomas that arise medial to the vertical group remain medial to the vertical group and similarly gliomas that arise lateral to the vertical group remain lateral to it. Gliomas that arise from the commissural fibers have a predilection for bilateral extension. Tumors that arise in the region of the genu and the forceps minor extend bilaterally anteriorly into the medial frontal brain. Tumors of the body of the corpus callosum extend bilaterally into the fronto –parietal brain depending on the part of the corpus callosum that is involved. Tumors of the splenium and forceps major extend bilaterally into the medial parieto-occipital lobes. As a general rule corpus callosal gliomas do not cross the vertical group of fibers and remain confined medial to this group of fibers [18–20].

4 Tumors of the Short Arcuate Fibers

4.1 Frontal Lobe

Tumors that arise from the short arcuate fibers of the frontal lobe can be divided into medial, lateral and basal groups. Medial frontal group of tumors involve the short arcuate fibers of the superior and middle frontal gyri. As the tumors grow they expand the involved gyrus. Inferior expansion of the gyri causes these tumors to reach upto the ependymal lining of the frontal horn of the ventricle, which is usually not breached (Figs. 2 and 3). By virtue of the short arcuate connection to the adjacent gyrus the tumors of the middle frontal gyrus can involve the superior frontal gyrus. Lateral frontal tumors are those arising from the short arcuate fibers of the inferior frontal gyrus. As they grow, these tumors expand the inferior frontal gyrus but do not reach the ventricle generally due to a barrier of a multitude of long association fibers preventing its extension towards the ventricle. Tumors arising from the short arcuate fibers of the lateral basal frontal brain or the orbital gyri are localized to the orbital gyri. Growth of these tumors posteriorly is towards the floor of the frontal horn. Forceps minor prevents the medial extension of these tumors. Tumors in the medial basal frontal region are largely those that arise from the forceps minor.

4.2 Central Lobe

Gliomas of the central lobe are those arising from the short arcuate fibers in the region of the precentral and postcentral gyrus (Figs. 4 and 5).

4.3 Parietal Lobe

Tumors of the parietal lobe can be divided into medial and lateral groups. Tumors arising from the short arcuate fibers of the inferior parietal lobule are the more common variety. They either involve the supramarginal gyrus or the angular gyrus singly or in combination (Fig. 6). Expansion of these tumors causes them to reach the lateral wall of the trigone of the ventricle. However, the tumors do not cross the lateral wall of the ventricle due to the presence of the optic radiations. Tumors arising from the short arcuate fibers of the superior parietal lobe form the medial group (Figs. 7 and 8). Some tumors that are seen to involve the medial parietal lobe in the region of

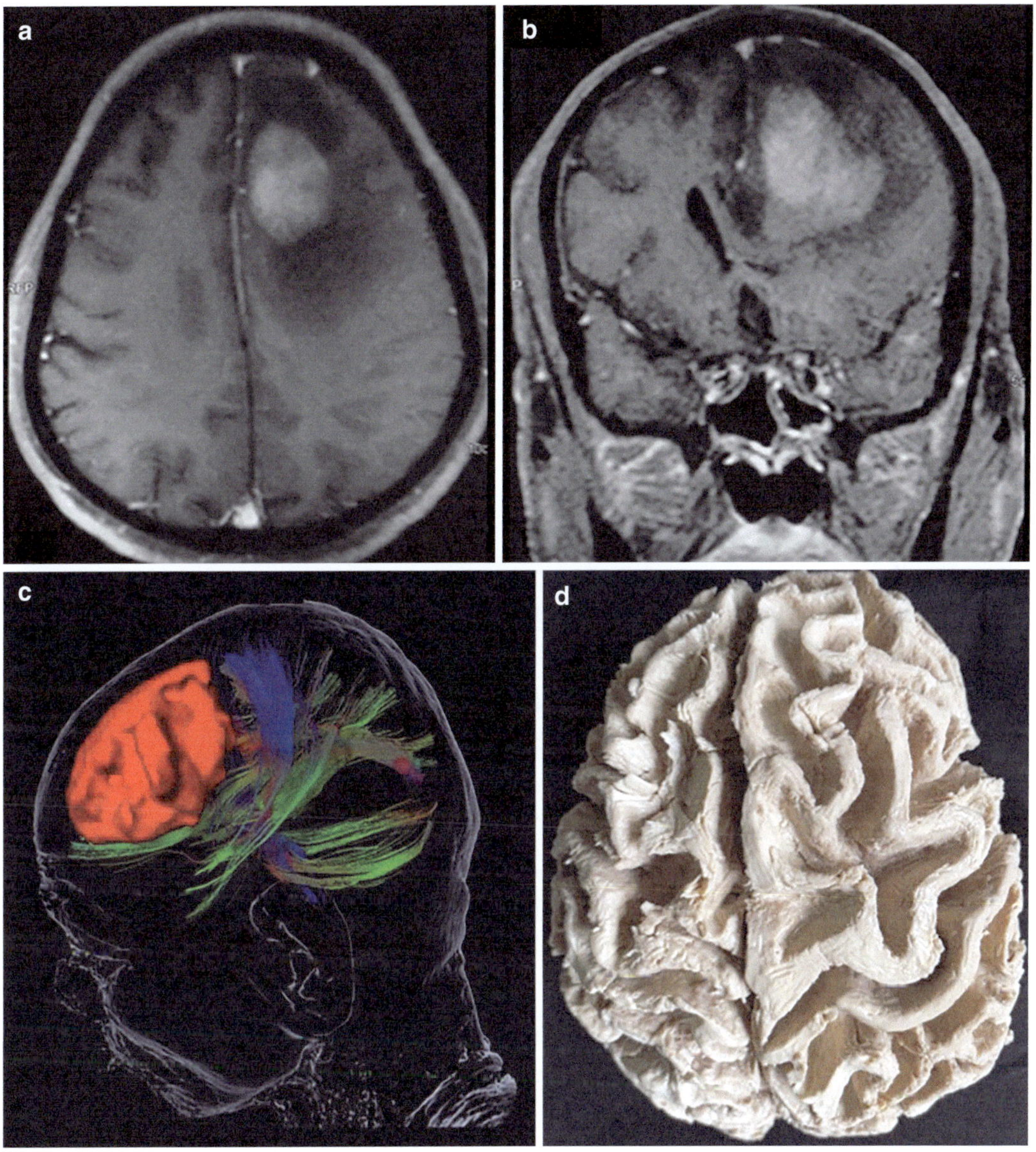

Fig. 2 Images of a 58 year old male patient with a medial frontal glioma. (**a**) T1 weighted axial image showing a contrast enhancing lesion in the region of the superior frontal gyrus. (**b**) Contrast enhanced coronal image showing the tumor. (**c**) Tractography images showing the relation of the lesion to the fiber tracts. The lesion involves the short arcuate fibers in the region of the superior frontal gyrus, lies medial to the vertical group of fibers and displaces the superior longitudinal fasciculus inferiorly. (**d**) Image of the superior surface of the cerebral hemispheres. The gray matter has been removed. The short arcuate fibers and their connections to adjoining gyri can be visualized

the precuneus are actually those that arise from the cingulum and traverse upwards by virtue of the short arcuate connection of the cingulum to the medial parietal lobe.

4.4 Temporal Lobe

Tumors of the temporal lobe can be divided into medial, basal and lateral groups. Tumors in the

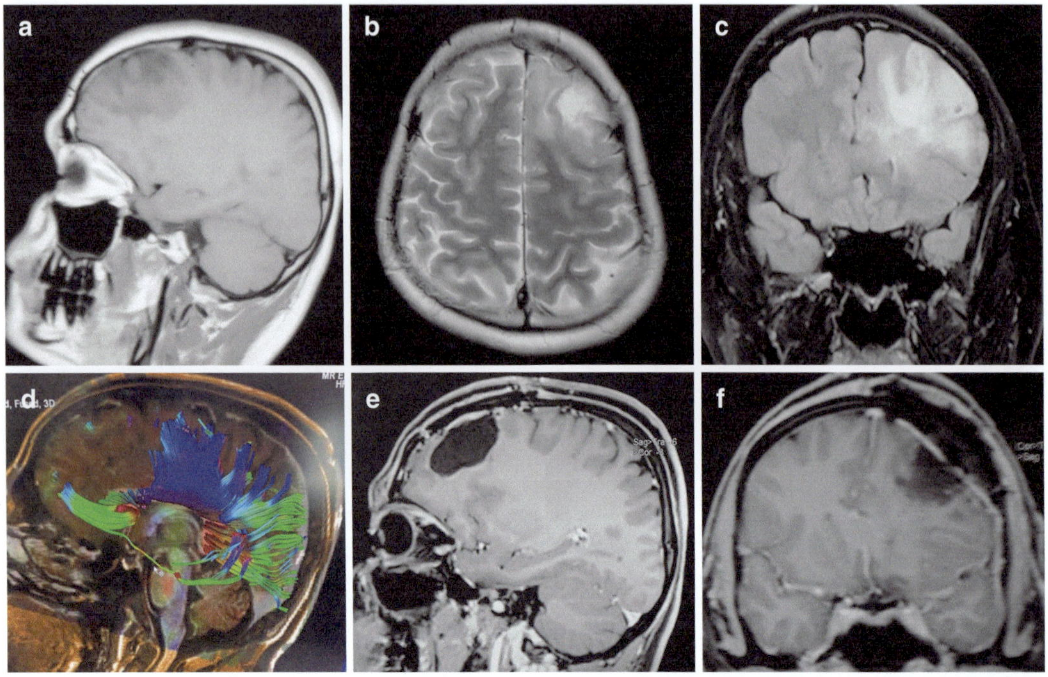

Fig. 3 Images of a – 37 year old male patient. (**a**) T1 weighted sagittal MRI showing a low grade gliomas in the region of the left middle frontal gyrus. (**b**) T2 weighted axial image showing the tumor. (**c**) FLAIR coronal image showing the tumor. (**d**) Tractography image showing the tumor medial to the projection group of fibers. (**e**) Post-operative T1 weighted sagittal contrast image showing the radical tumor resection. (**f**) Post-operative T1 weighted contrast coronal image showing the resection cavity

lateral group arise from the short arcuate fibers of the superior, middle and inferior temporal gyri. These tumors lie lateral to all the association fibers that run into the temporal lobe from the frontal lobe (Fig. 9). They extend upto the lateral wall of the temporal horn but do not breech the ependymal layer. Tumors of the medial temporal group involve the parahippocampal gyrus and in contrast to the lateral group are in close relation to the medial aspect of the temporal horn or the choroidal fissure (Fig. 10). These tumors project in to the medial portion of the ventricle do not breech the lateral wall of the temporal horn. Tumors of the basal group arise from short arcuate fibers of the fusiform gyrus (Fig. 11). These tumors can grow medially or laterally. There is no major long association fiber bundle preventing its extension either medially or laterally. These tumors are limited superiorly by the floor of the temporal horn.

4.5 Insular Lobe

The insula is a complex region where a number of fiber tracts run in close proximity. We have divided tumors of the insula into 4 types. These include gliomas arising from the extreme capsule or the short and long insular gyri, tumors of the uncinate fasciculus, tumors of the inferior fronto-occipital fasciculus and tumors of the anterior commissure. Tumors of the extreme capsule or those of the insular gyri remain confined to the gyri of the insula. These form the localized group of gliomas of the insula (Figs. 12 and 13). Tumors that arise from the uncinate fasciculus involve the basal frontal brain and the anterior medial temporal lobe. Tumors arising from the inferior fronto-occipital fasciculus involve the basal portion of the frontal lobe and then spread across to the posterior temporal lobe following the direction of its fibers. Tumors that involve the anterior commissure spread across the

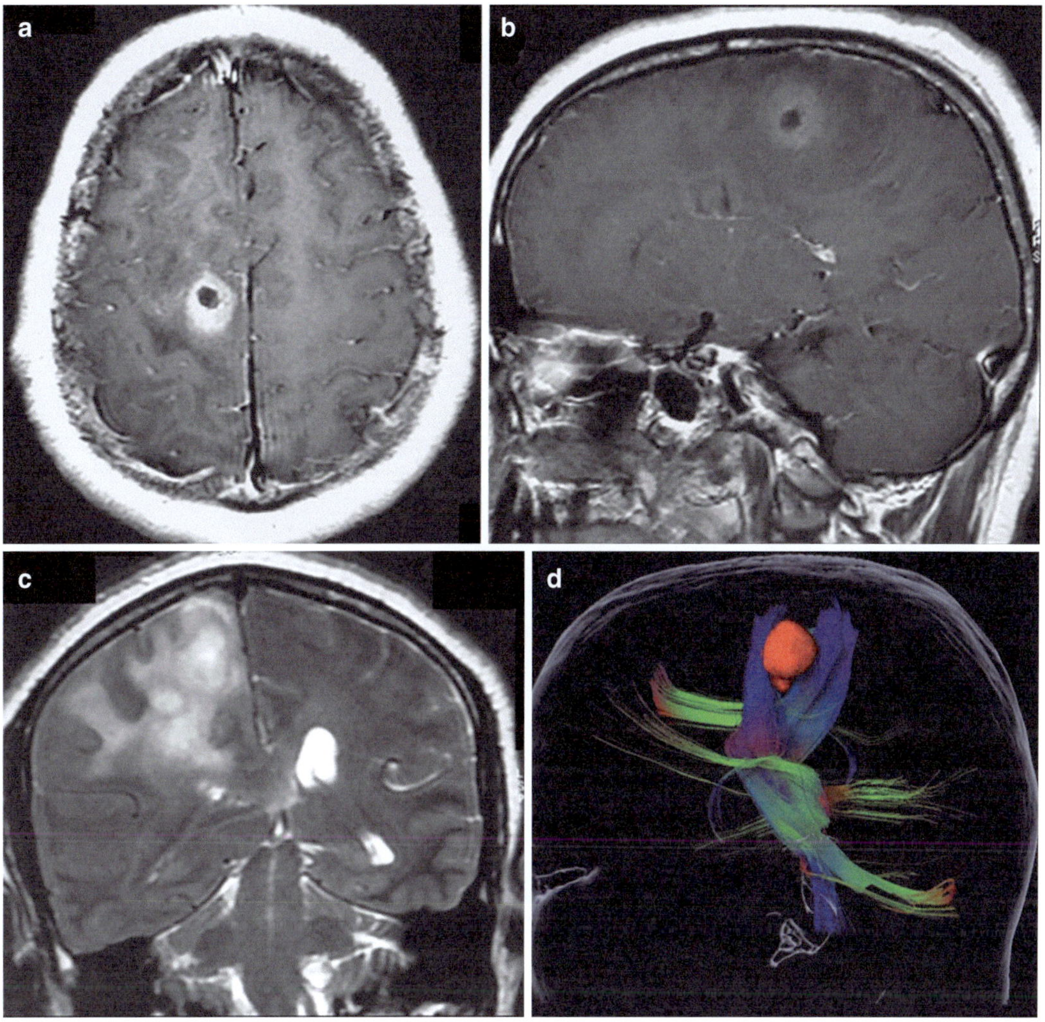

Fig. 4 Images of a 35 year old male patient with a pre-central tumor. (**a**) T1 weighted axial image showing a contrast enhancing glioma in the region of the pre-central gyrus. (**b**) Sagittal image showing the tumor. (**c**) T2 weighted image showing the tumor. The edematous short arcuate fibers can be visualized. (**d**) Tractography images showing the tumor arising from the short arcuate fibers in the precentral gyrus. The corticospinal tracts are seen draped around the tumor

midline to the contralateral side and also can have a frontal and temporal extension. These gliomas are further discussed in the section of tumors arising from long association and commissural fibers.

4.6 Occipital Lobe

Tumors of the occipital lobe can be divided into medial and lateral groups. These tumors were divided into medial and lateral groups based on their relation to the occipital horn of the lateral ventricle. Tumors that arose from short arcuate fibers lateral to the occipital horn formed the lateral group. These tumors were bounded laterally by the optic radiations. Tumors arising from the short arcuate fibers medial to the occipital horn formed the medial group. They were bounded laterally by the fibers of the forceps major.

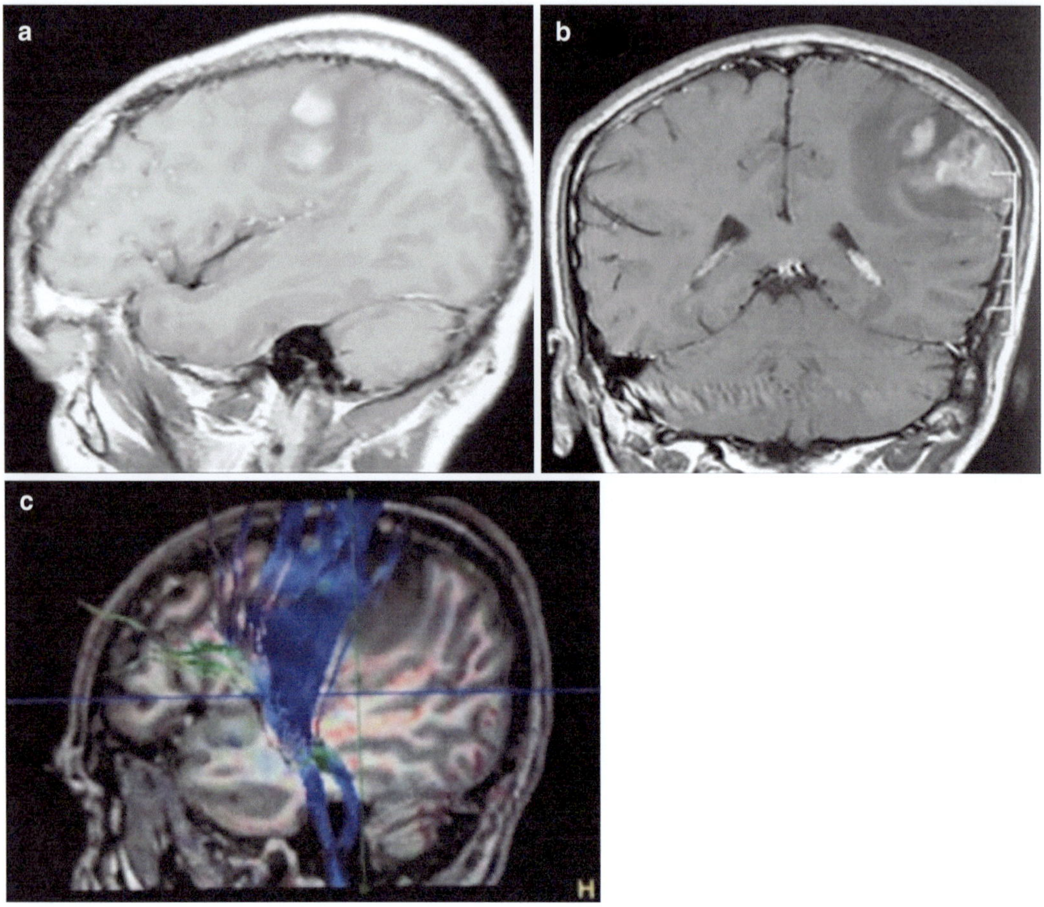

Fig. 5 Images of a 27 year old male with a post-central tumor. (**a**) T1 weighted sagittal contrast image showing the lesion in the region of the post-central gyrus. (**b**) Coronal image showing the tumor. (**c**) Tractography images showing the glioma arising from short arcuate fibers in the region of the post central gyrus. The vertical group of projection fibers are displaced anteriorly by the tumor

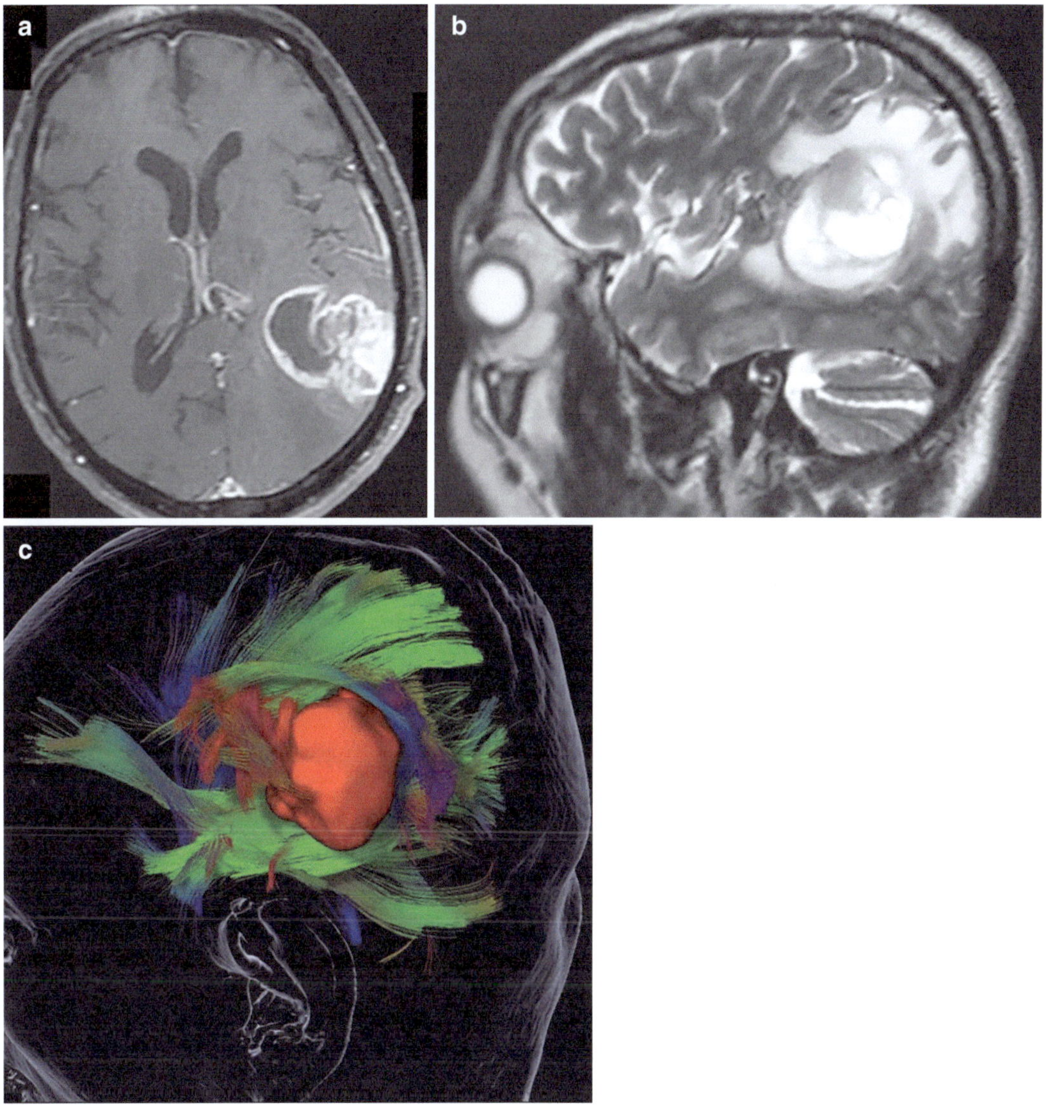

Fig. 6 Images of a 60 year old patient with a lateral parietal tumor. (**a**) T1 weighted axial contrast image showing the tumor in the region of the inferior parietal lobule. (**b**) T2 weighted sagittal image showing the lesion in the region of the angular gyrus. (**c**) Tractography images showing the tumor arising from short arcuate fibers in the region of the angular gyrus and displacing the arcuate fasciculus and the inferior fronto-occipital fasciculus medially. The tumor adjoining short arcuate fibers are edematous

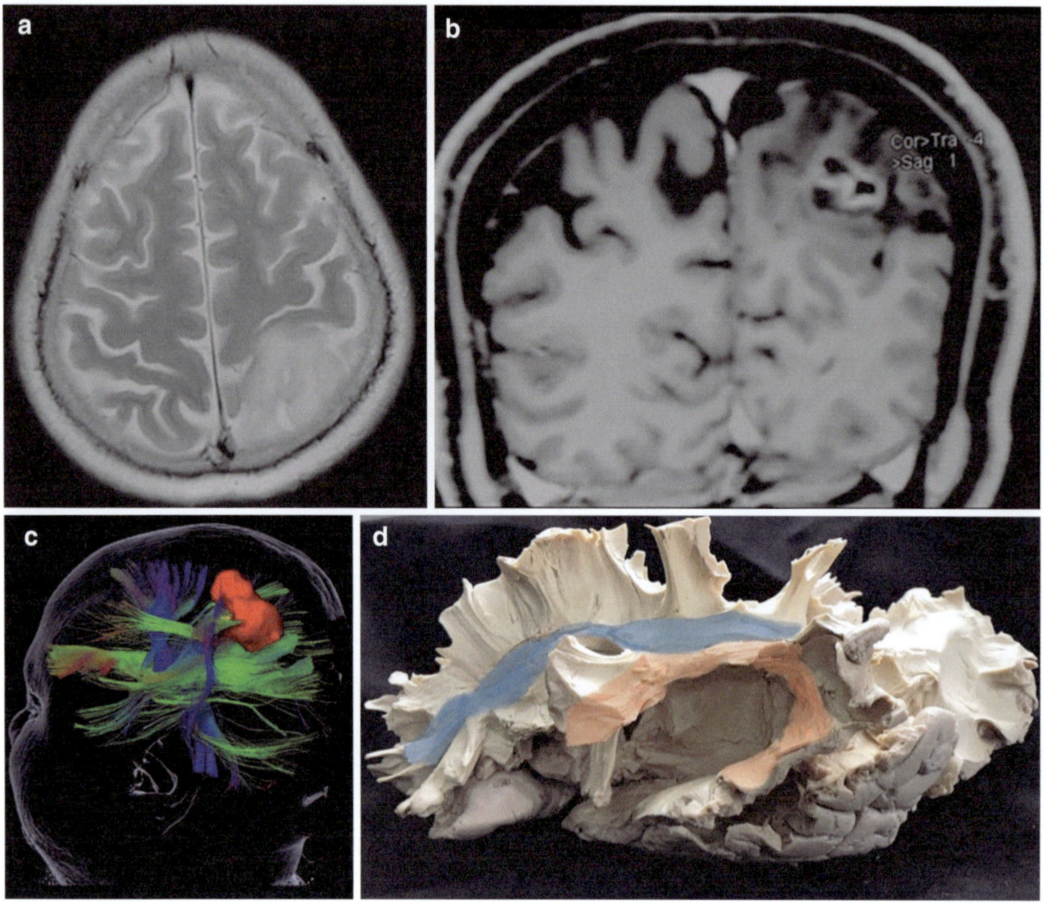

Fig. 7 Images of a 45 year old male patient with a medial parietal tumor. (**a**) T2 weighted axial image showing the tumor in the region of the superior parietal lobule. (**b**) Coronal contrast enhanced image showing the tumor in the region of the superior parietal lobule. (**c**) Tractography images showing the tumor involving the short arcuate fibers in the region of the superior parietal lobule. The normal course of tumor adjoining arcuate fasciculus, the inferior fronto-occipital fasciculus and the vertical group of fibers is visualized. (**d**) Image of lateral aspect of the cerebral hemisphere. The frontal, parietal and temporal operculum has been removed. The normal course of the arcuate fasciculus (orange) and the lateral portion of the superior longitudinal fasciculus (blue) and its relation to the superior and inferior parietal lobule can be visualized

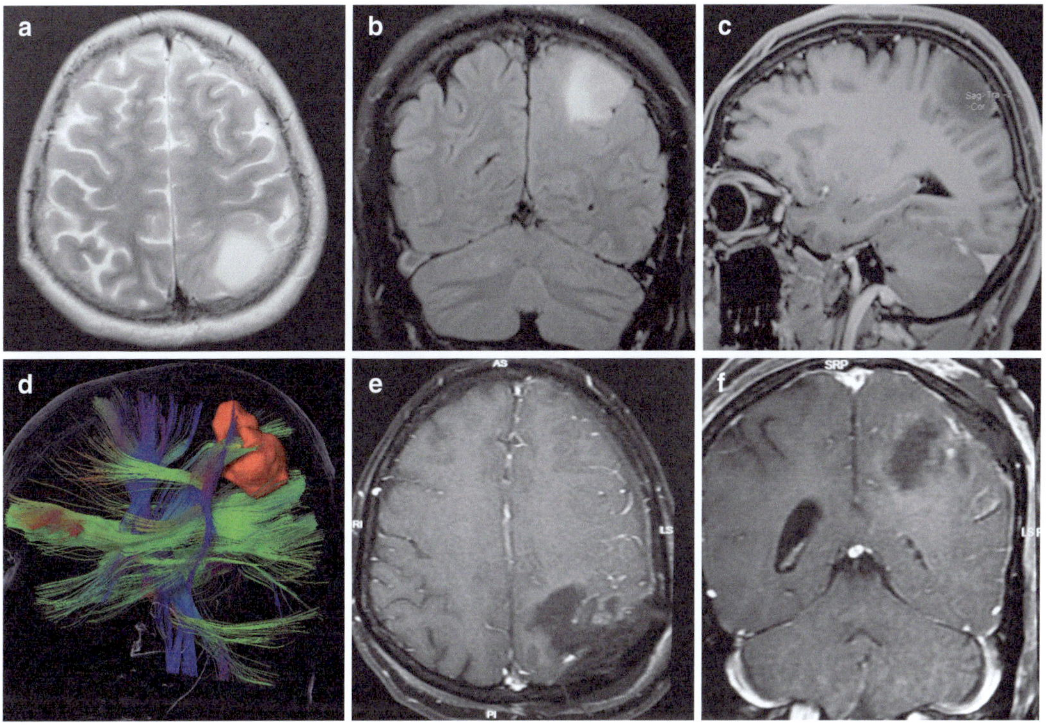

Fig. 8 Images of a 34 year old male patient. (**a**) T2 weighted axial image showing a low grade glioma in the region of the superior parietal lobule. (**b**) FLAIR coronal image showing the tumor. (**c**) T1 weighted sagittal contrast image showing the tumor in the superior parietal lobule. (**d**) Tractography image showing the tumor involving the short arcuate fibers in the region of the superior parietal lobule. The normal course of tumor adjoining arcuate fasciculus, the inferior fronto-occipital fasciculus and the vertical group of fibers is visualized. (**e**) Post-operative T1 weighted axial contrast image showing the radical tumor resection. (**f**) Post-operative coronal contrast image showing the resection cavity

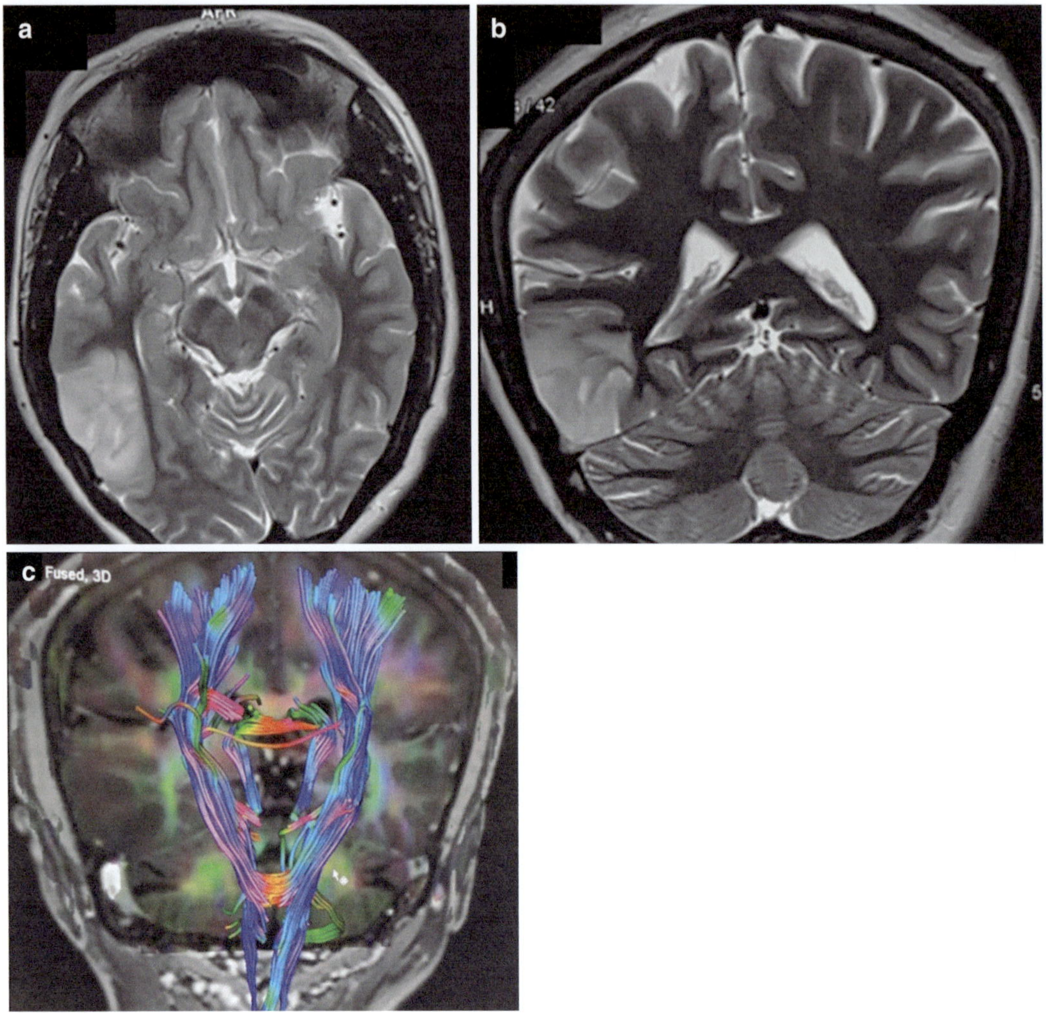

Fig. 9 Images of a 30 year old female patient with a lateral temporal glioma. (**a**) T2 weighted axial image showing the tumor in the inferior temporal gyrus. (**b**) Coronal image showing the tumor in the inferior temporal gyrus. (**c**) Tractography images showing the glioma arising from short aruate fibers in theregion of the inferior temporal gyrus. No major vertical or horizontal fiber tracts are in the vicinity

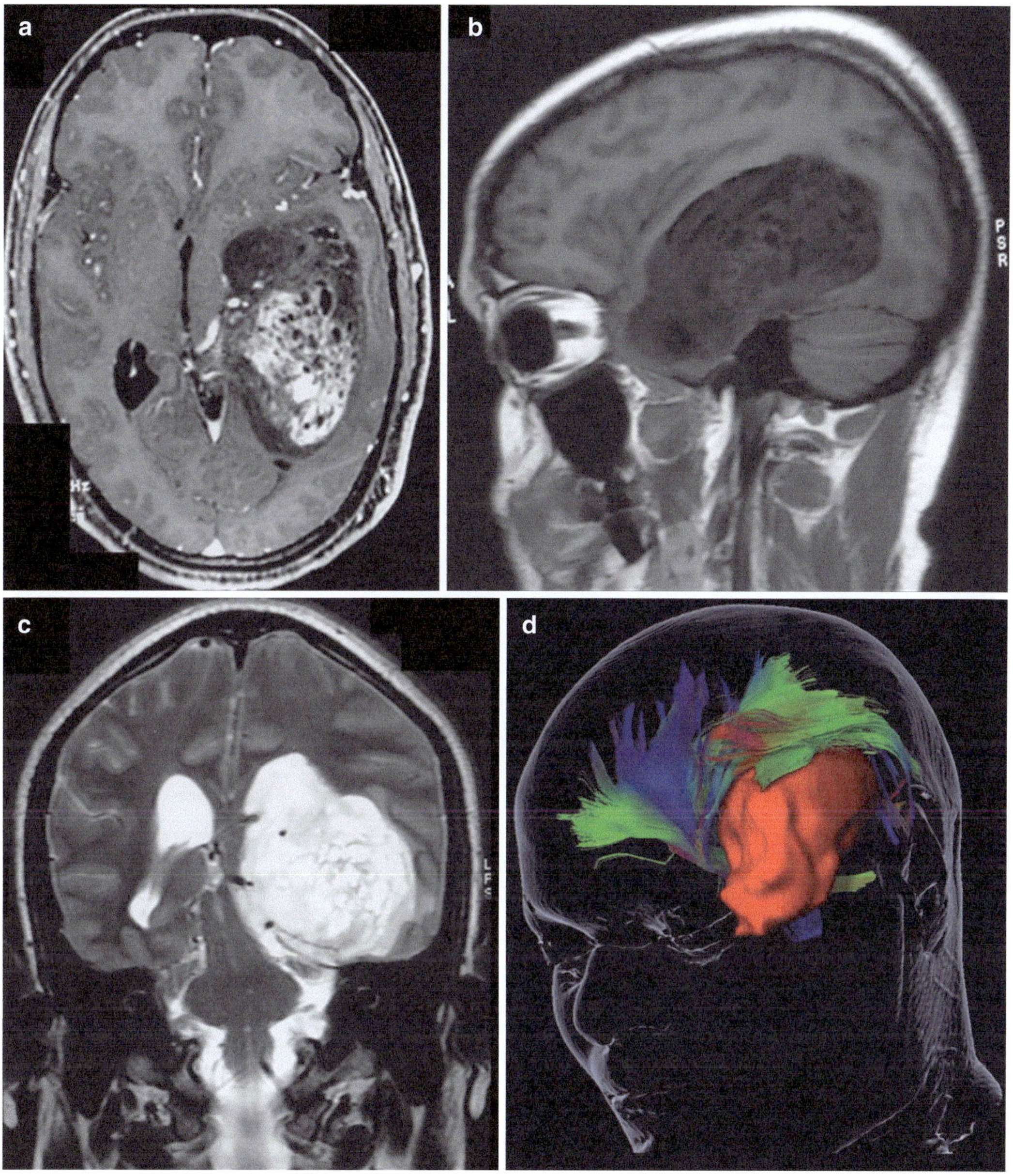

Fig. 10 Images of a 29 year old male patient with a medial temporal tumor. (**a**) T1 weighted contrast image showing a large glioma lesion arising from the region of the parahippocampal gyrus. (**b**) T1 weighted image showing the tumor. (**c**) T2 weighted coronal image showing the glioma in the medial temporal lobe. The tumor involves the short arcuate fibers of the parahippocampal gyrus and displaces all adjoining structures including the temporal horn. (**d**) Tractography images showing the inferior fronto-occipital fasciculus displaced laterally and superiorly and the projection fibers pushed medially by the tumor

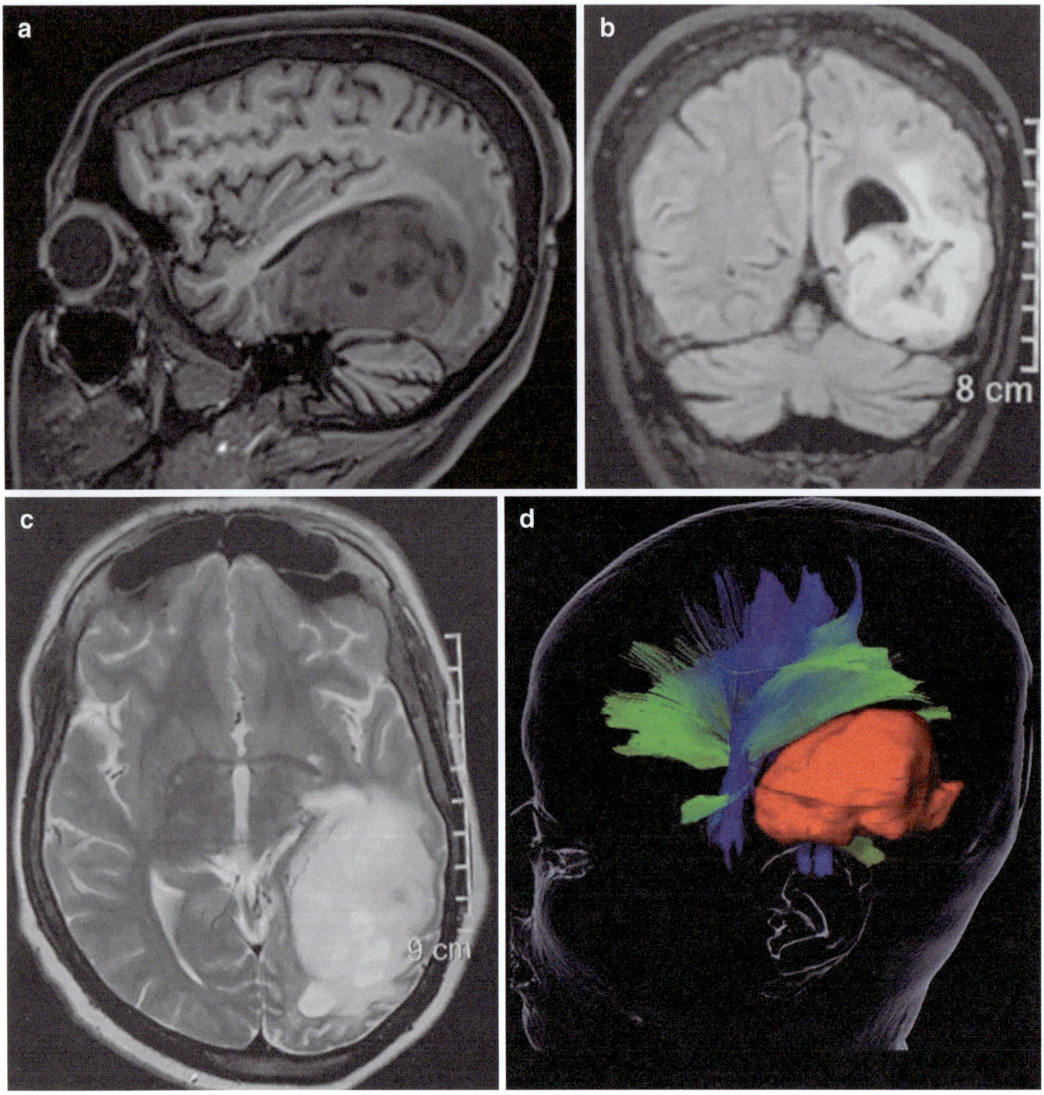

Fig. 11 Images of a 32 year old female patient with a basal temporal tumor. (**a**) T1 weighted image showing a lesion in the region of the fusiform gyrus displacing the floor of the temporal horn superiorly. (**b**) FLAIR image showing the glioma in the basal temporal region. (**c**) T2 weighted axial image showing the well defined glioma. (**d**) Tractography images showing that the tumor has involved the short arcuate fibers of the fusiform gyrus and displaced the inferior fronto-occipital fasciculus displaced superiorly and the arcuate fasciculus anteriorly

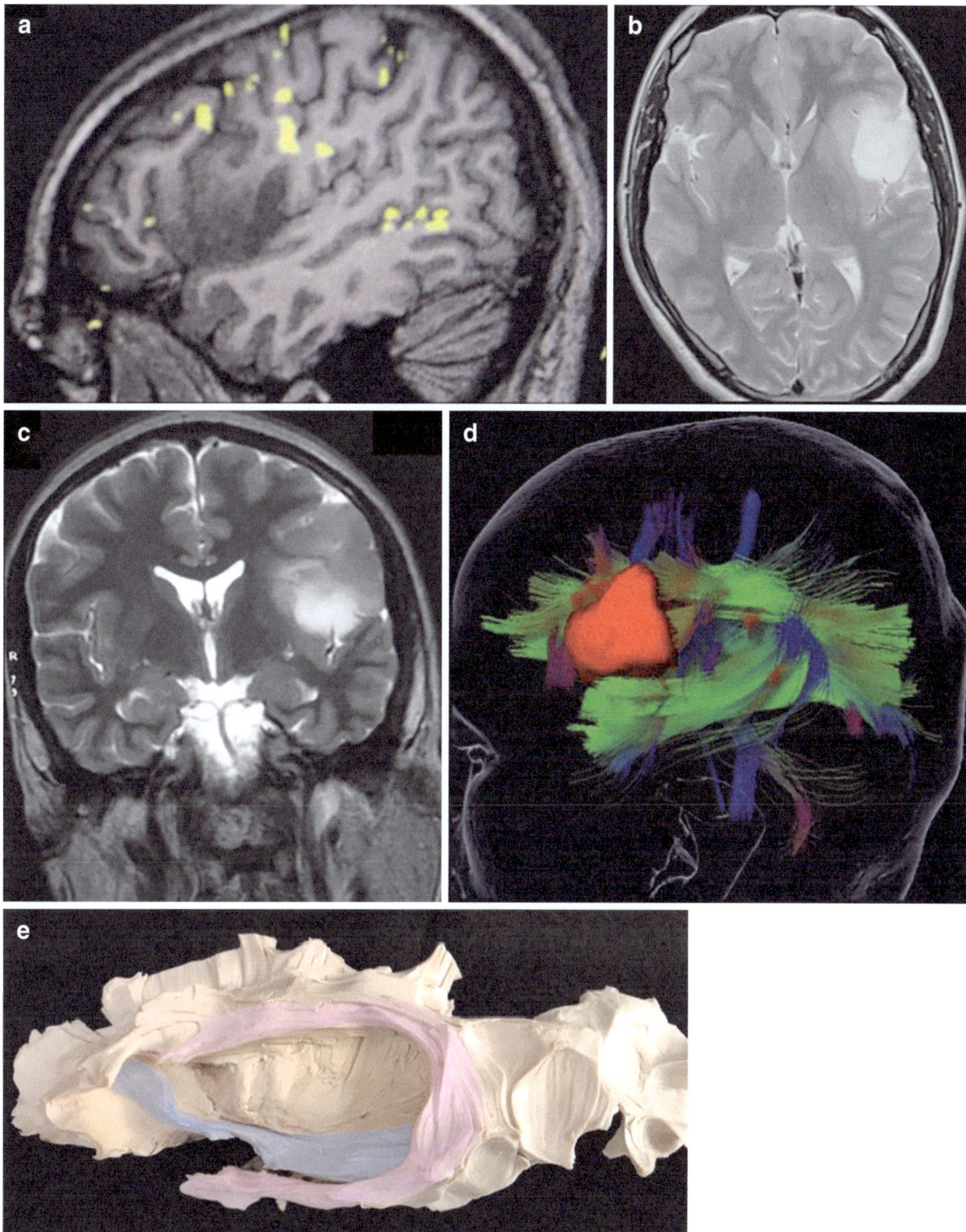

Fig. 12 Images of a localized insular glioma. (**a**) Functional MRI showing a localized insular glioma. The speech area has been marked out in yellow. (**b**) T2 weighted axial image showing the insular glioma. (**c**) Coronal image showing the tumor. The edematous tumor adjoining hyperintensity involves the course of the short arcuate fibers. (**d**) Tractography images showing the arcuate fasciculus around the tumor and the inferior fronto-occipital fasciculus running infero-medially in relation to the tumor. (**e**) Dissected specimen showing the course of the arcuate fasciculus (light pink) and inferior fronto-occipital fasciculus (blue) around the insula

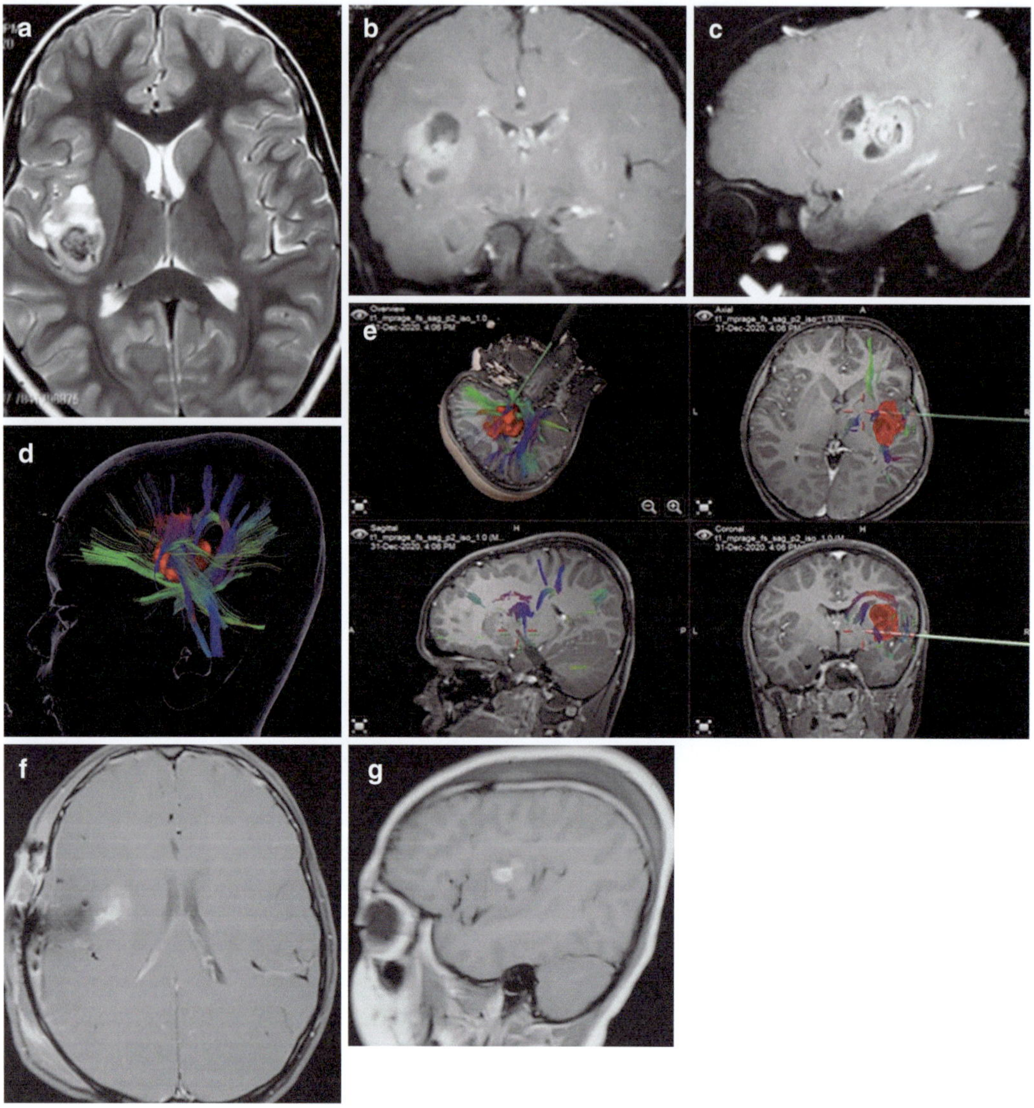

Fig. 13 Images of a 14 year old male child. (**a**) T2 weighted axial image showing a localized insular gyri in the region of the long insular gyri. (**b**) Coronal contrast image showing the enhancing tumor. (**c**) Sagittal contrast image showing the high grade glioma. (**d**) Tractography images showing the arcuate fasciculus around the tumor and the inferior fronto-occipital fasciculus running inferior to the tumor. (**e**) Fused tractography and neuronavigation images showing the trajectory used to approach the tumor. (**f**) Post-operative axial contrast image showing the tumor resection. (**g**) Post-operative sagittal contrast image showing the tumor resection

5 Tumors of the Long Association Fibers

5.1 Cingulum

Gliomas arising from the cingulum are fairly common and often mistaken as corpus callosal gliomas. These tumors are located in the medial frontal and parietal region rostral to the corpus callosum. They may extend superiorly to the frontal or parietal region respectively by virtue of their connections or they may extend inferiorly towards the corpus callosum, and lastly they may extend along the antero-posterior length of the cingulum itself (Figs. 14 and 15). The vertical group of fibers, which they do not cross, limits the lateral extension of these tumors. As a general rule gliomas of the corpus callosum tend to spread into both the hemispheres whereas gliomas of the cingulum tend to be unilateral.

5.2 Uncinate Fasciculus

Tumors arising from the uncinate fasciculus are one of the diffuse types of insular gliomas. When the insular glioma extends into the basal frontal and anterior temporal region it can be said to arise from the uncinate fasciculus (Fig. 16).

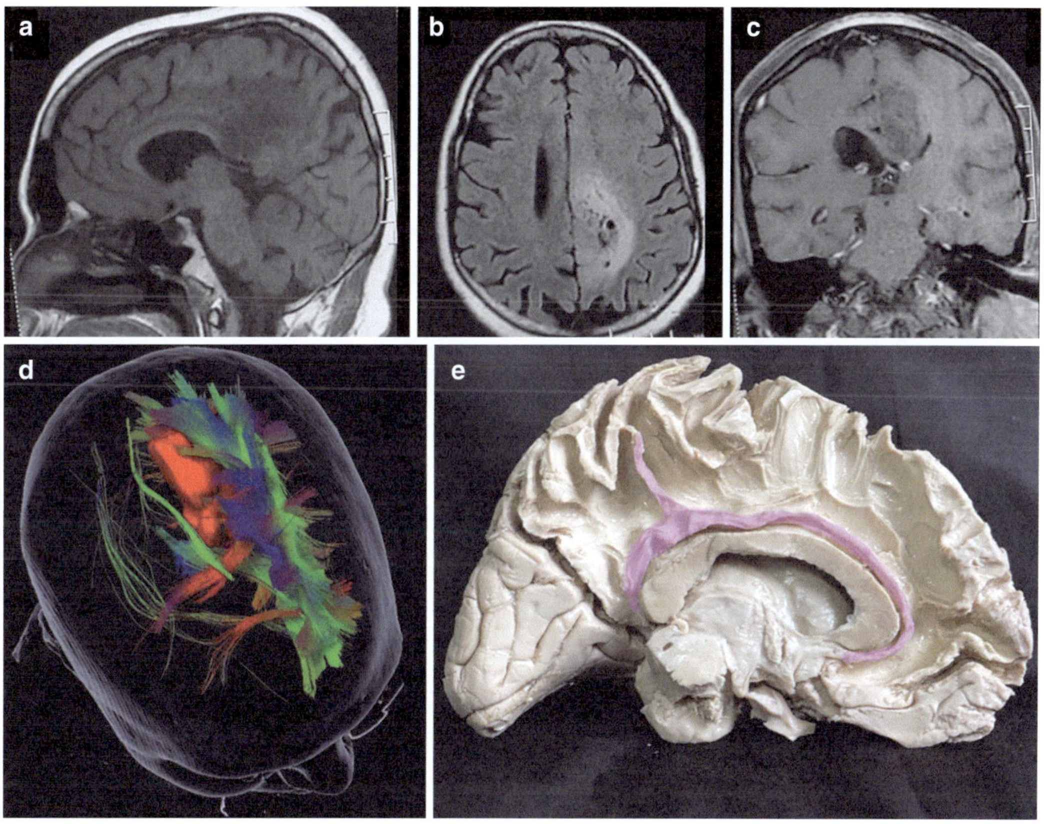

Fig. 14 Images of a 49 year old female patient with a lesion arising from the cingulum. (**a**) T1 weighted sagittal image showing the diffuse lesion arising from the cingulum and extending inferiorly towards the corpus callosum and superiorly in the medial parietal region. (**b**) FLAIR image showing the tumor arising from the cingulum. (**c**) Coronal image showing the tumor in the cingulum displacing the corpus callosum inferiorly. (**d**) Superior view of tractographic images showing the tumor. The tumor is limited laterally by the vertical group of fibers. (**e**) Dissected medial aspect of the cerebral hemisphere showing the course of the cingulum (pink). Note the parietal connection of the cingulum

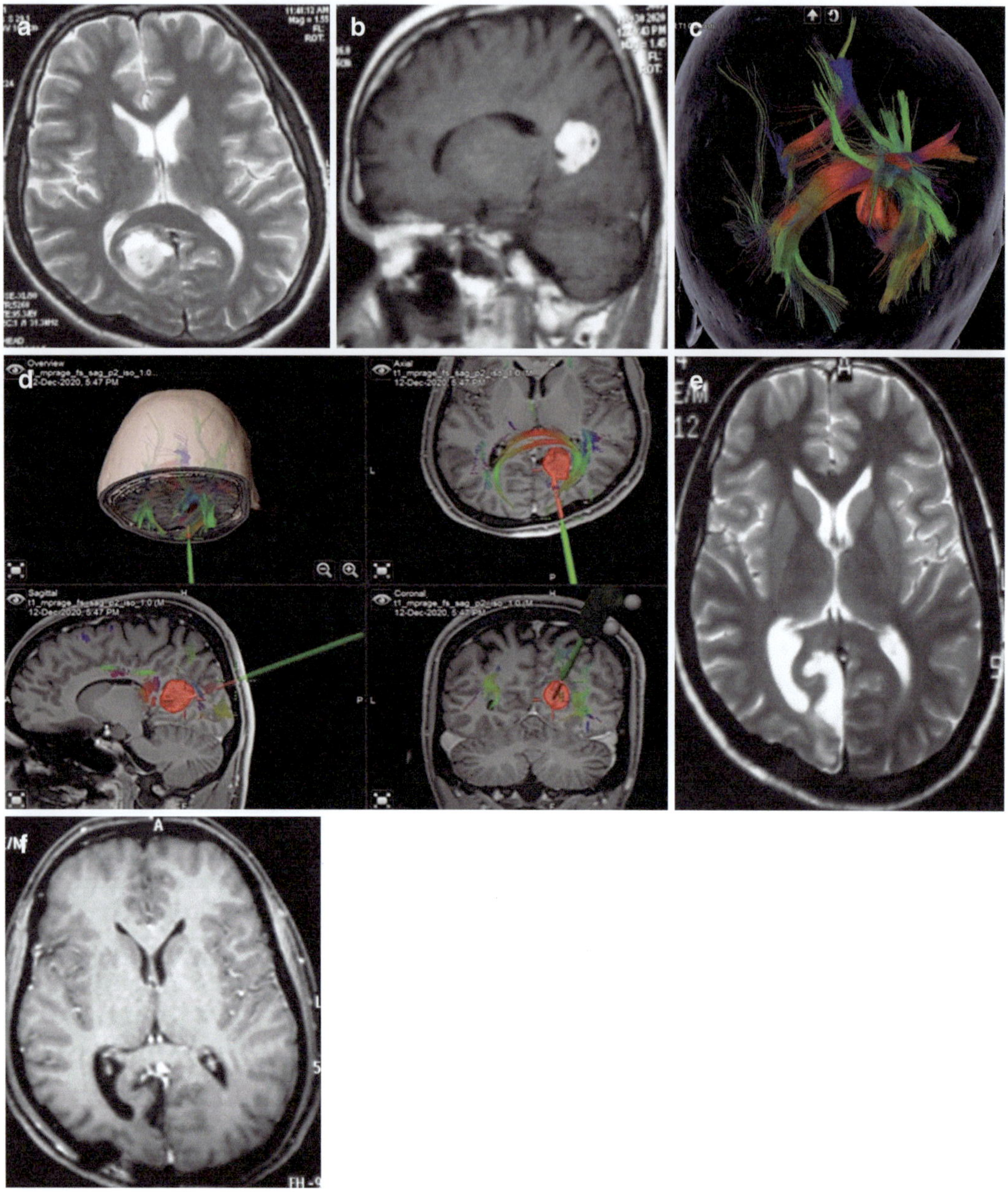

Fig. 15 Images of a 17 year old male patient. (**a**) T2 weighted axial image showing a lesion in the region of the isthmus of the cingulum. (**b**) T1 weighted contrast sagittal image showing a contrast enhancing tumor arising from the isthmus of the cingulum. (**c**) Tractography images showing the fibers of the forceps major displaced laterally by the tumor. (**d**) Fused tractograply and neuronavigation images showing the tumor with the fibers displaced around it. An interhemispheric approach was used to resect the tumor. (**e**) Post-operative T2 weighted axial image showing the supra-total tumor resection. (**f**) Post-operative T1 weighted axial contrast image showing the supra-total tumor resection

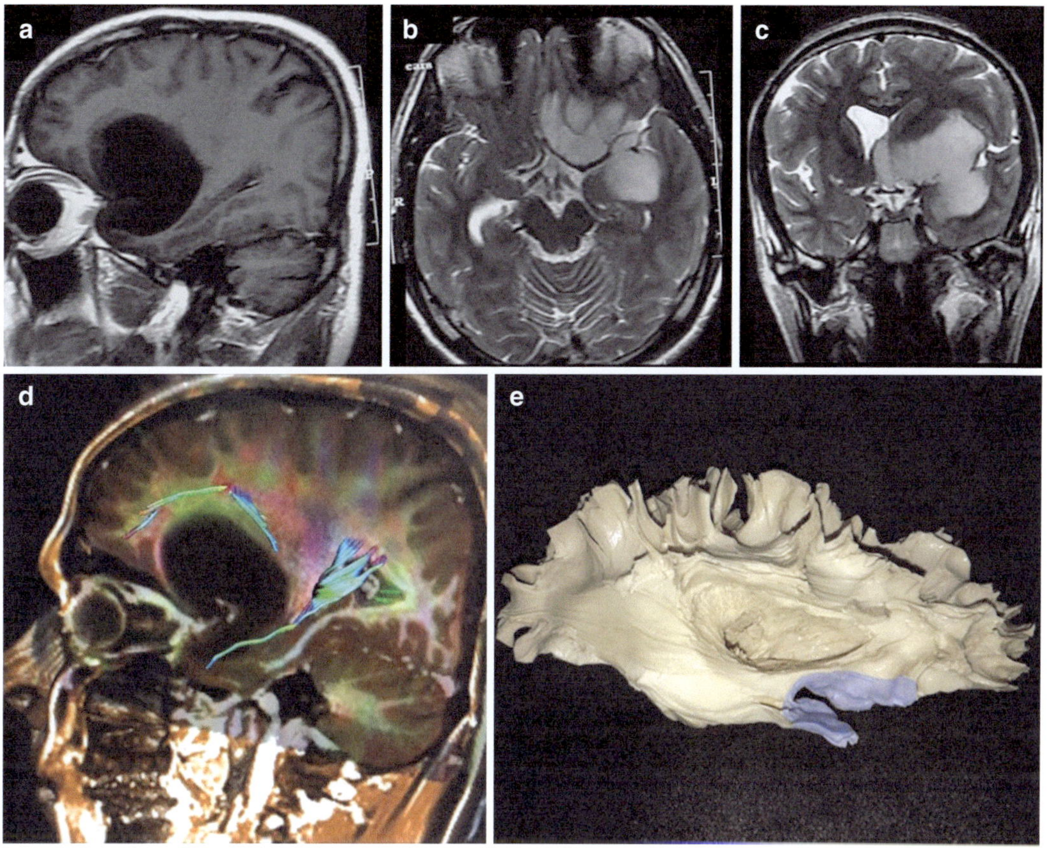

Fig. 16 Images of a 40 year old male patient with a glioma arising from the uncinate fasciculus. (**a**) T1 weighted sagittal image showing a characteristic tumor arising from the uncinate fasciculus. (**b**) T2 weighted axial image showing the tumor following the course of the uncinate fasciculus. (**c**) T2 weighted coronal image showing the tumor extension along the uncinate fasciculus. (**d**) Tractographic images showing some fibers of the uncinate fasciculus around the tumor. (**e**) Dissection of the lateral aspect of the cerebral hemisphere showing the white fibers. The uncinate fasciculus has been shaded purple

5.3 Inferior Fronto-Occipital Fasciculus

Gliomas of the inferior fronto-occipital fasciculus are included in the diffuse type of insular gliomas (Fig. 17). These tumors are those that extend into the frontal region and then via the fasciculus grow towards the posterior temporal region lateral to the temporal horn. These tumors have a typical inferior frontal and posterior temporal extension.

5.4 Superior Longitudinal Fasciculus, Middle Longitudinal Fasciculus, Inferior Longitudinal Fasciculus

Tumors of the superior longitudinal fasciculus, middle longitudinal fasciculus and inferior longitudinal fasciculus are relatively rare. Tumors of the superior longitudinal fasciculus could be divided into medial and lateral groups. Medial

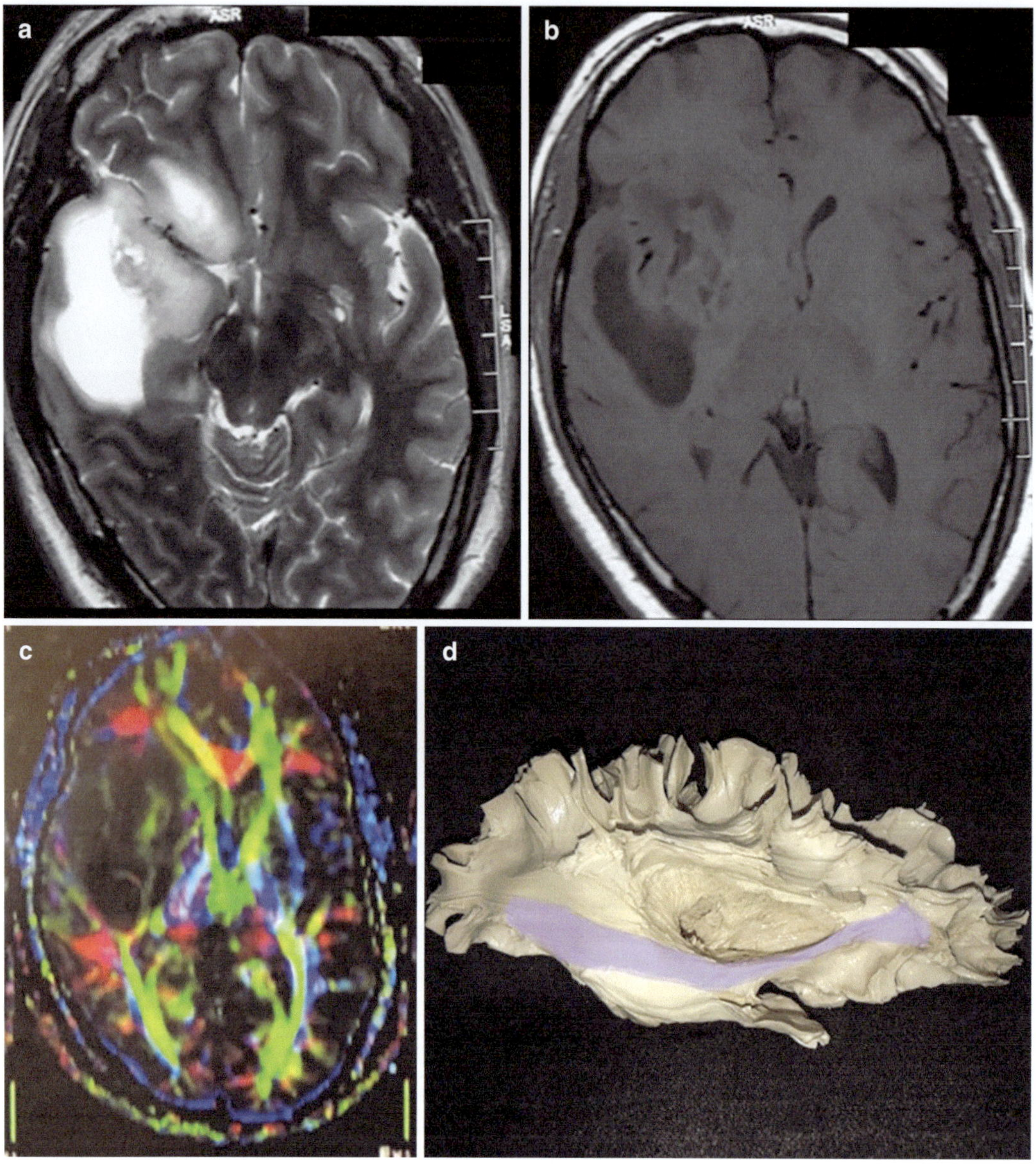

Fig. 17 Images of a diffuse insular glioma arising from the inferior fronto-occipital fasciculus. (**a**) T2 weighted image showing the diffuse glioma having a characteristic extension along the inferior fronto-occipital fasciculus. (**b**) T1 weighted image showing the tumor extension along the lateral aspect of the temporal horn. (**c**) Diffusion tensor images showing the tumor along the course of the inferior fronto-occipital fasciculus. (**d**) Dissected specimen showing the course of the inferior fronto-occipital fasciculus (purple)

tumors of the superior longitudinal fasciculus are identified by their location inferior to the superior and middle frontal gyri and superior to the cingulum (Fig. 18). They extend along an antero-posterior direction and do not specifically involve a single gyrus. Tumors of the lateral superior longitudinal fasciculus are difficult to differentiate from tumors of the arcuate fasciculus as these two bundles run close to each other [9]. Tumors of the inferior longitudinal fasciculus diffusely involve the basal part of the temporal lobe in an antero-posterior direction. In our series we could not identify any tumor that arose from the middle longitudinal fasciculus.

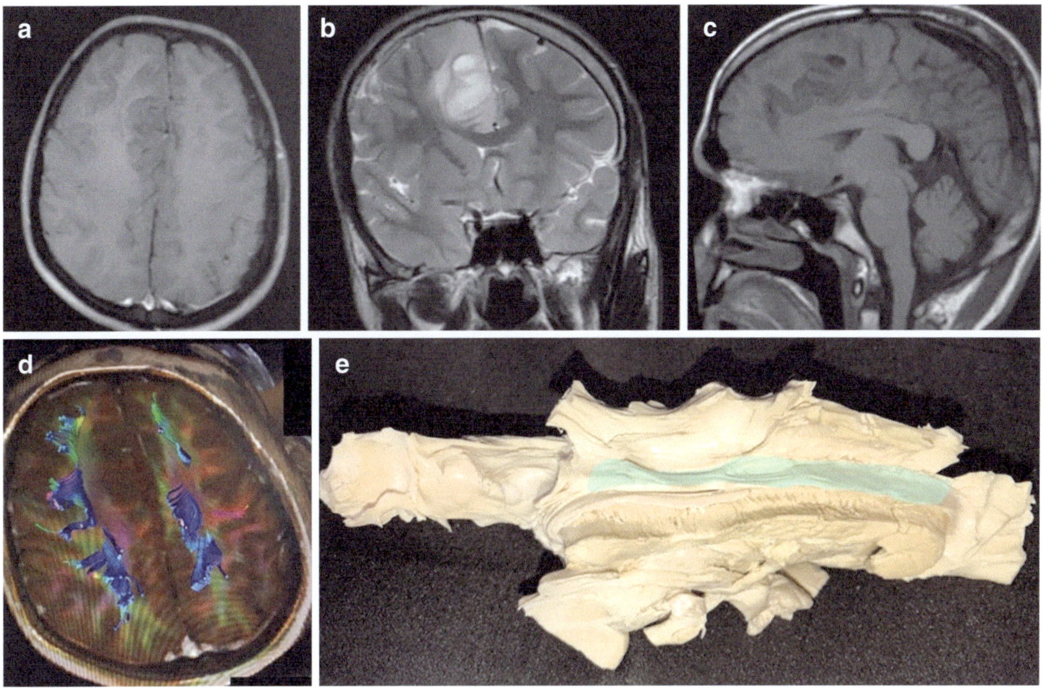

Fig. 18 Images of a 28 year old male patient with a diffuse lesion in the region of the medial portion of the superior longitudinal fasciculus. (**a**) T1 weighted axial image showing the tumor arising in the region of the medial portion of the superior longitudinal fasciculus. (**b**) T2 weighted coronal image showing the lesion just above the cingulum. (**c**) T1 weighted sagittal image showing the diffuse tumor along the superior longitudinal fasciculus and just superior to the cingulum. (**d**) Tractographic images showing the tumor lying medial to the vertical group of fibers and limited by it. (**e**) Dissected specimen of the medial aspect of the cerebral hemisphere showing the medial portion of the superior longitudinal fasciculus (green) above the cingulum

5.5 Arcuate Fasciculus

These tumors are also relatively less common. These tumors were found to grow around the insula just beneath the frontal, parietal and temporal operculum.

6 Tumors of the Commissural fibers

6.1 Corpus Callosum

Tumors of the corpus callosum are common and are generally bilateral. They can be further subdivided into tumors arising from the genu and forceps minor, tumors of the body and tumors of the splenium and forceps major. Tumors arising from the genu or the forceps minor (anterior callosal radiations) grow anteriorly into the bilateral medial frontal lobes, medial to the frontal horn of the lateral ventricle. Tumors of the body of the corpus callosum (dorsal callosal radiation) begin in the midline or eccentrically but ultimately extend bilaterally into both the cerebral hemispheres. They grow into the hemispheres but again are limited bilaterally by the vertical group of fibers, which form a firm barrier that is not crossed by these tumors (Fig. 19). Tumors of the splenium or of the forceps major (posterior callosal radiations) grow bilaterally into the medial parieto-occipital lobes. They lie medial to the occipital horn of the lateral ventricle. Tumors of the forceps minor and forceps major are often referred to as "butterfly gliomas".

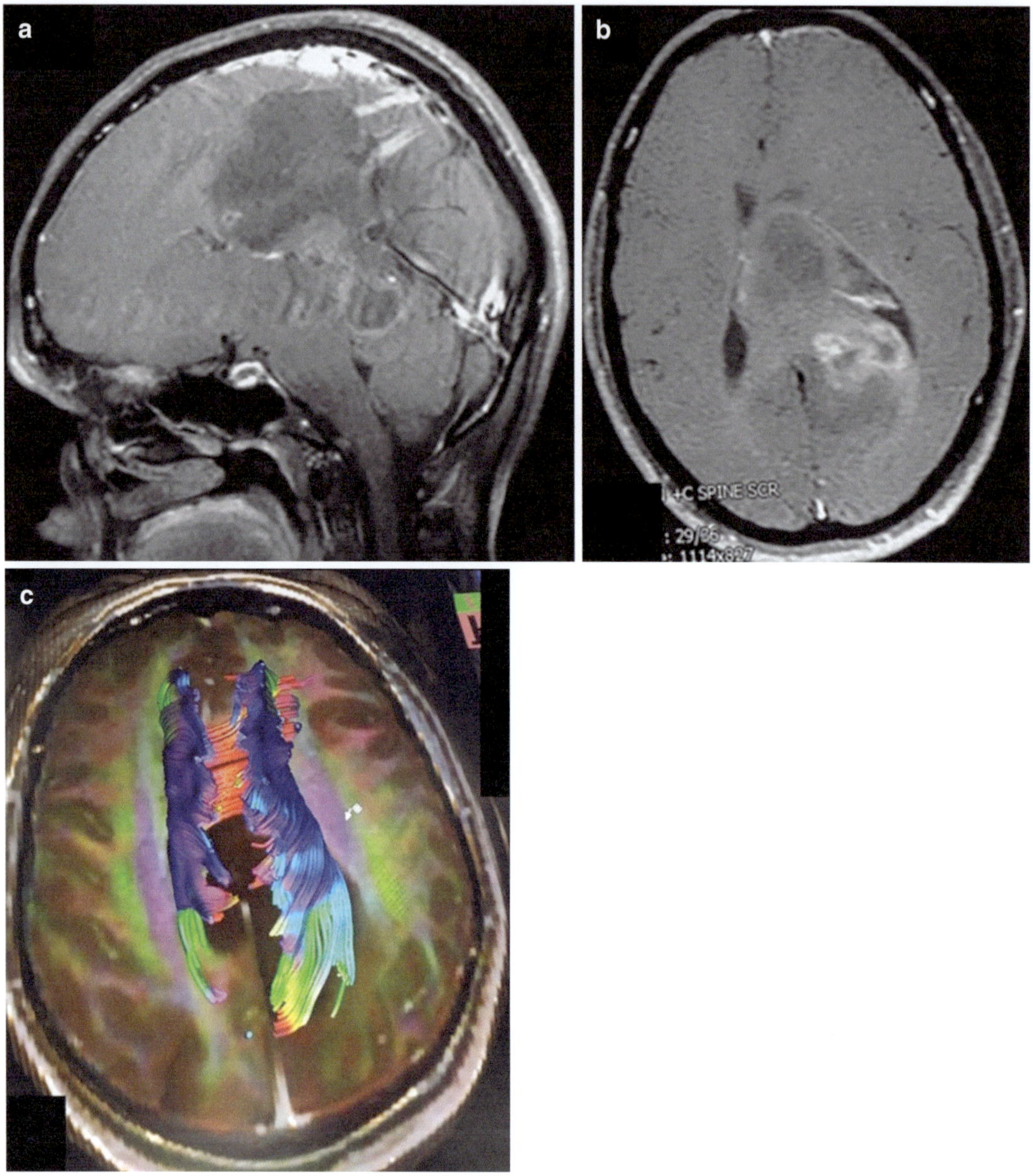

Fig. 19 Images of a 42 year old female patient with a glioma arising from the corpus callosum. (**a**) T1 weighted sagittal contrast image showing the diffuse tumor arising from the body of the corpus callosum. (**b**) T1 weighted axial contrast image showing the bilateral extension of the tumor along the corpus callosum. (**c**) Tractographic images showing the bilateral extension of the tumor along the fibers of the corpus callosum. Some fibers of the corpus callosum are seen to be pushed anteriorly. The tumor is limited on both the sides by the vertical group of projection fibers

6.2 Anterior Commissure

Tumors arising from the anterior commissure also form a subset of diffuse type of insular gliomas. These tumors may grow in multiple directions. They may involve bilateral basal frontal regions by its frontal extension or they can grow in the posterior temporal region as the temporal extension of the anterior commissure forms a component of the sagittal stratum [10]. The temporal extension of the anterior commissure is the only fasciculus to cross the vertical group of projection fibers and it does so at its inferior edge between the anterior limb and genu of the internal capsule.

6.3 Hippocampal Commissure

These gliomas are rare. When they do occur they lie beneath the corpus callosum and above the thalamus and spread bilaterally. They are difficult to distinguish between tumors of the ventral portion of the corpus callosum and of the thalamus.

6.4 Posterior Commissure

Tumors arising from the posterior commissure are rare. We did not encounter any case in our series.

7 Surgical Implications

White fibers of the brain have a distinct architecture with fixed and defined routes. The vertical group of projection fibers forms a firm barrier with no intermingling (or crossing) of fibers on either side of this barrier. Based on our radiological observations and anatomical understanding we speculate that gliomas originate in short arcuate fibers, long association fibers or commissural fibers [18–20]. Tumors on the lateral aspect of the vertical group did not involve the corpus callosum even at late stages. Similarly, not all gliomas reach the ventricle even after they have grown significantly in their sizes. Tumors of the cingulum and those arising from the commissural fibers do not grow laterally beyond the vertical group of fibers. Similarly, tumors of the insula and arcuate fasciculus do not grow medially beyond the vertical group. Tumors of the short arcuate fibers grow and expand the corresponding gyrus and remain localized. Tumors of the long association fibers spread along the anatomical extent of the tract more commonly in an antero-posterior direction. Tumors of the commissural fibers grow bilaterally.

We speculate that the origin of gliomas is from a tract rather than a cell unlike benign extraaxial tumors where the tumor arises from a single nerve fiber and grows by displacement of normal nerves. The growth of glioma is along the tract that is involved in the form of a bulk. The uninvolved adjoining tracts are compressed and deformed and form a preventive barrier against tumor extension [18–20].

We also observed that there was no difference in the pattern of growth of a low or a high-grade glioma. Both low and high grade gliomas arising from short arcuate fibers were localized to the corresponding gyrus. Tumors arising from long association fibers and commissural fibers were diffuse and had a complex shape matching the anatomy of the fiber tract.

Based on this anatomical knowledge and our surgical experience we discuss novel surgical strategies for glioma resection. We advocate an en-masse surgical resection for both low and high grade localized gliomas arising from the short arcuate fibers [19]. A clear plain of dissection is available for most of these gliomas during surgery. This strategy of resection is actually a supra-marginal resection as the glioma is only localized to that region. This kind of resection is possible without using any kind of fluorescence as the complete lesion is removed en masse. This holds true for all localized gliomas except for those in the insula where an enmasse resection is difficult due to the presence of the branches of the middle cerebral arteries.

Even tumors that arise from short arcuate fibers in eloquent cortical locations are localized and can be excised by identification of a well-defined plain of resection. The eloquent cortex was seen to be displaced by the growth of the tumor in low-grade tumors and in majority of high-grade tumors. A tailored resection was used for tumors when the eloquent cortex was intra-operatively mapped to be in close vicinity to the tumor. For diffuse tumors from the long tracts we advocate a safe radical piece meal excision as a total en masse resection may not be anatomically possible. This classification proposed could also have implications for planning adjuvant therapy following surgery where the entire involved white fiber tract can be included in the treatment field.

8 Conclusions

The anatomical classification of white fibers proposed by us was found to have implication in understanding the origin and extension of both low and high-grade gliomas. Understanding the trajectory of the fibers makes it possible to evaluate the direction of growth of the tumors and also provides us with defined limits which the tumor will not breech. This understanding greatly aids in safe radical excision of the glial tumors.

References

1. Louis DN, Perry A, Reifenberger G, von Deimling A, Figarella-Branger D, Cavenee WK, Ohgaki H, Wiestler OD, Kleihues P, Ellison DW. The 2016 World Health Organization classification of tumors of the central nervous system: a summary. Acta Neuropathol. 2016;131:803–20.
2. Sanai N, Berger MS. Surgical oncology for gliomas: the state of the art. Nat Rev Clin Oncol. 2018;15(2):112–25.
3. McKhann GM, Duffau H. Low-grade glioma epidemiology, pathophysiology, clinical features, and treatment. Neurosurg Clin N Am. 2019;30(1):xiii–xiv.
4. Duffau H. Higher-order surgical questions for diffuse low-grade gliomas: supramaximal resection, neuroplasticity, and screening. Neurosurg Clin N Am. 2019;30(1):119–28.
5. Yordanova YN, Duffau H. Supratotal resection of diffuse gliomas - an overview of its multifaceted implications. Neurochirurgie. 2017;63(3):243–9.
6. Yasargil MG. Microneurosurgery volume IVA: CNS tumors: surgical anatomy, neuropathology, neuroradiology, neurophysiology, clinical considerations, operability, treatment options. Stuttgart: Georg Thieme; 1994.
7. Giese A, Westphal M. Glioma invasion in the central nervous system. Neurosurgery. 1996;39(2):235–50; discussion 250-2.
8. Mandonnet E, Capelle L, Duffau H. Extension of paralimbic low grade gliomas: toward an anatomical classification based on white matter invasion patterns. J Neuro-Oncol. 2006;78(2):179–85.
9. Swanson KR, Bridge C, Murray JD, Alvord EC Jr. Virtual and real brain tumors: using mathematical modeling to quantify glioma growth and invasion. J Neurol Sci. 2003;216:1–10.
10. Swanson KR, Alvord EC Jr, Murray JD. A quantitative model for differential motility of gliomas in grey and white matter. Cell Prolif. 2000;33:317–29.
11. Burgess PK, Kulesa PM, Murray JD, Alvord EC Jr. The interaction of growth rates and diffusion coefficients in a three dimensional mathematical model of gliomas. J Neuropathol Exp Neurol. 1997;56:704–13.
12. Sawaya R, Hammoud M, Schoppa D, Hess KR, Wu SZ, Shi WM, Wildrick DM. Neurosurgical outcomes in a modern series of 400 craniotomies for treatment of parenchymal tumors. Neurosurgery. 1998;42(5):1044–55; discussion 1055-6.
13. Friedlein K, Bozhkov Y, Hore N, Merkel A, Sommer B, Brandner S, Buchfelder M, Savaskan NE, Eyüpoglu IY. A new functional classification system (FGA/B) with prognostic value for glioma patients. Sci Rep. 2015;21(5):12373.
14. Shah A, Goel A, Jhawar SS, Patil A, Rangnekar R, Goel A. Neural circuitry: architecture and function - a fiber dissection study. World Neurosurg. 2019;125:e620–38.
15. Shah A, Jhawar SS, Goel A. Analysis of the anatomy of the Papez circuit and adjoining limbic system by fiber dissection techniques. J Clin Neurosci. 2012;19(2):289–98.
16. Shah A, Goel A, Patil A, Goel A. Letter to the Editor. Superior longitudinal fasciculus. J Neurosurg. 2019;6:1–2.
17. Shah A, Jhawar SS, Goel A. Letter to the Editor. Optic radiations and anterior commissure. J Neurosurg. 2015;123(3):824–6.
18. Goel A, Shah A, Dandpat S, Rai S, Prasad A. Gliomas are 'confined' to a named white matter tract: a revolution in understanding gliomas. J Craniovert Junction Spine. 2020;11:252–3.
19. Goel A, Shah A, Vutha R, Dandpat S, Hawaldar A. Is "En masse" tumor resection a safe surgical strategy for low-grade gliomas? Feasibility report on 74 patients treated over four years. Neurol India. 2021;69(2):406–13.
20. Shah A, Vutha R, Goel A. Radiological evaluation of anatomical extensions of gliomas based on white fiber tracts: proposal of a novel classification. Neurol India. 2022;70:1492.

Surgical Strategy for Low-Grade Gliomas Arising from Short Arcuate Fibers

Atul Goel, Ravikiran Vutha, and Abhidha Shah

1 Introduction

The general understanding is that low-grade gliomas are by nature 'diffuse', 'invasive' or 'infiltrating' into the normal brain parenchyma. Considering their frequently identified wide extension and location in critical, functional or eloquent brain areas, injury to subcortical pathways can be responsible for morbidity encountered during surgical resection [1–17]. Safe resection of tumors near functional areas necessarily involves intraoperative identification of functional sites both at the cortical and the subcortical levels. For several years, the generally agreed upon or gold standard treatment strategy for low grade glioma was either symptomatic drug treatment and clinical observation or a biopsy for histological confirmation that was then followed up with adjuvant treatment. Current view is that radical resection of low grade gliomas is 'safely' possible and is associated with recovery from symptoms, improved longevity and quality of life and decreased potential for 'malignant' transformation. Moreover, adjuvant chemotherapy/radiotherapy treatment can be avoided. Technical adjuncts such as awake surgery and use of 5-ALA, real-time ultrasound, cortical and subcortical mapping and intra-operative MRI are currently in vogue to achieve the goal of maximal tumor resection with minimum or no morbidity.

Based on anatomical understanding of the three dimensional architecture of the white fibers of the brain and our surgical experience in dealing with low grade gliomas we speculate that the point of origin of glioma is from the white fiber tracts of the brain [18–20]. The propagation of the glioma occurs in the direction of the involved fiber bundle. From our study of gliomas over the years, we realised that like the benign brain tumors *viz.* meningiomas and neurinomas, gliomas grow by expansion into the adjoining gyri and displace the adjoining fibre tracts and provide a well-defined plane of surgical dissection around its circumference [18, 19]. This concept is in variance with the generally agreed viewpoint that gliomas merge into the adjoining normal brain parenchyma and encase normal blood vessels.

On the basis of our recent anatomical understanding of the architecture of white fibers we speculate that gliomas are 'confined' tumors

Dr. Ravikiran Vutha was involved in this chapter preparation

A. Goel (✉)
Department of Neurosurgery, Lilavati Hospital and Research Center, Bandra, Mumbai, India

Department of Neurosurgery, R.N. Cooper Hospital, Mumbai, India

Bombay Hospital Institute of Medical Sciences and Research Center, Mumbai, India

R. Vutha
Consultant Neurosurgeon, Apollo Hospitals, Mumbai, India

A. Shah
Department of Neurosurgery, K.E.M. Hospital and Seth G.S. Medical College, Parel, Mumbai, India

A. Shah et al. (eds.), *Functional Anatomy of the Brain: A View from the Surgeon's Eye*,
https://doi.org/10.1007/978-981-99-3412-6_12

[18, 19]. They arise from white matter tracts and grow in an expansile and 'disciplined' fashion along the involved tract and gyrus whilst displacing the adjacent tracts around it [18–24]. The glioma arises from a named white fiber tract involving it and displaces the adjacent uninvolved fiber bundles. The gliomas tend to be 'localized' when they arise from short arcuate fibers and remain confined to the affected gyrus/gyri [19, 20]. Gliomas arising from long association and commissural fibers follow a similar pattern of growth and expansion but are more 'diffuse' as they spread along the length of the tract in the hemisphere [18–20]. When the tumor involves the commissural fibers the glioma propagation tends to be bilateral. Both low and high-grade gliomas have a similar pattern of expansion and well-defined boundaries and are different only in their growth pattern. The surgical strategy was based on the concept that gliomas arise from and extend along a named white fibre tract and grow by circumferential expansion. The adjoining fiber tracts are displaced, limit and 'contain' the spread of the tumor and are functionally involved by virtue of compression and deformation and not by invasion, involvement or destruction [18–20]. We recently reported our surgical strategy for treatment of low-grade gliomas involving the short arcuate white fibres based on this concept, which is summarized in this chapter [19].

2 Pre-operative Planning

In this chapter we discuss our surgical strategy for gliomas arising from the short arcuate fibers or the localized group of gliomas. WHO Grade I tumors like pilocytic astrocytoma are not included here. Table 1 identifies the location of the gyrus harbouring the tumor affected short arcuate fibres in our series of low grade gliomas which we recently published [19].

3 Clinical Evaluation

All patients undergo a Mini- mental state examination (MMSE) to assess neurological and cognitive status. The patients are scored in the immediate post-operative period and at every follow up.

4 Imaging

Preoperative investigations include a plain, FLAIR and gadolinium enhanced MRI with MR spectroscopy, MR perfusion and diffusion tensor imaging. Functional MRI is performed for tumors in eloquent locations. Tractography was performed in all patients on the Brainlab planning software. The tumors are classified according to their fiber tract of origin and on the basis of the gyrus involved.

5 Adjuncts During Surgery

Neuronavigation and Tractography: Neuronavigation is performed in all patients using a Brainlab Vector Vision planning station. Tractography was performed using the iPlan Fiber Tracking software. By using several seed points located in one or more ROIs (region of interest) the connecting fiber bundles were tracked using fiber assignment by continuous tracking (FACT) approach [1].

Ultrasound: Real time ultrasound monitoring is done using a Samsung real time portable scanner prior to handling of the brain and after the completion of the surgical tumor resection in all patients.

Intraoperative neurophysiology: Intraoperative neuromonitoring is used only in select patients as discussed below. After induction of total intravenous anesthesia, corkscrew electrodes are applied over the ipsilateral motor area and subdermal electrodes are placed on the contralateral face, upper and lower limbs. Motor mapping is performed mainly for tumors in the vicinity of the motor cortex. Multimodality neurophysiologic monitoring is done using direct and subcortical monopolar and bipolar stimulation, transcranial motor evoked potentials (MEPs) and somatosensory evoked potentials (SSEPs) (NIM - Eclipse, Medtronic, Dublin, Ireland) [19]. High frequency transcranial MEPs were measured at intervals of 3–5 minutes during the tumor resection using a stimulation intensity of 300–500 volts. Bipolar cortical mapping was performed using a 5 mm bipolar probe using a biphasic square wave pulse of 1 ms duration at 60 Hz. Stimulation was begun at 4 mA and increased to a maximum of 16 mA sequentially. Subcortical mapping with monopolar probe is performed in patients with tumors in vicinity of the motor pathways. Awake craniotomy for language testing is performed when the tumor is in proximity to the motor speech areas, in vicinity of the arcuate fasciculus or the inferior fronto-occipital fasciculus on the dominant side. Stimulation was performed with 60 Hz using a bipolar probe starting at 1 mA and then increasing up to 6 mA in 0.5 mA increments. The stimulation was performed on the cortex at 10 mm intervals. Intra-operative fluorescence to evaluate the extent of tumor resection was not used in any case. Intra-operative MRI is not available at our institute and hence was not used.

6 Surgical Strategy

The surgical strategy essentially involved the following three steps: (1) To identify the gyrus/gyri associated with the tumor affected short arcuate fibers, (2) Understanding the location and displacement of long association, projection and commissural fibers in the vicinity of the tumor and (3) Performing an *en-masse* and radical surgical resection by developing a plane of dissection between the tumor and adjoining normal brain and keeping in perspective the subcortical anatomical boundaries formed by the displaced tracts and cortical eloquence [19]. Although the methodology of surgical procedure was essentially individual case based and cannot be generalised, the surgical principle was common. The location of the tumor was identified by its color, consistency and vascularity. Intraoperative ultrasound and navigation tools were used to confirm the location and extent of the tumor [19]. Low-grade gliomas generally appear 'whiter' and were visually distinguishable from normal adjoining normal brain. The tumors were firmer and gritty in consistency and were relatively avascular. A plane of dissection of the tumor from the adjoining brain was identified and developed using blunt surgical dissection. The sulcal blood vessels were displaced by the tumor and could be protected. Experience in identification of normal from abnormal and developing the dissection plane appeared key surgical steps. The dissection plane between tumor and normal brain was reliable and safe and allowed radical tumor resection. For tumors that were in proximity to 'functional' cortical motor area, a tailored customised resection was performed whenever there was even moderate difficulty in identifying the plane of dissection or when there were negative indicators by neurophysiological monitoring.

7 Post-operative Imaging

Post-operative MRI is performed in all patients within 48 h of surgery and again at an interval of 3 months to gauge the extent of resection. The pre-operative and post-operative FLAIR images are compared to quantify the resection. The resection is labelled as subtotal (more than 15 mL of tumor residue) when some portion of the tumor is left behind to preserve function or is left behind inadvertently [1, 2]. A supra-maximal resection was defined as resection of all signal abnormality and when the immediate post-operative resection cavity is bigger than the pre-operative tumor cavity [2]. A gross total resection is defined as resection of all signal abnormality and when the tumor resection cavity is same as the pre-operative tumor volume (Figs. 1, 2, 3, 4, 5, 6, 7, 8, 9, and 10).

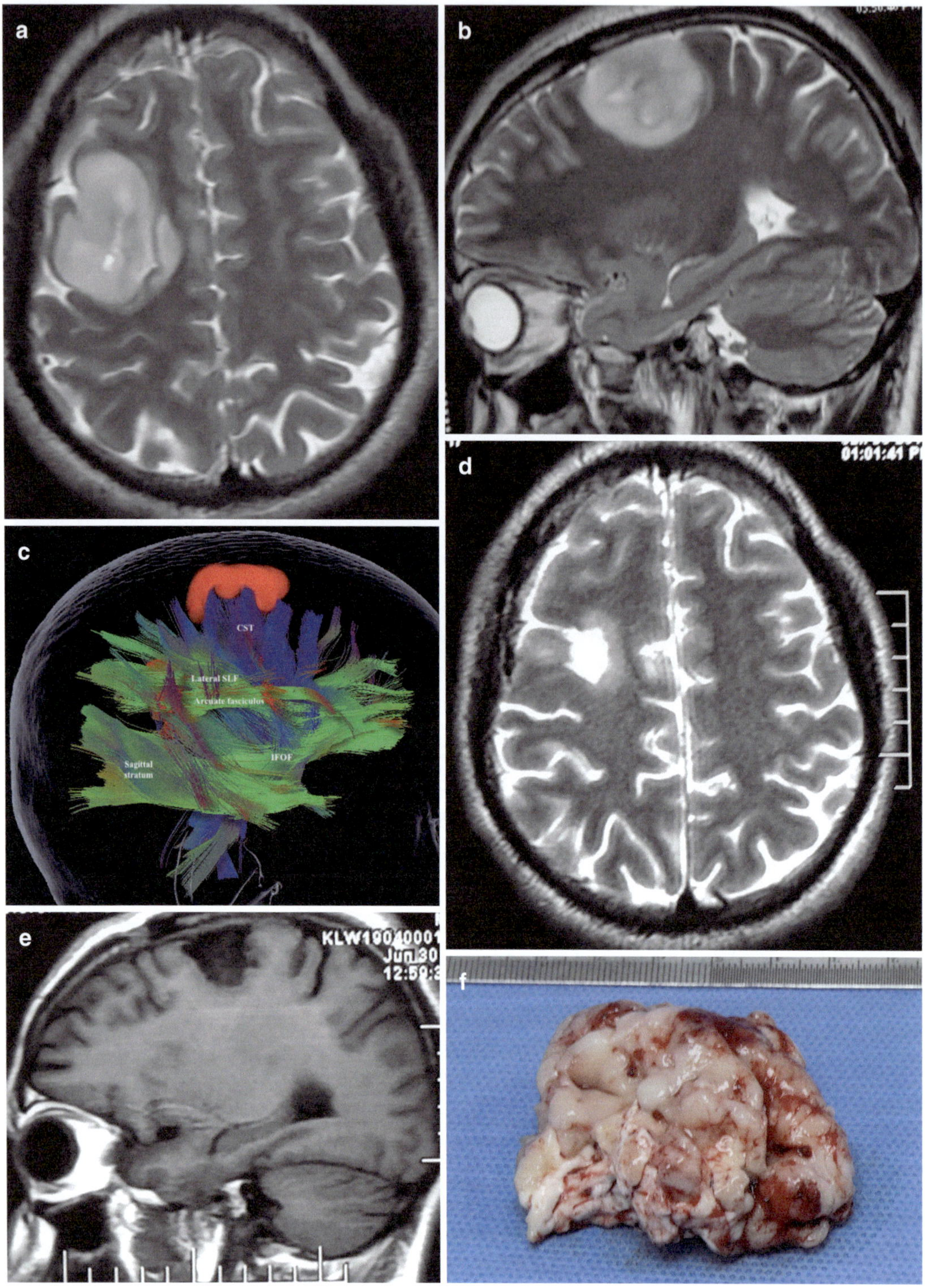

Fig. 1 Images of a 33 - year old male patient. (**a**) T2 weighted axial image showing a right medial frontal low grade glioma arising from short arcuate fibers in the region of the middle frontal gyrus, anterior to the precentral gyrus. (**b**) T2 weighted sagittal image showing the tumor. (**c**) Tractography image showing the tumor medial to the vertical group of projection fibers. (**d**) Post-operative T2 weighted axial image showing a resection of the tumor. (**e**) T1 weighted sagittal image showing the resection cavity. (**f**) Post-operative specimen showing en-masse resection of the tumor

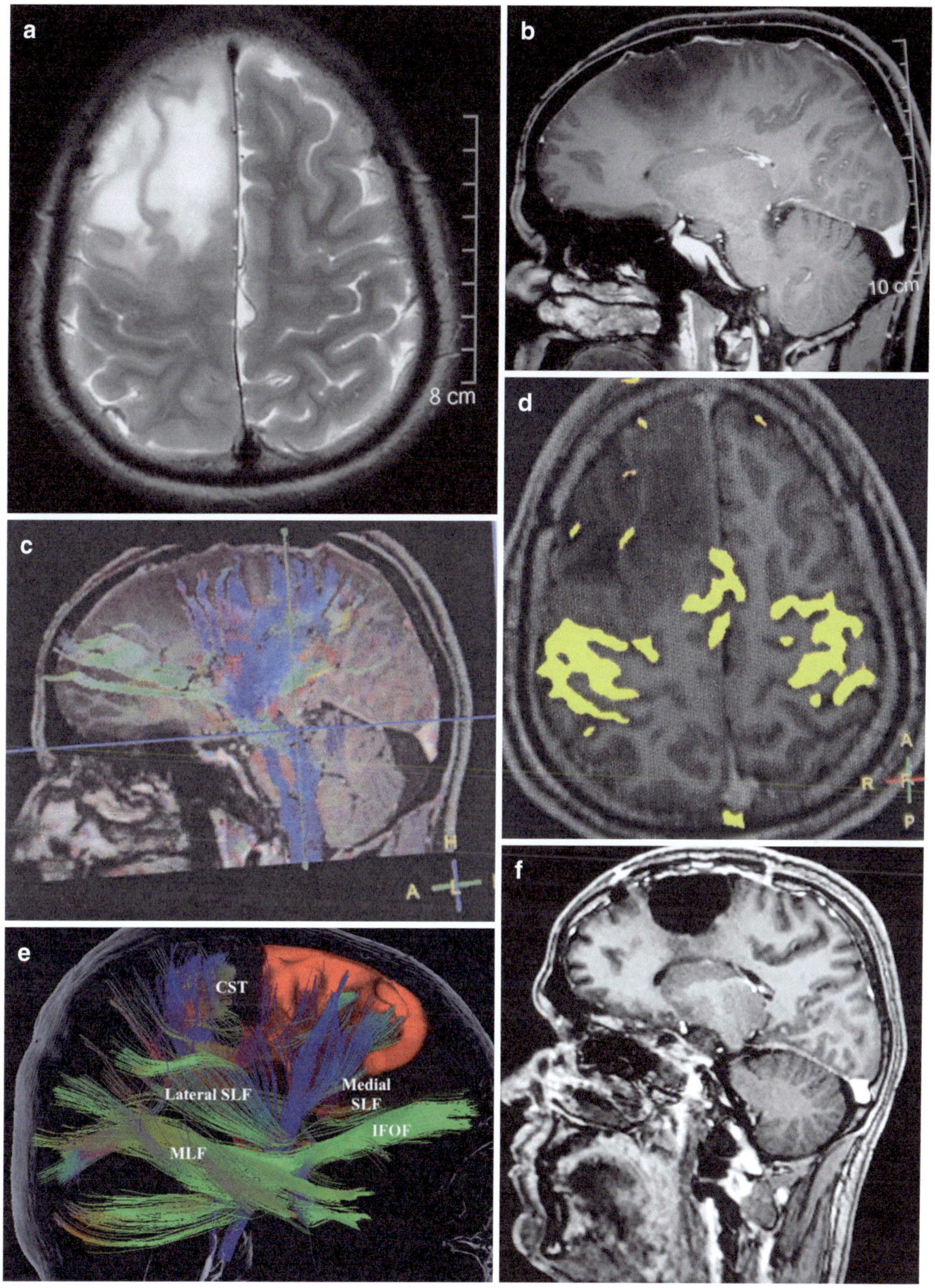

Fig. 2 Images of a 35-year-old male patient. (**a**) T2 weighted axial MRI image showing a right medial frontal low grade glioma arising from the short arcuate fibers in the region of the superior and middle frontal gyrus. (**b**) T1 weighted contrast sagittal image showing the hypointense tumor without any contrast enhancement. (**c**) Tractography image showing the tumor medial to the vertical group of projection fibers. (**d**) Functional MRI image showing the motor area abutting the posterior margin of the tumor. (**e**) Tractography images showing the tumor displacing the vertical group of projection fibers laterally and the medial SLF inferiorly. (**f**) Post-operative T1 weighted contrast-enhanced image showing the resection cavity. (**g**) Post-operative T2 weighted axial image showing the resection cavity. (**h**) Post-operative FLAIR image showing the tumor resection. (**i**) Post-operative specimen showing the en-masse resection

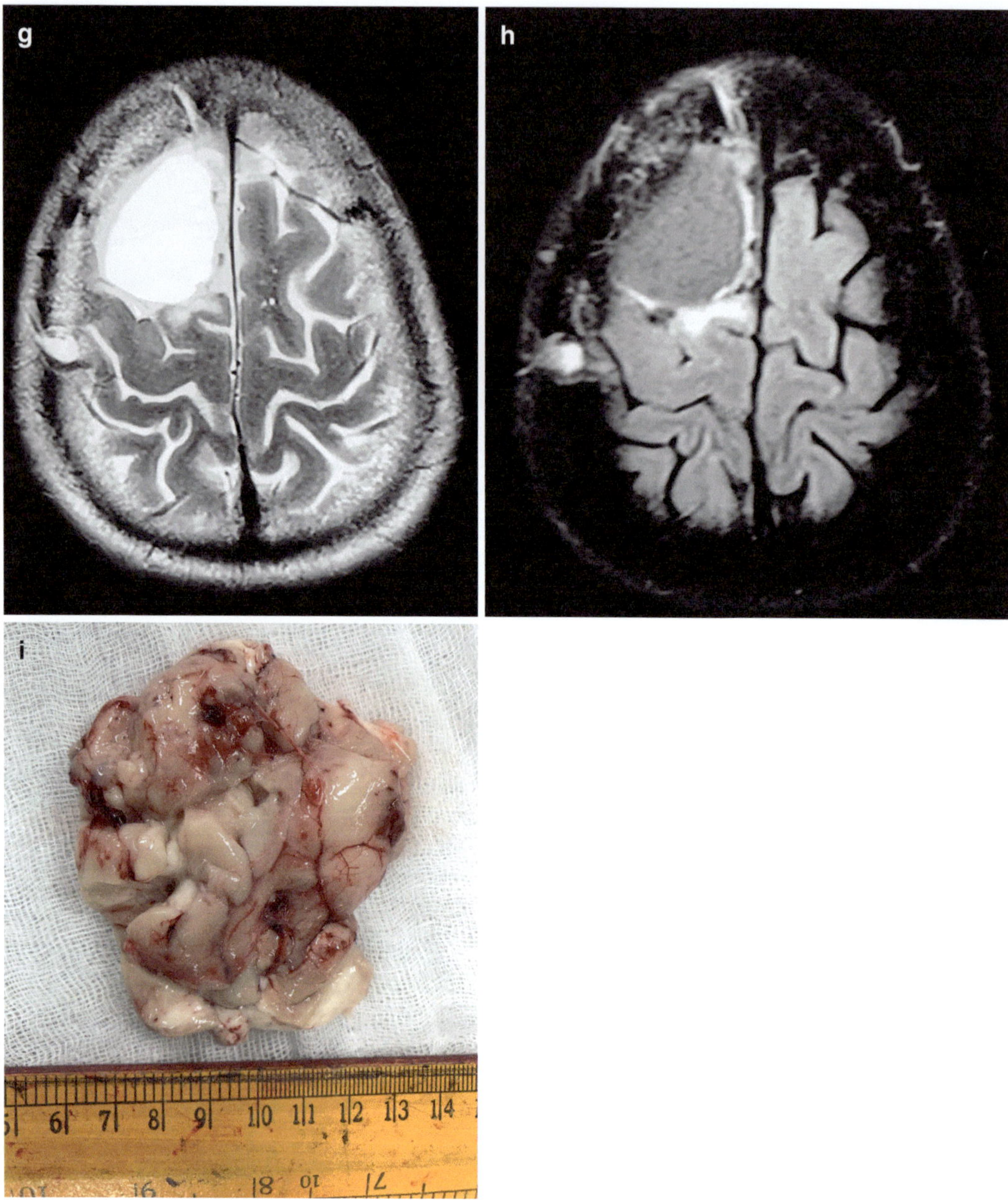

Fig. 2 (continued)

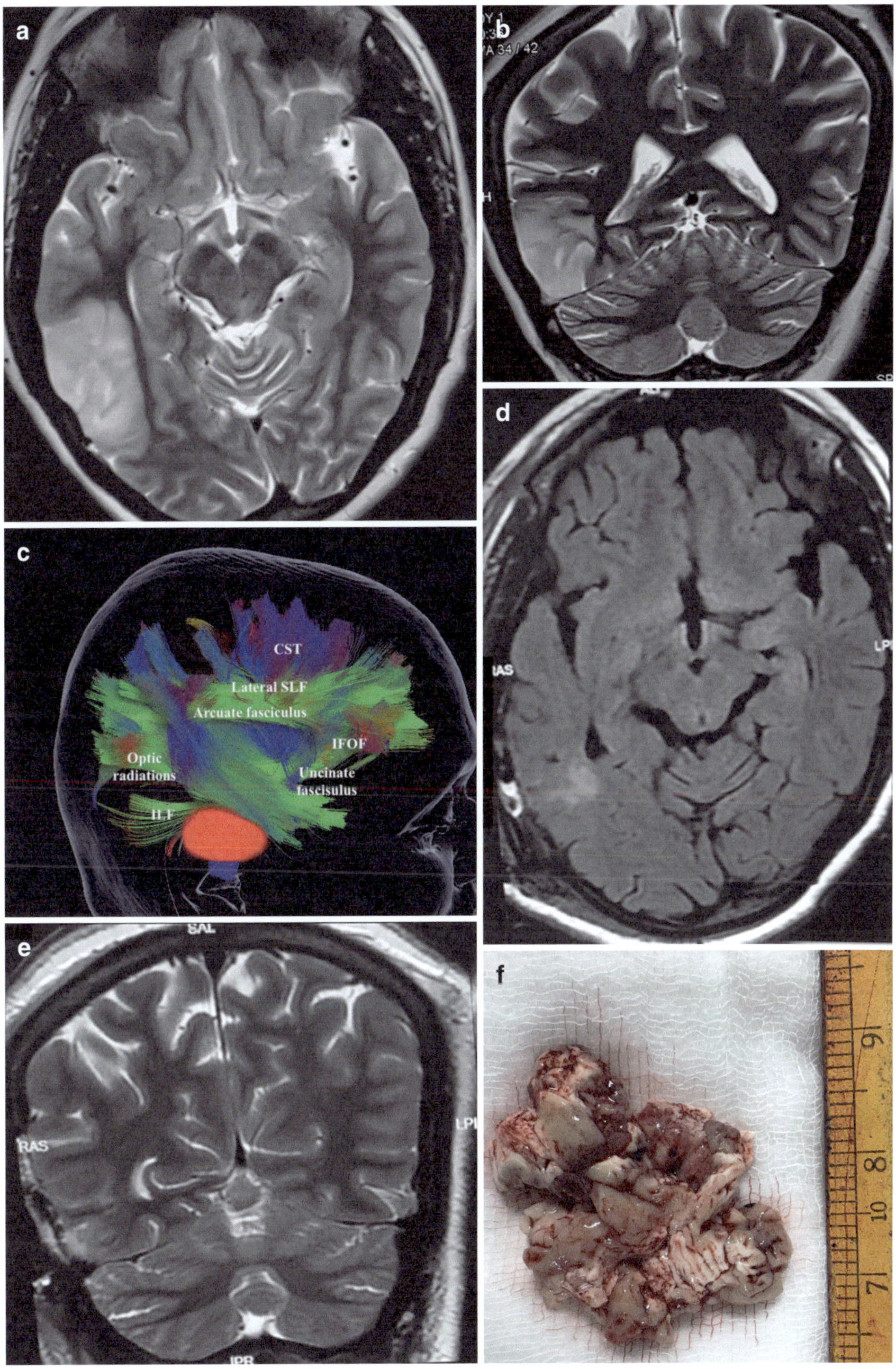

Fig. 3 Images of a 30-year-old female patient. (**a**) T2 weighted axial image showing a right lateral temporal low-grade glioma arising from the short arcuate fibers of the inferior temporal gyrus. (**b**) T2 weighted coronal image showing the tumor. (**c**) Tractography images showing no major white fibers in the vicinity of the tumor. All the fibers are present superior and medial to the tumor. (**d**) Post-operative T2 weighted axial image showing the tumor resection. (**e**) Post-operative coronal image showing the tumor resection. (**f**) Post-operative specimen showing the en-masse resection

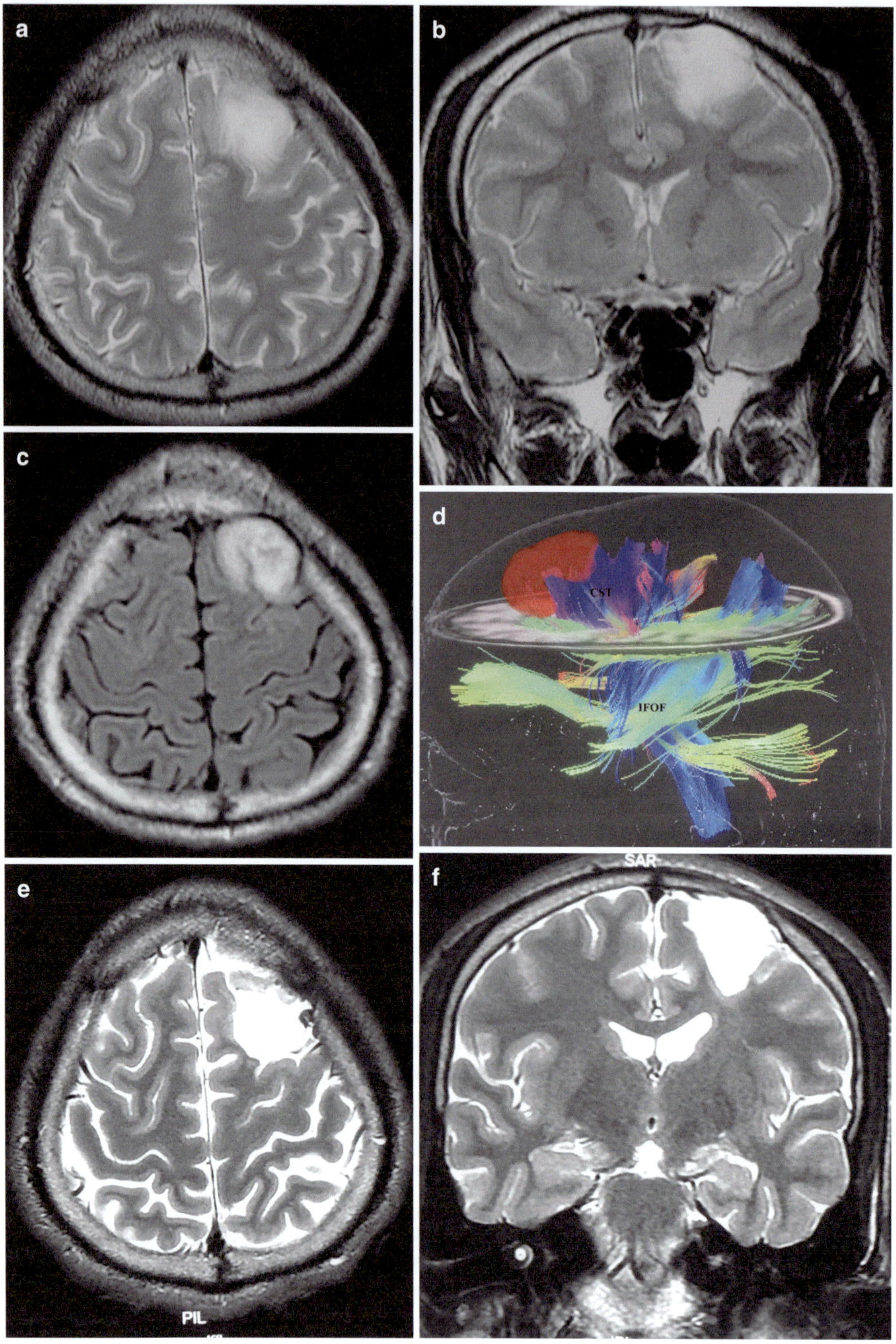

Fig. 4 Images of a 21-year-old male patient. (**a**) T2 weighted axial image showing a left medial frontal low grade glioma arising from short arcuate fibers in the region of the middle frontal gyrus. (**b**) T2 weighted coronal image showing the tumor. (**c**) FLAIR image showing the tumor. (**d**) Tractography images showing the tumor medial to the vertical group of projection fibers. (**e**) Post-operative T2 weighted axial image showing the supra-total tumor resection. (**f**) Post-operative coronal image showing the resection cavity. (**g**) FLAIR image showing the supra-total resection. (**h**) Post-operative specimen showing the en-masse resection

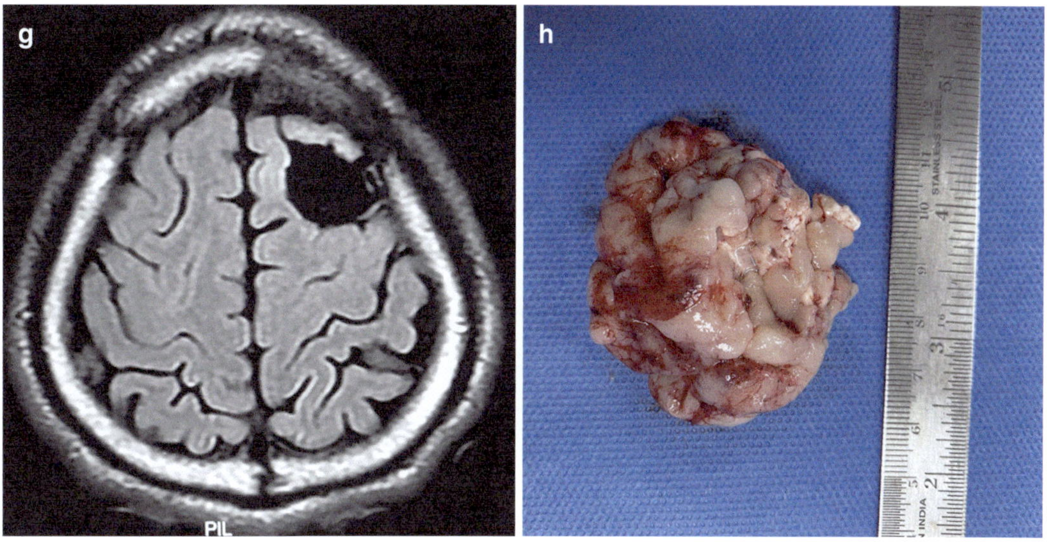

Fig. 4 (continued)

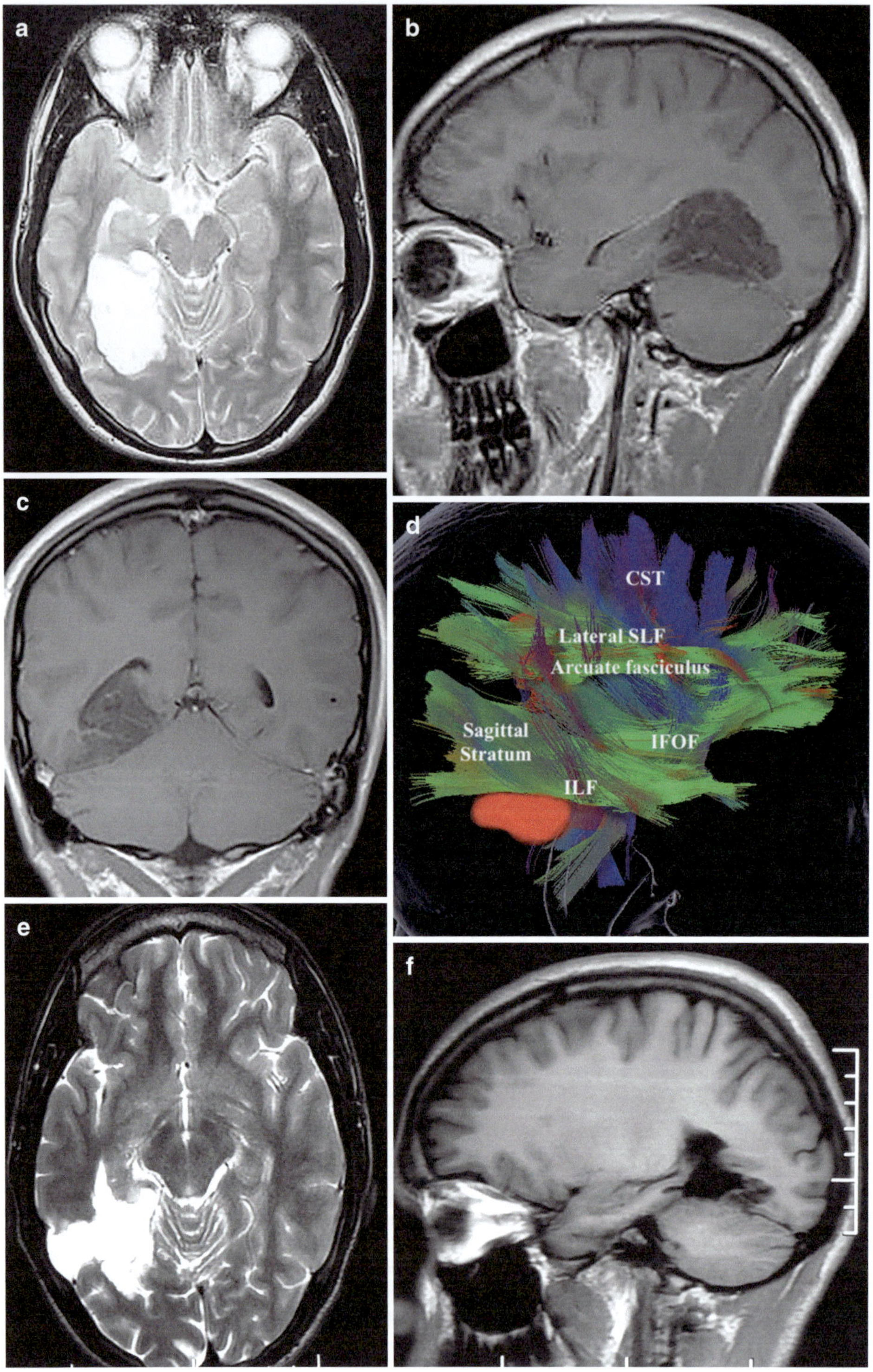

Fig. 5 Images of a 39-year male patient. (**a**) T2 weighted axial image showing a right medial temporal low grade glioma. (**b**) T1 weighted sagittal image showing the hypointense tumor. (**c**) T1 weighted coronal image showing the medial temporal glioma. (**d**) Tractography images showing the medial temporal tumor lying inferior and medial to the sagittal stratum. (**e**) Post-operative T2 weighted axial image showing the supra-total resection. (**f**) Post-operative T1 weighted sagittal image showing the resection cavity. (**g**) Post-operative T1 weighted coronal image showing the supra-total resection

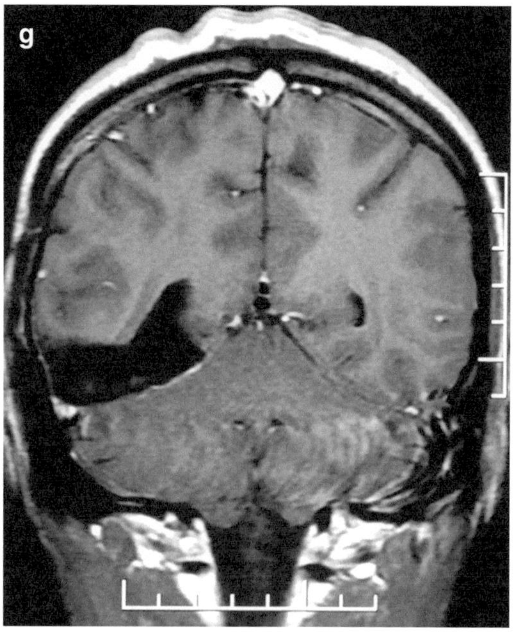

Fig. 5 (continued)

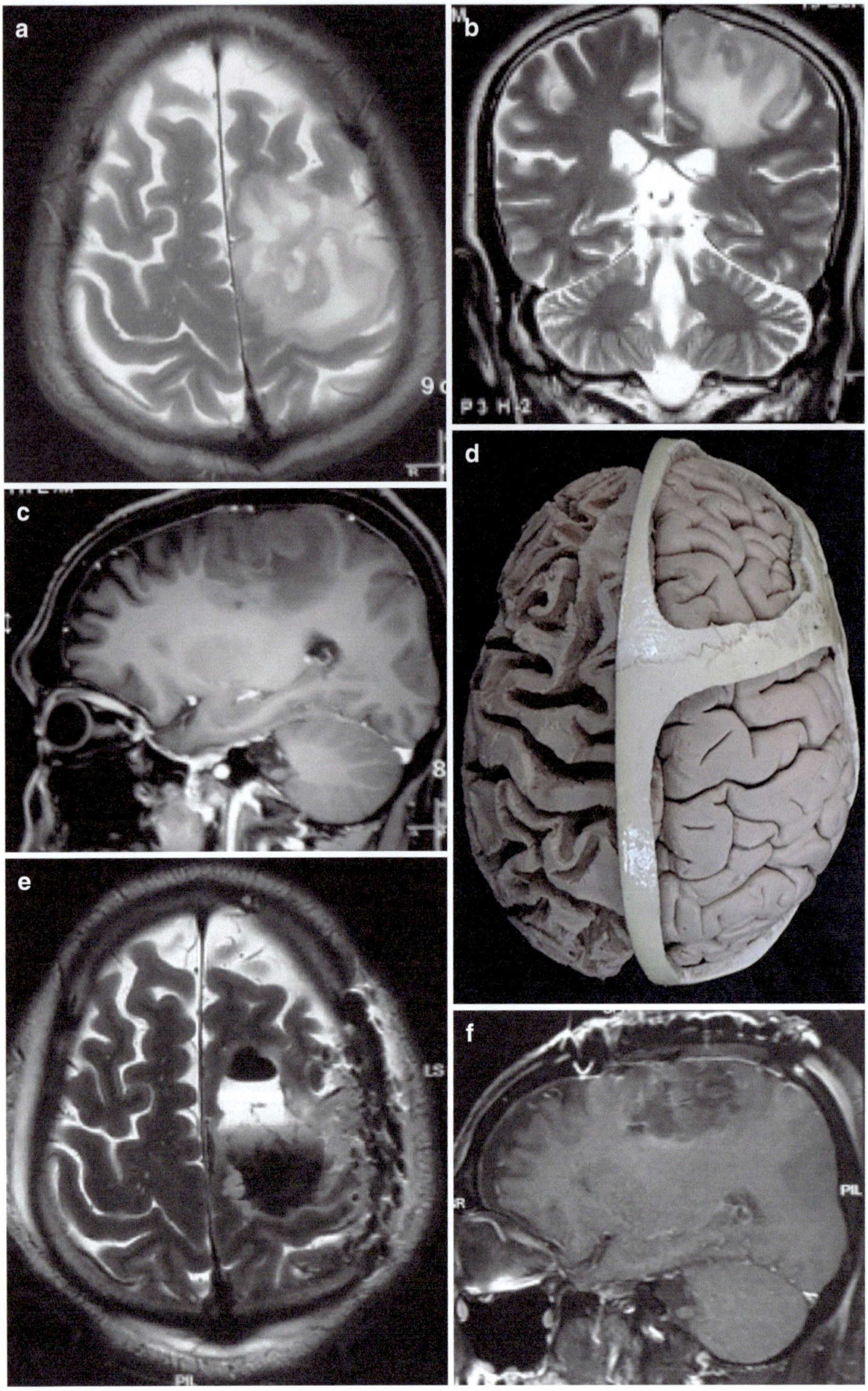

Fig. 6 Images of a 39 year old male patient. (**a**) T2 weighted axial image showing a low grade glioma involving the left precentral gyrus. (**b**) T2 weighted coronal image showing the tumor. (**c**) T1 weighted contrast sagittal image showing the tumor in the region of the precentral gyrus. (**d**) Cadaveric image showing the relation of the precentral gyrus to the coronal suture. (**e**) Post-operative T2 weighted axial image showing the resection. (**f**) Post-operative T1 weighted sagittal image showing the resection cavity. (**g**) Post-operative T2 weighted coronal image showing the resection. (**h**) Post-operative specimen showing the en – masse resection

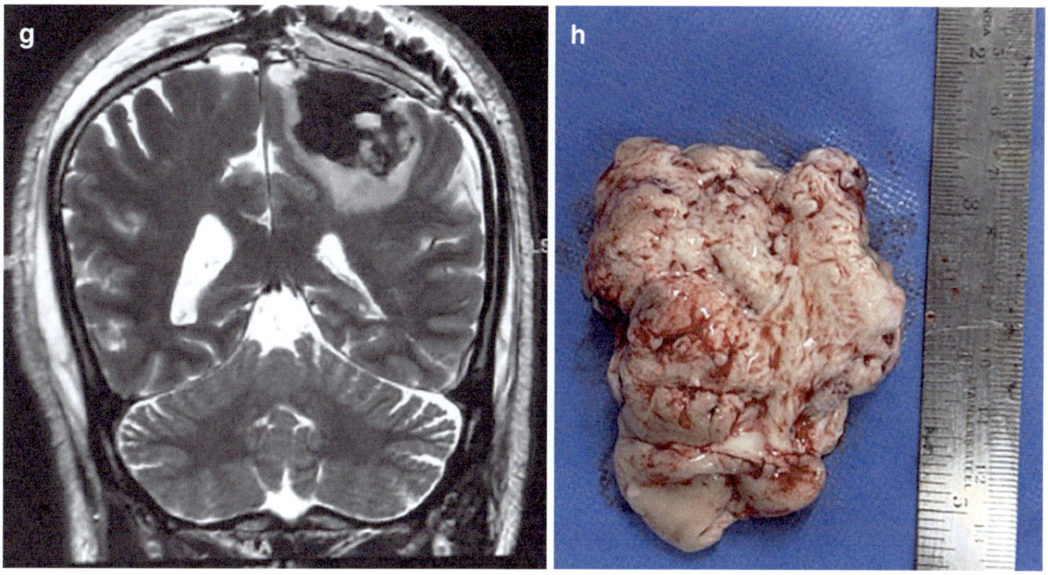

Fig. 6 (continued)

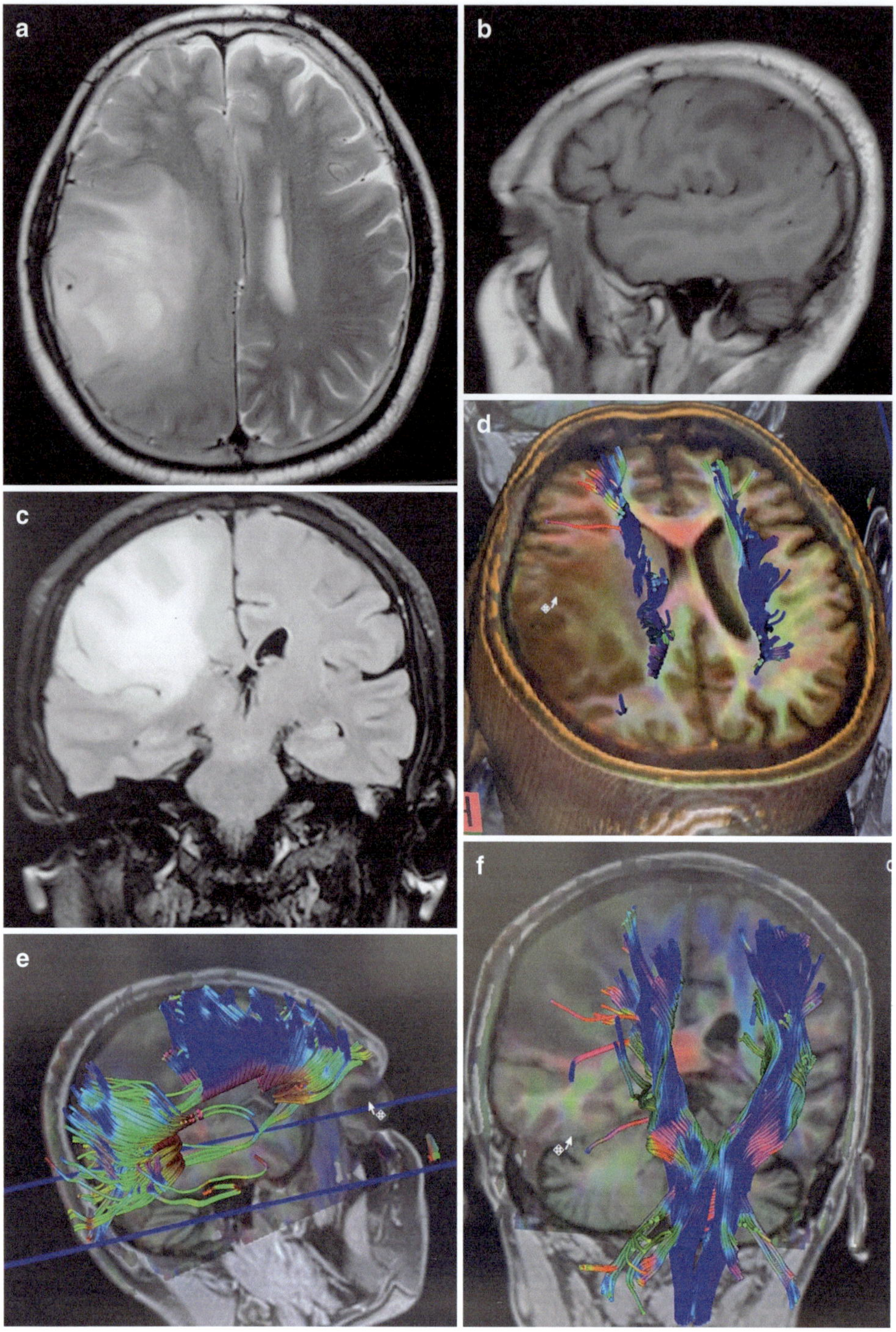

Fig. 7 Images of a 48 year old male patient. (**a**) T2 weighted axial image showing a low grade glioma arising from short arcuate fibers in the region of the right postcentral gyrus. (**b**) T1 weighted sagittal image showing the tumor in the region of the post-central gyrus. (**c**) T2 weighted coronal image showing the tumor. (**d**) Tractography image showing the tumor lateral to the vertical group of projection fibers. (**e**) Tractography image showing the medially displaced projection fibers. (**f**) Tractography image showing the uninvolved and displaced projection fibers. (**g**) Post-operative T1 weighted axial image showing the tumor resection. (**h**) Post-operative sagittal image showing the resection cavity. (**i**) Coronal FLAIR image showing the resection. (**j**) Post-operative sagittal tractography image showing the intact projection fibers. (**k**) Post-operative axial tractography image showing the resection cavity and the projection group of fibers. (**l**) Post-operative specimen showing the en-masse resection

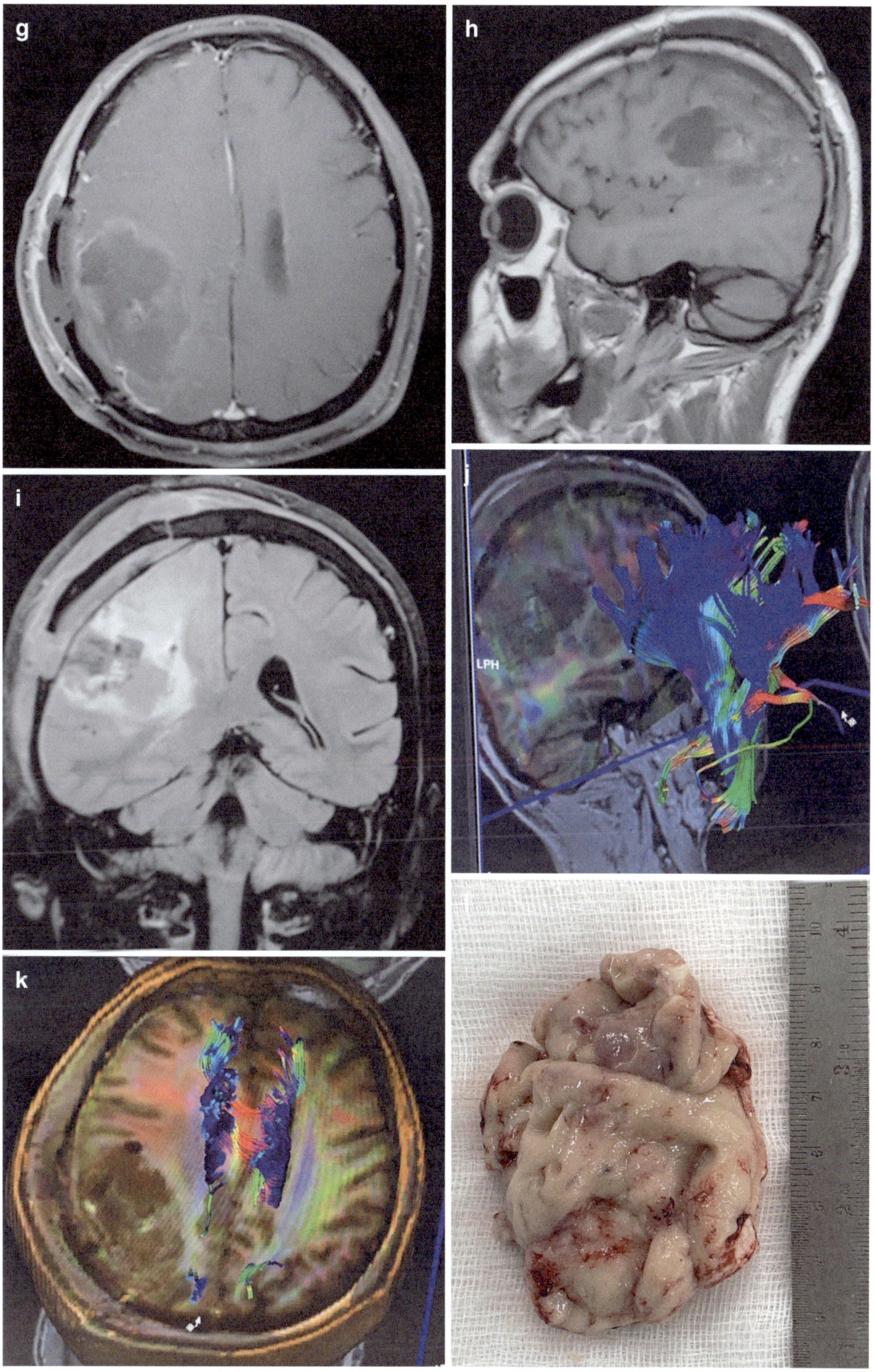

Fig. 7 (continued)

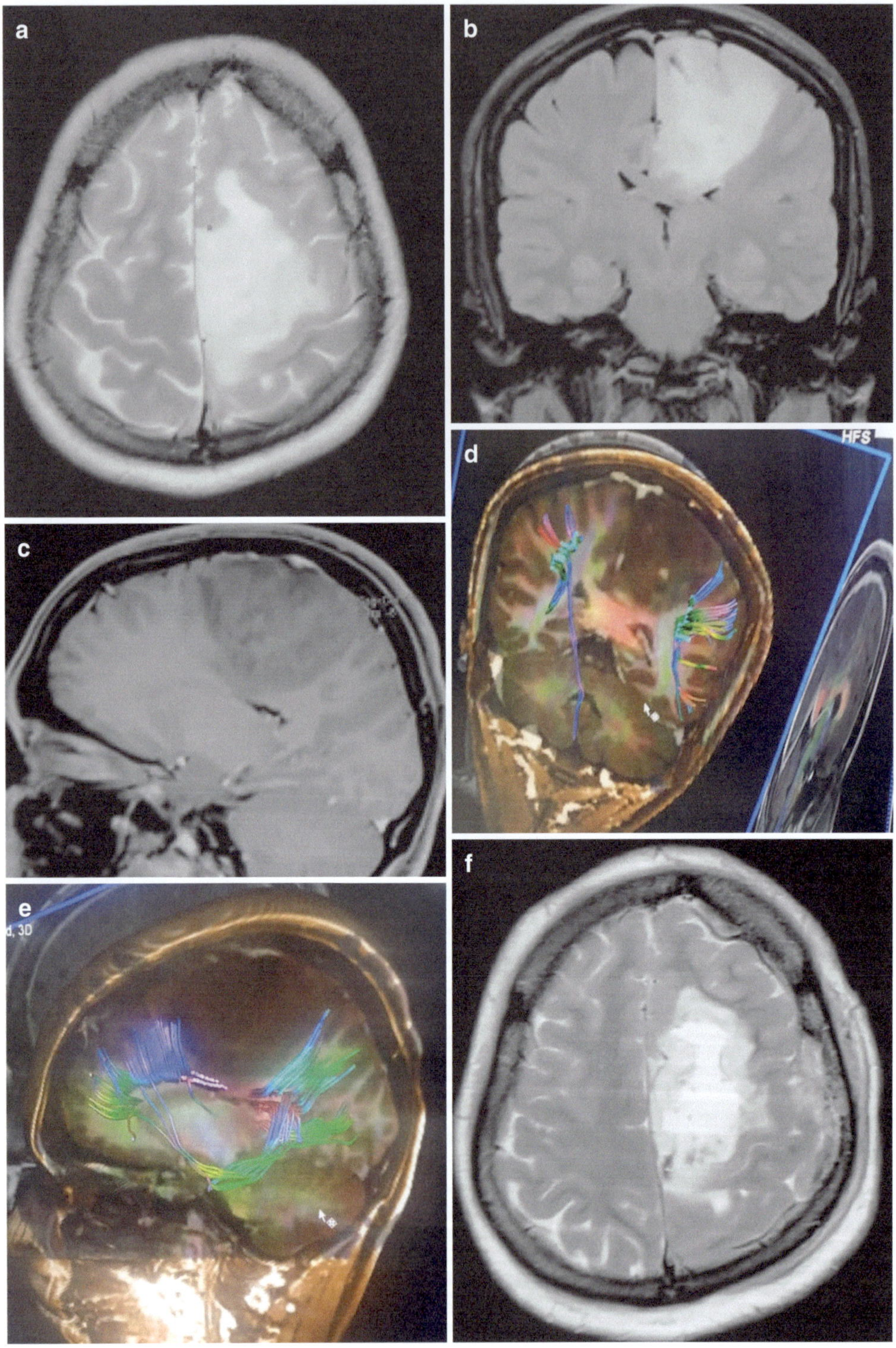

Fig. 8 Images of a 31 year old female patient. (**a**) T2 weighted axial image showing a low grade glioma involving the left precentral gyrus. (**b**) T2 weighted coronal image showing the tumor. (**c**) T1 weighted contrast sagittal image showing the tumor in the region of the precentral gyrus. (**d**) Tractography image showing the displacement of the projection fibers. (**e**) Tractography image showing the medial and inferior displacement of the projection fibers. (**f**) Post-operative T2 weighted axial image showing the resection. (**g**) Post-operative T1 weighted axial contrast image showing the resection. (**h**) Post-operative T1 weighted sagittal image showing the resection cavity. (**i**) Post-operative specimen showing the en – masse resection

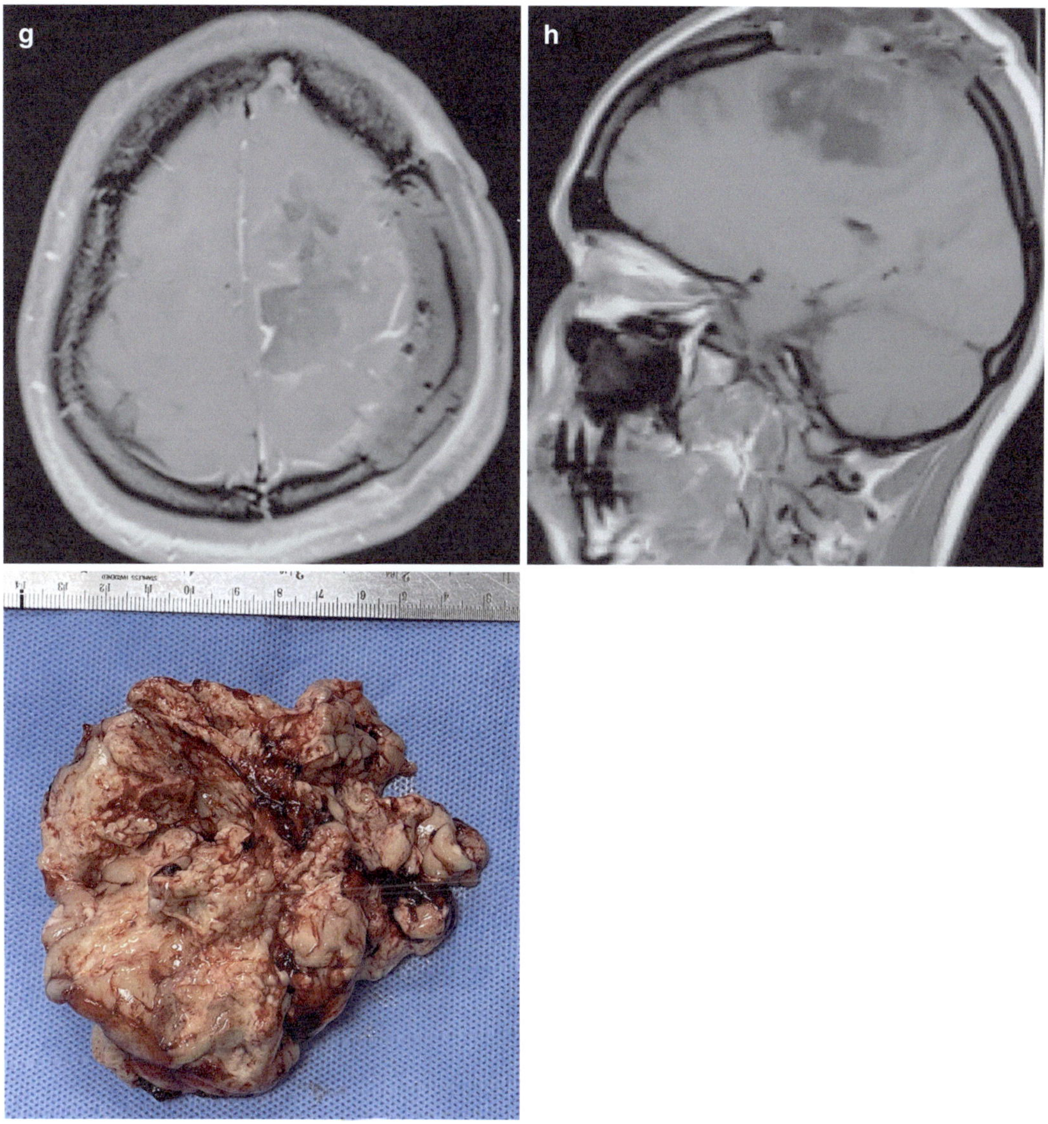

Fig. 8 (continued)

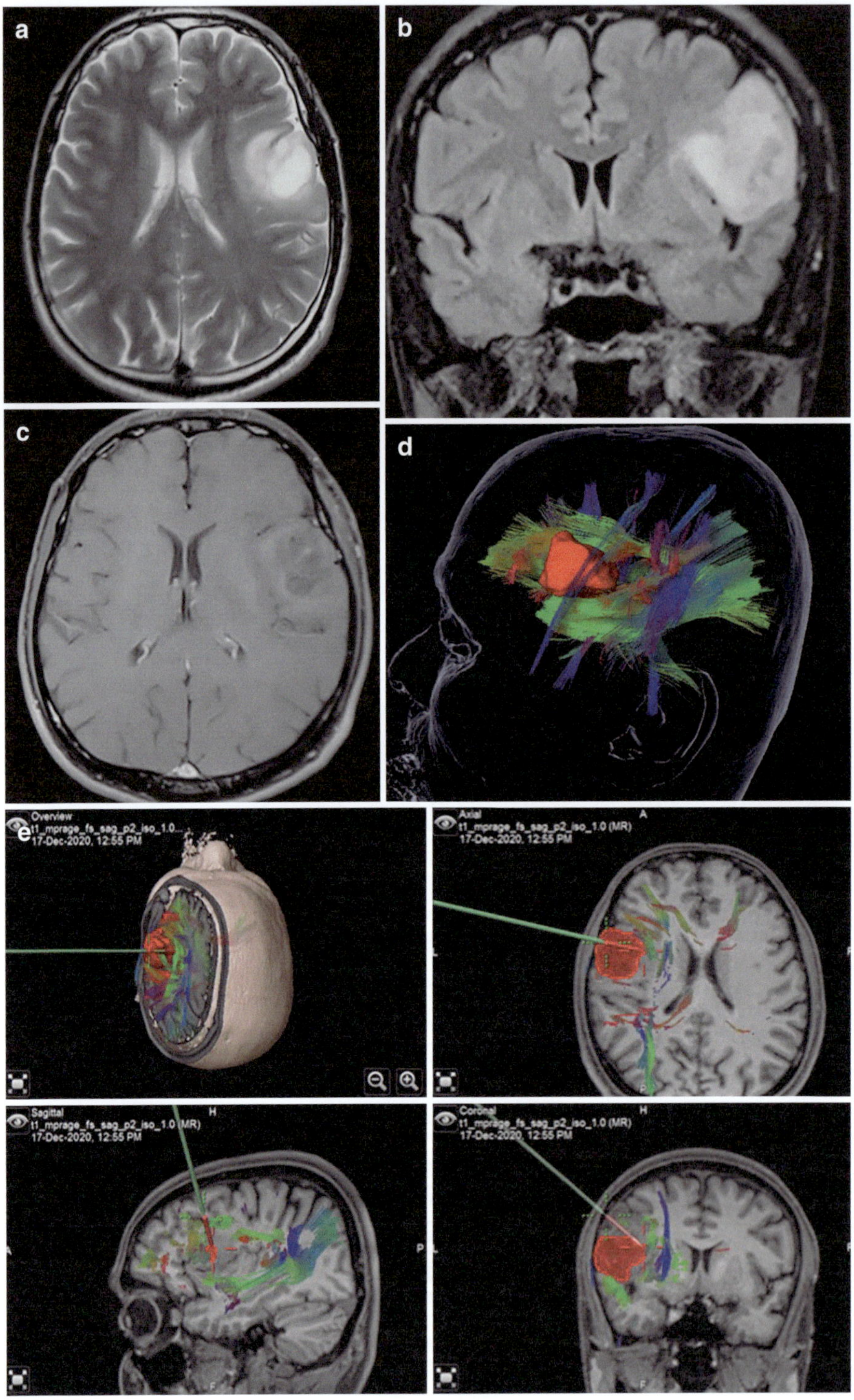

Fig. 9 Images of a 35 year old male patient. (**a**) T2 weighted axial image showing a left lateral frontal low grade glioma arising from short arcuate fibers in the region of the inferior frontal gyrus. (**b**) FLAIR coronal image showing the tumor. (**c**) T1 weighted axial image showing the tumor. (**d**) Tractography images showing the tumor lateral to the vertical group of projection fibers. The arcuate fasciculus lies medial to the tumor. (**e**) Fused neuronavigation and tractography images showing the tumor and the fibers displaced around it. (**f**) Post-operative specimen showing the en-masse resection. (**g**) Post-operative axial image with tractography showing the tumor resection. (**h**) Post-operative coronal image with tractography showing the resection cavity

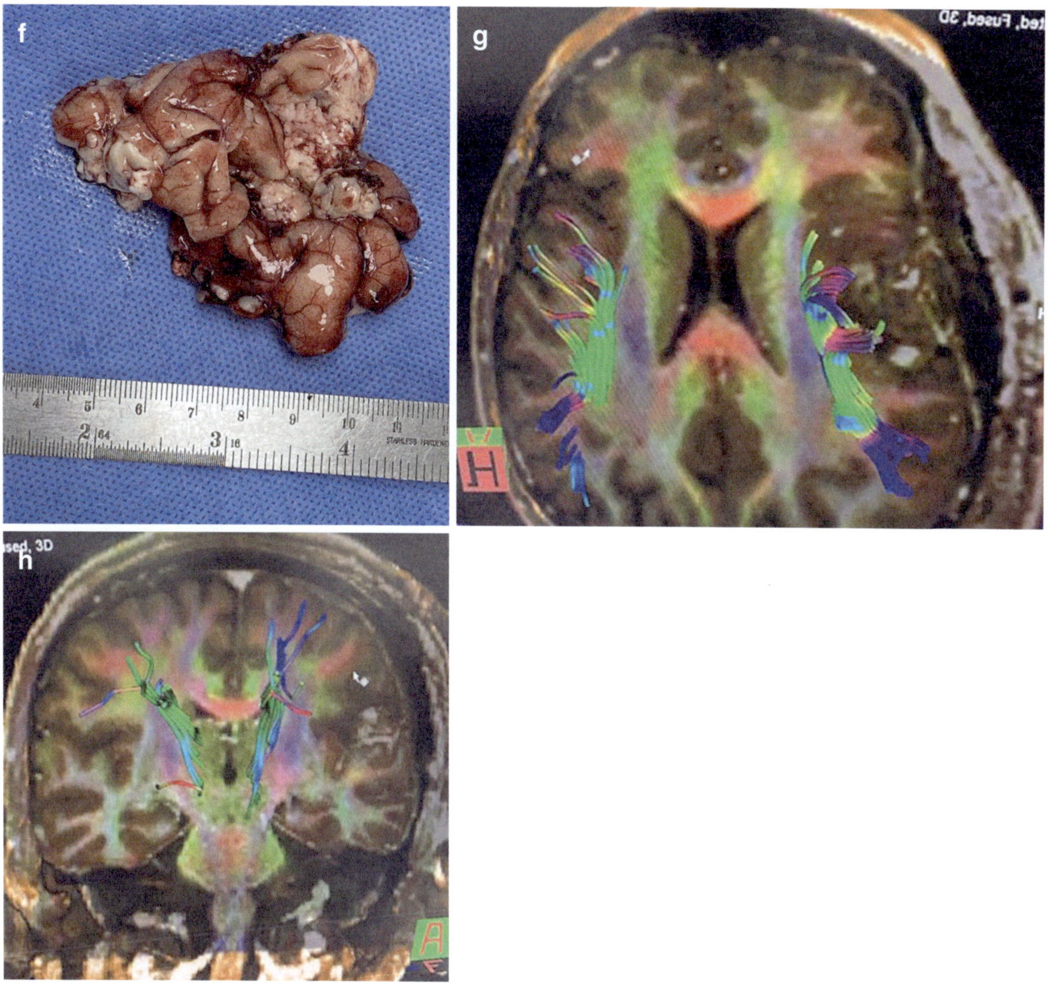

Fig. 9 (continued)

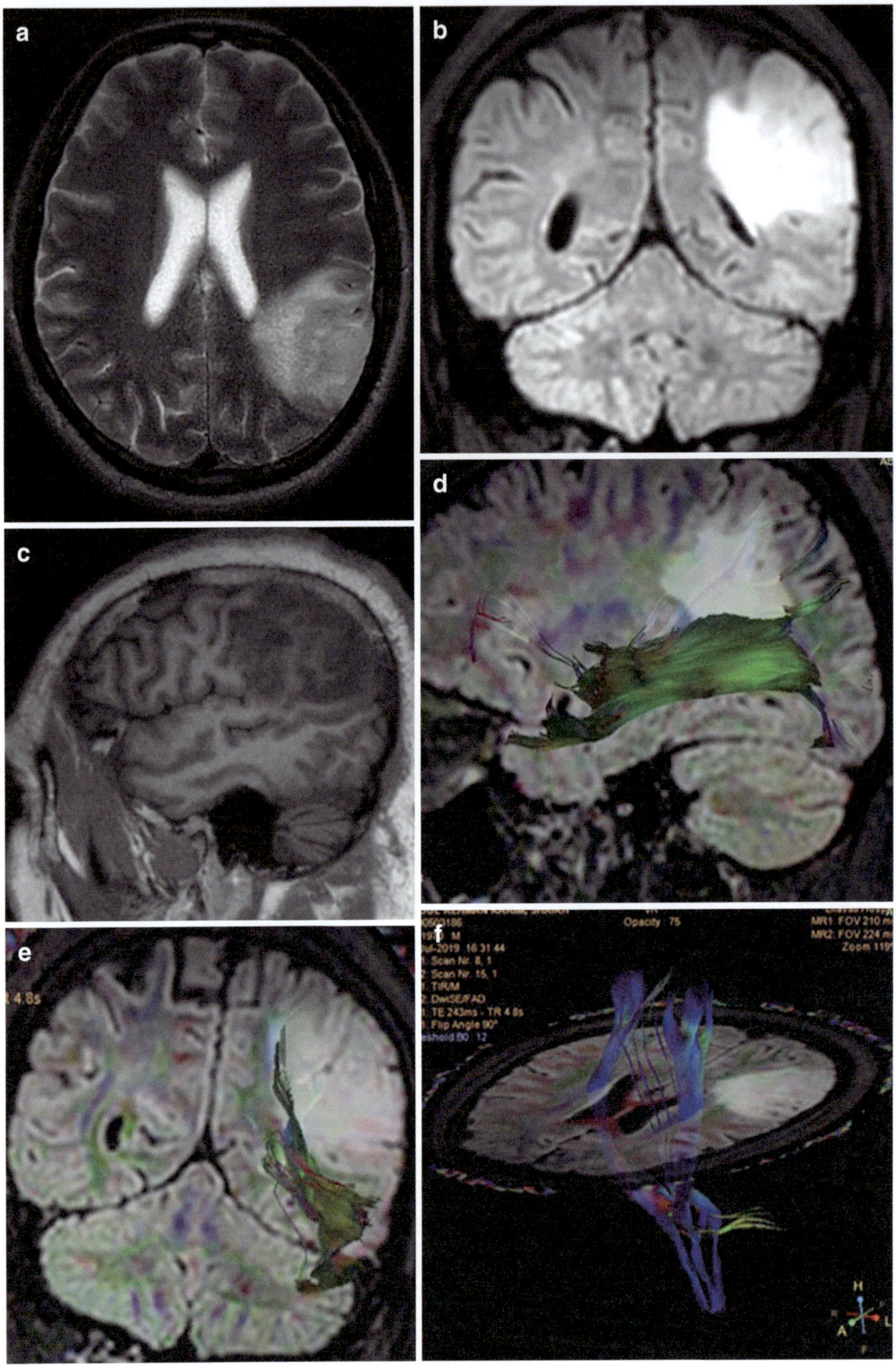

Fig. 10 Images of a 39 year old male patient. (**a**) T2 weighted axial image showing a low grade glioma involving the left superior parietal lobule. (**b**) FLAIR coronal image showing the tumor. (**c**) T1 weighted sagittal image showing the tumor in the region of the superior parietal lobule. (**d**) Tractography image showing the tumor medial and superior to the optic radiations and the sagittal stratum. (**e**) Tractography image showing the tumor medial to the optic radiations. (**f**) Tractography image showing the projection fibers anterior to the tumor. (**g**) Post-operative T2 weighted axial image showing the resection. (**h**) Post-operative axial FLAIR image showing the resection. (**i**) Post-operative T1 weighted sagittal contrast image showing the resection cavity. (**j**) Post-operative specimen showing the en – masse resection

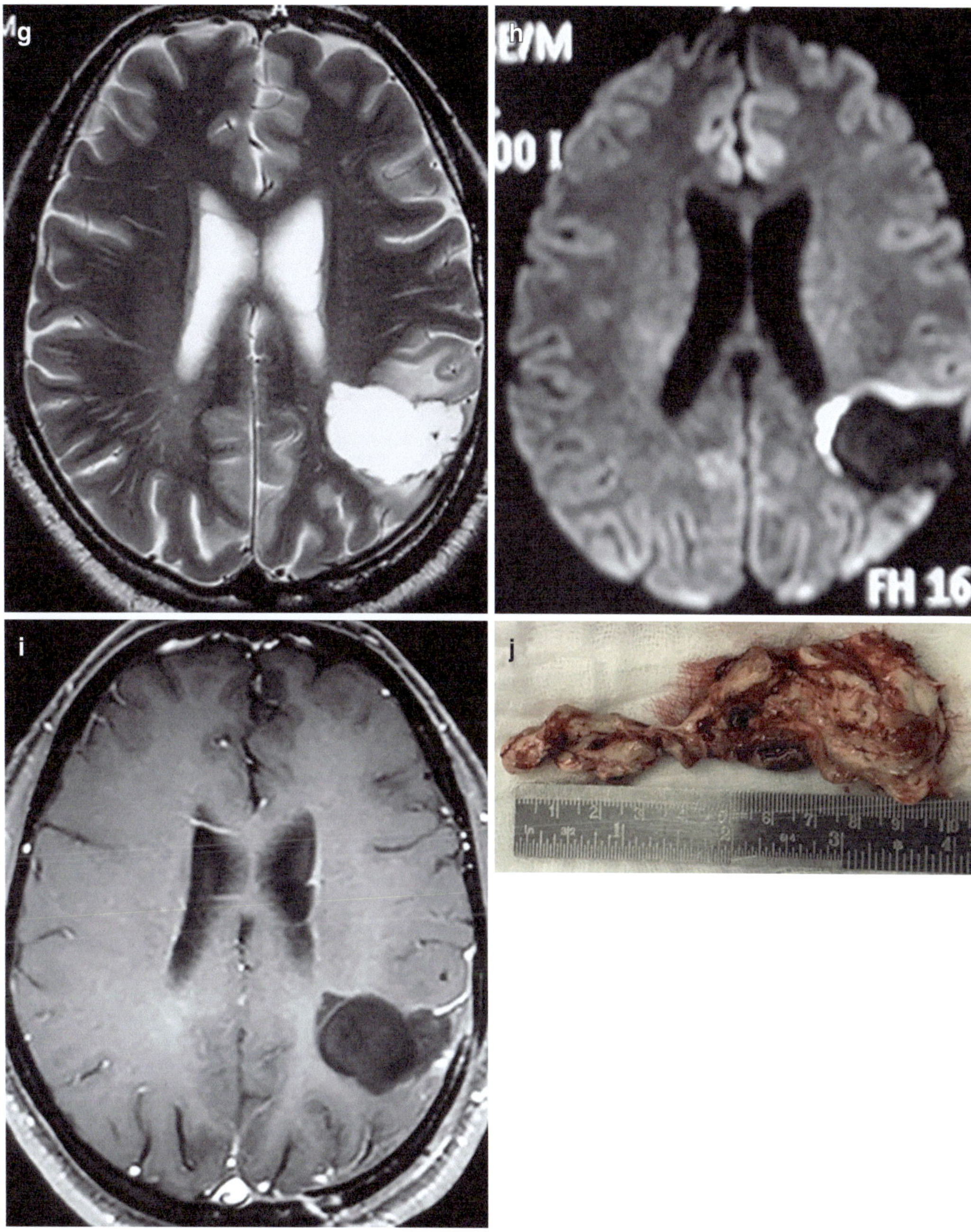

Fig. 10 (continued)

8 Surgical Results

We recently reported our series of 74 patients with low-grade gliomas involving the short arcuate fibers that were operated between the years 2016 to June 2019 [19]. There were 46 males and 28 females. The ages of the patients ranged from 21 to 55 years (average 33 years). The duration of symptoms ranged from 10 days to 5 years. The location of the tumors, the fiber tracts in vicinity and the nature of tumor resection are summarized in Table 1. Table 2 summarises the clinical features at the time of presentation. Visual impression and consistency of tumor allowed

Table 1 Table showing the location of the gliomas

Location of glioma with respect to location of the short arcuate fiber and the gyrus involved	Number of patients	Fiber tracts in vicinity	Need for intra-operative neurophysiological monitoring	Preferred mode of resection	Number of patients with en- masse resection
Frontal Medial (SFG, MFG) Lateral (IFG) Basal (OG, GR)	15 9 2	Corona radiata, projection fibers, medial SLF AF, Lateral SLF, IFOF, Frontal aslant tract IFOF, UF, forceps minor	No Yes on dominant side No	En - masse Tailored En - masse	15 5 2
Temporal Medial (PHG) Lateral (STG, MTG, ITG) Basal (FG)	10 6 3	IFOF, OR, UF, AC, Cerebral peduncle MLF, ILF, IFOF, UF, OR ILF	No Yes on dominant side especially STG No	En-masse in small tumors En-masse for MTG, ITG) En-masse	2 5 3
Parietal Medial (SPL) Lateral (IPL)	2 9	Corona radiata, projection fibers, medial SLF AF, MLF, IFOF, Lateral SLF	No Yes on dominant side	En-masse Tailored	2 2
Occipital Medial (SOG, MOG) Lateral (IOG)	3 1	Forceps major OR	No No	En- masse En-masse	3 1
Central Pre-central/Premotor/ Supplementary motor Post-central	12 2	Projection fibers, Frontal aslant tract, SLF Projection fibers, SLF	Yes Yes	Tailored Tailored	6 1

Abbreviations: *SLF* superior longitudinal fasciculus, *AF* arcuate fasciculus, *IFOF* inferior fronto-occipital fasciculus, *OR* optic radiations, *UF* uncinate fasciculus, *AC* anterior commissure, *MLF* middle longitudinal fasciculus, *ILF* inferior longitudinal fasciculus, *SFG* superior frontal gyrus, *MFG* middle frontal gyrus, *IFG inferior frontal gyrus*, *OG* orbital gyri, *GR* gyrus rectus, *PHG* parahippocampal gyrus, *STG* superior temporal gyrus, *MTG* middle temporal gyrus, *ITG* inferior temporal gyrus, *FG* fusiform gyrus, *SPL* superior parietal lobule, *IPL* inferior parietal lobule, *SOG* superior occipital gyrus, *MOG* middle occipital gyrus, *IOG* inferior occipital gyrus

Table 2 Table showing the presenting symptoms

Presenting symptoms	Number of patients
Headache	65
Seizures	70
Hemiparesis	3
Language abnormalities	
Fluency	2
Naming	1
Forgetfulness	5

development of a plane and radical tumor resection. The strategy of *en-masse* resection entailed removal of the entire affected gyrus in the region of the involved short arcuate fibers (Figs. 1, 2, 3, 4, 5, 6, 7, 8, 9, and 10). Essentially this surgical strategy of developing a surgical plane between normal brain and glioma led to a "supramarginal" resection in 25 patients. An essentially *en-masse* tumor resection was possible in 47 cases. Twelve patients had subtotal resection. Thirty-seven patients had gross total tumor resection and 25 patients had supra-maximal resection. At follow up that ranged from 11 to 56 months (mean 21 months) all patients improved in their preoperative complaints of headache and seizures following surgery. Three patients with motor weakness improved in their neurological deficits following surgery. None of the patients had any decline in their cognitive functions and were able to lead full functional lives.

Out of the three patients who had preoperative speech deficit, in 2 patients the tumor was located in the left inferior frontal gyrus and in 1 patient the tumor was in the fusiform gyrus. All these three patients improved in their speech function following surgery. One patient with a lesion in the pars opercularis and without any preoperative speech deficit developed anarthria with stimulation during awake surgery and underwent a subtotal tumor resection. Immediately following surgery the patient had dysarthria, which improved at follow up and the patient had normal speech at follow up of 2 years. Two patients with tumors located in the posterior superior frontal gyrus developed SMA (supplementary motor area) syndrome after surgery but recovered completely within 3 months. Post-operative MRI performed in the patients with the new deficits did not reveal presence of any ischemic injury or haemorrhage. Post-operative MRI scans were performed 48 hours after surgery, at 3 months and at 6 monthly intervals thereafter. Patients with complete tumor excision on FLAIR imaging were clinically observed without any use of adjuvant treatment. The histology of the patients was WHO Grade II astrocytoma (33 patients) or oligodendroglioma (41 patients). Twelve patients who had undergone a subtotal resection were given adjuvant radiotherapy and chemotherapy. At follow up of 11–52 months (average 30 months) the tumor residue was found to be stable in this group of patients. One patient with a medial temporal tumor developed a recurrence of the tumor 26 months after the first surgery and was re-operated. The histology was Grade II diffuse astrocytoma. Post-operative MRI showed a total resection of the recurrent tumor. None of the other patients had any recurrence or growth of residual tumor. Surgery was essentially based on the understanding of anatomical framework of white fiber tracts. By using the anatomical fibre tract model and information obtained from tractography and functional MRI it was possible to perform *en-masse* tumor resection in 47 (63.5%) patients with 'minimal' or no neurological morbidity. The resection was supra marginal in 25 (33.8%) patients (Table 3).

Table 3 Table showing the differences in clinical and radiological characteristics of patients and extent of resection

Variable	Total number of patients	Supratotal resection (No. of patients)	Total resection (No. of patients)	Subtotal resection (No. of patients)
Number of patients	74	25	37	12
Cerebral hemisphere				
Dominant	33	9	16	8
Non dominant	41	16	21	4
Tumor location by our classification				
Frontal				
Medial	15	8	7	–
Lateral	9	–	5	4
Basal	2	2	–	–
Temporal				
Medial	10	2	7	1
Lateral	6	2	3	1
Basal	3	3	–	–
Parietal				
Medial	2	2	–	–
Lateral	9	2	5	2
Occipital				
Medial	3	3	–	–
Lateral	1	1	–	–
Central				
Pre-central + Premotor + Supplementary motor	12	–	8	4
Post-central	2	–	2	–

9 Surgical Pearls

The location of the tumor is the most important factor in determining the feasibility of *"en masse"* supra total resection. It is possible to achieve an *"en-masse"* total or supra-total resection in all the patients with medial and basal frontal tumors, in basal temporal tumors and in medial parietal and occipital tumors [19]. Tumors involving or in vicinity of eloquent cortex are operated with the same surgical strategy of en-masse resection with more liberal use of neurophysiological monitoring. Awake surgery is used for patients with tumors in the left hemisphere in the region of the inferior frontal gyrus, superior temporal gyrus and angular gyrus [19].

Identification of normal from abnormal, dissection around the surface of the tumor, preservation of blood vessels coursing in the vicinity are important prerequisites of successful *en-masse* resection. The plane of dissection between normal brain and abnormal tumor involved brain was identified to be well-defined and reliable. In all the patients tractography showed that the long association fibers, the commissural fibers and the projection fibers were essentially displaced by the tumor [19, 20].

The sulcul blood vessels are located around the dome of the tumor and encased within the tumor lobulations and need to be preserved as they are not 'tumor vessels' but are normal coursing, *en-passage* or transit blood vessels. Whilst neurological deficits related to brain handling and manipulation usually recover over time, deficits due to a vascular injury are generally long-standing.

Improvement in neurological symptoms (95.9% patients in our series) following surgical tumor resection is indicative of the fact that the displaced tracts were functionally compromised due to pressure and not due to destruction, invasion or involvement. There was improvement in motor deficits and speech in 6 patients in our series. This may even be attributed to normalization, re-organization and plasticity of functional cortical areas around the dome of the tumor. None of our patients had any recognisable deterioration in cognitive and psychological function and were able to regain active and functional lives. This suggests that the tumor affected brain was essentially 'non- functional'. Histological examination of the resected tumor specimen at multiple areas within the tumor bulk showed tumor in all components and absence of normal brain configuration. We have recently started performing post-operative tractography to analyse the status and re-alignment of the tracts following surgery.

Intraoperative navigation and real time ultrasound form important tools to identify the location, extent and depth of the tumor. Although these technological tools are useful and are used, the color and consistency of the tumor is identified to be more reliable to locate the tumor. As our experience with en-masse tumor resection matured and the dissection technique was perfected, low-grade gliomas even in eloquent motor and speech areas are resected using the same technique. Despite the fact that an attempt was made to resect the tumor radically and by *en-masse* technique, postoperative imaging occasionally shows presence of residual tumor. This suggests that either the tumor extended into adjoining parenchyma with tail like protrusions or the plane of surgical dissection was lost during surgery. From our initial experience in our series it appears that the fear of affecting neural damage to critical adjoining tracts, a conservative surgical resection was probably carried out and the plane of dissection was lost at some points.

Some authors have advocated a subpial en-bloc – sulcal to sulcal resection when possible as it maintains margins throughout the tumor removal without interfering with tumor vascularity. In tumors near eloquent cortex a piece meal resection is advocated [6].

Although intraoperative neurophysiological monitoring and imaging are being used currently in some cases, it appears that with more experience in dissecting normal from abnormal and by technique of *en-masse* resection, both these intraoperative procedures are dispensable. It also appears that with more experience in developing a plane between the tumor and adjoining normal structures, the need of awake craniotomy can be minimized or avoided.

10 Conclusion

The technique of 'en-masse' resection of low grade glioma is possible, safe and ensures radical tumor resection. Real-time anatomical information of the involved and displaced white fibre tracts is crucial for adopting such surgical strategy.

References

1. Duffau H. Higher-order surgical questions for diffuse low-grade gliomas: supramaximal resection, neuroplasticity, and screening. Neurosurg Clin N Am. 2019;30(1):119–28.
2. Rossi M, Ambrogi F, Gay L, et al. Is supratotal resection achievable in low-grade gliomas? Feasibility, putative factors, safety, and functional outcome. J Neurosurg. 2019;2019:1–14.
3. Duffau H. Long-term outcomes after supratotal resection of diffuse low-grade gliomas: a consecutive series with 11-year follow-up. Acta Neurochir. 2016;158(1):51–8.
4. Buchfelder M, Zhao Y. Is awake surgery for supratentorial adult low-grade gliomas the gold standard? Neurosurg Rev. 2018;41(1):1–2.
5. Bush NA, Chang SM, Berger MS. Current and future strategies for treatment of glioma. Neurosurg Rev. 2017;40(1):1–14.
6. Krivoshcya D, Prabhu SS, Weinberg JS, Sawaya R. Technical principles in glioma surgery and preoperative considerations. J Neuro-Oncol. 2016;130(2):243–52.
7. Watts C, Sanai N. Surgical approaches for the gliomas. Handb Clin Neurol. 2016;134:51–69.
8. de Leeuw CN, Vogelbaum MA. Supratotal resection in glioma: a systematic review. Neuro-Oncology. 2019;21(2):179–88.
9. Sanai N, Berger MS. Surgical oncology for gliomas: the state of the art. Nat Rev Clin Oncol. 2018;15(2):112–25.
10. Yordanova YN, Duffau H. Supratotal resection of diffusegliomas - an overview of its multifaceted implications. Neurochirurgie. 2017;63(3):243–9.
11. Hervey-Jumper SL, Berger MS. Evidence for improving outcome through extent of resection. Neurosurg Clin N Am. 2019;30(1):85–93.

12. Lau D, Hervey-Jumper SL, Han SJ, Berger MS. Intraoperative perception and estimates on extent of resection during awake glioma surgery: overcoming the learning curve. J Neurosurg. 2018;128(5):1410–8.
13. Han SJ, Morshed RA, Troncon I, et al. Subcortical stimulation mapping of descending motor pathways for perirolandicgliomas: assessment of morbidity and functional outcome in 702 cases. J Neurosurg. 2018:1–8.
14. Castellano A, Bello L, Michelozzi C, Gallucci M, Fava E, Iadanza A, et al. Role of diffusion tensor magnetic resonance tractography in predicting the extent of resection in glioma surgery. Neuro-Oncology. 2012;14:192–202.
15. Raco A, Pesce A, Fraschetti F, et al. Motor outcomes after surgical resection of lesions involving the motor pathway: a prognostic evaluation scale. World Neurosurg. 2017;103:748–56.
16. Trinh VT, Fahim DK, Shah K, Tummala S, McCutcheon IE, Sawaya R, et al. Subcortical injury is an independent predictor of worsening neurological deficits following awake craniotomy procedures. Neurosurgery. 2013;72:160–9.
17. Keshri V, Deshpande RP, Chandrasekhar YB, Panigrahi M, Rao IS, Babu PP. Risk stratification in low grade glioma: a single institutional experience. Neurol India. 2020;68:803–12.
18. Goel A, Shah A, Dandpat S, Rai S, Prasad A. Gliomas are 'confined' to a named white matter tract: a revolution in understanding gliomas. J Craniovertebr Junction Spine. 2020;11:252–3.
19. Goel A, Shah A, Vutha R, Dandpat S, Hawaldar A. Is "en masse" tumor resection a safe surgical strategy for low-grade gliomas? Feasibility report on 74 patients treated over four years. Neurol India. 2021;69(2):406–13.
20. Shah A, Vutha R, Goel A. Radiological evaluation of anatomical extensions of gliomas based on white fiber tracts: proposal of a novel classification. Neurol India. 2022;70:1492.
21. Shah A, Goel A, Jhawar SS, Patil A, Rangnekar R, Goel A. Neural circuitry: architecture and function - a fiber dissection study. World Neurosurg. 2019;125:e620–38.
22. Shah A, Jhawar SS, Goel A. Analysis of the anatomy of the Papez circuit and adjoining limbic system by fiber dissection techniques. J Clin Neurosci. 2012;19(2):289–98.
23. Shah A, Goel A, Patil A, Goel A. Letter to the Editor. Superior longitudinal fasciculus. J Neurosurg. 2019;132:1309–11.
24. Shah A, Jhawar SS, Goel A. Letter to the Editor: Optic radiations and anterior commissure. J Neurosurg. 2015;123(3):824–6.

Anatomy of the Limbic System and Surgery for Medial Temporal Lesions

Abhidha Shah, Sukhdeep Singh Jhawar, and Atul Goel

1 Introduction

The limbic system is phylogenetically the oldest and functionally the most complex and least understood parts of the human brain [1]. The limbic system developed when primitive animals developed an olfactory system with a sense of smell to explore its surroundings. The sense of smell was the most important sense for survival in these primitive animals. Along with the development of the olfactory system, on the medial and lateral aspect of the primitive brainstem, thalamus and striatum the early limbic structures appeared. These constituted of the hippocampus and amygdala medially and the olfactory structures and piriform cortex laterally. Around this three layered allocortex developed the six layered mesocortex becoming the parahippocampus and the cingulate gyrus medially and the insula laterally thus forming the limbic and paralimbic structures [2, 3]. It is essential to understand that functionally one of the major fundamentals of the limbic system is that its components project to the hypothalamus and brainstem directly and not to the basal ganglia as does the remainder of the telencephalon [4, 5].

The circular bending of the upper part of the neural tube around the thalami led the developing telencephalic structures, the lateral ventricular cavities and deeper structures in close relation to the ventricular cavities to assume a C shaped configuration. The term "limbus" means 'border' or 'margin' in Latin and it refers to the limbic cortical areas that form a ring around the hilus of the hemispheres [1]. The various components of the limbic system are arranged at the epicenter of the hemispheres in a characteristic 'C' shaped architecture forming major circuits, which will be discussed below.

The gray matter components of the limbic system are the hippocampal formation, amygdaloid nuclei, hypothalamus, nucleus accumbens, cingulate cortex, major areas of the prefrontal cortex, and nuclei of the midbrain. The fiber bundles that are components of the limbic system include the cingulum, longitudinal striae, fornix, anterior commissure, mamillothalamic and mamillotegmental tracts, stria terminalis, striae medullaris thalami, diagonal band of Broca, fasciolus retroflexus, ansa peduncularis, uncinate fasciculus, dorsal longitudinal fasciculus and the medial forebrain bundle [1]. The anatomy of these

A. Shah (✉) · S. S. Jhawar
Department of Neurosurgery, K.E.M. Hospital, Pune, India

Seth G.S. Medical College, Parel, Mumbai, India

A. Goel
Department of Neurosurgery, Lilavati Hospital and Research Center, Bandra, Mumbai, India

Department of Neurosurgery, R.N. Cooper Hospital, Mumbai, India

Bombay Hospital Institute of Medical Sciences and Research Center, Mumbai, India

A. Shah et al. (eds.), *Functional Anatomy of the Brain: A View from the Surgeon's Eye*,
https://doi.org/10.1007/978-981-99-3412-6_13

structures has already been covered in an earlier chapter. Here we discuss the connectivity of the limbic system.

2 Delineation of the Limbic Connections - Medial to Lateral Dissection of the White Matter

Commencing the dissection from the medial aspect of the hemisphere can anatomically delineate the limbic system best [1]. The cerebral hemispheres are bisected in the midline through the corpus callosum. The indusium griseum and the medial and the lateral longitudinal striae can be identified on the surface of the corpus callosum before splitting the brain. The medial surface of the cerebral hemisphere is visualized and major anatomical landmarks such as the corpus callosum, the cingulate gyrus, the thalamus and the hypothalamus are identified (Figs. 1 and 2). The ependymal coverings over medial surface of thalamus and hypothalamus are then peeled off carefully [1]. The dissection begins with coring of the gray matter of the thalamus. The mamillothalamic tract can be identified emerging from the mammillary body and running upward till it reaches the anterior nucleus of the thalamus (Figs. 3, 4, 5, 6, and 7). The mamillotegmental tract can be identified as a discrete bundle emerging from the mamillary body and going inferiorly upto the midbrain tegmental nuclei ventral to the aqueduct. The dissection is then commenced above the corpus callosum in the region of the cingulate gyrus. The cingulate gyrus begins in the subcallosal region then continues above the corpus callosum and in the region of the splenium then bends inferiorly to continue as the parahippocampus. The gray matter of the cingulate gyrus is cored out and the cingulum is identified. It can be traced in its entirety anteriorly till its termination in the subcallosal region. Posteriorly, the cingulum narrows above the splenium into the isthumus of the cingulum and then continues into the medial temporal lobe to end as the radiation of the cingulum, which can be identified after removal of the parahippocampal gray matter

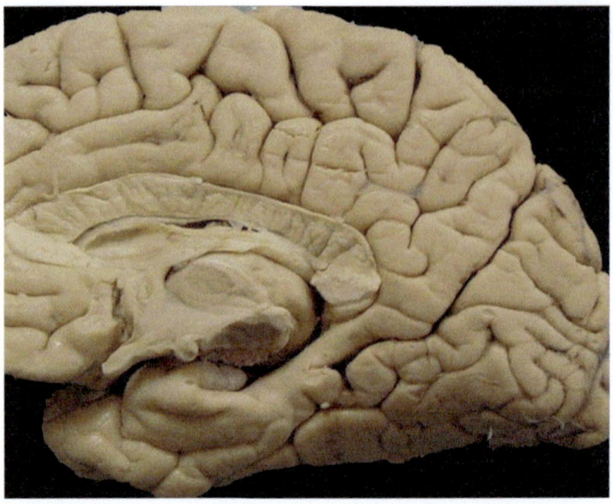

Fig. 1 Medial view of the cut cerebral hemisphere showing the limbic lobe encircling the diencephalon

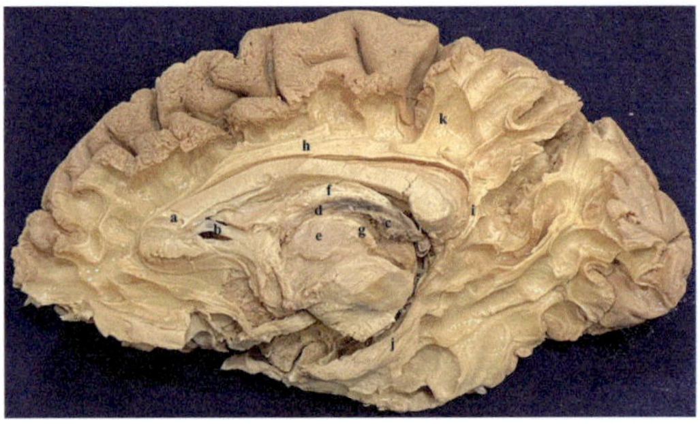

Fig. 2 Medial view of the cerebral hemisphere showing the various anatomical landmarks. The gray matter of the cingulate and parahippocampal gyrus has been cored out showing the entire extent of cingulum. (**a**) corpus callosum (**b**) frontal horn of lateral ventricle (**c**) choroid plexus in the choroidal fissure (**d**) choroidal fissure running between the fornix superiorly and thalamus inferiorly (**e**) thalamus (**f**) fornix (**g**) stria medullaris thalami, (**h**) Cingulum (**i**) isthmus of the cingulum (**j**) radiation of the cingulum (**k**) Fibers of the cingulum entering the parietal cortex

(Fig. 2). After tracing the whole length of the cingulum the fibers of the cingulum entering into the frontal and parietal cortices are traced. Attention is now turned to the mesial temporal brain (Figs. 8 and 9). The gray matter of the parahippocampal gyrus is removed to visualize, the subiculum, dentate gyrus and fimbria of fornix. The fimbria is traced posteriorly and superiorly till it continues into the crus of the fornix [1]. The body of the fornix is separated from the undersurface of the corpus callosum by blunt dissec-

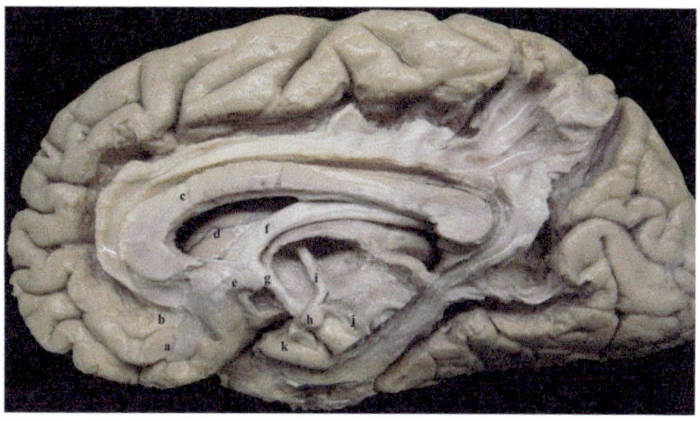

Fig. 3 Dissection of the medial aspect of the hemisphere to show the mamillothalamic and mamillotegemental tracts. (**a**) paraterminal gyrus (**b**) subcallosal gyrus (**c**) corpus callosum (**d**) caudate head in frontal horn of lateral ventricle (**e**) anterior commissure (**f**) body of fornix (**g**) columns of fornix (**h**) mamillary body (**i**) mamillothalamic tract (**j**) mamillotegmental tract (**k**) uncus

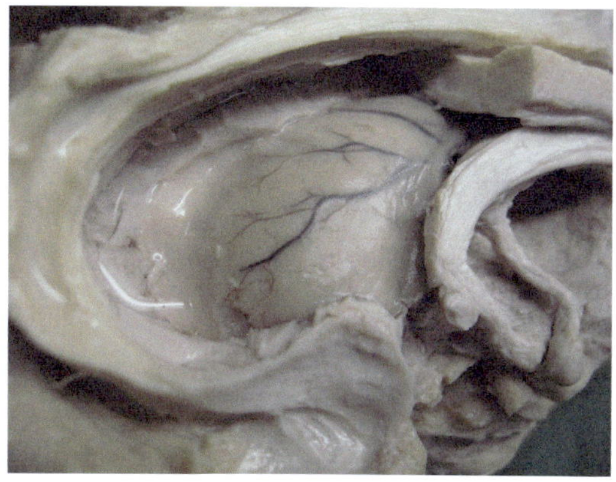

Fig. 5 Superior view showing the caudate head forming the floor of the lateral ventricle

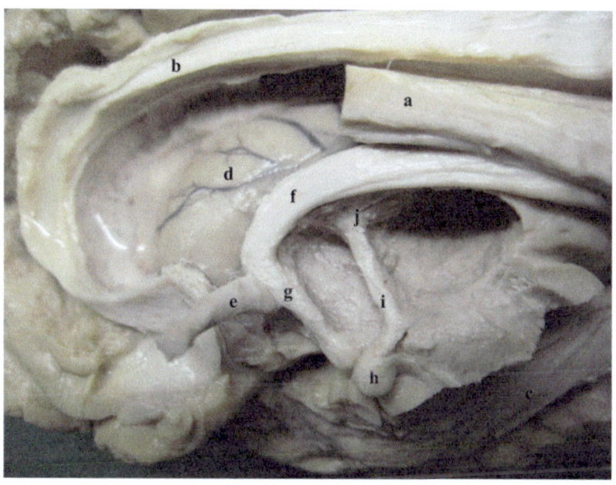

Fig. 4 Image showing a closer view of the medial aspect of the hemisphere. The anterior half of the corpus callosum has been cut. (**a**) corpus callosum, (**b**) cingulum, (**c**) radiation of the cingulum, (**d**) caudate head in frontal horn of lateral ventricle (**e**) anterior commissure (**f**) body of fornix (**g**) columns of fornix (**h**) mamillary body (**i**) mamillothalamic tract (**j**) anterior nucleus of the thalamus

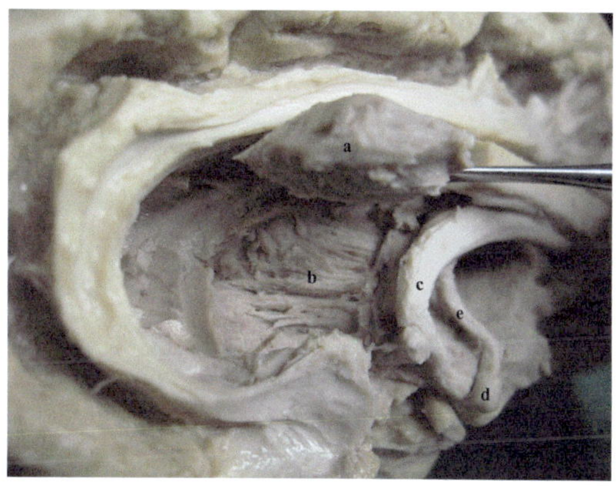

Fig. 6 The caudate head has been lifted off in one piece demonstrating the anterior thalamic radiation. The fibers from the anterior nucleus of thalamus are seen running towards the cingulum. (**a**) caudate head (**b**) anterior thalamic radiation (**c**) fornix (**d**) mamillary body (**e**) mamillothalamic tract

tion. The body is then traced anteriorly into the columns of the fornix, which end in the mamillary body. The anterior commisure is then identified anterior to the column of fornix. The dissection is then continued to trace the continuation of the mamillothalamic tract to the cingulum. The corpus callosum is removed carefully avoiding damage to the cingulum. The caudate head and body in the floor of the lateral ventricle are visualized (Figs. 3, 4, 5, 6, and 7). The stria terminalis can be seen running between the thalamus and caudate. The ependyma over the caudate is then removed. The whole caudate head is then lifted off in one piece to reveal the underlying fibers radiating from the thalamus and entering into the undersurface of the cingulum. These fibers intermingle with the fibers of the anterior limb of the internal capsule. This completes the medial dissection. The basal forebrain region is then visualized (Fig. 10). The olfactory tract, the medial and lateral olfactory striae, the medial, anterior, posterior and lateral orbital gyri on the basal surface of the frontal lobe are identified

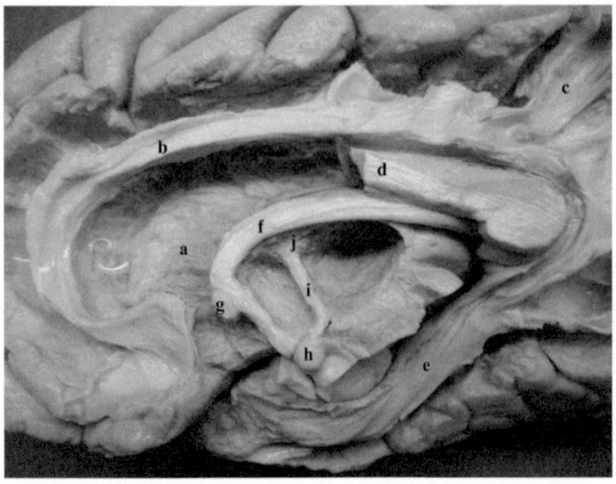

Fig. 7 After removal of the caudate head, the anterior thalamic radiation projecting to the cingulum can be seen. Fibers form the cingulum can also be seen radiating to the parietal cortex. (**a**) anterior thalamic radiation (**b**) cingulum (**c**) Fibers from the cingulum entering the parietal cortex (**d**) cut end of the corpus callosum (**e**) radiation of the cingulum (**f**) fornix (**g**) anterior commisure (**h**) mamillary body (**i**) mamillothalamic tract (**j**) anterior nucleus of thalamus

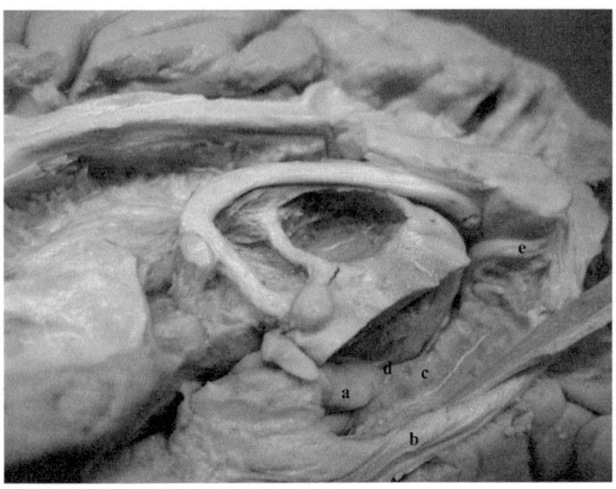

Fig. 8 Dissection of the mesial temporal structures. (**a**) Hippocampus (**b**) radiation of the cingulum (**c**) dentate gyrus (**d**) fimbria (**e**). fasciolar/subsplenial gyrus

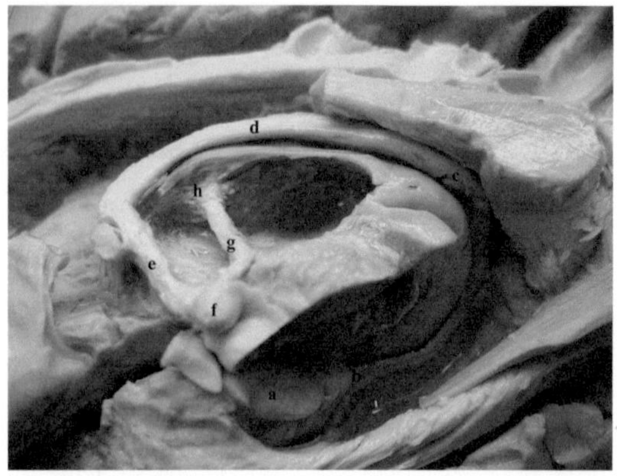

Fig. 9 The dentate gyrus has been removed to visualize the fimbria continuing as the fornix. (**a**) hippocampus (**b**) fimbria (**c**) crus of fornix (**d**) body of fornix (**e**) column of fornix (**f**) mamillary body (**g**) mamillothalamic tract (**h**) anterior nucleus of thalamus

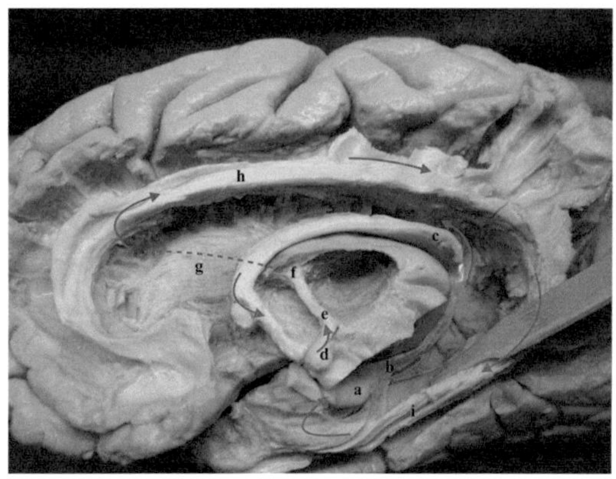

Fig. 10 Image showing an in situ dissection of the Papez circuit. (**a**) Hippocampus, (**b**) Fimbria, (**c**) Fornix, (**d**) Mammillary body, (**e**) Mammillothalamic tract, (**f**) Anterior nucleus of thalamus, (**g**) Anterior thalamic radiations, (**h**) Cingulum, (**i**) Radiation of the cingulum. The purple arrows show the course of the Papez circuit

(Fig. 11) [6–8]. The superficial gray matter in the region of the posterior orbital gyrus is cored out. Initially the ventromedial fibers of the uncinate fasciculus are seen. Once these are removed the deep gray matter of the substantia innominata can be visualized. Within the gray matter of the basal forebrain two distinct white fiber bundles can be identified, the anterior commissure and the diagonal band of Broca. The anterior commissure is identified as a discrete bundle of fibers that run transversely connecting the hemispheres across the midline anterior to the columns of the fornix (Fig. 12). On the basal surface of the brain it divides into a frontal extension and a temporal extension. The frontal extension can be identified under the gray matter of the medial orbital gyrus within the basal forebrain structures. The temporal extension traverses laterally towards the limen insulae and then inferior to the globus pallidus

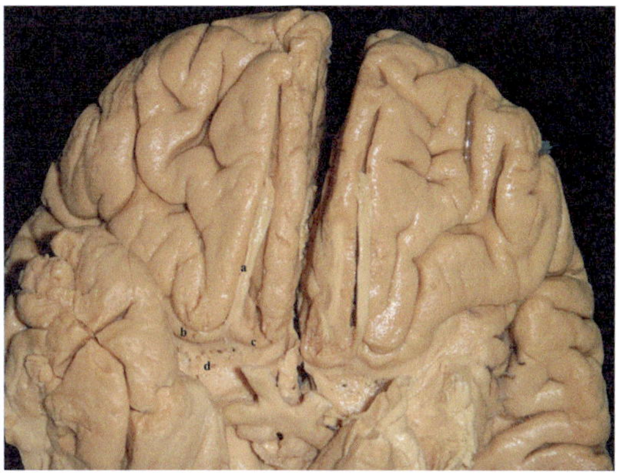

Fig. 11 View of the inferior surface of the brain showing the basal forebrain region. The lateral olfactory stria can be seen running along the lateral margin of the anterior perforated substance to reach the piriform region. The medial olfactory stria becomes continuous with the subcallosal region and the paraterminal gyrus. (**a**) olfactory tract (**b**) lateral olfactory stria (**c**) medial olfactory stria (**d**) anterior perforated substance

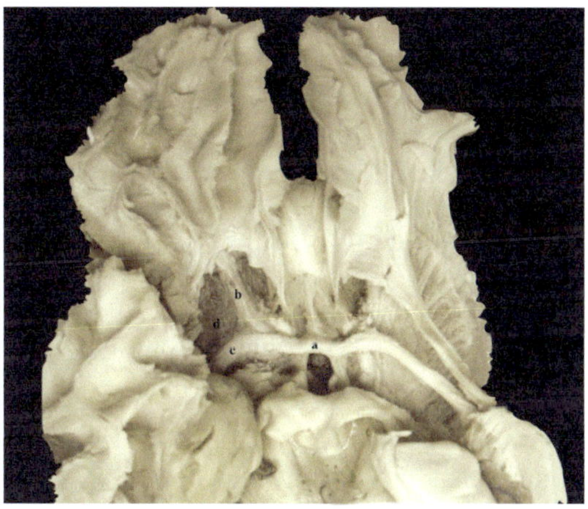

Fig. 12 Deeper dissection of the inferior surface of the basal forebrain. The anterior commissure can be seen. It is a compact fiber bundle, which crosses the midline. Laterally it splits into two, a smaller anterior or olfactory portion and the larger posterior portion. Also seen is the diagonal band of Broca. (**a**) anterior commissure (**b**) anterior portion of anterior commissure (**c**) posterior portion of anterior commissure (**d**) sunstantia innominata

and joins the inferior fronto-occipital fasciculus to become a component of the sagittal stratum (Figs. 12 and 13). The diagonal band of Broca is visualized as a bundle of fibers that runs anterior to the optic tract from the medial part of the uncus (amygdala) to the septal region [6–8].

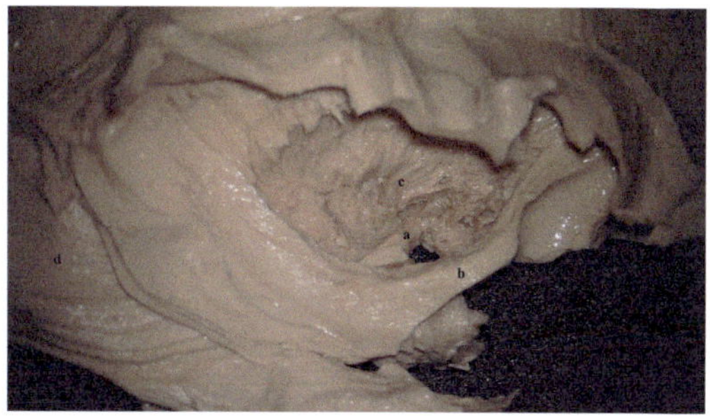

Fig. 13 Lateral view of the cerebral hemisphere. The insular region has been dissected to show the course of the posterior portion of the anterior commissure. The continuation of its fibers into the sagittal stratum can be seen. (**a**). anterior commissure (**b**). inferior fronto-occipital fasciculus c. internal capsule (**d**) sagittal stratum and optic radiations

3 Anatomical Correlates and Limbic Connections

The hippocampus and the amygdala form the key components of the limbic system. The amygdala and hippocampus participate in distinct circuits and connect the limbic system with the rest of the brain. The hippocampal formation is principally related to telencephalic and diencephalic structures via circuits whose basic purpose is to convert short-term memory into long term memory. The circuits that include the amygdala are more strictly related to the emotions and ultimately influence the effector systems (autonomic, neuroendocrine, and motor), primarily via the hypothalamus [1, 2].

Embryologically, the hippocampal formation is laid down along the hippocampal fissure. This fissure lies above and parallel to the choroidal fissure. As the temporal lobe develops, both these fissures are carried downwards and forwards to form a C shaped configuration. The hippocampal formation does not develop to the same extent across the whole length of the hippocampal fissure. The dorsal part of the fissure is invaded by fibers of the corpus callosum and the hippocampal formation in this portion becomes a vestigeal band – the indusium griseum [1]. The inferior portion of the hippocampal fissure differentiates into the hippocampal formation, which consists of the hippocampus proper, the dentate gyrus and the subiculum [1].

The parahippocampal gyrus extends from the inferior surface medially towards the hippocampal fissure where it curves inwards to form the hippocampus, which in turn again curves inwards to form the dentate gyrus. The portion of the parahippocampal gyrus at the edge of the hippocampal fissure is known as the subiculum and the area between the subiculum and the parahippocampal gyrus is known as the presubiculum [1].

When viewed from the medial aspect, the dentate gyrus is visualized as a thin band of cortex with a toothed appearance that lies between the fimbria above and the hippocampal fissure below (Fig. 8). The fimbria and the dentate gyrus run together backward till the splenium. Here, the dentate gyrus loses its denticulations and becomes the fasciolar gyrus, which then runs superiorly over the corpus callosum [1]. On the superior surface of the corpus callosum, the fasciolar gyrus becomes a thin lamina of gray matter, namely the indusium griseum, also known as the supracallosal gyrus. Running within the indusium griseum are two bands of myelinated fibers, the medial and lateral longitudinal striae of Lancisii. The indusium griseum and the longitudinal striae course over the entire corpus callosum and terminate in the paraterminal gyrus. The paraterminal gyrus then continues as the diagonal band of Broca.

The hippocampus (Gr. Seahorse) is where the Papez circuit begins and ends [1]. The hippocampal formation forms the medial part of the floor of the temporal horn. It rests on the subiculum (L. support), which forms the superior surface of the parahippocampal gyrus [1]. The hippocampus is separated from the parahippocampal gyrus and the subiculum by the hippocampal fissure. From the hippocampus fibers form a band known as the alveus (L. Trough or hollow). The alveus is made up of axons of the pyramidal cells in the subiculum and the hippocampus. It continues as the fimbria. Lateral to the fimbria and separated from it by the fimbrio-dentate sulcus, lies the dentate gyrus. This gyrus is named for its tooth like impressions, which can be clearly appreciated in the Fig. 7. This gyrus continues as the fasciolar gyrus, then as the subsplenial gyrus in the subsplenial region and then curves over the corpus callosum as the induseum griseum following its curve to end in the paraterminal gyrus. This then merges with the diagonal band of Broca. The fimbria then forms the crus of the fornix (L. vault, arch), which curve superolaterally over the pulvinar of the thalamus [1]. Further anteriorly, the two crura meet in the midline to form the body of the fornix. The two crura of the fornices are connected by the hippocampal commissure. Information is transmitted from one hemisphere to the other through this pathway. The body of the fornix then turns forward and downward ventral to the corpus callosum, as the column of the fornix to end in the mamillary body (Figs. 2, 3, 4, 5, 6, and 7). The fibers of the fornix on approaching the anterior commissure divide into a pre commissural and a post commissural portion. Majority of the fibers form the post commissural portion. These fibers arise mainly from the subiculum and terminate into the mamillary body. The pre-commisural fornix is made up of fibers from the hippocampus proper and end in the septal area. This forms the efferent pathway from the hippocampus. The post commissural fornix terminates into the medial mammillary nucleus. Fibers then arise from here and ascend upwards to the anterior nucleus of the thalamus forming the mamillothalamic tract of Vicq d'Azyr. Near its origin the mamillothalamic tract gives rise to another smaller bundle known as the mamillotegmental tract which terminates according to many authors into the tegmental nucleus of Gudden [1, 9–11]. From the anterior nucleus of the thalamus fibers radiate upward and mingle with the fibers of the anterior limb of the internal capsule to reach the cingulum (L. girdle) and the cingulate gyrus [1]. The cingulum is an association bundle, which runs superior and parallel to the copus callosum. Its fibers terminate anteriorly below the rostrum of the corpus callosum in the subcallosal gyrus. Posteriorly the cingulum narrows above the splenium, where it forms the isthumus of the cingulum and this continues below as the radiation of the cingulum to end in the anterior parahippocampal region adjacent to the hippocampus.

Thus, the hippocampus receives afferent fibers from the supramamillary regions via the fornix, from the cortical areas and the nuclei of the thalamus via the cingulum and also some fibers from the amygdala. The efferent fibers from the hippocampus travel via the fornix, terminating in the septal region (pre-commisural fibers); in the mamillary body and thalamus (post-commisural fibers) and in the entorhinal cortex and splenial regions (non-forniceal fibers) [1, 12].

The entorhinal area constitutes the anterior one-third portion of the parahippocampal gyrus. It is bounded laterally by the rhinal sulcus and posteriorly by the collateral sulcus. This area has many afferent and efferent connections with the hippocampus and the association areas of the cortex and forms an important relay center [1, 13].

Connections also exist between the subiculum/entorhinal cortex and the nucleus accumbens, caudate nucleus and putamen. The entorhinal cortex and the subiculum is in turn connected to various cortical areas of the brain. Apart from these connections, the hippocampal formation also has reciprocal connections with the amygdala, claustrum, septal area, thalamus and the supramamillary areas. The connections between the supramammillary area and the hippocampal formation constitute the hypothalamic input to the hippocampus. The neuronal activity in the hippocampal formation is also influenced by brainstem nuclei such as the ventral tegemental nuclei, the locus coeruleus and the midbrain raphe nuclei [10, 11].

The dentate gyrus receives input mainly from the entorrhinal area. The input is mainly by means of two pathways, the perforant path and the alvear path. The fibers of the perforant path originate in the lateral part of the entorrhinal cortex, traverse through the subiculum and terminate in the dentate gyrus. The alvear path originates in the medial part of the entorrhinal cortex and enters the hippocampus from the ventricular surface after traversing the alveus [14].

The *amygdala* (Gr. Almond like) lies entirely within the confines of the uncus [1]. The uncus forms the external marker of the amygdala. Superiorly, the amygdala blends with the globus pallidus and there is no clear demarcation between this transition.

The amygdala consists of the lateral, centeromedial and basal nuclei, which lie underneath the semilunar gyrus, ambient gyrus and the uncinate gyrus along the medial aspect of the uncus. Among its heterogeneous structure containing many different cell groups, a distinction can be made between a cortical or olfactory amygdala (related to olfactory structures), a basolateral amygdala (more closely related to many cortical areas), and a larger centromedial amygdala (more related to behavior and emotions) [2].

The amygdala gives rise to two major pathways, the stria terminalis and the ventral amygdalofugal pathway. The stria terminalis is a C-shaped structure, which exits from the caudal end of the amygdala (centromedial portion) and curves around the lateral ventricle in close proximity to the caudate nucleus, terminating in the bed nucleus of the stria terminalis and the hypothalamus Some fibers also terminate in the hypothalamus and some join the medial forebrain bundle [1, 14].

The bed nucleus of the stria terminalis, runs along with the stria terminalis and is largest rostrally close to the base of the head of the caudate nucleus. It lies lateral to the columns of fornix and dorsal to the anterior commissure. It receives input from all nuclei of the amygdala. The bed nucleus of the stria terminalis has widespread connections to the rest of the limbic system and therefore provides an important route by which the amygdala can affect structures beyond the connections of the stria terminalis itself [1]. The ventral amygdalofugal pathway courses medially from the amygdala to the basal nucleus of Meynert and the dorsal medial nucleus of the thalamus. It is a diffuse, myelinated bundle lying ventral to the globus pallidus (medial pole) [1]. It courses medially and rostrally beneath the lentiform nucleus, passes through the substantia innominata to reach the hypothalamus, septal areas, lateral preoptic area and the nucleus of the diagonal band of Broca. Some fibers proceed via the inferior thalamic peduncle to reach the dorsomedial nucleus of the thalamus.

There are many interconnections between the amygdala and the various sensory areas and the limbic association areas including the cingulate gyrus, temporal pole, insula, perirhinal and frontal cortices. Thus the amygdala receives an enormous amount of sensory information from various cortical regions. In turn, the amygdala has reciprocal connections with the olfactory bulbs, basal forebrain, striatum, hippocampus, thalamus, hypothalamus and the brainstem. The amygdala represents an important relay whereby external stimuli can influence and activate emotions via these connections [1, 14].

The mediobasal frontal cortical area of each cerebral hemisphere, composed of the paraterminal gyrus and the paraolfactory gyri, is also considered a limbic cortical area [2–5] (Fig. 3). It extends from the olfactory tubercle rostrally to the hypothalamus caudally. It lies over the anterior perforated substance and does not have any definite boundaries. The anterior and posterior paraolfactory gyri are oriented vertically and lie anterior to the paraterminal gyrus. They are separated from each other by the anterior paraolfactory sulcus and this area is known as the subcallosal area. The basal most portion of the cingulate gyrus connects by a fold to the gyrus rectus anterior to this subcallosal region. This area is known as the cingulate pole. The paraterminal gyrus is situated posterior to the paraolfactory gyri and separated from them by the posterior olfactory sulcus. It lies on the medial surface of the hemisphere immediately facing and almost continuous with the lamina terminalis [2]. The small anterior curvature of the paraterminal gyrus is called the prehippocampal rudiment and connects with the indusium griseum. Inferiorly the paraterminal gyrus is continuous with the diagonal band of Broca and the lateral olfactory stria [2].

The generic term olfactory cortical areas refers to the olfactory nerves, bulb, tract, trigone, lateral, and medial striae, as well as the anterior perforated substance, the diagonal band of Broca, and the piriform lobe, in each cerebral hemisphere [2]. The piriform lobe comprises: (1) the prepiriform cortical area; (2) the lateral olfactory stria which extends to the gyrus semilunaris; (3) the uncus of the parahippocampal gyrus and its small gyri (uncinate gyrus, caudal portion of the dentate gyrus or band of Giacomini, and intralimbic gyrus); and (4) the entorhinal area [2].

The basal forebrain includes the septal area, the olfactory tubercle, parts of the amygdala and the substantia innominata. The term substantia innominata is also used synonymously with the term basal nucleus [6–8]. The lateral olfactory stria and gyrus pass along the lateral margin of the anterior perforated substance to reach the piriform region. These fibers terminate in the piriform cortex and in the corticomedial part of the amygdaloid nuclear complex. The medial olfactory stria becomes continuous with the subcallosal and the paraterminal gyrus (Figs. 11, 12, and 13). The subcallosal area and the paraterminal gyrus constitute the septal area, beneath which are the septal nuclei. The septal region is situated on the medial surface of the cerebral hemisphere, immediately facing the anterior commissure. This region is also known as the *septum verum* ("true septum"), or *precommissural septum*, in counterpoint to the septum pellucidum, which does not contain neuronal cells and is situated posterior and superior to the anterior commissure, morphologically constituting the medial walls of the frontal horns and the ventricular bodies [1, 6–8]. The medial septal nucleus becomes continuous with the nucleus and tract of the diagonal band of Broca which then connects with the hippocampal formation [1]. The septal nuclei receive afferents principally from the hippocampus and subiculum of the parahippocampal gyrus via the indusium griseum as well as from the precommissural fornix. The efferents of the septal nuclei project to the area surrounding the hippocampus via the fornix itself and to the hypothalamus, the mesencephalon [via the fasciculus retroflexus], and the habenula [via the stria medullaris thalami], which in turn projects to the mesencephalon. Functionally, the septal nuclei are responsible for connecting limbic structures with the hypothalamus and the brainstem, principally via the hippocampal formation [1]. The so-called septal syndrome is clinically characterized by exaggerated reactions to environmental stimuli and the consequent behavioral alterations, principally related to eating and drinking habits, as well as by episodes of rage and disorders in the sexual sphere [1]. Afferents to the basal forebrain arise from the amygdala, the insular, temporal, piriform and entorhinal cortices. From the basal forebrain, neurons project to various regions of the cerebral cortex.

The stria medullaris, fasciolus retroflexus, stria terminalis and the medial forebrain bundle are the limbic pathways, which connect the cerebral hemispheres, the diencephalon and the midbrain [1].

The stria medullaris is a fiber bundle that runs from the septal region to the habenular (L. diminutive of habena – a strap) nuclei. The impulses from the habenular nuclei are then conveyed via the fasciculus retroflexus to the midbrain [1].

The medial forebrain bundle projects to the midbrain tegmentum, after originating from the septal and lateral preoptic region. It also carries ascending fibers from the midbrain. Ascending fibers to the mamillary body run in the mamillary peduncle. They arise from the dorsal and ventral nuclei of Gudden and traverse to the mamillary body. Some fibers continue rostrally to the lateral hypothalamic regions, medial septal nuclei and preoptic areas via the medial forebrain bundle [1].

4 Limbic Circuits

Morphologically the limbic system can be viewed as connections between the mediobasal frontal cortical areas, mesial temporal region, olfactory cortical areas and the ventral striato-pallidal region. These basal areas are connected ventrally by a network of fibers underneath the anterior portion of the central core with its posterior border being roughly delineated by the anterior commissure and dorsally by the cingulate gyrus along the medial cerebral surface and the parahippocampal gyrus along the basal surface with the two gyri being continuous around the splenium. This configuration of the connections along the C shaped limbic lobe and the basal forebrain of each hemisphere resemble a slightly tilted circle, with its superior portion medial and its inferior portion lateral, encircling the diencephalic structures.

The various intricate connections outlined above can be understood by dividing them into limbic circuits. Nieuwenhuys et al. outlines five main circular circuits (medial to lateral) [15]: (1) stria medullaris thalami, habenular nuclei, and habenulointerpeduncular tract; (2) amygdala, posterior extension of the amygdala (stria terminalis), and anterior extension of the amygdala which also project to the bed nucleus of the stria terminalis, and that together constitute the so-called extended amygdala; (3) fimbria, crura, body, and column of fornix, which connect the hippocampus to the mammillary body; (4) hippocampus and indusium griseum, which connect the hippocampus with the paraterminal gyrus; and (5) cingulate gyrus and parahippocampal gyrus [4, 5, 15].

We have classified the limbic connections into five main circuits. These are the Papez circuit, the supracallosal circuit, the dorsal amygdalofugal pathway, the ventral amygdalofugal pathway and the limbic connections to the brainstem.

4.1 Papez Circuit

Papez circuit is the most well known and established limbic circuit (Figs. 10 and 14). It begins at the hippocampus, continuing as the fimbria, fornix and ending at the mamillary body [1, 16]. From the mamillary body the mamillothalamic tract emerges and reaches the anterior nucleus of the thalamus. From the thalamus fibers of the anterior thalamic radiations emerge, traverse in the anterior limb of the internal capsule to reach the cingulum, which turns around the splenium

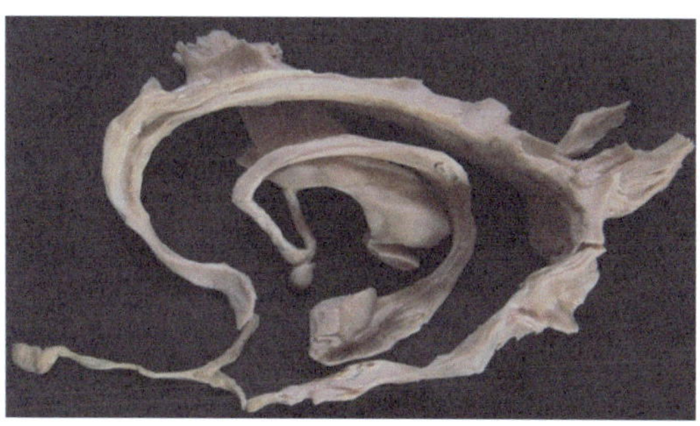

Fig. 14 Image showing the dissected Papez circuit

of the corpus callosum to end as the radiation of the cingulum into the hippocampus, thus completing the loop. The cingulum by means of its frontal and parietal connections along the superior frontal gyrus and the precuneus connects the circuit to the neocortex.

4.2 Supracallosal Circuit

The supracallosal circuit of the limbic system connects the dentate gyrus to the septal nuclei. It courses above the corpus callosum and connects the mediobasal temporal region to the medial frontal lobe (Fig. 15). The circuit is made up of a few fibers of the fimbria along with the subsplenial gyrus, which is the continuation of the dentate gyrus traversing above the corpus callosum as the indusium grisium and the medial and longitudinal striae to the subcallosal region. This loop connects the hippocampal formation to the septal nuclei thus forming an outer supracallosal limbic ring.

4.3 Dorsal Amygdalo-Fugal Pathway

The dorsal amygdalofugal pathway is represented by the stria terminalis (Fig. 15). The stria terminalis is a C-shaped structure, which exits from the caudal end of the amygdala and curves around the lateral ventricle in close proximity to the caudate nucleus, terminating in the bed nucleus of the stria terminalis and the hypothalamus thus connecting the amygdala to the septal area and the hypothalamus.

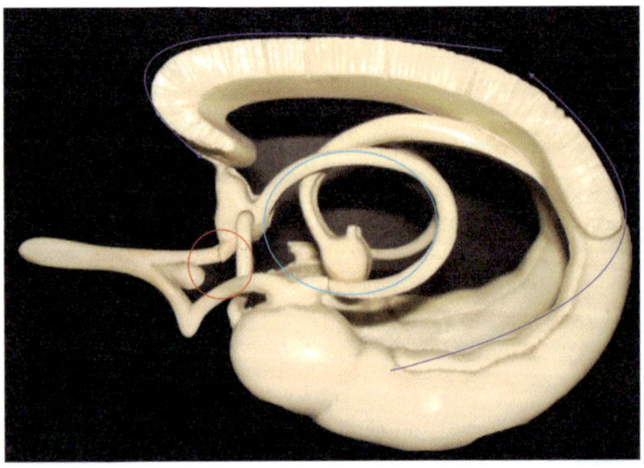

Fig. 15 Image of a model of the limbic system. The other circuits of the limbic system have been delineated. Purple arrows: Supracallosal circuit, Red circle: Ventral amygdalofugal pathway, Blue circle: Dorsal amygdalofugal pathway

4.4 Ventral Amygdalo – Fugal Pathway

The centromedial part of the amygdala gives rise to a ventral bundle of fibers known as the ansa peduncularis (Fig. 15). This bundle forms the ventral amygdalofugal pathway and consists of amygdalofugal fibers connecting the amygdala to the septal region, thalamus and hypothalamus. It runs ventral to the lentiform nucleus along the basal forebrain region. Another bundle of fibers runs posterior and superior to the ansa peduncularis and is known as the extracpasular thalamic peduncle as it traverses in the ventral striatopallidal region and not through the internal capsule. It consists of fibers connecting the amygdala and anterior temporal lobe to the medial thalamic nucleus and the hypothalamus.

5 White Fiber Correlates of Amygdalo-Hippocampectomy Through the Middle Temporal Gyrus Approach

Various approaches like the transylvian approach, the middle temporal gyrus approach, the temporopolar approach, keyhole subtemporal approach and the inferior temporal gyrus approach, have been described for approaching medial temporal lesions and for performing amygdalo-hippocampectomy (Fig. 16) [17–30]. Each of these approaches involves varying degrees of damage to the long association bundles in the vicinity of the temporal lobe and the temporal stem. We recently described the anatomical correlates of a variation of the middle temporal gyrus approach used by us to approach medial temporal lesions and to perform a combined tem-

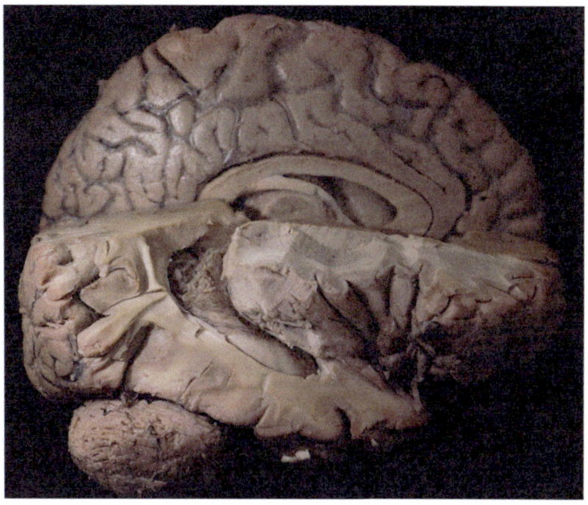

Fig. 16 An oblique axial cut has been made to demonstrate the lateral and superior view of the temporal horn and the choroidal fissure. The hippocampus is seen occupying the medial part of floor of the temporal horn. The fimbria can be seen on the surface of the hippocampus as they become the fornix. The choroidal fissure is seen between the fornix and the thalamus. The relation of the hippocampus to the temporal horn can be clearly visualized

poral lobe resection and amygdalo-hippocampectomy for mesial temporal lobe sclerosis [31]. The advantage of the approach is that it avoids/minimizes the damage to the fibers of the sagittal stratum i.e. the inferior fronto-occipital fasciculus (IFOF), the temporal extension of the anterior commissure (AC) and the optic radiations.

5.1 Anatomical Correlates *Safe Zone to Enter the Temporal Horn*

The long association fibers on the lateral aspect of the temporal lobe in the region of the posterior superior temporal gyrus are the fibers of the arcuate fasciculus superiorly and laterally and the fibers of the middle longitudinal fasciculus (MLF), the IFOF and AC inferiorly and medially. The fibers in the region of the anterior superior temporal gyrus are the temporal fibers of the uncinate fasciculus (UF). The fibers in the region of the inferior temporal gyrus are the fibers of the inferior longitudinal fasciculus (ILF). The fibers of the UF can be separately identified from the IFOF as they curve anteriorly to reach the anterior temporal lobe. The fibers of the ILF can be separately identified in the anterior and inferior aspect of the temporal lobe and then traced posteriorly as a distinct bundle separate from the bulk of the sagittal stratum. The fibers encountered in the region of the middle temporal gyrus are the middle longitudinal fasciculus, the IFOF, temporal fibers of the AC and the Meyer's loop of the optic radiation from lateral to medial. The fibers of the IFOF, temporal extension of the AC and the optic radiations form the major component of the sagittal stratum in the region of the temporal horn.The fibers of the IFOF, AC and the optic radiation lie in the roof of the temporal horn in the middle portion of the middle temporal gyrus and in the lateral wall of the temporal horn in the posterior portion of the gyrus (Fig. 17).

We described a safe area in the anterior portion of the lateral wall of the temporal horn (an average of 20 mm from the anterior wall of the temporal horn), which was devoid of any major fibers (Goel's safe entry zone) (Fig. 18) [31]. The amygdala, which lies at the junction of the anterior wall and the roof of the temporal horn, lies anterior to the sagittal stratum in this region and hence can be resected at this junction (Fig. 16). We advocate that the temporal horn should be entered through this safe area (Figs. 17 and 18). Further discon-

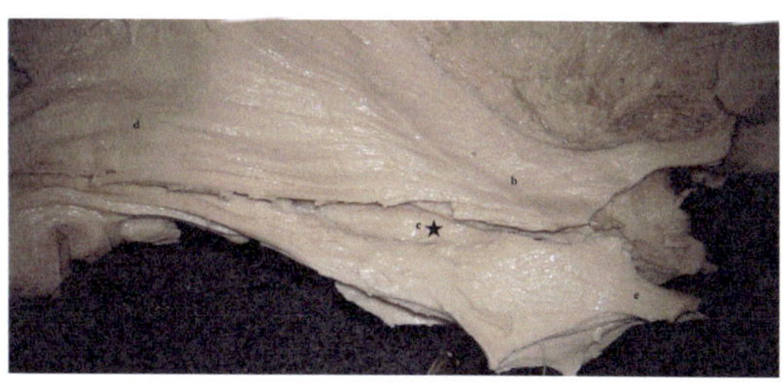

Fig. 17 Lateral view of the dissected temporal lobe showing the hippocampus in the temporal horn and the safe area to enter the temporal horn. In the anterior part of the temporal horn the fibers of the sagittal stratum are in the roof of the temporal horn and in the posterior part the fibers lie in the lateral wall. (**a**) Internal capsule, (**b**) Inferior fronto-occipital fasciculus, (**c**) Head of the hippocampus, (**d**) Sagittal stratum, (**e**) Uncinate fasciculus. The black star shows the safe area to enter the temporal horn

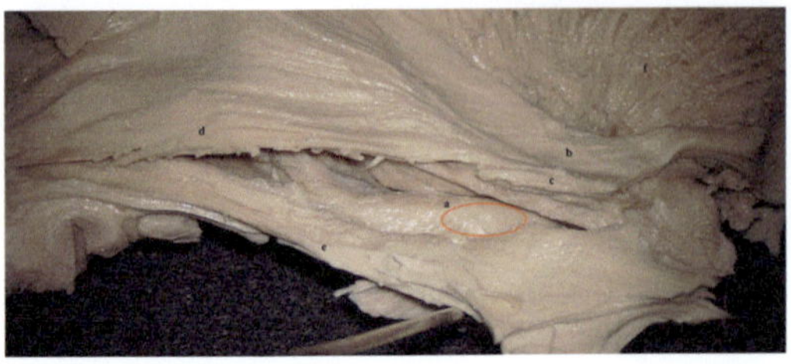

Fig. 18 The temporal horn has been opened to show the relations of the fibers more clearly. (**a**) Hippocampus, (**b**) Inferior fronto-occipital fasciculus, (**c**). Fibers of the Meyer's loop (sagittal stratum) in the roof of the temporal horn, (**d**) Fibers of the sagittal stratum forming the lateral wall of the posterior part of temporal horn and atrium, (**e**). Inferior longitudinal fasciculus, (**f**). Internal capsule. The orange circle shows the safe area to enter the temporal horn

nection and dissection should be carried out close to the hippocampus in the floor of the temporal horn to avoid injury to the sagittal stratum in the roof and lateral wall of the temporal horn.

5.2 En Bloc Amygdalo-Hippocampectomy and Anterior Temporal Lobectomy Through the Middle Temporal Gyrus: Surgical Technique [31, 32]

A slightly anteriorly curving incision is taken starting 1 cm below the zygoma and extending superiorly about 1.5 cm anterior to the tragus. The temporalis fascia and muscle are dissected and reflected with due protection to the frontalis branches of the facial nerve and the neurovascular supply of the temporalis muscle. The temporalis muscle is reflected anteriorly or inferiorly or split to obtain adequate exposure as has been discussed by us previously [33, 34]. A temporal craniotomy is performed remaining below the superior temporal line. The basal extent of the exposure is till the root of the zygoma and till the floor of the middle cranial fossa. The dura is opened in a cruciate manner. The dural opening exposes the inferior portion of the superior temporal gyrus, the middle and the inferior temporal gyri. A corticectomy is then performed in the middle temporal gyrus superior to the inferior temporal sulcus about 25 mm from the temporal pole. The corticectomy is deepened till the safe entry zone (25–30 mm) anterior portion of the temporal horn is entered and the hippocampal head is visualized. From this point all further dissection is carried out near the hippocampus inferior to the roof of the temporal horn. The amygdala is disconnected first anteriorly and superiorly, then the incision is extended into the floor of the temporal horn superior to the hippocampal head till the choroidal fissure. The incision is deepened across the entire extent of the hippocampus in the floor, including the parahippocampal gyrus till the arachnoid of the tentorial incisura is visualized. The arachnoid layer is not breached to protect the posterior cerebral artery, the third nerve and the cerebral peduncle. The incision is then extended inferiorly in the region where the tail of the hippocampus turns in the temporal horn and the tail is disconnected. This completes the disconnection of the amygdala and the hippocampus. Then the whole lateral neocortex lateral to the disconnected amygdala and hippocampus is disconnected from the temporal lobe. Then the whole specimen consisting of the amygdala, hippocampus, uncus, parahippocampus and the lateral neocortex is excised en bloc thus completing the resection.

We recently described a safe entry zone through the middle temporal gyrus that was devoid of fibers of the IFOF and optic radiations (sagittal stratum) [31]. This area was approximately 24 mm posterior to the temporal pole and extended for an average of 20 mm posteriorly from the anterior wall of the temporal horn. Once the temporal horn is entered at this point working under the roof of the temporal horn whilst performing the mesiotemporal resection will protect the sagittal stratum. We prefer to first enter the temporal horn then perform an enbloc resection of the amygdala, hippocampus, parahippocampus and the mesial temporal structures till the arachnoid boundary is reached at the tentorial edge. Following this the lateral temporal neocortex that is inferior, anterior and lateral to our entry point is resected. This technique of resection decreases the chances of injury to the IFOF and the Meyer's loop. Unlike current ATL procedures which involve first performing a temporal lobectomy and then disconnection of the amygdala and hippocampus our technique involves first entering the temporal horn through the desribed safe point. The amygdala and hippoacampus are disconnected and then the incision continues to the lateral neocortex and the whole specimen is removed "en bloc" [31, 32]. As the disscetion is performed below the level of the roof of the temporal horn the fibers of the optic radiations can be preserved.

References

1. Shah A, Jhawar SS, Goel A. Analysis of the anatomy of the Papez circuit and adjoining limbic system by fiber dissection techniques. J Clin Neurosci. 2012;19(2):289–98.
2. Ribas G. The cerebral architecture. In: Applied cranial-cerebral anatomy: brain architecture and anatomically oriented microneurosurgery. Cambridge: Cambridge University Press; 2018. p. 15–61. https://doi.org/10.1017/9781316661567.004.
3. MacLean PD. A triune concept of the brain and behavior. Toronto: University of Toronto Press; 1973.
4. Heimer L, Switzer RC, Van Hoesen GV. Ventral striatum and ventral pallidum. Trends Neurosci. 1982;5:83–7.
5. Heimer L, VanHoesen GW, Trimble M, Zahm DS. Anatomy of neuropsychiatry: the new anatomy of the basal forebrain and its implications for neuropsychiatric illness. Amsterdam: Elsevier; 2008.
6. Goel A, Shah A, Ramdasi R, Patni N. Orbital cortical approach to lesions around the frontal horn of the lateral ventricle: indication and surgical parameters. Acta Neurochir. 2014;156(4):825–30.
7. Shah A, Goel A, Jhawar SS, Patil A, Rangnekar R, Goel A. Neural circuitry: architecture and function - a fiber dissection study. World Neurosurg. 2019;125:e620–38.
8. Shah A, Dandpat S, Goel A. Anatomic correlates and surgical experience with Goel orbital cortical approach to caudate head tumors. World Neurosurg. 2021;148:e674–9.
9. Morgane PJ, Galler JR, Mokler DJ. A review of systems and networks of the limbic forebrain/limbic midbrain. Prog Neurobiol. 2005;75(2):143–60.
10. Nauta WJH. Hippocampal projections and related neuronal pathways to the mid-brain in the cat. Brain. 1958;81:319–40.
11. Nauta WJH, Domesick VB. Crossroads of limbic and striatal circuitry: hypothalamo-nigral connections. In: Livingston KE, Hornykiewicz O, editors. Limbic mechanisms. New York: Plenum Press; 1976. p. 75–93.
12. Mark LP, Daniels DL, Naidich TP, Hendrix LE. Limbic connections. AJNR Am J Neuroradiol. 1995;16(6):1303–6.
13. Ribas GC. The cerebral sulci and gyri. Neurosurg Focus. 2010;28(2):E2.
14. Parent A. The limbic system. In: Carpenter's human neuroanatomy. Philadelphia: Lippincott Williams & Wilkins; 1996. p. 744–86.
15. Nieuwenhuys R, Voogt J, Van Huijzen C. The human central nervous system. 3rd ed. Berlin: Springer-Verlag; 1988.
16. Papez JW. A proposed mechanism for emotion. Arch Neurol Psychiatr. 1937;38:725–43.
17. Yasargil MG, Türe U, Boop FA. Selective amygdalohippocampectomy. In: Kaye AH, Black PM, editors. Operative neurosurgery, vol. 2. London: Harcourt Publishers Limited; 2000. p. 1275–83.
18. Yasargil MG, Krayenbuhl N, Roth P, et al. The selective amygdalohippocampectomy for intractable temporal limbic seizures. J Neurosurg. 2010;112(1):168–85.
19. Miyamoto S, Kataoka H, Ikeda A, et al. A combined subtemporal and transventricular/transchoroidal fissure approach to medial temporal lesions. Neurosurgery. 2004;54(5):1162–7. [discussion: 1167–9]
20. Niemeyer P, Bello H. Amygdalohippocampectomy in temporal lobe epilepsy: microsurgical technique. Excerpta Med. 1973;293:20.
21. Spencer D, Burchiel K. Selective amygdalohippocampectomy. Epilepsy Res Treat. 2012;2012:382095.
22. Wheatley BM. Selective amygdalohippocampectomy: the trans-middle temporal gyrus approach. Neurosurg Focus. 2008;25(3):E4.
23. Olivier A. Transcortical selective amygdalohippocampectomy in temporal lobe epilepsy. Can J Neurol Sci. 2000;27(Suppl 1):S68–76. [discussion: S92–6]
24. Miyagi Y, Shima F, Ishido K, et al. Inferior temporal sulcus approach for amygdalohippocampectomy guided by a laser beam of stereotactic navigator. Neurosurgery. 2003;52(5).1117–23. [discussion:1123–4]
25. Park TS, Bourgeois BF, Silbergeld DL, et al. Subtemporal transparahippocampal amygdalohippocampectomy for surgical treatment of mesial temporal lobe epilepsy. Technical note. J Neurosurg. 1996;85(6):1172–6.
26. Takaya S, Mikuni N, Mitsueda T, et al. Improved cerebral function in mesial temporal lobe epilepsy after subtemporal amygdalohippocampectomy. Brain. 2009;132(Pt 1):185–94.
27. Hori T, Yamane F, Ochiai T, Kondo S, Shimizu S, Ishii K, Miyata H. Selective subtemporal amygdalohippocampectomy for refractory temporal lobe epilepsy: operative and neuropsychological outcomes. J Neurosurg. 2007;106(1):134–41.
28. Little AS, Smith KA, Kirlin K, et al. Modifications to the subtemporal selective amygdalohippocampectomy using a minimal-access technique: seizure and neuropsychological outcomes. J Neurosurg. 2009;111(6):1263–74.

29. Whiting AC, Chen T, Swanson KI, Walker CT, Godzik J, Catapano JS, Smith KA. Seizure and neuropsychological outcomes in a large series of selective amygdalohippocampectomies with a minimally invasive subtemporal approach. J Neurosurg. 2020;12:1–9.
30. Harput MV, Türe U. The Paramedian Supracerebellar-Transtentorial Selective Amygdalohippocampectomy for Mediobasal Temporal Epilepsy. Oper Neurosurg (Hagerstown). 2018;14(4):462.
31. Shah A, Lunawat A, Jhawar SS, Goel A, Goel A. White fiber correlates of amygdalohippocampectomy through the middle temporal gyrus approach. World Neurosurg. 2022;157:e156–65.
32. Muzumdar D, Patil M, Goel A, Ravat S, Sawant N, Shah U. Mesial temporal lobe epilepsy - an overview of surgical techniques. Int J Surg. 2016;36(Pt B):411–9.
33. Goel A, Nadkarni T. Basal lateral subtemporal approach for trigeminal neurinomas: report of an experience with 18 cases. Acta Neurochir. 1999;141(7):711–9.
34. Goel A. Splitting of temporalis muscle for basal exposure—technical note. Neurol Med Chir (Tokyo). 1997;37(1):59–62.

Medial Temporal Lobe Tumors: Surgical Anatomy and Technique

Yuanzhi Xu, Maximiliano Alberto Nunez, Vera Vigo, and Juan C. Fernandez-Miranda

Abbreviations

AChA	anterior choroidal artery
AHA	anterior hippocampal artery
AHPA	anterior hippocampal-parahippocampal artery
AHV	anterior hippocampal vein
AITA	anterior-inferior temporal artery
ALHV	anterior longitudinal hippocampal vein
ITA	inferior temporal artery
LPChA	lateral posterior choroidal artery
MHA	middle hippocampal artery
MITA	middle inferior temporal artery
MTLE	mesial temporal lobe epilepsy
MTR	medial temporal region
PCA	posterior cerebral artery
PHA	posterior hippocampal artery
PHPA	posterior hippocampal-parahippocampal artery
PITA	posterior-inferior temporal artery
PLHV	posterior longitudinal hippocampal vein
PLU	posterior limit of the uncus
SA	splenial artery
SAH	selective amygdalohippocampectomy
UHA	uncohippocampal artery

Y. Xu
Department of Neurosurgery, Huashan Hospital, Shanghai Medical College, Fudan University, Shanghai, China

Department of Neurosurgery, Stanford Hospital, Stanford, California, USA

M. A. Nunez · V. Vigo · J. C. Fernandez-Miranda (✉)
Department of Neurosurgery, Stanford Hospital, Stanford, California, USA
e-mail: drjfm@stanford.edu

1 Introduction

The medial temporal lobe (MTL), also termed as the temporomesial or mediobasal temporal region has the most intricate cortical anatomy. The MTL has a unique histological architecture with different layers of mesocortex and isocortex (Fig. 1a). Thus, the MTL is highly related to the pathogenesis of epilepsy and tumors. The MTL is located in the medial and inferior surface of the temporal lobe, and is separated from the thalamus by the choroidal fissure. Since the MTL is located deep in the brain and is near vital neural and vascular elements, surgical removal carries considerable risk. It must be preserved during surgery unless it is necessary for treatment [1]. Nevertheless, safe resection can be achieved with high knowledge of the microsurgical anatomy of this region and choosing the right surgical approach to avoid postoperative complications.

Here we give a comprehensive review on the cortical and subcortical features, arterial distribution, and venous drainage of MTL based on our

A. Shah et al. (eds.), *Functional Anatomy of the Brain: A View from the Surgeon's Eye*,
https://doi.org/10.1007/978-981-99-3412-6_14

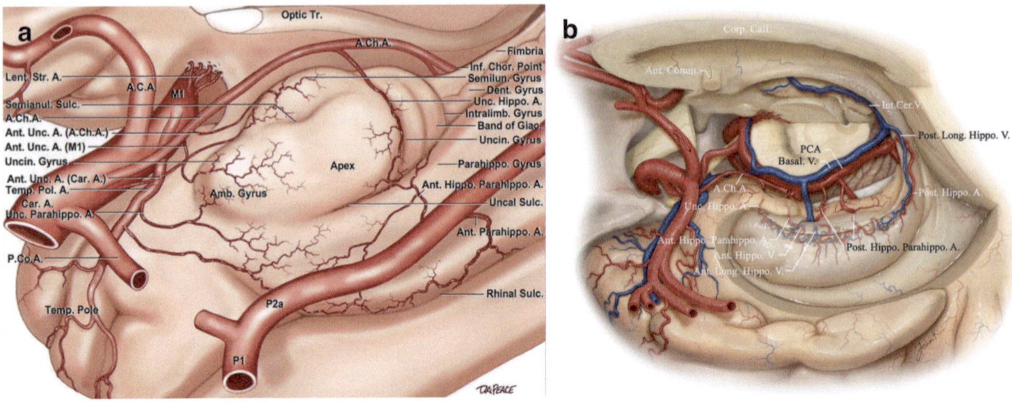

Fig. 1 Artistic illustration of MTL vascular pattern. (**a**) Medial view of anterior segment of MTL. The AChA gives off an anterior uncal artery that irrigates the semilunar gyrus and an unco-hippocampal artery that irrigates the uncinate gyrus and band of Giacomini and penetrates the uncal sulcus to vascularize the extra-ventricular hippocampal head. The ICA gives off an anterior uncal artery that supplies the semilunar gyrus. This branch usually is present when the anterior uncal artery of the AChA is absent. An anterior uncal artery also arises from the M1 segment of the MCA and supplies the ambient gyrus. An unco-parahippocampal artery arises from the temporopolar artery and irrigates both the ambient gyrus and the anterior parahippocampal area. Branches from the P2a segment of the PCA irrigate the anterior parahippocampal region (anterior parahippocampal artery) or both the anterior para-hippocampal gyrus and hippocampal head (anterior hippocampal-parahippocampal artery). (**b**) Superior view of ventricular vascular pattern of MTL, which composed of one uncohippocampal artery and one anterior hippocampal-parahippocampal artery in anterior segment of MTL; two posterior hippocampal parahippocampal arteries in the middle segment of MTL; and one posterior hippocampal artery in the posterior segment of MTL. A. Ch.A, anterior choroidal artery; Ant. Hippo. Parahippo. A., anterior hippocampal-parahippocampal artery; Ant. Hippo. V., anterior hippocampal vein; Ant. Inf. Temp. A., anterior-inferior temporal artery; Ant. Long. Hippo. V., anterior longitudinal hippocampal vein; Ant. Parahippo. A., anterior parahippocampal artery; AHA, anterior hippocampal artery; Calc. A., calcarine artery; Calc. Sulc., calcarine sulcus; Calc. V., calcarine vein; CN, cranial nerve; Coll. Sulc., collateral sulcus; Hippo. Head, hippocampal head; ICA, internal cerebral artery; Int.Cer.V., internal cerebral vein; L.P.Ch.A, lateral posterior choroidal artery; MHA, middle hippocampal artery; Mid. Inf. Temp. A., middle-inferior temporal artery; M.P.Ch.A, medial posterior choroidal artery; P.Comm.A., posterior communicating artery; PHA, posterior hippocampal artery; Post. Comm., posterior communicating; Post. Hippo. Parahippo. A., posterior hippocampal-parahippocampal artery; Post. Hippo. V., posterior hippocampal vein; Post. Inf. Temp. A., posterior-inferior temporal artery; Post. Long. Hippo. V., posterior longitudinal hippocampal vein; Pt., pituitary; Rec. A., recurrent artery; Splen. A., splenial artery; Sup. Coll., superior colliculus; Sup. Temp., superior temporal; Unc. Hippo. A., uncal hippocampal artery; V., vein. (With permission from Juan Carlos Fernandez-Miranda, MD)

previous sequel publication [1, 2] (Fig. 1b), as well as surgical implications when approaching the MTL via different approaches.

2 Neural Features of the MTL

The MTL can be classified into three segments: anterior, middle, and posterior [1]. There are cisternal and ventricular components in each segment. The anterior segment starts from the superior end of the rhinal sulcus and terminates at the posterior end of the uncus. The middle segment starts from the posterior margin of the uncus and terminates at approximate level of the quadrigeminal plate. The posterior segment begins at the quadrigeminal plate and terminates at the calcarine point, which is the intersection of the calcarine sulcus and parieto-occipital sulcus (Fig. 2a).

2.1 Anterior Segment

Uncus and entorhinal areas form the cisternal elements of the anterior segment of the MTL. The

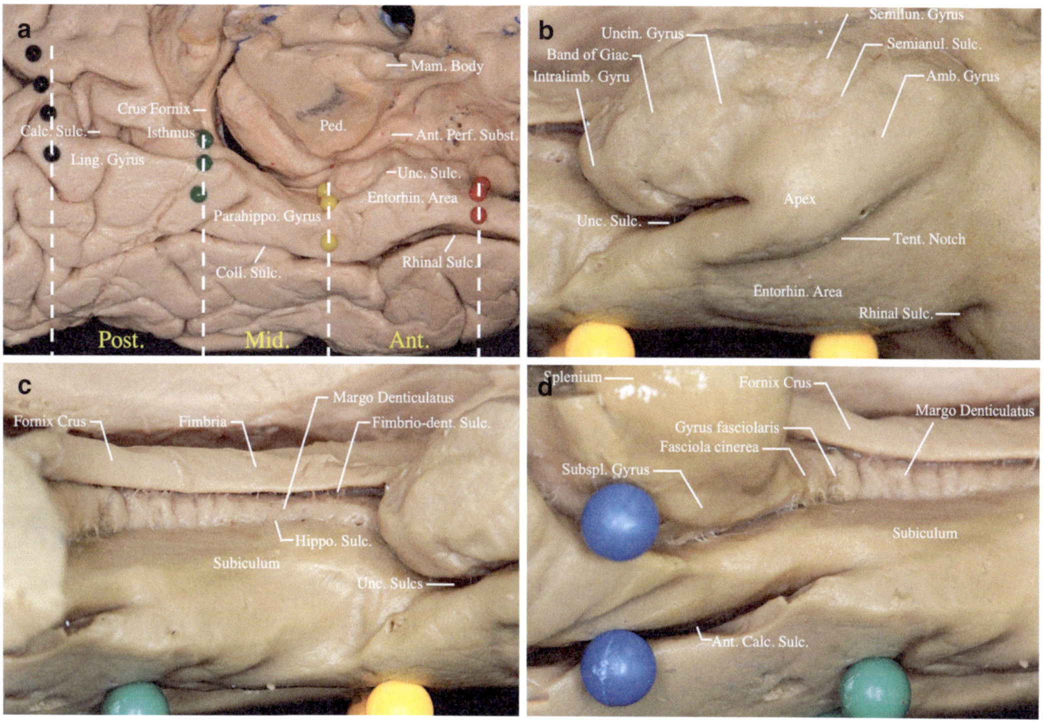

Fig. 2 Neural features and subdivision of MTL. (**a**) Mediobasal view of the left cerebral hemisphere. The MTR can be divided into 3 segments: anterior, middle, and posterior. The anterior segment extends posteriorly from where the rhinal sulcus turns upward at the posterior edge of the temporal pole to a vertical line crossing the posterior edge of the uncus. The middle segment extends posteriorly from the posterior edge of the uncus to a vertical line at the level of the quadrigeminal plate. The posterior segment extends from the quadrigeminal plate to a vertical line passing through the calcarine point located at the junction of the parieto-occipital and calcarine sulci. (**b**) Enlarged view of anterior segment of MTL. Hippocampus head is running rostro-medially from the uncal recess to posterior limit of uncus (PLU) according its contour of digitations in the floor of temporal horn. Posterior limit of uncus is the posterior end of uncus which located at the site of anterior attachment end of the fimbria and posterior to intralimbic gyrus), served as a landmark to divid hippocampal head and body. The uncal sulcus, uncinate gyrus, band of Giacomini, and intralimbic gyrus was involved in extraventricular hippocampal head. (**c**) Enlarged view of middle segment of MTL. The extraventricular hippocampal body is made of several structures, namely from superior to inferior: the fimbria; the fibrio-dentate sulcus; the margo denticulatus, subiculum of parahippocampus and subsplenial gyrus. (**d**) Enlarged view of posterior segment of MTL. Hippocampus tail is running caudo-medially in coronal, from the end of body to the terminal of subsplenial gyrus. The extraventricular hippocampal tail consists of fasciola cinereal (FC) which is the smooth caudal extension of the margo denticulatus; gyrus fasciolaris (GF), the enlargement of fibrio-dentate sulcus posteriorly extended; and the subsplenial gyrus, a prolongation of the gyrus fasciolaris, forms the end of the hippocampal tail, underneath the splenium, and continually extend with the indusium griseum above the splenium. (With permission from Juan Carlos Fernandez-Miranda, MD)

uncus is separated into anterior and posterior parts, which meet at uncal apex. Two minor gyri: the ***ambient*** and ***semilunar gyrus***, demarcated by the semiannular or amygdaloid sulcus, make up the anterior part of the uncus. The posterior part of uncus is separated into upper and lower portions according to the uncal sulcus. There are two surfaces at the upper part: one superior and one inferior, which is composed by the extra-ventricular elements of the hippocampal head. The superior surface is bounded by the posterior continuation of the entorhinal sulcus superiorly semiannular sulcus anteriorly, the anterior-most part of the fimbria posteriorly, and the uncal sulcus inferiorly. Three small gyri, from ventral to dorsal, ***uncinate gyrus***, ***band of Giacomini*** (terminal segment), and ***intralimbic gyrus*** are identified on upper part of posterior uncus [3, 4]. The parahippocampal gyrus constitutes the area below the uncal sulcus. The ***entorhinal area*** occupies the anterior parahippocampal gyrus, lying medially to the rhinal sulcus (Fig. 2b).

Semilunar Gyrus

The semilunar gyrus, which appeared as a spherical bulge on the upper aspect of the anterior uncus, covers the cortical amygdaloid nucleus. It is separated by entorhinal sulcus from the anterior perforated substance medially (Fig. 2b).

Ambient Gyrus

The ambient gyrus, composed mostly of entorhinal cortex, occupies the anterior, medial and inferior portions of the anterior uncus. It is confined by the semilunar gyrus superiorly, the anterior end of the uncal sulcus posteriorly and the tentorial edge inferiorly. The ambient gyrus joins the entorhinal area of the parahippocampal gyrus anteroinferiorly and the limen insula and lateral olfactory gyrus anterolaterally (Fig. 2b).

Uncinate Gyrus

The uncinate gyrus, considered as the anterior-most part of posterior uncus, is a protrusion situated superiorly to anterior end of the uncal sulcus, anteriorly to the band of Giacomini, and posteriorly to the ambient gyrus (Fig. 2b).

Band of Giacomini

The band of Giacomini is the second prominence of posterior uncus above the uncal sulcus, which located between the uncinate gyrus and intralimbic gyrus. It is regarded as the extension of dentate gyrus that travels vertically at the surface of the posterior uncus.

Intralimbic Gyrus

The intralimbic gyrus comprises the posterior-most part of the posterior uncus, which functioned as the anchor of anterior fimbria. The CA3 and CA4 sectors of hippocampus are located in intralimbic gyrus.

Entorhinal Area

The entorhinal area, also known as Brodmann's area 28, is a region of the allocortex found in the anterior parahippocampal gyrus. It is located inferiorly to the uncal apex and ambient gyrus, medially to the rhinal sulcus and anteriorly to the level of the posterior end of the uncus (as the posterior demarcation), which is considered correspondently to the extra-ventricular bottom of hippocampal head.

The ventricular elements of anterior segment of the MTL is formed predominantly by amygdala and hippocampus head. The hippocampus head running rostro-medially from the uncal recess to posterior limit of uncus (PLU) according to its contour of digitations in the floor of temporal horn. Posterior limit of uncus is the posterior end of uncus which is located at the site of anterior attachment end of the fimbria and posterior to intralimbic gyrus, served as a landmark to divide hippocampal head and body. The character of intraventricular surface included hippocampal digitations (also known as pes hippocampus) with the shallow longitudinal sulci which forms 2 or 3 small lobules on hippocampus head.

2.2 Middle Segment

The middle segment of MTL is composed of cisternal and ventricular elements, and the former is built by posterior parahippocampal gyrus, which has 3 surfaces: superior, medial, and inferior [5]. The ***fimbria*** of the fornix, the ***dentate gyrus*** and the ***subiculum***, running from medial to lateral, constitute the superior surface of middle segment. The fimbriodentate sulcus separates the fimbria from the dentate gyrus, whereas the hippocampal sulcus separates the dentate gyrus from the subiculum. The superior and medial (constituted mainly by the ***presubiculum***) surface occupies the upper and lower part of ambient cistern, respectively. The inferior surface of middle segment, limited by the collateral sulcus laterally, faces the cerebellar tentorium, where the trajectory of superior petrosal sinus serves as a demarcation line for anterior and middle segments of MTL (Fig. 2c).

The ventricular elements of middle segment of the MTL is built by choroidal fissure, hippocampus body and collateral eminence, from medial to lateral. The hippocampus body is a swiss roll appearance with two interlocking U-shaped structures (dentate gyrus and cornu ammonis), running sagittally from the PLU to the point the hippocampus turns caudo-medially in coronal. The intraventricular surface is the posterior continuation of hippocampal digitations with choroidal plexus above.

2.3 Posterior Segment

There are three cisternal surfaces in the posterior segment: medial, inferior and anterior. The calcarine sulcus, the most important landmark on the cisternal element splits the medial surface into a superior portion, which is formed by parasplenial region including isthmus of the cingulate gyrus anteriorly and the inferior pole of the precuneus gyrus posteriorly, and an inferior portion, which is composed by the anterior part of the ***lingual gyrus***. The anterior surface is hidden in the cleft between splenium of the corpus callosum and the isthmus of the cingulate gyrus, which is composed of the fornix crus and extraventricular portion of the hippocampal tail. Three anatomic structures including ***fasciola cinereal*** (prolongation of dentate gyrus), ***fasciolaris gyrus*** (prolongation of fibrio-dentate sulcus) and ***subsplenial gyrus*** (prolongation of fasciolaris gyrus), located from rostral to caudal, made up the extraventricular portion of hippocampal tail. The crus of the fornix curves superomedially toward the inner surface of the splenium, wrapping around the posterior surface of the pulvinar to form choroidal fissure (Fig. 2d).

The ventricular elements of posterior segment of the MTL is composed by hippocampal tail, the collateral trigone and calcar avis, rostrally to caudally. The calcar avis, posteriorly to collateral trigone and inferiorly to the bulb of the corpus callosum, is the corresponding ventricle prominent structure of the ending calcarine sulcus, traditionally known as the hippocampus minor.

3 Arterial Distribution of the MTL

The AChA, ICA, MCA, and PCA were the arteries most significantly implicated in the supply of the MTL. Rich anastomoses among branches of the main supplying arteries of the MTL were noted [1]. Significantly, anastomoses form a complete arcade at hippocampal sulcus, and with the perforated branch into hippocampus proper it forms a rake-like appearance (Figs. 1, 3, and 4) [2].

3.1 Anterior Choroidal Artery (AChA)

The anterior choroidal artery (AChA) originates from the posterior wall of the ICA as a single trunk (77%) in most cases. Two types of pseudoduplicated AChA are presented: one, consisting of 2 separate arteries arising from the ICA, and the other, consisting of a single artery that immediately divides into 2 trunks. Two segments comprise the course of AChA: the cisternal and plexal segment. The cisternal segment of AChA, which starts from its ICA origin and ends at the inferior choroidal point, runs in the carotid cistern firstly then reaches the crural cistern between crus cerebri and uncus. The average length of cisternal segment is 23 mm (16–30 mm), and it gives an average of 9.5 branches (5–15 branches), which are divided into three groups depending on their distribution: (1) medial temporal group; (2) perforating group; and (3) choroidal branches group [1]. The medial temporal group can be subdivided into 3 subgroups according to the destination: (1) anterosuperior subgroup to

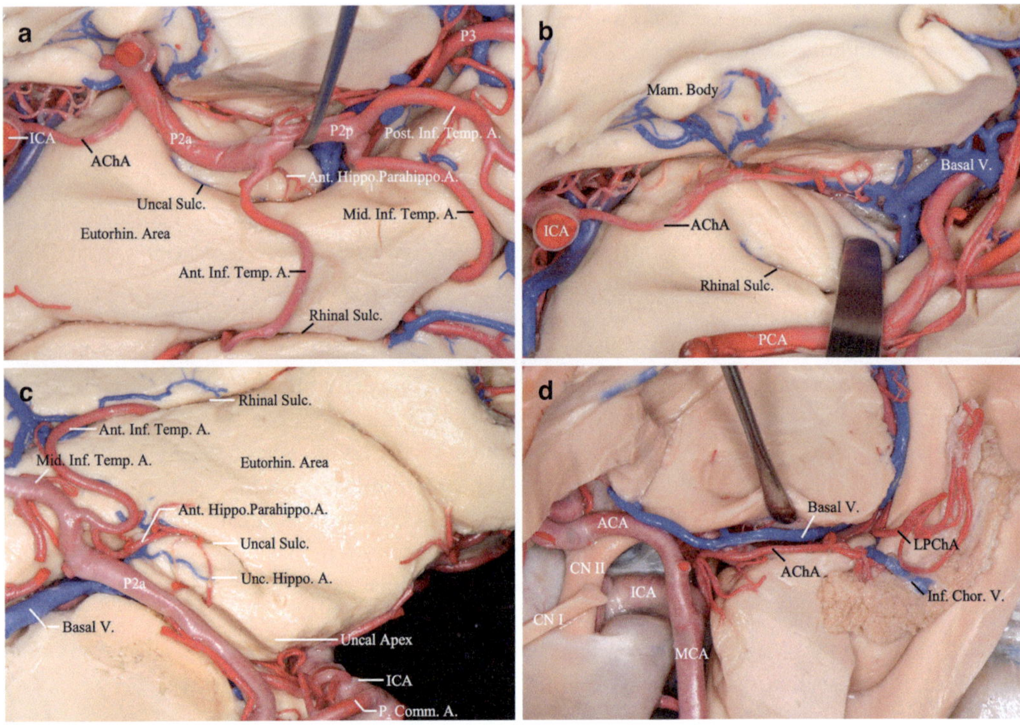

Fig. 3 Arterial Supply of anterior segment of MTL. (**a**) Medioinferior view of the uncal sulcus shows 1 anterior hippocampal-parahippocampal artery (AHPA), originating from the AITA, penetrating the uncal sulcus to irrigate the extraventricular part of the hippocampal head. (**b**) Inferiorly retraction of PCA showing the entry point of the AChA in the choroidal fissure at the inferior choroidal point. (**c**) Inferior view of anther specimen shows 1 AHPA and 1 UHA, entering the uncal sulcus. (**d**) Superior view of another dissected temporal lobe showing the ventricular features of MTL. The AChA and LPChA enter the lateral ventricle in the posterior limit of the uncus, supply the choroidal plexus. A.Ch.A, anterior choroidal artery; Ant. Hippo. Parahippo. A., anterior hippocampal-parahippocampal artery; Ant. Hippo. V., anterior hippocampal vein; Ant. Inf. Temp. A., anterior-inferior temporal artery; Ant. Long. Hippo. V., anterior longitudinal hippocampal vein; Ant. Parahippo. A., anterior parahippocampal artery; AHA, anterior hippocampal artery; Basal V., basal vein; Chor. Plex., choroid plexus; Coll. Sulc., collateral sulcus; Entorhin. Area, entorhinal area; Mam. Body, mamillary body; MHA, middle hippocampal artery; Mid. Inf. Temp. A., middle-inferior temporal artery; Parahippo. Gyrus, parahippocampal gyrus; PCA, posterior cerebral artery; P. Co. A., posterior communicating artery; PHA, posterior hippocampal artery; Post. Hippo. Parahippo. A., posterior hippocampal-parahippocampal artery; Post. Inf. Temp. A., posterior-inferior temporal artery; Splen. A., splenial artery; Tent. Notch, tentorial notch; Uncal Sulc., uncal sulcus; Unc. Hippo. A., uncal hippocampal artery. (With permission from Juan Carlos Fernandez-Miranda, MD)

anterior uncus, (2) medial subgroup to posterior uncus, and (3) inferior subgroup to the entorhinal area of parahippocampus. The plexal segment of the AChA starts from the inferior choroidal point at choroidal fissure, courses posteromedially at the medial choroidal plexus, and ends at the choroidal glomus. It usually supplies the plexus coordinated with LPChA in the temporal horn, but on occasion, it could feed the choroidal plexus alone in atrium, even arriving in foramen of Monro. According to our previous research, 38% AChA sent out perforating neural branches at the periphery of lateral geniculate body in the plexal segment, named as capsulothalamic artery. It gives blood supply to posterior limb (caudodorsal and ventral part), the retrolenticular segment (containing Meyer's loop) of the internal capsule, the tail of the caudate nucleus, and the lateral thalamic nucleus [1, 6, 7].

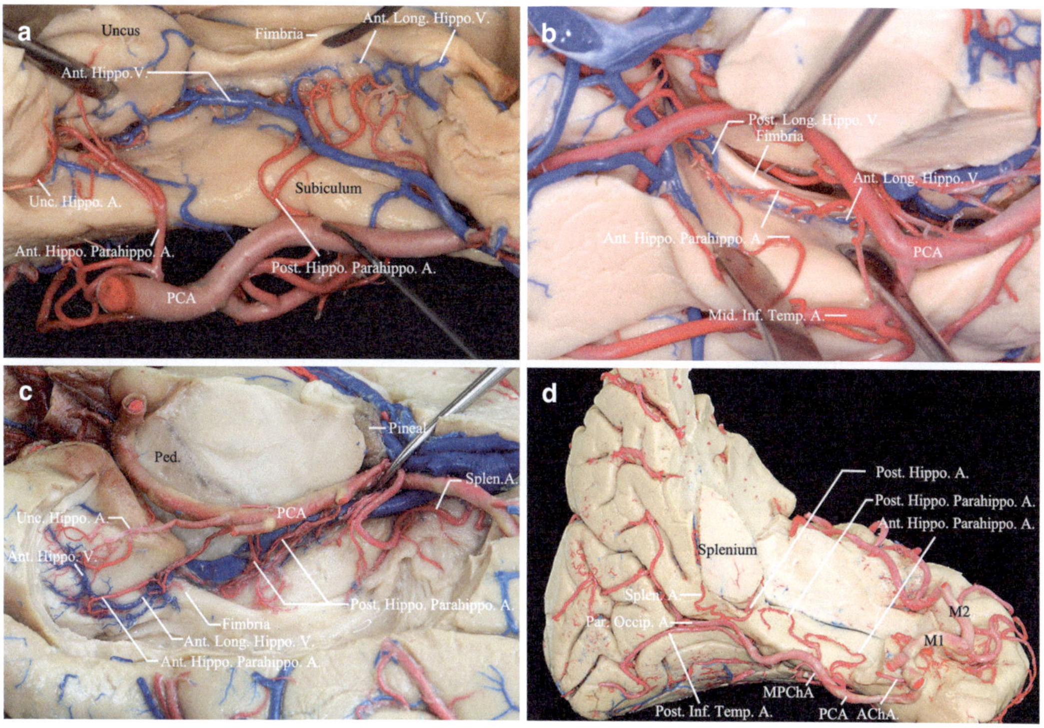

Fig. 4 Arterial Supply of the middle and posterior segment of MTL. (**a**) Medial view of the uncal sulcus in another specimen. One AHPA and 1 posterior hippocampal parahippocampal artery (PHPA), both originating from the PCA, serve to irrigate the hippocampal head and body. No UHA was found in this specimen. (**b**) Medioinferior view of the fimbriodentate sulcus showing 2 AHPAs, originating from the MITA, running above the flat surface of the subiculum of the parahippocampus with a slightly curved contour. It ultimately reaches the hippocampal sulcus and fimbriodentate sulcus to feed the hippocampal body. (**c**) Inferior view of another dissected MTL shows the arterial network in this sample, which is formed by 1 UHA and 1 AHPA supplying the anterior segment of MTL; 4 PHPAs supplying the middle segment of MTL; and 1 splenial artery (SA) originating from the posterior inferior temporal artery, distributing into the posterior segment of MTL. (**d**) Medial view of the another temporal lobe showing vascular network in this specimen, which is formed by 1 AHPA supplying the anterior segment of MTL, 2 PHPAs supplying the middle segment of MTL, and 1 splenial artery (SA) and 1 posterior hippocampal artery supplying the posterior segment of MTL. A. Ch.A, anterior choroidal artery; Ant. Hippo. Parahippo. A., anterior hippocampal-parahippocampal artery; Ant. Hippo. V., anterior hippocampal vein; Ant. Inf. Temp. A., anterior-inferior temporal artery; Ant. Long. Hippo. V., anterior longitudinal hippocampal vein; Ant. Parahippo. A., anterior parahippocampal artery; AHA, anterior hippocampal artery; Basal V., basal vein; Chor. Plex., choroid plexus; Coll. Sulc., collateral sulcus; Entorhin. Area, entorhinal area; Mam. Body, mamillary body; MHA, middle hippocampal artery; Mid. Inf. Temp. A., middle-inferior temporal artery; Parahippo. Gyrus, parahippocampal gyrus; PCA, posterior cerebral artery; P. Co. A., posterior communicating artery, PHA, posterior hippocampal artery; Post. Hippo. Parahippo. A., posterior hippocampal-parahippocampal artery; Post. Inf. Temp. A., posterior-inferior temporal artery; Splen. A., splenial artery; Tent. Notch, tentorial notch; Uncal Sulc., uncal sulcus; Unc. Hippo. A., uncal hippocampal artery. (With permission from Juan Carlos Fernandez-Miranda, MD)

Anterior Uncal Arteries (AUA)

The anterior uncal arteries (AUA) originate from the AChA, ICA or MCA to the anterosuperior surface of the uncus and irrigate the anterior uncus. Usually, AChA generates an averagely of 1.2 (0–3) anterior uncal arteries. The anterior uncal artery is not present in only 17% of cases in our previous study. The AUA normally appear in the first one-third of the AChA, right before it hits the lateral margin of the optic tract, at a distance of 5 mm (0–15 mm) on average, from the origin of the AChA. It gives blood supply to semilunar gyrus in 60% of cases, semilunar and ambient gyri in 30% of cases and ambient gyrus in 10% of cases [1].

Posterior Uncal Artery (PUA)

The posterior uncal artery (PUA) originates from the AChA and irrigates the posteromedial uncus which is formed by the cisternal surface of the head of hippocampus. Usually AChA gives off 0 - 3 posterior uncal arteries (average 1.3). They emerges normally in the middle third of the AChA, 13 mm on average (6–20 mm) distal to the origin of AChA before it arrives at the ending of band of Giacomini. The PUA descends down the band of Giacomini and accesses the uncal sulcus. When more than 1 PUA is present, they were dispersed to the posteromedial uncus, where they send out a number of minuscule branches to uncinate or intralimbic gyri without entering uncal sulcus [1].

Uncohippocampal Artery (UHA)

The uncohippocampal artery (UHA) originates from PUA or AChA directly in most cases, and penetrates into the uncal sulcus and is distributed in the anterior uncal sulcus, or the superior surface of entorhinal area which is hidden in the uncal sulcus. The entry point at uncal sulcus, in relation to the uncal apex, is averagely 8.9 mm (7-13 mm) posteriorly. During the hidden course in uncal sulcus, it runs serpentinely to the deepth of uncal sulcus and gives off several perforated ramies to supply the hippocampal head, uncinate gyrus superjacently and the anterior parahippocampus gyrus subjacently. The UHA is commenly anastomosed with the AHPA at the uncal sulcus (76%) [1, 2].

Unco-Parahippocampal Artery (UPHA)

The unco-parahippocampal artery (UPHA) usually originates from the branches of AChA (AUA or UHA) to the inferior surface of the anterior MTL and gives blood supply to the entorhinal area [1].

3.2 Middle Cerebral Artery (MCA)

The M1 segment of MCA is intimately related to the MTL. The proximal part of M1 is located posteroinferiorly to the ambient gyri, superiorly to anterior perforated substance. The horizonal M1 wraps around the limen insula as it crosses from internal to lateral, where it typically splits into two trunks. Early branches are cortical branches that originate from the ventromedial side of M1 segment. Lenticulostriate arteries are the perforating branches that enter the anterior perforated substance and arise from the dorsolateral side of M1 segment (rarely from M2) [8, 9].

The irrigation is highly variable at the anterior MTL, which showed 3 differentiated patterns in our previous study: anterior pattern in 41% of cases, anteroinferior pattern in 51% of cases and inferior pattern in 8% of cases. In the anterior pattern, the anterior uncus, especially anterior ambient gyrus is irrigated by the ***anterior uncal arteries (AUA).*** In the anteroinferior pattern, the feeding artery also descended to the inferior surface of MTL and supplied anterior entorhinal area after passing the ambient gyrus, named as the same with ***unco-parahippocampal arteries (UPHA)***. The arteries in the third inferior pattern, which are known as the ***anterior parahippocampal arteries (APHA)***, irrigated the anterior portion of the entorhinal area by reaching the inferior surface of the anterior MTL [1].

3.3 Internal Carotid Artery (ICA)

The choroidal segment of the ICA, exactly the section between the AChA and the bifurcation, is intimately related to the MTL. Remarkably, these branches from ICA to MTL are present when the anterior uncal branch of the AChA or MCA is absent, which shows a reciprocal relationship between the uncal branches of the AChA, ICA, or MCA. According to their distribution, 96% of branches were ***AUA***, and 4% were ***APHA.*** It's interesting to note that the distribution of the AUAs of the ICA mimicked the pattern of the AUAs of the AChA when their origin was closer to AChA than to the end segment of ICA. The distribution of the AUAs of the ICA, on the other hand, mimicked the pattern of the AUAs of the MCA when their origin was closer to the end segment of ICA than the origin of the AChA [10, 11].

3.4 Posterior Cerebral Artery (PCA)

The PCA and its branches are the main feeding vessels of the MTL. It has a close connection to the anterior, middle, and posterior MTL. The PCA is divided into four segments based on constant anatomic landmarks: precommunicating segment (P1), postcommunicating segment (P2), which can be subdivided into anterior or crural (P2a) segment and posterior or ambient (P2p) segment, quadrigeminal segment (P3), and calcarine segment (P4) [12].

The anterior MTL is intimately related to the P2a segment. As it travels from the interpeduncular cistern to the crural cistern and turns at the crus cerebri, it has a close spatial relationship with uncal sulcus and posterior uncus. The middle segment of the MTL is related to P2p segment, which courses in the ambient cistern at this level and lies on the medial aspect of the posterior parahippocampal gyrus, then moves to the medial margin of the parahippocampal gyrus. The posterior inferior temporal artery (PITA) and the parieto-occipital arterial trunk (P2p) comprise the most prevalent pattern of PCA proximal bifurcation [13].

The P3 segment, which passes via the quadrigeminal cistern, the splenium of corpus callosum, and the anteroinferior aspect of the cingulate isthmus sequentially before entering the anterior calcarine sulcus, is associated with the posterior MTL [1]. Before exiting the quadrigeminal cistern, all branches of the inferior temporal arteries have already been raised from PCA. The distal bifurcation of the PCA was seen in 86% of cases in our previous research [1], where the parieto-occipital arterial trunk divided into the calcarine artery and parieto-occipital artery. The branches arising from the PCA to vascularize the MTL are described below.

Inferior Temporal Arteries (ITA)

The group of ITAs originates from the PCA, and can be divided into three vessels according to the location: ***anteroinferior temporal arteries (AITA), middle inferior temporal artery (MITA) and posterioinferior temporal artery (PITA).*** The AITA originates from P2a in 39% of hemispheres or P2p in 61% of hemispheres [1], typically located near the level of the posterior uncus (anterior-middle segment of MTL transition), and the distance is always less than 1 cm to that level. The AITA traverses the inferior surface of anterior parahippocampus to reach rhinal sulcus, and in 69% of cases sends out 1–2 branches of anterior parahippocampal artery that vascularizes the entorhinal region and anterolateral parahippocampus.

Anterior Hippocampal-Parahippocampal Artery (AHPA)

The AHPA origins from PCA (42%), in which P2a hold 52% and P2p hold 48%, AITA(56%), LPChA (2%). The enter point of AHPA at uncal sulcus is at an average of 13 mm posterior to the uncal apex and at an average of 1 mm anterior to the posterior limit of the uncus (PLU), which usually lies hidden underneath the anterior hippocampus vein. Frequently it penetrates the posterior uncal sulcus and produces several perforated rami to supply the posterior uncal sulcus, the head, and anterior body of hippocampus. The AHPA is regularly anastomosed with the UHA and PHPA in the uncal sulcus (74% and 56%), respectively.

Posterior Hippocampal-Parahippocampal Artery (PHPA)

There is a large of variability of PHPA origin, among which were PCA (33%), MITA(20%), PITA(33%), Splenial artery (9%), AITA(5%), LPChA(2%). The PHPA travels above the horizonal plane of the subiculum with a liner or slightly curved shape and provides dispersed rami to surrounding structures. It curves posteriorly upon reaching the hippocampal sulcus and gives rise to several perforated branches into the hippocampal proper.

Lateral Posterior Choroidal Artery (LPChA)

Typically, the LPChA originates from the posterior segment of the P2 with double trunks as a C-shaped trajectory. The superior trunk of

LPChA generally travels in the ambient cistern to reach the choroidal fissure and feed the medial side of choroid plexus, then extends posteriorly to the fimbria and crus of fornix and terminates at the posterolateral surface of pulvinar. The inferior trunk of LPChA normally reaches the fimbrio-dentate sulcus passing beneath the fimbria, penetrating the choroidal fissure seldom, and finally terminates at the posteromedial surface of pulvinar. The LPChA is divided into three segments according to our previous research: the first or cisternal segment, from PCA to choroidal fissure with a length of 10.6 ± 2.5mm; the second or forniceal segment, from the penetrating site at choroidal fissure, exactly posterior to the inferior choroidal point with a distance of 8.2 ± 5.7mm, to the posterior choroidal fissure with a length of 28.7 ± 6.8mm; and the last or pulvinar segment, from the posterior choroidal fissure to pulvinar with a length of 5.9 ± 2.2mm [14].

Medial Posterior Choroidal Artery (MPChA)
The MPChA is commonly derived from the P2 segment of the PCA or PCA branches. The course of MPChA can be divided into 2 segments: cisternal and plexal segment. The cisternal segment ascends deep to PCA and gives tiny branches to the tegmentum, midbrain, pulvinar and pineal gland. The plexal segment is formed once it penetrates the velum interpositum. It then feeds the ipsilateral choroid plexus at the roof of the third ventricle. It passes via the foramen of Monro and anastamoses with branches from LPChA.

Splenial Artery (SA)
The SA, also termed as *posterior pericallosal artery (PPA)*, originates from parieto-occipital branch of the PCA in quadrigeminal cistern, then runs superiorly toward the splenium of corpus callosum, enters into the callosum sulcus between isthmus of the cingulate gyrus and the splenium of corpus callosum. It continues to ascend to the roof of splenium and anastomoses with the pericallosal artery. In our precious research, the SA produces a curved loop at medial surface of splenium giving out multiple branches penetrating the subiculum to feed the hippocampal tail in 49% hemispheres [2].

Parieto-Occipital Artery (POA)
The POA is the terminal branch of PCA, which normally travels into the parieto-occipital sulcus and nourishes the cuneus and precuneus.

Calcarine Artery (CA)
The CA, Named according to its course in the calcarine fissure, originates usually from the P3 segment of PCA. It courses deep in the fissure, supplies to the cuneussuperiorly and the lingual gyrusinferiorly.

4 Venous Drainage

The basal vein of Rosenthal is the primary venous pathway that drains the MTL. It is divided into three segments: the first (also known as striate or anterior) segment; the second (also known as peduncular or middle) segment, which is further subdivided into two subsegments: the anterior peduncular segment and posterior peduncular segment; and the third (also known as mesencephalic or posterior) segment (Fig. 5) [15].

4.1 Anterior Segment of the MTL

The anterior segment of the MTL is composed of striate segment and the anterior peduncular segment of the basal vein of Rosenthal. The striate segment of basal vein travels parallel to the anterosuperior part of uncus, while the anterior peduncular segment of basal vein runs parallel to posterosuperior part of uncus. Both veins converge at the site of the uncal apex, the anterior boundary of the crus cerebri, below the optic tract, which corresponds to the origins of the peduncular vein [1]. The striate segment of basal vein is typically created by the junction of the ***anterior cerebral vein (ACV)*** medially and the deep ***middle cerebral vein (MCV)*** laterally. The ***inferior striate veins***, ***olfactory***, and ***fronto-orbital veins*** served as branches draining into MCV or ACV [16]. Generally, the anterior region of the uncus is drained by anterior uncal vein or a cortical

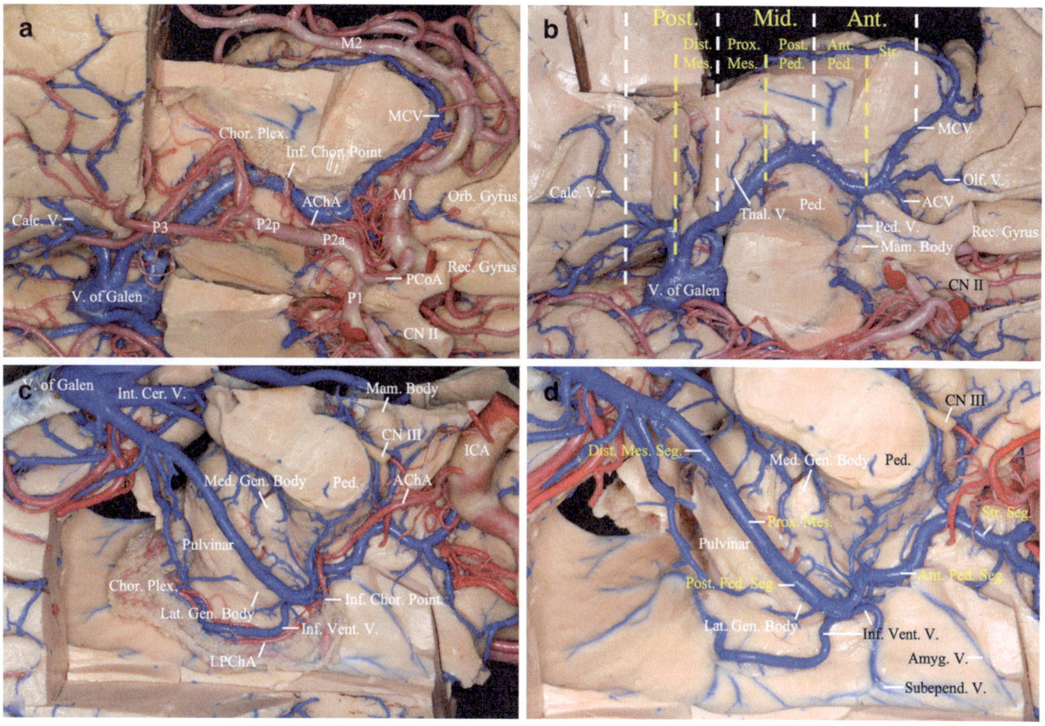

Fig. 5 Venous drainage pattern of the MTL. (**a**) Inferior view of right MTL and the PCA is the main feeding vessel. The P2 is divided into 2 portions: the P2a courses in the crural cistern and the P2p courses in the ambient cistern. (**b**) Removal of PCA to reveal the basal vein and its tributaries in the MTL. The corresponding relationship between segmental classification of the basal vein and the segmental classification of the MTR are shown in yellow font. The striate and anterior peduncular segments of the basal vein are located along the anterior segment of the MTR. The posterior peduncular and proximal mesencephalic segments of the basal vein are situated in the middle segment of the MTR. The distal mesencephalic segment of the basal vein belongs to the posterior segment of the MTR. (**c**) Inferior view of the roof of another right temporal horn. The junctional point between the inferior ventricular vein and the peduncular segment marks the transition between the anterior and middle segments of the MTR, and corresponds to the cisternal level of the inferior choroidal point and the junction of the crural and ambient cisterns. (**d**) Further removal of choroidal plexus reveal the subependymal veins draining the roof and lateral wall of the temporal horn join to form the inferior ventricular vein. In this case, the amygdalar vein drains separately from the inferior ventricular vein into the anterior basal anastomotic vein. (With permission from Juan Carlos Fernandez-Miranda, MD)

vein, which may drain into the deep MCV, striate segment, or anterior peduncular segment of the basal vein. The beginning of the anterior peduncular segment is the junction of basal vein and peduncular vein. Frequently, the peduncular vein interacts with its counterpart on the other side via the posterior communicating vein, which may connect with the anterior pontomesencephalic vein inferiorly. The ***premammillary vein*** arises from the region anterior to the mammillary body and often connects with the peduncular vein, however it could also drain straight into the basal vein.

4.2 Middle Segment of the MTL

The middle segment of the MTL includes the posterior peduncular and the proximal mesencephalic segment of the basal vein. The posterior peduncular segment begins at the junction of the ***inferior ventricular vein (IVV)*** and basal vein and ends at the lateral mesencephalic sulcus. The confluence point of IVV and basal veins is located at the inferior choroidal point, which runs at the transition between the crural and ambient cisterns. The typical IVV drains the superior and lateral surface of the temporal horn of ventricle, as well as the fiber of Meyer's loop. There are several tributaries of IVV, named according to draining territory, among which are the ***amygdalar vein*** anteriorly, the ***subependymal vein*** and the ***inferior choroid vein*** posteriorly. Important tributaries of posterior peduncular segment of the basal vein include ***anterior hippocampal vein (AHV), anterior longitudinal hippocampal vein (ALHV)* and *transverse hippocampal veins (THV)*** [17].

Anterior Hippocampal Vein (AHV)
The AHV runs along the uncal sulcus to collect veins from the hippocampal head and drain into the inferior ventricular or basal veins.

Anterior Longitudinal Hippocampal Vein (ALHV)
The ALHV travels in the anterior fibriodentate sulcus or hippocampal sulcus and drains hippocampal body typically into the inferior ventricular vein or basal vein. There is substantial anastomosis between each other. The AHV, ALHV, together with inferior ventricular vein drain into the basal vein at the site of posterior limit of uncus, which presented as "paw" shape. In 68% of specimen, there three veins (AHV, ALHV, PLHV) appeared concurrently, forming a continuous longitudinal arcade [2].

Transverse Hippocampal Veins (THV)
The THVs are a series of extremely thin veins that travel medially over the collateral eminence and the alveus of the hippocampus. They penetrate the fimbriodentate sulcus then enter the ambient cistern and finally drain into the anterior longitudinal hippocampal vein. The proximal mesencephalic segment of basal vein courses under the pulvinar into the ambient cistern and the main tributaries of proximal mesencephalic segment of the basal vein are the ***lateral mesencephalic (LMV)*** and ***posterior thalamic vein (PTV)***. The PTVs, also termed as pulvinar veins are a collection of veins that empty into the proximal portion of the posterior basal vein at the ambient cistern.

4.3 Posterior Segment of MTL

The distal mesencephalic segment of basal vein travels in the quadrigeminal cistern, terminates at the internal cerebral vein or the vein of Galen, and the main tributaries are ***posterior longitudinal hippocampal (PLHV)***, ***medial temporal (MTV)*** and ***lateral atrial vein (LAV)***. Other uncommon tributaries are the ***internal occipital (IOV)***, ***medial atrial (MAV)***, ***splenial*** (posterior pericallosal), ***pre-central cerebellar***, and ***superior vermian vein*** [18].

Posterior Longitudinal Hippocampal Vein (PLHV)
As the tributary of internal cerebral vein in most case (not basal vein), the PLHV runs in the posterior hippocampal sulcus, and collects the transverse hippocampal veins that drain the hippocampal tail. Frequently, it joins to the medial atrial vein, which pass through the fornix crus inferiorly to access the posterior fibriodentate sulcus, draining into the internal cerebral vein.

Medial Temporal Vein (MTV)
The MTVs, also known as occipitotemporal cortical veins, drain the posteroinferior parahippocampus. As an alternative, this area is drained by the occipitotemporal basal veins and the temporobasal veins, which typically empty into the lateral tentorial sinus laterally rather than the basal vein medially.

Lateral Atrial Vein (LAV)
Multiple branches on the lateral atrium of the ventricle give birth to the LAV. It travels via the crus of the fornix and choroid fissure to access the quadrigeminal cistern and meet the posterior basal vein finally, and otherwise, rarely, drains into the calcarine veins or internal cerebral [1].

Medial Atrial Vein (MAV)
Multiple branches on the medial atrium and occipital horn of the ventricle give birth to the MAV, which perforates the medial portion of the

crus of the fornix in order to get access to quadrigeminal cistern ultimately.

Internal Occipital Vein (IOV)
The IOV, also termed as calcarine vein, collects the tributary vein in the anterior cuneus and lingual gyrus, specifically the anterior portion of the calcarine sulcus, and travels in the anteromedial calcarine sulcus to access the quadrigeminal cistern finally. Additionally, it may collect the veins from the isthmus part of cingulate gyrus.

5 Surgical Approach

The surgical approaches to the medial temporal lobe can be divided into 4 groups: superior, lateral, inferior, and posterior, according to the direction of exposure. The superior approach, also known as the transsylvian-transinsular approach, is conducted through the sylvian fissure to access the anteroinferior limiting sulcus of the insula (temporal stem). The lateral approach is entered through middle temporal gyrus in the lateral surface of the temporal lobe. The inferior approach is performed through the fusiform gyrus and collateral sulcus in the inferior surface of the temporal lobe. The posterior approach is conducted through occipital interhemispheric approach or supracerebellar transtentorial approach to access the medial temporal lobe from posterior to anterior [19].

5.1 Transsylvian Translimen Insula Approach

The pterional craniotomy is most frequently performed to expose the sylvian fissure. The sylvian fissure is extensively dissected from lateral to internal, typically starting 2 cm posterior to the pars triangularis and dissecting forward to the anterior sylvian point, then deep to chiasmatic, carotid, and crural cisterns. The uncus, limen insula, inferior limiting sulcus, M2 are revealed gradually; then the amygdala, which is situated in the anterior part of uncus and comprised the anterior border of the temporal horn was removed to get access to the head of hippocampus. Further enlargement of this corridor involves a 1.5 cm incision directed inferomedially at a 45-degrees toward the midpoint of the lateral wall of cavernous sinus in the anteriormost portion of the inferior limiting sulcus and limen insula. After entering the temporal horn, the choroidal plexus lies directly posterior to the head of hippocampus. The retraction of choroidal plexus is critical to perform toward the superior temporal horn to facilitate the exposure of choroidal fissure and the entry of AChA and inferior ventricular vein at the inferior choroidal point [20]. According to our previous study, the choroidal segment of the LPChA is then revealed at the choroidal fissure with an 8 mm distance posterior to inferior choroidal point, and it runs along the medial portion of the choroidal plexus and the fimbria [14]. Then opening the anterior choroidal fissure through at the tenia fimbriae should be conducted with carefulness to protect the AChA and inferior ventricular vein at the inferior choroidal point, and the choroidal fissure is dissected on the forniceal side instead of the thalamic side to reduce the injury of inferior ventricular vein. Next, hippocampal dissociation develops with an incision from the hippocampal head and uncal recess downhill to the collateral and rhinal sulci. The posterior margin of disconnection of the hippocampal and parahippocampal gyrus is roughly 1 cm posterior to the inferior choroidal point, which equates to the estimated level of the lateral geniculate body, lateral mesencephalic sulcus and posterior border of the cerebral peduncle. The head and body of the hippocampus, as well as a portion of the parahippocampal gyrus, are removed as one piece after dissociation (Fig. 6).

5.2 Lateral Transtemporal Approach

A pretemporal craniotomy is performed to reveal the sylvian fissure superiorly, temporal pole and anterior two-thirds of the temporal lobe, and the

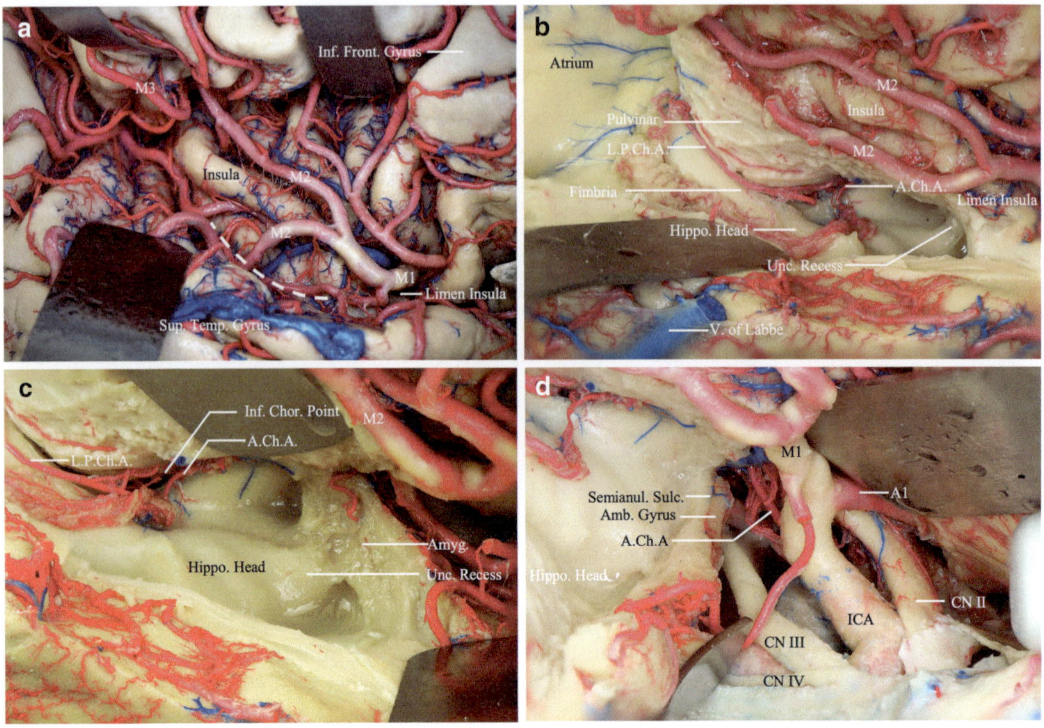

Fig. 6 Stepwise dissection via the transsylvian translimen insula approach to MTL. (**a**) In the transsylvian translimen insula approach, after wide splitting of the Sylvian fissure, the uncus, limen insula, and inferior limiting sulcus are exposed. Corticectomy of the temporal stem (white dotted line) is performed from the limen insula and extends posteriorly along the inferior limiting sulcus of the insula. (**b**) The temporal horn and hippocampal head are accessed. The choroid plexus, anterior choroidal artery (AChA), and lateral posterior choroidal artery are identified on the hippocampus. (**c**) Enlarged view of hippocampal digitations. The uncal recess, located in the anterior part of the temporal horn between the hippocampal head and ventricular surface of the amygdala. The AChA enters the temporal horn and choroid plexus at the inferior choroidal point. (**d**) When the entire amygdala is carefully removed, the basal cisterns are exposed, which brings the internal cerebral artery and its branches, the oculomotor nerve, and the optic nerve into view. The temporopolar artery arising from M1 has been sectioned near its origin for better exposure. The semilunar gyrus, which represents the cortical projection of the amygdala, was resected, and the semiannular and ambient gyri can be seen below. (**e**) Further resection of the uncal gyrus reveals the uncohippocampal artery (UHA) and lateral surface of the mesencephalon. (**f**) Resection of the hippocampal head reveals the uncal sulcus, where the vessels of the hippocampal head are hidden. The UHA arises from the AChA, distributes into the uncinate gyrus and band of Giacomini, and penetrates the anterior uncal sulcus to vascularize the extraventricular hippocampal head. The anterior hippocampal-parahippocampal artery originates from the PCA, then enters into the posterior uncal sulcus, and distributes into the hippocampal head. In this specimen, the anterior hippocampal artery anastomoses with the UHA of the AChA. The anterior hippocampal vein, which distributes into the uncal sulcus, and the anterior longitudinal hippocampal vein, which distributes into the hippocampal sulcus, drain the hippocampal head and body, respectively, into the basal vein at the site of the posterior limit of the uncus, which is paw shaped. (**g**) Resection of the hippocampal body reveals that 2 posterior hippocampal parahippocampal arteries, originating from the PCA, run above the flat surface of the subiculum of the parahippocampus with a curved contour, supplying the adjacent structures with scattered rami. (**h**) Final surgical field for amygdalohippocampectomy, showing the PCA and its branches. A.Ch.A, anterior choroidal artery; Amb. Gyrus, ambient gyrus; Amyg., amygdala; Ant. Hippo. Parahippo. A., anterior hippocampal-parahippocampal artery; Ant. Hippo. V., anterior hippocampal vein; Ant. Long. Hippo. V., anterior longitudinal hippocampal vein; Basal A., basal vein; CN, cranial nerve; Hippo. Head, hippocampal head; ICA, internal cerebral artery; Inf. Chor. Point, inferior choroidal point; Inf. Front. Gyrus, inferior frontal gyrus; L.P.Ch.A, lateral posterior choroidal artery; Post. Hippo. Parahippo. A., posterior hippocampal-parahippocampal artery; Semianul. Sulc., semiannular sulcus; Sup. Temp. Gyrus, superior temporal gyrus; Unc. Hippo. A., uncal hippocampal artery; Unc. Recess, uncal recess; Unc. Sulc., uncinate sulcus; V., vein. (With permission from Juan Carlos Fernandez-Miranda, MD)

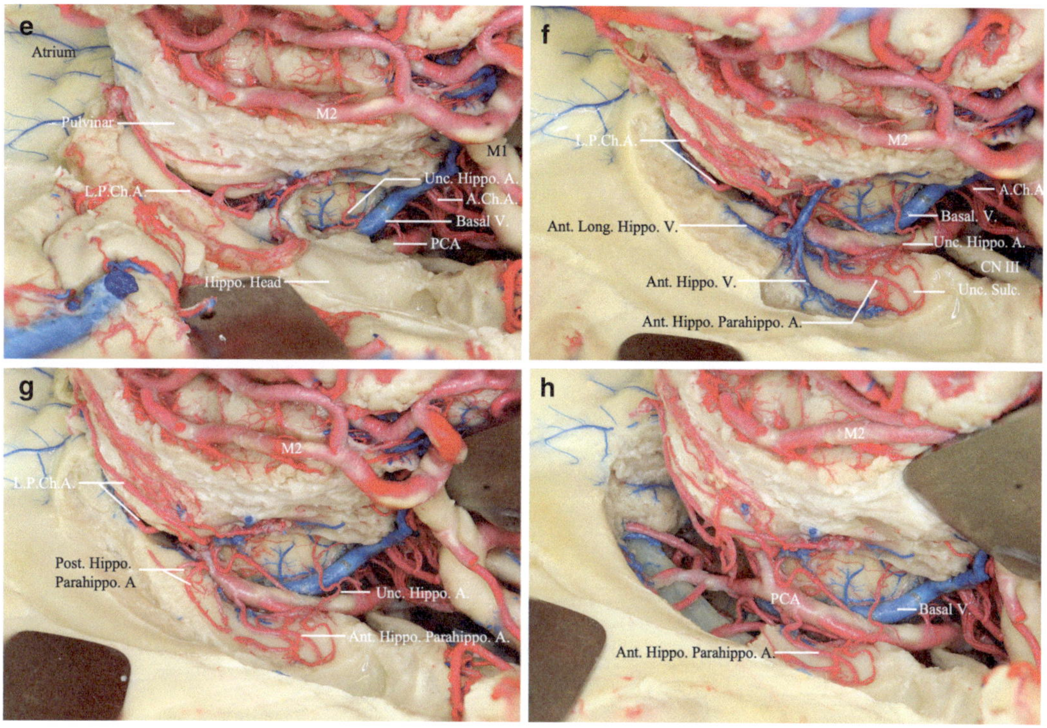

Fig. 6 (continued)

floor of the middle fossa inferiorly. When the temporal lobe is accessible, a corticectomy is performed along the middle temporal gyrus, often beginning 2 cm posterior to the temporal pole and continuing 3–4 cm posteriorly (5 cm on the dominant hemisphere and 6 cm on the nondominant hemisphere posterior to the temporal pole). The inferior temporal sulcus could be utilized for a transsulcal approach, whereas the superior temporal gyrus usually is avoided due to Heschl's gyrus. After entering the temporal horn, the choroid plexus, collateral eminence, and head and body of the hippocampi are distinguished. To expose and open the choroidal fissure, the choroid plexus is elevated. The medial temporal resection is accomplished by employing the same techniques as translimen insula approach (Fig. 7) [20].

5.3 Subtemporal Approach

A linear, horseshoe-shaped or small question mark incision is recommended to expose the region of temporal squama and the root of the zygomatic arch, then the craniotomy is performed with substantial middle fossa floor exposure to prevent brain retraction (protect the Labbé's vein) and increase the angle of surgical view through the inferior surface of temporal lobe. The temporal lobe is elevated once the dura is opened to reveal the inferior surface. To improve exposure, the basal cisterns can be accessed and cerebrospinal fluid released. The collateral and rhinal sulcus between the parahippocampal gurus and the fusiform gyrus is the first recognized anatomic struture during operation, which served as the landmark for accessing the temporal horn. The seperation between the fimbria and the choroid plexus is performed once within the temporal horn, exposing the ambient cistern. When the temporal horn is accessed by the lower temporal surface in the basal approaches, the lateral dissociation at approximate level of the lateral edge of the hippocampus and the collateral eminence is accomplished. The anterior and posterior dissociation of the hippocampus, as well as the excision of the amygdala are similar to transsylvian translimen insula approach [20, 21].

5.4 Supracerebellar Transtentorial Approach

The patient is placed in a park bench position with head rotated contralaterally, thus creating a

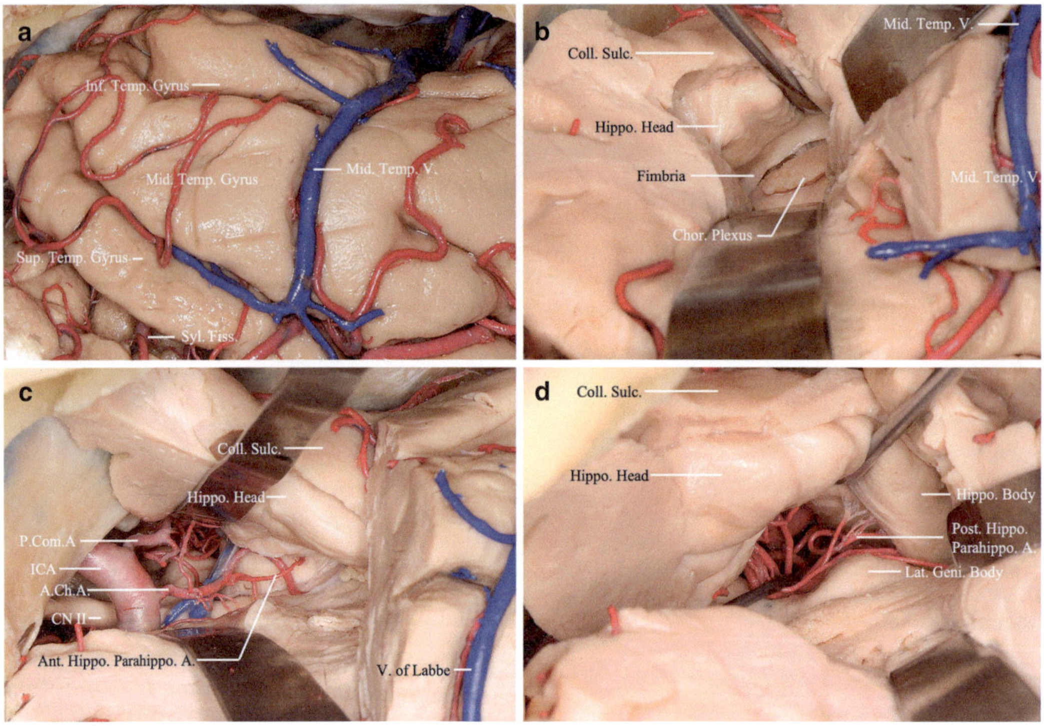

Fig. 7 Stepwise dissection of lateral transtemporal approach to MTL. (**a**) The lateral transtemporal approach via right frontotemporal craniotomy was completed to expose the temporal lobe. (**b**) An anterior temporal lobectomy was performed step by step. The half-lateral portion was resected to expose the temporal horn and hippocampal head. A retractor is used to reveal the fimbria and choroidal plexus. (**c**) Connection to the medial temporal lobe was cut to reveal the carotidal cistern and the spatial relationship between the internal cerebral artery, anterior choroidal artery, and hippocampus. (**d**) Further resection of the anterior temporal lobe revealed the hippocampal body. Lateral retraction of the hippocampal body shows 2 posterior hippocampal parahippocampal arteries (PHPAs) serving as the feeding arteries in this portion. (**e**) Medial view of the resected anterior temporal lobe, in which the vessels were preserved. In this specimen, 1 uncohippocampal artery and 1 anterior hippocampal-parahippocampal artery (AHPA) serve as the anterior hippocampal artery complex supplying the hippocampal head, and 3 PHPAs serve as the middle hippocampal artery (MHA) complex supplying the hippocampal body. (**f**) The surgical field after the anterior temporal lobectomy was completed. The posterior ambient cistern and quadrigeminal cistern were exposed, where the posterior cerebral artery and its branches to the temporal lobe, the basal vein and the lateral surface of the pedunculus can be seen. A.Ch.A, anterior choroidal artery; AHA, anterior hippocampal artery; Ant. Hippo. Parahippo. A., anterior hippocampal-parahippocampal artery; Ant. Inf. Temp. A., anterior-inferior temporal artery; Band of Giac., band of Giacomini; Chor. Plexus, choroid plexus; CN, cranial nerve; Coll. Sulc., collateral sulcus; Hippo. Body, hippocampal body; Hippo. Head, hippocampal head; ICA, internal cerebral artery; Inf. Temp. Gyrus, inferior temporal gyrus; Intralimb. Gyrus, intralimbic gyrus; Lat. Geni. Body, lateral geniculate body; MHA, middle hippocampal artery; Mid. Inf. Temp. A., middle-inferior temporal artery; Mid. Temp. Gyrus, middle temporal gyrus; Mid. Temp. Sulc, middle temporal sulcus; Mid. Temp. V., middle temporal vein; PCA, posterior cerebral artery; P.Com.A., posterior communicating artery; Ped., peduncle; Post. Hippo. Parahippo. A., posterior hippocampal-parahippocampal artery; Sup. Temp. Gyrus, superior temporal gyrus; Syl. Fiss., sylvian fissure; Temp. Horn, temporal horn; Uncal Hippo. A., uncal hippocampal artery; Uncal Sulc., uncal sulcus; Uncin. Gyrus, uncinate gyrus; V., vein. (With permission from Juan Carlos Fernandez-Miranda, MD)

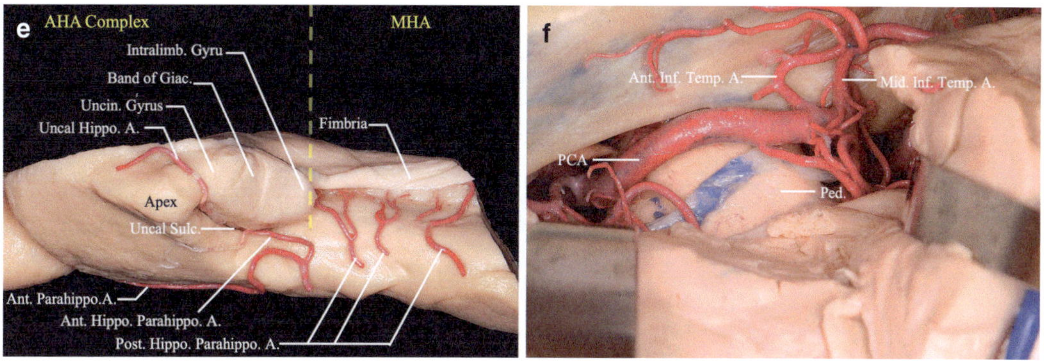

Fig. 7 (continued)

direct access to the tentorium and medial posterior temporal lobe. A linear incision was performed from the superior nuchal line (superior to the transverse sinus) to foramen magnum (CSF in cisterna magna drainage) to expose the suboccipital region. The dura was opened in a U-shaped manner with the base superiorly and as wide as possible, then bridging veins between tentorium and cerebellum could be obliterated and cut to increase the working space between tentorium and superior surface of cerebellum. The Galenic venous complex, superior cerebellar artery, trochlear nerve and the superior and inferior colliculus are distinguished when the arachnoid membrane of the quadrigeminal cistern is opened. The tentorium was cut, starting at the medial free margin to lateral to expose the MTL [22].

6 Illustrative Cases

Case 1. Uncal Dysembryoplastic Neuroepithelial Tumor (DNET)
A 30-years old man presented intermittent déjà vu seizures for 8 years, and was confirmed as left temporal lobe seizures by EEG. MRI following seizure revealed a left temporal lobe uncohippocampal mass with cystic features. He developed cognitive difficulties, and the executive functioning neuropsycological test results indicated relatively intact cognitive function with the exception of reduced naming suggestive of dominant lateral temporal dysfunction. We performed a left frontotemporal craniotomy and transsylvian approach to the medial temporal lobe for microsurgical resection of tumor. Six weeks after surgery, the patient's speech and memory exhibited considerable improvement, and postoperative MRI revealed a gross complete resection (Fig. 8).

Case 2. Panhippocampal Low-Grade Glioma
A 51-years old male presented seizure, and MRI revealed a mass occupy hippocampal head, body and tail which was named after panhippocampal leision (Fig. 9). The fiber tracting image showed arcuate fasciculus (AF) superiorly, inferior fronto-occipital (IFOF) laterally, optic radiation superiorly and laterally encased the tumor (Fig. 10). The patient underwent a two-stage operation, which required maximize resection while minimizing morbidity.

In the first stage, The transsylvian approach was performed to the uncus and hippocampus via the inferior insular sulcus and temporal horn. Debulking is from the anterior segment of uncus to hippocampal body with piecemeal fashion by ultrasonic aspirator, where several landmarks were encountered by sequence including amygdala, temporal horn, hippocampal head, choroid plexus, choroid fissure, rhinal sulcus, collateral sulcus. Post-first-stage radiology presented an intact cortex and well-protected fibers (Fig. 10). The anterior middle portion of the tumor has been resected and residual tumor was left between quadrigeminal plate and the temporal lobe which could not be accessed through transsylvian approach. In the second stage, the supracerebellar transtentorial approach was performed with sitting position. The tumor, located in the quadrigeminal cistern, was

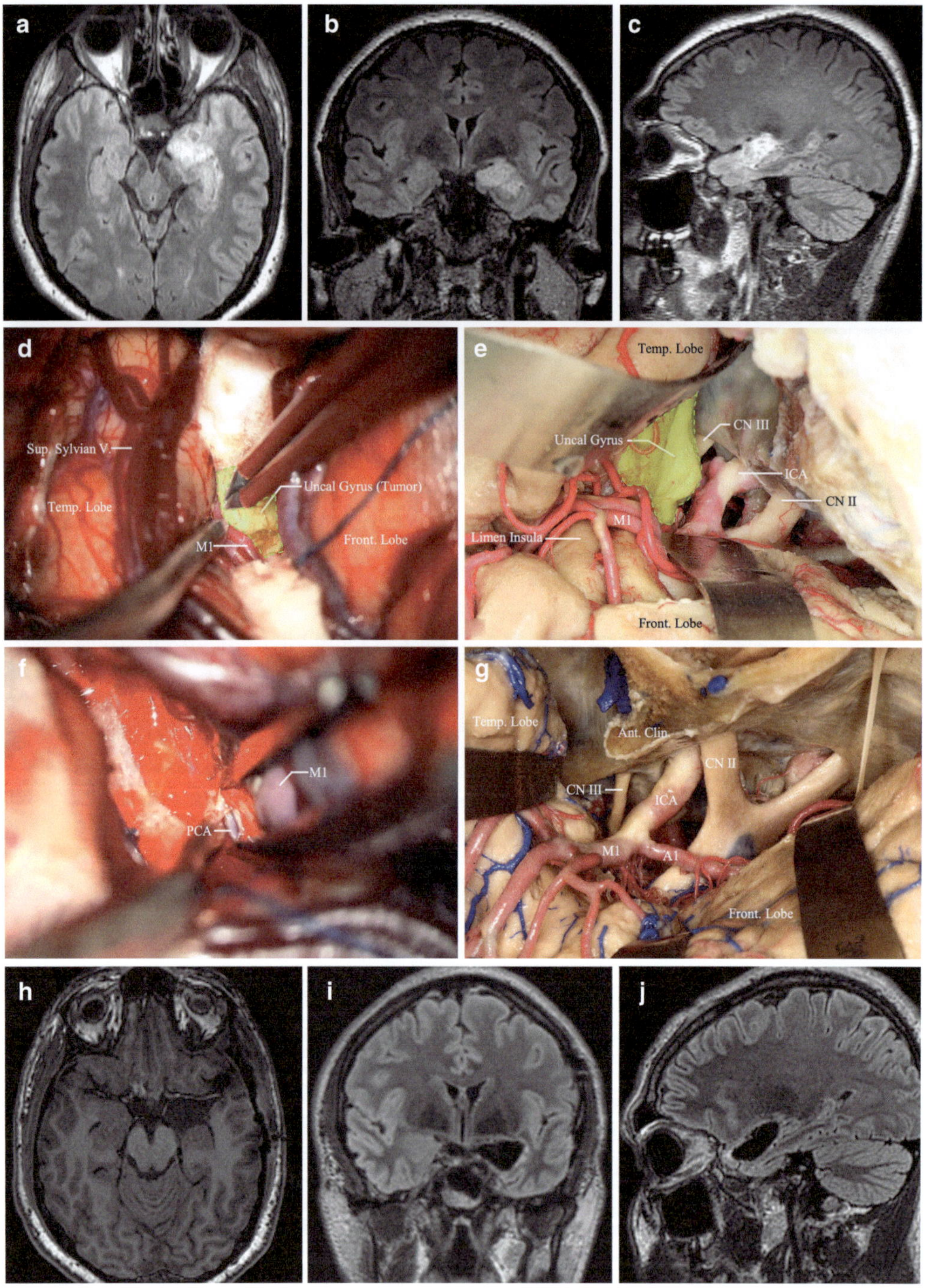

Fig. 8 Illustrative Case 1 of left uncal dysembryoplastic neuroepithelial tumors (DNET). (**a–c**) Preoperative contrast-enhanced MRI images show a left temporal lobe unco-hippocampal mass with cystic features. (**d**) A left frontotemporal craniotomy and transsylvian approach was performed to the uncal gyrus for microsurgical resection of tumor. (**e**) The same microsurgical perspective to left uncal gyrus on a specimen. The location of the lesion has been highlighted (green shaded area) to facilitate the understanding of its anatomic relationships. (**f**) When tumor was resected totally, the PCA appeared in the deep surgical field. (**g**) The similar anatomic view with (**f**) was acquisited on specimen. (**h–j**) The postoperative MRI showed gross total resection. (With permission from Juan Carlos Fernandez-Miranda, MD)

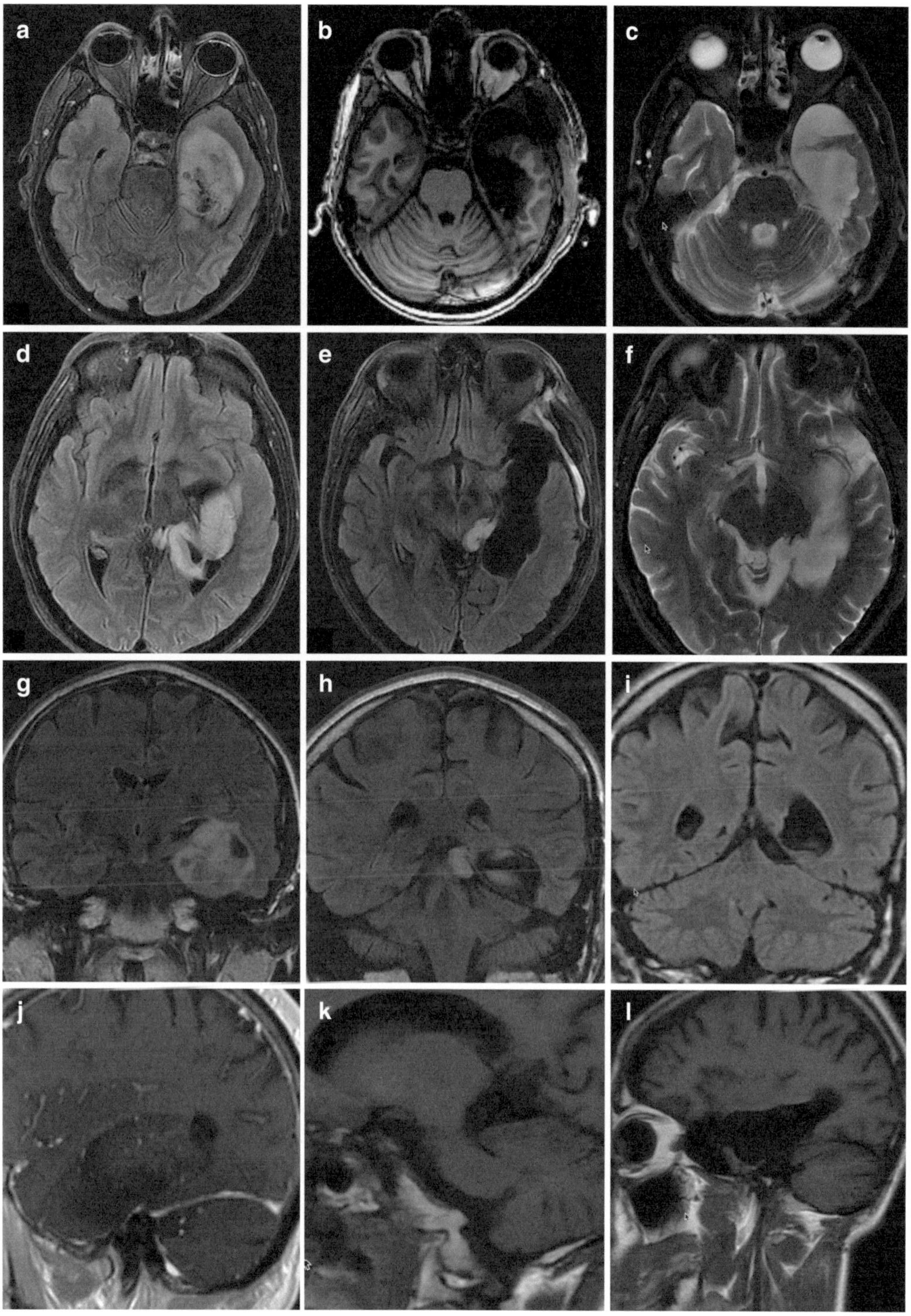

Fig. 9 Illustrative Case 2 of left panhippocampal low-grade glioma treated by two-stage microsurgical resection. (**a**, **d**, **g**, **j**) Preoperative MRI images revealed a mass occupy hippocampal head, body and tail suggestive of a panhippocampal leision. (**b**, **e**, **h**, **k**) The post first stage MRI showed the anterior middle portion of the tumor has been resected and the residual tumor was left between quadrigeminal plate and the temporal lobe which could not be accessed through transsylvian approach (**c**, **f**, **i**, **l**) The post second stage MRI showed gross total resection for panhippocampal glioma. (With permission from Juan Carlos Fernandez-Miranda, MD)

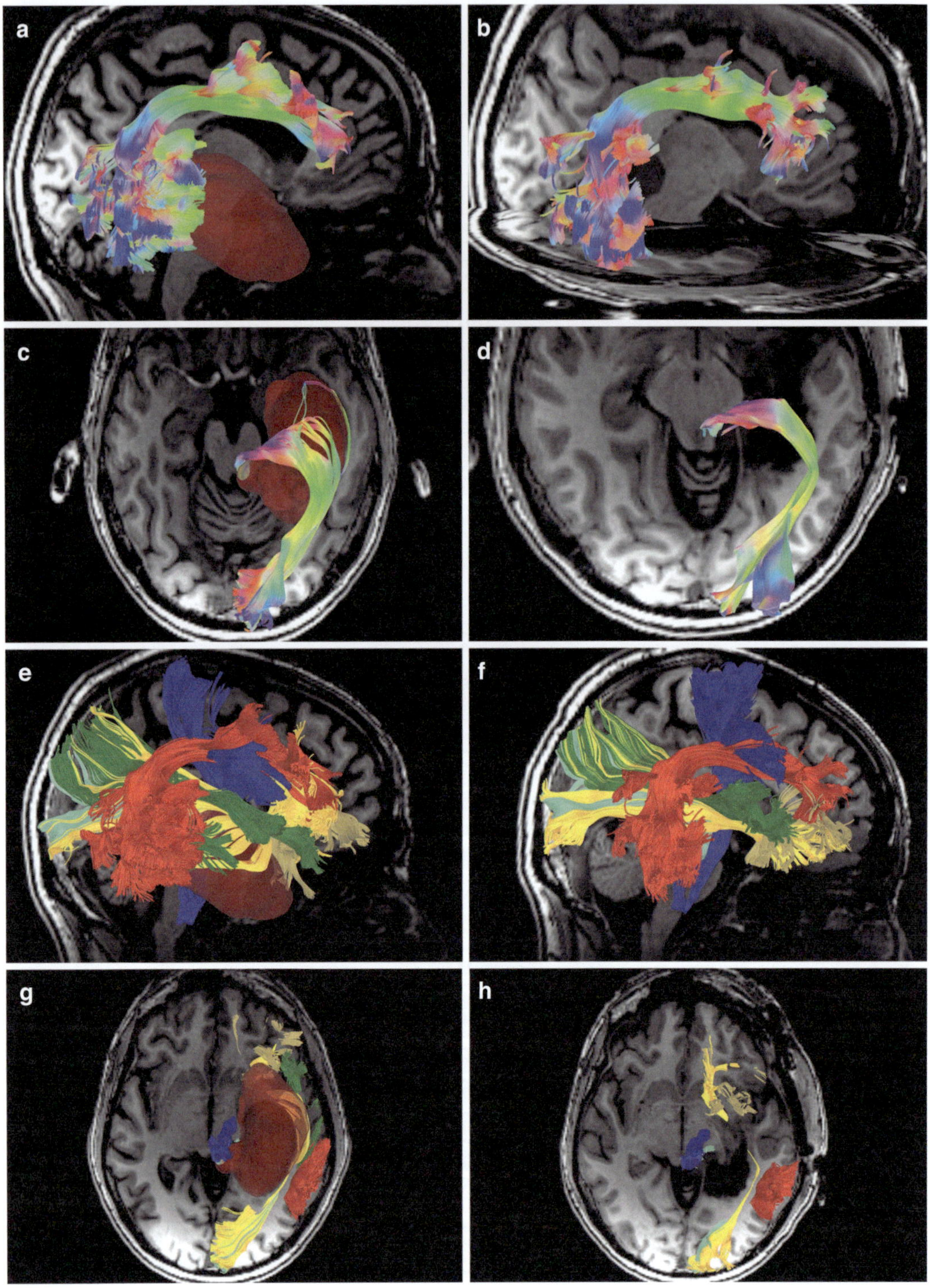

Fig. 10 Fiber tractography of pre-operation and post-operation in panhippocampal low-grade glioma. (**a, c, e, g**) The fiber tracting image showed arcuate fasciculus (**a**) superiorly, inferior fronto-occipital laterally, optic radiation (**c**) superiorly and laterally encased the tumor. (**b, d, f, h**) Post-first-stage radiology presented an intact cortex and well-protected fiber. (With permission from Juan Carlos Fernandez-Miranda, MD)

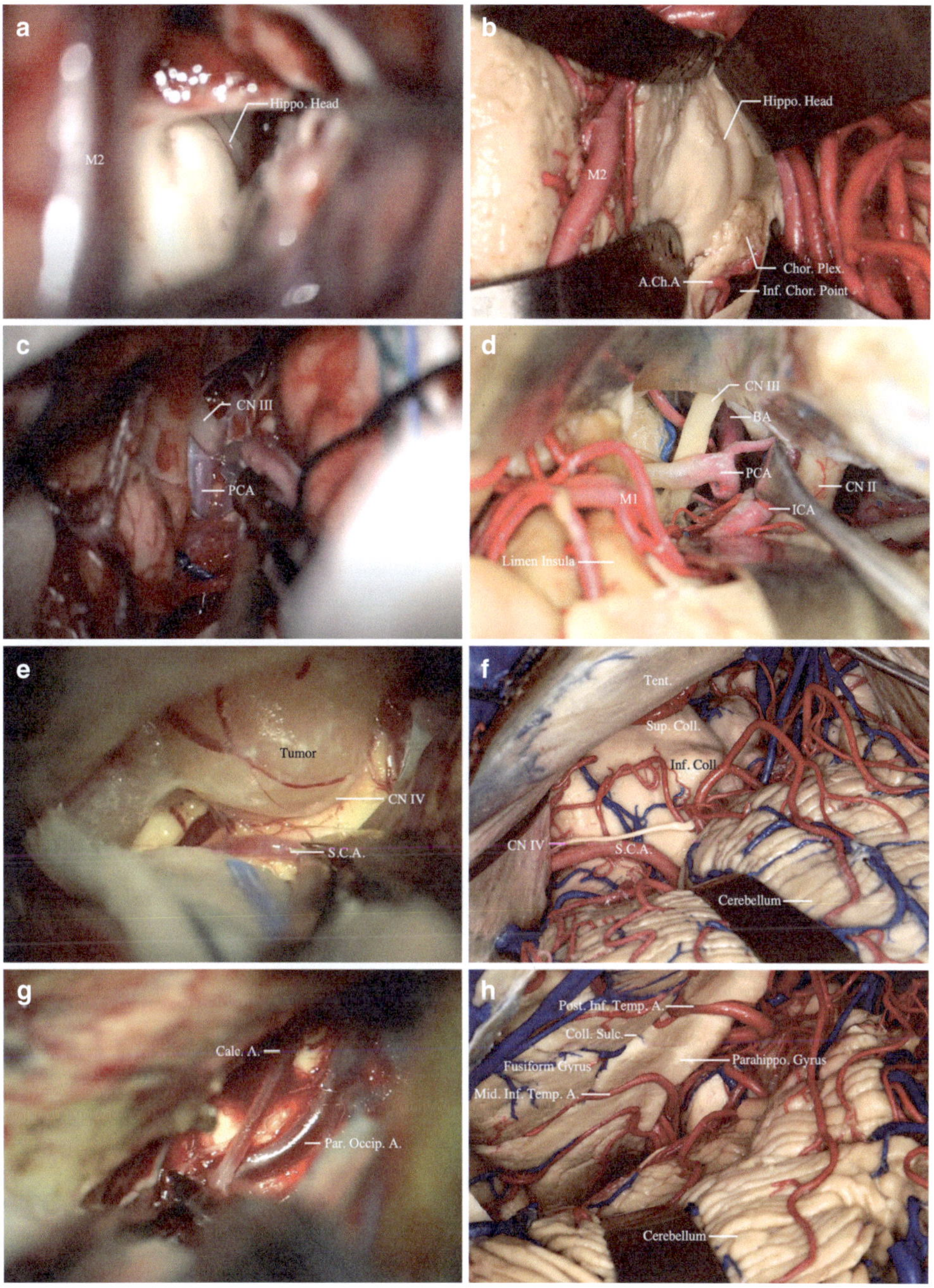

Fig. 11 Trans-sylvian (Stage 1) and supra-cerebellum trans-tentorium approach (Stage 2) for panhippocampal low-grade glioma. (**a**) In the first stage, the head and the body of hippocampus was exposed by transsylvian approach through the inferior insular sulcus and temporal horn. (**b**) The same microsurgical perspective to left hippocampal head on a specimen. (**c**) After resection the oculomotor nerve and the posterior cerebral artery were revealed. (**d**) The similar anatomic view with (**c**) was illustrated on dissected specimen. (**e**) In the second stage, the supracerebellar transtentorial approach was performed and the tumor, located in the quadrigeminal cistern, was observed a close relationship with trochlear nerve. (**f**) The similar anatomic view with (**e**) to present the trochlea nerve and surrounding structures. (**g**) After total resection of tumor, the calcarine artery and the parietal occipital trunk, as the branches of posterior cerebral artery, were revealed. (**h**) similar anatomic view with (**g**) to parahippocampal gyrus through supra-cerebellum trans-tentorium approach. (With permission from Juan Carlos Fernandez-Miranda, MD)

separated carefully from surrounding neurovascular structures including trochlear nerve, SCA, PCA with its branches and brainstem (Fig. 11). Postoperative MRI shows the tumor was resected totally, and the patient returned to work eight weeks after the second stage operation with exception of superior quadrantanopia.

7 Conclusion

The MTL is the location of the most complicated cortical anatomy, involved in key elements of limbic system, intricate irrigation and drainage system composed by AChA, ICA, MCA, PCA and the basal vein of Rosenthal, which pose a great challenge to neurosurgeon. Four operative approaches including the transsylvian-transinsular approach, the lateral transtemporal approach, the subtemporal approach, and the supracerebellar transtentorial approach are the main surgical corridors to the MTL. A comprehensive knowledge of the microvascular anatomy of MTL is required for surgery of medial temporal tumor, contributing to surgical planning, execution and successful outcomes.

References

1. Fernandez-Miranda JC, de Oliveira E, Rubino PA, et al. Microvascular anatomy of the medial temporal region: part 1: its application to arteriovenous malformation surgery. Neurosurgery. 2010;67(3 Suppl Operative):ons237-276; discussion ons276.
2. Xu Y, Mohyeldin A, Nunez MA, et al. Microvascular anatomy of the medial temporal region. J Neurosurg. 2021;1(aop):1–13.
3. Duvernoy HM, Cattin F, Risold P-Y. The human hippocampus. 4th ed. Berlin: Springer; 2013.
4. Wen HT, Rhoton AL Jr, de Oliveira E, et al. Microsurgical anatomy of the temporal lobe: part 1: mesial temporal lobe anatomy and its vascular relationships as applied to amygdalohippocampectomy. Neurosurgery. 1999;45(3):549–91.
5. Ulm AJ, Tanriover N, Kawashima M, Campero A, Bova FJ, Rhoton AL Jr. Microsurgical approaches to the perimesencephalic cisterns and related segments of the posterior cerebral artery: comparison using a novel application of image guidance. Neurosurgery. 2004;54(6):1313–27.
6. Rhoton AL Jr, Fujii K, Fradd B. Microsurgical anatomy of the anterior choroidal artery. Surg Neurol. 1979;12(2):171–87.
7. Goldberg HI. The anterior choroidal artery. In: Newton TH, Potts DG, editors. Radiology of the skull and brain, vol. 2, Book 2. St. Louis: CV Mosby; 1974. p. 1628–58.
8. Gibo H, Carver CC, Rhoton AL Jr. Microsurgical anatomy of the middle cerebral artery. J Neurosurg. 1981;54(2):151–69.
9. Rhoton AL Jr. The supratentorial arteries. Neurosurgery. 2002;51(4 Suppl):S53–S120.
10. Tanriover N, Kawashima M, Rhoton AL Jr, Ulm AJ, Mericle RA. Microsurgical anatomy of the early branches of the middle cerebral artery: morphometric analysis and classification with angiographic correlation. J Neurosurg. 2003;98(6):1277–90.
11. Gibo H, Rhoton AL Jr. Microsurgical anatomy of the supraclinoid portion of the internal carotid artery. J Neurosurg. 1981;55(4):560–74.
12. Zeal AA, Rhoton AL Jr. Microsurgical anatomy of the posterior cerebral artery. J Neurosurg. 1978;48(4):534–59.
13. Seoane ER, Tedeschi H, de Oliveira E, Siqueira MG, Calderón GA, Rhoton AL Jr. Management strategies for posterior cerebral artery aneurysms: a proposed new surgical classification. Acta Neurochir. 1997;139(4):325–31.
14. Xu Y, Mohyeldin A, Doniz-Gonzalez A, et al. Microsurgical anatomy of the lateral posterior choroidal artery: implications for intraventricular surgery involving the choroid plexus. J Neurosurg. 2021;135(5):1534–49.
15. Huang YP, Wolf BS. The basal cerebral vein and its tributaries. In: Newton TH, Potts DG, editors. Radiology of the skull and brain, vol. 2, Book 3. St. Louis: CV Mosby; 1971. p. 2111–54.
16. Wolf BS, Huang YP, Newton CM. The lateral anastomotic mesencephalic vein and other variants in drainage of the basal vein. AJR Am J Roentgenol. 1963;89:411.
17. Rhoton AL Jr. The cerebral veins. Neurosurgery. 2002;51(4 Suppl):S159–205.
18. Chaynes P. Microsurgical anatomy of the great cerebral vein of Galen and its tributaries. J Neurosurg. 2003;99(6):1028–38.
19. Kucukyuruk B, Richardson RM, Wen HT, et al. Microsurgical anatomy of the temporal lobe and its implications on temporal lobe epilepsy surgery. Epilepsy Res Treat. 2012;2012:769825.
20. Campero A, Tróccoli G, Martins C, et al. Microsurgical approaches to the medial temporal region: an anatomical study. Neurosurgery. 2006;59(suppl_4):ONS-279-ONS-308.
21. Shah A, Lunawat A, Jhawar SS, Goel A, Goel A. White finer correlates of amygdalohippocampectomy through the middle temporal gyrus approach. World Neurosurg 2022;157:e156–e165.
22. Weil AG, Middleton AL, Niazi TN, et al. The supracerebellar-transtentorial approach to posteromedial temporal lesions in children with refractory epilepsy. J Neurosurg Pediatr. 2015;15(1):45–54.

Anatomy of Basal Forebrain and Orbital Cortical Approach to Caudate Head Tumors

Atul Goel, Jash Doshi, and Abhidha Shah

1 Introduction

The mediobasal frontal region harbors prominent gray matter masses and white matter connections and is an important component of the limbic system. The mediobasal frontal cortical areas, the olfactory cortical areas and the ventral- striato-pallidal region form a continuum on the ventral surface of the cerebral hemisphere connecting the medial part of the temporal brain to the basal and medial parts of the frontal brain [1]. The caudate nucleus is a prominent gray matter nucleus in the basal portion of the frontal lobe. Tumors arising in or around the vicinity of the caudate nucleus form a rare subset of lesions. The location of the caudate nucleus in the central core of the brain amidst a myriad of critical white fiber bundles makes it a difficult location to access surgically. Lesions located in the caudate head are challenging to access surgically owing to their location in the depth of the cerebral hemisphere and proximity to critical eloquent gray matter [2]. Traditional avenues for access to the caudate head include transcortical approaches through the superior or middle frontal gyrus and the transcallosal approaches [3–6]. The ipsilateral and contralateral interhemispheric approaches have been abundantly evaluated to access this region [3]. The contralateral interhemispheric approach has been recommended to approach lesions in the lateral portion of the frontal horn and caudate nucleus owing to the possibility of a direct angle of surgical trajectory [3, 7]. However all these approaches have the disadvantage of a long trajectory, possibility of venous injury due to cerebral retraction, a steep angle of working, an incision through the corpus callosum and ependyma of the lateral ventricle and potential of injury to the pericallosal arteries [2]. Furthermore the transcallosal approaches have the disadvantage of difficulty or inability to access the depths of a larger lesion. For selected larger size lesions in the region, long surgical trajectory, narrow operative corridor and more critical neural fibers particularly the genu of the internal capsule located in the depth of the surgical field makes interhemispheric approach cumbersome and sometimes dangerous [1, 3]. Transcortical cere-

Dr Jash Doshi was involved in this chapter preparation

A. Goel (✉)
Department of Neurosurgery, Lilavati Hospital and Research Center, Bandra, Mumbai, India

Department of Neurosurgery, R.N. Cooper Hospital, Mumbai, India

Bombay Hospital Institute of Medical Sciences and Research Center, Mumbai, India

J. Doshi · A. Shah
Department of Neurosurgery, K.E.M Hospital and Seth, G.S. Medical College, Mumbai, India

A. Shah et al. (eds.), *Functional Anatomy of the Brain: A View from the Surgeon's Eye*,
https://doi.org/10.1007/978-981-99-3412-6_15

bral convexity approach is a possible alternative but the location of critical association fibers in the surgical trajectory can make access to the lesion limited or impossible [4]. Goel et al earlier described an orbital cortical or basal frontal approach through the medial orbital surface of the frontal lobe to access lesions located in and lateral to the frontal horn of the lateral ventricle [2, 8]. In this chapter we describe the anatomy of the basal forebrain, subcallosal region and the relevant white fiber tracts that are in the vicinity of the surgical track when our orbital cortical approach is adopted. We have divided basal orbital approach into two avenues namely the medial and lateral orbital gyri approaches [2].

2 Anatomical Correlates

2.1 Cortical Anatomy

The orbital surface of the frontal lobe is divided by the deep olfactory sulcus into the medial gyrus rectus and the lateral orbital gyri (Fig. 1). The olfactory bulb and tract are located in the olfactory sulcus. The orbital gyri make up most of the basal portion of the frontal lobe. The H shaped cruciform sulcus of Rolando divides the orbital gyri into anterior, posterior, medial and lateral orbital gyri [2, 9, 10]. The posterior orbital gyrus is situated just anterior to the anterior perforated substance and connects medially to the medial orbital gyrus and this connection is named as the posteromedial orbital lobule. The medial, anterior and lateral orbital gyri are connected to the superior, middle and inferior frontal gyri along the frontal pole.

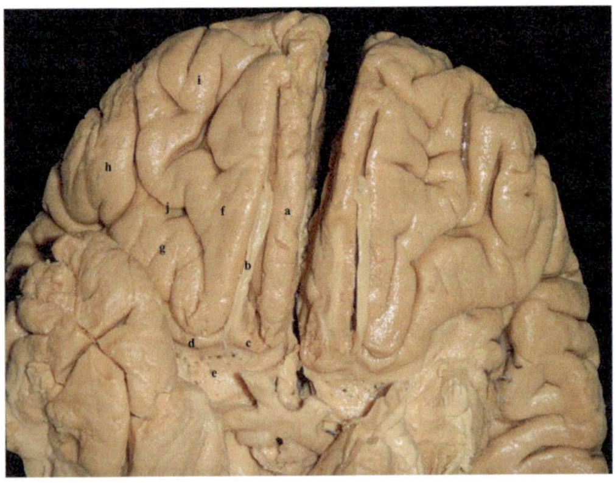

Fig. 1 Inferior view of the brain showing the basal/orbital surface of the frontal lobes. a. Gyrus rectus, b. Olfactory tract, c. Medial olfactory stria, d. Lateral olfactory stria, e. Anterior perforated substance, f. Medial orbital gyrus, g. Posterior orbital gyrus, h. Lateral orbital gyrus, i. Anterior orbital gyrus, j. Cruciform sulcus of Rolando

The olfactory tract runs in the olfactory sulcus on the orbital surface of the frontal lobe. Further posteriorly the olfactory tract divides into the medial and lateral olfactory striae. The medial olfactory stria becomes continuous with the subcallosal and the paraterminal gyrus on the medial surface of the frontal brain. The lateral olfactory stria passes along the lateral and anterior margin of the anterior perforated substance hugging the posterior margin of the posterior orbital gyri to reach the piriform region in the temporal lobe [2, 8–10].

2.2 Subcortical Anatomy

2.2.1 Deep Gray Matter

Basal Forebrain

The basal forebrain includes the septal area, the olfactory tubercle, parts of the amygdala and the substantia innominata [8–10]. It extends from the olfactory tubercle rostrally to the hypothalamus caudally (Figs. 2, 3, and 4). It lies over the anterior perforated substance and does not have any defined boundaries [1]. The anterior perforated substance is located just above the bifurcation of the internal carotid artery and forms the roof of the space,

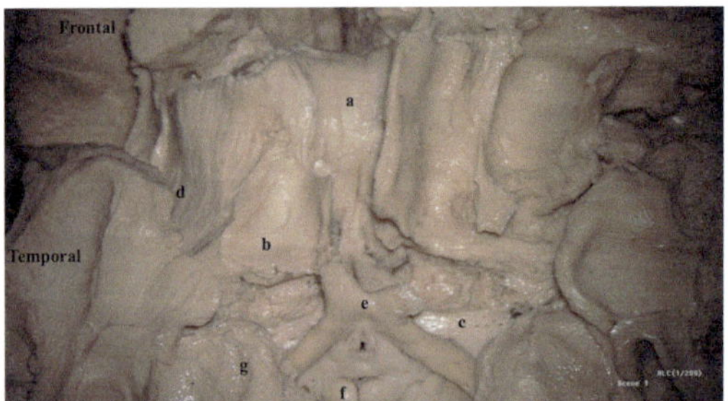

Fig. 2 Inferior view of the fiber dissected specimen of the orbital surface of the frontal lobe. a. Genu of corpus callosum, b.Fibres of the uncinate fasciculus, c. Anterior perforated substance, d. Fibers of the IFOF, e. Optic chiasm, f. Mammillary body, g. Medial temporal brain

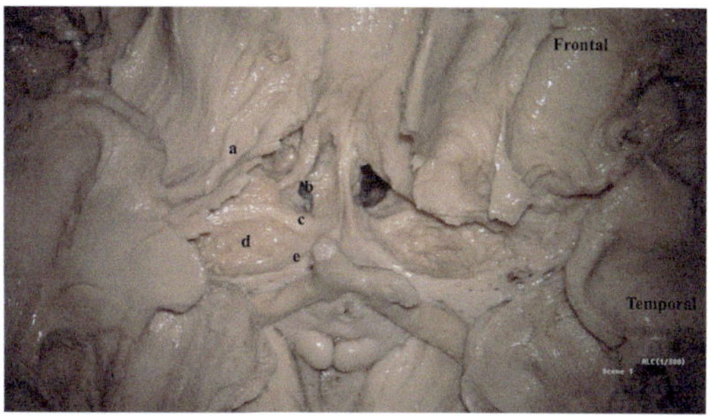

Fig. 3 Deeper dissection of the basal forebrain. a. Fibers of the uncinate fasciculus, b. Frontal horn, c. Anterior commissure, d. Substantia innominata, e. Diagonal band of Broca

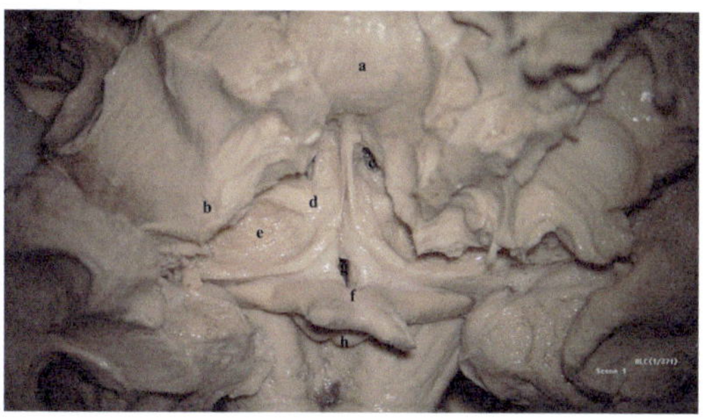

Fig. 4 View of the basal brain from an anterior perspective. a. Genus of corpus callosum, b. fibers of the uncinate fasciculus, c. Frontal horn, d. Anterior commissure, e. Substantia innominata or ventral striatum, f. Optic chiasm, g. Third ventricle, h. Mammillary bodies

which carries the distal portion of the internal carotid artery and its bifurcation into the anterior and middle cerebral artery. The anterior perforated substance acquires its name due to the multiple perforations seen in the area that are caused by the lenticulostriate arteries which are the perforating branches of these vessels. The boundaries of the anterior perforated substance are the medial and lateral olfactory striae anteriorly, the interhemispheric fissure medially, the optic tracts posteriorly and the limen insulae and the uncus laterally [1]. The anterior perforated substance is closely related to the anterior limb of the internal capsule. Laterally it is continuous with the temporal stem and the limen insulae where it extends along the prepiriform area. Posteriorly it is continuous with the semilumar gyrus in the periamygdaloid area. Medially it is continuous with the septal area and the hypothalamus. The most posterior portion of the anterior perforated substance just in front of the optic tract traverses the diagonal band of Broca, which is made up of amygdalo- septal -fugal fibers connecting the amygdala to the mediobasal forebrain [1].

The term substantia innominata is often used synonymously with the term basal nucleus. However the correct terminology for this region would be the ventral-striatal- pallidal region [1]. It is made up of the nucleus accumbens, which is the basal connection of the most anterior and inferior portion of the caudate head and the putamen (ventral striatum), the ventral portion of the globus pallidus, the nucleus basalis of Meynert and the amygdalofugal fibers of the ansa peduncularis and the ventral portion of the amygdala. It lies deep to the ventromedial frontal fibers of the uncinate fasciculus between the anterior perforated substance and the anterior commissure. (Figs. 2, 3, and 4) Superiorly it is related to the anterior limb of the internal capsule, medially it is continuous with the septal region and hypothalamus and laterally it merges with the temporal stem.

2.2.2 Caudate Nucleus

The caudate nucleus is a C shaped structure that is closely related to the lateral ventricle. (Fig. 5) It consists of three parts: the head, body and tail. The head of the caudate nucleus forms the lateral wall of the frontal horn and is the part most commonly affected by tumors [2]. The body of the caudate nucleus forms the lateral wall of the body of the lateral ventricle and the tail forms a part of the lateral wall of the atrium and of the roof of the temporal horn of the lateral ventricle. The tail of the caudate nucleus merges with the amygdala. The caudate nucleus wraps around the lateral edge of the thalamus and is separated from it by the stria terminalis. The nucleus accumbens is located below and medial to the head of the caudate nucleus. There are direct connections between all parts of the caudate nucleus and the lentiform nucleus and these are known as the transcapsular gray matter [11]. The head of the caudate nucleus is related to the frontal horn medially. The white fibers in close proximity to the head of the caudate nucleus are the forceps minor anteriorly and medially, the inferior fronto-occipital fas-

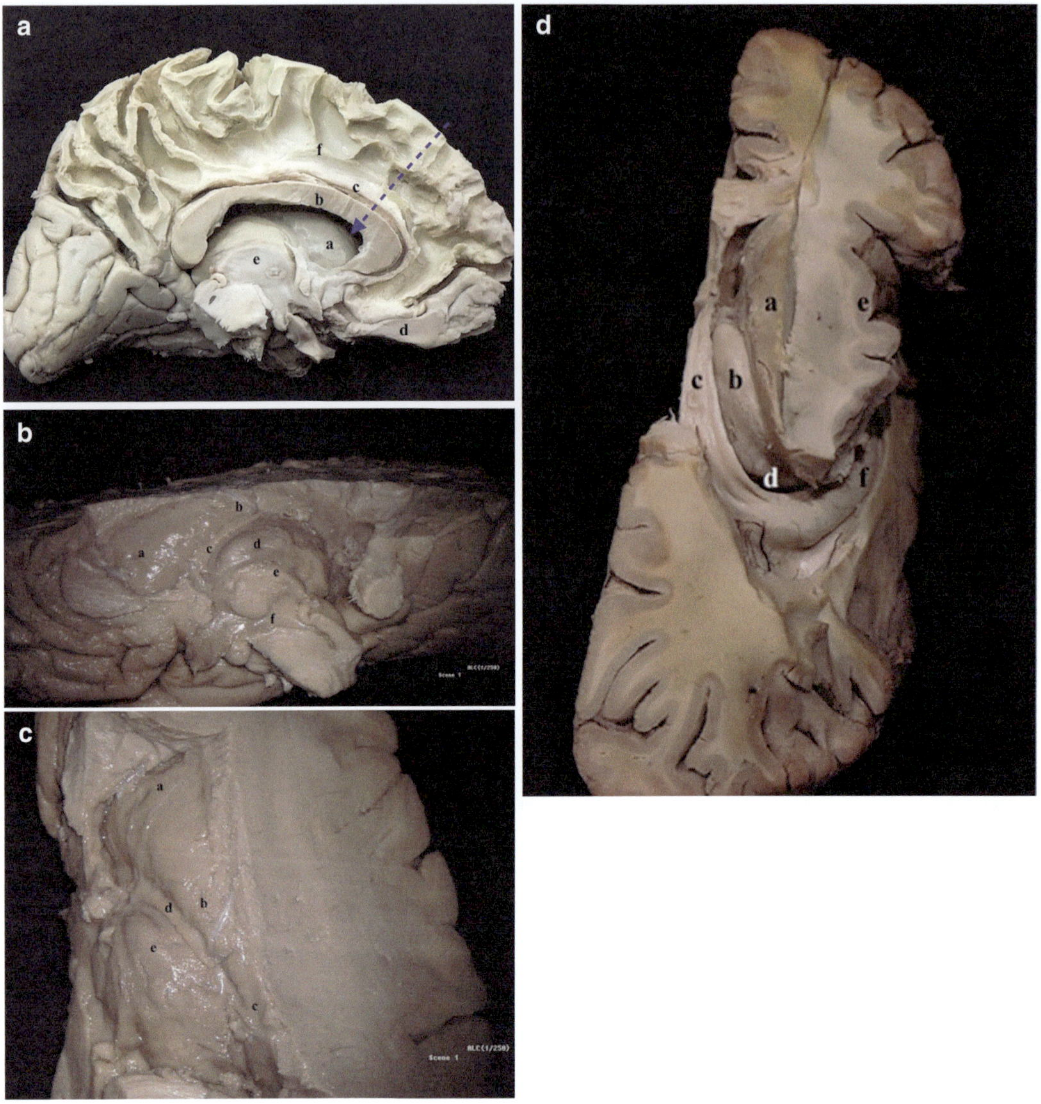

Fig. 5 Medial view of the cerebral hemisphere. (**a**) Medial view of the cerberal hemisphere with fiber dissection performed in the frontal and parietal lobes. a. Caudate nucleus, b. Corpus callosum, c. Cingulum, d. Gyrus rectus, e. Thalamus, f. Medial portion of the superior longitudinal fasciculus. Purple arrow indicates the trajectory of the transcallosal approach to the caudate nucleus. (**b**) Medial view of the cerebral hemisphere with the corpus callosum cut to show the caudate nucleus. a. Head of the caudate nucleus, b. Body of the caudate nucleus, c. Stria terminalis, d. Thalamus, e. Stria medullaris thalami, f. Third ventricle. (**c**) Supeior view of the caudate nucleus after cutting of the corpus callosum. a. Head of the caudate nucleus, b. Body of the caudate nucleus, c. Tail of the caudate nucleus, d. Stria terminalis, e. Thalamus. (**d**) Superior view of the lateral ventricle. a. Caudate nucleus as it runs from the frontal to the temporal horn, b. Thalamus, c. Fornix, d. Choroid fissure, e. Insula, f. Hippocampus

ciculus laterally and the uncinate fasciculus postero-inferiorly and laterally (Figs. 2, 3, and 4). The caudate head lies at an average distance of 34 mm (range 32–36 mm) from the tip of the frontal pole, 24 mm (range 23–26 mm) from the cortex of the medial basal orbital surface of the frontal lobe, 35 mm (range 33–37 mm) from the cortex of the basal lateral orbital surface and 37 mm (range 34–41 mm) from the superior surface of the frontal lobe. The mean depth of the caudate nucleus when approached from the contralateral interhemispheric approach was 41 mm (range 39–45 mm). The shortest trajectory to approach the caudate head is when it is approached from the medial or lateral basal orbital surface.

2.3 White Fibers

The first fibers that are visualized after removing the superficial gray matter of the orbital surface of the frontal brain are the short arcuate fibers. These are short association fibers that interconnect neighboring gyri. Removing the short association fibers in the region of the medial gyrus rectus will reveal the anteriorly curving fibers of the forceps minor [2]. Inferior and lateral to the fibers of the forceps minor are association fiber bundles that connect the orbitofrontal gyri to the supero-medial portion of the thalamus [11]. These fibers run lateral to the gyrus rectus, inferior to the frontal horn and further posteriorly run lateral to the caudate nucleus. Deeper dissection in the region of the medial orbital gyrus will expose the frontal horn. The caudate head lies lateral to the frontal horn and roughly corresponded to the transverse limb of the H shaped cruciform sulcus [2]. Fiber bundles running lateral to the caudate connect the orbitofrontal gyri to the midbrain via the crus cerebri. The medial portion of the orbitofrontal region connects to the thalamus and the brainstem. The lateral and posterior portion of the orbitofrontal gyri mainly connects with the ipsilateral temporal, parietal and occipital cortices by two major association bundles namely the inferior occipitofrontal fasciculus and the uncinate fasciculus.

2.3.1 Inferior Occipitofrontal Fasciculus (IFOF)

The exact frontal terminations of the IFOF have not been clearly defined in literature [12, 13]. Our dissection studies identified fibers originating from the lateral and posterior orbital gyri that joined the fibers from the middle and inferior frontal gyri at a level lateral and just posterior to

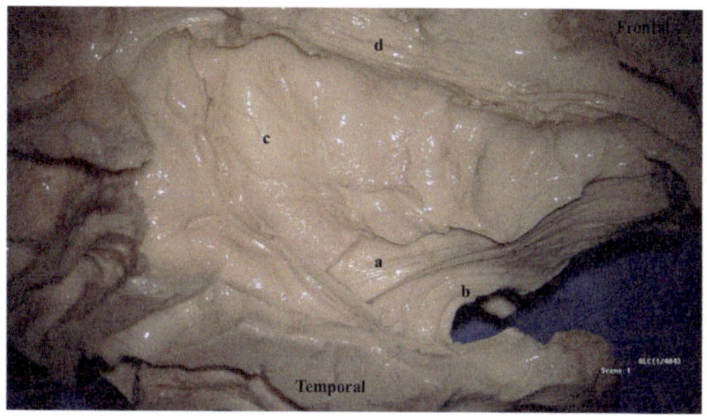

Fig. 6 Lateral view of a fiber dissected brain specimen showing the fibers of the IFOF and the UF as they run between the frontal and temporal lobes in the region of the limen insula. a. Fibers of the IFOF, b. Fibers of the UF, c. Extreme Capsule, d. Arcuate fasciculus

the caudate nucleus to form the IFOF [2, 8–10]. (Figs. 2, 3, 4, and 6) This region was found to be at a depth of 26 mm (23–27 mm) from the basal surface of the frontal brain. All the fibers aggregate and pass lateral and anterior to the corona radiata fibers and medial and inferior to the superior longitudinal and arcuate fasciculus towards the temporal lobe as a constituent of the temporal stem. In the region of the insula the inferior fronto-occipital fasciculus and the uncinate fasciculus form the ventral portion of the external capsule with the inferior fronto-occipital fasciculus lying superiorly and the uncinte fasciculus lying inferiorly. After exiting the temporal stem the majority of the IFOF passes through the superior and middle temporal gyri to reach the parietal and occipital gyri. Some fibers of the IFOF turn anteriorly along with the uncinate fasciculus to end in the superior and middle temporal gyri [2]. The fibers of the IFOF running posteriorly form a component of the sagittal stratum and lie lateral to the optic radiations. Fibers running medially in the IFOF project to the parietal lobe and fibers running laterally project to the occipital lobe.

2.3.2 Uncinate Fasciculus

The uncinate fasciculus is a hook shaped tract running in the temporal stem connecting the orbitofrontal and temporal regions (Figs. 2, 3, 4, and 6) The uncinate fasciculus originates in the orbitofrontal gyri runs laterally deep to the IFOF, anteroinferior to it in the temporal stem and ends in the superior and middle temporal gyri. The dorsolateral portion of the uncinate fasciculus connects the temporal pole to the lateral orbitofrontal gyri and its ventromedial portion connects the temporal pole to the medial orbitofrontal cortex and the septal area [2]. (Figs. 2, 3, and 4) The dorsolateral portion of the uncinate fasciculus runs together with the frontal portion of the IFOF and could not be delineated from it [2]. The ventromedial portion of the UF was distinctly identified on the basal surface of the frontal brain coursing just anterior to the anterior perforated substance. It covers the inferior and medial sides of the nucleus accumbens and reaches the region below the genu of the corpus callosum. This portion of the UF was found to be 62 mm (range 59–64 mm) from the frontal pole and 50 mm (48–52 mm) from the lateral orbital surface.

2.3.3 Anterior Commissure

The anterior commissure is a horizontal bow shaped fasciculus, which connects the frontal and temporal lobes of one hemisphere to the other. It runs in a canal of gray matter (canal of Gratiolet) parallel and deep to the uncinate fasciculus and the occipitofrontal fasciculus [2, 8–10]. It consists of a frontal and temporal extension. The anterior commissure was found to run roughly 3–7 mm (mean 5 mm) behind and parallel to the ventromedial portion of the UF. The substantia inominata lies in the area between the UF and the anterior commissure deep to the anterior perforated substance. (Figs. 2, 3, 4, 7, and 8) The frontal extension of the anterior commissure was found to lie just beneath the junction of the medial gyrus rectus with the posterior orbital gyrus. The temporal portion of the anterior commissure runs deep to the lentiform nucleus crosses the internal capsule to end as part of the sagittal stratum running beneath the superior and middle temporal gyri to the occipital brain. The distance of the anterior commissure from the frontal pole was 72 mm (range - 68–75 mm) and from the lateral orbital surface it was 68 mm (range - 63–70 mm).

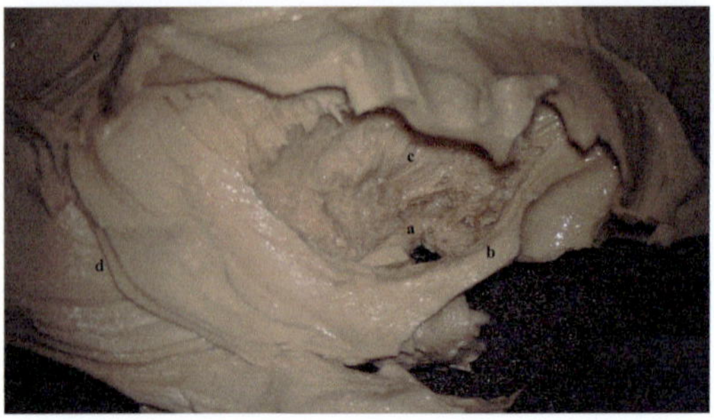

Fig. 7 Lateral view of the cerebral hemisphere. The insular region has been dissected to show the course of the temporal portion of the anterior commissure. The continuation of its fibers into the sagittal stratum can be seen. a. anterior commissure b. Fibers of the IFOF c. internal capsule d. sagittal stratum and optic radiations, e. Arcuate fasciculus

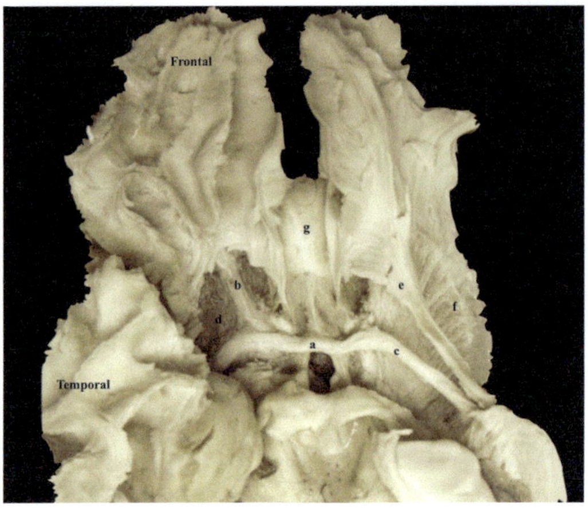

Fig. 8 Deeper dissection of the inferior surface of the basal forebrain. The anterior commissure can be seen. It is a compact fiber bundle which crosses the midline. Laterally it splits into two, a smaller anterior or olfactory portion and the larger posterior portion. a. anterior commissure b. frontal portion of anterior commissure c. temporal portion of anterior commissure d. Ventral striatum, e. Fibers of the IFOF, f. Internal capsule, g. Corpus callosum

2.3.4 Forceps Minor

The forceps minor consists of the anteriorly curving fibers of the genu of the corpus callosum into the medial frontal lobe. The fibers pass superior and anterior to the frontal horn forming the anterior wall of the frontal horn. The distance of the forceps minor from the frontal pole was found to be an average of 37 mm (range - 35–38 mm) [2].

2.4 Surgical Correlates

The principle surgical landmarks related to the approach have been described in our earlier publication [2, 8] and are summarized here. The patients were placed in a supine position with the head extended and turned to the contralateral side. A hairline fronto–temporal skin incision was marked for a small fronto-temporal craniotomy. After opening the dura the frontal lobe was retracted off the orbital roof and the cerebrospinal fluid released from the Sylvian and optico-carotid cisterns. The anterior cerebral artery, the recurrent artery of Heubner and the olfactory tract were visualized and safeguarded. The Heubner's artery traverses posteriorly in the gyrus rectus and the orbitofrontal cortex and should be avoided and protected. It passes inferiorly and laterally to the origin of the lateral olfactory striae before reaching the anterior perforated substance. To have a larger access, the cortical incision and corticetomy can be widened circumferentially. Then depending on the projection of the tumor either a medial or a lateral cortical approach can be used to approach the tumor. (Fig. 9).

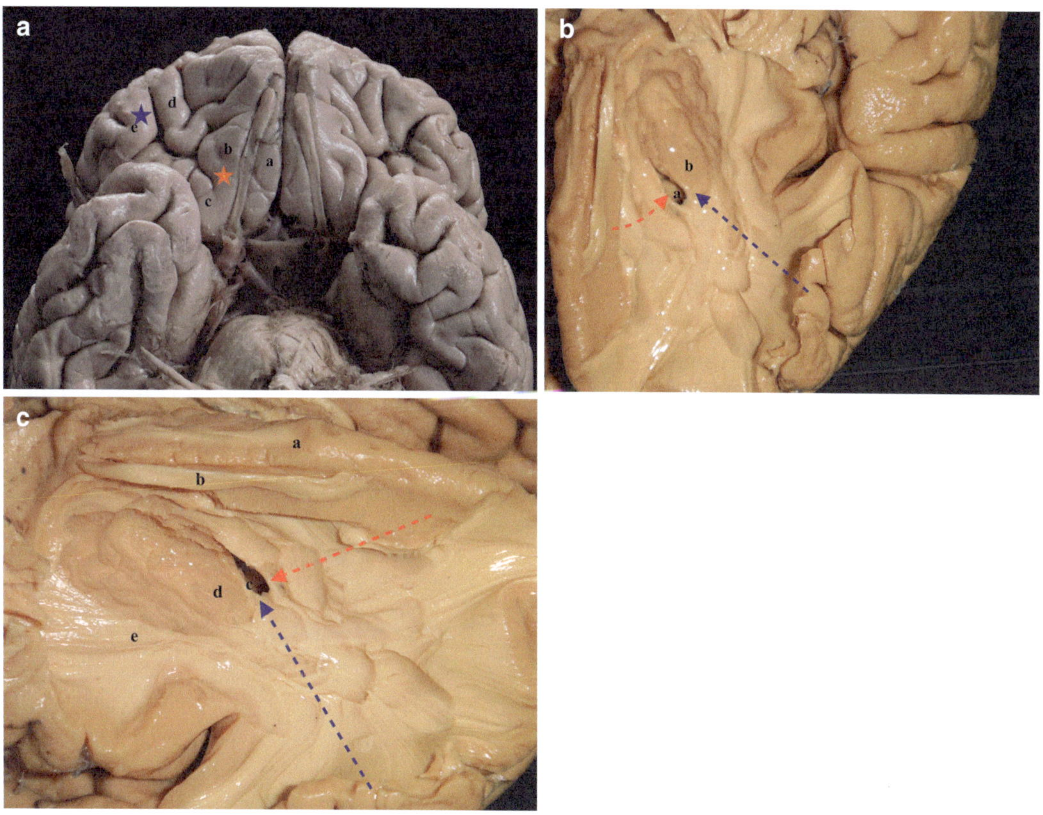

Fig. 9 Cadaveric dissection images showing the trajectory and relevant anatomy of the orbital cortical approach. (**a**) Inferior view of the brain showing the basal/orbital surface of the frontal lobes. a. Gyrus rectus, b. Medial orbital gyrus, c. Posterior orbital gyrus, d. Anterior orbital gyrus, e. Lateral orbital gyrus. The orange star indicates the point of corticectomy whilst using the medial orbital approach. The purple star indicates the point of corticectomy for the lateral orbital approach. (**b**) Inferior view of the brain with diisection to show the frontal horn and the caudate head. a. Frontal horn, b. Caudate head. Red arrow shows the trajectory of the medial orbital cortical approach. Purple arrow shows the trajectory of the lateral orbital cortical approach. (**c**) View from an inferior and lateral perspective of the orbital surface of the frontal lobe. a. Gyrus rectus, b. Olfactory tract, c. Frontal horn, d. Caudate head, e. Frontal fibers of the IFOF. Red arrow shows the trajectory of the medial orbital cortical approach. Purple arrow shows the trajectory of the lateral orbital cortical approach

2.4.1 Medial Orbital Cortical Approach

The medial orbital cortical approach to approach the head of the caudate nucleus involves a corticectomy through the medial orbital gyrus [2]. In a lateral perspective, the tip of the frontal horn is in line with the tip of the temporal pole. For the inferior approach to the frontal horn the incision on the medial orbital gyrus is made about 0.5 cm lateral to the olfactory tract [2]. The cortical incision was taken 0.5 cm lateral to the olfactory tract on the medial orbital gyrus just posterior to a line from the tip of the temporal pole. The corticectomy was widened circumferentially and the angle of attack was approximately 70 degrees from the horizontal and the direction of the approach was laterally and inferiorly. The incision travelled through the most lateral aspect of the frontal horn to enter the caudate head. This trajectory avoids injury to the bulk of the fibers of the forceps minor, which would be displaced medially and anteriorly.

2.4.2 Lateral Orbital Cortical Approach

The lateral orbital cortical approach involves a corticectomy in the lateral orbital gyrus. The corticectomy is made 1 cm lateral and superior to the transverse limb of the cruciform sulcus of Rolando [2]. The corticectomy is deepened in a 45-degree angle from the horizontal directed medially and inferiorly. The most inferior fibers of the IFOF lie at a depth of about 26 mm (range 23–27 mm) from the basal surface of the frontal lobe. Keeping this in mind damage to the inferior most fibers of the IFOF can be avoided. Wide opening of the Sylvian fissure, relaxation of the brain and lateral basal frontal exposure can be used effectively to obtain a suitable angulation for conduct of surgery. Use of neuronavigation can assist in directing the incision.

Depending on the site and extent of the lesion, either medial or lateral orbital gyral approach can be selected. Both the trajectories directly provid access to caudate head lesions without traversing any major white fiber bundle. The approach permits a radical excision of the caudate head tumor without violating any eloquent structure.

3 Conclusions

The transcallosal approaches have been preferred over transcortical approaches in general as the former avoids injury to the cortex and to important white fiber bundles. The major white fiber bundles on the inferior surface of the frontal lobe are the uncinate fasciculus, the frontal fibers of the anterior commissure, the inferior fronto-occipital fasciculus and medially the fibers of the forceps minor. The orbital cortical approach as described avoids injury to all these major fasciculi making it an ideal surgical route for these lesions located in midst of maze of critical neural structures. It also provides a short trajectory to reach the lesion and a comfortable working angle, avoids damage to major neurovascular structures and enhances the possibility of radical tumor resection (Figs. 10 and 11).

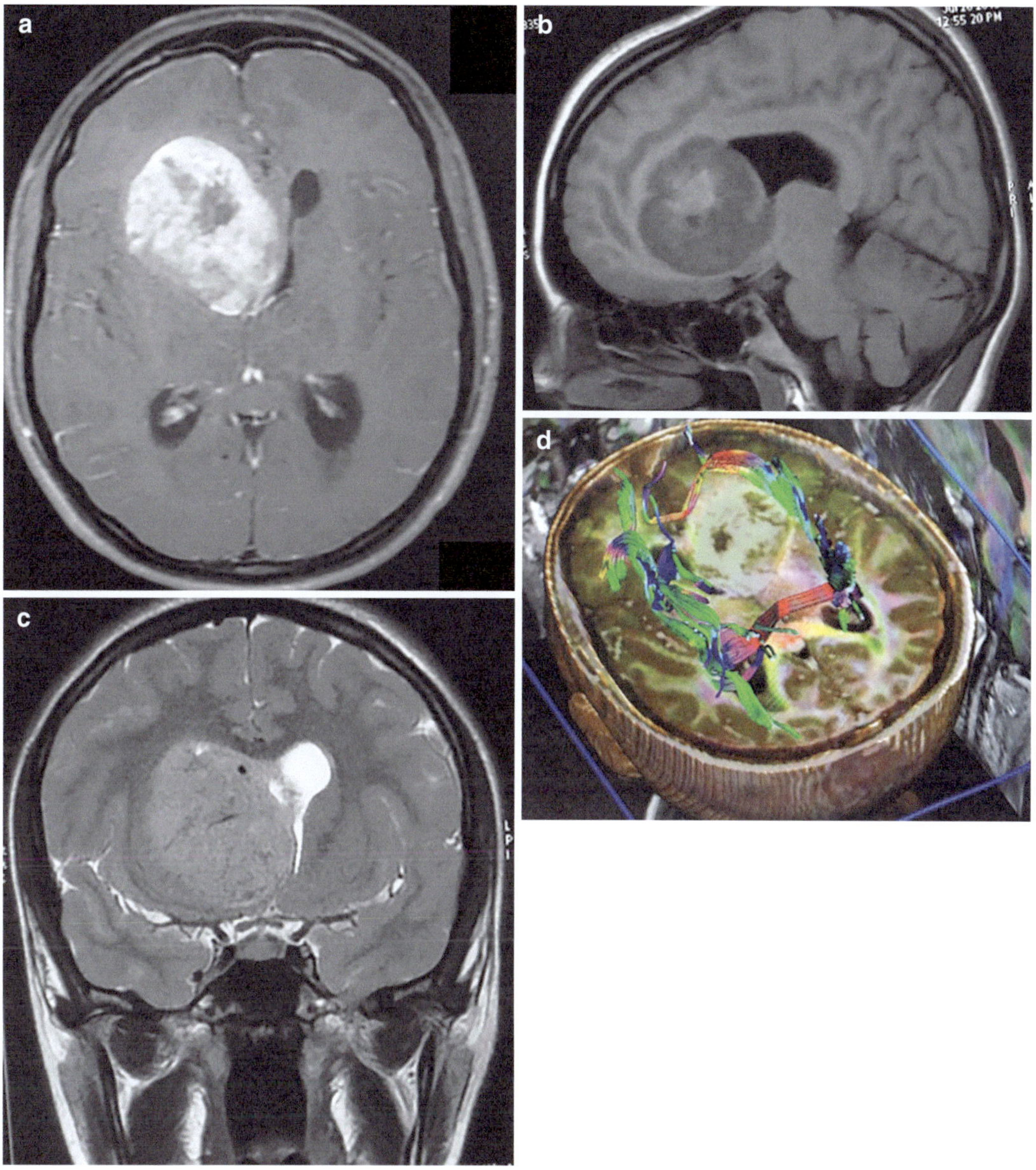

Fig. 10 Images of a 19 year old female patient. (**a**) Axial contrast enhanced MRI image showing a large enhancing tumor in the region of the right caudate head. (**b**) T1 weighted sagittal image showing the tumor. (**c**) T2 weighted coronal image showing the tumor in the region of the caudate head. (**d**) Tractography image showing the fibers of the forceps minor pushed anteriorly by the tumor and the fibers of the IFOF displaced laterally. (**e**) Tractography images showing the projection fibers pushed laterally by the tumor. (**f**) Post-operative image showing the tumor resection

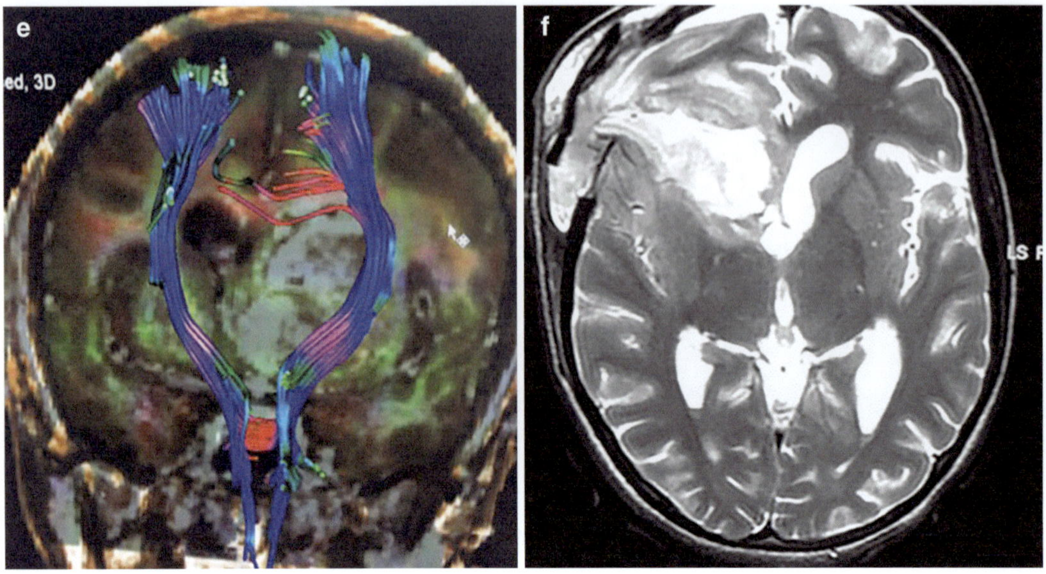

Fig. 10 (continued)

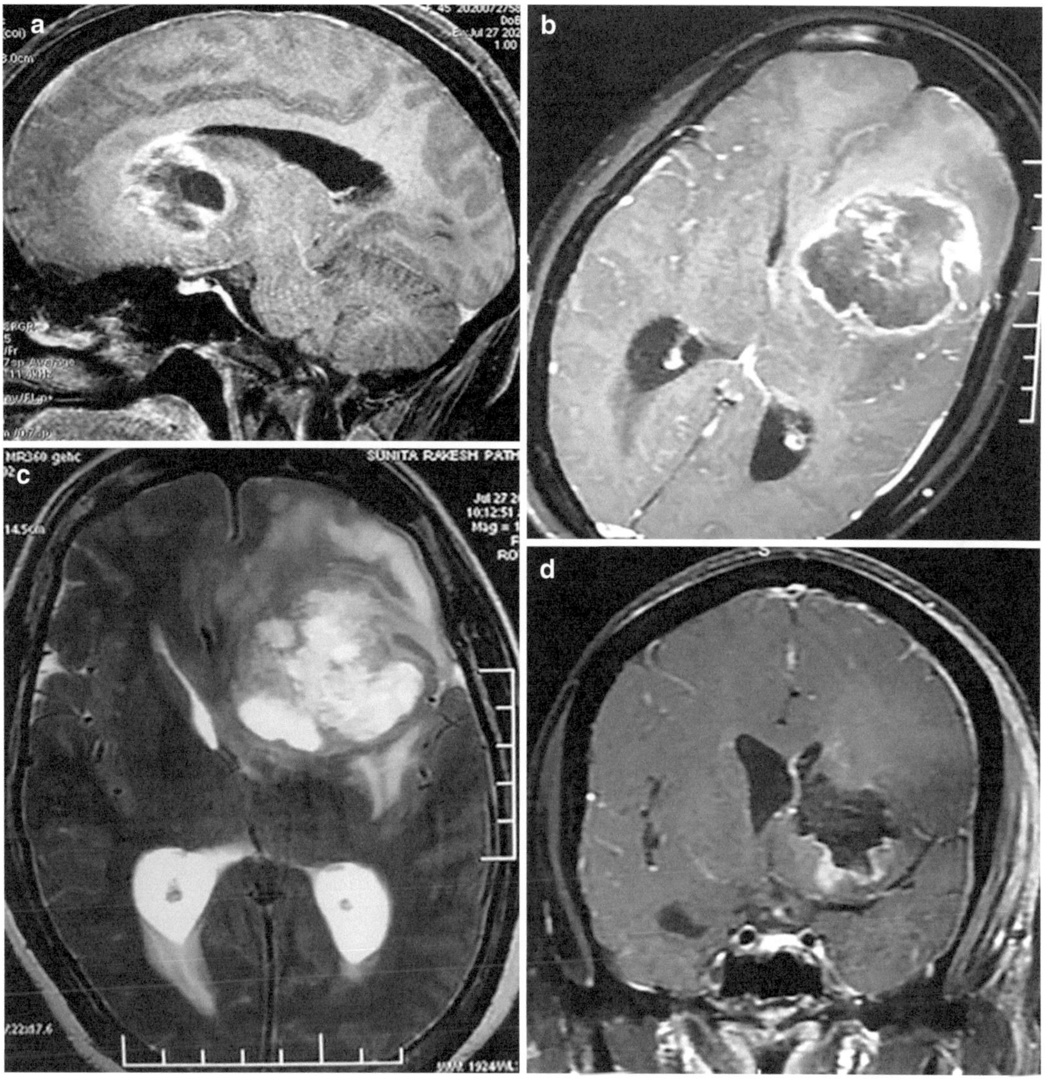

Fig. 11 Images of a 44 year old female. (**a**) T1 weighted sagittal image showing a large isointense tumor in the region of the caudate head. Few streaks of hyperintensity are visualized in the periphery of the tumor. (**b**) Axial contrast image showing the tumor. (**c**) T2 weighted axial image showing the tumor. (**d**) Post-operative contrast enhanced coronal image showing the tumor resection. (**e**) Post-operative contrast enhanced axial image showing the tumor resection

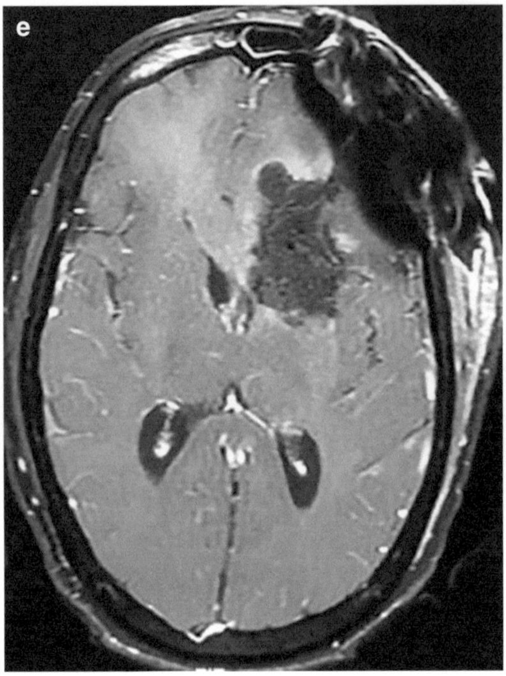

Fig. 11 (continued)

References

1. Ribas G. The cerebral architecture. In: Applied cranial-cerebral anatomy: brain architecture and anatomically oriented microneurosurgery. Cambridge: Cambridge University Press; 2018. p. 15–61. https://doi.org/10.1017/9781316661567.004.
2. Shah A, Dandpat S, Goel A. Anatomic correlates and surgical experience with Goel orbital cortical approach to caudate head Tumors. World Neurosurg. 2021 Apr;148:e674–9.
3. Belykh E, Yağmurlu K, Lei T, Safavi-Abbasi S, Oppenlander ME, Martirosyan NL, Byvaltsev VA, Spetzler RF, Nakaji P, Preul MC. Quantitative anatomical comparison of the ipsilateral and contralateral interhemispheric transcallosal approaches to the lateral ventricle. J Neurosurg. 2018;128(5):1492–502.
4. Ellenbogen RG. Transcortical surgery for lateral ventricular tumors. Neurosurg Focus. 2001;10(6):E2.
5. Milligan BD, Meyer FB. Morbidity of transcallosal and transcortical approaches to lesions in and around the lateral and third ventricles: a single-institution experience. Neurosurgery. 2010;67:1483–96.
6. Rhoton AL Jr. The lateral and third ventricles. Neurosurgery. 2002;51(4 Suppl):S207–71.
7. Briggs RG, Nix CE, Conner AK, Palejwala AH, Smitherman AD, Teo C, Sughrue ME. An awake contralateral, transcallosal approach for deep-seated gliomas of the basal ganglia. World Neurosurg. 2019;130:e880–7.
8. Goel A, Shah A, Ramdasi R, Patni N. Orbital cortical approach to lesions around the frontal horn of the lateral ventricle: indication and surgical parameters. Acta Neurochir. 2014;156(4):825–30.
9. Shah A, Goel A, Jhawar SS, Patil A, Rangnekar R, Goel A. Neural circuitry: architecture and function-a Fiber dissection study. World Neurosurg. 2019;125:e620–38.
10. Shah A, Jhawar SS, Goel A. Analysis of the anatomy of the Papez circuit and adjoining limbic system by fiber dissection techniques. J Clin Neurosci. 2012;19(2):289–98.
11. Çırak M, Yağmurlu K, Kearns KN, Ribas EC, Urgun K, Shaffrey ME, Kalani MYS. The caudate nucleus: its connections, surgical implications, and related complications. World Neurosurg. 2020;139:e428–38.
12. De Benedictis A, Sarubbo S, Duffau H. Subcortical surgical anatomy of the lateral frontal region: human white matter dissection and correlations with functional insights provided by intraoperative direct brain stimulation: laboratory investigation. J Neurosurg. 2012;117(6):1053–69.
13. Baydin S, Gungor A, Tanriover N, Baran O, Middlebrooks EH, Rhoton AL Jr. Fiber tracts of the medial and inferior surfaces of the cerebrum. World Neurosurg. 2017;98:34–49.

Pineal Region Anatomy with Clinical Correlates

Viktoria Sefcikova, Zabina Satar, Manni Waraich, and George Samandouras

1 Introduction

The deep-seated neural and vascular structures of the pineal region remain formidable obstacles during tumour resection, requiring an accurate understanding of their topography, relationships, and functional importance. Surgical approaches and anatomical complications are largely independent of the wide neuropathological range of tumours and lesions affecting the pineal region. Therefore, the current chapter will focus mainly on the embryology and surgically-relevant complex anatomy of the pineal region, without extending into the diagnostic work-up, neuroradiological investigations, or neuropathological aspects of pineal region tumours.

The term "pineal" is derived from the Latin term *pinealis* or *pinea*, referring to a pine cone [1]. Early reports of the pineal region were provided sequentially by the notable anatomists Herophilus, Galen, and Vesalius, each extensively elaborating on pineal topography and presumed imaginative functions. Later, French philosopher, mathematician, and scientist Descartes considered the pineal gland to be the *"seat of the soul"* [1–3].

2 Embryology

The rudimentary pineal gland (also known as the *pineal anlage*) is formed in weeks 8 to 9 of gestation from a midline evagination of the roof of the third ventricle, specifically at the junction between the diencephalon and mesencephalon (Fig. 1a, b) [4]. This evagination occurs immediately anterior to the already developing posterior commissure (i.e., *posterior commissure anlage*). Unlike the pineal anlage, which projects posteriorly, the posterior commissure anlage projects anteriorly and is larger in size (Fig. 1b). This size discrepancy enables the posterior commissure anlage to serve as the primary factor orienting the pineal gland [4].

A well-differentiated subcommissural organ (SCO) is also situated in the diencephalon-mesencephalon junction, posterior to the developing pineal gland. The SCO is a well-conserved

V. Sefcikova
University College London, Queen Square Institute of Neurology, London, UK

Faculty of Medicine, The University of Queensland, Brisbane, QLD, Australia

Z. Satar · G. Samandouras (✉)
University College London, Queen Square Institute of Neurology, London, UK

University College London Hospitals NHS Trust, Victor Horsley Department of Neurosurgery, The National Hospital for Neurology and Neurosurgery, London, UK

M. Waraich
Department of Neuroanaesthesia, The National Hospital for Neurology and Neurosurgery, London, UK

A. Shah et al. (eds.), *Functional Anatomy of the Brain: A View from the Surgeon's Eye*,
https://doi.org/10.1007/978-981-99-3412-6_16

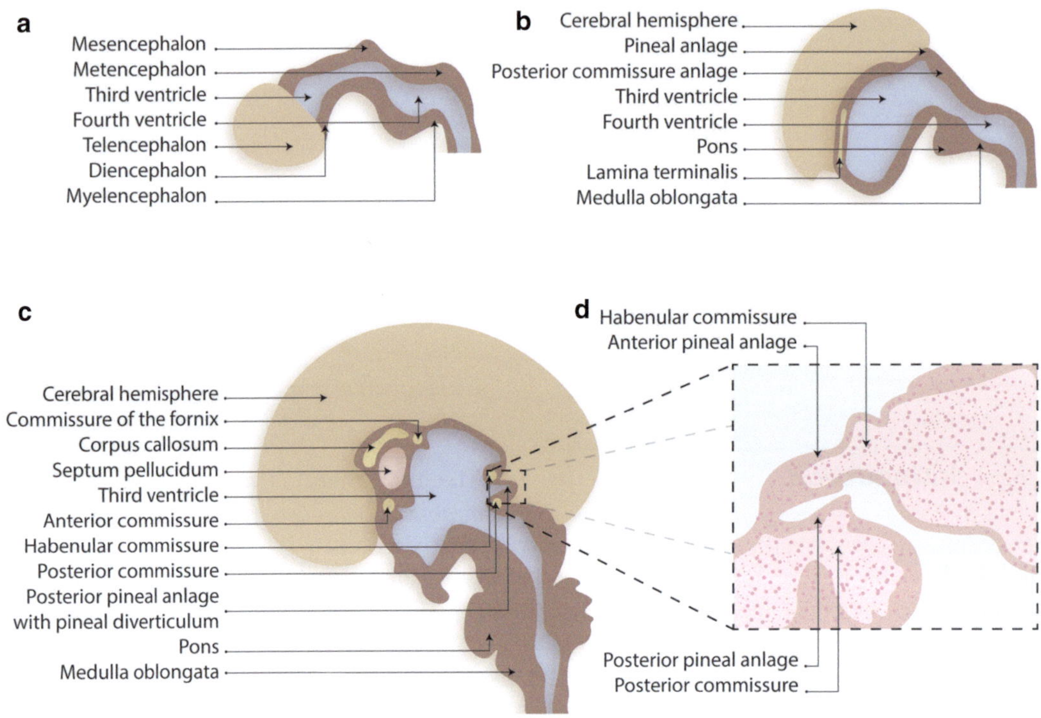

Fig. 1 Sagittal illustrations of the brain at gestational ages of approximately (**a**) 5 weeks; (**b**) 8 weeks; and (**c**) 3–4 months. (**d**) depicts an illustration of histology at approximately 3–4 months gestation

structure in vertebrates. In humans, it maintains high activity during embryogenesis, but becomes a vestigial structure in the adult form [5, 6]. The SCO secretes a glycoprotein called SCO-spondin into the embryonic cerebrospinal fluid (CSF), which proceeds in several directions, including rostrally, towards the developing pineal gland [6]. This secretion pathway promotes neurodifferentiation and proper morphogenesis of the pineal gland [6, 7].

During weeks 9 to 10 of gestation, the pineal anlage, which originally began its development as an extension of the third ventricle roof, further develops to become the pineal diverticulum—a bulging pouch continuous with the third ventricle [4]. Around the same time, the anlage of the habenular commissure also appears, situated above the pineal anlage, while the posterior commissure anlage continues to develop beneath the pineal anlage [4]. Throughout pineal embryogenesis, the cell types present in the pineal anlage change in accordance with the stage in development. At this stage, the pineal anlage is composed of undifferentiated neuroepithelium, consistent with the cells making up the roof plate of the diencephalon.

Between weeks 10 to 11 of gestation, the pineal anlage begins to develop a mass of cells which end up creating two segments, the anterior and posterior [4, 8]. At this stage, the anterior and posterior segments can be distinguished not only based on their connections, but also morphology—the anterior cells are arranged in a net-like manner, while the posterior cells are arranged in palisade layers [9]. Innervation also commences as a nerve bundle can be identified at the distal pole of the pineal anlage, which proceeds to branch out into smaller bundles with some fibres entering the pineal anlage via the habenular commissure [9].

The habenular commissure engages with the anterior pineal anlage after the anterior segment increases in size, and the posterior commissure embraces the posterior segment [9]. This proliferative stage occurs between months 2–4 of gestation (Fig. 1c, d) [4, 9]. At the start, neuroepithelial cells of the pineal anlage express transcription factor Pax6+ and intermediate filament protein vimentin [10]. Pax6+ is an important regulator of this proliferation with all diving cells that express the factor generating either pinealocytes (hormone-producing cells) or astrocytes. The latter cell type is formed when cells express vimentin in addition to Pax6+ [10]. This means Pax6+ contributes heavily to the formation of all pineal cell lineages except for microglia, which are of mesodermal origin [10–12].

While blood vessels are evident in week 11 of gestation, the bulk of the vascular network of the pineal gland develops alongside the proliferative stage [9]. Vascular connective tissue enters the pineal tissue through the periphery [4, 9]. The anterior pineal anlage is the first to obtain its vascular network from its connections to the habenular commissure, while the posterior pineal anlage lags behind and retains its palisade arrangement of cells [9]. Blood vessels have been reported in the posterior pineal anlage in specimens sampled 2-3 weeks after the vasculature in the anterior pineal anlage had already formed [9], although these changes possibly occurred soon.

Additional architectural changes take place in the pineal anlage during months 3–4 of gestation, including the pineal diverticulum dividing into the pineal recess and cavum [4]. It has been suggested this division of the pineal diverticulum is initiated by spongioblasts lining the diverticulum [4]. Spongioblasts can differentiate into either ependymal cells or neuroglia, and the differentiation into neuroglial precipitates the reduction of the diverticulum [4]. The innervation begins to take shape, as around this time the nervus conarii appears at the distal pole of the pineal anlage [9]. The nervus conarii carries sympathetic fibres, important in coordinating pineal gland activity with the circadian rhythm [13–15].

In subsequent months, the entrance of the nervus conarii becomes more diffuse [9]. The pineal gland develops three main forms of innervation. This includes the sympathetic and parasympathetic innervation, stemming from distant ganglia. Specifically, the superior cervical ganglia for the sympathetic, and the sphenopalatine and otic ganglia for the parasympathetic. The third mode of innervation is termed "central innervation" as fibres enter the pineal stalk from the brain, directly [15]. Central innervation includes connections from the medial paraventricular nucleus of the hypothalamus [15, 16], lateral hypothalamus [17], lateral geniculate body of the thalamus [18, 19], and dorsal raphe nucleus [20].

The last set of architectural changes occur in the subsequent months, including projection of the pineal recess, lined by ependymal cells, into the pineal gland anteriorly and enclosure by neuroglial peduncles [4]. The vascular connective tissue continues to develop, and a connective tissue sheath forms a septum separating the anterior and posterior pineal anlagen [4, 21].

From months 5 to 8 of gestation, the cellular structure of the pineal gland notably changes to include: pineal parenchyma, which is made up of (a) pinealocytes (approximately 95% of cells), and (b) interstitial cells including astrocytes and microglial/perivascular phagocytes (the remaining 5%) [4, 9, 10, 15]. The microglia have been implicated in synaptic pruning [10]. The generation of this final pineal parenchyma is evident by the presence of large cell size, pale nuclei, presence of 1–2 nucleolar masses, granular cytoplasm, and surrounding hyaline [4].

Following these changes, the most notable anatomical changes in-utero have completed. The pineal gland's architecture remains consistent in the post-natal period. However, the gland continues to grow in size until around age 2 and then increases in weight during puberty to reach adult dimensions [4, 22–24].

3 Vascular Anatomy

3.1 Arteries

The two major arterial trunks interfacing with the pineal gland region are the posterior cerebral (PCA) and superior cerebellar artery (SCA), which emerge deep to the incisural space and remain mostly visible from posterior surgical approaches, (Fig. 2). The PCA arises at the bifurcation of the basilar artery and divides topographically into four segments along its course through the inter-peduncular cistern (P1), ambient cistern (P2), quadrigeminal cistern (P3), and its last, post-cisternal segment (P4), where it usually bifurcates into the calcarine and parieto-occipital branches above the free tentorial edge [25].

The SCA, typically originating 2–2.5 mm from the basilar artery, extends below CN III, encircles the upper part of the pons and traverses almost parallel to the basal vein of Rosenthal and the PCA [25, 26]. The P2 and P3 segments of the PCA lie between the pulvinar of the thalami and the superior colliculi of the quadrigeminal plate, while the relatively small calibre loops of the SCA are often seen anterior and lateral to the pineal gland, as the midline lacks major arterial structures [27].

Additional arterial branches of the deep incisural space include the medial posterior choroidal artery (MPChA), the collicular (or quadrigeminal) artery, and the rare, posterior pericallosal artery [2]. The MPChA, typically a branch of the PCA, supplies the pineal body itself, in addition to the quadrigeminal plate and the choroidal plexus of the third ventricle, extending anteriorly until the foramen of Monro [1, 2]. The MPChA usually has three segments; the first lies in the ambient cistern, the second lies in the quadrigeminal cistern, and the third is in the cistern of the velum interpositum, reaching the choroid plexus of the third ventricle [2]. The MPChA can originate from either the P2 (~70%), P1 (~13.5%), P3 (~8%), or the cortical branches (~8%) [26]. Descriptions of variations in the anatomy of the MPChA include differences in quantity, with the artery being single in 54–62% of patients, and multiple in 38–46% [26, 28]. The collicular artery, also originating from the PCA, usually divides into superior and inferior branches over the quadrigeminal plate [2]. Finally, the posterior pericallosal artery, a rare terminal branch of the anterior cerebral artery (ACA), encircles the splenium and courses the superior aspect of the pineal gland before penetrating the cistern of the velum interpositum [2]. The lateral posterior choroidal artery (LPChA) has cisternal and plexal segments [29] and supplies the choroid plexus, the pulvinars and, more anteriorly, the parahippocampal gyri [30]. The LPChA can frequently be seen as a double trunk (superior and inferior) rather than a single trunk [28, 30].

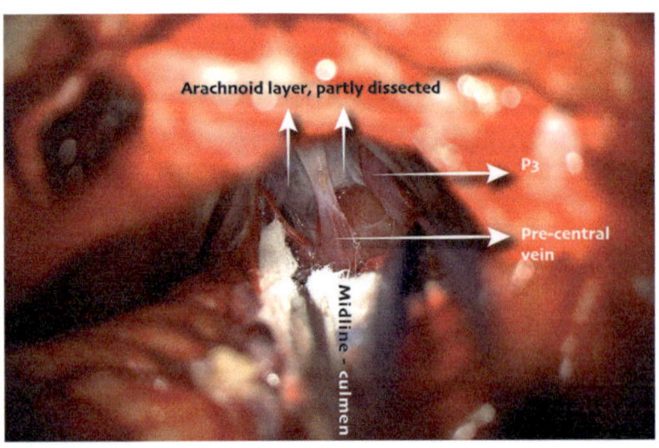

Fig. 2 Following division of the surface bridging veins in the sitting position, the culmen is depressed and the arachnoid is dissected exposing the third segment of the PCA, and the midline pre-central vein (or vein of the cerebello-mesencephalic fissure). Copyright © George Samandouras. Used with permission

3.2 Veins

The surgical venous anatomy of the pineal region is complex and a potential source of major complications, including stroke, if inadvertently injured or coagulated. In addition, they are covered by a thick, opaque arachnoidal layer, requiring meticulous dissection before the major venous branches are visualised [27]. The vein of Galen (VG) is placed most dorsally in the venous complex, with a length ranging from under 10 mm to up to 25 mm, and joins the inferior sagittal sinus to form the straight sinus (SS) at the level of the tentorial apex (Fig. 3) [26].

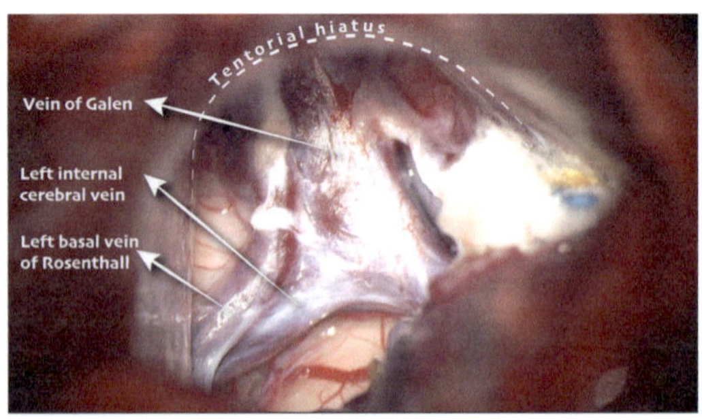

Fig. 3 Exposure of the vein of Galen and internal cerebral veins following dissection of the overlying opaque arachnoid membrane (same patient as in Fig. 9). Copyright © George Samandouras. Used with permission

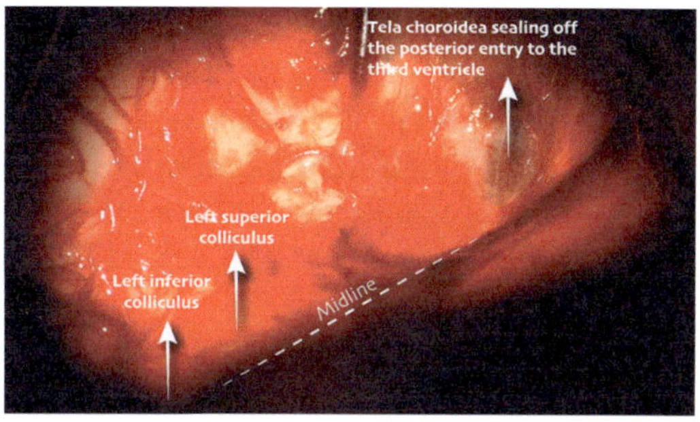

Fig. 4 Paramedian panoramic exposure of dorsal midbrain and the tela choroidea following tumour resection (same patient as in Fig. 9). Copyright © George Samandouras. Used with permission

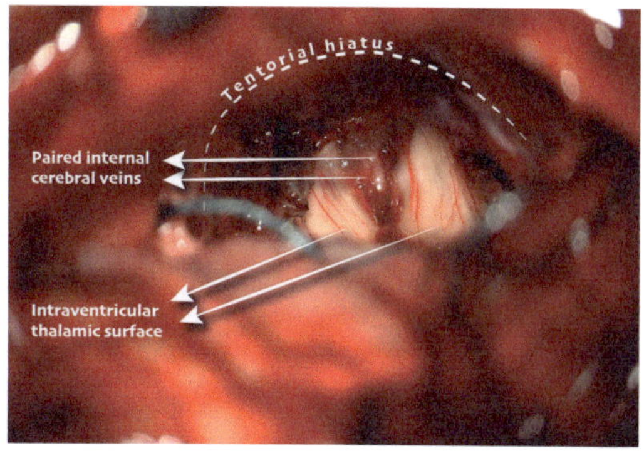

Fig. 5 Exposure of the intraventricular part of the internal cerebral veins' entry into the posterior third ventricle. This entry is sealed by the tela choroidea in normal subjects but tumours often invade this thin membrane. Copyright © George Samandouras. Used with permission

The paired basal veins of Rosenthal (BvRo) are placed more laterally and superficially at a paramedian plane, exiting the ambient cistern converging to the VG, while the internal cerebral veins (ICV) lie at a deeper plane, more medially and also underneath the BvRo and exit through the cistern of the velum interpositum, before converging to the VG (Figs. 3, 4, and 5) [27].

Both ICVs run parallel in close proximity to one another, and converge at the apex of the pineal gland, draining into the VG, which also receives flow from the midline precentral cerebellar vein, or the vein of the cerebellomesencephalic fissure, representing a key, midline venous landmark (Fig. 5).

4 Neural Anatomy

The pineal region lies posterior to the midbrain, in the area designated by Rhoton as the posterior incisural space (Fig. 4). The pineal body, normally measuring between 5-10 mm, is rarely, if ever, seen in pineal region tumours and is attached to the posterior part of the third ventricle with two commissures: the habenular commissure superiorly and the posterior commissure inferiorly. Superior to the pineal gland lies the splenium of the corpus callosum (Fig. 3), and more anterior and deeper, as seen from a posterior exposure, the curving crura of the fornices are encountered and, between them, the hippocampal commissure runs longitudinally. Laterally and above the superior colliculi lie the pulvinar of the thalami (Figs. 2 and 3).

The trochlear nerve is not visible through a midline approach as the prominent culmen obstructs its view, while a paramedian approach enables trochlear nerve visualisation underneath the free edge of the tentorium (Fig. 6).

Cortical parcels reaching the pineal region include the posterior parts of the parahippocampal gyri, located above the tentorium. These parcels, along with the dentate gyri, can be reached by dividing the tentorium. More superficially, cerebellar parcels related to the pineal region include the culmen of the vermis, specifically its uppermost segment, and the quadrangular lobules of the cerebellar hemispheres.

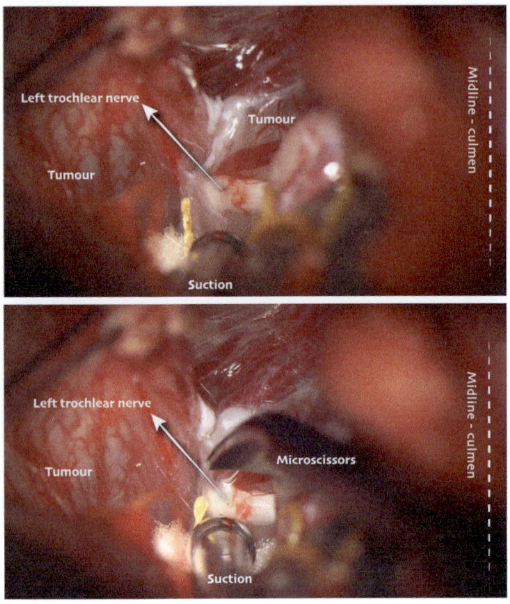

Fig. 6 Identification and dissection of the left trochlear nerve in a supra-cerebellar/infratentorial approach. Top figure, micro-dissection; bottom figure, complete exposure and preservation. Observe that the trochlear nerve is visible through a paramedian view as the bulky culmen (midline) obstructs its exposure (same patient as in Fig. 8). Copyright © George Samandouras. Used with permission

5 Anaesthetic Considerations for the Sitting Position

The sitting position is advantageous for pineal tumour removal as it provides good surgical visualisation. However, there are a few anaesthetic complications associated with the position, including venous air embolism (VAE) and haemodynamic instability [31, 32].

In the sitting position, a VAE occurs when air is entrained from an open vein by negative intravenous pressure due to the gravitational effect of a low central venous pressure (Fig. 7). Historically, the reported incidence of VAE has been around 40% [33]. However, with the use of transoesophageal echocardiography (TOE), the true incidence may be closer to 70% in posterior fossa surgery [34]. It is important to note that TOE captures a multitude of small VAEs, which may have limited clinical significance. Nonetheless, VAE always has the potential to become an arterial air embolism if a right-to-left pressure gradient exists in the heart, allowing the gas to travel from the venous to the arterial circulation through a patent foramen ovale (PFO). This is known as a paradoxical air embolism. Approximately thirty percent of the population have a PFO [35]. The physiological effects of any VAE depend on the volume of air entrained (Table 1).

In addition to the standard neuroanaesthetic preoperative assessment, a bubble echocardiography study is ordered to evaluate for the presence of a PFO, as it is a contraindication to the sitting position. For a bubble echocardiography study, either an apical 4 chamber view or a subcostal view on transthoracic echo (TTE) is used whilst a swift injection of agitated saline/air mixture is given into a peripheral vein. The patient is asked to perform a Valsalva manoeuvre to transiently raise the pressures in the right side of the heart, so if a PFO is present, the bubbles can be seen entering the left atrium. The appearance of micro-bubbles on the left side of the heart, after their appearance in the right heart chambers, is considered a positive bubble echocardiography study.

Intraoperative routine neuroanaesthesia monitoring includes ECG, non-invasive blood pressure (NIBP), continuous end-tidal CO_2 ($ETCO_2$), oxygen saturations, and spirometry (Fig. 7). Additional invasive monitoring includes arterial cannulation for continuous invasive pressure monitoring and arterial blood gases (ABG), as well as central venous cannulation (CVC; i.e., a 20–24 cm femoral line) which allows for the use of vasopressors in the event of cardiovascular collapse. An internal jugular line adds no further benefit in the event of a VAE, as aspiration of air from the CVC is not routinely performed, as the CVC tip needs to be positioned correctly at the superior vena cava-right atrial junction to facilitate aspiration of any air bubbles. The most important point about the anaesthetic technique, whether it be total intravenous or volatile anaesthesia, is to maintain cardiovascular stability by preventing hypotension, which might increase the likelihood of a VAE intraoperatively. Most anaesthetic monitors have pulse pressure variation (PPV) monitoring when invasive arterial monitoring is instituted. In the mechanically ventilated patient, there is a difference in pulse pressure between inspiration

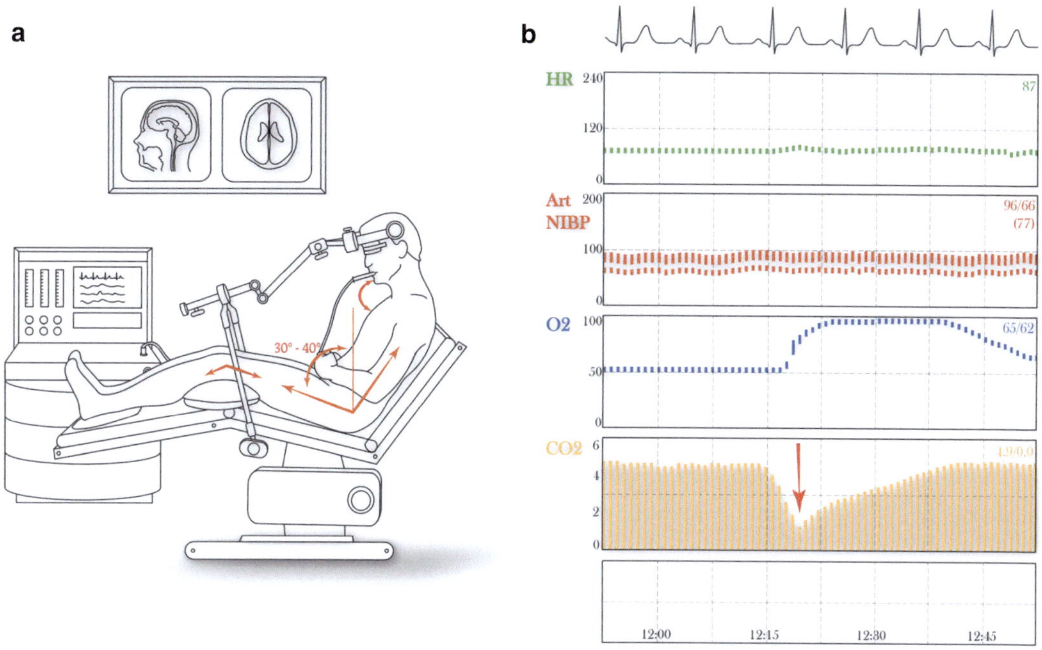

Fig. 7 (**a**) The patient begins supine on the operating table. The Mayfield skull clamp is applied, after which the patient is gradually (over the course of 1–2 minutes) brought to Fowler's (sitting) position. The superior back support is removed after the Mayfield skull clamp is secured anteriorly. The head is positioned to remain midline, and the neck is made to assume a flexed position 2–3 cm from the suprasternal/jugular notch. Elevation of the legs is achieved by placing pillows below the knees and flexing the hips. Hip flexion is limited to 45° maximum to (i) trade-off between raised intrathoracic pressure and reduced risk of VAE, attributable to reduced diaphragmatic movement; and (ii) limit sciatic nerve compression, femoral vein compression beneath the inguinal ligament, and venous pooling and hypotension. (**b**) An example of real anaesthetic monitor values during the sitting position. A $\geq$ 3 mmHg drop in ETCO2 (illustrated in yellow) is commonly seen and not indicative of VAE, and classification schemes commonly report visible air bubbles, as well as cardiovascular and haemodynamic instability [37]. HR, heart rate; Art NIBP, arterial non-invasive blood pressure. Reprinted from Sharma et al. [27]

Table 1 The volume of embolised air and the corresponding clinical signs under anaesthesia

Amount of air entrained	Likely signs under anaesthesia
Small <0.5 mL/kg	Sudden decrease in end tidal CO2 ($ETCO_2$) as seen on capnography with no change in ventilator settings Difference > 1.33 kPa (10 mmHg) $PaCO_2$-$ETCO_2$ on ABG done at the same time as capnography change Sudden desaturation Sudden decrease in SBP
Moderate 0.5–2.0 mL/kg	Unexplained persistent hypotension Increase in peak airway pressures secondary to pulmonary oedema Ischaemic ST changes seen on ECG
Large >2 mL/kg	Cardiovascular collapse Hypoxaemia
Fatal 3–4 mL/kg	Cardiac arrest

and expiration. This is usually about 10% in the euvolaemic patient. Volume optimisation of the patient by ensuring adequate preload, and hence PPV <10%, reduces the hypotension associated with the sitting position.

The patient is ventilated with an inspired concentration of oxygen of 40–50%. The higher partial pressure of inspired oxygen decreases the partial pressure of nitrogen in the blood. This produces a positive pressure gradient for the diffusion of nitrogen from any air bubbles entrained in the circulation thus accelerating bubble resorption.

The $ETCO_2$ will alert the anaesthetist to a suspected VAE by a sudden drop in $ETCO_2$ without a change in ventilation settings. An ABG done at this time will show a higher $PaCO_2$.The difference between arterial and $ETCO_2$ in a normal healthy adult is about 0.25–0.5 kPa (2–5 mmHg) due to physiological dead space. A difference greater than 1 kPa (10 mmHg) usually indicates a VAE. In the event of a VAE, the neurosurgeon must be alerted while the inspired oxygen concentration is increased to 100% and the surgical field is irrigated with saline. The shunt caused by the air bubbles in the pulmonary circulation takes about 15–30 min to resolve. Serial ABG will show normalisation of arterial-end tidal CO_2 difference.

In the event of cardiovascular collapse secondary to a paradoxical air embolism, the operating table is tilted head down to prevent further entrainment of air and aid with the hypotension. Hypotension is treated with vasopressors and fluid resuscitation [27, 36].

At the end of surgery, a swift emergence from anaesthesia using an awake remifentanil technique allows for clinical assessment of any neurosurgical deficit not explained by the surgery that may be due to a VAE.

6 Surgical Techniques

The complex pineal region can be accessed by four major approaches: (a) the supracerebellar/infratentorial (SC/IT), either midline or paramedian, (b) the occipital transtentorial, (c) the posterior interhemispheric transcallosal, and (d) the combined supra-/IT approach for very large tumours such as giant falcotentorial meningiomas. However, the SC/IT approach appears to be the most versatile (Table 2), and will be described in detail. After positioning the patient in the sitting position, a midline skin incision is made, and muscle dissection is carried out from the external occipital protuberance to the level of C2, without detaching its muscular attachments from C2.

The authors' preference is to perform a craniectomy, although a craniotomy is a good alternative. Emphasis is placed on exposing the inferior half of the transverse sinus completely, initially by thinning down the bone using a diamond drill and completing the removal with Kerrison rongeurs. Adequate lateral bony exposure is critical, as a narrow bony opening will not allow the cerebellum to relax and descend inferiorly, resulting in a narrow and crowded surgical field. The foramen magnum does not require opening, and of course, the posterior arch of C1 does not require removal.

Table 2 Advantages and disadvantages of using the sitting position in pineal region tumour surgery

Advantages	Disadvantages
Operative/positional	
Gravity facilitates culmen/cerebellar descent and tumour dissection from splenium Facilitates drainage of blood, clear operative field Better anatomical orientation of a complex neurovascular region Natural corridor with no normal tissue violation during approach	Fatigue of operator's hands Risk of postoperative haematoma when patient returns to horizontal position (0.4%) Risk of brachial plexus injury if arms allowed at sides rather than folded on abdomen [38] Cervical quadriplegia has been reported (neck flexion + anterior spinal artery compression + systemic hypotension) [38] Sciatic nerve injury from nerve tension (piriformis syndrome) prevented by gentle knee flexion [38]
Anaesthetic	
Easy ventilation (unencumbered chest)	Risk of VAE (%) Invasive anaesthetic monitoring with CVP catheter (risk of pneumothorax or subclavian vein thrombosis)
Physiological	
CSF drainage for surgical field, reduces ICP/brain oedema Carotid/vertebral system unobstructed, reduces ICP	Decreased cerebral blood flow due to lower haemodynamic arterial pressure Venous blood pooling in lower extremities Risk of post-op or tension pneumocephalus (0.2%) [39] Risk of subdural haematoma (1.3%) [40]

VAE venous air embolism. Adapted from Sharma et al. [27]

The dura is opened in a standard "Y" fashion with ligation of the occipital sinus, as occasionally it can be of significant calibre and venous blood flow. There are numerous cortical veins suspending the superior cerebellar surface from the tentorium, which are sequentially diathermised on their tentorial attachment, rather than the cerebellar surface, and divided by microscissors. This manoeuvre will allow the gradual inferior desent of the cerebellum.

The superior part of the vermis, known as the culmen, requires a gentle depression with cottonoid patties to allow further cerebellar descent, but no retractors are used. The first landmark encountered is the tentorial hiatus (Figs. 2 and 3), along with a typically thick, opaque arachnoidal membrane covering the pineal region and the tectal plate. In larger tumours, the arachnoid membrane may be stretched and significantly thin (Fig. 2). Although surgical videos usually demonstrate a sharp dissection with micro-scissors, our practice is to use a microhook to gently detach the arachnoid from the underlying vulnerable veins, as inadvertent venous injury may be difficult to control and can result in a deep-seated venous infarction.

Prior to embarking on tumour resection, it is important for the surgeon to understand the position of the splenium, pulvinar of the thalami, superior colliculi (Fig. 4), PCAs and SCAs when visible (Fig. 2), and the posterior third ventricle (Fig. 4). The pineal gland is rarely identifiable during tumour surgery (Fig. 4).

Smaller tumours, up to three centimetres, can be removed *en bloc*. A Rhoton #8 microdissector (*Integra, Plainsboro, NJ*) is used to develop a plane between the tumour and surrounding neural structures, only placing traction and gentle pressure on the side of the tumour. It is easier to develop the initial dissection plane from the visible splenium, and then proceed from the thalami, leaving the midbrain for last, as the angle of view makes visualisation difficult without significant prior dissection.

The authors do not use bipolar diathermy unless for coagulating very small bleeding vessels. Frequent use of diathermy carries the risk of ischemic stroke and makes the tumour-brain interface indistinct. Larger tumours require debulking, with repeated cycles of further dissection and debulking after initial planes have been developed, until the tumour is removed following release of its last attachments. A routine postoperative MRI is performed within 24 hours of the surgery (Figs. 8 and 9).

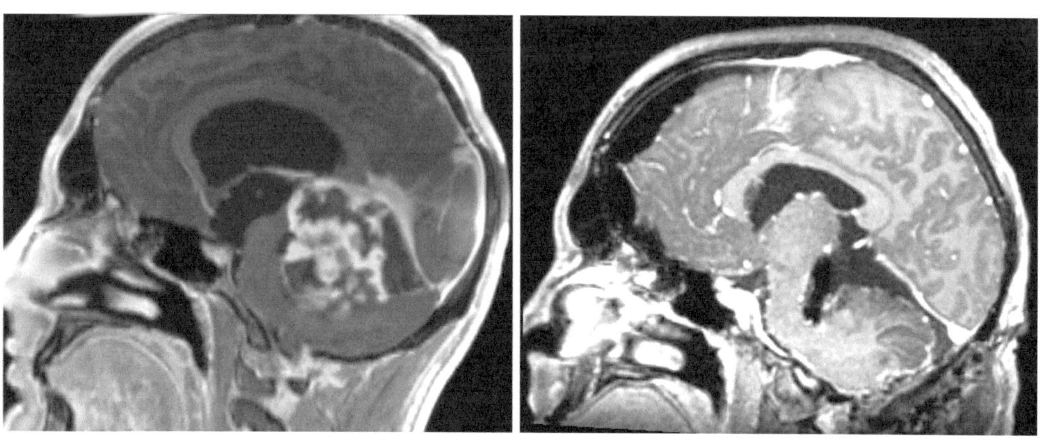

Fig. 8 Pre-operative (left) and post-operative (right) T1W sagittal MR images of a large posterior fossa tumour extending from the pineal region to the obex, which was operated on in the sitting position (same patient as in Fig. 6)

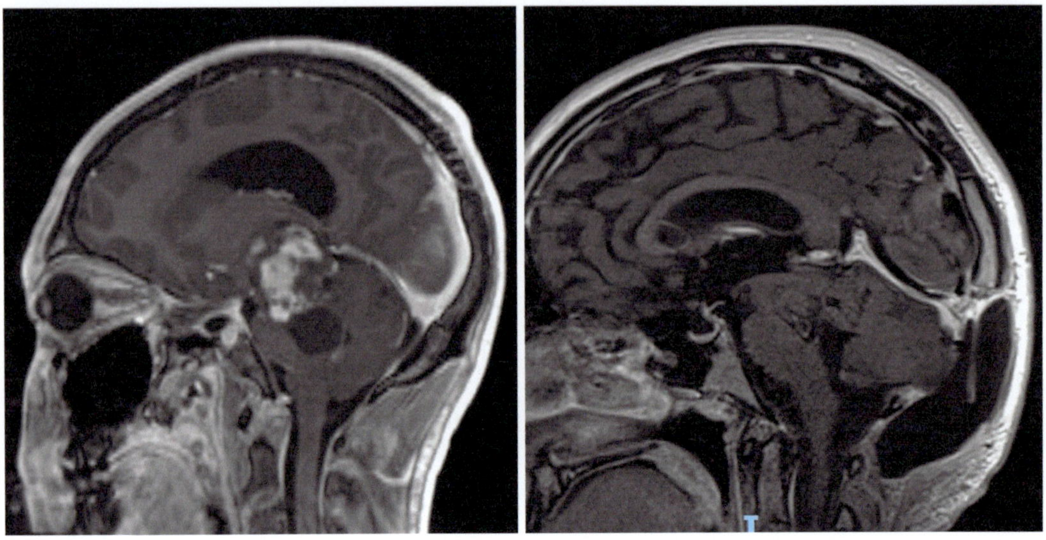

Fig. 9 Pre-operative (left) and post-operative (right) T1W sagittal MR images of a midbrain and thalamic tumour with extensive involvement of the pineal region, which was operated on in the sitting position (same patient as in Figs. 3 and 4)

References

1. Simon E, Afif A, M'Baye M, Mertens P. Anatomy of the pineal region applied to its surgical approach. Neurochirurgie. 2015;61(2):70–6. https://doi.org/10.1016/j.neuchi.2013.11.008.
2. Duvernoy HM, Parratte B, Tatu L, Vuillier F. The human pineal gland: Relationships with surrounding structures and blood supply. Neurol Res. 2000;22(8):747–90. https://doi.org/10.1080/01616412.2000.11740753.
3. Erlich SS, Apuzzo MLJ. The pineal gland: anatomy, physiology, and clinical significance. J Neurosurg. 1985;63(3):321–41. https://doi.org/10.3171/jns.1985.63.3.0321.
4. Cooper E. The human pineal gland and pineal cysts. J Anat. 1932;67(1):28–46.
5. Rodriguez EM, Oksche A, Montecinos H. Human subcommissural organ, with particular emphasis on its secretory activity during the fetal life. Microsc Res Tech. 2001;52(5):573–90. https://doi.org/10.1002/1097-0029(20010301)52:5<573::AID-JEMT1042>3.0.CO;2-6.
6. Vera A, Stanic K, Montecinos H, Torrejon M, Marcellini S, Caprile T. SCO-spondin from embryonic cerebrospinal fluid is required for neurogenesis during early brain development. Front Cell Neurosci. 2013;7:80. https://doi.org/10.3389/fncel.2013.00080.
7. Olsson R. Subcommissural ependyma and pineal organ development in human fetuses. Gen Comp Endocrinol. 1961;1:117–23. https://doi.org/10.1016/0016-6480(61)90039-9.
8. Krabbe K. Histologische und embryologische Untersuchungen über die Zirbeldrüse des Menschen. Anat Hefte. 1916;54:190–319.
9. Hulsemann M. Development of the innervation in the human pineal organ. Light and electron microscopic investigations. Z Zellforsch Mikrosk Anat. 1971;115(3):396–415. https://doi.org/10.1007/BF00324942.
10. Ibanez Rodriguez MP, Noctor SC, Munoz EM. Cellular basis of pineal gland development: emerging role of microglia as phenotype regulator. PLoS One. 2016;11(11):e0167063. https://doi.org/10.1371/journal.pone.0167063.
11. Mitchell TN, Free SL, Williamson KA, Stevens JM, Churchill AJ, Hanson IM, et al. Polymicrogyria and absence of pineal gland due to PAX6 mutation. Ann Neurol. 2003;53(5):658–63. https://doi.org/10.1002/ana.10576.
12. Abouzeid H, Youssef MA, ElShakankiri N, Hauser P, Munier FL, Schorderet DF. PAX6 aniridia and interhemispheric brain anomalies. Mol Vis. 2009;15:2074–83.
13. Bowers CW, Dahm LM, Zigmond RE. The number and distribution of sympathetic neurons that innervate the rat pineal gland. Neuroscience. 1984;13(1):87–96. https://doi.org/10.1016/0306-4522(84)90261-6.
14. Borjigin J, Zhang LS, Calinescu AA. Circadian regulation of pineal gland rhythmicity. Mol Cell Endocrinol. 2012;349(1):13–9. https://doi.org/10.1016/j.mce.2011.07.009.
15. Moller M, Baeres FM. The anatomy and innervation of the mammalian pineal gland. Cell Tissue Res. 2002;309(1):139–50. https://doi.org/10.1007/s00441-002-0580-5.

16. Korf HW, Wagner U. Evidence for a nervous connection between the brain and the pineal organ in the guinea pig. Cell Tissue Res. 1980;209(3):505–10. https://doi.org/10.1007/BF00234762.
17. Fink-Jensen A, Moller M. Direct projections from the anterior and tuberal regions of the lateral hypothalamus to the rostral part of the pineal complex of the rat. An anterograde neuron-tracing study by using Phaseolus vulgaris leucoagglutinin. Brain Res. 1990;522(2):337–41. https://doi.org/10.1016/0006-8993(90)91480-5.
18. Moller M, Korf HW. The origin of central pinealopetal nerve fibers in the Mongolian gerbil as demonstrated by the retrograde transport of horseradish peroxidase. Cell Tissue Res. 1983;230(2):273–87. https://doi.org/10.1007/BF00213805.
19. Mikkelsen JD, Cozzi B, Moller M. Efferent projections from the lateral geniculate nucleus to the pineal complex of the Mongolian gerbil (Meriones unguiculatus). Cell Tissue Res. 1991;264(1):95–102. https://doi.org/10.1007/BF00305726.
20. Leander P, Vrang N, Moller M. Neuronal projections from the mesencephalic raphe nuclear complex to the suprachiasmatic nucleus and the deep pineal gland of the golden hamster (Mesocricetus auratus). J Comp Neurol. 1998;399(1):73–93.
21. Tapp E. The histology and pathology of the human pineal gland. Prog Brain Res. 1979;52:481–500. https://doi.org/10.1016/S0079-6123(08)62955-6.
22. Sumida M, Barkovich AJ, Newton TH. Development of the pineal gland: measurement with MR. AJNR Am J Neuroradiol. 1996;17(2):233–6.
23. Tapp E, Huxley M. The weight and degree of calcification of the pineal gland. J Pathol. 1971;105(1):31–9. https://doi.org/10.1002/path.1711050105.
24. Tapp E, Huxley M. The histological appearance of the human pineal gland from puberty to old age. J Pathol. 1972;108(2):137–44. https://doi.org/10.1002/path.1711080207.
25. Matsuo S, Baydin S, Gungor A, Miki K, Komune N, Kurogi R, et al. Midline and off-midline infratentorial supracerebellar approaches to the pineal gland. J Neurosurg. 2017;126(6):1984–94. https://doi.org/10.3171/2016.7.JNS16277.
26. Yamamoto I, Kageyama N. Microsurgical anatomy of the pineal region. J Neurosurg. 1980;53(2):205–21. https://doi.org/10.3171/jns.1980.53.2.0205.
27. Sharma P, Abdul M, Waraich M, Samandouras G. Pineal region tumours in the sitting position: how I do it. Acta Neurochir. 2022;164(1):79–85. https://doi.org/10.1007/s00701-021-04821-3.
28. Zeal AA, Rhoton AL Jr. Microsurgical anatomy of the posterior cerebral artery. J Neurosurg. 1978;48(4):534–59. https://doi.org/10.3171/jns.1978.48.4.0534.
29. Fujii K, Lenkey C, Rhoton AL Jr. Microsurgical anatomy of the choroidal arteries: lateral and third ventricles. J Neurosurg. 1980;52(2):165–88. https://doi.org/10.3171/jns.1980.52.2.0165.
30. Xu Y, Mohyeldin A, Doniz-Gonzalez A, Vigo V, Pastor-Escartin F, Meng L, et al. Microsurgical anatomy of the lateral posterior choroidal artery: implications for intraventricular surgery involving the choroid plexus. J Neurosurg. 2021:1–16. https://doi.org/10.3171/2020.8.JNS202230.
31. Domaingue CM. Anaesthesia for neurosurgery in the sitting position: a practical approach. Anaesth Intensive Care. 2005;33(3):323–31. https://doi.org/10.1177/0310057X0503300307.
32. Dallier F, Di Roio C. Sitting position for pineal surgery: some anaesthetic considerations. Neurochirurgie. 2015;61(2-3):164–7. https://doi.org/10.1016/j.neuchi.2014.10.110.
33. Bedford R. Venous air embolism: An historical perspective. Semin Anesth. 1983;2:169–76.
34. Fathi AR, Eshtehardi P, Meier B. Patent foramen ovale and neurosurgery in sitting position: a systematic review. Br J Anaesth. 2009;102(5):588–96. https://doi.org/10.1093/bja/aep063.
35. Koutroulou I, Tsivgoulis G, Tsalikakis D, Karacostas D, Grigoriadis N, Karapanayiotides T. Epidemiology of patent foramen ovale in general population and in stroke patients: a narrative review. Front Neurol. 2020;11:281. https://doi.org/10.3389/fneur.2020.00281.
36. Mirski MA, Lele AV, Fitzsimmons L, Toung TJ. Diagnosis and treatment of vascular air embolism. Anesthesiology. 2007;106(1):164–77. https://doi.org/10.1097/00000542-200701000-00026.
37. Gunther F, Frank P, Nakamura M, Hermann EJ, Palmaers T. Venous air embolism in the sitting position in cranial neurosurgery: incidence and severity according to the used monitoring. Acta Neurochir. 2017;159(2):339–46. https://doi.org/10.1007/s00701-016-3034-7.
38. Wilder BL. Hypothesis: the etiology of midcervical quadriplegia after operation with the patient in the sitting position. Neurosurgery. 1982;11(4):530–1.
39. Lunsford LD, Maroon JC, Sheptak PE, Albin MS. Subdural tension pneumocephalus. Report of two cases. J Neurosurg. 1979;50(4):525–7. https://doi.org/10.3171/jns.1979.50.4.0525.
40. Standefer M, Bay JW, Trusso R. The sitting position in neurosurgery: a retrospective analysis of 488 cases. Neurosurgery. 1984;14(6):649–58. https://doi.org/10.1227/00006123-198406000-00001.

Anatomy of Corpus Callosum and Its Connections and Surgery for Corpus Callosal Tumors

Abhidha Shah, Dhiraj Parihar, and Atul Goel

1 Introduction

The appearance of the corpus callosum is one of the most distinct modifications seen in placental mammals and is a diagnostic character just like the placenta itself. It signified a major leap in evolution and its presence maybe linked to the binding mechanisms related to midline fusion and bimanual coordination [1]. Prior to the corpus callosum interhemispheric transfer amongst vertebrates was via the anterior commissure and the hippocampal commissure. It is believed that the corpus callosum originated in placental mammals as a shortcut to reduce interhemispheric transmission times for fibers connecting the motor and sensory areas which otherwise would have to take the longer route through the anterior commissure [2, 3]. In present day placental mammals however only a small contingent of callosal fibers are seen to connect the primary sensory and motor cortices and most callosal fibers are seen to connect higher order centers as has been demonstrated in this chapter. Callosal axons have been found to terminate in diverse regions of the brain apart from the topographically equivalent areas thus participating in spreading corticocortical synchronous activation to multiple homotropic and heterotopic regions in the hemispheres. This synchronous activity can explain the callosal propagation of an epileptic focus and the effectiveness of a callosotomy to prevent expansion of epileptic activity [1].

The word commissure derives its origin from the Latin "commissura" which means joining or connecting together. In the brain a "commissure" is defined as the junction where two anatomical parts meet. There are five major commissures in the brain: the anterior commissure, the posterior commissure, the hippocampal commissure, the habenular commissure and the corpus callosum [3]. Though the massa intermedia connects the medial surfaces of the two thalami it is not a true commissure as it is made up of glial tissue and not neurons. The corpus callosum is the largest commissure and connects the two cerebral hemispheres to each other in the depth of the interhemispheric fissure. (Fig. 1) We recently classified the white fibers of the brain into five groups [4]. The first or superficial layer is the

A. Shah (✉)
Department of Neurosurgery, K.E.M. Hospital and Seth G.S. Medical College, Mumbai, India

D. Parihar
Seth G.S. Medical College and K.E.M Hospital, Mumbai, India

A. Goel
Department of Neurosurgery, Lilavati Hospital and Research Center, Bandra, Mumbai, India

Department of Neurosurgery, R.N. Cooper Hospital, Mumbai, India

Bombay Hospital Institute of Medical Sciences and Research Center, Mumbai, India

A. Shah et al. (eds.), *Functional Anatomy of the Brain: A View from the Surgeon's Eye*,
https://doi.org/10.1007/978-981-99-3412-6_17

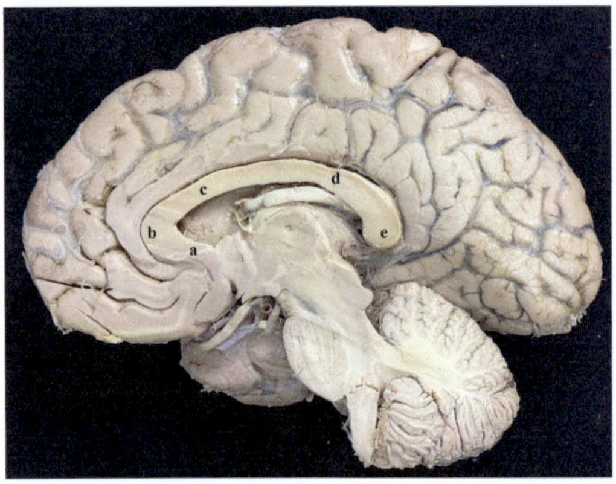

Fig. 1 Image of the medial surface of the cerebral hemisphere showing the parts of the corpus callosum. a. Rostrum, b. Genu, c. Body, d. Isthmus and e. Splenium

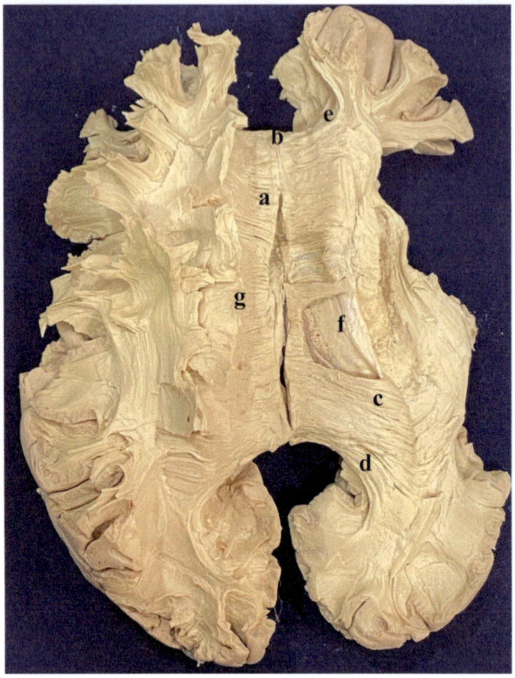

Fig. 2 Superior view of the dissected cerebral hemispheres showing the corpus callosum and its radiations. The right half of the hemisphere shows the ventral callosal radiations. The left half of the hemisphere shows the dorsal callosal radiations. a. Dorsal callosal radiations, b. Genu of the corpus callosum, c. Tapetum or the posterior ventral callosal radiations, d. Forceps major, e. Forceps minor, f. Lateral ventricle, g. Dorsal callosal radiations as they curve upwards to join the corona radiata

short arcuate fibers. The second layer or the middle layer is that of the long association fibers. The deep layer consists of long association fibers that are in proximity to the ventricles. The central group consists of the commissural fibers. The vertical group consists of the projection fibers. The commissural group of fibers or the central group is bounded on either side by the vertical group of projection fibers. The corpus callosum connects the frontal, parietal, temporal, occipital, insular and limbic lobes and the basal ganglia of the hemispheres to each other. We have recently published a classification that divides the fiber connections of the corpus callosum into four types: dorsal radiations, ventral radiations, anterior radiations and posterior radiations [3]. This chapter discusses the course and connections of each of these radiations and its surgical implications.

2 Gross Anatomy

The corpus callosum lies in between the two medial surfaces of the hemisphere and forms a bridge connecting the two halves of the brain to each other. The parts of the corpus callosum are rostrum, genu, body, isthmus and splenium. (Fig. 1) On the superior surface of the corpus callosum, a thin layer of grey matter the indusium griseum is visualized [3]. The indusium griseum is the supracallosal remnant of the hippocampal formation. It connects the dentate gyrus to the subcallosal region [2]. Within the indusium grisium two thin white fiber bundles can be identified. These are the medial and lateral longitudinal striae of Lancisi. These white fibers are considered to be the aberrant fibers of the fornix. Once this thin layer of grey matter is removed the transversely running fibers of the corpus callosum can be identified.

2.1 The Callosal Radiations

The corpus callosum comprises of transverse running fibers extending from the rostrum to the splenium connecting both the cerebral hemispheres. (Figs. 2, 3, 4, 5, 6, 7, and 8) The white fiber connections or radiations of the corpus callosum can be divided into four types viz.: superior/dorsal connections, ventral/inferior

Fig. 3 Lateral view of the cerebral hemispheres showing the various components of the callosal radiations. a. anterior ventral callosal radiations showing fibers from the genu ending in the region of the caudate nucleus, b. posterior ventral callosal radiations or tapetal fibers, c. dorsal callosal radiations, d. cavity of the lateral ventricle, e. Dorsal callosal radiations as they join fibers of the corona radiata, f. fibers of the inferior fronto-occipital fasciculus as they form part of the sagittal stratum, g. fibers of the forceps minor, h. Fibers of the forceps major, i. Putamen

Fig. 4 Anterior view of the dissected corpus callosum showing the anterior callosal radiations or the forceps minor. a. Genu of the corpus callosum, b. Forceps minor or anterior callosal radiations, c. Dorsal callosal radiations

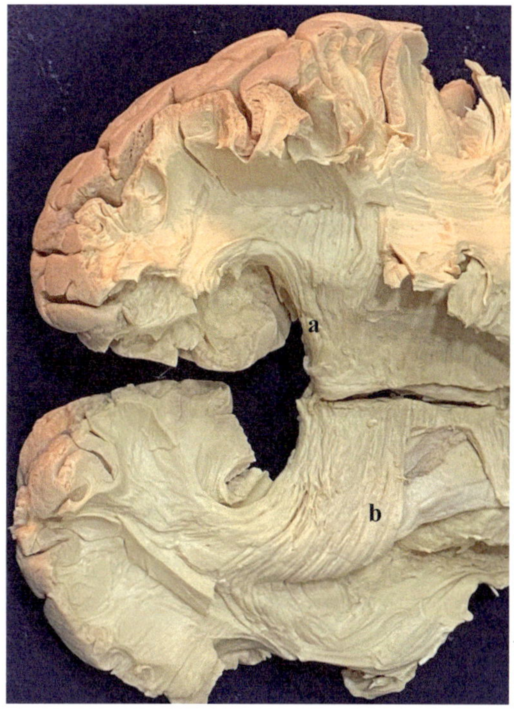

Fig. 5 Posterior view of the dissected corpus callosum. a. Posterior callosal radiations or the forceps major can be seen arising from the splenium and connecting the medial occipital regions of the two hemispheres, b. Posterior component of the ventral callosal radiation or the tapetal fibers turning inferiorly as the roof and lateral wall of the atrium and temporal horn

connections, anterior connections and posterior connections [3]. Fibers emanating from the body of the corpus callosum curve either superiorly or inferiorly. The superior or the dorsal radiations are those fibers that traverse laterally and enter the medial surface of the hemisphere and then curve upwards. The inferior or the ventral callosal radiations are those that traverse laterally from the genu and the body of the corpus callosum and curve inferiorly. (Figs. 2 and 3) The anterior radiations or the forceps minor originate from the genu and rostrum of the corpus callosum and curve anteriorly into the medial and basal frontal lobes. (Fig. 4) The posterior radiations or the forceps major traverse posteriorly from the splenium of the corpus callosum and enter the medial aspect of the occipital lobes. (Fig. 5) The dorsal callosal radiations are limited by the vertical fibers of the corona radiata on both the sides. The ventral callosal radiations are closely related to the lateral ventricle and envelop the lateral ventricle within its fibers.

2.2 Dorsal Callosal Radiations

The superior or the dorsal callosal radiations arise from the entire extent of the body of the corpus callosum and run transversely bilaterally into the medial surface of the hemispheres [3]. (Figs. 2 and 3) They traverse below the cingulum and the medial portion of the superior longitudinal fasciculus (SLF) and then turn upward lateral to the medial SLF (SLF—I) to join the fibers of the

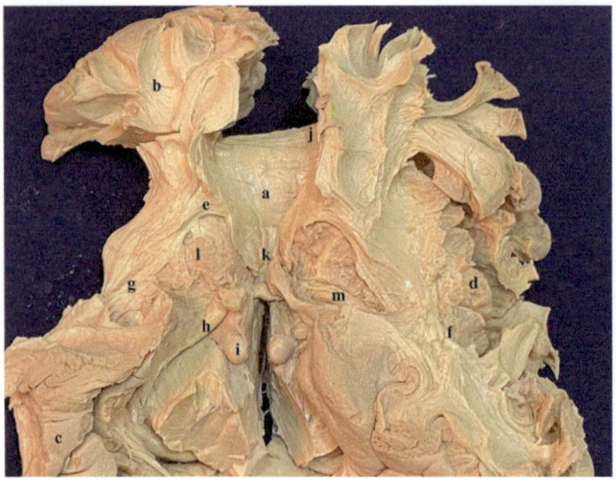

Fig. 6 Inferior view of the cerebral hemisphere. The brainstem and the mammillary bodies have been bisected in the midline.a. Rostrum of the corpus callosum, b. Frontal lobe, c. Temporal lobe, d. Insular gyri, e. Fibers from the rostrum of the corpus callosum going towards the insula, f. Fibers of the inferior fronto-occipital fasciculus, g. Fibers of the uncinate fasciculus, h. Optic tract, i. Mamillary body, j. Forceps minor. k. Longitudinal stria of Lancisi ending in the subcallosal region, l. Substantia innominate, m. Anterior commissure

corona radiata. Fibers in the anterior part of the body, connect the medial portion of the superior frontal gyrus of both the hemisperes. Fibers in the posterior part of the body at a distance of 3.5 cm (range 3.2–3.7 cm) to 7 cm (range 6.9–7.1 cm) from the genu were found to connect the posterior portion of the medial superior frontal gyrus (supplementary motor area), precentral and postcentral gyri or the paracentral lobule of the two halves of the hemispheres to each other [3]. Fibers in the region of the isthmus, 7 cm (range 6.9–7.1 cm) from the genu and just anterior to the splenium connected the region of the precuneus of the 2 hemispheres to each other [3]. (Fig. 8) Thus the dorsal callosal radiations essentially connect the motor and supplementary motor areas.

2.3 Ventral Callosal Radiations

The ventral callosal radiations were found to have an anterior component and a posterior component. Anteriorly in the region of the genu, fibers traversed laterally and inferiorly and then

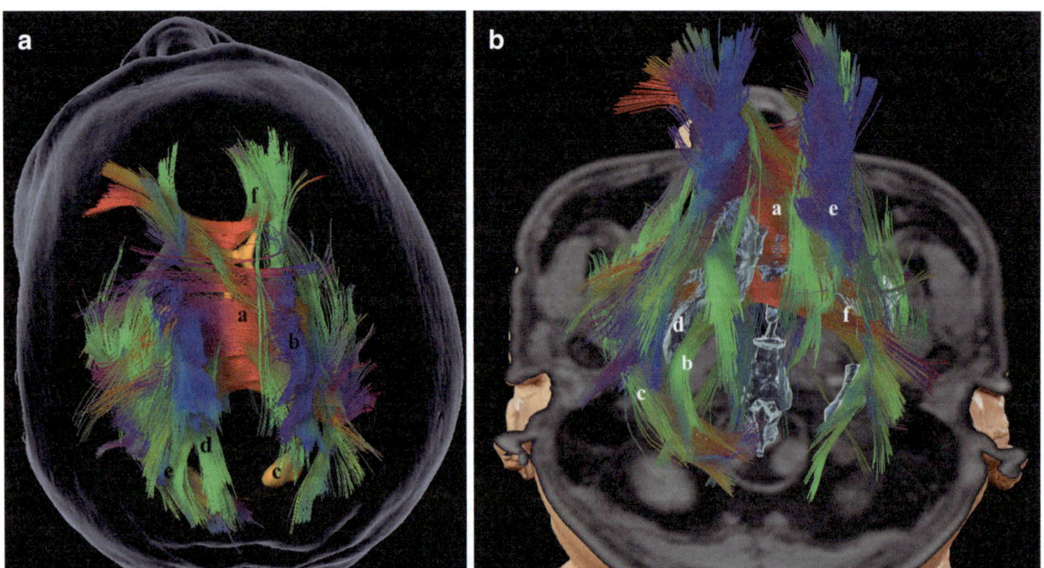

Fig. 7 Tractography images. (**a**) Superior view of tractography image. a. Callosal fibers, b. Projection fibers, c. Lateral ventricle, d. Forceps major, e. Optic radiations, f. Forceps minor. (**b**) Posterior view of tractography image showing the corpus callosal fibers and the relation to the lateral ventricle. a. Callosal fibers, b. Forceps major, c. Optic radiations, d. Atrium of the lateral ventricle, e. Projection fibers, f. Tapetal fibers

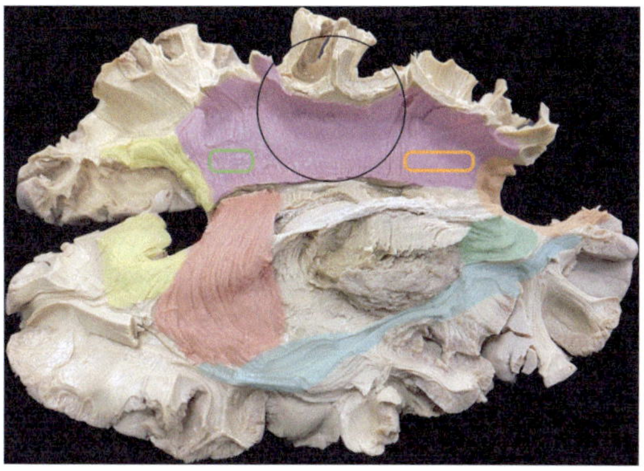

Fig. 8 Image showing the callosal radiations and the safe areas for callosotomy. Orange color—Anterior callosal radiations—forceps minor, Green color—Anterior component of the ventral callosal radiations, Purple color—Dorsal callosal radiations, Red color—Posterior component of ventral callosal radiation/tapetum, Yellow color—Posterior callosal radiations—Forceps major, Blue color—Inferior fronto-occipital fasciculus. The black circle indicates the region of the precentral and postcentral gyri and the part of the dorsal callosal radiations emanating from this region. The ornage rectangle indicates the safe area for transverse callosotomy for the anterior interhemispheric approach. The green rectangle indicates the safe area for transverse callostomy for the posterior interhemispheric approach

turned posteriorly to terminate in the region of the caudate head. We have called these fibers as the anterior component of the ventral callosal radiations [3]. (Fig. 3) These anterior ventral callosal radiations are closely related to the fibers of the IFOF inferiorly. The posterior component of the ventral callosal radiations run laterally forming the roof of the body, atrium and occipital horn of the lateral ventricle and then turn inferiorly to form the innermost layer of the lateral wall of the temporal and occipital horns of the lateral ventricle. These fibers are better known as the tapetum. (Fig. 3) The fibers of the tapetum are the medial most fibers in the layers of fibers that constitute the sagittal stratum. The tapetal fibers are related laterally to the optic radiations, temporal extension of the anterior commissure, the fibers of the IFOF and the arcuate fasciculus. (Fig. 8) The ventral callosal radiations connect the temporal lobes and the basal ganglia of both the hemispheres to each other.

2.4 Anterior Callosal Radiations

The anterior callosal radiations are those that originate from the genu and rostrum of the corpus callosum [3]. Majority of the fibers curve forwards to enter the medial orbitofrontal region. These fibers formed the medial wall of the frontal horn of the lateral ventricle and are better known as the forceps minor. (Fig. 4) The forceps minor connect the medial orbitofrontal regions of the hemispheres to each other. Some fibers originating in the rostrum instead of curving anteriorly as the forceps minor curved laterally on the basal surface of the frontal lobe and ran along the medial portion of the uncinate fasciculus to reach the temporal stem in close association with the ventral external capsule thus connecting the insular regions of the hemispheres to each other. (Fig. 6).

2.5 Posterior Callosal Radiations/ Forceps Major

Fibers originating from the splenium of the corpus callosum were seen to curve posteriorly to enter the medial occipital region as the forceps major or the posterior callosal radiations [3]. (Fig. 5) The forceps major form a bulge in the medial wall of the atrium of the lateral ventricle known as the bulb of the atrium and connect the medial occipital lobes of the hemispheres to each other.

It was observed that by virtue of its callosal radiations the corpus callosum connects almost all regions of one hemisphere to the other either directly or indirectly. (Fig. 8).

3 Surgical Implications

This classification proposed by us has surgical implications in understanding the spread of primary corpus callosal gliomas and in devising the most appropriate surgical strategy. Understanding the anatomy gives an insight into the direction and extent of callosotomy to be performed whilst performing epilepsy surgery and whilst performing a callosotomy to gain intraventricular access.

The corpus callosum envelops the entire lateral ventricle within its callosal radiations [4]. The rostrum of the corpus callosum forms the floor of the frontal horn and the genu forms its anteromedial boundary [5]. The fibers of the forceps minor curve forward and form the medial wall of the frontal horn. The forceps major forms the medial wall of the occipital horn of the lateral ventricle. The body of the corpus callosum forms the roof of the body of the lateral ventricle. The ventral callosal radiations of the anterior portion of the body travel laterally in the roof the frontal horn to reach the caudate head. The ventral callosal radiations from the posterior part of the body of corpus callosum (tapetal fibers) travel laterally and inferiorly and form the lateral wall of the temporal horn, atrium and the occipital horn thus enveloping these portions of the ventricle. Peltier et al. defined ventral callosal radiations as those that arose from the rostrum and ended at the junction of the first and second third of the corpus callosum [6]. They found that some of these fibers reached the temporal pole and the amygdaloid nucleus. We have defined all the inferior connections of the corpus callosum to the temporal lobe as the ventral callosal radiations [3]. In our recent study we identified connections from the rostrum of the corpus callosum traversing laterally in close association with the fibers of the uncinate fasciculus and inferior to the IFOF to reach the temporal stem and thereby the temporal lobe [3]. We also identified fibers arising from the genu of the corpus callosum traversing laterally to reach the caudate nucleus [3]. These fibers were found superior to the IFOF. (Fig. 8).

3.1 Callosotomy

The Witelson classification divides the corpus callosum into 5 vertical segments topographically: the anterior third, the anterior and posterior midbody, the posterior third and the posterior one fifth [7]. According to their study, the motor fibers were found to traverse in the anterior part of the midbody of the corpus callosum [7]. In a recent study Kucukyuruk et al. found that motor fibers traversed in the posterior mid body and isthmus of the corpus callosum. They suggested extending the callosotomy to the posterior midbody till the anterior edge of the splenium to have effective disconnection for epilepsy surgery [8]. The mean distance of the most anterior motor fiber to the tip of the genu as measured by them was an average of 41.7 ± 5.72 mm (min: 31 mm, max: 57 mm) [8]. In the study conducted by us we found the mean distance of the most anterior motor fiber as measured form the tip of the genu was 3.5 cm [3]. This difference in measurement may be due to the inclusion of the most posterior fibers connecting the posterior supplementary motor area as the anterior limit of the measurement in our study.

The technique of surgical disconnection of the corpus callosum was first proposed by Van Wagenen and Herren with the aim of interruption of interhemispheric transfer of impulses to control epilepsy. The technique consisted of a median longitudinal incision perpendicular to the transversely running fibers of the corpus callosum [9]. Mazza et al. proposed performing a transverse incision in the corpus callosum to access lesions of the third ventricle based on the miscrosurgical anatomy of the neuronal fibers and the vascular supply [10]. This led to better neuropsychological outcomes when compared to the longitudinal callosotomy procedure.

Based on our anatomical dissections and our experience we recommend that the incision in the corpus callosum should be in a transverse direction making it parallel to the line of the transversely running fibers when performing a callosotomy for approaching ventricular lesions. This transverse incision will cause a splitting of the fibers without damaging them. A longitudinal incision in the antero-posterior direction will transect the transversely running fibers of the corpus callosum. This longitudinal incision would be most appropriate whilst performing a callosotomy for disconnection as in epilepsy surgery.

Corpus callosotomy has been found to be effective in treating drop attacks and generalized tonic clonic seizures in children. However the extent of callosotomy that leads to a favorable seizure outcome is not very clear. Traditionally an anterior callosotomy was preferred to decrease the morbidity associated with sectioning of the splenium but this often requires a second stage surgery and a complete callosotomy for persistent seizures and drop attacks. A complete callosotomy was associated with better seizure outcomes [11]. Recent anatomical dissections and DTI studies have shown that the fibers connecting the premotor and motor areas traverse in the mid-body and posterior third of the corpus callosum [11]. The selective posterior callosotomy as described by Paglioli et al. involves a disconnection of the posterior corpus callosum with sectioning of the premotor and motor fibers and sparing of prefrontal connectivity [12, 13]. The procedure led to comparable seizure outcomes with historical series of complete callosotomy and avoided the transient post-operative apathy that occurred due to disconnection of the anterior callosal fibers [13].

In our anatomical dissections we found that the fibers connecting the posterior portion of the supplementary motor area, primary motor and sensory cortices were located in the posterior part of the body of the corpus callosum beginning 3.5 cms behind the genu [3]. Some fibers were seen in the region of the isthmus but not in the splenium. Thus for an effective disconnection we recommend that the callosotomy should include the posterior part of the body of the callosum to disconnect the supplementary motor, primary motor and sensory areas.

For anterior interhemispheric transcallosal approaches we recommend making a transverse callosotomy in the anterior portion of the body of the corpus callosum [3]. (Fig. 8) The site of callosotomy should not extend beyond 3.5 cms from the genu. While performing a posterior interhemispheric approach we recommend a transverse callosotomy in the region of the isthmus in front of the splenium [3]. This avoids majority of the motor—sensory fibers in the posterior part of the body and also the visual and auditory fibers that are in the splenium.

3.2 Corpus Callosal Gliomas

Primary corpus callosal gliomas are not very common. We have recently classified gliomas based on their white fiber origin into localized and diffuse group of gliomas [14–16]. Corpus callosal gliomas belong to the diffuse group of gliomas. Even though the radiological picture of corpus callosal gliomas can sometimes be bizzare, the anatomical extent of these lesions can be gauged by understanding the anatomy of the callosal radiations. Our recent understanding is that gliomas are 'confined' tumors. They grow along the named fiber tracts and displace the adjoining normal fiber tracts that are functionally involved by virtue of pressure and deformation and not due to destruction or invasion [14]. Gliomas of the corpus callosum can be subdivided into tumors arising from the genu and forceps minor, tumors of the body, tumors of the splenium and forceps major and tumors of the ventral callosal radiations. Generally callosal gliomas tend to have a bilateral hemispheric involvement. (Figs. 9 and 10) Tumors arising from the genu of the corpus callosum grow along the forceps minor into the medial orbitobasal frontal lobes. The tumors are limited laterally by the frontal horn of the lateral ventricles. Gliomas arising in the region of the body of the corpus callosum (dorsal callosal radiations) grow and expand superiorly towards the cingulum and can extend bilaterally into the medial fronto-parietal regions. However they are always limited laterally by the projection fibers, which they do not cross. Tumors of the splenium or the forceps major expand bilaterally into the medial occipital lobes and are limited laterally by the occipital horn of the lateral ventricle, which they do not cross. Tumors of the anterior ventral callosal radiations extend from the genu towards the caudate head and can ultimately reach the temporal stem due to the close association with the IFOF as seen in our study. Tumors of the posterior ventral callosal radiations extend along the tapetum and cross over to the lateral aspect of the temporal and occipital horns.

We observed that like in other locations, gliomas of the corpus callosum remain 'confined' to callosal fibres and grow by displacement of

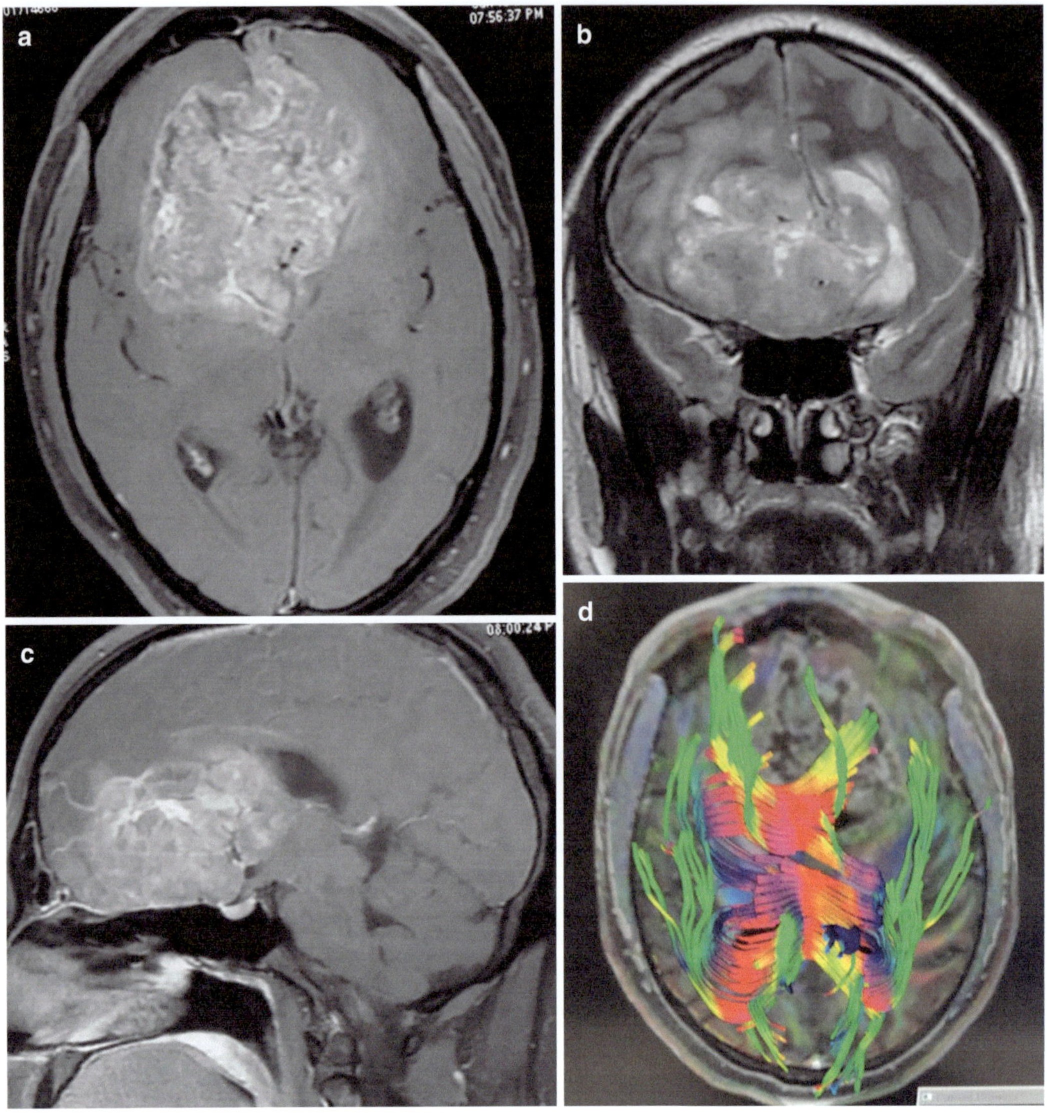

Fig. 9 Images of a 43 year old male patient. (**a**) T1 weighted axial contrast image showing a large corpus callosal glioma arising from the anterior callosal radiations. (**b**) T2 weighted coronal image showing the tumor. The tumor spreads along the fibers of the forceps major and is limited laterally by the vertical group of projection fibers. (**c**) Sagittal contrast enhanced image showing the tumor. (**d**) Tractography images showing the tumor from the forceps minor and displacing the adjoining fibers. (**e**) Post-operative image showing the radical tumor excision with post-operative changes

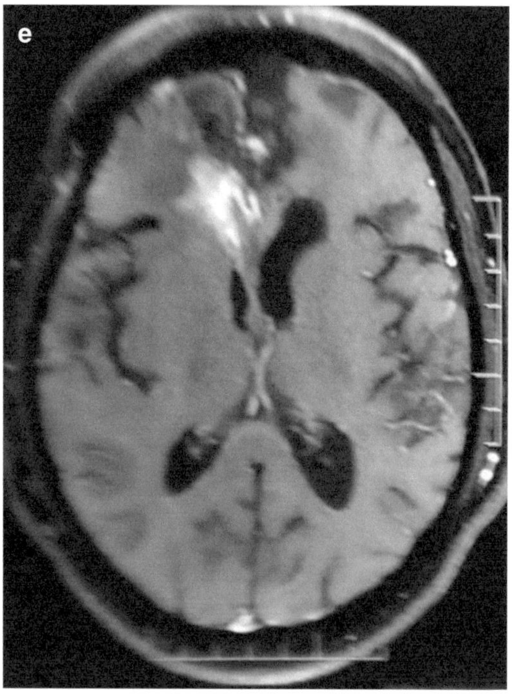

Fig. 9 (continued)

adjoining fibre tracts that are involved by virtue of pressure distortion and not by invasion or destruction. A well-defined surgical plain of dissection can be developed between the tumor and the adjoining normal brain parenchyma [16]. Though gliomas of the corpus callosum have the potential to involve different components of the callosal radiations, we have observed from our experience that the tumors initially remain confined to the component of the radiation that they arise from. When they enlarge the gliomas involve the component of the callosal radiation that is the most adjacent. Tumors of the anterior, superior and posterior radiations can be accessed surgically by an inter-hemispheric approach. Tumors of the ventral callosal radiation generally have an unusual morphology and the approach has to be individualized based on the region of maximum extent of the tumor.

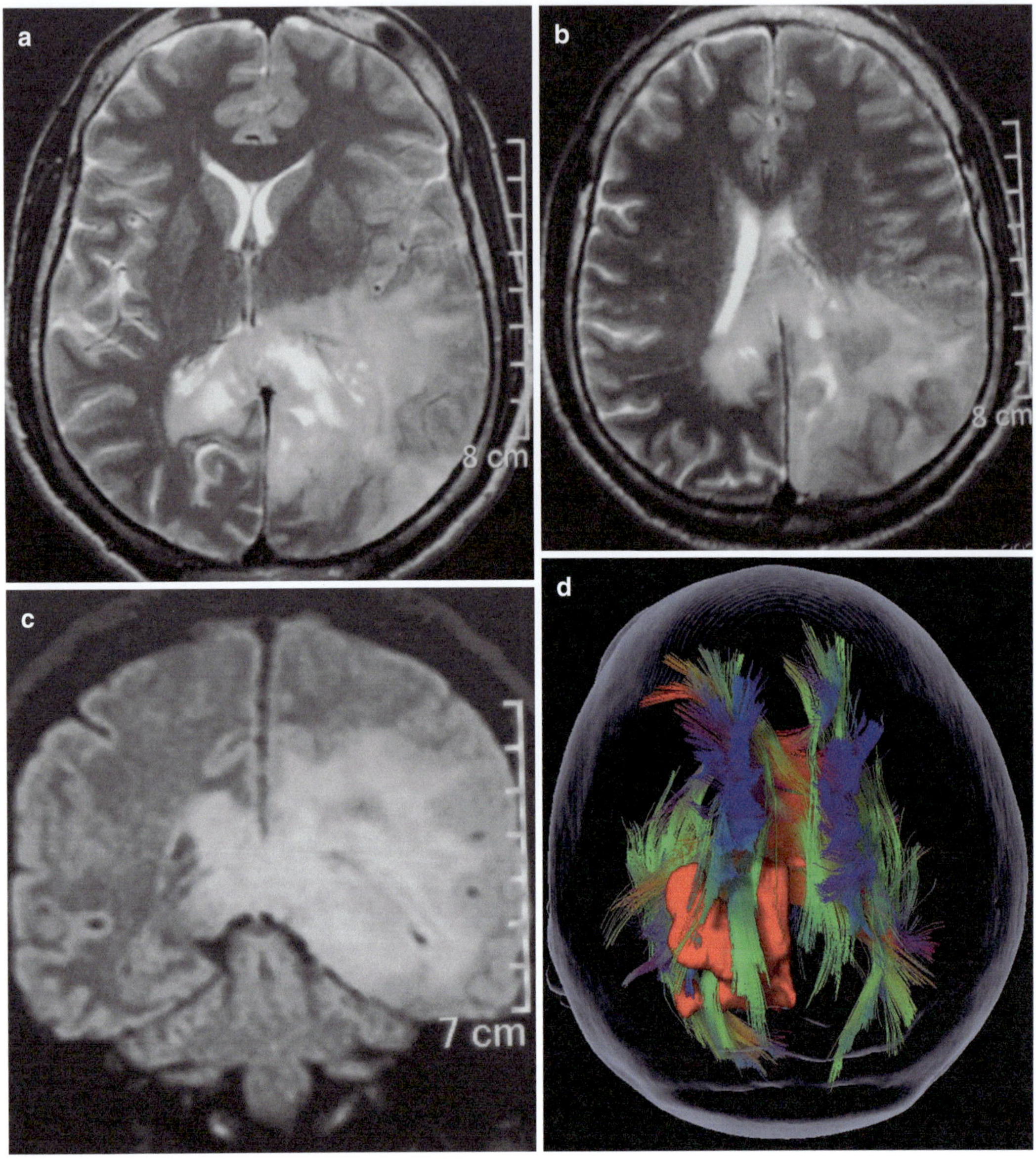

Fig. 10 Images of a 39 year old male patient. (**a**) T2 weighted axial image showing a glioma arising from the forceps major and also involving the ventral callosal radiations. (**b**) T2 weighted axial image showing the involvement of the fibers of the body of the corpus callosum and the posterior component of the ventral callosal radiations. (**c**) Coronal FLAIR image showing the bilateral spread of the tumor. (**d**) Tractography image showing the spread of the tumor along the forceps major and the ventral callosal radiations. (**e**) T1 weighted axial contrast image showing the tumor excision. (**f**) T2 weighted axial image showing the tumor excision. (**g**) Post-operative coronal FLAIR image

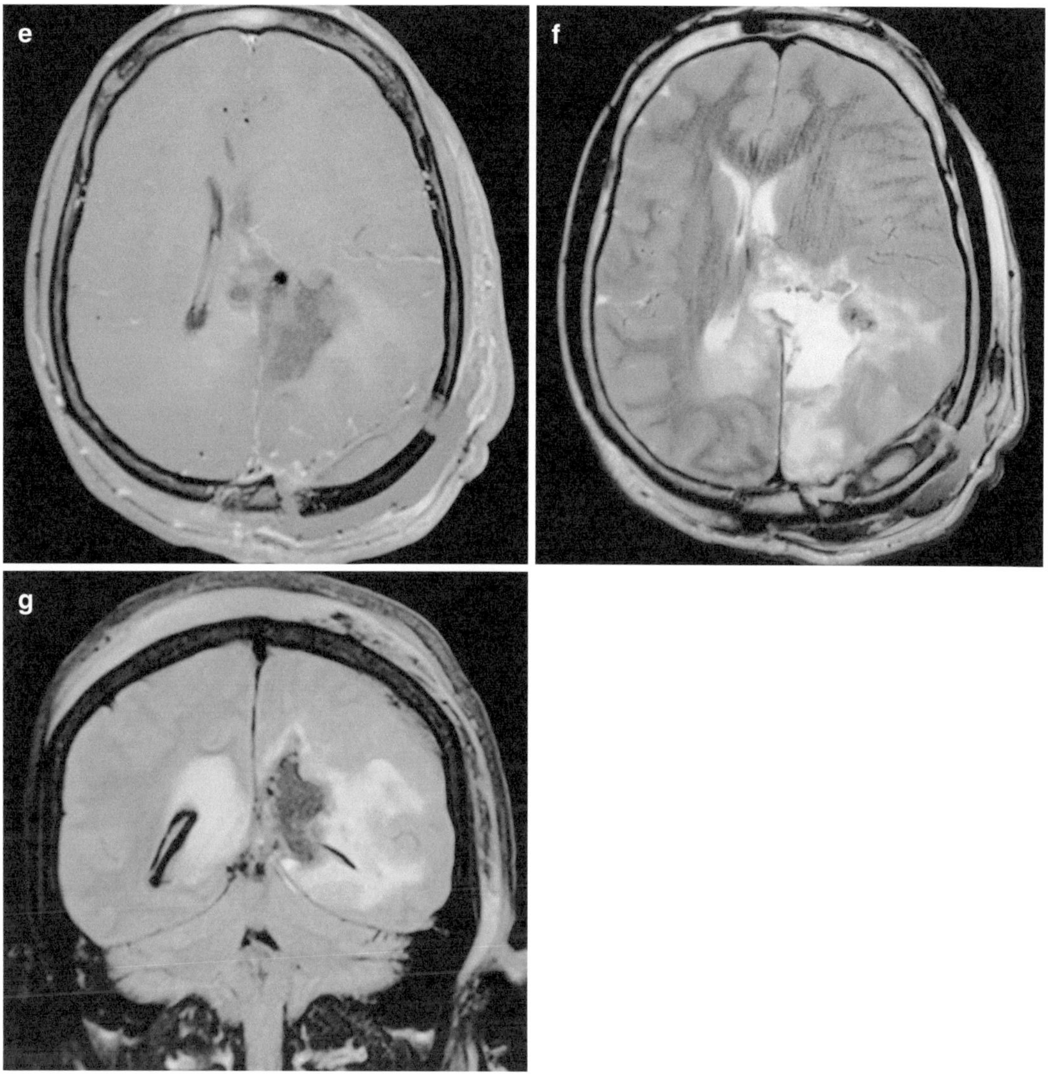

Fig. 10 (continued)

4 Conclusions

Understanding the fiber anatomy of the corpus callosum is helpful when operating on primary corpus callosal gliomas. A transverse incision should be performed when using the corpus callosum as a route to access intraventricular lesions and a longitudinal incision when performing a callosotomy for epilepsy surgery.

References

1. Aboitiz F, Montiel J. One hundred million years of interhemispheric communication: the history of the corpus callosum. Braz J Med Biol Res. 2003;36(4):409–20.
2. Shah A, Jhawar SS, Goel A. Analysis of the anatomy of the Papez circuit and adjoining limbic system by fiber dissection techniques. J Clin Neurosci. 2012;19(2):289–98.

3. Shah A, Jhawar S, Goel A, Goel A. Corpus callosum and its connections: a Fiber dissection study. World Neurosurg. 2021 Jul;151:e1024–35.
4. Shah A, Goel A, Jhawar SS, Patil A, Rangnekar R, Goel A. Neural circuitry: architecture and function-a Fiber dissection study. World Neurosurg. 2019;125:e620–38.
5. Rhoton AL Jr. The lateral and third ventricles. Neurosurgery. 2002;51(4 Suppl):S207–71.
6. Peltier J, Verclytte S, Delmaire C, Deramond H, Pruvo JP, Le Gars D, Godefroy O. Microsurgical anatomy of the ventral callosal radiations: new destination, correlations with diffusion tensor imaging fiber-tracking, and clinical relevance. J Neurosurg. 2010;112(3):512–9.
7. Witelson SF. Hand and sex differences in the isthmus and genu of the human corpus callosum. A postmortem morphological study. Brain. 1989;112(Pt 3):799–835.
8. Küçükyürük B, Uzan M, Avyasov R, Tahmazoğlu B, İşler C, Sanus GZ, Tanrıöver N. Evaluation of ideal extent of corpus callosotomy based on the location of Intracallosal motor Fibers. World Neurosurg. 2020;144:e568–75.
9. Van Wagenen WP, Herren RY. Surgical division of commissural pathways in the corpus callosum. Relation to spread of an epileptic attack. Arch Neurol Psychiatr. 1940;44:740–59.
10. Mazza M, Di Rienzo A, Costagliola C, Roncone R, Casacchia M, Ricci A, Galzio RJ. The interhemispheric transcallosal-transversal approach to the lesions of the anterior and middle third ventricle: surgical validity and neuropsychological evaluation of the outcome. Brain Cogn. 2004;55(3):525–34.
11. Mirza FA. Commentary: selective posterior callosotomy for treatment of epileptic drop attacks: video documentation of the surgical technique: 2-dimensional operative video. Oper Neurosurg (Hagerstown). 2020;19(5):E516–7.
12. Paglioli E, Martins WA, Azambuja N, Portuguez M, Frigeri TM, Pinos L, Saute R, Salles C, Hoefel JR, Soder RB, da Costa JC, Hemb M, Theys T, Palmini A. Selective posterior callosotomy for drop attacks: a new approach sparing prefrontal connectivity. Neurology. 2016;87(19):1968–74.
13. Frigeri T, Paglioli E, Soder RB, Martins WA, Paglioli R, Mattiello R, Paganin R, Palmini A. Control of drop attacks with selective posterior callosotomy: anatomical and prognostic data. Epilepsy Res. 2021;171:106544.
14. Shah A, Vutha R, Goel A. Radiological Evaluation of Anatomical Extensions of Gliomas Based on White Fiber Tracts: Proposal of a Novel Classification. Neurol India. 2022;70(4):1492–99.
15. Goel A, Shah A, Dandpat S, Rai S, Prasad A. Gliomas are 'confined' to a named white matter tract: a revolution in understanding gliomas. J Craniovert Jun Spine. 2020;11:252–3.
16. Goel A, Shah A, Vutha R, Dandpat S, Hawaldar A. Is "En Masse" Tumor Resection a Safe Surgical Strategy for Low-Grade Gliomas? Feasibility Report on 74 Patients Treated Over Four Years. Neurol India. 2021;69(2):406–13.

The Central Core of the Brain

Eduardo Carvalhal Ribas and Hung Tzu Wen

Abbreviations

ACA	Anterior Cerebral Artery
Acc.	Accessory
AChA	Anterior Choroidal Artery
AComA	Anterior Communicating Artery
ALS	anterior limiting sulcus
Amygd.	Amygdala
Ant. Comm.	Anterior Commissure
Ant.	Anterior
APS	Anterior Perforated Substance
BA	Basilar Artery
Caud. Nucl.	Caudate Nucleus
Chor. Fiss.	Choroidal Fissure
Chor. Plex.	Choroidal Plexus
Cing.	Cingulate
Clauscort.	Claustrocortical fibers
Cor. Rad	Corona Radiata
Corp. Cal.	Corpus Callosum
Cu	Cuneus
DLPFC	dorsolateral prefrontal cortex
FLP	frontal limen point
Fus.	Fusiform
G.	Gyrus
Gl.	Gland
Glob. Pall.	Globus Pallidus
GPe	external segment of globus pallidus
GPi	internal segment of globus pallidus
Hipothal.	Hypothal.Hypothalamic
Int. Caps.	Internal Capsule
ICA	Internal Carotid Artery
IFOF	Inferior Fronto-occipital Fascicle
IFOF	inferior fronto-occipital fascicle
ILS	inferior limiting sulcus
Inf. Chor. Point	Inferior Choroidal Point
Ins.	Insular
Int. Caps.	Internal Capsule
Isth.	Isthmus of Cingulate Gyrus
ITG	Inferior Temporal Gyrus
Lat.	Lateral
Lent. Nucl.	Lentiform Nucleus
LGB	Lateral Geniculate Body
LSA	Lenticulostriate Arteries
MCA	Middle Cerebral Artery
Mesenc.	Mesencephalon
Olf.	Olfactory
OMPFC	orbital and medial prefrontal cortex
Opt. Rad.	Optic Radiation
Opt.	Optic
Parac.	Paracentral Lobule

E. C. Ribas
Division of Neurosurgery, Hospital das Clínicas, University of São Paulo Medical School, São Paulo, Brazil

Hospital Israelita Albert Einstein, São Paulo, Brazil

H. T. Wen (✉)
Division of Neurosurgery, Hospital das Clínicas, University of São Paulo Medical School, São Paulo, Brazil

A. Shah et al. (eds.), *Functional Anatomy of the Brain: A View from the Surgeon's Eye*,
https://doi.org/10.1007/978-981-99-3412-6_18

Parahip.	Parahippocampal
PCA	Posterior Cerebral Artery
PChorA	Posterior Choroidal Arteries
PComA	posterior communicating artery
PMChorA	Postero-medial Choroidal Artery
Post.	Posterior
PreCu	Precuneus
RAH	Recurrent Artery of Heubner
Rect.	Rectus
S.	Sulcus
Sept. Pel.	Septum Pellucidum
Sept.	Septal Region
SFG	Superior Frontal Gyrus
SLF	Superior Longitudinal Fascicle
SLS	superior limiting sulcus
SNc	pars compacta of substantia nigra
SNr	pars reticulata of substantia nigra
STN	subthalamic nucleus
Subic.	Subiculum
Thal.	Thalamus
TLP	temporal limen point
TPA	Thalamoperforating Arteries
Transv.	Transverse
UF	Uncinate Fascicle
Vent.	Ventricle.
VP	ventral pallidum
VS	ventral striatum

1 Introduction

Most of the brain has a uniform pattern of organization, with grey matter on its surface and white matter underneath. The first represents the cortex, where bodies and dendrites of neurons are found, and the second is composed by long neuronal projections called axons which are surrounded by myelin nodes, a lipid substance that confers the white color to this substance.

The phylogenetic and embryological development of the brain surface is based on its invagination processes, forming fissures and sulci which are natural extensions of the subarachnoid space. These processes led to a typical superficial organization of sulci and gyri. Each mammalian species has a highly characteristic cerebral surface, with humans having a higher complexity. Sulci depths are variable, ranging from 1 to 3 cm approximately, and almost two thirds of the human neocortex are hidden away in the depth of the sulci. There's a considerable intersubject variability if individual sulci are seen in detail, but the cerebral gyri constitute a continuum of the cerebral surface and their appearance as distinct gyri is only superficial, since they are continuous along the entire sulcal depth and along its extremities [1, 2].

However, deep to the subcortical thick layer of white matter, also named centrum semi-ovale, a distinct region stands out: the central core of the brain. The central core is an ovoid shaped region at the morphological center of each cerebral hemisphere and its structures display a different pattern of organization. Whereas the cerebral lobes have a monotonous arrangement with grey matter on its surface and white matter underneath, the central core is formed by collections of grey matter that appear to be encapsulated by layers of white matter. Also, the disposition of grey and white matter roughly alternate sequentially from lateral to medial at this region. (Fig. 1a, b).

The central core is basically composed of the insular surface, basal ganglia, and thalamus, connected to the rest of the supratentorial compartment by the cerebral isthmus, represented by the continuation of the internal, external, and extreme capsules toward the cerebral lobes. It also harbors part of the anterior commissure, part of the amygdalofugal pathways, and the substantia innominata region [3]. During its phylogenetic and embryological development, the brain undergoes a process of circular curvature, which places each thalamus at the morphological center of each cerebral hemisphere and confers a

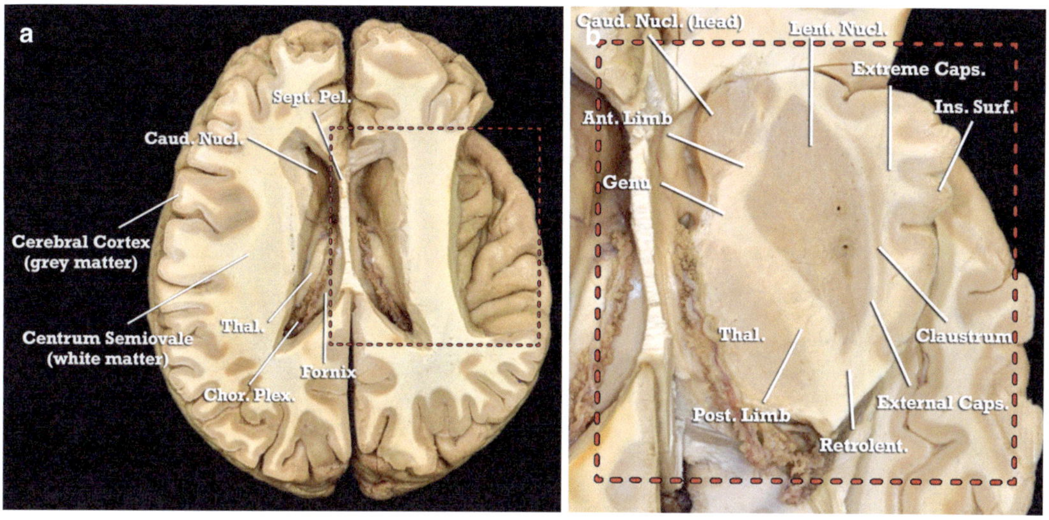

Fig. 1 The central core of the brain. (**a**) At this superior view, an axial cut was made inferior to corpus callosum and the right frontoparietal operculum was removed. Both lateral ventricular cavities can be seen, separated by the septum pellucidum at the midline. Each cerebral hemisphere harbors a central core at its morphological center, represented by the region between the caudate nucleus and the thalamus medially and the insular surface laterally. (**b**) an axial cut through the central core reveals its composition resembling layers, with grey matter structures and white matter capsules roughly alternating sequentially from lateral to medial at this region

C-shape form to nearby structures around the thalamus [4].

The central core is composed by structures with very different origins and functions, but are found topographically together forming this solid and well-delimited block between the brainstem and the cerebral lobes. From an organizational standpoint, the concept of a central core facilitates to understand the anatomical disposition of other encephalic regions: 1- the brainstem, formed by the medulla, midbrain and mesencephalon, continues superiorly with two central cores, being one at each side; 2- each central core is at the center of each cerebral hemisphere and the cerebral lobes are placed around it, being the frontal lobe anteriorly, the parietal lobe superiorly, the occipital lobe posteriorly, the temporal lobe inferiorly, the limbic lobe encircling the central core and the insular surface on its lateral surface (Figs. 2a, b and 3a–c). From a neurosurgical perspective, the central core concept of the brain is particularly justified since the complexity and possible morbidities of surgery in this region are very relevant [3].

a

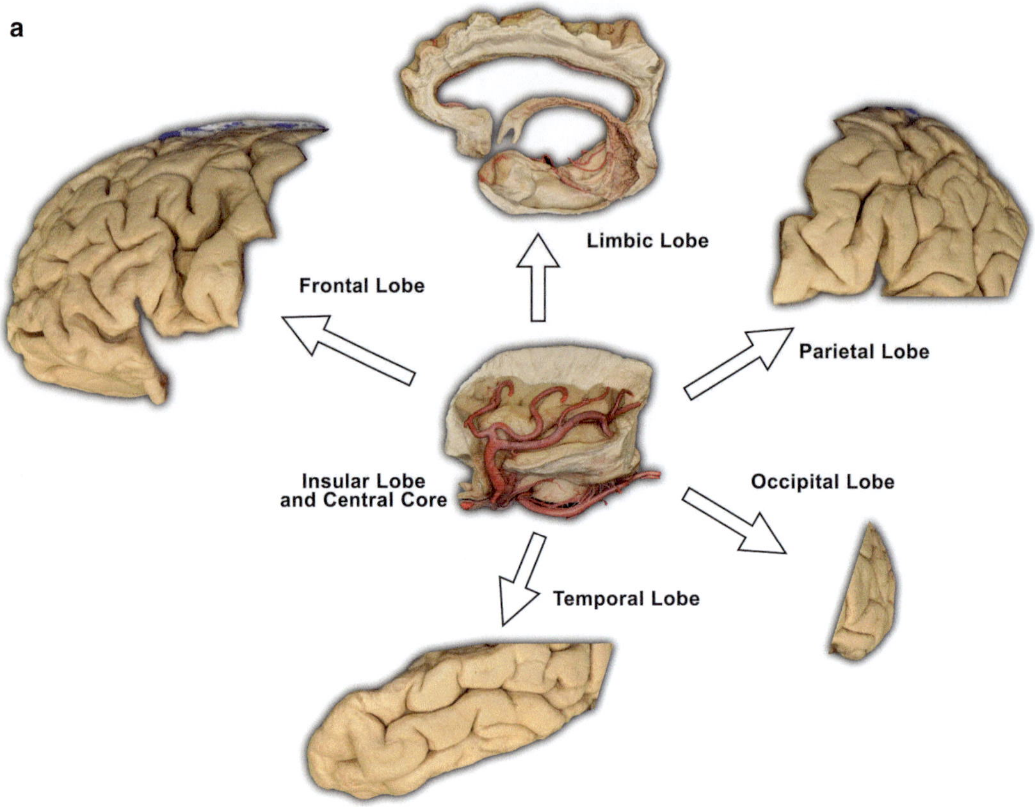

b

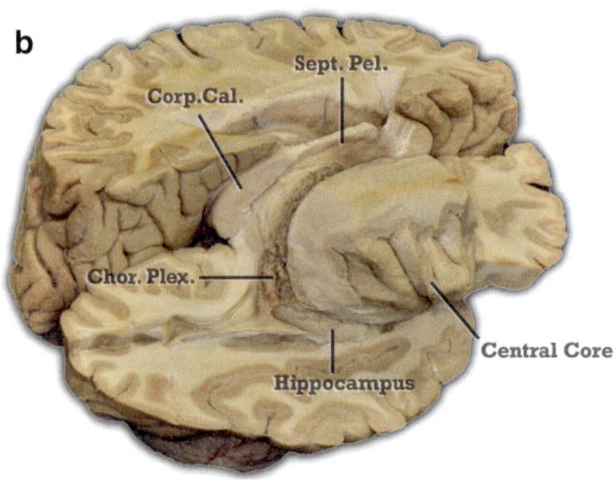

Fig. 2 The central core at the center of the brain. (**a**) Each central core is at the center of a cerebral hemisphere and the cerebral lobes are placed around it, being the frontal lobe anteriorly, the parietal lobe superiorly, the occipital lobe posteriorly, the temporal lobe inferiorly, the limbic lobe encircling the central core and the insular surface on its lateral surface. (**b**) The lateral ventricle also encircles the central core, with its extensions or horns dispositioned in relation to the thalamus: the frontal horn anteriorly, the body superiorly, the atrium posteriorly, the temporal horn inferiorly, and the occipital horn being a posterior extension of the atrium

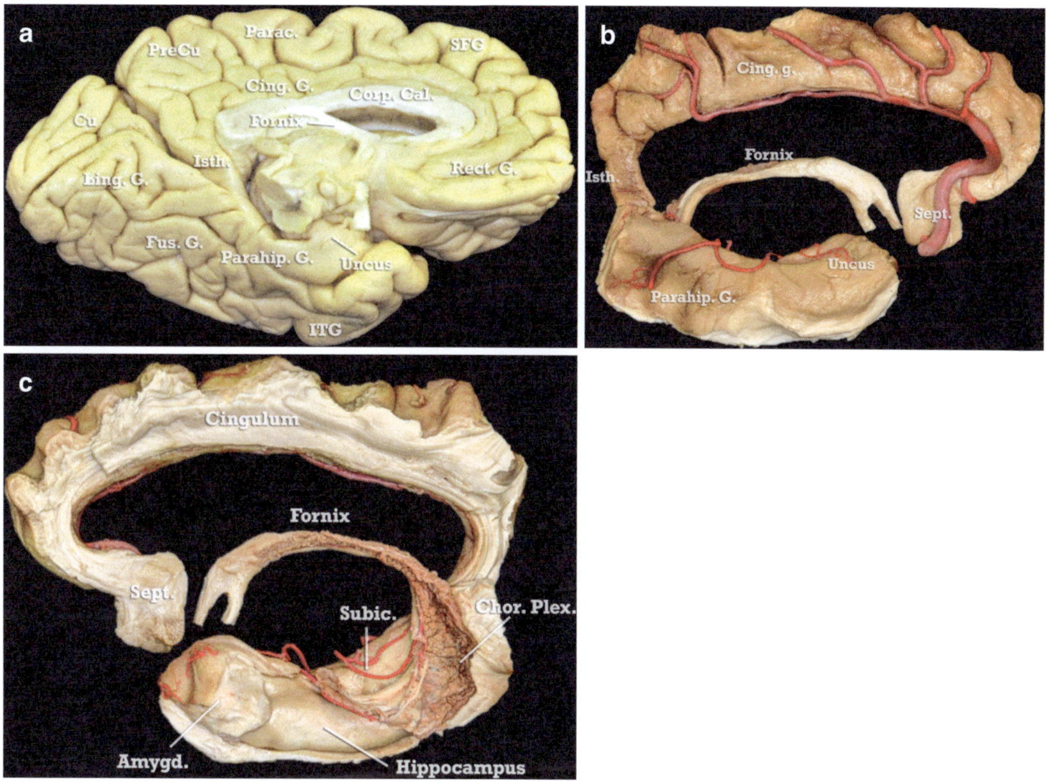

Fig. 3 The limbic lobe. (**a**) At the medial and basal surfaces of the brain, a group of structures form a ring around the central core. Due to their marginal arrangement around the central core, particularly the thalamus, these structures are called "limbic". (**b**) At the surface, a cortical limbic ring is formed by the septal region, the cingulate gyrus, the isthmus of the cingulate gyrus and the parahippocampal gyrus. (**c**) If this cortical ring is turned around, inner subcortical cortical structures can be seen also forming a limbic ring: amygdala, hippocampus, fornix and the cingulum

2 Defining the Central Core

The insular surface is the most lateral aspect of the central core, and its limits are also landmarks to define the central core region. A peri-insular, or circular, sulcus circumscribes the insular surface and can be further subdivided into an anterior, superior and inferior insular limiting sulci [5]. The limen insula is a hook-like structure found at the antero-inferior vertex of the insula and is connected to the anterior and inferior limiting sulci (ALS and ILS, respectively). This region is promptly and easily seen as the neurosurgeon splits the sylvian fissure and faces the insular surface, serving as a helpful intraoperative landmark. The middle cerebral artery (MCA) is found directly anterior to the limen insula and its trajectory bends at this point, transitioning from the M1 segment underneath the anterior perforated substance to the M2 segment spreading over the insular surface. As the limen insula joins the ALS and ILS, it defines other two relevant landmarks: the frontal and temporal limen points, respectively. The superior limiting sulcus (SLS) meets the ALS anteriorly and the ILS posteriorly, defining the anterior and posterior insular points, respectively, but since these points are underneath the fronto-parietal operculum and their surgical exposure requires extensive and laborious dissection of the sylvian fissure, their intraoperative use as landmarks is not routinely employed [3] (Fig. 4a).

The limits of the central core region can be explained by three cutting planes, which arise at

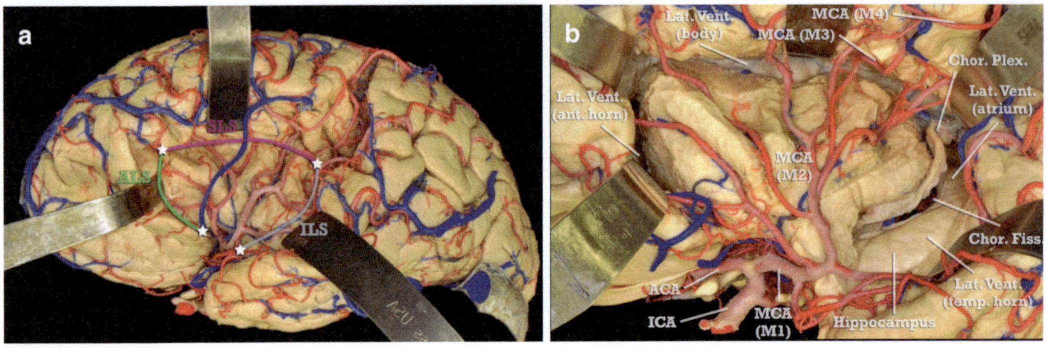

Fig. 4 The limits of the central core. (**a**) The insular surface is the most lateral aspect of the central core. A peri-insular, or circular, sulcus circumscribes the insular surface and can be further subdivided into an anterior, superior and inferior insular limiting sulci. The limen insula is a hook-like structure found at the antero-inferior vertex of the insula and is connected to the anterior and inferior limiting sulci (ALS and ILS, respectively. As the limen insula joins the ALS and ILS, it defines other two relevant landmarks: the frontal and temporal limen points, respectively. The superior limiting sulcus (SLS) meets the ALS anteriorly and the ILS posteriorly, defining the anterior and posterior insular points, respectively. (**b**) The limits of the central core region can be explained by three cutting planes, which arise at the insular limiting sulci laterally and meet the lateral ventricle medially. The anterior plane extends medially from the ALS to the frontal horn of the lateral ventricle superiorly and to the anterior limit of the septal region inferiorly. This anterior plane also extends inferiorly, terminating at the anterior limit of the anterior perforated substance (APS) at the base of the brain, defined as the medial and lateral olfactory striae. The superior cutting plane arises at the SLS and reaches the body of the lateral ventricle medially. From the ILS, another cutting plane arises: its posterior aspect reaches the atrium of the lateral ventricle medially and its inferior aspect reaches the temporal horn of the lateral ventricle inferiorly

the insular limiting sulci laterally and meet the lateral ventricle medially. The anterior plane extends medially from the ALS to the frontal horn of the lateral ventricle superiorly and to the anterior limit of the septal region inferiorly. With this plane, the head of the caudate nucleus and the septal region are included inside the central core region, while the orbitofrontal cortex, cingulate gyrus, and genu of the corpus callosum are excluded anteriorly. This anterior plane also extends inferiorly, terminating at the anterior limit of the anterior perforated substance (APS) at the base of the brain, defined as the medial and lateral olfactory striae. As result, the anterior perforated substance region, corresponding to the APS and structures located above and posterior to the APS (i.e., the nucleus accumbens, which corresponds to the ventral striatum; the Meynert nucleus; the ventral amygdalofugal pathways; and the substantia innominata, which roughly corresponds to the ventral globus pallidus) are included at the ventral aspect of the central core (Fig. 4b).

The superior cutting plane arises at the SLS and reaches the body of the lateral ventricle medially, including the body of the caudate nucleus inferiorly and excluding the corpus callosum superiorly. From the ILS, another cutting plane arises: its posterior aspect reaches the atrium of the lateral ventricle medially and its inferior aspect reaches the temporal horn of the lateral ventricle inferiorly, both including the tail of the caudate nucleus inside the central core.

From the lateral ventricle, these cutting planes continue further medially, passing through the choroidal fissure. This fissure is a natural cleft between the thalamus and the fornix in the lateral ventricle, extending in a C-shaped arc from the foramen of Monro and coursing around the superior, posterior, and inferior surfaces of the thalamus to its inferior termination, called the inferior choroidal point, which is located just behind the head of the hippocampus. Use of this fissure excludes most limbic structures (the parahippocampal gyrus, amygdala, hippocampus, and fornix) from the central core concept, but they have an intimate relation with it as they encircle this region, and the caudate nucleus and stria terminalis (running in the groove between the thalamus and caudate nucleus) are included inside the central core.

The anterior plane first reaches the midline just in front of the septal region and then turns posteriorly toward the foramen of Monro to enter the third ventricle, with its course being lateral to the columns of the fornix in order to keep this limbic structure outside the central core. The superior plane passes through the most superior aspect of the choroidal fissure and curves inferiorly, crossing the velum interpositum cistern and entering the roof of the third ventricle. The inferior plane connects the atrium and temporal horn of the lateral ventricle to the quadrigeminal and ambient cisterns, respectively, posteriorly and inferiorly. Since the choroidal fissure terminates at the inferior choroidal point, another cutting plane have to be chosen to exclude the temporo-limbic medial structures that are anterior to this point, mainly the amygdala the temporal pole. Since the internal carotid artery or the proximal segment of M1 are routinely exposed in vascular surgeries and most neurosurgeons are familiar to this anatomy, the use of the carotid-choroidal line might be more convenient [6]. This line is traced from the bifurcation of the internal carotid artery to the inferior choroidal point, just lateral to the optic tract and to the anterior choroidal artery course, and a cutting plane at this line will detach the temporo-limbic medial structures laterally and safeguard the central core medially (Fig. 5a–c).

The medial aspects of the central core can be seen inside the ventricular cavities. The head of

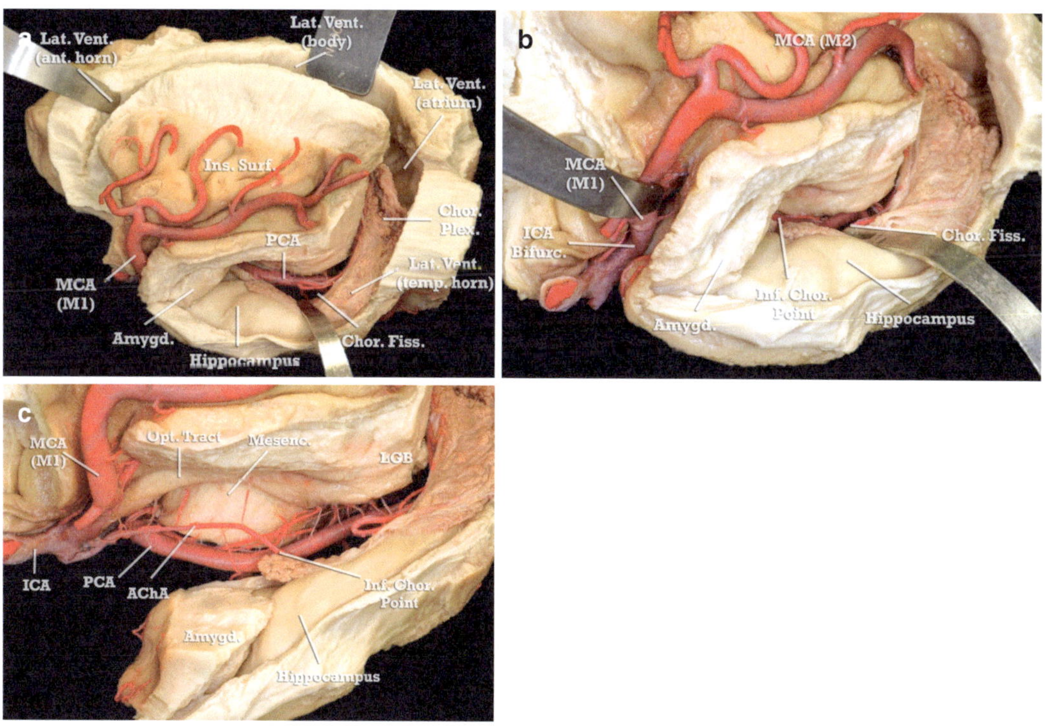

Fig. 5 The ventral aspect of the central core. (**a**) The choroidal fissure extends from the foramen of Monro to the inferior choroidal point, located inside the temporal horn and having the head of the hippocampus and amygdala anterior to this point. The hippocampus and fornix are found inside the lateral ventricle and can be detached from the central core by enlarging the choroidal fissure. (**b**) The amygdala continues superior and medially with the central core, particularly with the globus pallidus, and a cut needs to be make in order to separate both structures. (**c**) A line traced from the bifurcation of the internal carotid artery to the inferior choroidal point, named carotid-choroidal line, just lateral to the optic tract and to the anterior choroidal artery course, can be used a reference to make this cut. This plane finishes to detach the temporo-limbic medial structures laterally and safeguard the central core medially

the caudate nucleus is the anterior most medial structure of the core, and bulges inside the anterior horn of the lateral ventricle. It is encircled superior, anterior and inferiorly by a natural cleft that separates it from the trunk, genu and rostrum of the corpus callosum, respectively. The thalamus is the posterior most medial structure of the core, and its medial surface is the lateral wall of the third ventricle. The hypothalamic sulcus, present at this surface, is a shallow sulcus that extends from the foramen of Monro to the opening of the cerebral aqueduct and marks the transition from the thalamus, located superiorly, to the hypothalamus, located inferiorly and medially. The epithalamus consists of the pineal body, the right and left habenular nuclei, and the habenular and posterior commissures. Both the hypothalamus and the epithalamus are singular regions and are best understood as being separated from the central core, which is present in both hemispheres.

The thalamus is also the most inferior structure of the central core, but there is not a natural division between the diencephalon and mesencephalon. This transition is usually represented by a line drawn inferior to the optic tract and the lateral geniculate body (LGB). The metathalamus, which corresponds to the medial and lateral geniculate bodies, is the most inferolateral substructure of the central core.

3 Deep Structures

An axial cut of the central core easily reveals its composition roughly made of an alternating sequence of grey and white matters. The insular cortex is the most lateral and superficial aspect of the central core, followed medially by the extreme capsule underneath. This cortex is marked by a central sulcus, which divides into anterior and posterior regions. The first is composed by short insular gyri, usually three or four, that converge inferiorly forming the insular apex. The second is made by long insular gyri, usually two or three, that will not meet the insular apex. (Fig. 6a)

A careful dissection of this cortical layer and the extreme capsule exposes a thin sheet of the gray matter named claustrum (Fig. 6b). This collection of grey matter is more evident dorsally, where it's more densely packed, and ventrally has a loser arrangement that resembles islands grey matter embedded together with the white fibers of the extreme and external capsules conferring a more brownish appearance to this region. These portions are known as dorsal and ventral claustrum, respectively. Microscopic and histochemical investigation indicates the basolateral amygdaloid nuclei are reciprocally connected with the claustrum and develop from the same pallial sectors, mounting that these amygdaloid nuclei can be considered as a ventromedial extension of the claustrum by some authors. (Fig. 6c) [7].

The insula is connected to dorsal thalamus, amygdala and with a large number of cortical areas. Due to this vast connections and knowledge gathered from stimulation and ablation studies in animals and humans, the insular cortex has been implicated in olfactory, gustatory, viscerosensory, visceromotor, somatosensory and vestibular functions [7]. On the other hand, the functional importance of the claustrum remains unclear. Anatomically, it's improbable that the claustrum is related to low-level sensory processing, given the lack of inputs from sensory thalamic nuclei and sparse inputs from primary sensory cortices, or to direct generation and/or modification of motor output, given lack of connectivity with striatum. Crick and Koch proposed that this area could play a key role in consciousness due to its reciprocally widespread connectivity, including the frontal eye field, visual cortex and other frontal, posterior parietal and inferior temporal cortices [8].

The lentiform nucleus is found between the external and internal capsules, located laterally and medially respectively (Fig. 6d–f). It is entirely contained inside the margins of the central core, being not present under the depths of the insular limiting sulci. Its name derives from the its biconvex shape that resembles a glass lens, and is composed by the putamen nucleus and globus pallidus. The putamen is darker, more friable and larger, present underneath almost all insular area. The globus pallidus is medial to the putamen, and its identification is favored by its pale appearance and its harder consistency, as well as by its posterior (and superior) position in relation to the anterior commissure. Both are separated by an external medullary lamina, and the globus pallidus is subdivided by an internal medullary lamina into external and internal portions.

The caudate nucleus and the thalamus are the most medial structures of the central core and are bounded laterally by the internal capsule. The caudate has a head, which is larger, and becomes progressively thinner as it continues posteriorly encircling the thalamus and becoming its body and tail. Also, the caudate changes progressively its position in relation to the thalamus during its course, being anteriorly, supero-laterally, postero-laterally and infero-laterally to the anterior, superior, posterior and inferior thalamic surfaces, respectively.

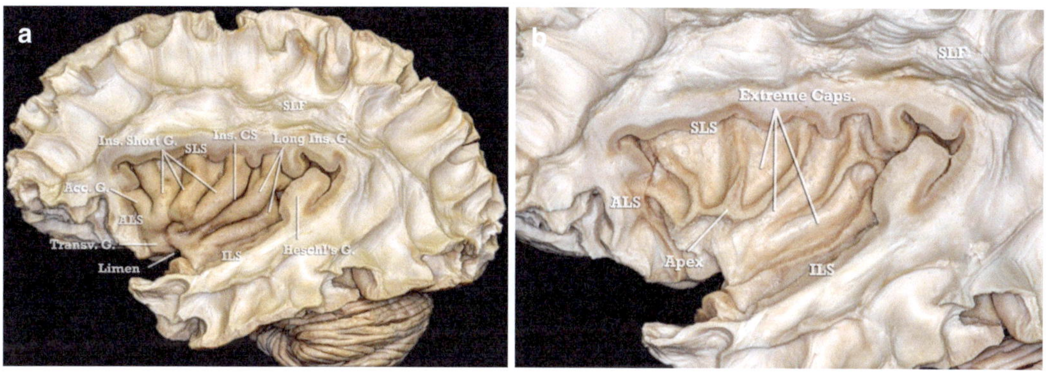

Fig. 6 Deep grey matter structures and white matter pathways related to the central core. (**a**) The insular cortex is the most lateral and superficial aspect of the central core, followed medially by the extreme capsule underneath. This cortex is marked by a central sulcus, which divides the insular cortex into anterior and posterior regions. The first is composed by short insular gyri, usually three or four, that converge inferiorly forming the insular apex. The second is made by long insular gyri, usually two or three, that will not meet the insular apex. (**b**) The extreme capsule is the first subcortical layer, found directly underneath the insular cortex. It is mainly composed of the short association fibers, interconnecting the insular gyri, and continues spreading under the insular limiting sulci to the opercula, linking the insular and opercular cortices. (**c**) Dissection deep to the extreme capsule exposes the claustrum. This a thin sheet of the grey matter is more evident dorsally, where it's more densely packed, and ventrally has a loser arrangement that resembles islands grey matter embedded together with the white fibers of the extreme and external capsules conferring a more brownish appearance to this region. The external capsule harbors the white matter pathways between the claustrum and the lentiform nucleus, located laterally and medially respectively. It is formed by 3 fiber bundles, all at the same depth, but arranged in a sequential anterior to posterior disposition: the uncinate fascicle, the inferior fronto-occipital fascicle, and the claustrocortical fibers. (**d**) The removal of IFOF fibers, exposes the lentiform nucleus. This nucleus is found between the external and internal capsules, and is formed by the putamen and the globus pallidus. The putamen is darker, more friable, larger and present underneath almost all insular area. (**e**) The removal of the putamen exposes the globus pallidus, whose identification is favored by its pale appearance and its harder consistency, as well as by its posterior (and superior) position in relation to the anterior commissure. The anterior commissure has a cylindrical appearance along the midline, and as it progresses laterally it divides into a very tiny and involuted ventral extension directed anteriorly toward each olfactory region and a lateral extension directed to the temporal lobe. The lateral extension passes through a tunnel (Gratiollet's canal) at the ventral aspect of the globus pallidus, continues under the putamen and underneath the ILS, and then progressively fans out along the dorsal aspect of the temporal lobe. The nucleus accumbens represents the ventral portions of the putamen and the caudate nucleus located anterior and inferior to the anterior commissure (*) (**f**) The internal capsule is exposed medial to the lentiform nucleus and represents most of the projection system of the brain. Its fibers have a radial disposition that, as they reach the outer margin of the central core represented by the caudate nucleus, fan out towards all cerebral cortices and receive the name "corona radiata". Among its fibers, the visual radiation can be demonstrated connecting LGB to the calcarine fissure. The removal of the nucleus accumbens demonstrates that IC fibers are not found inside this region (**)

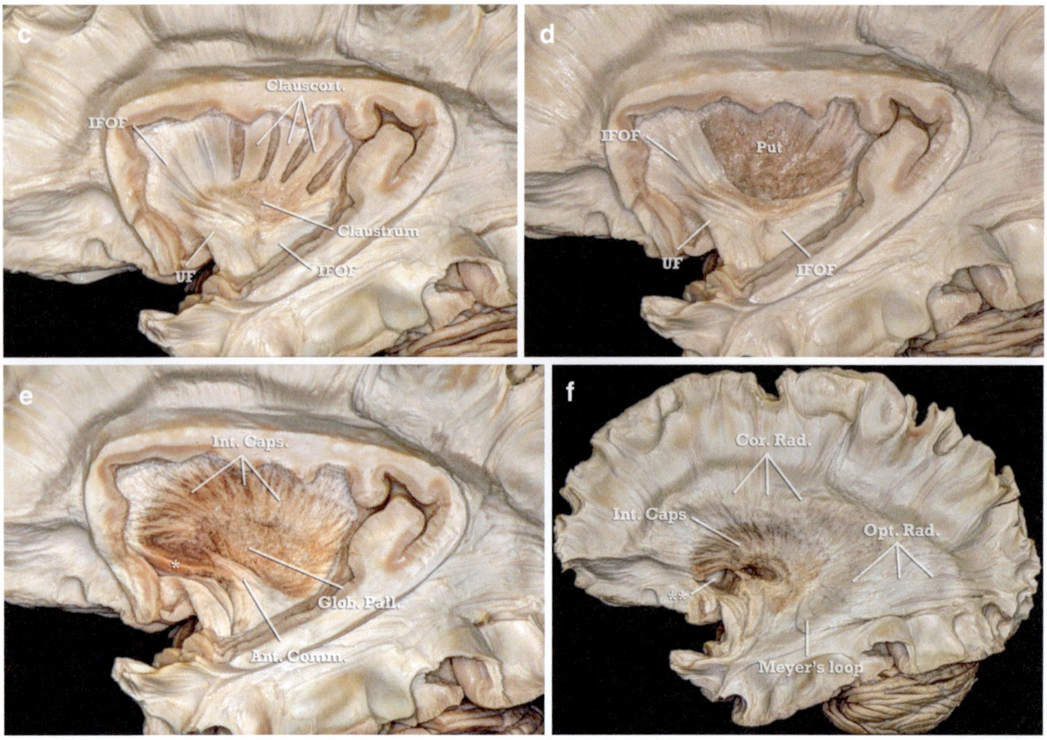

Fig. 6 (continued)

The thalamus is a well-defined ovoid structure, having its name derived from the ancient Greek meaning "chamber," and at that time referred to the innermost room of Greek mansions, usually reserved for the bridal couple. Six thalamic surfaces are described according to standard anatomic axes. Of these, the superior, medial, and posterior are free surfaces adjoining the ventricular cavities or cisternal spaces, whereas the anterior, lateral, and most of the inferior surface are covered by the internal capsule, the prethalamus and the mesencephalon [9].

The *lateral ventricle surface* of the thalamus is the free surface exposed inside the body, atrium and temporal horn of the lateral ventricle. This surface is bounded laterally by the strio-thalamic sulcus, which is found between the caudate nucleus and the thalamus and has the stria terminalis inside, and medially by the choroidal fissure, where the taenia choroidea attaches the choroid plexus to the thalamus [9].

The choroidal fissure is a natural cleft between the fornix and the thalamus, and holds the choroid plexus attached laterally to the thalamus by the taenia thalami and medially to the fornix by the taenia fornicis. This membrane continues further medially as the superior tela choroidea. On the medial aspect of the thalamus and inferior to the choroidal fissure, the inferior tela choroidea is found attached to the stria medullaris thalami. Both superior and inferior tela choroidea are thin membranes that stretches from one thalamus to the other, constituting the roof of the third ventricle and having a space in between called velum interpositum cistern, which harbors both internal cerebral veins and the postero-medial choroidal arteries distal segments [10]. The *velar surface* of the thalamus is exposed inside the vellum interpositum cistern.

The *third ventricle surface* of the thalamus makes the lateral wall of this ventricular cavity, and is bounded superiorly by the stria medullaris thalami and inferiorly by the hypothalamic sulcus, which is the inferior-medial limit of the central core. The *cisternal surface* of the thalamus is exposed at the posterior and inferior aspects of the thalamus, medial to the fornix, having the pulvinar and the geniculate bodies at this surface and opening into the quadrigeminal and ambient cisterns [9].

The thalamus is the major structure of the diencephalon, and stands on top the the mesechephalon without a natural division between them. This large ovoid mass has an external medullary lamina that divides the main thalamus, also known as the dorsal thalamus, from the reticular nucleus of the thalamus, which is named as the ventral thalamus and is a thin sheet of grey matter located between the lateral and ventrolateral surface of the dorsal thalamus medially and the internal capsule laterally. The dorsal thalamus is divided by an internal medullary lamina into a medial portion, which harbors the medial and dorsal thalamic nuclei, and a lateral portion, related to the lateral and ventral nuclei. This lamina also bifurcates anteriorly and separates the anterior thalamic nucleus from the rest of the dorsal thalamus [7].

Most thalamic nuclei are designated *specific nuclei* because they are reciprocally connected with specific areas or regions of the cerebral cortex and can be related to a particular cerebral function: (1) sensory relay, intermediating ascending sensory projections to the primary sensory fields of the neocortex, (2) motor relay, intermediating projections from the cerebellar nuclei and centres belonging to the extrapyramidal motor system, such as the substantia nigra and globus pallidus, to the motor and premotor cortical fields, (3) association, having strong reciprocal connections to cortical association areas, and (4) limbic nuclei, having connections to limbic cortices such the cingulate and entorhinal, presubiculum and parasubiculum. Other nuclei are called *non-specific thalamic nuclei* due to their diffuse cortical projections, and include the intralaminar, midline and posterior nuclei. Caudally, the ventral nuclei continue with the medial geniculate body which, together with the lateral geniculate body (LGB), is termed metathalamus [7]. If the inferior limit of the central core is established as the plane where the optic tract encircles the top of the mesechephalon to reach the LGB, the subthalamic region can be understood as partially included in the central core. This region is located between the dorsal thalamus and the hypothalamus and harbors the subthalamic nucleus and the zona incerta, located at the bottom of the reticular nucleus of the thalamus and also known as the thalamic fascicle or field H2 of Forel [3].

The *epithalamus* is formed by the stria medullaris, habenular complex and the pineal gland. The first is bundle of fibers that forms a ridge at the superior aspect of the medial surface of the thalamus, whose fibers originate at the septal region and the lateral preoptico-hypothalamic continuum, assemble posterior to the anterior commissure and terminate posteriorly at the habenular complex (some at the ipsilateral and others at the contralateral habenular complex by decussating through the habenular commissure). The habenular complex is a small collection of grey matter posterior and medial to both thalami that receives afferents mainly via the stria medullaris and project through the habenulo-interpeduncular tract (or fasciculus retroflexus) to the interpeduncular nucleus, located in the floor of the interpeduncular fossa. The pineal gland is regarded in mammals as a photosensitive neuroendocrine organ, which secrets mainly serotonin, norepinephrine, and melatonin in a circadian rhythm, suggesting that this gland functions as a biologic clock [7].

The hypothalamus is a diencephalic region located ventral to the hypothalamic sulcus, which is shallow sulcus found at the inferior aspect of the *third ventricle surface* of the thalamus. This sulcus extends from the foramen of Monro to the opening of the aqueduct, including at the hypothalamic region the lamina terminalis, optic chiasm, infundibulum, mammillary bodies and superior aspect of the mesencephalon. The third ventricle, hypothalamus, and epithalamus are grouped together in the midline, between the central cores of the left and right hemispheres.

The term basal ganglia includes the caudate nucleus, putamen, and the globus pallidus and three closely related structures: the substantia nigra, the ventral tegmental area and the subthalamic nucleus. Together with the cerebral cortex, the basal ganglia orchestrate the planned, motivated goal-directed behaviors, since the processes that lead to movement, including the elements that drive actions, such as emotions, motivation, and cognition, and the execution of these movements. Its circuitry can be separated into parallel and integrative networks, known as motor, cognitive, and limbic circuits, and each related to different parts of the basal ganglia. Its ventral regions are involved in reward and reinforcement behaviors, that culminate in habit formation and addictions; its more central areas are associated to cognitive functions such as procedural learning and working memory tasks; and the dorsolateral portion of the striatum, caudal to the anterior commissure, is related to movements control [11, 12].

The striatum, which comprises the caudate nucleus, putamen and accumbens, is the main input structure to the basal ganglia and have been subdivided into ventral and dorsal portions by Heimer et al. based on connectivity and functional considerations [13]. The concept of the ventral striatum (VS) includes the nucleus accumbens and the broad continuity between the caudate nucleus and putamen ventral to the rostral internal capsule and the anterior commissure, the olfactory tubercle and the rostrolateral portion of the anterior perforated space adjacent to the lateral olfactory tract. The ventral and medial borders of the VS are more clearly defined, but its dorsal and lateral aspects merge imperceptibly with the dorsal striatum, without a clear separation. The VS has been related to different aspects of reward processing and, while both the dorsal and ventral striatum receive input from the cortex, thalamus, and brainstem, the VS alone also receives a dense projection from the amygdala and hippocampus.

The substantia innominata should be understood as part of the ventral striatum and ventral pallidum, which are located inside a region defined by the APS and ansa peduncularis anteriorly and inferiorly, the anterior commissure posteriorly, and the anterior limb of the internal capsule superiorly. The fibers of the internal capsule converge from all cortical surfaces toward the central core but are not found anterior and inferior to the anterior commissure, resulting in an absence of clear separation between the grey matter nuclei at this region. Although different nuclei have been individualized inside the substantia innominata by means of different methods, [14] the lack of reliable macroscopic subdivisions makes this region better understood as a macroscopic continuous layer of gray matter, from lateral to medial, where all these nuclei are embedded and may blend together [3].

4 White Matter Fibers

The central core is connected to other encephalic regions by an extensive set of white matter pathways, including association, projection and commissural fibers. These fibers have been grouped and termed as white matter capsules because, at an axial cut, they seem to encapsulate the basal ganglia, but in fact they are not confined to the central core and continue to other regions. These capsules should rather be seen as compartments between the grey matter collections of the central core, each composed of different white matter fibers. Anatomical careful dissection, and radiological tractography, is able to follow the trajectory and cortical terminations of these fibers with detail and now each capsule can be understood as composite of different fiber bundles, each connecting different encephalic regions and having different neurological functions.

The extreme capsule is the first subcortical layer, found directly underneath the insular cortex. It is mainly composed of the short association fibers, interconnecting the insular gyri, and continues spreading under the insular limiting sulci to the opercula, linking the insular and opercular cortices (Figs. 6b and 7a, b)

The external capsule harbors the white matter pathways between the claustrum and the lentiform nucleus, located laterally and medially respectively. It is formed by 3 fiber bundles, all at the same depth, but arranged in a sequential anterior to posterior disposition: the uncinate fascicle (UF), the inferior fronto-occipital fascicle (IFOF), and the claustrocortical fibers. The UF is found directly posterior to the limen insula, assuming its same hook-like shape, and its fibers connect the fronto-orbital region to the anterior temporal region. As the UF fibers leave the central core region, they pass deep to the ALS inferior aspect and the ILS anterior aspect. The IFOF is located directly posterior to the UF, and its fibers pass superiorly, under the ALS and the most anterior part of the SLS, to reach the middle and superior frontal gyri, and inferiorly, under the ILS, to reach the posterior temporal, parietal, and

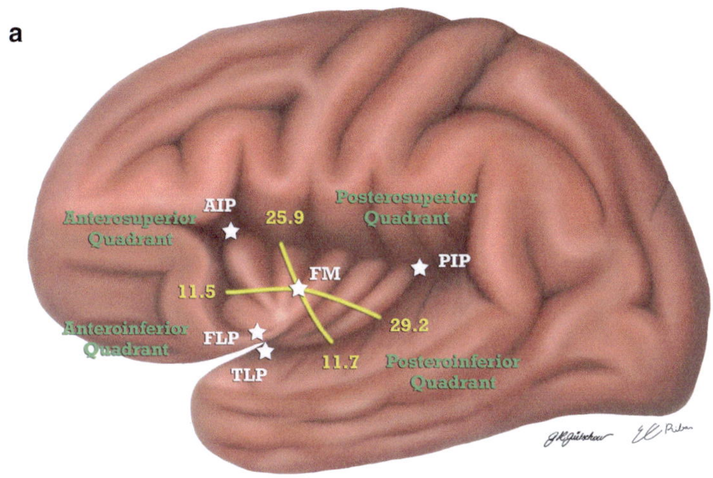

Fig. 7 Illustration of main white matter pathways present deep to the insular limiting sulci. (**a**) A subdivision of the central core into quadrants represents an attempt to better understand this region and to possibly lead to more precise and safer surgeries. A vertical line informs the neurosurgeon that posteriorly lies the genu and the posterior limb of the internal capsule, comprising not only the pyramidal tract but also the superior thalamic radiations related to the somatosensory information. The anterior limit of these fibers inside the central core can be used to separate this region into an anterior compartment, related to the anterior limb of the internal capsule, and a posterior compartment, related to the more eloquent parts of the internal capsule. This limit can be drawn over the insular surface by a vertical line that connects the most anterior aspect of the precentral sulcus (found 25.9 mm posterior to the AIP along the SLS) to the genu of the internal capsule and to the anterior limit of the cerebral peduncle (found 11.7 mm posterior to the TLP along the ILS). Since the foramen of Monro, is at the same coronal plane as the genu of the internal capsule, its lateral projection over the insular surface can be used as the intermediate point of this vertical line. A horizontal line represents the floor of the lateral ventricle (found 11.5 mm superior to the FLP along the ALS) and the inferior margin of the thalamus, represented by the LGB (found 29.2 mm posterior to the TLP along the ILS). (**b**) The extreme capsule is mainly composed of the short association fibers, interconnecting the insular gyri, and continues spreading under the insular limiting sulci to the opercula, linking the insular and opercular cortices. Its fibers are seen deep to all insular limiting sulci. (**c**) The external capsule is formed by 3 fiber bundles, all at the same depth, but arranged in a sequential anterior to posterior disposition: the uncinate fascicle, the inferior fronto-occipital fascicle, and the claustrocortical fibers. Deep to the ALS, a transition point located 11.3 mm superior FLP marks the transition between the uncinate fascicle and the inferior fronto-occipital fascicle, with the first present from the FLP to this transition point and the second from this point to the AIP. Deep to the SLS, a transition point located 9.8 mm posterior to the AIP marks the transition between the inferior fronto-occipital fascicle and the claustrocortical fibers, with the first present from the AIP to this transition point and the second from this point to the PIP. Deep to the ILS, the uncinate fascicle is seen from the TLP to 10.0 mm posterior, followed by the inferior fronto-occipital fascicle from this point until 35.5 mm posterior to the TLP, and later claustro-cortical fibers are found posteriorly until the PIP. (**d**) The anterior commissure is found deep to the external capsule fibers and its fibers fan towards the temporal lobe, passing underneath the ILS. Deep to the ILS, anterior commissure fibers are located between 8.4 mm and 22.0 mm posterior to the TLP. (**e**) The internal capsule fibers are found deep to most insular limiting sulci extension. The fibers related to its anterior limb are found deep to the entire ALS length and to the SLS (from the AIP to 25.9 mm posterior). The optic radiation fibers can the distinguished from the other for Int. Caps. fibers due to its trajectory from the LBG to calcarine fissure; these fibers are located deep to the ILS between 10.6 mm and 34.5 mm posterior to the TLP. The other for Int. Caps. fibers (posterior to the anterior limb and superior to the optic radiation) are found deep to SLS (between 25.9 mm posterior to AIP and the PIP) and to the ILS (between 34.5 mm posterior to TLP and the PIP)

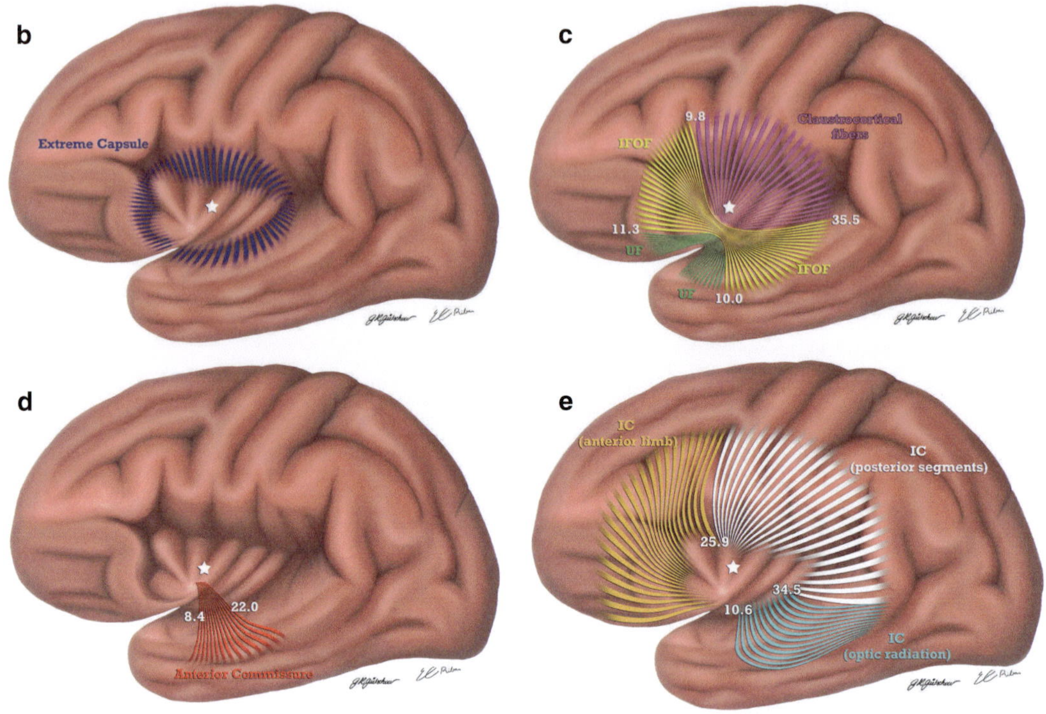

Fig. 7 (continued)

occipital regions. The claustrocortical fibers emerge from the claustrum in a radial pattern, more clearly identified arising from its dorsal portion, and pass under the SLS and the most posterior part of the ILS toward the brain convexity. (Figs. 6c, d and 7c)

The UF and IFOF pass through the antero-inferior aspect of the central core, at the ventral claustrum. Since the ventral claustrum is formed by a loose pattern of gray matter islands, the UF and IFOF are found passing superficially and underneath these islands in this region. The claustrocortical fibers, as they arise from the dorsal claustrum, are also seen lateral and medial to the claustrum. As a result, the UF, IFOF and claustrocortical fibers can be understood as contributing to form both the extreme and external capsules.

The anterior commissure connects mostly temporal mesial regions of both hemispheres that are not connected by the corpus callosum. It crosses the interhemispheric fissure perpendicular to it, dividing the columns of the fornix of each side into a major postcommissural component (directed to each mammillary body) and a smaller precommissural component (directed to each septal region). The anterior commissure has a cylindrical appearance along the midline, and as it progresses laterally it divides into a very tiny and involuted ventral extension directed anteriorly toward each olfactory region and a lateral extension directed to the temporal lobe. The lateral extension passes through a tunnel (Gratiollet's canal) at the ventral aspect of the globus pallidus, continues under the putamen and underneath the ILS, and then progressively fans out along the dorsal aspect of the temporal lobe, joining the sagittal stratum as it courses laterally (Figs. 6e and 7d).

The internal capsule Int. Caps. lies medial to the lentiform nucleus and represents most of the projection system of the brain—most importantly, the pyramidal tract, corticopontocerebellar tracts, cortico-striatal circuits and thalamic radiations. The emergence of the internal capsule separates the striatum into the putamen laterally and the caudate nucleus medially, and leaves the thalamus medial and posterior. This thick group of fibers is classically divided into an anterior limb, genu, posterior limb, and retrolenticular and sublenticular portions. At an axial cut, the internal capsule has V-shape, with its vertex pointing medially to the foramen of Monro, the anterior limb bounded laterally by the putamen and medially by the head of the caudate nucleus, the posterior limb bounded laterally also by the putamen but medially by the thalamus, and the retrolenticular and sublenticular portions at the posterior and inferior aspects of the putamen, respectively (Figs. 6f and 7e).

The Int. Caps. fibers have a radial disposition that, as they reach the outer margin of the central core represented by the caudate nucleus, fan out towards all cerebral cortices and receive the name "corona radiata". Among all Int. Caps. fibers, the pyramidal tract can be traced by anatomical careful dissection and radiological tractography due to its more known trajectory, including the precentral gyrus, genu and posterior limb of Int. Caps., and medullary pyramids [15]. The visual radiation can be demonstrated connecting LGB to the calcarine fissure, and further subdivided into 3 different 3 bundles: an anterior bundle (FOR Meyer's loop) that first has an anterolateral direction and later bends posteriorly to end along the inferior lip of the calcarine fissure; a central bundle that initially follows a lateral direction and then turns posteriorly to end at the lateral aspect of the occipital pole; and a posterior bundle that courses posteriorly and ends along the superior lip of the calcarine fissure [16]. The auditory radiation, present at the sublenticular portion of the internal capsule, emerges from the medial geniculate body and projects to the most anterior transverse temporal gyrus, also called Heschl's gyrus.

Basal ganglia circuitry is topographically and functionally organized into parallel and integrative networks, operating to execute goal directed behaviors. Its afferent projections are derived from three major sources: [1] cerebral cortex; [2] thalamus; and [3] brainstem. Although several cortical region project to the basal ganglia, the frontal cortex and the striatum stand out having massive projections from the first to the second, making the striatum the main input structure to the basal ganglia.

The frontal cortex in primates can be divided into several functional regions: the orbital and medial prefrontal cortex (OMPFC), involved in emotions and motivation; the dorsolateral prefrontal cortex (DLPFC), involved in higher cognitive processes or 'executive functions'; the premotor and motor areas, involved in motor planning and the execution of those plans. Furthermore, prefrontal, premotor, and motor cortices are thought to be organized in a hierarchical manner, with motor cortex being the final step to action. Inputs from the cortex and thalamus terminate in specific regions of the striatum: the dorsolateral striatum receives cortical input from sensory-motor areas, the central striatum receives input from associative cortical areas (DLPFC), and the ventromedial striatum receives input from limbic areas (OMPFC).

The striatum, in turn, projects topographically to the pallidal complex (which includes external and internal segments of the globus pallidus and the ventral pallidum) (GPe, GPi and VP, respectively) and the substantia nigra (including both the pars reticulata and pars compacta) (SNr and SNc, respectively). The GPi and GPe receive inputs from the dorsal striatum (from both the caudate nucleus and putamen) and the VP, defined at the globus pallidum portion located ventral to the anterior commissure, receives its input from the VS. Three pathways at these connections can be considered:

(1) a direct "cortico–basal ganglia pathway" projecting from the GPi and SNr to the thalamus and then to the cortex; (2) an indirect pathway with projections from the GPe and parts of the VP to the subthalamic nucleus (STN) and then to GPi, to later continue to thalamus and then to the cortex; and (3) an hyperdirect pathway comprising frontal cortex direct projections to the STN [11].

The thalamic peduncles comprise the corticothalamic and thalamocortical fibers that belong to the internal capsule and enter the thalami. They are located medial to the long corticofugal pathways that descend to the brain stem and spinal cord, and each can be found at the different internal capsule's segments. The anterior thalamic peduncle is found at the anterior limb of the internal capsule and has reciprocal connecting fibers to prefrontal and orbitofrontal parts of the cortex and cingulate gyrus. The genu of the internal capsule contains the most anterior fibers of the superior thalamic peduncle, while the posterior limb has the superior and posterior thalamic peduncles, with fibers connecting reciprocally the thalamus to the central, parietal and occipitotemporal areas. The retrolenticular portion of the internal capsule also has posterior thalamic peduncle fibers [7, 17, 18].

The inferior thalamic peduncle, named in the literature as an extracapsular thalamic peduncle since it is not placed between the caudate nucleus and the putamen like the fibers of the internal capsule, contains fibers connecting the thalamus to the orbitofrontal, insular and temporal cortices and the amygdaloid complex [3]. Other fibers arising from the amygdala are found at this region, and are collectively named as ventral amygdalofugal pathway system which includes amygdaloseptal fibers connecting the amygdala and anterior temporal cortex to the septal region and amygdalohypothalamic fibers project from the amygdala to the hypothalamus. Together with the amygdalothalamic fibers that project to the medial thalamic region, these fibers have been termed "ansa peduncularis" in the literature, because they sweep around the cerebral peduncle and occupy the space between the anterior perforated substance inferiorly and anterior commissure superiorly and posteriorly [19].

The globus pallidus is lateral to the internal capsule, and most of its projections terminate at the substantia nigra, in the midbrain. Other projections from the globus pallidus are represented by the lentiform fascicle and "ansa lenticularis", the first passing superiorly and the latter passing inferiorly to the subthalamic nucleus, both reaching the zona incerta medially and coming together as the thalamic fascicle. The subthalamic fascicle is a small connection from the globus pallidus directly to the subthalamic nucleus. The "ansa lenticularis" is another concept, that refers to the connections between the thalamo-subthalamic region and the globus pallidus within the ansa peduncularis. Its fibers can be individually recognized in tractography, showing connections between the globus pallidus and red nucleus, substantia nigra, subthalamic nucleus, and thalamus [20].

The mammillothalamic tracts represent an important connection between the hypothalamus and the central core. They are located within the lateral wall of the third ventricle, projecting from each mammillary body to the anterior thalamic nuclei (limbic thalamic nucleus). The stria terminalis is another small bundle of fibers that connects to the limbic system. It originates at the amygdala, continues posteriorly in a groove between the caudate nucleus and the thalamus, and ends under the head of the caudate at the bed nucleus of the stria terminalis [21, 22].

The corticopontocerebellar pathways arise from all cortical surfaces of the brain and come together as they course toward the cerebral peduncle, as well to constitute the transverse pontine fibers, which project to the cerebellum along the middle cerebellar peduncle [7]. Among them, the frontopontine fibers arise from the frontal cortex and are found at the anterior limb of the internal capsule (inside the central core) and later the medial third of the cerebral peduncle at the midbrain, while the other cortical corticopontine fibers that arise from other cerebral cortices occupy the more posterior segments of the internal capsule and the lateral third of the cerebral peduncle. The superior cerebellar peduncle is the main cerebellar output, and most of these fibers run from the dentate nucleus of the cerebellum toward the red nucleus at the midbrain and then the ventral lateral nuclei of the thalamus.

The medial lemniscus is formed by the spinothalamic tracts, carrying sensory information from the spinal cord to the medial and lateral ventral posterior nuclei of the thalamus. The fasciculus retroflexus connects the interpeduncular fossa to the habenular nucleus, and although these regions are not inside the central core per se, its course is at the inferomedial aspect of the central core [3].

The term "temporal stem" refers to all white matter fibers of the temporal lobe that converge and pass underneath the ILS toward the central core. The temporal stem can be divided into an anterior and medial component, which resembles an anterior temporal peduncle (formed by the ventral amygdalofugal fibers, UF, anterior aspects of the IFOF and of the anterior commissure, and the superior extension of the amygdala toward the globus pallidus), and a posterior and lateral component, constituted by the fibers that pass underneath the ILS between the tip of the temporal horn and LBG (posterior aspects of IFOF and anterior commissure, and optic radiation fibers) [5].

5 Vascularization

The arterial supply and venous drainage of the cerebral lobes are represented by major vessels over their cortical and ventricular surfaces, which send smaller vessels into the brain parenchyma. This typical pattern of vascularization of the brain is also present at the central core, related to the middle cerebral artery branches that supply the lateral aspect of the central core. However, it is interesting to note that, in addition to this pattern of vascularization, the central core also has deep and penetrating vessels related to its deep structures that are not found in other parts of the brain. Lenticulostriate arteries and striate veins related to the anterior and deep part of the central core converge at the anterior perforated substance (APS), and thalamoperforating arteries and inferior thalamic veins related to the posterior and deep part of the central core converge at the posterior perforated substance (Fig. 8).

The anterior perforating arteries, also named lenticulostriate arteries (LSA), are the group of numerous and small arteries that arise from the internal carotid, anterior choroidal, and anterior and middle cerebral arteries and enter the brain through the APS. This region is a rhomboid-shaped area at the basal surface of the brain, buried deep in the sylvian fissure stem, bounded anteriorly by the lateral and medial olfactory striae, posteriorly by the optic tract and the temporal lobe, laterally by the limen insulae; medially, it extends above the optic chiasm to the interhemispheric fissure. The APS can be divided into medial and lateral territories by an anterior to posterior oriented line projecting posteriorly from the olfactory tract, and further into anterior, middle, and posterior zones by diving the APS by other lines perpendicular to the first. Anatomical studies have associated each group of LSA to particular APS regions and supply of central core structures (Table 1) [23].

As these LSA ascend, they pass both anterior and posterior to the lateral extension of the anterior commissure, forming 2 peduncles (Fig. 8g). The peduncle anterior to the anterior commissure reaches the putamen and continues medially to the anterior part of the internal capsule, while the peduncle posterior to the anterior commissure supplies the globus pallidus and continues medially to the posterior part of the internal capsule. The caudate nucleus is supplied by the same arteries, which continue medially inside the internal capsule toward this nucleus. The most anterior and medial of these lenticulostriate arteries, which include the recurrent artery of Heubner (RAH), was found to ascend directly into the head of the caudate nucleus [3].

After penetrating the APS, the medial and lateral LSA and RAH have similar branching pattern. The primary vessels give rise to two to four secondary branches, each of which leads to tertiary branches, at the terminals of which are small pre-capillary arteriole envelopes. Their course is arched medially: first running through the most lateral components of the basal ganglia (putamen and globus pallidus), to later continue towards the IC and then to the caudate nucleus. Few anastomoses occur between these penetrating vessels or their branches. Also, penetrating vessels supplying the dorsal striatum contribute very little to the basal forebrain, despite passing through it. These primary, secondary and most of the tertiary branches of the medial and lateral LSA and the RAH remain within a particular zone of the striatum, being possible to associate each group to a region within the striatum and its related functional role: (1) the lateral lenticulostriate arteries (LSA) and sensorimotor striatum; (2) the medial LSA and associative striatum; and (3) the RAH and limbic striatum [24] (Fig. 9).

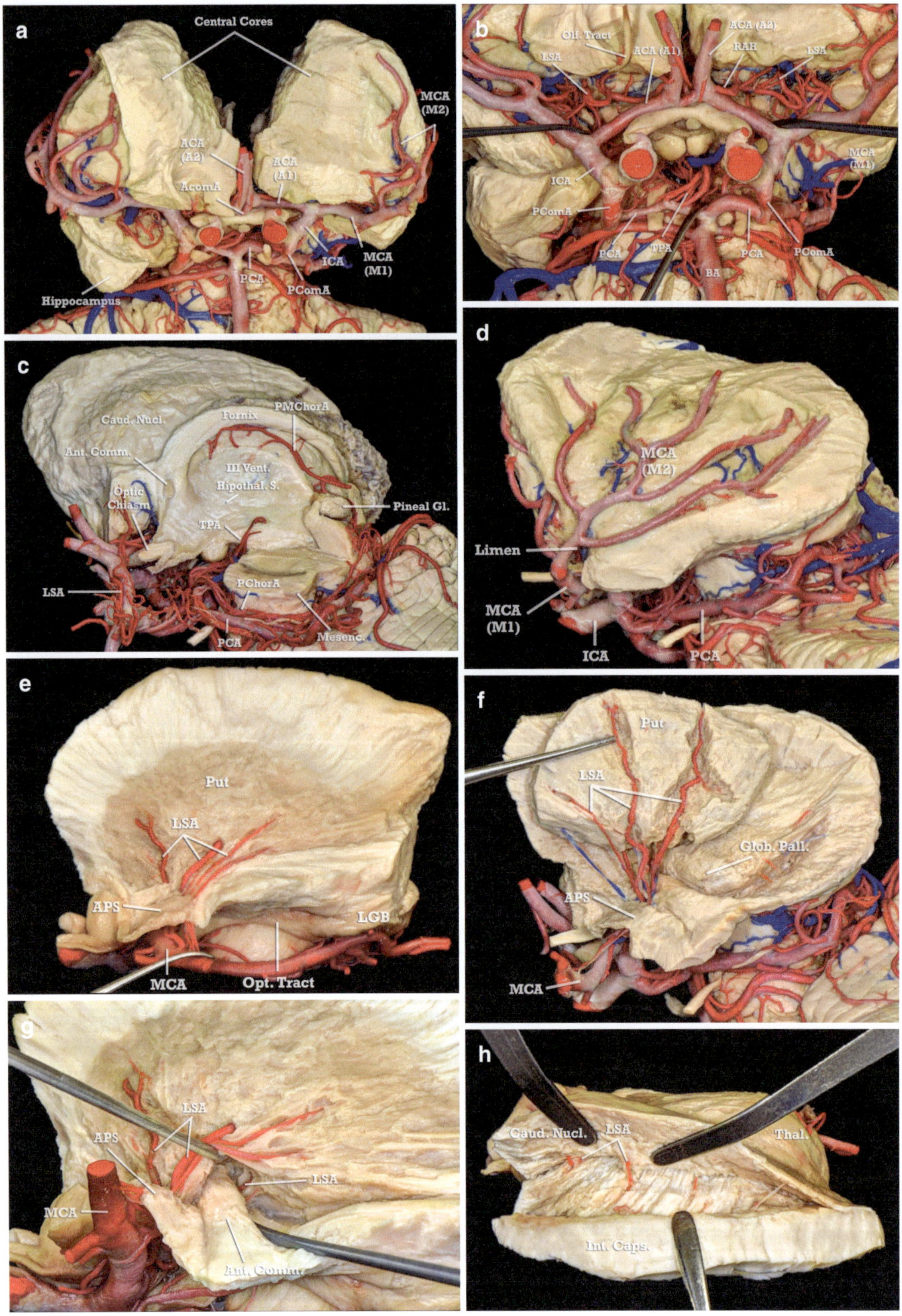
a
Central Cores
MCA (M2)
ACA (A2)
ACA (A1)
AcomA
ICA
MCA (M1)
PCA
PComA
Hippocampus
b
Olf. Tract
ACA (A1)
ACA (A2)
RAH
LSA
LSA
MCA (M1)
ICA
PComA
PCA
TPA
PCA
PComA
BA
c
Caud. Nucl.
Fornix
PMChorA
Ant. Comm.
III Vent.
Hipothal. S.
Optic Chiasm
Pineal Gl.
TPA
LSA
PChorA
PCA
Mesenc.
d
MCA (M2)
Limen
MCA (M1)
ICA
PCA
e
Put
LSA
APS
LGB
MCA
Opt. Tract
f
Put
LSA
Glob. Pall.
APS
MCA
g
LSA
APS
LSA
MCA
Ant. Comm.
h
Caud. Nucl.
LSA
Thal.
Int. Caps.

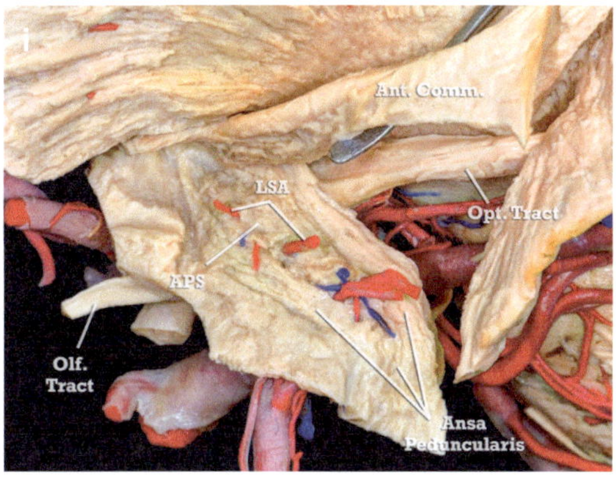

Fig. 18.8 (continued)

Small branches from the posterior communicating artery (PComA) penetrate the floor of the third ventricle between the optic chiasm and carotid peduncle to reach the hypothalamus, thalamus, subthalamus, and internal capsule. The premamillary artery is the largest branch of PComA, which penetrates the diencephalon just rostolateral to the mamillary body, and may irrigate the caudal part of the hypothalamus (the lateral mamillary nucleus, the posterior hypothalamic area and the caudal portion of the lateral hypothalamic area); the medial part of the subthalamus (the substantia nigra, Forel's field and the zona incerta); and the rostral portion of the thalamus. The premamillary artery may also originate optic branches, to supply a small ventromedial part of the optic tract, and peduncular branches to supply ventromedial part of the cerebral peduncle [25].

The small thalamoperforating arteries that arise from the basilar tip and proximal segments of the posterior cerebral artery pierce the posterior perforated substance, located behind the optic tracts and along the roof of the interpeduncular fossa, and are directed into the thalamus and posterior parts of the internal capsule. The medial posterior choroidal artery encircles the

Fig. 8 Arteries of the Central Core. (**a**) The central core has two patters of vascularization. The first is related to the middle cerebral artery branches that supply the lateral aspect of the central core, mainly branches arising the M2 and entering the insular surface. The second is composed by several small arteries that originate from the polygon of Willis, located inferior to central core, that penetrate the anterior and posterior perforated substances to supply deep portions of the central core. (**b**) The anterior perforating arteries, also named lenticulostriate arteries (LSA), are the group of numerous and small arteries that arise from the internal carotid, anterior choroidal, and anterior and middle cerebral arteries and enter the brain through the APS to supply the anterior portions of the central core. The thalamoperforating arteries arise from the basilar tip and proximal segments of the posterior cerebral artery pierce the posterior perforated substance, located behind the optic tracts and along the roof of the interpeduncular fossa, and are directed into the thalamus and posterior parts of the internal capsule. (**c**) The medial posterior choroidal artery encircles the midbrain, ascending posterior to it, and enters the velum interpositum cistern; along its course, it supplies the inferolateral and finally the medial aspects of the thalamus. (**d**) The middle cerebral artery, first running under the APS and sending lenticulostriate branches, reaches the limen insula and bends posteriorly over the insular surface. At the limen, or close to it, the middle cerebral artery bifurcates (or trifurcates) into trunks, having the inferior trunk coursing close to the inferior limiting sulcus and the superior trunk has a posterosuperior course directed toward the superior limiting sulcus. These trunks, as well as their several main branches, send small arteries into the insular surface to supply its cortex and whose course usually do not reach too deep into the insular subcortical white matter. (**e**, **f**) After penetrating the APS, the LSA first runs through the most lateral components of the basal ganglia (putamen and globus pallidus). (**g**) As these LSA ascend, they pass both anterior and posterior to the lateral extension of the anterior commissure, forming 2 peduncles. The peduncle anterior to the anterior commissure reaches the putamen and continues medially to the anterior part of the internal capsule, while the peduncle posterior to the anterior commissure supplies the globus pallidus and continues medially to the posterior part of the internal capsule. (**h**) Superior view of the central core, with caudate nucles being separated from the internal capsule. The LSA course is arched medially: first running through the most lateral components of the basal ganglia (putamen and globus pallidus), to later continue towards the IC and then to the caudate nucleus. (**i**) The APS is seen from superiorly, and the anterior commissure is displaced by the instrument. Ventral amygdalofugal pathways, named collectively as "ansa peduncularis, occupy the space between the anterior perforated substance inferiorly and anterior commissure superiorly and posteriorly midbrain, ascending posterior to it, and enters the velum interpositum cistern; along its course, it supplies the inferolateral and finally the medial aspects of the thalamus [3, 23].

The middle cerebral artery, first running under the APS and sending lenticulostriate branches, reaches the limen insula and bends posteriorly over the insular surface. At the limen, or close to it, the middle cerebral artery bifurcates (or trifurcates) into trunks, having the inferior trunk coursing close to the inferior limiting sulcus and the superior trunk having a posterosuperior course

Table 1 Each group of LSA has been related to most often pierce a particular APS regions and supply some central core structures

Origin	APS region	Supply
A1	Medial half	IC genu, and the adjacent part of the GP, IC posterior limb and thalamus
ICA	Posterior part of the central region	
AComA		GP medial two segments, IC posterior limb (inferior part) and thalamus (anterior and ventrolateral nuclei)
MCA - medial LSA	Middle and posterior portions of the lateral half	GP lateral part, IC anterior limb (superior part) and CN head (anterosuperior part)
MCA - intermediate LSA		Pass through the put and the adjacent GP, then arch medially and posteriorly to supply almost the entire IC anterior-to-posterior length (superior part), CN body and head
MCA - lateral LSA		
RAH	Anterior half of the full mediolateral region	CN head (most anterior and inferior part), put, and the adjacent part of IC anterior limb

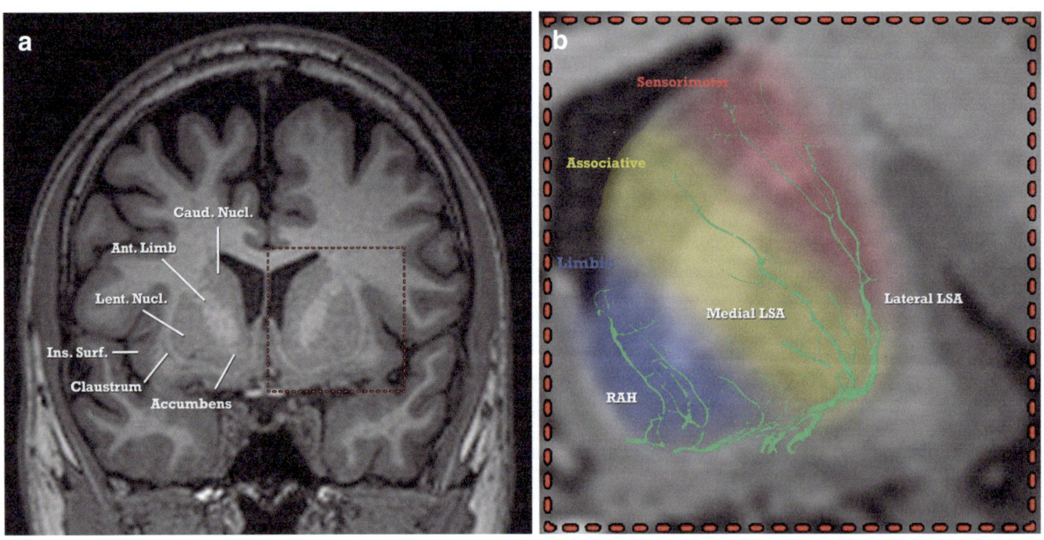

Fig. 9 Coronal view of an MRI study at the central core level. (**a**) The anterior limb of the internal capsule divides the striatum, with the putamen laterally and the caudate nucleus medially. The nucleus accumbens in found inferior to the anterior limb, representing a striatal nucleus that have not been divided by internal capsule fibers. (**b**) Basal ganglia circuitry is topographically and functionally organized into parallel and integrative networks, with inputs from the cortex and thalamus terminating in specific regions of the striatum: the dorsolateral striatum receives cortical input from sensory-motor areas, the central striatum receives input from associative cortical areas (DLPFC), and the ventromedial striatum receives input from limbic areas (OMPFC). Also, anterior perforating arteries remain within a particular zone of the striatum, being possible to associate each group to a region within the striatum and its related functional role: (1) the lateral lenticulostriate arteries (LSA) and sensorimotor striatum; (2) the medial LSA and associative striatum; and (3) the RAH and limbic striatum

directed toward the superior limiting sulcus. These trunks, as well as their several main branches, send small arteries into the insular surface to supply its cortex and whose course usually do not reach too deep into the insular subcortical white matter, mostly stopping before the claustrum [3]. Ture et al. reported that approximately 85–90% of insular arteries are short and supply the insular cortex and extreme capsule; 10% are medium sized and also supply the claustrum and external capsule; and the remaining 3–5% are long and extend as far as the corona radiata [26]. These long insular arteries are perforator-like and mostly located in the posterior region of the insula.

The venous drainage of the central core can be subdivided into a cisternal, a deep, and a ventricular group. The cisternal group is composed of the insular veins that drain the insular cortical surface and are directed to the limen insula, where they join the deep sylvian vein or the deep middle cerebral vein to later empty into the anterior part of the basal vein of Rosenthal. Occasionally, this group terminates at the sphenoparietal or cavernous sinuses. Posterior thalamic and epithalamic veins drain the posterior inferolateral and the posteromedial portions of the thalamus, respectively, and empty into the posterior part of the basal vein in its course along the perimesencephalic cisterns [3, 27].

The deep group is related to the drainage of the central core regions above the anterior and posterior perforated substance regions. The inferior striate veins drain the putamen, caudate nucleus, and internal capsule and converge to exit through the APS, where they join the deep middle cerebral and basal veins. The anteroinferior part of the thalamus is drained by the inferior thalamic veins, which exit through the posterior perforated substance to empty into the posterior communicating or peduncular veins, to also end at the basal vein [3, 27].

The ventricular group of veins drains the superior and medial aspects of the central core. These veins course along the medial and lateral walls of the ventricles in a subependymal location and exit these cavities through the choroidal fissure to join the internal cerebral, basal, and great veins. At the frontal horn of the lateral ventricle, anterior thalamic and anterior caudate veins drain the anterior portion of the thalamus and the head of the caudate nucleus, respectively; at the body of the ventricle, the thalamostriate, thalamocaudate, and posterior caudate veins drain the body of the caudate nucleus; at the atrium, lateral atrial veins drain the pulvinar of the thalamus and the tail of the caudate nucleus; and at the temporal horn, the inferior aspect of the thalamus and the tail of the caudate nucleus are drained by the inferior ventricular vein located at the roof of this horn [3, 27].

6 Central Core Segmentation

Detailed description of the cortical and subcortical anatomy of the central core, including the trajectory of the major white matter tracts inside this region, has been the objective of some articles and is important for neurosurgeons aiming to operate central core pathologies [3, 28, 29].

The corticobulbar and cortico-spinal tracts course inside the central core and their injuries usually lead to disabling motor deficits. These tracts arise mostly from the precentral gyrus and project through the genu and the posterior limb of the internal capsule to the cerebral peduncle at the midbrain, progressing to the motor nuclei of the brainstem and spine. The anterior limit of these fibers inside the central core can be used to separate this region into an anterior compartment, related to the anterior limb of the internal capsule, and a posterior compartment, related to the more eloquent parts of the internal capsule. This limit can be drawn over the insular surface by a vertical line that connects the most anterior aspect of the precentral sulcus to the genu of the internal capsule and to the anterior limit of the cerebral peduncle. Since the foramen of Monro, is at the same coronal plane as the genu of the internal capsule, its lateral projection over the insular surface can be used as the intermediate point of this vertical line. The lateral projection of the foramen of Monro is more frequently found at the third short insular gyrus (71%), but can also be found more posteriorly at the fourth (21%) or more anteriorly at the second (8%) short insular gyri.

A horizontal line can be drawn over the insular surface using the lateral projection of the foramen of Monro also as an intermediate point, but continuing to the lateral projection of the floor of the anterior horn anteriorly and of the LGB posteriorly. By using the vertical and horizontal lines together, the central core can be subdivided into four quadrants, each related to different structures.

Additionally, the central core can divided into lateral and medial portions by a parasagittal plane formed by the internal capsule and its vascularization coming from its anteroinferior aspects. Due to the high function importance of these structures, this plane forms a barrier that should not be crossed or damaged during surgeries. The lateral portion contains the insular surface, extreme capsule, claustrum, external capsule, putamen, and globus pallidus, whereas the medial one contains the caudate nucleus and the thalamus.

6.1 Anteroinferior Quadrant

The "anterior perforated substance region" and its contents are included at this quadrant since the anterior commissure is always slightly anterior and inferior to the foramen of Monro. All UF and anterior commissural fibers running inside the central core are also located here. Although only the inferior portion of the anterior limb of the Int. Caps. is at this quadrant, most of the vascularization to the internal capsule and basal ganglia runs inside this region. Medially, the anteroinferior quadrant continues into the septal area, inferior to the anterior horn of the lateral ventricle.

6.2 Anterosuperior Quadrant

The IFOF fibers are found inside this quadrant and the posteroinferior quadrant. Also, some claustrocortical fibers are present in the anterosuperior quadrant anterior to the vertical plane. The anterior limb of the internal capsule is found at this region, with the head of the caudate nucleus and the frontal horn medially.

6.3 Posterosuperior Quadrant

Most claustrocortical fibers are posterior the vertical line and some IFOF fibers are superior to the horizontal line, including them at this quadrant. Deep at the region, the genu, posterior limb, retrolenticular and some of the sublenticular portions of the Int. Caps. are found. Since the optic radiation fibers directed to the superior lip of the calcarine fissure have a slightly superior course after arising from the LGB, they are also located here. More medially, the body of the caudate and the thalamus are almost completely included inside this quadrant. The body of the lateral ventricle is deep to these structures, and its atrium is posterior.

6.4 Posteroinferior Quadrant

The IFOF fibers and the pyramidal tract, located within the genu and posterior limb of the internal capsule, are located under the insular surface of this quadrant. Also, the fibers of the optic radiation directed to the inferior lip of the calcarine fissure and to the occipital pole are present here. The third ventricle lies deep to this quadrant.

7 Surgical Approaches at the Central Core

A subdivision of the central core into quadrants represents an attempt to better understand this region and to possibly lead to more precise and safer surgeries.

The vertical line informs the neurosurgeon that posteriorly lies the genu and the posterior limb of the internal capsule, comprising not only the pyramidal tract but also the superior thalamic radiations related to the somatosensory information [30]. The horizontal line represents the floor of the lateral ventricle and the inferior margin of the thalamus, informing the neurosurgeon to go superiorly for lesions lateral to the lateral ventricle or to the thalamus, or inferiorly for lesions below the lateral ventricle or lateral to the third ventricle.

Additionally, the high functional importance of internal capsule and its vascularization makes this parasagittal plane act as a surgical barrier. Medial operative approaches may be the best to reach the caudate nucleus, thalamus, and medial anterior perforated substance region, whereas transinsular approaches are more suitable for lateral structures and lesions.

All transinsular approaches will initially cross the insular surface, extreme capsule and the claustrum. Although these regions have widespread connections to several other sites and have been related to important cerebral functions, Duffau et al. reported a series of 42 patients operated for a cerebral glioma involving the claustrum, with absence of response during its electrical stimulation and with functional recovery in all cases but three related to deep stroke due to perforating arteries damage. They conclude that removal of the claustrum combined to insular resection does not induce permanent sensorimotor nor cognitive disorders, particularly when affected by low-grade gliomas [31].

Transinsular approaches at the anteroinferior quadrant can damage the UF, while the IFOF is most likely harmed at anterosuperior and posteroinferior quadrants approaches. The existence of a ventral and a dorsal stream associated with language have been debated in the literature [32]. Findings based on subcortical mapping by electrical stimulation during awake neurosurgical procedures point that UF and the IFOF forms a ventral semantic stream while the superior longitudinal fascicle and, most importantly, its arcuate fasciculus component forms a dorsal phonemic stream. The ventral stream is present inside the central core, while the dorsal stream is a perisylvian pathway, outside the central core. The UF has an indirect role in the ventral semantic net work and surgical data reveals that can be partially removed without leading to permanent deficits, but damages to the IFOF at the dominant hemisphere can and may lead to language deficits. The IFOF also seems to play a role in awareness, elaboration of visual information for motor planning, reading, and attention, [33] and other studies have supposed the importance of the UF in category-specific naming tasks for objects and actions [34] or famous faces [35].

More deeply, lenticulostriate arteries at the anteroinferior quadrant restrain the neurosurgeon to advance further through lateral approaches. Lesions located medial to these arteries at this quadrant can be approached inferiorly through a supracarotid triangle formed by the proximal segments of the anterior and middle cerebral arteries at the orbitofrontal surface, a route described as the supracarotid-infrafrontal approach, or superiorly by an anterior transcallosal approach that extends through the floor of the lateral ventricle [36].

The anterior limb of the internal capsule is found deep at the anterosuperior quadrant. It carries anterior thalamic radiations and most of the corticostriatopallidal fibers, which have been related to behavioral disorders, but its resection superior to the anterior commissure might not lead to an important cognitive impairment [37, 38]. The other segments of the Int. Caps. are more eloquent and are located deep to the posterior quadrants, restraining the neurosurgeon to perform deep posterior transinsular approach. Also, cortical insular arteries, mostly located over the posterior surface of the insula, can be long and extend as far as the internal capsule, limiting their coagulation.

Lesions medial to the Int. Caps. at the posterosuperior quadrant will likely be related to the thalamus and can be addressed by a posterior interhemispheric or transparietal approach directed to the body or atrium of the lateral ventricle [39]. Lesions at the posterior and inferior aspects of the thalamus can be approached by a posterior interhemispheric or supracerebellar-infratentorial route, respectively [36, 39]. The third ventricle, medial to the posteroinferior quadrant, can be reached by an interhemispheric transchoroidal approach [10].

Acknowledgments We thank the medical student and artist Giovanna Hespanhol Gütschow (Pontifícia Universidade Católica do Paraná - PUCPR - Brazil) for drawing Fig. 7 with great talent and effort.

References

1. Braak H. Architectonics of the human telencephalic cortex [Internet]. Berlin, Heidelberg: Springer; 1980. [Cited 2021 Jul 13]. https://doi.org/10.1007/978-3-642-81522-5.
2. Ribas GC. The cerebral sulci and gyri. Neurosurg Focus. 2010;28(2):E2.
3. Ribas EC, Yağmurlu K, de Oliveira E, Ribas GC, Rhoton A. Microsurgical anatomy of the central core of the brain. J Neurosurg. 2018;129(3):752–69.
4. Basma J, Guley N, Michael II LM, Arnautovic K, Boop F, Sorenson J. The evolutionary development of the brain as it pertains to neurosurgery. Cureus [Internet] 2020 [cited 2021 Jul 14]; Available from: https://www.cureus.com/articles/22454-

the-evolutionary-development-of-the-brain-as-it-pertains-to-neurosurgery
5. Ribas EC, Yagmurlu K, Wen HT, Rhoton AL. Microsurgical anatomy of the inferior limiting insular sulcus and the temporal stem. JNS. 2015;122(6):1263–73.
6. Wen HT, Rhoton AL, de Oliveira E, Castro LHM, Figueiredo EG, Teixeira MJ. Microsurgical anatomy of the temporal lobe: part 2—Sylvian fissure region and its clinical application. Neurosurgery. 2009;65(6):ons1–ons36.
7. Nieuwenhuys R, Voogd J, van Huijzen C. The human central nervous system. 4th ed. New York: Springer; 2008. p. 967.
8. Crick FC, Koch C. What is the function of the claustrum? Phil Trans R Soc B. 2005;360(1458):1271–9.
9. Serra C, Türe U, Krayenbühl N, Şengül G, Yaşargil DCH, Yaşargil MG. Topographic classification of the thalamus surfaces related to microneurosurgery: a white matter fiber microdissection study. World Neurosurg. 2017;97:438–52.
10. Wen HT, Rhoton AL, de Oliveira E. Transchoroidal approach to the third ventricle: an anatomic study of the choroidal fissure and its clinical application. Neurosurgery. 1998;42(6):1205–17.
11. Haber SN. Corticostriatal circuitry. Dialogues Clin Neurosci. 2016;18(1):7–21.
12. Haber SN. The primate basal ganglia: parallel and integrative networks. J Chem Neuroanat. 2003;26(4):317–30.
13. Heimer L. The olfactory cortex and the ventral striatum. In: Livingston KE, Hornykiewicz O, editors. Limbic mechanisms. New York: Plenum Press; 1978. p. 95–187.
14. Olmos JS, Heimer L. The concepts of the ventral Striatopallidal system and extended amygdala. Annals NY Acad Sci. 1999;877:1–32.
15. Ebeling U, Reulen HJ. Subcortical topography and proportions of the pyramidal tract. Acta Neurochir. 1992;118(3–4):164–71.
16. Párraga RG, Ribas GC, Welling LC, Alves RV, de Oliveira E. Microsurgical anatomy of the optic radiation and related fibers in 3-dimensional images. Oper Neurosurg. 2012;71:ons160–72.
17. Fernández-Miranda JC, Rhoton AL, Álvarez-Linera J, Kakizawa Y, Choi C, de Oliveira EP. Three-dimensional microsurgical and tractographic anatomy of the white matter of the human brain. Neurosurgery. 2008;62(Supplement 3):SHC989–SHC1028.
18. Rhoton AL. The cerebrum. Neurosurgery. 2002;51(suppl_4):S1-1–S1-52.
19. Li M, Ribas EC, Wei P, Li M, Zhang H, Guo Q. The ansa peduncularis in the human brain: a tractography and fiber dissection study. Brain Res. 2020;1746:146978.
20. Li M, Ribas EC, Zhang Z, Wu X, Wang X, Liu X, et al. Tractography of the ansa lenticularis in the human brain. Clin Anat. 2022;35(3):269–79.
21. Shah A, Goel A, Jhawar SS, Patil A, Rangnelar R, Goel A. Neural Circuitry: Architecture and Function - A fiver dissection study. World Neurosurg. 2019;25:e620–e638.
22. Shah A, Jhawar SS, Goel A. Analysis of the anatomy of the Papez circuit and adjoining limbic system by fibre dissection techniques. J Clin Neurosci. 2012;19(2):289–98.
23. Rhoton AL. The supratentorial arteries. Neurosurgery. 2002;51(4 Suppl):S53–120.
24. Feekes JA. The vascular supply of the functional compartments of the human striatum. Brain. 2006;129(8):2189–201.
25. Gibo H, Marinkovic S, Brigante L. The microsurgical anatomy of the premamillary artery. J Clin Neurosci. 2001;8(3):256–60.
26. Türe U, Yaşargil MG, Al-Mefty O, Yaşargil DCH. Arteries of the insula. J Neurosurg. 2000;92(4):676–87.
27. Rhoton AL. The cerebral veins. Neurosurgery. 2002;51(4 Suppl):S159–205.
28. Tanriover N, Rhoton AL, Kawashima M, Ulm AJ, Yasuda A. Microsurgical anatomy of the insula and the sylvian fissure. J Neurosurg. 2004;100(5):891–922.
29. Türe U, Yaşargil DCH, Al-Mefty O, Yaşargil MG. Topographic anatomy of the insular region. J Neurosurg. 1999;90(4):720–33.
30. Gray H, Standring S, Ellis H, Berkovitz BKB, editors. Gray's anatomy: the anatomical basis of clinical practice. 39th ed. Edinburgh; New York: Elsevier Churchill Livingstone; 2005. p. 1627.
31. Duffau H, Mandonnet E, Gatignol P, Capelle L. Functional compensation of the claustrum: lessons from low-grade glioma surgery. J Neuro-Oncol. 2007;81(3):327–9.
32. Duffau H. The anatomo-functional connectivity of language revisited. Neuropsychologia. 2008;46(4):927–34.
33. Sarubbo S, De Benedictis A, Maldonado IL, Basso G, Duffau H. Frontal terminations for the inferior fronto-occipital fascicle: anatomical dissection, DTI study and functional considerations on a multi-component bundle. Brain Struct Funct. 2013;218(1):21–37.
34. Lu LH, Crosson B, Nadeau SE, Heilman KM, Gonzalez-Rothi LJ, Raymer A, et al. Category-specific naming deficits for objects and actions: semantic attribute and grammatical role hypotheses. Neuropsychologia. 2002;40(9):1608–21.
35. Glosser G, Salvucci AE, Chiaravalloti ND. Naming and recognizing famous faces in temporal lobe epilepsy. Neurology. 2003;61(1):81–6.
36. Chang EF, Gabriel RA, Potts MB, Berger MS, Lawton MT. Supratentorial cavernous malformations in eloquent and deep locations: surgical approaches and outcomes: clinical article. JNS. 2011;114(3):814–27.
37. Machado A, Haber S, Sears N, Greenberg B, Malone D, Rezai A. Functional topography of the ventral striatum and anterior limb of the internal capsule determined by electrical stimulation of awake patients. Clin Neurophysiol. 2009;120(11):1941–8.
38. Duffau H, Denvil D, Capelle L. Absence of movement disorders after resection of glioma invading the right striatum. J Neurosurg. 2002;97(2):363–9.
39. Rhoton AL. The lateral and third ventricles. Neurosurgery. 2002;51(4 Suppl):S207–71.

Fiber Anatomy of the Brainstem and Implications for Surgery of Brainstem Tumors

Abhidha Shah, Sukhdeep Singh Jhawar, Maximiliano Nunez, and Atul Goel

1 Introduction

The anatomical design of the brainstem is one of the most fascinating and intriguing components of the human brain. The brainstem is composed of both gray and white matter arranged in a complex three-dimensional framework. The intricate neural networks of the brainstem are responsible for most of the vital functions of the body that are necessary for survival. Thus understanding the complex anatomy of the brainstem is crucial as there can be no margin for error when operating in this region. Along with the native gray matter of the brainstem a multitude of efferent and afferent fibers traverse through the brainstem from forebrain to spinal cord. Most of these fiber bundles undergo a systemic rearrangement at various levels of brainstem. This chapter deals with the intrinsic anatomy of the brainstem. A novel system of division of fiber tracts of the brainstem into zones is presented which will aid in the understanding of the exquisite anatomy. [1]

The brainstem forms a conduit between the forebrain and the spinal cord and consists of the midbrain, pons and medulla from above downwards. (Figs. 1, 2, and 3). The anatomy of the brainstem can be best understood by describing it via three corridors: the ventral (anterior), the lateral and the dorsal (posterior) corridors. (Figs. 1, 2, and 3).

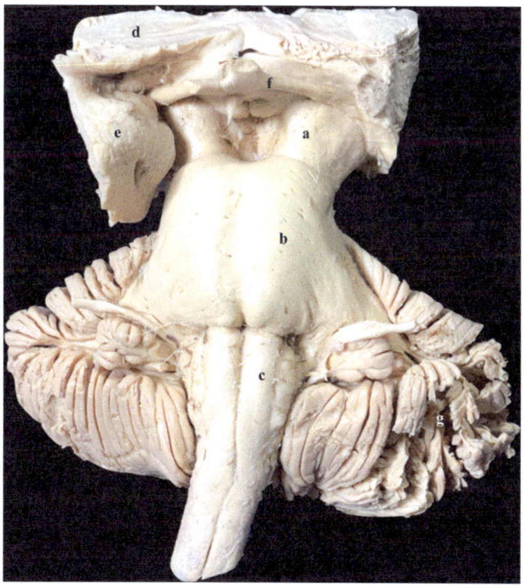

Fig. 1 Ventral view of the brainstem with its parts. a. Midbrain, b. Pons, c. Medulla, d. Thalamus, e. Temporal lobe, f. Optic chiasm, g, Cerebellum

A. Shah (✉) · S. S. Jhawar
Department of Neurosurgery, K.E.M. Hospital and Seth, G.S. Medical College, Mumbai, India

M. Nunez
Department of Neurosurgery, Stanford Hospital, Stanford, CA, USA
e-mail: maxi1971@stanford.edu

A. Goel
Department of Neurosurgery, Lilavati Hospital and Research Center, Bandra, Mumbai, India

Department of Neurosurgery, R.N. Cooper Hospital, Mumbai, India

Bombay Hospital Institute of Medical Sciences and Research Center, Mumbai, India

A. Shah et al. (eds.), *Functional Anatomy of the Brain: A View from the Surgeon's Eye*,
https://doi.org/10.1007/978-981-99-3412-6_19

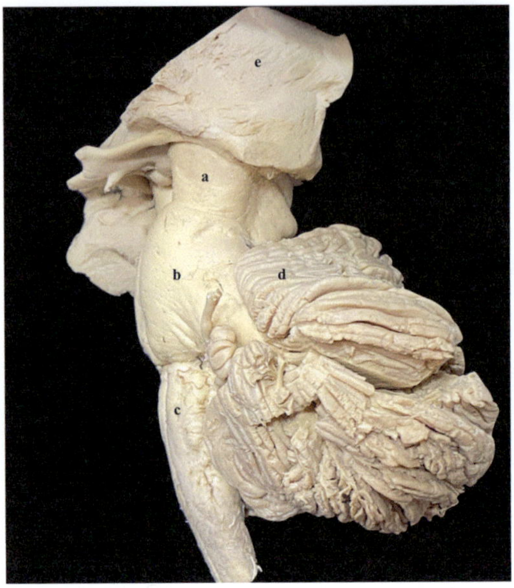

Fig. 2 Lateral view of the brainstem. a. Midbrain, b. Pons, c. Medulla, d. Cerebellum, e. Thalamus

Fig. 3 Dorsal view of the brainstem and cerebellum as it lies in the cranial cavity. a. Posterior surface of the midbrain, b. Cerebellum, c. Thalamus, d. Superior colliculus, e. Inferior colliculus, f. Pineal body, g. Third ventricle, h. Massa intermedia

2 Mid Brain

2.1 Extent

The midbrain or mesencephalon connects the pons and cerebellum with the diencephalon and cerebral hemispheres (Figs. 4 and 5). It extends from the optic tracts superiorly to the pontomesencephalic sulcus inferiorly [1]. The traditional boundary between the diencephalon and the mesencephalon is an imaginary line between the mammillary bodies and the posterior commissure [5].

2.2 Parts/Divisions

For descriptive purposes the midbrain can be divided into right and left cerebral peduncles ventrally and the tectum dorsally. Each cerebral peduncle is composed of the crus cerebri, which lie ventrally, and the tegmentum, which lies dorsally [1]. The substantia nigra can be identified as a dark pigmented lamina and separates the ventral crus and the dorsal tegmentum. (Fig. 4b). An oblique plane that crosses the cerebral aqueduct defines the division between the cerebral peduncle and the tectum. The tectum consists of dorsal portion consisting of four rounded eminences, named the corpora quadrigemina and a pretectal area.

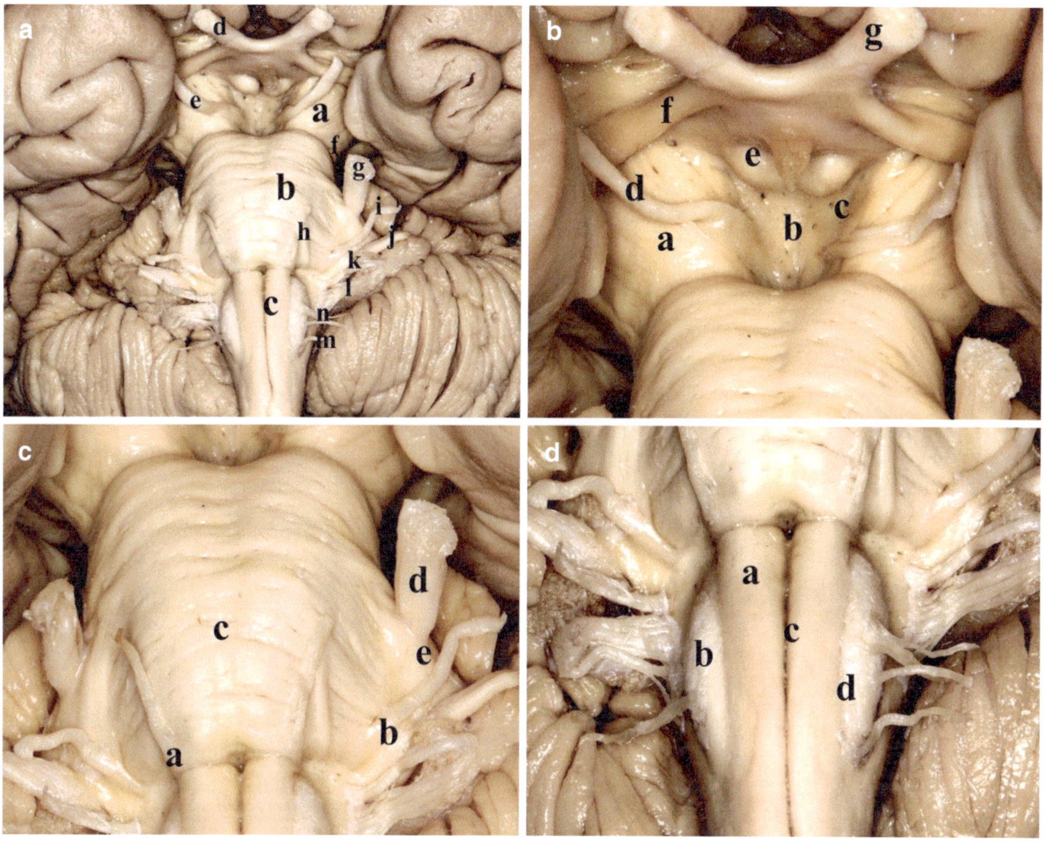

Fig. 4 Ventral view of brainstem. (**a**) Brainstem can be divided into three distinctive morphological parts consisting midbrain, pons and medulla from above downwards. Superiorly, the brainstem (midbrain)is continuous with structures forming thalamus, hypothalamus and cerebral hemispheres. Inferiorly, it is continuous with spinal cord (medulla). a. Midbrain, b. pons, c medulla, d. Optic nerve, e. Oculomotor nerve, f. Trochlear nerve, g. Trigeminal nerve, h. Abducens nerve, i. Facial nerve, j. Vestibulocochlear nerve, k. Glossopharyngeal nerve, l. Vagus nerve, m. Cranial part of accessory nerve, n. Hypoglossal nerve. (**b**) Ventral view of midbrain. When viewed from anterior aspect midbrain has two prominent converging fiber bundles called cerebral peduncles, which are separated by a deep fissure known as interpeduncular fossa. This fissure is wider superiorly than inferiorly. The medial surface of the peduncle forms the lateral boundary of the interpeduncular fossa, and is marked by a longitudinal furrow, the oculomotor sulcus. Oculomotor nerve can be seen emerging from medial aspect of both peduncles-from depth of interpeduncular fossa. Paired mammillary bodies can be seen forming roof of this fossa. a. Cerebral peduncles, b. Interpeduncular fossa, c. Oculomotor sulcus, d. Oculomotor nerve, e. Mammillary bodies, f. Optic tract, g, Optic nerve. (**c**) Pons is the post prominent part of brainstem ventrally. It extends frompontomesencephalic sulcus superiorly to pontomedullary sulcus inferiorly. Abducens, facial and vestibulocochlear nerve emerge from pontomedullary sulcus inferiorly. The anterior surface has a shallow vertical median groove for basilar artery called sulucs basilaris. Near the mid pontine level trigeminal nerve emerges which is taken as a landmark to define the plane of junction between pons and middle cerebellar peduncle. The transverse pontine fibers collect together to form middle cerebellar peduncle. a. Abducens nerve, b. Facial and vestibulo-cochlear nerves, c. Sulcus basilaris, d. Trigeminal nerve, e. Middle cerebellar peduncle. (**d**) The medulla oblongata is pyramidal in shape, with broad upper end seprated from pons by pontomedullary sulcus and narrow lower end continous with spinal cord. A plane passing transversely below the pyramidal decussation and above the first pair of cervical nerves is taken as lower limit of medulla oblogata. The Anterior median fissure extends along the entire length of the medulla oblongata ventrally. Upper part of the lateral compartment of the medulla contain oval prominences called as olive. Its lower part is only partly continuous with spinal lateral funiculus. Last four cranial nerves emerge from medulla as two set of rootlets from anterolateral and posterolateral sulcus respectively. a. Pyraminds, b. Olive, c. Anterior median fissure, d. Anterolateral sulcus

2.3 External Features

2.3.1 Anterior Corridor: Cerebral Peduncles

The crus cerebri occupy a prominent portion of the ventral surface of the midbrain. The crus cerebri or the cerebral peduncles are stationed like columns on either side of the midbrain resting on the upper portion of the pons. The peduncles are shielded anteriorly by the medial temporal lobe structures and can be seen only on retracting the medial temporal lobes laterally. (Fig. 4a, b). The structure of the cerebral peduncles can be said to resemble the greek mythological figure "Atlas" as he is holding up the celestial heavens with his two arms; the arms being the cerebral peduncles and the heaven being the cerebrum [1]. The cerebral peduncles extend from the optic chiasm superiorly, converging as they pass downward and backward towards the pontomesencephalic junction inferiorly. Between the two peduncles is a deep cleft known as the interpeduncular fossa. The interpeduncular fossa is a triangular region that has its apex inferiorly and is formed by the foramen caecum and base superiorly which is formed by the mammillary bodies. The posterior perforated substance lies in the interpeduncular fossa. The lower portion of cerebral peduncles, contains a nucleus named the interpeduncular ganglion and its upper part assists in forming the floor of the third ventricle (Fig. 4c, d) [1–3]. The medial surface of the peduncle forms the lateral limit of the triangular interpeduncular fossa. The lateral surface of cerebral peduncles is covered by the parahippocampal gyrus of the cerebral hemisphere. The trochlear nerve crosses the cerebral peduncle from behind forward, after exiting from the dorsal surface of the brainstem. The medial surface of the cerebral peduncle has a linear vertical groove known as the oculomotor sulcus. The oculomotor nerve exits the brainstem surface from this groove (Fig. 4e, f).

2.3.2 Lateral Corridor

The lateral surface of each peduncle is marked by a longitudinal groove, termed the lateral mesencephalic sulcus, which divides the lateral aspect of the midbrain into an anterior and posterior portion. The anterior portion is the lateral most part of the cerebral peduncle. The posterior portion is triangular in shape and is known as the leminiscal triangle. It is bounded anteriorly by the lateral mesencephalic sulcus, postero- superiorly by the inferior brachium and postero-inferiorly by the superior cerebellar peduncle. It is made up of the fibers of the lateral lemniscus as they pass backward and upward, to disappear under the inferior colliculus. (Fig. 4h, 5b).

2.3.3 Posterior Corridor: Corpora Quadrigemina

The tectum or dorsal midbrain exhibits four rounded elevations termed corpora quadrigemina or the superior and inferior colliculi (Fig. 5c). They lie under cover of the splenium of the corpus callosum, and are partly overlapped on either side laterally by pulvinar of the thalamus (Fig. 5c). They are situated above and in front of the superior medullary velum and superior cerebellar peduncle, and below and behind the third ventricle and posterior commissure [1–3]. They can be said to occupy a position behind the third and the fourth ventricles. The corpora quadrigemina are arranged in pairs (superior and inferior colliculi), and are separated from one another by a cruciate sulcus. The longitudinal part of this sulcus expands superiorly to form a slight depression that supports the pineal body that partly covers the superior colliculi. From the inferior end of the longitudinal sulcus, a white band, termed the frenulum veli, is prolonged downward to the superior medullary velum. On either side of this band below the inferior colliculi, the trochlear nerve emerges, and passes forward on the lateral aspect of the cerebral peduncle (Fig. 5c). The inferior colliculi are hemispherical, and appear more prominent than the superior even though the superior colliculi are darker in appearance [1–3]. The superior colliculi are associated with the visual pathway and the inferior colliculi are associated with auditory pathway.

From the lateral aspect of each colliculus a white band emerges and is prolonged upward and

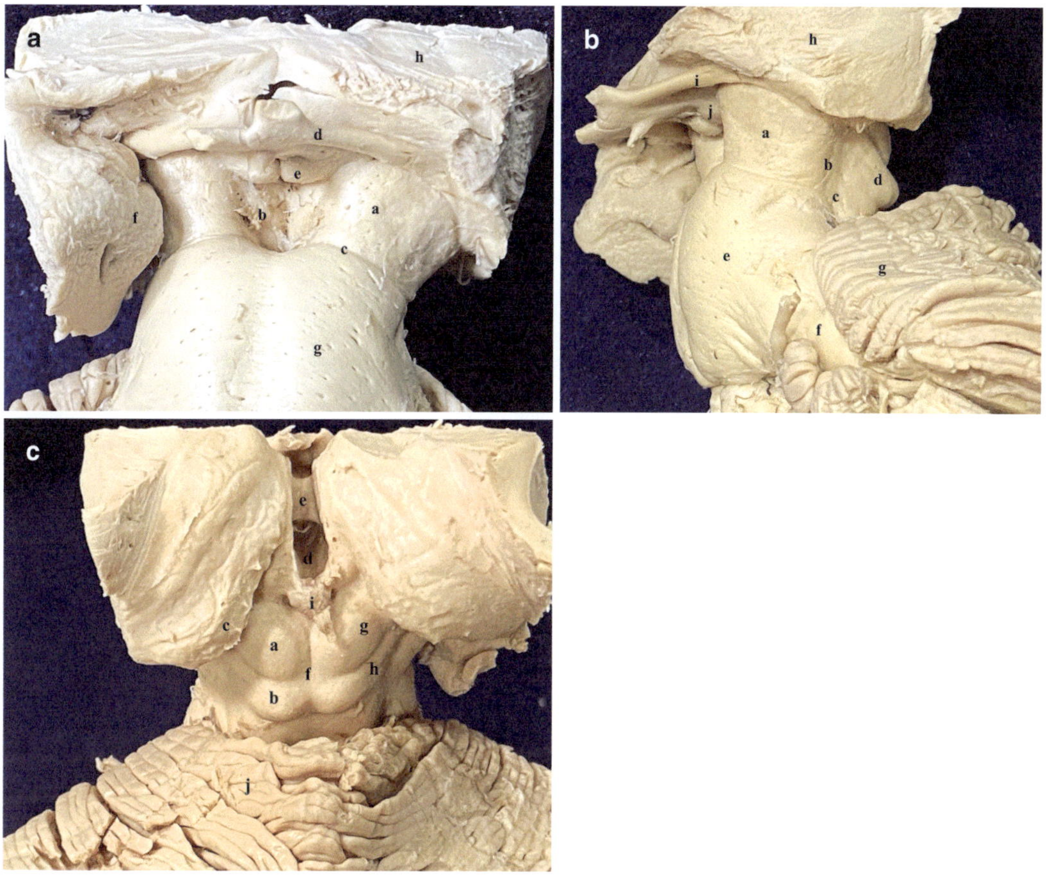

Fig. 5 Midbrain. (**a**) Close up view of the ventral part of the midbrain. a. Cerebral peduncle, b. Interpeduncular fossa and posterior perforated substance, c. Pontomesencephalic sulcus, d. Optic tract, e. Mammillary body, f. Parahippocampal gyrus of temporal lobe, g. Pons, h. Thalamus. (**b**) Close up view of the lateral part of the midbrain. a. Cerebral peduncle, b. Lateral mesencephalic sulcus, c. Lemniscal triangle, d. Tectal plate, e. Pons, f. Middle cerebellar peduncle, g. Cerebellum, h. Thalamus, i. Optic tract, j. Mammillary body. (**c**) Close up view of the dorsal part of the midbrain. a. Superior colliculus, b. Inferior colliculus, c. Pulvinar of thalamus, d. Third ventricle, e. Massa intermedia, f. Cruciate sulcus, g. Superior brachium, h. Inferior brachium, i. Pineal body, j. Superior surface of cerebellum

forward and are termed as the superior and inferior brachium (Fig. 5c). The superior brachium ascends upwards and laterally towards the thalamus. It passes between the pulvinar and the medial geniculate body to enter the lateral geniculate body. Some of its fibers enter the optic tract directly. The inferior brachium passes forward and upward from the inferior colliculus and disappears under cover of the medial geniculate body.

2.4 Internal Features

The internal structure of the midbrain can be understood by dividing it into three parts, the crus cerebri, the tegmentum and the tectum. (Fig. 6). The tegmentum (red nucleus) and tectum (quadrigeminal plate) constitute the dorsal midbrain and the cerebral peduncle and substantia nigra constitute the ventral midbrain (Fig. 6f–h).

Fig. 6 Fiber dissection of midbrain. (**a**) Midbrain traverses the hiatus in tentorium cerebelli, connecting pons and cerebellum with forebrain. Ventrally the cerebral peduncles can be seen as dominating cylinderical structures connverging towards the pons as they desecnd. These bundles are seprated by a midline fissure called as interpeduncular fossa which is wider superiorly. The oculomotor nerve can be seen emerging from medial aspect of both peduncles from depth of interpeduncular fossa. Upper limit is demarcated laterally by optic tracts and medially by floor of third ventricle. Uncus of mesial temporal lobe can be seen covering cerebral peduncles on both sides. (**b**) Each cerebral peduncleis semilunar in cross section and consists entirely of longitudinal bundles of efferent fibers, which arise from different parts of cerebral cortex. After removal of some part of cerebral peduncle deeply pigmented semilunar lamina can be seen which is called substantianigra. Corticospinal and corticonuclearfibres occupy middle two third of peduncle. Corticopontinefibers arise in cerebral cortex and end in nuclei pontis. They can be grouped into frontopontine from frontal lobesituated in medial fifth of crusandtemporopontine fibers from the temporal lobe situated in lateral part of the peduncle. a. Frontopontine fibers in crus cerebri, b. Substantianigra, c. Corticospinal tract continuing into the pons, d. Optic tract, e. Medial lemniscus, f. Interpeduncular fossa, g. Oculomotor nerve, h. Cerebral peduncle, i. Temporopontine fibers in crus cerebri. (**c**) Corticospinal fibers can be traced down into pons as well defined fiber bundle more so in lower pons before entering into medullary pyramids. a. Corticospinal tract entering into the pons from the midbrain. (**d**) Anterior parts of temporal lobes have been removed on both sides to show upper limit of midbrain. Fiber bundles of crus cerebri can be seen in relation to optic tract before entering midbrain. Left side corticospinal fibers are traced downwards into pons on way to medullary pyramids. a. Optic tract, b. Corticospinal tract in the cerebral peduncle, c. Mammillary bodies, d. Corticospinal tract entering the pons, e. Anterior perforated substance, f. Fornix, g, Optic nerve, h. Olfactory tract. (**e**) Ventral view of brainstem in another specimen showing corticospinal fiber tract extending from midbrain to upper medulla on both sides. Oculomotor nerve can be seen emerging from interpeduncular sulcus. a. Corticospinal tract as it traverses the midbrain, pons and medulla. (**f**) Close up view of same specimen showingcruscerebri with substantianigra. Oculomotor nerve emergesfrominterpeduncular fossa on medial aspect of substantianigra. Part of red nucleus is visible dorsomedial to substantianigra. a. Corticospinal fibers, b. Substantianigra, c, Occulomotor nerve, d. Substantianigra, e. Frontopontine fibers, f. Temporopontine fibers, g. Fornix, h. Mammillary body, i. Red nucleus, j. Interpeduncular nucleus, k. Internal capsule. G. Corticospinal tract is removed on right side and partially on left side exposing deep midbrain structures. Substantianigra convex ventral surface is visible. It is thicker medially and reaches upto lateral crural sulcus. a. Corticospinal tract, b. Substantianigra, c. Oculomotor nerve, d. Cut frontopontine fibers, e. Temporopontine fibers, f. Fasciculus retroflexus (**h**) Midsagittal section of brainstem showing fiber dissection of midbrain, pons and medulla. Fibers of superior cerebellar peduncle can be seen going from dentate nucleus to red nucleus and thalamus and forms lateral wall of superior half of roof of fourth ventricle. Medial longitudinal fasciculus runs just ventral to cerebral aqueduct and close to midline near floor of fourth ventricle. The medial longitudinal fasciculus passes lateral and ventral to oculomotor nucleus and ascends to end in *interstitial nucleus* (of Cajal) in the lateral wall of third ventricle, just above the cerebral aqueduct. The medial lemniscus when viewed from lateral aspect ascends through the pons forming a concave shape facing ventrally. In the midbrain it ascends in dorsolateral part of tegmentum in the angle between red nucleus and substantianigra. a. Superior cerebellar peduncle, b. Medial longitudinal fasciculus, c. Lateral lemniscus, d. Medial lemniscus, e. Corticopontine fibers, f. Corticospinal fibers, g. Pyramidal decussation, h. Central tegmental tract, i. Middle cerebellar peduncle

2.4.1 Ventral Midbrain

White Fibers

Each peduncle is semilunar in section and contains the corticospinal, corticonuclear and corticopontine fibers. The corticonuclear fibres end in nuclei of cranial nerves and other brainstem nuclei. Corticospinal fibres which is mainly the pyramidal tract continues into the pons and the medulla. Corticopontine fibers arise in cerebral cortex and end in the pontine nuclei. They form two groups viz. frontopontine from frontal lobe and, temporopontine fibers from the temporal and occipital lobes. The organization of the white fibers in the peduncle can be understood by dividing the peduncle into five parts. The medial most one – fifth portion of the peduncle is made up of the corticonuclear (Geniculate fasciculus) and the corticopontine fibers (fronto-pontine fibers – Arnold's fasciculus). The middle three portions of the peduncle comprises of the corticospinal tract [1–3]. The lateral most one fifth portion of the crus is made up of the corticopontine fibers (temporo-pontine fibers – Turk's fasciculus) (Figs. 6c, d) [1–3].

Gray Matter

The most prominent gray matter mass in the ventral midbrain is the substantia nigra and extends from the oculomotor sulcus medially to the lat-

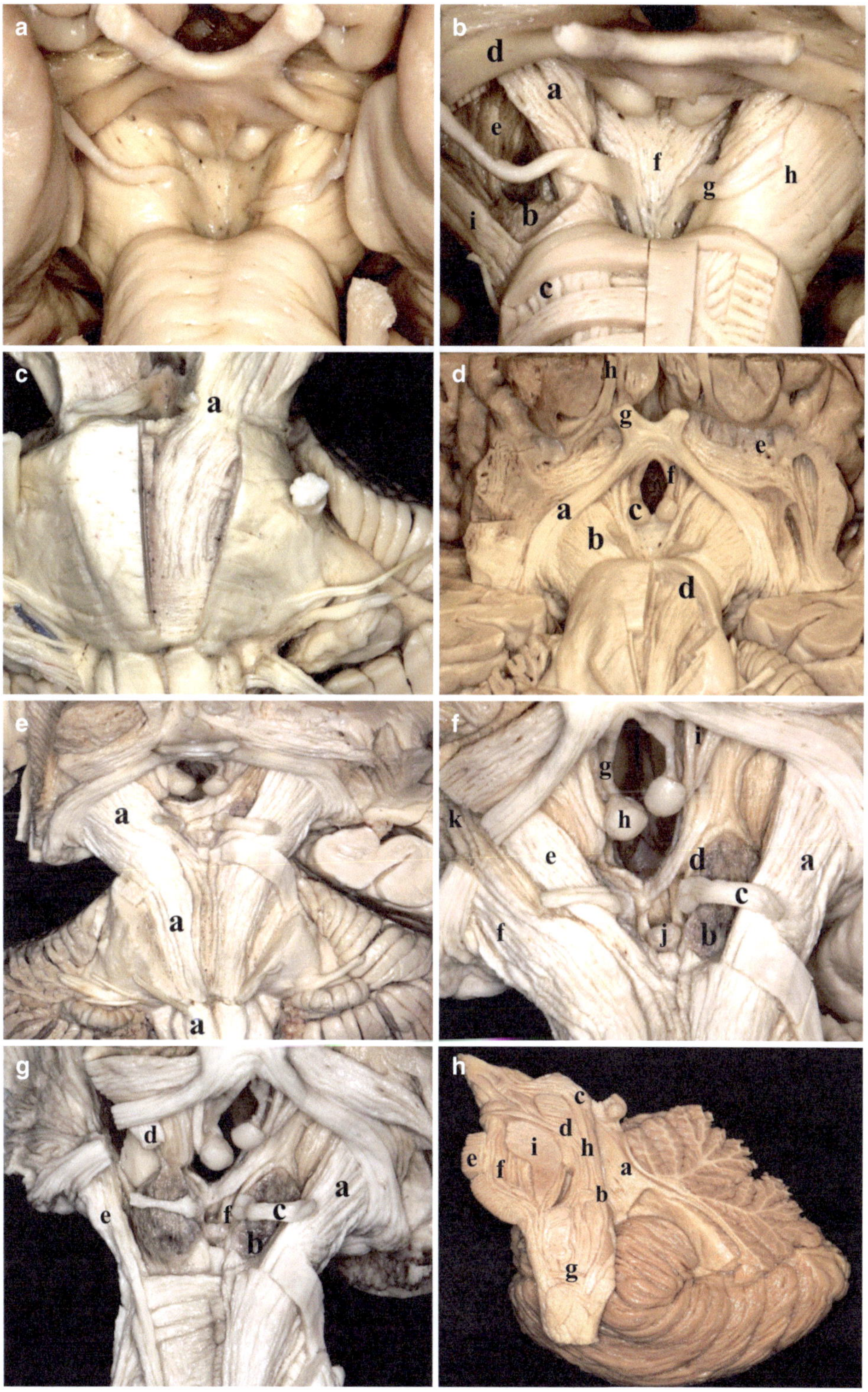
a
b
d
a
e
f
g
h
i
b
c
c
a
d
h
g
e
f
a
c
b
d
e
a
a
a
f
i
g
h
k
e
d
a
c
f
j
b
g
d
a
e
f
c
b
h
c
d
i
h
e
f
a
b
g

eral mesencephalic sulcus laterally. In a craniocaudal perspective it extends from the upper surface of the pons inferiorly to the subthalamic region superiorly. Its color is due to the presence of pigmented neurons. (Figs. 6f, g). On transverse section the substantia nigra appears as a semilumar transverse band which is concave dorsally which is more thicker medially than laterally.

2.4.2 Dorsal Midbrain

White Fibers

White fibers of the dorsal midbrain mainly consist of superior cerebellar peduncle, the tectobulbar fibers, the rubrospinal tract, the rubrothalamic tract, the MLF, the medial lemniscus, the spinal lemniscus, the trigeminal lemniscus and the lateral lemniscus.The tracts which are in the midline are the medial longitudinal fasciculus, the tectospinal tract, the decussation of the superior cerebellar peduncle, the dorsal and ventral tegmental decussation and the rubrospinal tract from dorsal to ventral. Further laterally is the central tegmental tract and the lemisncal system. The leminiscal system is made up of the medial lemniscus, the trigeminal lemniscus, the spinal lemniscus and the lateral lemniscus from medial to lateral.

Decussations

The midbrain harbours a series of three decussations. At the level of the superior colliculus are the dorsal and ventral tegmental decussations. The dorsal tegemental decussation is formed by the crossing of the tectobulbar fibers. The tectobulbar tract connects the superior colliculus to the medulla. Further ventrally is the ventral tegmental decussation which is made up of the intersecting fibers of the rubrospinal fasciculus. The rubrospinal tract connects the red nucleus to various spinal cord nuclei. At the level of the inferior colliculus the superior cerebellar peduncle decussates. The superior cerebellar peduncle carries efferent fibers from the dentate nucleus to the opposite red nucleus and the ventral spinocerebellar tracts.

Gray Matter

The gray matter of the midbrain is made up of the nuclei of the third, fourth and fifth cranial nerves and gray matter inherent to the brainstem. The native gray mater of the midbrain consists of the red nucleus, the interpeduncular ganglion, the substantia nigra and the reticular formation.

The red nucleus and the interpeduncular ganglion as the major gray matter masses of the tegmentum. The red nucleus is situated in the anterior part of the tegmentum, and is continues supeirorly into the posterior part of the subthalamic region. It extends upto the inferior one third of the cerebral peduncle. The interpeduncular ganglion occupies the lower thirds of the ceberal peduncle. Dorsal to the red nucleus are the oculomotor nucleus superiorly and the trochlear nucleus inferiorly. The oculomotor nucleus consists of the main motor nucleus, the Perlia's nucleus and the Edingar – Westphal's nucleus [2]. Oculomotor nucleus is located next to the midline at the level of the lower half of the superior colliculus and the upper half of the inferior colliculus, and between the cerebral aqueduct dorsally and the decussation of the superior cerebellar peduncle ventrally. The intramesencephalic segment of the oculomotor nerve passes inside and along the medial side of the red nucleus to exit the midbrain in the lateral wall of the interpeduncular fossa from the oculomotor sulcus. (Figs. 6f, g) The trochlear nucleus is located adjacent to the midline in the midbrain at the level of the lower half of the inferior colliculi. The fibers of the trochlear nerve exit the nucleus, traverse around the aqueduct and emerge dorsally from the sides of the frenulum of the superior medullary velum. Thus the midbrain contains the intramesencephalic segments of both the oculomotor and trochlear nerves. Lateral to the aqueduct and dorsally lies the mesencephalic tract and nucleus of the fifth nerve continuing up from the pons. Dispersed between the red nucleus and the central gray matter extending posteriorly upto the colliculi and the posterior commissure are the neurons of the reticular formation [1–3].

3 Pons

3.1 Extent

The pons or as its name suggests "bridge" serves as the main conduit between the cerebrum and the cerebellum and lies between the mesencephalon and rhombencephalon [1–3]. It extends from the pontomesencephalic sulcus superiorly to the pontomedullary sulcus inferiorly. The inferior site of transition from pons to medulla is demarcated superficially by the pontomedullary sulcus in which abducens, facial and vestibulocochlear nerves emerge.

3.2 Parts/Divisions

The pons is divided into two parts: a ventral base and a dorsal tegmentum. The trapezoid body and the medial lemniscus demarcate the two parts from each other.

3.3 External Features

3.3.1 Anterior Corridor

The ventral surface of the pons is rounded and very prominent and is often referred to as the basis pontis. (Figs. 1, 2, 5, and 7). The anterior surface has a shallow vertical median furrow, the sulcus basilaris, which is occupied by the basilar artery. The bulk of the prominenence of the pons is due to the descending CSTs. It merges laterally with the middle cerebellar peduncle (brachium pontis) The point of emergance of trigeminal nerve defines the plane of tranistion between pons and middle cerebellar peduncle. Facial nerve can be seen to emerge at lower border of pons in the pontomedullary sulcus in the interval between olive and inferior cerebellar peduncle of medulla oblongeta. The fibers of vestibular nerve enter the medulla oblongeta on medial side of those of cochlear fibers, and pass between inferior peduncle and spinal tract of trigeminal nerve. (Fig. 7).

3.3.2 Lateral Corridor

The two main structures of the lateral surface of the pons are the middle cerebellar peduncle and the trigeminal nerve. The pontine nuclei which are present in the basis pontis give rise to transverse pontine fibers, which traverse transversly to make the middle cerbellar peduncle. The trigeminal nerve emerges laterally on each side on the lateral surface of the pons at the mid pontine level. It has a small superomedial motor root and a large inferolateral sensory root (Figs. 7a, b and 8).On

Fig. 7 Fiber dissection of ventral pons. (**a**) Pons is most prominent part of brainstem on ventral aspect and is located between pontomesencephalic sulcus superiorly and pontomedullary sulcus inferiorly. The inferior site of transition from pons to medulla is demarcated superficially by ventrolateral sulcus in which abducens, facial and vestibulocochlear nerve emerge. Near the mid pontine level trigeminal nerve emerges on each side with a small superomedial motor root and a large inferolateral sensory root. a. Pontomesencephalic sulcus, b. Pontomedullary sulcus, c. Sulcus basilaris, d. Trigeminal nerve. (**b**) Ventral pons has been dissected to demonstrate longitudinal and transverse fibers tracts. Transverse fibers form three bundles of superficial, middle and deep transverse fiber projections based on relation to longitudnal descending corticospinal fibres. Superficial transverse fibers can be seen as thick layer of fibers on ventral surface of pons ventral to the descending fibers of the corticospinal tract. The middle fibers run between the fibers of the corticospinal tract. The transverse pontine fibers collect togather to form middle cerebellar peduncle. a. Superficial transverse pontine fibers (Stratum superficiale pontis), b. Corticospinal tract, c. Middle transverse pontine fibers (Stratum complexum), d. Middle cerebellar peduncle, e. Trigeminal nerve, f. Vestibulocochlear nerve, g. Facial nerve, h. Glossopharyngeal nerve, i. Vagus nerve. (**c**) The deep transverse fibers lie on the dorsal aspect of the corticospinal fibers. They course to the lateral border of the pons, where they form part of the middle cerebellar peduncle. a. Deep transverse pontine fibers (Stratum profundum pontis), b. Corticospinal tract, c.Superficial transverse pontine fibers (Stratum superficiale pontis), d. Trigeminal nerve, e. Middle cerebellar peduncle, f. Medial lemniscus. (**d**) The dorsal or tegmental part of pons is exposed after dissecting deep transverse fibers. It is chiefly composed of upward continution of reticular formation and gray matter of medulla oblengata. More dorsal and adjacent to deep transverse fibers is flattened lemniscus band of ascending longitudnal fiber tracts. Medial lemniscus lies on either side as as flattend oval bundle, related laterally to trigeminal and spinal lemniscus. a. Superficial transverse pontine fibers (Stratum superficiale pontis), b. Corticospinal tract, c.Deep transverse pontine fibers (Stratum profundum pontis), d. Medial Lemniscus, e. Medial longitudinal fasciculus. (**e**) On disseting deep to medial lemniscus, the next longitudnal fiber bundle is the medial longitudnal fasciculus which is paramedian, ventral to fourth ventricle, near the abducens nucleus withfacial fibers ascending lateral to this. Other longitudnal tract related laterally and ventrally is the central tegmental tract which connects red nucleus with inferior olivary nucleus. a. Superficial transverse pontine fibers (Stratum superficiale pontis), b. Corticospinal tract, c. Central tegmental tract, d. Medial Longitudinal fasciculus, e. Tectospinal tract, f. Locus coerulus. (**f**) Fourth ventricle is visible here as cavity of pons. Upper portion of roof of fourth ventricle is formed by superior cerebellar peduncle and superior medullary velum. a. Fourth ventricle, b. Superficial transverse pontine fibers (Stratum superficiale pontis), c. Corticospinal tract, d. Trigeminal nerve (motor root medial, sensory root is lateral), e. Middle cerebellar peduncle. (**g**) Left oblique lateral view of fiber dissection of pons. Corticospinal fibers can be seen converging to medullary pyramids as a compact bundle after traversing ventral pons and transverse fibers. Behind this is medial lemniscus which form flat lamina of fibers extending from medial raphe. Medial lemniscus, when viewed from lateral aspect, ascend through the pons forming a concave shape facing ventrally. Trigeminal nerve can be seen entering upper margin of anterior aspect of pons, where its fibers can be seen entring perpendicular to transverse pontine fibers going into cerebellun via middle cerebellar peduncle. a. Corticospinal tract, b. medullary pyramid, c. Medial lemniscus, d. Trigeminal nerve, e. Olive. (**h**) Close up view of same specimen shows transverse running fibers of middle cerebllar peduncle. Facial nerve can be seen to emerge at lower border of pons in the interval between olive and inferior peduncle of medulla oblongeta. The fibers of vestibular nerve enter the medulla oblongeta on medial side of those of cochlear fibers, and pass between inferior peduncle and spinal tract of trigeminal nerve. a. Corticospinal tract, b. Middle cerebellar peduncle, c. Trigeminal nerve, d. Facial and vestibulocochlear nerves, e. Medullary pyramid

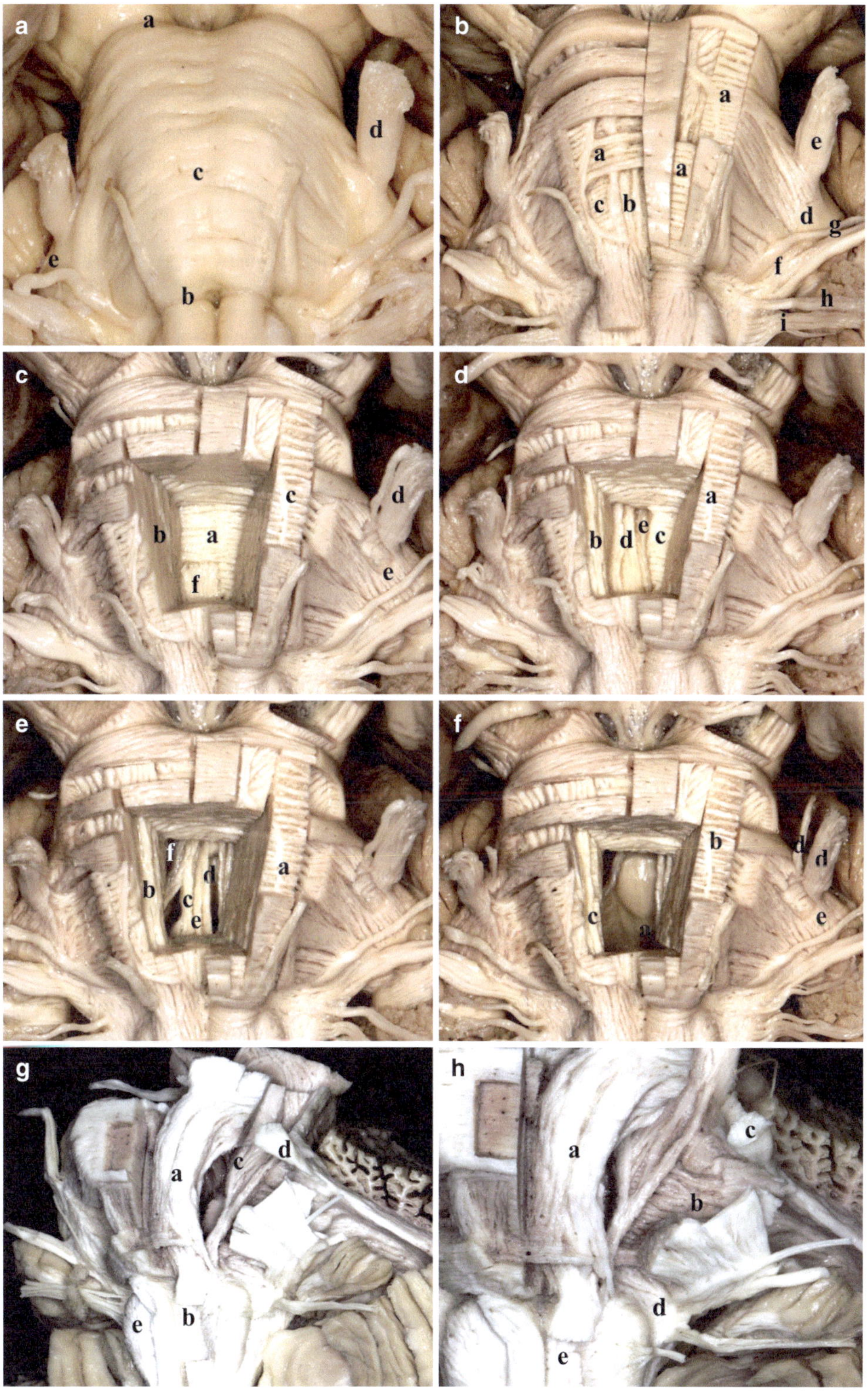
a
a
d
c
e
b
b
a
e
a
a
c b
d g
f
h
i
c
c
d
b a
f
e
d
a
b d e c
e
f
a
b
d
c
e
f
b
d d
c
e
a
g
a
c
d
e
b
h
a
c
b
d
e

Fig. 8 Fiber dissection of midsagittal section of pons. (**a**) Lateral view of cerebellopontine angle. Trigeminal nerve can be seen arising from upper margin of ventral aspect of pons. Below this is flocculus of cerbellum which is irregular lobular extension from nodule of vermis situated between middle cerebellar peduncle and biventral lobule of cerebellum. a. Trigeminal nerve, b. Flocculus. (**b**) Ventral view of flocculus and relation to cranial nerves. Cerebellopontine and cerebellomedullary fissure can be seen converging behind flocculus. Fibers of facial and vestibulocochlear can be seen arising from pontomedullary junction medial to flocculus. Glossopharyngeal, accessory and vagus enters posterolateral sulcus of medulla medial and inferior to flocculus. a. Cerebellopontine fissure, b. Cerebellomedullary fissure, c. Flocculus, d. Facial and Vestibulocochlear nerves, e. Abducens nerve, f.Glossopharyngeal nerve, g, Vagus nerve, h, Cranial part of accessory nerve. (**c**) Mid sagittal view showing roof of fourth ventricle. Upper portion of roof is formed by superior peduncle and superior medullary velum. Lower portion, by the inferior medullary velum, epithelial lining of ventricle covered by tela choroidea and tenia of fourth ventricle. a. Superior cerebellar peduncle, b. Tela choroidea, c. Medial longitudinal fasciculus, d. Fourth ventricle. (**d**) Fiber dissection of mid sagittal section of pons. Fibers of abducens nerve can be seen arising from abducens nucleus located near floor of fourth ventricle in paramedian location in lower half of pons. Fibers after emerging pass ventrally descend through reticular formation and medial lemniscus and transverse the ventral pons to emerge at its inferior border. a. Abducens nerve fibers, b. Corticospinal tract, c. Central tegmental tract, d. Medial longitudinal fasciculus, e. Medial Lemniscus. (**e**) Sagittal view of brainstem showing fiber dissection of medial lemniscus and medial longitudinal fasciculus. Medial longitudinal fasciculus extend close to midline and ventral to floor of fourth ventricle and ends in interstitial nucleus of cajal in lateral wall of third ventricle. Medial lemniscus ascends from sensory decussation on each side as a flattened tract, near the median raphe. The tract increases in size as it ascends to pons. When viewed from side as seen here, medial lemniscus can be seen here ascending through pons forming a concave shape facing ventrally. In the midbrain it forms a curved lamina dorsolateral to substantia nigra to terminate in thalmus. a. Corticospinal tract, b. Central tegmental tract, c. Superior cerebellar peduncle, d. Medial Lemniscus, e. Medial longitudinal fasciculus. (**f**) Midsagittal view of motor and sensory root of trigeminal nerve passing through substance of pons after removal of portion of dorsal pons. Nerve can be seen going towards motor and main sensory nucleus close to its posterior surface and along the line of lateral margin of fourth ventricle. The sensory nucleus is lateral to motor one and beneth the superior cerebellar peduncle. a. Trigeminal nerve, b. Central tegmental tract, c. Corticospinal tract, d. Medial lemniscus, e. Medial longitudinal fasciculus

either side of lower part of pons is the region called as the cerebellopontine angle (Fig. 7g, h). This region lies near the foramen of Luschka which is the lateral opening of the fourth ventricle. This area is intimately related to facial, vestibulocochlear, and glossopharyngeal nerves.

3.3.3 Posterior Corridor

The posterior or dorsal aspect of pons forms the upper aspect of floor of fourth ventricle which will be described in the section on fourth ventricle.The upper portion of roof of fourth ventricle is formed by superior cerebellar peduncle and superior medullary velum (Fig. 9).

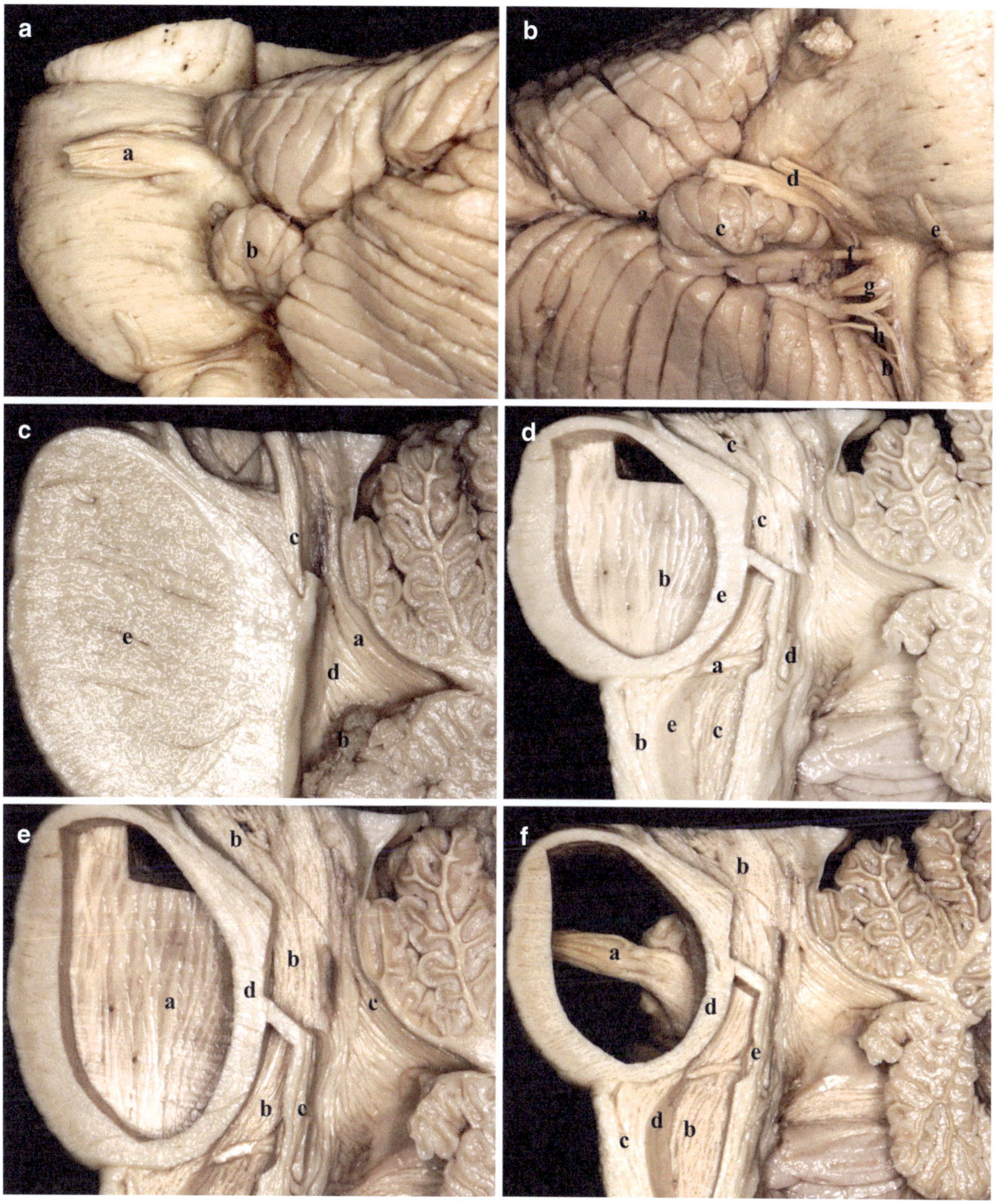
a
a
b
b
a
d
c
e
f
g
h
b
c
c
e
a
d
b
d
c
c
b
e
a
d
b
e
c
e
b
b
a
d
c
b
c
f
b
a
d
e
c
d
b

Fig. 9 Superficial fiber dissection of medulla oblongata. (**a**) The medulla oblongata is pyramidal in shape with its broad extremity being directed upward toward the pons, while its narrow, lower end is continuous with the spinal cord. The medulla oblongata extends from the lower margin of the pons to a plane passing transversely below the pyramidal decussation and above the first pair of cervical nerves. The Anterior median fissure ends above at lower border of ponsinto a small triagular depression called faramen caecum and its lower end is interrupted by pyramidal decussation. The region between anterolateral sulcus and posterolateral sulcus is lateral compartment. Its upper part contain oval prominence called Olive, and lower part is only partly continus with spinal lateral funiculus. Hypoglossal rootlets can be seen, emerging from anterolateral sulcus whereas accessory, vagus and glossopharyngeal nerves are visible entering through posterolateral sulcus. a. Anterior median fissure, b. Foramen caecum, c. Pyramid, d. Anterolateral sulcus, e. Olive f. Hypoglossal nerve rootlets. (**b**) Oblique lateral view of medulla showing lateral region of medulla. The lateral district is limited in front by the anterolateral sulcus and the roots of the hypoglossal nerve, and behind by the posterolateral sulcus and the roots of the accessory, vagus, and glossopharyngeal nerves. Olive can be seen in upper part, while its lower part is of the same width as the lateral funiculus of the spinal cord, and appears on the surface to be a direct continuation of it. a. Anterolateral sulcus, b. Posterolateral sulcus, c. Olive, d. Trigeminal nerve, e. Abducens nerve, f. Facial nerve, g, Vestibulocochlear nerve, h, Rootlets of the glossopharyngeal nerve, i. Rootlets of vagus nerve. (**c**) Obilque lateral view of olive with superficial fiber dissection. Some fibers, termed the anterior external arcuate fibers, emerge from the fissure curve lateralward and upward over the surface of the medulla oblongata to join the inferior peduncle.The ventral spinocerebellar fasciculus is continued upward on the lateral surface of the medulla oblongata in the same relative position it occupies in the spinal cord until it passes under cover of the external arcuate fibers. It passes beneath these fibers just dorsal to the olive and ventral to the roots of the vagus and glossopharyngeal nerves and it continues upward through the pons along the dorso-lateral edge of the lateral lemniscus. Dorsal spinocerebellar fasciculus is carried into the inferior peduncle in the posterior district. a. Pyramid, b. Anterior external arcuate fibers, c. Ventral spinocerebellar fasciculus. (**d**) Fiber dissection of anterior and lateral district of medulla. The two pyramids contain the motor fibers which pass from the brain to the medulla oblongata and spinal cord. Pyrmidal fibers can be seen decussating in lower medulla. a. Pyramid, b. Pyramidal decussation, c. Medial lemniscus, d. Olive, e. olivo-olivary fibers, f. Uncrossed fibers of the corticospinal tract. (**e**) Close up view of upper end of medulla at pontomedullary junction. Corticospinal fibers can be seen here getting into compact bundle at lower end of pons as they continue along either side of anterior median fissure into medullary pyramids. Its upper end is slightly constricted, and between it and the pons the fibers of the abducens nerve emerge. a. Corticospinal tract, b. Abducens nerve, c. Medial lemniscus, d. Foramen caecum. The asterisk shows the area where the corticospinal tracts have been cut. (**f**) View of lower end of medulla showing pyrmidal decussation. Approximately two third fiber of corticospinal tract leave the pyramids in successive bundles, and decussate in the anterior median fissure, forming what is termed the pyramidal decussation. a. Pyramidal decussation

3.4 Internal Features

The internal sections of the pons shows it to be composed of two parts which differ in appearance and structure [1]. The ventral portion mainly consists of an abundance of white fibers arranged in transverse and longitudinal bundles, together with a small amount of gray substance. The dorsal tegmental portion is a continuation of the reticular formation of the medulla oblongata with fiber tracts, and most of its constituents are continued into the tegmentum of the midbrain.

3.4.1 Ventral Pons

White Fibers

The white fibers of the ventral pons are arranged in three transverse bundles and one longitudinal bundle. The longitudinal fiber bundle is the corticospinal tract. The transverse fibers form three bundles based on their relation to the longitudinal descending CST (Fig. 7b, c). The most superficial transverse fibers constitute a thick layer of fibers on ventral surface of pons. This superficial layer runs anterior to the corticospinal tract and is

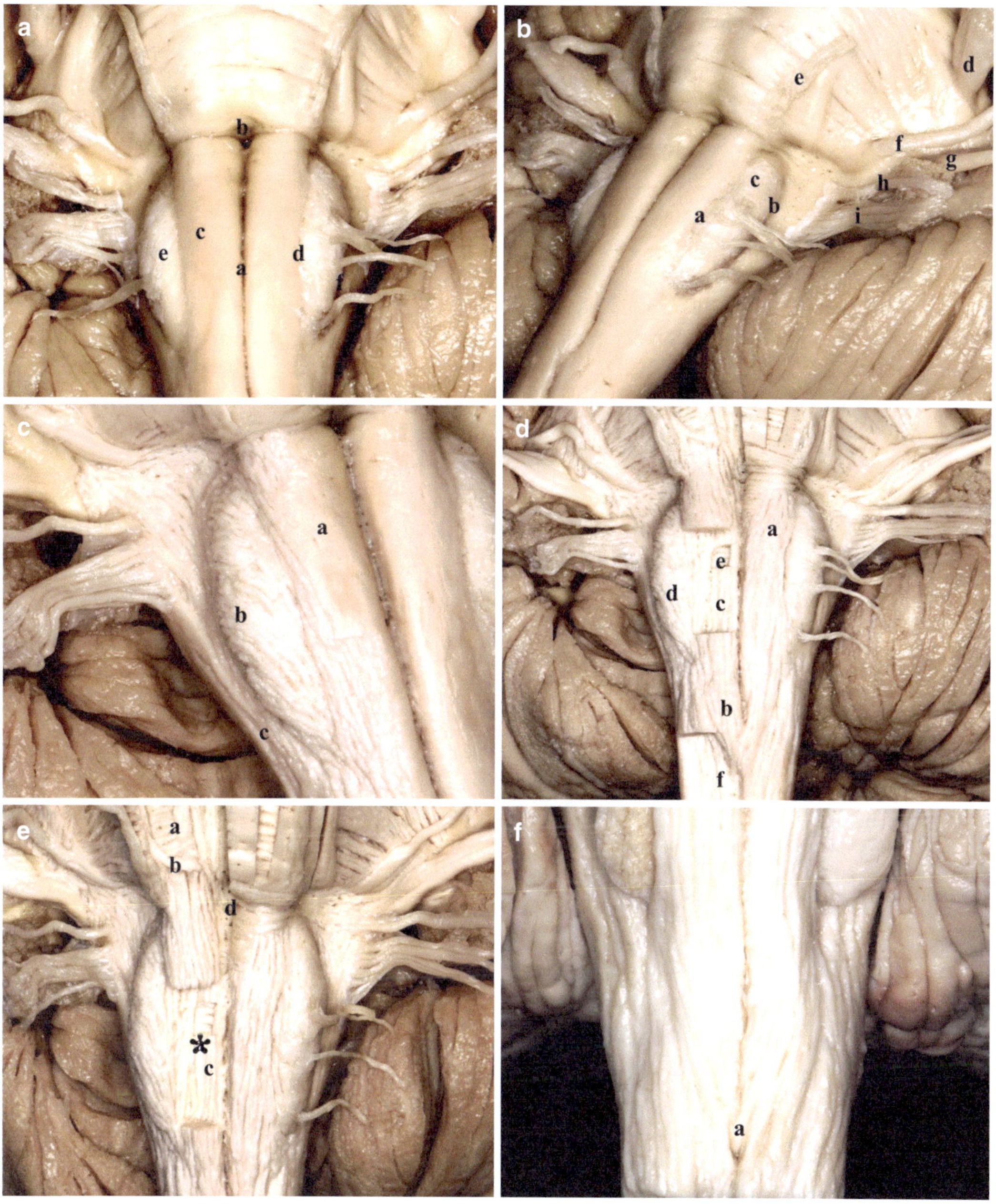
a
b
c
e
a
d
b
e
d
f
g
c
h
a
b
i
c
a
b
c
d
a
e
d
c
b
f
e
a
b
d
*
c
f
a

known as the stratum superficiale pontis. The middle layer of transverse fibers is known as the stratum complexum and it separates the longitudinal running fibers of the pyramidal tract [1]. The third or deep layer is called as the stratum profundum pontis and it traverses dorsal to the pyramidal pathway (CST) [1, 2]. The longitudinal fibers of corticopontine, cortinuclear and CST descend from the crus cerebri of the midbrain and enter the pons compactly but rapidly disperse into fascicles, seperated by pontine nuclei and transverse pontine fibers (Fig. 7). Corticospinal fibers traverse the pons in multiple discrete fascicles to converge again into compact tracts, to emerge as the medullary pyramids.These longitudinal fasciculi stream downward on either side of the midline and cause a forward projection of the superficial transverse fibers, and thus give rise to the eminences on the ventral surface of the pons. The corticopontine fibers originating from the cerebrum terminate on the pontine nuclei which then give rise to the pontocerebellar fibers. The corticonuclear fibers of the pons end in the trigeminal, abducens and facial nuclei. Thus the medial half of pons mainly constitute longitudinal CST where as lateral half or peritrigeminal area is mainly composed of tranverse fibers from the pontine nuclei. The transverse pontine fibers collect together to form the middle cerebellar peduncle.

Gray Matter

The gray matter of the base of the pons is made up of the scattered pontine nucle which give rise to the pontocerebellar pathway.

3.4.2 Dorsal Pons

The dorsal or tegmental part of the pons is an upward continuation of the reticular formation and gray substance of the medulla oblongata. It is composed of white fibers, native gray matter masses of the brainstem and the cranial nerve nuclei. It is subdivided by a median raphé, which, however, does not extend into the basilar part, being obliterated by the transverse pontine fibers.

White Fibers

The trapezoid body lies dorsal to the ventral part of the lower pons. It consists of transverse running fibers, which can be difficult to dissect due to intermingling of crossing fibers [1, 4]. The trapezoid body is formed by the intersecting fibers which originate in the ventral cochlear nuclei and then ascend laterally and supriorly as the lateral lemniscus. The region behind trapezoid body and adjoining ventral pons is occupied by important ascending tracts. The medial lemniscus occupies a transversly elongated flat lamina next to midline, adjacent and dorsal to the trapezoid body and when viewed from lateral aspect ascends through the pons forming a concave shape facing dorsally (Fig. 7g, h). Lateral to this are the trigeminal and spinal lemniscus. Still more laterally is the lateral lemniscus in upper pons. Further dorsally and close to the median raphe/midline lie the medial longitudinal fasciculus, the tectospinal tract, the vestibulospinal tract and the rubrospinal tract. Lateral to this is the central tegmental tract and the spinal tract of the trigeminal nerve.

Gray Matter

The rest of the dorsal part of the pons is a continuation upward of the reticular formation of the medulla oblongata. Within these scattered nerve cells, there are some prominent masses of gray substance, viz., the superior olivary nucleus and the nuclei of the trigeminal, abducens, facial, and vestibulocochlear nerves. From lateral to medial at varying levels in the pons, these are the dorsal and ventral cochlear nuclei, main sensory nucleus of the trigeminal nerve, the motor nucleus of the trigeminal nerve, the nuclei of the lateral lemniscus and the facial nucleus. Further medially are the vestibular nuclei and the abducens nucleus. These nuclei will be discussed in detail in the section on floor of the fourth ventricle.

4 Medulla

4.1 Extent

The medulla oblongata extends from the pons superiorly to the spinal cord inferiorly. The line of demarcation between the medulla and the spinal cord is a horizontal plane that passes below the pyramidal decussation.

4.2 Parts/Divisions

The medulla oblongata is pyramidal in shape, and has its broad extremity directed upward towards the pons, ant its narrow, lower end is continuous with the spinal cord (Fig. 9). The central canal of the spinal cord is prolonged into its lower half, and then opens into the cavity of the fourth ventricle in its upper half. The medulla oblongata may therefore be divided into a lower closed part containing the central canal, and an upper open part corresponding with the lower portion of the fourth ventricle.

4.3 External Features

The surface of medulla is marked by series of fissures and sulci that divide it into three regions (Fig. 9a, b). In the midline ventrally, the anterior median fissure and dorsally, the posterior median sulcus are upward continuations of corresponding fissures seen in spinal cord. These two sulci divide the medulla into two halves. The anterior median fissure is present along the whole extent of the ventral medulla. It terminates at the lower border of the pons as an enlargement, which is triangular in shape and is known as foramen caecum (Fig. 9d, e). The lower end of the anterior median fissure is interrupted by the obliquely criss crossing fibers of the pyramidal decussation (Fig. 9f). The posterior median fissure exists only in the closed part of the medulla oblongata and gradually becomes shallower from below upward. It finally ends at about the middle of the medulla oblongata, where the central canal expands into the cavity of the fourth ventricle. Each half of medulla is marked by two sulci, the anterolateral and posterolateral sulci, which are also continuations of the corresponding sulci of spinal cord [1]. The anterolateral sulcus is in line with ventral roots of spinal nerve from where fibers of the hypoglossal nerve emerge in linear series. Similarly, the posterolateral sulcus is in line with dorsal nerve roots of spinal cord, from where the accessory, vagus, and glossopharyngeal nerve rootlets emerge. Based on these sulci each half of the medulla oblongata is subdivided into three districts, anterior, middle, and posterior.

4.3.1 Anterior Corridor

The anterior district lies between the anterior median fissure and the anterolateral sulcus and is occupied by the pyramids. (Fig. 9a–c) Having crossed the midline in the region of the inferior portion of the anterior medial fissure, they pass down in the posterior part of the lateral funiculus of the spinal cord as the lateral corticospinal fasciculus [1]. The remaining fibers, which occupy the lateral part of the pyramid do not cross the midline, but are carried downward as the anterior corticospinal fasciculus into the anterior funiculus of the same side. The sixth cranial nerve can be seen emerging from the ponto-medullary sulcus anterior to the pyramids.

4.3.2 Lateral Corridor

The lateral district is limited in front by the anterolateral sulcus and behind by the posterolateral sulcus. The roots of the hypoglossal nerve emerge in the anaterolateral sulcus and the roots of the accessory, vagus, and glossopharyngeal nerves emerge from the posterolateral sulcus (Fig. 9b). The upper portion of the lateral district is made up of a prominent oval mass that is named the olive. The olive does not extend all the way upto the pons. Just before reaching the pons there is a small depression on the superior end of the olive. This depression is known as the supraolivary fossa and is the region of emergence of the facial nerve and the nervus intermedius. There is another depression posterior to the supraolivary fossa known as the lateral fossa and from this region the vestibulocochlear nerves emerge. The lower part of the lateral district is of the same width as the lateral funiculus of the spinal cord, and appears on the surface to be a direct continuation of it (Fig. 9c). Only a few fiber bundles that constitute the lateral column of the spinal cord continue upwards in the lateral district. The lateral corticospinal tract from the lateral funiculus of the spinal cord curves medially as the pyramidal decussation and then occupies its position in the anterior district of the opposite side [1]. The dorsal spinocerebellar fasciculus continues upwards in the inferior cerebellar peduncle. So only the ventral spinocerebellar fasciculus continues in the lateral district of the medulla underneath the external arcuate fibers from the lateral spinal funiculus. (Fig. 9c)

4.3.3 Posterior Corridor

The posterior district lies between the posterior median sulcus and posterolateral sulcus and contains the tracts that enter from posterior funiculus of spinal cord (Fig. 9c). They are the fasciculus gracilis medially and fasciculus cuneatus laterally. These fasciculi end in rounded elevations called the gracile and cuneate nuclei [1]. Lateral to the cuneate fasciculus is an elevation, which is formed by the spinal nucleus of the trigeminal nerve and its tract. This elevation is knows as the tuberculum cinereum or the trigeminal tubercle (not to be confused with tuber cinereum in the floor of the third ventricle) [1]. The posterior surface of upper medulla forms the lower part of floor of fourth ventricle. This fossa is bound on either side by inferior cerebellar peduncles. The features of this part of medulla are described with floor of fourth ventricle.

4.4 Internal Features

The internal arrangement of grey and white matter in the lower closed part of medulla is similar to that of spinal cord. However above this it changes gradually due to presence of fourth ventricle. The gray substance, which in the spinal cord forms a continuous H-shaped column, becomes greatly modified and subdivided in the medulla oblongata and on account of the opening out of the central canal of the spinal cord, certain parts of the gray substance, which in the spinal cord were more or less centrally situated, are displaced in floor of fourth ventricle.

4.4.1 White Fibers

A deep dissection of fiber tracts is very difficult in this region but a general overview of relationships of pyramidal tracts, central grey matter and dorsal ascending tracts has been studied by us in our publication and is elaborated here [1]. (Figs. 9 and 10) Starting dorsally the gracile and cuneate fasciculi or the dorsal columns are prolonged upward into the lower part of the medulla oblongata, where they end respectively in the gracile and cuneate nuclei. Second order neurons originate from these nuclei (internal arcuate fibers) pass forward through the neck of the posterior column, curving forward, they decussate in the midline with the corresponding fibers of the opposite side, and run upward immediately behind the CST, as a flattened band, named medial lemniscus. This decussation is also known as the sensory decussation. It lies superior to the pyramidal decussation of the motor fibers. Further laterally is the medial lemniscus which is joined by the ipsilateral anterior and lateral spinothalamic tracts, the fibers of which are derived from the cells of the gray substance of the opposite side of the spinal cord. The anterior and lateral spinothalamic tracts run upwards as the spinal lemniscus.

Lateral to the pyramids and the olive the white matter of the medulla is made up of the spinocerebellar tracts. The ventral and dorsal spinocerebellar tracts ascend upwards from the spinal cord to occupy the lateral district of the medulla between the olivary nucleus and the inferior cerebellar peduncle. Both the ventral and dorsal spinocerebellar pathways are destined for the cerebellar vermis ultimately which they reach via the superior and inferior cerebellar peduncle respectively. Lateral to the spinocerebellar tracts is the inferior cerebellar peduncle [1].

The white matter on the ventral aspect of the medulla consists of the corticospinal and the corticonuclear fibers. The corticospinal tracts form the bulk of the pyramids. The corticonuclear fibers occupy the posterior and medial portion of the pyramid and are destined to reach the ambiguous and hypoglossal nuclei. On the surface of the pyramids and the olive transverse fibers or the anterior external arcuate fibers are visualized. These fibers are said to originate in the arcuate nucleus traverse over the surface of the pyramids and the olive and then enter the inferior cerebellar peduncle. The arcuate nucleus is present on the anterior and medial aspect of the pyramid close to the midline. The other white matter tracts that run within the substance of the medulla are the medial longitudinal fasciculus and the tectospinal tracts medially and the vestibulospinal, rubrospinal and reticulospinal tracts laterally. Further laterally is the spinal tract of the trigeminal nerve.

4.4.2 Gray Matter

The gray matter of the medulla oblongata consists of nuclei of the cranial nerves and gray matter native to the medulla itself. The cranial nerve nuclei in the medulla are the hypoglossal nucleus, the dorsal motor nucleus of the vagus, the nucleus of the solitary tract, the spinal nucleus of the trigeminal nerve and the nucleus ambiguous. The gray matter masses of the medulla are the inferior olivary nucleus, the dorsal and medial accessory olivary nuclei and the gracile and cuneate nuclei. Some of these nuclei are traceable from the gray substance of the spinal cord, while others are represented in only the medulla.

Motor Nuclei

The hypoglossal nucleus represents the continuation of the base of the anterior horn of the spinal cord into the medulla. The hypoglossal nucleus can be seen anterolateral to the central canal in the closed part of the medulla but moves to the midline in the floor of the fourth ventricle where it forms an eminence known as the hypoglossal trigone (Fig. 9). The nerve roots leave the ventral side of the nucleus, pass forward between the reticular formation, and emerge from the anterolateral sulcus.

The nucleus ambiguus which is the somatic motor nucleus of the glossopharyngeal, vagus and cranial portion of the accessory nerves, is the continuation of the dorso-lateral cell group of the anterior column of the spinal cord into the medulla [1]. The fibers of the nucleus ambiguus pass dorsally and approach the floor of the fourth ventricle. Here they make an acute bend laterally and anteriorly to join the fibers of the dorsal nucleus.

Sensory Nuclei

The gray matter base of the posterior horn of the spinal cord continues on the dorsal portion of the central canal in the closed part of the central canal. In the upper portion the gray matter masses move laterally in the floor of the fourth ventricle. It constitutes the terminal nuclei of the sensory fibers of the vagus and glossopharyngeal nerves, and further laterally the vestibular part of the acoustic nerve and the sensory root of the facial nerve. Still higher, it forms a mass of pigmented cells—the locus coeruleus—in which some of the sensory fibers of the trigeminal nerve appear to end.

The dorsal nucleus is often called the sensory nucleus or the terminal nucleus of the sensory fibers of the glossopharyngeal and vagus nerves [1]. It is situated behind the hypoglossal nucleus in closed part of medulla oblongata, whereas in the upper, open part it lies lateral to that nucleus, and corresponds to an eminence, named the ala cinerea or vagal triangle, in the rhomboid fossa described later.

The nucleus of the solitary tract is a sensory nucleus which receives afferent visceral (taste) fibers from the facial, glossopharyngeal and the vagus nerves. It lies dorsal to the central canal in the closed portion of the medulla and moves posterolateral to the dorsal motor nucleus of the vagus in the upper open part of the medulla [1].

The nucleus of the spinal tract of the trigeminal nerve is the lateral most of the cranial nerve nuclei in the medulla. In the closed portion it lies lateral to the cuneate nucleus medial to the spinal tract and further upwards the tract and the nucleus are present medial to the inferior cerebellar peduncle.

The gracile, the cuneate nucleus and the inferior olivary nucleus are represented on the surface of the medulla by the gracile and cuneate tubercle and the olive respectively [1]. The dorsal and medial accessory olivary nuclei lie on the ventral and dorsal aspect of the inferior olivary nucleus.

5 Floor of Fourth Ventricle

5.1 External

The fourth ventricle portion of the brainstem is also known as the rhomboid fossa, based on its shape. Its ventral wall, is formed by the back of the pons and medulla oblongata, and it constitutes the floor of the fourth ventricle (Fig. 10).

Fig. 10 Fiber dissection of dorsal midbrain and floor of fourth ventricle. (**a**) Posterior view of brainstem with superficial fiber dissection of pulvinar, pineal gland, tectal plate and superior cerebellar peduncle. Corpora quadrigemina can be seen here, above and in front of the superior medullary velum and superior peduncle, and below and behind the third ventricle and posterior commissure. The superior and inferior colliculican be seen separated from one another by a crucial sulcus. Pineal body, a cone-like structure which projects backward from the thalamencephalon and partly obscures the superior colliculi can be seen here. From the inferior end of the longitudinal sulcus, a white band, termed the frenulum veli, is prolonged downward to the anterior medullary velum; on either side of this band the trochlear nerve emerges. The superior brachium extends lateralward from the superior colliculus, and, passing between the pulvinar and medial geniculate body, is partly continued into an eminence called the lateral geniculate body, and partly into the optic tract. The inferior brachium passes forward and upward from the inferior colliculus and disappears under cover of the medial geniculate body. a. Pulvinar of the thalamus, b. Pineal body, c. Superior colliculus, d. Inferior colloculus, e. Third ventricle, f. Cruciate sulcus, g. Frenulum veli, h. Trochlear nerve, i. Superior cerebellar peduncle, j. Superior brachium, k. Inferior brachium, l. Medial geniculate body, m. Lateral lemniscus.(**b**) Roof has been removed on both sides exposing floor of fourth ventricle also called rhomboid fossa. One side superficial fibre dissection is done while preserving ependymal floor on other side. The fossa is divided into three parts superior, intermediate and inferior. The superior part is triangular in shape and limited laterally by the superior cerebellar peduncle and its base is represented by an imaginary line at the level of the upper ends of the superior fovea.The intermediate part extends from this level to that of the horizontal portions of the taeniae of the ventricle. The inferior part is triangular, and its downwardly directed apex, named the calamus scriptorius, is continuous with the central canal of the closed part of the medulla oblongata. Facial colliculus, which overlies the nucleus of the abducens nerve and ascending portion of the root of the facial nervecan be seen on right side after fibre dissection. Hypoglossal trigoneis visible on left side as triagular region in lower part of median. Trigone is seen to consist of a medial and a lateral area separated by a series of oblique furrows. The medial area corresponds with the upper part of the nucleus of the hypoglossal nerve, the lateral with a small nucleus, the nucleus intercalatus. a.Mesencephalic tract of the trigeminal nerve, b. Medial sulcus, c. Sulcus limitans, d. Supeior fovea, e. Superior cerebellar peduncle, f. Abducens nucleus, g. Inferior cerebellar peduncle, h. Facial nerve, i. Middle cerebellar peduncle, j. Lateral lemniscus, k. Medial longitudinal fasciculus, l. Locus coerulus. (**c**) Superficial fibre dissection of floor of fouth ventricle and facial colliculus. The most superficial layer of the floor of fourth ventricle is ependymal lining and a subependymal layer of glia have been removed on left also. The abudcent nerve nucleusand the facial nerve fibers loping around it form elevation on floor of fouth ventricle called as facial colliculus. The fibers of the facial nerve pursue a tortuous course in the substance of the pons. At first they pass backward and medialward until they reach the rhomboid fossa, close to the median sulcus, where they are collected into a round bundle; this passes upward and forward, producing an elevation, in the rhomboid fossa, and then takes a sharp bend, and arches lateralward through the substance of the pons. Ascending part of internal genu of facial nerve is medially related to fibers of medial longitudinal fasciculus. Lateral to fovea is rounded elevation of vestibular area extending into lateral recess where it forms feebly marked swelling as auditory tubercle. White fibrous strands called striae medullares can be seen crossing vestibular area and medial eminence and disappear into median sulcus. Caudal to inferior fovea between hypoglossal triangle and lower part of vestibular area is the vagal triangle overlying dorsal vagal nucleus. a. Abducens nucleus, b. Inferior fibers of the facial nerve, c. Superior fibers of the facial nerve, d. Medial longitudinal fasciculus, e. Vestibular area, f. Striae medullaris, g. Hypoglossal trigone, h. Region of solitary nucleus and tract, i. Inferior cerebellar peduncle, j. Central tegmental tract, k. Spinal tract of trigeminal nerve, l. Area postrema, m. Fasciculus gracilis, n. Fasciculus cuneatus, o. Superior cerebellar peduncle, p. Middle cerebellar peduncle, q. Vagal trigone. (**d**) Close up view of rostral and intermediate floor of fourth ventricle. Fiber dissection done to demonstrate facial colliculcus with facial nerve fibers, medial longitudinal fasciculus, trigeminal mesencephalis tract and central tegmental tract. The trigeminal mesencephalic tract ascend deep to floor of fourth ventricle between superior cerebellar peduncle laterally, sulcus limitans medially, central tegmental tract ventrally and locus coeruleus dorsally. Central tegmental fasciculus is group of efferent fibers from red nucleus to inferior olivary nucleus. Central tegmental tract is located deep to locus coreuleus and trigeminal mesencephalic tract and dorsal to medial lemniscus. a. Abducens nucleus, b, Inferior fibers of the facial nerve, c. Superior fibers of the facial nerve, d. Medial Longitudinal fasciculus, e. Suprafacial safe entry zone, f. Central tegmental tract, g. Mesencephalic nucleus and tract of the trigeminal nerve, h. Striae medullares, i. Acoustic tubercle, j. Inferior cerebellar peduncle, k. Hypoglossal trigone, l. Region of nucleus and tract of solitarius, m- Infrafacial safe entry zone, n. Spinal tract of trigeminal

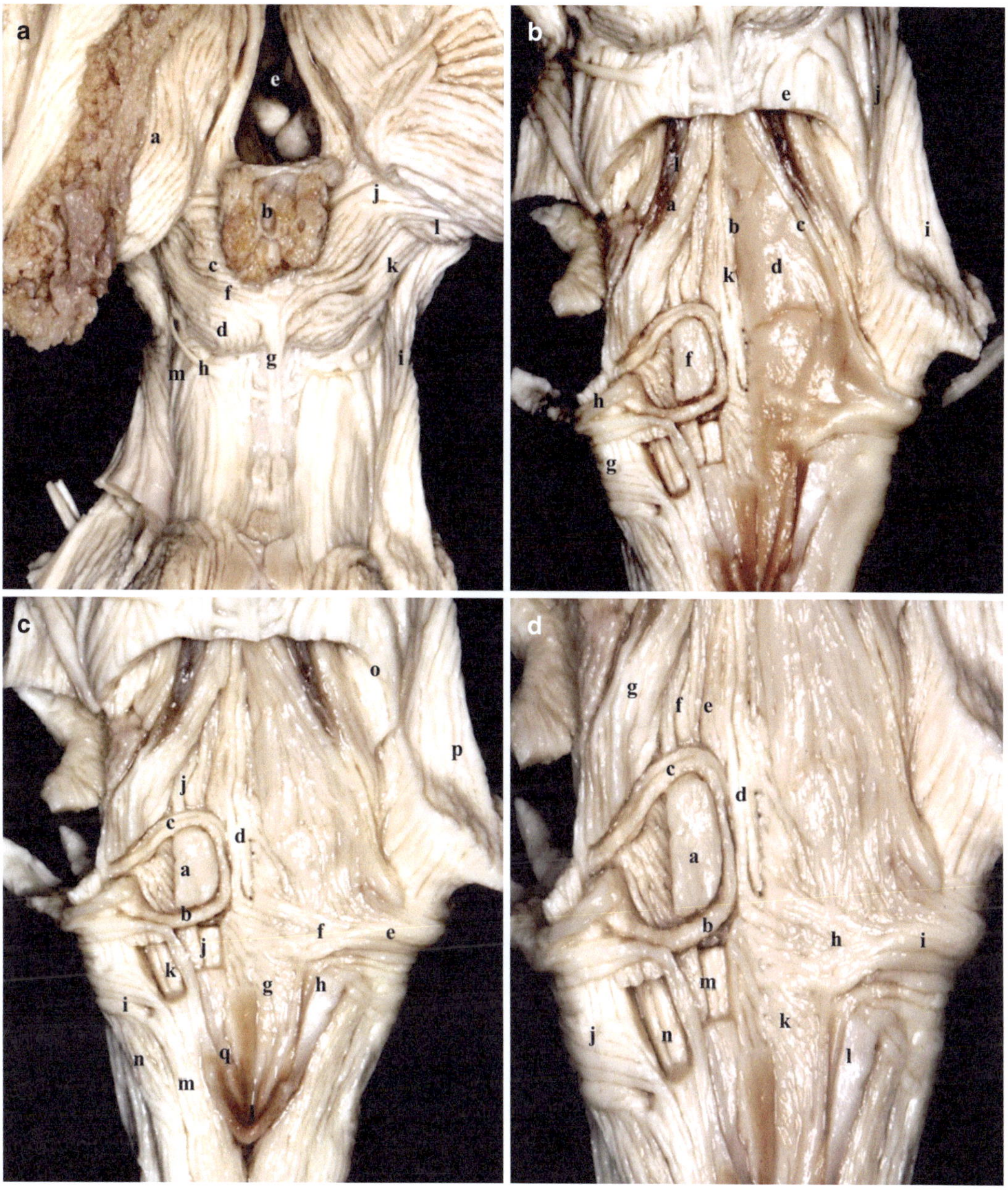
a
e
a
b
j
l
c
k
f
d
g
i
m
h
b
e
j
l
a
b
c
i
k
d
f
h
g
c
o
p
j
c
d
a
b
f
e
j
k
g
h
i
n
q
m
l
d
g
f
e
c
d
a
b
h
i
m
k
j
n
l

The most superficial layer of the floor of the fourth ventricle is a neuroglial layer which is the ependyma of the ventricle. Beneath this is a thin layer of gray matter. The fossa can be divided into three parts viz. superior, intermediate, and inferior (Figs. 10b, c). The superior and inferior portions are triangular in shape. The superior portion has its apex superiorly and its base inferiorly whereas the inferior portion is an inverted triangle with its apex inferiorly and its base superiorly [1]. The superior portion is bounded laterally by the superior cerebellar peduncles. The base is an imaginary line passing through the upper ends of the superior fovea. The inferior portion is bounded by the inferior cerebellar peduncles laterally and its apex is formed by a narrow recess known as the calamus scriptorius which continues with the central canal. The intermediate part of the floor lies between the two triangles. It extends from a line passing through the upper parts of the superior fovea superiorly to the horizontal portion of the taenia inferiorly. It is limited laterally by the middle cerebellar peduncle and the lateral recesses of the fourth ventricle inferiorly. (Fig. 10c, d).

The floor of the fourth ventricle is divided into two halves by the median sulcus.(Fig. 10b). On either side of this sulcus is an elevation, the medial eminence, bounded laterally by a sulcus, the sulcus limitans (Fig. 10b). The superior portion of the medial eminence is marked by an elevation known as the facial colliculus. The elevation is formed by the nucleus of the abducens nerve, and in part by the ascending portion of the fibers of the facial nerve. In the inferior part of the fossa the medial eminence assumes the form of a triangular area, the hypoglossal trigone. The trigone is seen to consist of a medial and a lateral area separated by a series of oblique furrows. The medial area is formed by the hypoglossal nucleus and the lateral area by the nucleus intercalatus.

The sulcus limitans forms the lateral boundary of the medial eminence. The sulcus limitans deepens at its superior and inferior ends to form the superior and inferior fovea. The superior fovea is triangular and its boundaries are the superior cerebellar peduncle superolaterally, the vestibular area inferolaterally and the sulcus limitans medially.The base of the superior fovea lies at the same level as the upper border of the facial coliculus. The main sensory nucleus and the motor nucleus of the trigeminal nerve lie in the superior fovea. Ventral to the trigeminal nuclei lies the locus coeruleus. The locus coeruleus is bluish gray in color owing to the presence underneath of deeply pigmented nerve cells, known as the substantia ferruginea. The inferior fovea constitutes the vagal trigone. (Fig. 10b, c) It is triangular in shape, greyish in color and it overlies the dorsal nucleus of the vagus nerve.

Lateral to the superior and the inferior fovea is the vestibular area which is seen to extend into the lateral recess. The prominence seen at the lateral portion of the vestibular area is the auditory tubercle. Winding around the inferior cerebellar peduncle and crossing the vestibular area and the medial eminence are seen white fibrous strands called striae medullares which forms a portion of cochlear division of the acoustic nerve and disappear into median sulcus. Caudal to inferior fovea between hypoglossal triangle and lower part of vestibular area is the vagal triangle overlying dorsal vagal nucleus as described earlier. This triangle is crossed below by narrow translucent ridge called funiculus separans, and between this and gracile tubercle is a small tongue shaped area called area postrema (Fig. 10).

5.2 Internal

Superficial fibre dissection of floor of fouth ventricle and facial colliculus was done to demonstrate relation of deep structures to surface. The most superficial layer of the floor of fourth ventricle is ependymal lining and a subependymal layer of glia. The abducens nerve nucleus and the facial nerve fibers loping around it form an elevation on floor of fouth ventricle known as facial colliculus (Fig. 10c, d). The nucleus of the facial nerve is situated deeply in the reticular formation of the pons, on the dorsal aspect of the superior olivary nucleus. Nerve fibers arising from facial nerve nucleus pursue a tortuous course in the substance of the pons. At first they pass dorsally and medially until they reach the lower pole of the abducens nucleus near rhomboid fossa, where they are collected into a round bundle. This bundle passes upward and forward on medial side of nucleus, producing an elevation, in the rhomboid fossa. Finally at upper pole of nucleus fibres takes a sharp bend, and arches lateralward through the substance of the pons (Fig. 10d). This ascending part of internal genu of facial nerve fibers is medially related to fibers of MLF. Finally these fibers travel ventrally between facial nucleus medially and spinal nucleus of trigeminal nerve laterally and to emerge at its lower border in the interval between the olive and the inferior peduncle of the medulla oblongata. The MLF courses adjacent to midline near the floor of forth ventricle and connects third fourth and sixth nerve nuclei (Fig. 10d). The trigeminal nerve after reaching its motor and main sensory nucleus in pons, divides into a rostrally directed trigeminal mesencephalic nucleus tract and caudally directed spinal tract.The trigeminal mesencephalic tract ascends deep to floor of fourth ventricle between superior cerebellar peduncle laterally, sulcus limitans medially, CTT ventrally and locus coeruleus dorsally (Fig. 10d). CTT is group of efferent fibers from red nucleus to inferior olivary nucleus. At the level of midbrain CTT originates from dorsomedial red nucleus and descends ipsilaterally, passing dorsally through superior cerebellar peduncle decussation and lateral to MLF. CTT is located deep to locus coreuleus and trigeminal mesencephalic tract and dorsal to medial lemniscus.

6 Organization of White Fibers of Brainstem (Figs. 11, 12, and 13)

The major bulk of the brainstem is made up of white fibers and within it is the interspersed gray matter. The gray matter of the brainstem consists of nuclei of the cranial nerves and some native brainstem nuclei. The white matter of the brainstem can be said to be composed of three types of fibers [1].

These are the short and long projection fibers, the traversing fibers and the association fibers. The arrangement of these fibers can be viewed as

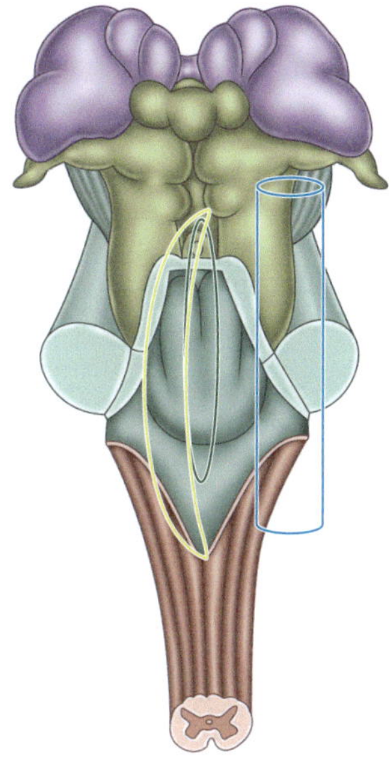

Fig. 11 Schematic dorsal view of the brainstem showing the division of white fibers into three zones. Blue cylinder – Zone 1 or outermost zone, Yellow crescent – Zone 2 or middle zone, Green Oval – Zone 3 or innermost zone

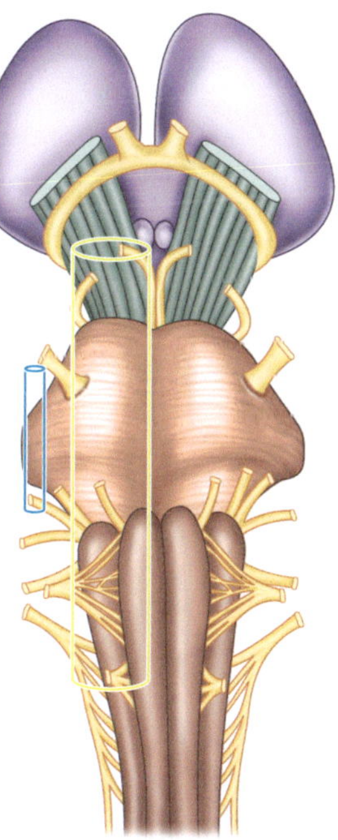

Fig. 12 Schematic ventral view of the brainstem showing the zones. Blue cylinder – Zone 1 or outermost zone. (Note: Most of the fibers in Zone 1 are seen on the dorsal aspect of the brainstem.) Yellow cylinder – Zone 2 or middle zone. Zone 3 has not been depicted in this iamge as it lies in the center of the brainstem and is better visualized from the dorsal aspect

arranged in three zones. The short projection fibers are mainly fibers that originate in the brainstem and project to the cerebellum or the adjacent geniculate bodies and fibers that originate in the cerebellum and project to the brainstem. These include the superior, middle and inferior cerebellar peduncle, the superior and inferior brachium and the lateral lemniscus. These bundles occupy the outermost zone of the brainstem circumference and we have classified them as Zone 1. The next zone or Zone 2 lies medial to Zone 1 and consists of traversing fibers and long projection fibers. Traversing fibers of the brainstem include those fibers that just pass the brainstem on their way to and from the cortex, diencephalon and the cerebellum. These include the corticospinal tract, the spinal lemniscus and the dorsal and ventral spinocerebellar tracts. The long projection fibers are those that connect the brainstem to the cerebral cortex, diencephalon and the spinal cord. This sect of fibers consists of both ascending and descending fibers and includes the medial lemniscus, the trigeminal lemniscus, the corticobulbar fibers, the corticopontine fibers, the reticulospinal, the rubrospinal and the vestibulospinal tracts. The third types of fibers of the brainstem are the association fibers. These fibers essentially connect the brainstem nuclei to each other and to the spinal cord. These include the medial longitudinal fasciculus, the tectobulbar tract and the central tegmental tract. These fibers are in Zone 3 and comprise the innermost zone of fibers. The fibers traverse in their respective zones throughout the brainstem except at the point of their decussation where they occupy the midline. Thus at the level of the midbrain the major decussations are the decussations of the rubrospinal tract, tectobulbar tract and the superior cerebellar peduncle. At the pons are the crossing fibers of the middle cerebellar peduncle and the trapezoid body. At the medulla there is the pyramidal decussation and the decussation of the medial lemniscus. All the three zones can be visualized as three circles occupying the brainstem. Zone 1 is the outermost circle, within which lies Zone 2 and then Zone 3. Zone 3 is in the centre of the brainstem around the midline. (Figs. 11, 12, and 13).

Fig. 13 Schematic representation of cut section at various levels in the brainstem showing the various zones. Zone 1– Blue (Color coding of tracts: Green color – Medial geniculate body and lateral lemniscus, Olive green color – Superior cerebellar peduncle, Pink color – Middle cerebellar peduncle, Purple color – Inferior cerebellar peduncle), Zone 2 – Yellow (Color coding of tracts: Light orange – Corticospinal and corticobulbar tracts, Light blue color – Medial lemniscus and gracile and cuneate fasciculi at the level of obex, Dark pink color – Spinal lemniscus, Brown color – Trigeminal lemniscus) Zone 3 – Green (Color coding: Yellow – Medial longitudinal fasciculus, Red – Tectospinal tract). (**a**) Cut section of the midbrain at the level of the superior colliculus. Zone 1 includes the superior cerebellar peduncle, the lateral lemniscus and the superior brachium. Zone 2 includes from ventral to dorsal the corticospinal and corticonuclear tracts, the medial lemniscus, the spinal lemniscus and the mesencephalic tract of the trigeminal. Zone 3 includes the medial longitudinal fasciculus, the central tegmental tract and the tectobulbar tracts. (**b**) Cut section of the midbrain at the level of the inferior colliculus. Zone 1 includes the superior cerebellar peduncle, the lateral lemniscus and the inferior brachium. Zone 2 includes from ventral to dorsal the corticospinal and corticonuclear tracts, the medial lemniscus, the spinal lemniscus and the mesencephalic tract of the trigeminal. Zone 3 includes the medial longitudinal fasciculus, the central tegmental tract and the tectobulbar tracts. (**c**) Cut section of the pons at the level of the trigeminal nuclei. Zone 1 includes the superior cerebellar peduncle and the middle cerebellar peduncle. Zone 2 includes from ventral to dorsal the corticopontine and the corticospinal fibers, the medial lemniscus, the spinal lemniscus and the trigeminal lemniscus. Zone 3 includes the medial longitudinal fasciculus, the central tegmental tract and the tectobulbar tracts. (**d**) Cut section of the pons at the level of the facial colliculus. Zone 1 includes the middle and the inferior cerebellar peduncle. Zone 2 includes from ventral to dorsal the corticopontine and the corticospinal fibers, the medial lemniscus, the spinal lemniscus and the trigeminal lemniscus. Zone 3 includes the medial longitudinal fasciculus, the central tegmental tract and the tectobulbar tracts. (**e**) Cut section of the medulla at the level of the olivary nuclei. Zone 1 includes the inferior cerebellar peduncle. Zone 2 includes from ventral to dorsal the corticopontine and the corticospinal fibers, the medial lemniscus, the spinal lemniscus and the trigeminal lemniscus. Laterally it includes the dorsal spinocerbellar tracts. Zone 3 includes the medial longitudinal fasciculus, the central tegmental tract and the tectobulbar tracts. (**f**) Cut section of the medulla at the level of the obex. Zone 1 includes the inferior cerebellar peduncle. Zone 2 includes from ventral to dorsal the corticopontine and the corticospinal fibers, the spinal lemniscus, the trigeminal lemniscus and the fasciculus gracilis and cuneatus. Laterally it includes the dorsal and ventral spinocerbellar tracts. Zone 3 includes the medial longitudinal fasciculus, the central tegmental tract and the tectobulbar tracts

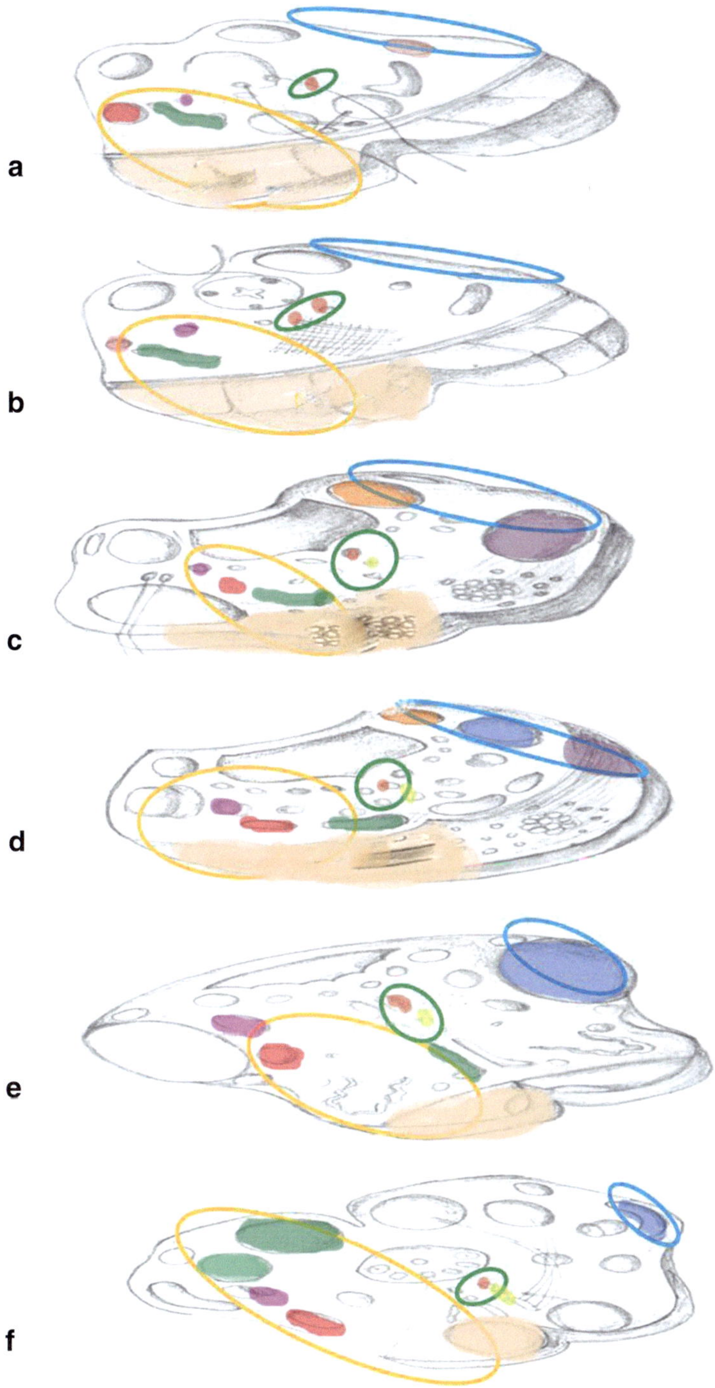
a
b
c
d
e
f

7 Zone 1

7.1 Cerebellar Peduncles

7.1.1 Superior Cerebellar Peduncle

The superior cerebellar peduncles emerge from the medial portion of each cerebellar hemisphere and ascend upwards towards the tectum [1]. The two peduncles are connected to each other by the superior medullary velum. Thus they form both the lateral wall and the roof of the fourth ventricle. (Fig. 10) The fibers of the superior cerebellar peduncle are mainly derived from the cells of the dentate nucleus of the cerebellum and emerge from the hilus of this nucleus. They are continued upward beneath the corpora quadrigemina, and the fibers of the two peduncles undergo a complete decussation ventral to the cerebral aqueduct. Having crossed the midline they divide into ascending and descending groups of fibers, the former ending in the red nucleus, the thalamus, and the nucleus of the oculomotor nerve, while the descending fibers can be traced as far as the dorsal part of the pons. The superior cerebellar peduncle is the main efferent pathway of the cerebellum.

7.1.2 Middle Cerebellar Peduncle

Middle cerebellar peduncles are composed entirely of fibers arising from the cells of the pontine nuclei of the opposite side that end in the cerebellar cortex (corticopontine fibers). (Fig. 7) The fibers of the middle cerebellar peduncle are arranged in superficial and deep fasciculus [1]. The superficial fasciculus is derived from the superficial and middle transverse fibers of the pons and it is directed backward and lateralward to be distributed mainly to the lobules on the inferior surface of the cerebellar hemisphere and to the folia on the under surface close to the vermis. The deep fasciculus comprises most of the deep transverse fibers of the pons. The middle cerebellar peduncle is the main afferent pathway of the cerebellum.

7.1.3 Inferior Cerebellar Peduncle

The inferior cerebellar peduncles pass at first upward and lateralward, forming part of the lateral walls of the fourth ventricle, and then bend abruptly backward to enter the cerebellum between the superior and middle peduncles. (Fig. 10) The inferior cerebellar peduncle consists of fibers of the dorsal spinocerebellar fasciculus of the spinal cord, fibers from the olivary nuclei, vestibular fibers, fibers from the reticular formation, some fibers from the gracile and cuneate nuclei and some fibers of the ventral spinocerebellar tract [1].

7.1.4 Lateral Lemniscus

The lateral lemniscus comes to the surface of the midbrain along its lateral sulcus, and disappears under the inferior colliculus. The lateral lemniscus consists of fibers arising from the cochlear nuclei (acoustic striae), the superior olivary nucleus and the trapezoid body [1]. The fibers ascend upwards to end in the ipsilateral or contralateral inferior colliculus. Some fibers traverse further onwards to reach the thalamus and the auditory area of the hemispheres.

7.1.5 Superior Brachium

The superior brachium consists of corticotectal fibers that come from the visual cortex, the adjacent parietal lobe and the frontal eye fields and terminate in the superior colliculus [1]. The fibers pass between the pulvinar of the thalamus and the medial geniculate body and reach the superior colliculus. En route to the lateral geniculate body the fibers cross the fibers of the inferior brachium.

7.1.6 Inferior Brachium

The inferior brachium consists of short projection fibers of the auditory pathway, which originate in the inferior colliculus and end in the medial geniculate body.

8 Zone 2

8.1 Corticospinal Tract (CST)

CSTs are made up of predominantly fibers of neurons of the motor area of the cortex which projects down to anterior horn cells of spinal cord. After passing through corona radiata and internal capsule they enter brainstem as discrete fiber bundle and traverse in the cerebral peduncles of midbrain, ventral pons and pyramidal region of medulla (Figs. 6 and 7). Just rostral to spino-medullary junction, majority of corticospinal fibers cross the median plane as pyramidal decussation to continue as lateral CST in the spinal cord (Fig. 9) [1]. Most of the fibers decussate and enter the lateral funiculus of the spinal cord as the lateral corticospinal tract. The fibers that pass downwards into the spinal cord without decussation form the anterior corticospinal tract, which occupies the anterior portion of the spinal cord close to the midline. They can be dissected easily and traced along their pathway, as they are located superficially in ventral brainstem. In the pons CST are a little difficult to define in view of crossing transverse fibers but can be defined if followed from either midbrain or medulla [1]. When dissected they are predominantly occupying medial half of ventral pons. In medulla, they again can be identified as discreet fiber bundles, as the medullary pyramids.

8.2 Medial Lemniscus

The fibers from the gracile and cuneate nuclei proceed towards the midline in a curved course as the internal arcuate fibers. These fibers then cross the midline forming the sensory decussation and turn upwards to form the medial lemniscus. In the medulla the medial lemniscus ascends from sensory decussation on each side as a flattened tract, between the median raphe and the inferior olivary nucleus. The tract then moves laterally and increases in size as it ascends to pons [1]. When viewed from the side as seen here, medial lemniscus ascends through the pons forming a concave shape facing dorsally (Fig. 7g, h). In the midbrain it ascends in dorsolateral part of tegmentum in the angle between red nucleus and substantia nigra. Fibers in medial and trigeminal lemnisci continue a rostral course to end in lateral and medial ventral posterior nuclei of thalamus respectively.

8.3 Trigeminal Tract

The trigeminal nerve enters the brainstem in the middle of the pons just superior to the middle cerebellar peduncle. (Figs. 7 and 8) It courses towards the fourth ventricle to reach trigeminal motor and sensory nuclei. Trigeminal nerve can be seen entering upper margin of anterior aspect of pons, where its fibers can be seen entering perpendicular to the transverse pontine fibers which are going into cerebellum via middle cerebellar peduncle. Two tracts originate from the pons from the trigeminal nuclei, one ascending towards the midbrain (mesencephalic tract) and one descending towards the spinal cord (spinal tract). The trigeminal mesencephalic tract ascends deep to floor of fourth ventricle between superior cerebellar peduncle laterally, sulcus limitans medially, CTT tract ventrally and locus coeruleus dorsally (Fig. 10). The spinal tract runs caudally with trigeminal spinal nucleus lying on its medial aspect. It is related laterally to the inferior cerbellar peduncle, medially to the medial lemniscus and dorsally to the gacile and cuneate nuclei [1]. The spinal tract runs caudally as far as the C3 level of the spinal cord and them merges with the Lissauer's tract in the dorsal horn of the spinal cord.

8.4 Dorsal and Ventral Spinocerebellar Tracts

The spinocerebellar tracts have their origin in the thoracic and lumbosacral segments of the spinal cord. The dorsal spinocerebellar tract originates in the nucleus dorsalis or Clarke's column in the thoracic and upper lumbar spinal cord. It consists of mainly uncrossed fibers that ascend to the brainstem [1]. The ventral spinocerebellar tract originates in the lumbosacral enlargement of the spinal cord. It comprises mainly of crossed fibers. The fibers first cross in the anterior commissure of the spinal cord and then double cross again in the cerebellum. In the brainstem they lie on the lateral aspect of the spinal cord medial to the cerebellar peduncles. The dorsal spinocerebellar tracts enter the cerebellum through the inferior cerebellar peduncle whilst the ventral tracts continue upwards to the pons and enter the cerebellum via the superior cerebellar peduncle [1]. The dorsal and ventral spinocerebellar tracts carry proprioceptive fibers from the lower limbs. The fibers from the upper limb carrying proprioception to the cerebellum travel via the cuneocerebellar tract. The fibers originate in the accessory cuneate nucleus and travel as posterior external arcuate fibers and the cuneocerebellar tracts and enter the cerebellum via the inferior cerebellar peduncle.

8.5 Spinal Lemniscus

The spinal lemniscus comprises of the spinothalamic tracts. The spinothalamic tracts originating in the spinal cord cross in the anterior commissure and ascend upwards. In the medulla the tracts lie dorsal to the inferior olivary nucleus and medial to the spinocrebellar tracts [1]. In the pons the spinal lemniscus lies lateral to the medial and trigeminal lemniscus and medial to the middle cerebellar peduncle. In the midbrain the tract lies dorsolateral to the red nucleus and dorsal to the medial lemniscus.

9 Zone 3

9.1 Medial Longitudinal Fasciculus (MLF)

MLF is a compact fiber bundle situated in a paramedian location close to floor of fourth ventricle which interconnects oculomotor, trochlear, abducens, vestibular, spinal accessory and reticular nuclei coordinating conjugate eye and neck movements (Figs. 7d, e, 8e, and 10d). At the level of the midbrain, it is found paramedian between the cerebral aqueduct dorsally and the decussation of the superior cerebellar peduncle ventrally [1]. The MLF passes ventrolateral to the oculomotor nuclei where it mixes with the intramesencephalic segment of the oculomotor nerve and ends in the interstitial nucleus of Cajal, in lateral wall of third ventricle. At the level of the lower half of the inferior colliculus, the MLF connects to the trochlear nucleus and travels near the floor of the fourth ventricle and adjacent to the midline of the pons (Fig. 10d). It curves ventrally at the lower edge of the facial colliculus, approaches the medial lemniscus, and passes ventral to the hypoglossal triangle. The MLF retains its position relative to central grey matter through midbrain, pons and upper medulla. In lower medulla it is displaced ventrally by successive decussation of medial lemnisci and lateral CSTs, to continue in anterior spinal fasciculus.

9.2 Central Tegmental Tract (CTT)

CTT is a group of efferent fibers from red nucleus to inferior olivary nucleus. At the level of midbrain CTT originates from dorsomedial red nucleus and descends ipsilaterally, passing dorsally through superior cerebellar peduncle decussation and lateral to MLF [1] (Fig. 10). CTT is located deep to locus coreuleus and trigeminal mesencephalic tract and dorsal to medial lemniscus.

9.3 Tectobulbar and Tectospinal Tract

Both of these tracts originate in the superior colliculus and terminate in the brainstem and the spinal cord. The fibers that are destined to reach the spinal cord curve ventrally around the periaqueductal gray matter, cross to the oppposite side as the dorsal tegmental decussation and then continue downwards in the midline close to the medial longitudinal fasciculus as the tectospinal tract [1]. The tectobulbar tract also runs close to the medial longitudinal fasciclus and terminates in the paramedian pontine reticular formation.

9.4 Brainstem Safe Entry Zones

Due to the compactness of the neural structures in the brainstem, tumors of the brainstem tend to distort the anatomy making it difficult to identify the anatomical safe entry zones as have been previously described and discussed.This configuration of fibers into zones can be used to understand the safe entry zones into the brainstem [1]. Zone 1 is essentially made up of white fibers that lie at the lateralmost perimeter of the brainstem. Entry into the brainstem through this zone is relatively safe and avoids major neurological deficits. Zone 2 comprises of both major gray matter masses of the brainstem and important ascending and descending white fibers. This zone provides narrow avenues through which tumor removal is possible. Zone 3 lies in the depth near the midline of the brainstem and consists of fibers connecting brainstem nuclei to each other and few motor cranial nerve nuclei.

10 Midbrain

Lateral Midbrain: Tumors on the lateral aspect of the midbrain can be approached through the lateral mesencephalic sulcus and the leminiscal triangle. (Zone 1). The boundaries are the lateral mesencephalic sulcus anteriorly, the inferior brachium postero-superiorly and the superior cerebellar peduncle posteroinferiorly. In the depth the approach is limited by the medial lemniscus dorsally and the red nucleus ventrally [1].

Ventral Midbrain: The ventral portion of the midbrain constitutes fibers of Zone 2 and consists of the corticospinal, corticobulbar, fronto-pontine and temporo-pontine fibers. The traditionally described safe entry zone for ventral midbrain tumors is through the sulcus lateral to the exit of the occulomotor nerve [1]. This approach traverses the fronto-pontine fibers in the medial one-fifth portion of the cerebral peduncle. The limits of the approach are the corticospinal tracts laterally and the red nucleus in the depth.

Dorsal Midbrain: The dorsal midbrain and tectal plate lesions can be approached through supra-collicular and infra-collicular areas. The supra-collicular avenue is through an incision just rostral to the superior colliculus. The approach is limited in the depth by the aqueduct medially and the mesencephalic tract of the trigeminal and the central tegmental tract laterally [1]. The infra-collicular approach is through the frenulum and above the trochlear nerve. The approach is limited by the aqueduct in the depth. If the entry is made lateral to the midline the approach is limited laterally by the superior cerebellar peduncle and the lateral lemniscus and in the depth by the mesencephalic tract of the trigeminal and the central tegmental tract.

Lateral Cerebral Peduncle Approach to Ventral Midbrain Tumors: We recently described a novel entry point for tumors located in the ventral midbrain [5] (Fig. 14). The lateral most one fifth portion of the crus is made up of the cortico-pontine fibers (parieto – occipito –temporal pontine fibers – Turk's fasciculus). These fibers after originating from the cortex traverse medial to the optic radiation in the retrolenticular portion of the internal capsule and then enter the midbrain in the lateral one-fifths of the cerebral peduncle [5]. The exact function of these fibers is not known but they are believed to be involved in pursuit eye movements. This avenue through the lateral one-fifth of the cerebral peduncle can be used to approach tumors located in the ventral midbrain in the cerebral peduncles.

The entry point is through the lateral one-fifth of the cerebral peduncle, 5 mm anterior to the lateral mesencephalic sulcus and approximately 5 mm superior to the fourth nerve. (Fig. 14c, d) The limits of the approach are the corticospinal tracts ventrally, the substantia nigra dorsally and the red nucleus and medial lemniscus in the depth. The trajectory of the approach is towards the interpeduncular cistern. The direction of the approach is approximately 60° from the horizontal plane when using a retrosigmoid avenue and approximately 10° from the horizontal plane when using a subtemporal approach. (Fig. 14b) In majority of the cases the tumor displaces the structures providing a natural plane of dissection. The incision in the brainstem is made vertically. (Fig. 14c, d) This incision is in the direction of the caudally and vertically running parieto-

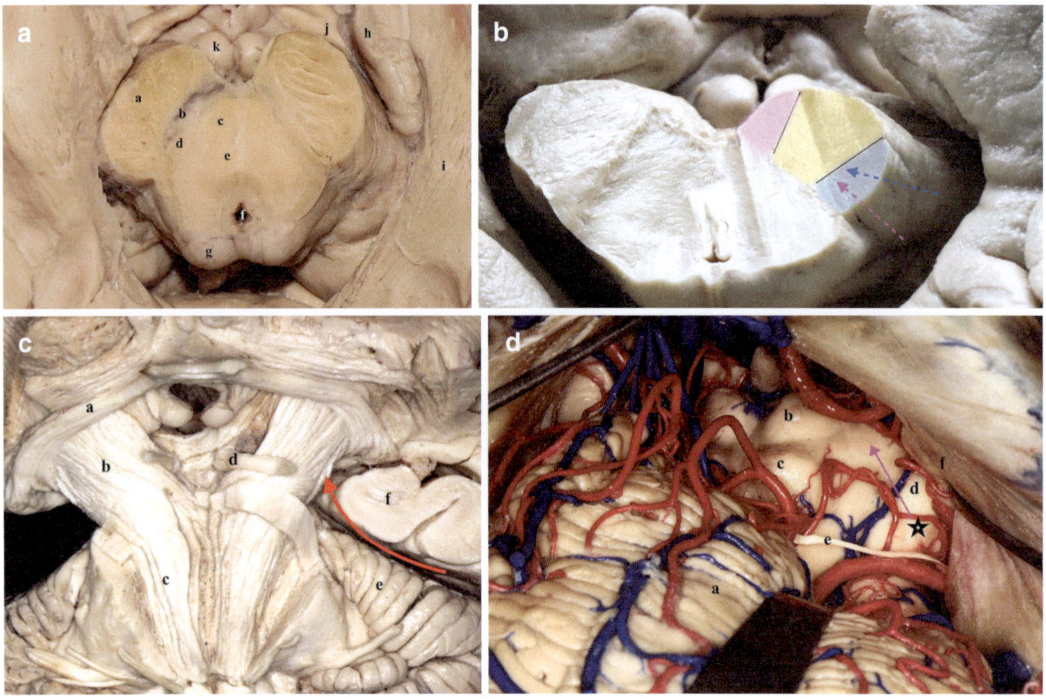

Fig. 14 Images showing the relevant surgical anatomy of the lateral cerebral peduncle approach to the midbrain. (**a**) Inferior view of the brain with cut-section of the brainstem at the level of the inferior colliculus. a. Cerebral peduncle, b. Substantia nigra, c. Red nucleus, d. Medial lemniscus, e. Medial longitudinal fasciculus, f. Cerebral aqueduct, g. Inferior colliculus, h. Uncus, i. Parahippocampus, j. Optic tract, k. Mamillary bodies. (**b**) Cut section of the midbrain of another specimen. The arrangement of fibers in the cerebral peduncle can be understood by dividing the peduncle into five parts. The medial most one – fifth portion of the peduncle (shaded pink) is made up of the corticonuclear and the frontopontine fibers. The middle three portions of the peduncle (shaded yellow) comprises of the corticospinal tract. The lateral most one fifth portion (shaded blue) of the crus is made up of the parieto – occipito –temporal pontine fibers. The dotted pink arrow shows the trajectory of the approach from a supracerebellar avenue. The dotted blue arrow shows the trajectory of the approach from a subtemporal avenue. (**c**) Frontal view of a dissected specimen of the brainstem showing the fiber bundles. a. Optic tract, b. Corticospinal tract in the midbrain, c. Corticospinal tract in the pons, d. Oculomotor nerve, e. Cerebellum, f. Temporal lobe. The red arrow shows the approach to the lateral cerebral peduncle that can be obtained from both the sub temporal and retrosigmoid supracerebellar routes. The incision in the cerebral peduncle is oriented vertically in the direction of the corticospinal and corticopontine fibers. (**d**) Image of an injected cadaveric specimen simulating the retrosigmoid lateral supracerebellar route. a. Cerebellum, b. Superior colliculus, c. Inferior colliculus, d. Lateral one-fifth of the cerebral peduncle, e. Trochlear nerve, f. Tentorium. The asterisk shows the entry point into the lateral cerebral peduncle. The pink arrow shows the trajectory of the approach

occipito-temporal pontine fibers and will thus split the fibers. A transverse incision will damage the vertically traversing fibers. The neural and vascular structures encountered during the approach will be posterior cerebral artery superiorly, the superior cerebellar artery inferiorly and the trochlear nerve as it traverses from the dorsal aspect of the midbrain towards the cavernous sinus. We found this approach to be safe as it provided a direct trajectory to the lesion. When operated using a lateral supracerebellar transtentorial approach or a subtemporal route the direction of the approach is in a straight line with the microscopic view. (Figs. 15, 16, and 17).

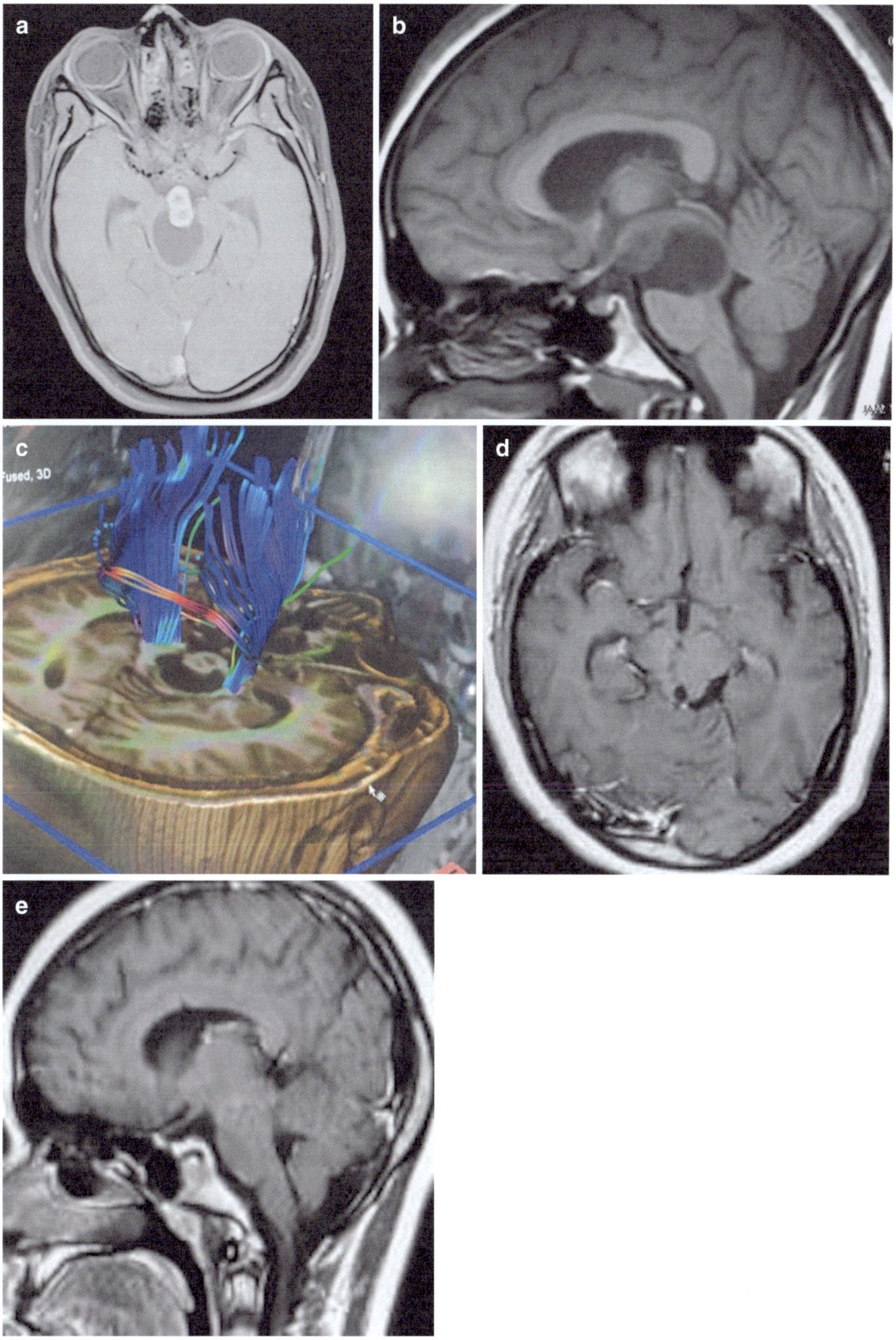

Fig. 15 Images of a 21 year old female patient. (**a**) T1 weighted contrast enhanced axial MRI showing a solid cystic tumor in the midbrain. The solid portion of the tumor is in the ventral midbrain posterior to the interpeduncular fossa. (**b**) T1 weighted sagittal image showing the tumor. (**c**) Tractography images showing the ventrally displaced corticospinal tracts. (**d**) Post-operative contrast enhanced axial image showing excsion of the tumor. (**e**) Post-operative contrast enhanced sagittal scan

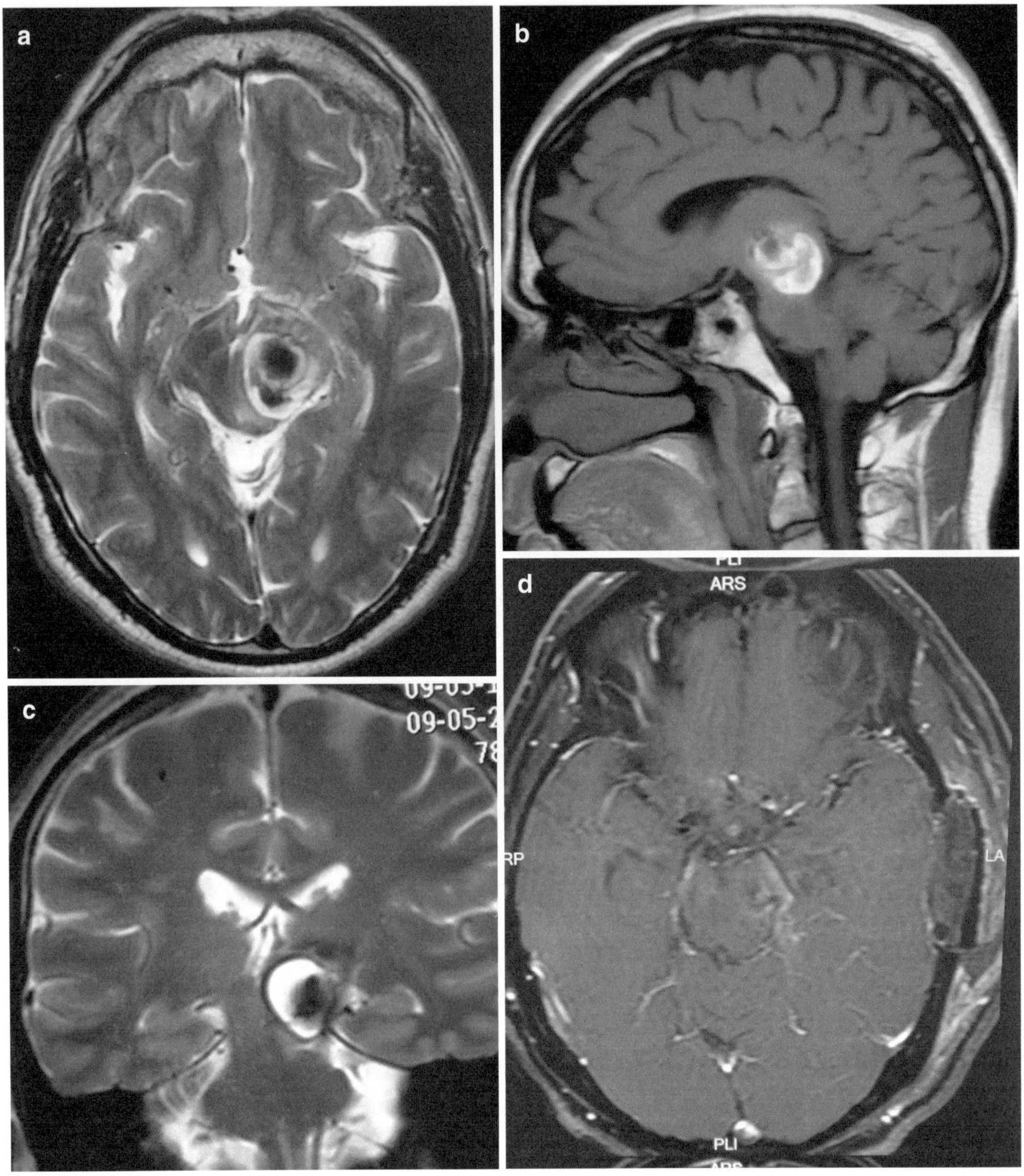

Fig. 16 Images of a 42 year old male patient. (**a**) T2 weighted axial MRI image showing the cavernoma with bleed in the left cerebral peduncle. (**b**) T1 weighted sagittal scan showing the midbrain cavernoma with bleed. (**c**) T2 weighted coronal image showing the cavernoma projecting on to the surface at the lateral aspect of the cerebral peduncle. (**d**) Post-operative axial image showing resection of the cavernoma

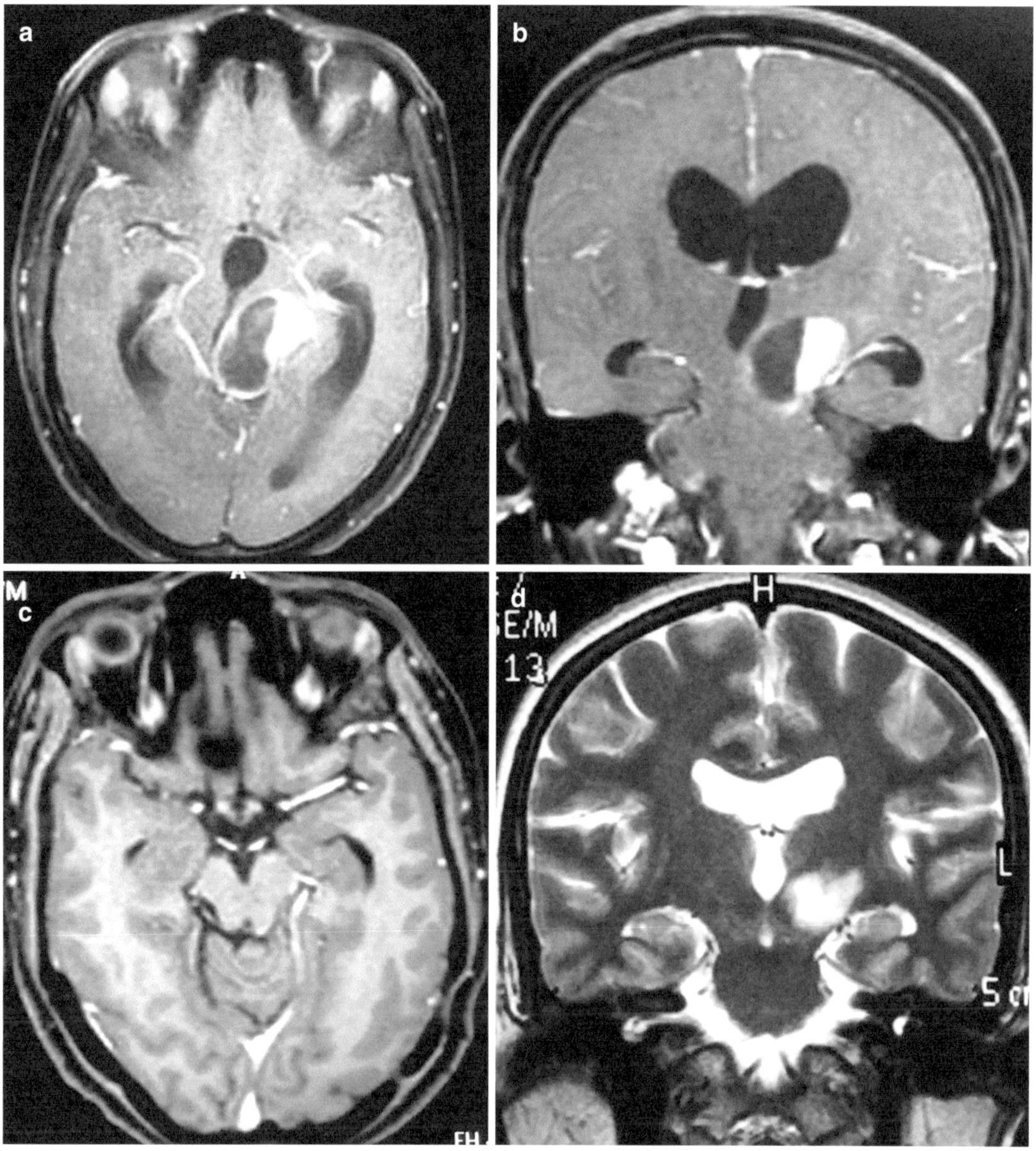

Fig. 17 Images of a 32 year old female patient. (**a**) Contrast enhanced axial MRI image showing a solid cystic tumor in the left half of the midbrain. The solid component of the tumor is in the cerebral peduncle. (**b**) Coronal image showing the tumor projecting onto the lateral aspect of the cerebral peduncle. (**c**) Post-operative contrast enhanced axial image showing complete excision of the tumor. (**d**) Post-operative T2 weighted coronal image showing the tumor resection

11 Pons

Ventro-lateral pons: Strictly speaking the pons cannot be entered from a purly ventral aspect due to the presence of the corticospinal tract. Tumors in the ventral pons anterior to the medial lemniscus can be approached via the middle cerebellar peduncle [1]. (Zone 1) The safest point of entry is the area anterior and inferior to the trigeminal nerve and above the facial nerve. The structures that will be encountered is the superficial and middle layer of transverse pontine fibers, the pontine nuclei and then the deep layer of the tranverse pontine fibers. The limits of the exposure are the corticospinal tract ventromedially and the medial and the spinal lemniscus in the depth. Entering the middle cerebellar peducle posterior and below the trigeminal nerve risks injury to the intrapontine fibers of the fifth cranial nerve. Making an incision in the area around the seventh cranial nerve carries risk of injury to the intrapontine fibers of the sixth and seventh cranial nerves.

Dorsal pons: Entry to the dorsal pons is through the floor of the fourth ventricle. Traditionally safe entry points have been described above and below the facial colliculus. The entry into the pons above the facial colliculus is bounded by the medial longitudinal fasciculus medially, the fibers of the facial nerve running above the facial colliculus inferiorly, the sulcus limitans laterally and the frenulum veli superiorly [1]. (Zone 2) The safe entry into the pons below the facial colliculus lies in the intermediate part of the fourth ventricle. It is demarcated by the intrapontine fibers of the facial nerve that lie below the facial colliculus superiorly, the tela choridea inferiorly, the medial longitudinal fasciculus medially and the facial nucleus and the nucleus ambiguus laterally. These areas are relatively devoid of major structures and can be used to appraoch lesions in the pons [1]. (Zone 2).

12 Medulla

*Ventral Medulla:*Just like the ventral portion of the pons the ventral portion of the medulla is occupied predominantly by the corticospinal tract. Hence safe approaches to the ventral medulla are through the pre-olivary and the post-olivary sulcus. The pre-olivary sulcus lies between the pyramid medially and the olive laterally. The entry should be made below the rootlets of the hypoglossal nerve and above the C1 nerve rootlets [1]. The post-olivary sulcus lies between the olive medially and the inferior cerebellar peduncle laterally. The entry is made ventral to the rootlets of the gloospharyngeal and vagal nerves. In the depth the approach is limited by the spinal lemniscus and the nucleus ambiguus.

Dorsal Medulla: The dorsal medulla can be entered through the posterior median sulcus, through the posterior intermediate sulcus and the posterior lateral sulcus. The entry is usually made below the obex.

References

1. Shah A, Jhawar SS, Nunez M, Goel A, Goel A. Brainstem anatomy: a study on the basis of the pattern of fiber organization. World Neurosurg. 2020;134:e826–46.
2. Párraga RG, Possatti LL, Alves RV, Ribas GC, Türe U, de Oliveira E. Microsurgical anatomy and internal architecture of the brainstem in 3D images:surgical considerations. J Neurosurg. 2016;124(5):1377–95.
3. Yagmurlu K, Rhoton AL Jr, Tanriover N, Bennett JA. Three-dimensionalmicrosurgical anatomy and the safe entry zones of the brainstem. Neurosurgery. 2014;10Suppl 4:602–19. discussion 619–20
4. Akiyama O, Matsushima K, Nunez M, Matsuo S, Kondo A, Arai H, Rhoton AL, Matsushima T. Microsurgical anatomy and approaches around the lateral recess with special reference to entry into the pons. J Neurosurg. 2018;129(3):740–51.
5. Shah A, Jhawar SS, Goel A. The lateral cerebral peduncle approach to ventrally placed intra-axial midbrain tumors: a technical note. J Clin Neurosci. 2021;89:226–31.

Virtual Reality Surgical Simulation and Planning

Taichi Kin

1 Image Processing

1.1 Images Used for Surgical Assessment

Medical fusion 3DCG, which fuses multiple medical image data, is used. A three-dimensional skull model is constructed from images obtained by computed tomography (CT) [1–4], and a three-dimensional vascular model is constructed from images obtained by CT angiography [5], magnetic resonance angiography (MRA) [6–8], or three-dimensional rotational angiography (3DRA) (Fig. 1) [9–11]. Some reports have indicated that more detailed vascular models can be constructed by simultaneously using several of these datasets [9, 12]. Balanced steady-state free precision (balanced-SSFP) imaging is often used to construct models of the brain cortices, cranial nerves, and cerebral ventricles [9, 10, 13], while source images for brain tumors are obtained by contrast-enhanced T1-weighted imaging, T2-weighted imaging, and positron emission tomography, which are considered to best visualize lesions [1, 2, 13–16].

To generate detailed medical fusion 3DCG images, the source images need to be taken at the highest resolution possible (512 pixels) and in thin slices (0.4–1 mm). These conditions may impose limits on imaging time and area [9, 13].

1.2 Image Analysis Software

Many types of image analysis software are used for virtual reality surgical simulation. However, because medical image analysis software programs are originally intended to be used for diagnosing diseases, medical software programs that allow virtual surgical operations, such as virtual deformation of tissues, are extremely limited. Thus, there are many reports that medical image analysis software programs are first used to reconstruct three-dimensional images, which are then output as shape data and used with other non-medical software programs to perform virtual reality surgical simulation. In terms of pricing, some types of software are free, while others are expensive. In addition, many types of image processing software are only used for research purposes. Recently, the GRID medical software was released in Japan. This software uses artificial intelligence technology to generate medical fusion 3DCG images and allows surgeons to perform surgical operations in virtual reality surgical simulation. (Fig. 2).

T. Kin (✉)
The University of Tokyo, Tokyo, Japan
e-mail: tkin-tky@g.ecc.u-tokyo.ac.jp

A. Shah et al. (eds.), *Functional Anatomy of the Brain: A View from the Surgeon's Eye*,
https://doi.org/10.1007/978-981-99-3412-6_20

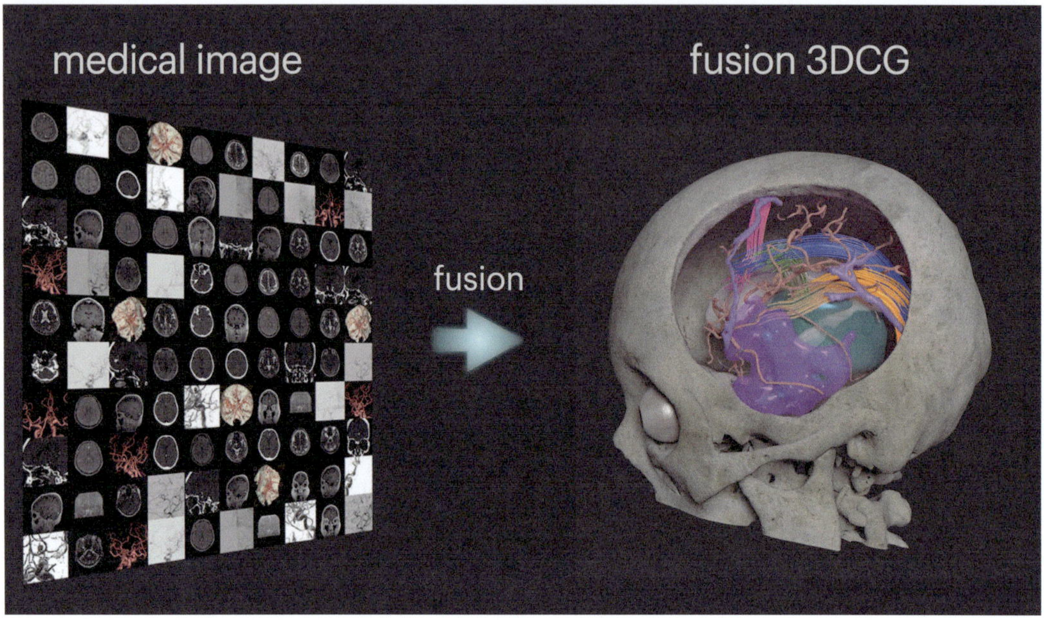

Fig. 1 Medical fusion 3DCG. 3DCG: three-dimensional computer graphics

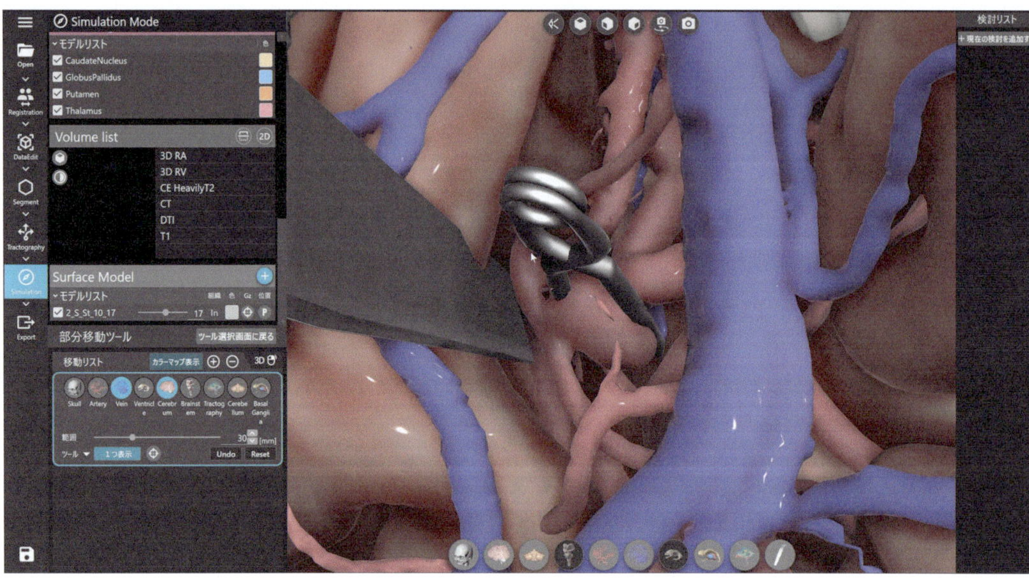

Fig. 2 Virtual reality surgical simulation using fusion 3DCG. 3DCG: three-dimensional computer graphics

1.2.1 Major Software Programs that Generate Medical Fusion 3DCG Images

They include Amira (Thermo Fisher Scientific, Waltham, Massachusetts) [1–3, 17, 18], Mimics (DICOM, Mimics 17.0, Materialise company) [19–21], 3D Slicer (Surgical Planning Laboratory, Harvard University, USA) [6, 16], iPlan (BrainLab, Westchester) [22], Horos (GNU General Public License, version 3) [15], OsiriX (Pixmeo Sarl) [4], and Dextroscope (Bracco Group, Kent Ridge Digital Labs, Singapore) [18].

1.2.2 Main Software Programs for Virtual Reality Surgical Simulation

The main software programs include CardinalSim (https://cardinalsim.stanford.edu/) [3] and Dextroscope (Bracco Group, Kent Ridge Digital Labs, Singapore) [18], which are used for interactive operations with virtual surgical devices and stereoscopic visualization; MATLAB [23], which is used for shape measurement; HyperMesh (Altair Engineering, Inc.) [21], which is used for the finite analysis of stress, strain, etc.; Maya (Autodesk, San Rafael, California), which is used for editing of three-dimensional computer graphics polygon models such as rigging; and Unity (Unity Technologies, San Francisco, California) [24], which is used as a development engine for virtual surgical simulators.

1.3 Image Processing Methods

Many articles on the clinical application do not describe imaging processing methods in detail. It is not easy to determine the accuracy of three-dimensional models, which are a tool for surgical simulation, from articles. Consequently, it may be difficult to determine the validity of the accuracy and clinical usefulness of surgical simulation. Typical image processing methods are described below.

1.3.1 Registration

Medical fusion 3DCG images are generated through registration, which entails aligning the coordinate systems of multiple different medical image datasets on the image processing software. (Fig. 3) Some reports specifically describe registration methods, such as normalized mutual information [1, 2, 8, 24]. However, because of the recent availability of many types of software that fully automate registration [1], many articles do not describe registration methods.

1.3.2 Segmentation

The process of extracting target areas from source images is referred to as segmentation. Source images that best visualize the target tissues are used. Coarse structures, such as the skull, blood vessels, and tumors, can be extracted by automated segmentation methods, such as single-threshold segmentation and region-growing segmentation [12]. (Fig. 4) Because complete automatic extraction is difficult for cranial nerves and microstructures [2, 3], many reports describe a combination of automated, semi-automated, and manual operations [9, 25]. For example, coarse structures are automatically extracted, whereas microstructures are manually extracted. Ohtani et al. proposed a method in which two-dimensional images of facial nerves are plotted at 3 mm intervals to align the center of the nerves and fused to visualize the nerves [8]. The segmentation methods often differ among software and facilities.

1.3.3 Rendering

Rendering is a process of generating (visualizing) three-dimensional graphics. Volume rendering and surface rendering are commonly used methods for medical three-dimensional images [26]. (Fig. 5) Volume rendering is frequently used to visualize entire source images in medical three-dimensional images for diagnosis. This method transforms poorly-delineated borders into well-delineated borders in images by adjusting color and transparency and is suitable for understanding the whole image and information on internal structures [27, 28]. Surface rendering is a process used to convert medical image data into a three-dimensional polygonal model with a selected threshold for the target tissues. Compared to volume rendering, surface rendering is superior in terms of clear visualization of microstructures and virtual editing, such as the exclusion of the brain. In addition, the polygonal models constructed by surface rendering can be output as realistic image data by processing the models for texture, lighting, etc. For the above reasons, many recent reports have described virtual reality surgical simulation using surface rendering.

1.3.4 Simulation Operations

Many neurosurgical procedures require craniotomy and brain deformation. If these procedures can be performed virtually, their virtual forms can

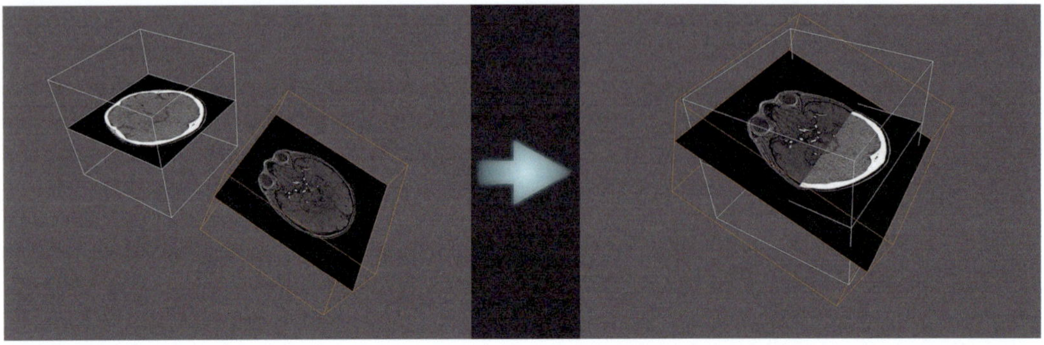

Fig. 3 Registration between CT and MRI using normalized mutual information method

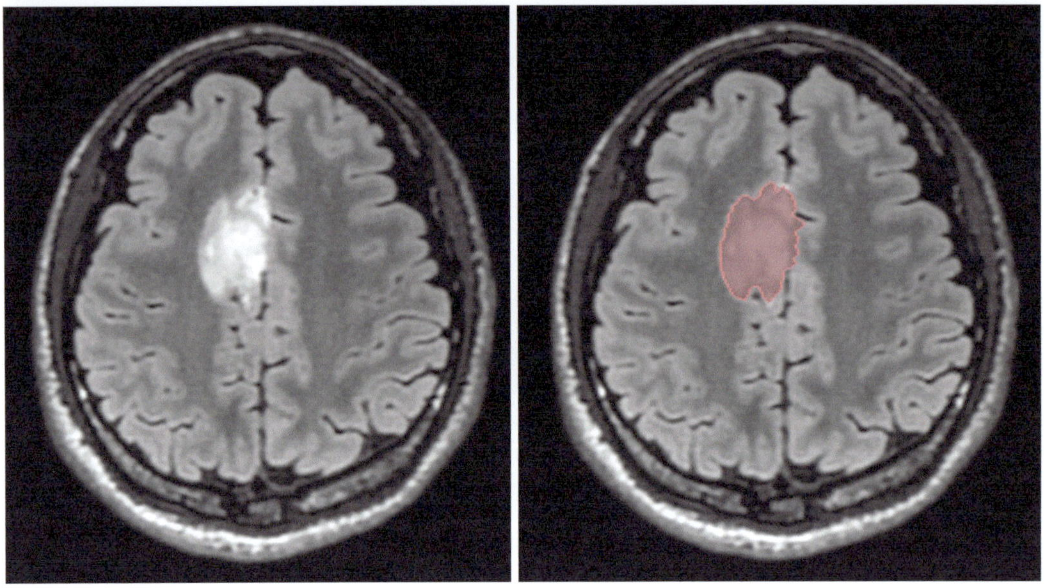

Fig. 4 Region growing

be used for many neurosurgical simulations. However, many reports on the clinical application of virtual reality surgical simulation have described the use of three-dimensional fusion images generated from preoperative images to plan for surgery. Specifically, these reports describe planning rather than simulation. Regarding simulation operations, there are many reports on the virtual exclusion of bones but only a few reports on virtual tissue deformation. Shono et al. added virtual arachnoid membrane to medical fusion 3DCG images and virtually simulated arachnoid incision, brain deformation, and clipping in patients undergoing craniotomy clipping of cerebral aneurysm [24]. Eastwood et al. reported the development of a platform for estimating trepanation sites without interference by surgical devices and the conditions of the devices on computers for third ventriculostomy and pineal tumor biopsy [23]. Li et al. compared surgical plans developed by surgeons with the surgical plans based on analysis of biomechanically calculated stress and strain in the surgical assessment of early synostosis of sagittal sutures. They reported that the latter was superior in terms of symmetry [21].

1.3.5 Application of Device

Combinations of various hardware devices have been attempted in virtual surgical simulation. Because virtual reality surgical assessment using

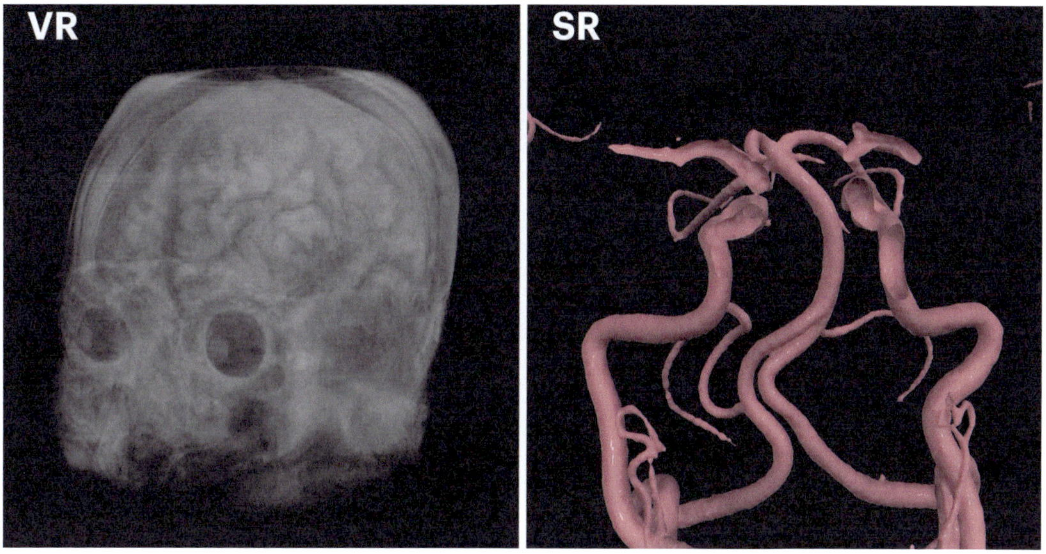

Fig. 5 Volume rendering and surface rendering. VR: volume rendering, SR: surface rendering

head mount displays improves a sense of immersion and provides better stereoscopic visualization, the displays are considered to contribute to a better understanding of patient-specific anatomy and to facilitate decision-making on surgical techniques [29–31]. Surgical assessment using three-dimensional print models has been reported for more than 10 years. In recent years, advances in printing technology have allowed the printing of complex shapes. Uzunoglu et al. used three-dimensional print models of cerebral arteriovenous malformation for preoperative assessment [32]. It has been reported that head mount displays and controllers are used with HTC Vice (HTC Corporation, Xindian District, New Taipei City, Taiwan) [33] and the SuRgical planning platform (Surgical Theater, Inc.) [11, 34] to improve user interface and operability in computer simulation requiring complex operations. These devices are considered to contribute to the improvement of a sense of immersion and usability in surgical simulation.

2 Clinical Evaluation

Various techniques have been used to evaluate the usefulness of surgical simulation.

2.1 Tool Accuracy Evaluation

Fernandes et al. used free software for surgical assessment of 23 patients with the supratentorial disease and reported the difference between medical fusion 3DCG images and surgical findings as 2.6 ± 1.0 mm [15]. Eastwood et al. constructed medical fusion 3DCG models in 16 children with brain tumors who underwent both third ventriculostomy and pineal tumor biopsy and developed a platform for estimating trepanation sites without interference by surgical devices and the conditions of the surgical devices by setting multiple regions of interest. Based on estimation obtained by their platform, they reported that the ideal trepanation sites where inserted parts of surgical devices were shorter and less curved were located on average at 22.5 mm from the coronal suture and 30 mm from the sagittal suture [23].

2.2 Subjective Evaluation of Users

Sato et al. conducted a questionnaire survey of neurosurgical residents on evaluation of 14 patients with deep-seated meningioma. They reported that medical fusion 3DCG was more useful for preoperative assessment than two-dimensional images in terms of anatomical structure, accurate understanding of surgical fields, methods of preserving important blood vessels, and understanding of how to achieve safe and extensive tumor resection [1]. Won et al. conducted a study on transnasal endoscopic surgery performed on 10 patients with skull base disease, in which three specialists ranked the correlation with intraoperative findings. Won et al. reported that medical fusion 3DCG was superior in identifying surgical findings, anatomical characteristics, and lesion sites [3]. Yoshino et al. evaluated the usefulness of medical fusion 3DCG for surgical assessment on a 4-point scale and reported that medical fusion 3DCG was more useful than conventional imaging in three of eight patients and far more useful in five patients than conventional imaging [9].

2.3 Improved Visibility of Tissues and Diagnostic Accuracy

Yoshino et al. performed preoperative assessment using medical fusion 3DCG in five patients with posterior fossa hemangioblastoma and analyzed the localization of feeding vessels and draining veins. Their results showed that medical fusion 3DCG was superior to cerebral angiography in three-dimensional identification [35]. Yoshino et al. performed a preoperative assessment using medical fusion 3DCG in eight patients with cerebellar arteriovenous malformation to examine whether the preoperative findings of abnormal blood vessels were consistent with the intraoperative findings. They reported that the abnormal vessels visualized by the proposed medical fusion 3DCG were consistent with the intraoperative findings in all patients [9]. Fujimoto et al. used visibility scores to evaluate abnormal blood vessels in three patients with an arteriovenous fistula at the craniocervical junction. They reported that medical fusion 3DCG (97%) was significantly superior to cerebral angiography (51%) in terms of visualization of vascular structures, particularly those of microvessels [2]. Yao et al. diagnosed the offending vessels and evaluated the severity of nerve compression in 42 patients with neurovascular compression syndrome. As for the diagnosis of the offending vessels, medical fusion 3DCG showed a sensitivity of 97.6% and a specificity of 100%. The κ coefficient for predicting the severity of compression was >0.75 ($P < 0.001$) [6]. For neurovascular compression syndrome, Ohtani et al. reported that the diagnostic rate of the offending vessels by medical fusion 3DCG was 100% in 15 patients with hemifacial spasm. Medical fusion 3DCG appears to be highly useful for the diagnosis of neurovascular compression syndrome. Li et al. performed surgical assessment using virtual endoscopy in 18 patients with suprasellar arachnoid cysts. When visibility scores were used to evaluate the reproducibility of preoperative findings in intraoperative findings, virtual endoscopy was associated with an extremely favorable reproducibility, measured as a score of 0.96 (the highest possible score of 1) [16].

2.4 Surgical Plans, Surgical Results, and Outcomes

Regarding aneurysms, several reports indicate that the clips used in preoperative assessment were used during surgery [19, 20, 24]. Nowell et al. performed automatic planning of electrode placement for epilepsy surgery in 35 patients and reported that automatic planning devised safer and more effective plans than manual planning [7]. Goertz et al. performed a surgical assessment for clipping of ruptured cerebral aneurysm in 157 patients and reported that the incidence of ischemic complications caused by intraoperative maneuvers was significantly lower after the surgical assessment of three-dimensional images than after that of two-dimensional images (16.2% vs 35%) [5]. Li et al. compared the strategies planned by surgeons and those developed from the integration of computer-assisted three-dimensional simulation and biomechanical calculation in 18 patients undergoing surgery for early synostosis of sagittal sutures. Li et al. have reported that the calvaria visualized in the latter strategies was significantly more symmetrical and that the latter strategies tended to reduce the operative time and blood loss [21].

3 Cost

For clinical application of preoperative simulation, attention should also be paid to costs, such as time of image generation and material expenses. Yao et al. generated medical fusion 3DCG images in 42 patients undergoing neurovascular decompression. When the free software was used, the mean time of image generation was 12.9 min. Yao et al. reported that medical fusion 3DCG is superior in terms of cost performance [6]. Likewise, Fernandes et al., who compared three-dimensional images generated for supratentorial lesions and intraoperative findings in 23 patients, reported that image generation using the free software took an average of 35 min.[15] In addition, Li et al. reported that the mean time for generating three-dimensional images for surgical assessment was 10.2 min in 18 patients with suprasellar arachnoid cysts [16]. Ohtani et al. reported that, although paid software was used, the generation of medical fusion 3DCG images took 10–20 min per patient in 15 patients with hemifacial spasm [8]. Based on all these reports, the time and costs of image generation appear to be sufficiently acceptable for clinical application. The reason for the reduced time of image generation may be that, in all reports, three-dimensional images were reconstructed only from the minimum source image data necessary for surgical assessment rather than the entire source image data. On the other hand, when any simulation operation or analysis was performed on a computer, the time of image generation ranged from 1 to 10 h. The time and effort for surgical assessment substantially increase [3, 24].

References

1. Mitsuru S, Kensuke T, Hidetoshi M, et al. Three-dimensional multimodality fusion imaging as an educational and planning tool for deep-seated meningiomas. Br J Neurosurg. 2018;32(5):509–15.
2. So F, Keisuke T, Taichi K, et al. Three-dimensional angioarchitecture and microsurgical treatment of arteriovenous fistulas at the craniocervical junction. J Clin Neurosci. 2018;53:140–6.
3. Tae-Bin W, Peter H, Hyun LJ, et al. Early experience with a patient-specific virtual surgical simulation for rehearsal of endoscopic skull-base surgery. Int Forum Allergy Rhinol. 2017;8(1):54–63.
4. Rotariu DI, Ziyad F, Budu A, Poeata I, et al. The role of OsiriX based virtual endoscopy in planning endoscopic Transsphenoidal surgery for pituitary adenoma. Turk Neurosurg. 2017;27(3):339–45.
5. Lukas G, Christoph K, Jan B, et al. Microsurgery simulator of cerebral aneurysm clipping with interactive cerebral deformation featuring a virtual arachnoid. World Neurosurg. 2018;120:1163–70.
6. Shujing Y, Jiashu Z, Yining Z, et al. Multimodal image-based virtual reality Presurgical simulation and evaluation for trigeminal neuralgia and Hemifacial spasm. World Neurosurg. 2018;113:499–507.
7. Mark N, Roman R, Gergely Z, et al. A pipeline for 3D multimodality image integration and computer-assisted planning in epilepsy surgery. J Visualized Exp. 2016;111:e53450.
8. Keisuke O, Toshihiro M, Keiji O, et al. Preoperative three-dimensional diagnosis of neurovascular relationships at the root exit zones during microvascular decompression for Hemifacial spasm. World Neurosurg. 2016;92:171–8.
9. Yoshino M, Kin T, Hara T. Usefulness of high-resolution three-dimensional multi-fusion medical imaging for preoperative planning in patients with cerebral arteriovenous malformation. World Neurosurg. 2019;124:e755–63.
10. Wang B, Chen Y, Gai S, et al. Preoperative evaluation of neurovascular relationships for microvascular decompression: visualization using Brainvis in patients with idiopathic trigeminal neuralgia. Clin Neurol Neurosurg. 2021;210:106957.
11. Steineke TC, Barbery D. Microsurgical clipping of middle cerebral artery aneurysms: preoperative planning using virtual reality to reduce procedure time. Neurosurg Focus. 2021;51(2):E12.
12. Hasegawa H, Shin M, Kin T, et al. Fully endoscopic minimally invasive tumor resection for cystic cerebellar Hemangioblastoma. World Neurosurg. 2019;126:484–90.
13. Uchida T, Kin T, Koike T, et al. Identification of the facial colliculus in two-dimensional and three-dimensional images. Neurol Med Chir (Tokyo). 2021;61(6):376–84.
14. Koike T, Tanaka S, Kin T, et al. Accurate preoperative identification of motor speech area as termination of arcuate fasciculus depicted by Q-ball imaging tractography. World Neurosurg. 2022;164:S1878–8750.
15. de Oliveira F, Bruno S, da Costa S, Devanir M, Silva CR, et al. Clinical application of an open-source 3D volume rendering software to neurosurgical approaches. World Neurosurg. 2017;110:864–72.
16. Ye L, Yining Z, Jiashu Z, et al. Low-cost interactive image-based virtual endoscopy for the diagnosis and surgical planning of Suprasellar arachnoid cysts. World Neurosurg. 2016;88:76–82.
17. Lukas A, Jan G, Andres Robert H, et al. Stereolithographic models in the interdisciplinary

planning of treatment for complex intracranial aneurysms. Acta Neurochir. 2016;158:1711–20.
18. Eleftherios A, Eike S, Michael K, et al. A modified microsurgical endoscopic-assisted Transpedicular Corpectomy of the thoracic spine based on virtual 3-dimensional planning. World Neurosurg. 2016;91:424–33.
19. Liang W, Xun Y, Qiang H, et al. Three-dimensional intracranial middle cerebral artery aneurysm models for aneurysm surgery and training. J Clin Neurosci. 2018;50:77–82.
20. Liang W, Xun Y, Qiang H, et al. Comparison of two three-dimensional printed models of complex intracranial aneurysms for surgical simulation. World Neurosurg. 2017;103:671–9.
21. Xiang L, Wanchun Z, Jintao H, et al. Application of computer assisted three-dimensional simulation operation and biomechanics analysis in the treatment of sagittal craniosynostosis. J Clin Neurosci. 2017;44:323–9.
22. Makoto I, Takafumi N, Norio I, et al. A surgical strategy using a fusion image constructed from 11C-methionine PET, 18F-FDG-PET and MRI for glioma with no or minimum contrast enhancement. J Neuro-Oncol. 2018;138(3):537–48.
23. Eastwood Kyle W, Bodani Vivek P, Drade JM. Three-dimensional simulation of collision-free paths for combined endoscopic third Ventriculostomy and pineal region tumor biopsy: implications for the design specifications of future flexible endoscopic instruments. Oper Neurosurg. 2016;12:231–8.
24. Naoyuki S, Taichi K, Seiji N. et al, Microsurgery simulator of cerebral aneurysm clipping with interactive cerebral deformation featuring a virtual arachnoid. Oper Neurosurg. 14(5):579–89.
25. Tel A, Bagatto D, Tuniz F, et al. The evolution of craniofacial resection: a new workflow for virtual planning in complex craniofacial procedures. J Craniomaxillofac Surg. 2019;47(9):1475–83.
26. Kin T(correspondence), Nakatomi H, Shono N, Nomura S, Saito T, Oyama H, Saito N. Neurosurgical virtual reality simulation for brain tumor using high-definition computer graphics: a review of the literature. Neurol Med Chir (Tokyo). 2017;57(10):513–20.
27. Gosal JS, Tiwari S, Sharma T, et al. Simulation of surgery for supratentorial gliomas in virtual reality using a 3D volume rendering technique: a poor man's neuronavigation. Neurosurg Focus. 2021;51(2):E23.
28. Alsofy SZ, Sakellaropoulou I. Stroop Ralf: evaluation of surgical approaches for tumor resection in the deep Infratentorial region and impact of virtual reality technique for the surgical planning and strategy. J Craniofac Surg. 2020;31(7):1865–9.
29. Sugiyama T, Clapp T, Nelson J, et al. Immersive 3-dimensional virtual reality modeling for case-specific Presurgical discussions in cerebrovascular neurosurgery. Oper Neurosurg. 2021;20(3):289–99.
30. Perin A, Galbiati TF, Roberta A, et al. Informed consent through 3D virtual reality: a randomized clinical trial. Acta Neurochir. 2020;163(2):301–8.
31. Alsofy SZ, Sakellaropoulou I, Nakamura M, et al. Impact of virtual reality in arterial anatomy detection and surgical planning in patients with Unruptured anterior communicating artery aneurysms. Brain Sci. 2020;10(12):963.
32. Uzunoglu I, Kizmazoglu C, Husemoglu RB, et al. Three-dimensional printing assisted preoperative surgical planning for cerebral arteriovenous malformation. J Korean Neurosurg Soc. 2021;64(6):882–90.
33. Alsofy SZ, Nakamura M, Suleiman A, et al. Cerebral anatomy detection and surgical planning in patients with anterior Skull Base Meningiomas using a virtual reality technique. J Clin Med. 2021;10(10):681–95.
34. Filimonov A, Zeiger J, Goldrich D, et al. Virtual reality surgical planning for endoscopic endonasal approaches to the craniovertebral junction. Am J Otolaryngol. 2022;43(1):103219.
35. Yoshino M, Nakatomi H, Kin T, Saito T, Shono N, Nomura S, Nakagawa D, Takayanagi S, Imai H, Oyama H, Saito N. Usefulness of high-resolution 3D multifusion medical imaging for preoperative planning in patients with posterior fossa hemangioblastoma: technical note. J Neurosurg. 2017;127:139–47.

Artificial Intelligence in the Management of Glial Tumors

Gleb Danilov, Michael Shifrin, Timur Ishankulov, Konstantin Kotik, and Alexander Potapov

Neurosurgery in the XXI century is a complex discipline that exploits and integrates the achievements of basic and applied neurosciences. Information technologies have become an essential component of modern neurosurgery, generating unique data that capture and reflect functional and abnormal regularities in nervous system states.

Today we face a scientific challenge: data are generated and grow in volume much faster than we can extract knowledge from it or get any other benefit. We reasonably deal with new topical questions. What is the extent to which routinely collected data in neurosurgery are useful in research and practice? How informative are they? Do they implicitly hide important patterns about the human brain and its pathology that we do not know? Is there a way to effectively leverage large and complex data in neurosurgery? Since the number of questions is high and no straightforward answers are provided, a new branch of neurosurgical research is arising. This area relates to the application of data science to neurosurgical data. Data science methodology is extensive. Today, a related area of technologies that imitate human intellectual capabilities called "Artificial Intelligence" has gained immense popularity. Following the timely trends, in this chapter, we are discussing artificial intelligence in neurosurgery to understand how these technologies can help us in answering the questions posed above.

This book addresses the issues of brain tumor surgery. To be consistent, this chapter focuses on possible applications of artificial intelligence and its benefits in neurooncology, emphasizing glial tumor surgery. However, the reflections and conclusions made can be extrapolated to all neurosurgery and far beyond.

We want to tribute this chapter to our coauthor, professor Alexander Potapov, who had passed away before we prepared the manuscript. Professor Potapov was an innovator to inspire many marvelous ideas in basic and applied neurosurgical research. In 1981, the first electronic computer SM-4 was installed at N.N. Burdenko Institute of Neurosurgery (Moscow, Russian Federation). Professor Potapov shortly appreciated the potential of computing for research in neurosurgery. In 1984, he initiated a Department of Computer Sciences at N.N. Burdenko Institute of Neurosurgery. In recent years professor Potapov strongly supported the advancement of biomedical informatics and artificial intelligence technologies in neurosurgery. Today, a dedicated laboratory of biomedical informatics and artificial intelligence is actively working at

G. Danilov (✉) · M. Shifrin · T. Ishankulov
K. Kotik · A. Potapov (Deceased)
Laboratory of Biomedical Informatics and Artificial Intelligence, National Medical Research Center for Neurosurgery named after N.N. Burdenko, Moscow, Russian Federation

A. Shah et al. (eds.), *Functional Anatomy of the Brain: A View from the Surgeon's Eye*,
https://doi.org/10.1007/978-981-99-3412-6_21

N.N. Burdenko Neurosurgery Center. Thus, some experience in "neurosurgical artificial intelligence" was accumulated in recent years, making it possible to write this chapter.

1 What Is Artificial Intelligence?

The reader may find different definitions of the term "Artificial intelligence" as well as the history of the concept on the Internet. However, we have to admit that at the time of this writing most of neurosurgeons are not that familiar with the concept of "artificial intelligence". This term may easily confuse people who are not involved in it technically.

Our intuition about "artificial intelligence" is an image of a computer that can think and learn like a human. The question about the ability of machine to think was raised by one of the pioneers of computer science Alan Turing in his work "Computing Machinery and Intelligence" [1]. Despite the romance and attractiveness of such an idea, in the context of this book, we will define the concept of "artificial intelligence" in a more pragmatic way.

A formal definition for artificial intelligence in a technical sense may be "a mathematical solution which automates intellectual tasks normally performed by humans". This definition does not claim to be universal and the best, but it reflects the narrow working meaning of "artificial intelligence" when used in the specific application context.

At the same time, the term "artificial intelligence" is used to name a branch of computer science that focuses on creating intelligent machines that act like human beings. In that way, it implies many smart technologies with human-like behavior. Here we refer to machine learning, natural language and speech processing, computer vision, robotics, decision support systems, etc.

So, a formal definition of artificial intelligence is probably not that strict today. However, we see the approximate boundaries of this scientific/technological/philosophical area.

Why does the idea of "artificial intelligence" become so widely popular nowadays? First of all, we are living in the era of data. Today we collect data much faster than we can extract solid knowledge from it. So, it would be great to have a tool to proceed with the data we generate to make them work more effectively for us and improve its secondary usage efficiency. Fortunately, plenty of mathematical methods have been developed for these problems already in the XX-XXI centuries. With the growth in computer power, we observe at the beginning of the XXI century, the implementation of these methods was brought to life. Today we benefit from a cost-efficient way to process and accumulate a considerable amount of data on an unprecedented scale.

However, if we are talking primarily about advanced mathematics, why is it called "intelligence", specifically "artificial"? Because it can deal with data usually proceeded by humans. Finally, what makes artificial intelligence promising in different areas, is success stories in business and science and in those areas where you can get enough data in data-dependent problem-solving.

2 The Basics of Artificial Intelligence in Brief

Artificial intelligence has gone through various historical periods of heyday and decline of interest. There are two main popular paradigms of artificial intelligence used to solve "intellectual" tasks. Historically the first paradigm in this field was symbolic artificial intelligence. It is the approach based on explicit representation of knowledge (see Fig. 1). In this approach, you need some predefined human-readable knowledge to build a solution and program it explicitly. In other words, in the symbolic paradigm, you apply the rules to the data through classic programming to get the answer to your question. The symbolic paradigm was dominant in artificial intelligence research from the mid-1950s until the late 1980s. As an example, these technologies were used to create software that plays chess.

Machine learning is the second and modern paradigm of artificial intelligence based on the idea that you can learn the solution from data (Fig. 1).

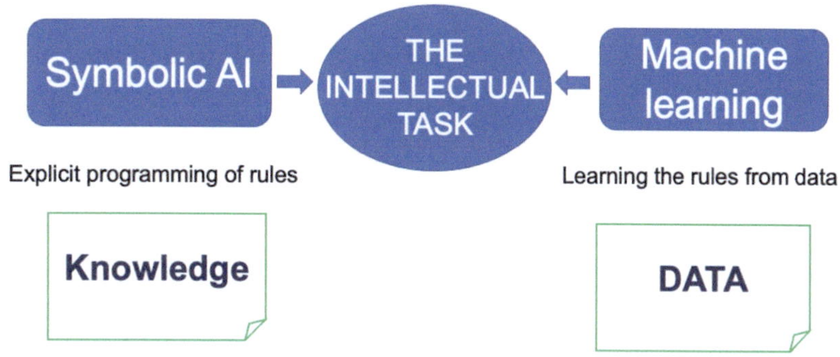

Fig. 1 Two paradigms of artificial intelligence

In machining learning, you feed the data into a computing machine that solves complex equations. In the output, you get the mathematical model that produces the answers to your questions.

Machine learning is not equal to human learning. It is much more primitive. Given data, for example, a set of predictors and a response, you choose some typical mathematical model that, in your opinion, may reflect the relationship between predictors and the answer. Then, solving equations, you learn the parameters of this model so that it works well on your data. This learning process is also called training. After the model is trained, its performance is tested on new data it has never seen before. No magic. It is all about computing.

There are two basic types of machine learning: supervised and unsupervised. Supervised learning produces a model that maps input data to output (target variable/response) based on example input-output pairs. Supervised machine learning is used to solve regression (predict quantity) and classification (predict category) tasks. Unsupervised learning does not exploit target variable and mines the intrinsic characteristics of data to find possible patterns and regularities.

Finally, all machine learning methods are divided into shallow and deep. This gradation relies on the architecture of models. So deep learning models are typically called artificial neural networks in which a learning process is organized in layers. In recent years deep learning showed significant advances in solving complex problems typically handled by humans earlier. The prominent examples are computer vision, speech recognition, natural language processing, machine translation, unmanned vehicle control. That is why artificial neural networks are strongly associated with the artificial intelligence concept.

Shallow learning is less strictly defined in artificial intelligence. However, let's assume that all the models opposed to the deep learning approach are referred to as shallow models. Exemplary "shallow" models are linear regression, logistic regression, support vector machine, decision trees, random forest, Naive Bayes, gradient boosting machine, etc.

Today, machine learning is more of an engineering field. The theory goes far behind the experiments with data. However, it also makes this area very exciting as a combination of science, engineering, programming, and art, which requires both mathematical and programming background, deep knowledge in the professional domain, and creativity.

3 Artificial Intelligence in Neurosurgery

What are the goals of using human-like computer behavior in medicine, in particular neurosurgery? The most obvious and general purpose is the automation (boosting up the speed and reliability) of medical processes: data-based multivariable diagnostics, decision reasoning, complication/outcome predicting. Imagine a computer that can read medical records, judge medical images, and suggest evidence-based clinical solutions much faster and more confidently than a doctor can do nowadays? That might be helpful.

Modern neurosurgery is grounded on the close integration of basic and applied neurosciences and technologies, including neuroimaging, optics (microneurosurgery, endoscopy), endovascular approaches, stereotaxis, neuronavigation, neuromonitoring, robotics, additive technologies, genetics, computer modeling, etc. At the same time, neurosurgery is the only clinical discipline that provides direct access to the living human brain for clinical research. Implementing a wide spectrum of technologies, neurosurgery routinely generates a lot of data that can potentially hide valuable information on how our nervous system functions and breaks. Therefore, the secondary use of these data is absolutely justified, nevertheless, a really challenging task. That is why I would like to draw the attention of the reader to the importance of using data science, artificial intelligence in the field of neurosurgery for the advancement of brain sciences including brain pathology.

What are the use cases of artificial intelligence in neurosurgery? Here we refer to our recent systematic review which can bring some insights into the trends obtained from scientific literature in more details [2, 3]. The majority of studies in artificial intelligence were conducted in neurooncology. A little less popular are the fields of functional, vascular, spinal neurosurgery and traumatic brain injury.

The most frequent topics in neurooncology we were able to distinguish using topic modeling were as follows:

1. Non-invasive tumor grading based on neuroimaging
2. Non-invasive molecular diagnostics based on neuroimaging
3. Segmentation of brain tumors based on neuroimaging
4. Predicting complications and treatment outcomes

Next, we will look at each topic with specific examples. A caveat is needed here: currently, artificial intelligence in neurosurgery is primarily an area of research. Now let us drill down into its current state and perspectives focusing on brain tumor surgery, particularly – into glioma management.

4 Non-invasive Glioma Subtyping

Gliomas are the most common neuroepithelial brain tumors, variable in histology and prognosis. The World Health Organization (WHO) classification proposed in 2007 and based on histogenesis divides all gliomas into four grades [4]. Experts often merge I-II grades into low-grade gliomas (LGG) and III-IV - to high-grade (HGG). The revision of this classification released in 2016 relied on molecular subtypes associated with treatment outcomes [5]. A new modified version of the WHO classification was released in 2021 [6]. Although the new system is much more advanced, its application worldwide is costly because of expensive biological tests. That is why the old WHO grade classification is still applicable, especially in low and middle-income countries.

Non-invasive molecular and histological diagnostics for nervous system tumors have the potential to dramatically improve the efficacy and safety of medical care and significantly reduce healthcare costs. The advances of artificial intelligence in neuroimaging open up new perspectives in that direction. In particular, radiomics and radiogenomics are promising technologies in diagnostic neurooncology today.

Radiomics is a term to designate a set of methods for extracting a large number of features from medical images [7]. These features may be informative to describe tumor and brain patterns that the naked eye cannot see. Thus, a proper image representation with radiomic features is hypothesized to help distinguish tumor types and relate them to clinical forms, prognosis, and best effective treatment. Although radiomics emerged from radiology and oncology, it can be applied to any research involving medical imaging.

The term "radiogenomics" in artificial intelligence is used to name technologies estimating the relations between tumor imaging features and gene expression. In other words, these methods aim to create imaging biomarkers that can identify the genomics of a disease without using a biopsy.

An application of radiomics and radiogenomics in neurooncology is driven by a momentum towards non-invasive diagnostics. Technically speaking, the tumor phenotype and molecular markers are reflected in and predicted by neuroimaging data. This is believed to be a reasonable way of implementing personalization in neurodiagnostics.

To the moment, glioma is one of the widely assessed tumor in neurooncology using radiomics and radiogenomics approaches. A systematic review by M. Tabatabaei et al. (2021) summarizes the most recent studies on glioma grading with radiomics and machine learning, and evaluates the clinical utility of these technologies, and demonstrates the weaknesses to be resolved in the future [8]. The authors selected 18 retrospective studies utilizing radiomics features from brain MRI published in 2017–2020. Multiple MR sequences appeared to be more efficient for feature extraction than a single sequence. Tl + C MR sequence was reported as the most powerful. Advanced modalities (MR perfusion and spectroscopy) could improve the performance of radiomics. Feature extraction from multiple regions (contrast-enhanced, intratumoral, edema) was more efficient than focusing on lesions only.

Dimensionality reduction and feature selection were universal approaches to increase the prediction accuracy. Gray-level co-occurrence matrix features were the most predictive. The deep neural network was used only in one study out of 18, and shallow learning (support vector machine, logistic regression, and random forest) was used in the overwhelming majority of research.

The authors were generally satisfied with the quality metrics of radiomics technologies dividing gliomas into grades (sensitivity, specificity, accuracy, and area under ROC curve exceeding 0.8–0.9 in the majority of cases). However, they noted that most of the results came from a single experiment on one training and testing sample split. This can artificially raise expectations for the quality of radiomics technologies in real clinical practice. The other limitations were related to outdated MRI archives, data heterogeneity, retrospective study design, class imbalance (predominance of high-grade gliomas), lack of external validation for generalizability, radiomics pipeline variations, studies selected before 2021. A serious disadvantage was binary glioma grading (low-grade vs. high-grade) in most of the studies.

In the series analyzed, the maximum number of patients did not exceed 285. The tens or hundreds of cases is another typical limitation of artificial intelligence research in neurosurgery. Surprisingly, deep neural networks were not common for glioma grade prediction.

The studies mentioned above focused on feature engineering primarily. Z. Ning et al. (2021) combined local and deep MRI features using a convolutional neural network (CNN) and developed a noninvasive glioma grading tool in 567 patients [9]. The authors report the AUC, sensitivity, and specificity in the testing cohort for this hybrid model as 0.88 (95% CI: 0.84,0.91), 88% (95% CI: 80%, 93%), and 81% (95% CI: 76%, 86%), respectively. The solution outperformed the models based solely on either radiomics or deep features ($p < 0.001$). The performance was comparable with that of radiologists.

In 2021 our group performed a pilot study of noninvasive glioma grading with deep learning [10]. In our study, we evaluated the diagnostic accuracy of subtyping gliomas by WHO grades (I-IV) with deep learning (DL) over preoperative magnetic resonance images (MRI) extracted from N.N. Burdenko Neurosurgery Center's database. We included 707 MRI studies in the project. A "3D classification" method of predicting tumor grade for a 3D patient's MRI demonstrated the best performance (accuracy = 0.83, ROC AUC = 0.95), comparable with other researchers exploiting different technologies. Our first results showed the separability of contrast-enhanced T1 axial MRI by WHO grades with deep learning.

5 Non-invasive Molecular Glioma Diagnostics

A number of molecular biomarkers migrated into the modern brain tumor WHO classification system has shown the importance for treatment planning, patient prognosis, and treatment response prediction. Those are isocitrate dehydrogenase (IDH) mutations, O6-methylguanine-DNA methyltransferase (MGMT) gene promoter methylation, p53 mutation, lp/19q codeletion, epidermal growth factor receptor (EGFR) amplification, phosphatase and tensin homolog (PTEN) loss, telomerase reverse transcriptase (TERT) mutations, and alpha thalassemia/mental retardation syndrome X-linked (ATRX). In a systematic review, A. Jian et al. (2021) evaluated the machine learning diagnostic accuracy in glioma molecular subtyping based on preoperative MRI [11]. A total of forty-four original articles were processed in the review. The overall sensitivity and specificity for predicting IDH mutation were 0.88 (95,901 T 0.83–0.91) and 0.86 (95% Cl 0.79–0.91) in training datasets and 0.83 to 0.85 in validation subsets, respectively. Both lp/19q codeletion and MGMT gene promoter methylation were predicted with a pooled sensitivity and specificity of 0.76 to 0.83 in training datasets. The sample size varied between 13 and 463 patients in the works reviewed. The majority of studies again relied on shallow learning. Neural networks were not used that often. However, the paper by A. Jian and co-authors demonstrates the potential of artificial intelligence technologies in predicting the molecular tumor profile. The quality of evidence for use of machine learning models was strongest for IDH mutation and MGMT methylation. That is a significant step towards faster, safer, less invasive, and less expensive personalized diagnostics based on neuroimaging.

The results of brain malignancy subtyping with radiomics are generally inspiring. The introduction of radiomics in neuroimaging led to novel diagnostic and predictive tools, justifying their application in personalized tumor diagnosis and treatment in neurooncology. Thus, the researchers focused on applying radiomics to brain MRI as the most common modality for brain tumor imaging. However, most data-driven studies in neurooncology are still limited by the relatively small amount of data. So, the stable reproducing "big data" and deep learning technologies is challenging. Most studies are subject for bias and limitations of retrospective research. Thus, the generalizability of solutions proposed in distinct publications is still questionable. External multicenter validation of machine learning models might be the most realistic and efficient way to improve generalizability.

6 Segmentation of Brain Tumors

Tumor segmentation with deep learning algorithms is typical for artificial intelligence in neurooncology due to the computational complexity of finding the visual object's boundaries. In Fig. 2 we demonstrate the autosegmentation of small tumors (metastases, meningiomas and acoustic schwannomas) with convolutional neural networks [12].

However, delineating glial tumors is a more challenging task. The standard MR1 sequences

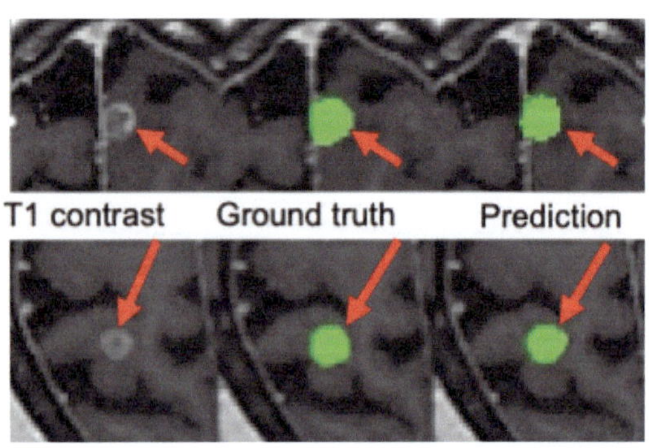

Fig. 2 Automatic delineation of small tumors for radiosurgery planning. A courtesy of Alexandra Dalechina

for glial tumor diagnostics are T2 weighted/fluid-attenuated inversion recovery (FLAIR) and T1 weighted contrast-enhanced. Contrast-enhanced MRI can help delineate tumor margins, but it is insufficient to detect low-density tumor cell infiltration around the contrast-enhanced area common for HGGs. Nevertheless, infiltrating tumor cells may be present in brain regions without any MRI abnormalities and might be related to the limited efficacy of glioma treatment. Perfusion-weighted MRI and diffusion tensor imaging (DTI) showed moderate accuracy for the detection of infiltration. Thus, the potential of artificial intelligence takes the stage.

S. Honore d'Este et al. (2021) systematically reviewed the literature on multimodal imaging integration with artificial intelligence to visualize tumor cell infiltration in gliomas [13]. The authors reviewed 14 studies. A reference for tumor infiltration in all studies was either the histological diagnosis after a stereotactic biopsy or tumor recurrence in the follow-up. The area under the ROC curve for detection reached 0.74–0.91 in six studies. The combination of multiparametric neuroimaging with artificial intelligence appeared to be promising for tumor cell infiltration detection in patients with brain gliomas. The existence of such effective models can optimize the extent of surgical resection and personalize radiotherapy planning.

E.J. van Kempen and co-authors (2021) provide a systematic literature review on automated segmentation of gliomas with machine learning based on 42 studies [14]. An overall dice similarity coefficient (the metric to evaluate segmentation quality) was 0.84 (95% CI: 0.82–0.86). Deep learning with convolutional neural networks predominated as a machine learning method in those studies. The majority of studies combined different MRI sequences for brain tumor segmentation. The lack of external model validation was again obvious. It is noticeable that despite rather good automated segmentation performance, there is still no wide acceptance and implementation of these methodologies in routine clinical practice.

An interesting overview of artificial intelligence applications in pediatric brain tumor imaging was performed by J. Huang and co-authors (2021) [15]. The most popular tasks in 22 selected studies were tumor diagnosis, followed by tumor segmentation and tumor detection. Almost half of the solutions investigated were based on neural networks. The majority of studies demonstrated the superiority of artificial intelligence for tumor diagnosis. The authors believe that automated interpretation of imaging studies may, to a certain extent, improve the performance of diagnostic radiologists. As the authors stated, artificial intelligence is best used to augment human expertise rather than replace it.

7 Predicting Complications and Treatment Outcomes in Neurooncology

Prognosis in medicine is a typical challenge addressed with computational methods for many decades. Supervised machine learning is intended for predictive analytics. Therefore, it is pretty natural that artificial intelligence is actively used to predict possible complications and outcomes of neurosurgical treatment. M. Kocher et al. (2020) reviewed the potential of radiomics and machine learning in neuroimaging and radiotherapy in patients with malignant gliomas and brain metastases [16]. The authors found that in glioma patients, important applications were a prediction of tumor recurrence location and differentiation of tumor progression from pseudoprogression (besides the WHO grade identification and molecular biomarkers determination in patients not eligible for biopsy or resection, as well as image auto-segmentation for target volume planning). Radiomics was also applied for additional detection of small brain metastases, accurate segmentation of multiple metastases, local response to radiosurgery prediction, and differentiation of local brain metastasis relapse from radiation injury. Again, the authors achieved high diagnostic accuracy (80%–90%) in most approaches. However, they exploited a large variety of neuroimaging and computational techniques.

Differentiating pseudoprogression from tumor progression might be a very important automation task. Although we did not find any literature review on the topic at the time of this writing and the number of studies is limited, some insights are brought by several authors. P. Lohmann and co-authors (2020) applied radiomics to O-(2-[^{18}F]fluoroethyl)-L-tyrosine positron emission tomography and reported 100% sensitivity and 100% negative predictive value in discriminating pseudoprogression and early progression in glioma patients post-chemoradiation [17]. J. Lee et al. (2020) differentiated true progression from pseudoprogression in diffuse infiltrating glioma with multi-parametric MRI and deep learning [18]. The authors combined convolutional neural networks (CNN) with long short-term memory LSTM) deep learning models. The input to the models included all available MRI sequences. The mean accuracy ranged from 0.62 to 0.75; the average ROC-AUC varied from 0.64 to 0.81. M. Patel and coauthors (2021) investigated machine learning processing radiomic, molecular and clinical data to distinguish between pseudoprogression and early progression in glioblastoma [19]. The authors achieved the area under the ROC curve of 0.80, sensitivity 78.2%, specificity 66.7%, and accuracy of 73.7% [19]. However, the typical limitation of all of these differential studies were a small sample size (less than 100 patients in each).

Q. Buchlak et al. (2021) addressed the scope of using machine learning applications to glioma MRI data in a systematic review [20]. They summed up, the primary goals of using artificial intelligence over MRI in glioma patients were diagnosis or classification of tumor grade, tumor detection, and automated segmentation, detection and prediction of tumor progression, noninvasive classification of glioma genetic and molecular characteristics, glioma survival prediction. The latest goal was specially addressed in a paper of I. Tewarie and co-authors [21]. These authors systematically reviewed the literature on predictive modeling in patients with glioblastoma. They performed a systematic literature search to identify all cases of overall survival prediction in glioblastoma with modelling. The predictive performance reported in the review varied widely by the ROC-AUC (O.5Q-O.98), C-index (0.66–0.70), and accuracy (0.69–0.98). The authors saw promising results using machine learning models trained on high-dimensional datasets. However, they admit that none of the models was implemented in clinical care.

8 Data in Neurosurgery Are Rarely Big, But Typically Complex

So far, we have provided examples of machine learning technologies mainly for image analysis. Indeed, neuroimaging is the most apparent area to apply artificial intelligence within neurosurgery, particularly in neurooncology. Why? On the one hand, images contain a large amount of data and are suitable for "big data" analysis methods. On the other hand, computer vision is one of the most advanced areas of artificial intelligence. In addition, the application of artificial intelligence in neuroimaging is best suited to the context of this book. However, there are other data types in neurosurgery and neurooncology to which artificial intelligence can be applied. We would like to briefly touch on possible applications of these methods beyond pure neuroimaging.

The typical sources of big data in neurosurgery (and neurooncology) apart from medical images are genome sequencing (terabytes of data), biosensors (especially those used in intensive care units or wearable devices), and medical records. The results of genome sequencing are within the scope of a science called bioinformatics. This is a distinct, specific area in which neurosurgeons, as a rule, do not have professional knowledge. However, it is very relevant for neurooncology since molecular biomarkers define the prognosis and treatment response in patients with brain tumors.

Monitors installed in neuro-intensive care units (neuro-ICU) or intraoperative neuromonitoring devices are providers of streaming signals reflecting patients' condition and vital signs online. Electroencephalography (EEG) and electrocorticography traditionally produce tons of data about the brain state and pathology. Any clinical or instrumental data accompanying neurocritical conditions are of exceptional value as they can reveal the reserves and limits of the human brain. That data might be really "big" in volume. Thus, for example, multivariable time series analysis to detect unfavorable events based on implanted or wearable biosensors is another area of artificial intelligence applications. F. Chaudhry et al. (2021) made a review on machine learning applications in neuro-ICU [22]. Artificial intelligence was used to detect cognitive-motor dissociation by EEG, predict mortality with intracranial pressure monitoring, triage suspected intracranial bleeds with computer tomography (CT), predict delayed cerebral ischemia and vasospasm after subarachnoid hemorrhage, detect seizure activity by EEG, predict hemorrhagic transformation following thrombolytic therapy for stroke with MRI and clinical data, predict and diagnose nosocomial intracranial infections, predict outcome.

At N.N. Burdenko Neurosurgery Center we have some personal experience in developing artificial neural networks to classify intracranial hemorrhage types with CT scans [23]. We also applied machine learning to study nosocomial meningitis with our academic partner [24]. We believe that data from neuro-ICU, especially in high-volume centers, might appear very fruitful for machine learning research and clinical applications.

Electronic health records (EHRs), including clinical narrative texts, are significant sources of data. However, they are poorly used in artificial intelligence studies and are underestimated, in our opinion. We want to draw the reader's attention to this data source and share our personal experience with it further.

Finally, it is noticeable, that the amount of data available in neurosurgical research is always limited. In the best scenario, there are hundreds of cases available. Rarely thousands of observations are included in the research. At the same time, modern machine learning, for example, neural networks, benefits from thousands, tens and hundreds of thousands of objects.

However, neurosurgery is rich in per-patient data. A multimodal set of clinical examination data, complex neuroimaging in a wide range of modalities, ultrasound, electroencephalography, laboratory tests, intraoperative studies, daily observation data, etc., is regularly obtained for every patient. In other words, the dimension of feature space for medical research objects significantly exceeds the number of such objects. That allows to state that data in neurosurgery is not as "big" as complex. That means developers need to invest efforts into creating methods to proceed with a limited set of large-dimensional data to apply artificial intelligence technologies in neurosurgery successfully.

9 Non-trivial Application of Artificial Intelligence in Neurooncology

The scope of artificial intelligence in neurosurgery and neurooncology is undoubtedly not limited to all the examples above. We discussed the mainstream in neurooncology, with a focus primarily on brain gliomas. At the same time, the range of tasks for using artificial intelligence is more expansive than we can imagine. For instance, X. Kong et al. (2018) proposed machine learning methods for automatic detection of acromegaly from facial photographs [25]. D. Johansson et al. (2019) conducted a study with accelerometry-based wearable sensors to detect tonic-clonic seizures under video-EEG [26]. M. Livne et al. (2019) proposed a deep-learning U-net framework for vessel segmentation in patients with cerebrovascular diseases showing good performance [27]. V. Staartjes et al. (2021) developed synthetic computed tomography of the lumbar spine based on MRI. The potential of artificial intelligence to recognize critical anatomic landmarks in real-time as a neuronavigation was shown in an animal study by T. Carson and co-authors (2021) [28]. We believe video processing with artificial intelligence is promising in neurosurgical research.

Artificial intelligence with natural language processing can be applied not only to medical records but also to scientific literature. In particular, T. Hana and co-authors (2019) were able to identify several dozen areas in neurooncology and predict trends in publication activity in these areas with acceptable accuracy [29].

10 What Is the Rationale for Using Artificial Intelligence in Neurosurgery?

We hope the reader has already got some idea of the possible application of artificial intelligence in neurosurgery by now. So, we will try to answer the question: what is the rationale for using artificial intelligence in neurosurgery? Are all the efforts in this new area worth it? Below are the arguments we subjectively see as essential.

1. Transition to the most non-invasive and safe technologies in neurosurgery

   As illustrated with the research in neurooncology, the first reason to adopt artificial intelligence in neurosurgery is to strive for the safest and most non-invasive care. We may imagine a non-invasive tumor biopsy even if it seems to be impossible today. Nevertheless, who knows what's coming on tomorrow?
2. Evidence-based risk assessment for adverse events

   The second rationale for artificial intelligence is a new evidence-based way to assess risks for adverse events. That is an essential component of decision support. Considering a vast spectrum of variables with machine learning might improve risk assessment.
3. Processing the maximum amount of information available

   Machine learning can account for much more factors than any single expert can handle. It can take hundreds of factors that describe the individual characteristics of the patient and his/her disease.
4. Personalization of care

   The fourth argument is related to the third. Considering patient details in a model is a way of evidence-based personalization of care.
5. Reducing time for clinical routine

   Saving much time for a doctor in clinical routine is a very rewarding automation task. Automated detecting, delineating and counting the number of brain lesions might be such an example.
6. Reduction of the interval between diagnostics and best treatment

   Digital technology have a potential to simplify diagnostic procedures and shorten the time from the disease manifestation to the best treatment administration.
7. Cost savings

   If we substitute some of the expensive technologies to digital, if we take some work burden off the doctor, this also leads to cost savings.
8. Any efficiency coming from automation in medicine...

   Today we may not imagine all the benefits artificial intelligence can bring to neurosurgery. It could be any efficiency one can imagine coming from automation. So, our current commitment is to figure out in which cases artificial intelligence is most valuable. That is one of the most important rationales for research in this field.

11 Is it Possible to Develop Artificial Intelligence at the Bedside?

In the final part of this chapter, we want to share our experience organizing artificial intelligence research in a neurosurgical facility. The authors of this chapter work at N.N. Burdenko Neurosurgery Center (Moscow, Russian Federation) - the biggest neurosurgical facility in Russia and one of the biggest in the world. The number of neurosurgical interventions performed annually reaches 10,000, with less than a 0.3% postoperative mortality rate. Over the past 23 years of medical information systems functioning in our Center, we have accumulated data on 150,000+ neurosurgical cases. The challenge to improve the efficiency of secondary data usage in research led to establishing the laboratory of bioinformatics and artificial intelligence at our Center in 2016.

Why create a specialized artificial intelligence laboratory within a high-volume clinical center? The first reason is obviously in organizing the Center's data, data access, and analysis. However, the other, no less important reasons exist. In our view, the applicability of artificial intelligence in neurosurgery can be achieved not only with effective models but primarily by choosing the correct automation problems critical for that domain. Probably, today there is a reasonably large area to determine these needs and tasks. Why do we think so? Because specialists in machine learning are not able to define a relevant task in the neurosurgery. In turn, neurosurgeons cannot formulate the problem in the way machine learning specialists expect it. Thus, the laboratory of biomedical informatics and artificial intelligence at N.N. Burdenko Neurosurgery Center is an example of a specialized structure that understands doctors' language and the conceptual apparatus of informaticians. Being "inside medicine," the laboratory conducts its research in artificial intelligence. The laboratory staff can understand the requirements of neurosurgery, which can be met through digital technologies. The lab also builds its prototypes for artificial intelligence solutions.

At the same time, the laboratory is the "interface" between the neurosurgical clinic and the basic science, providing effective connections between the National Center for Neurosurgery and external partners in digital technologies. Today the laboratory operates in the modern computing infrastructure possessing its own supercomputing facilities (a cluster of two supercomputers of 5 petaflops each). Thus, the laboratory might serve as a working model of a computing center inside a large medical institution.

Due to the accumulation of an extensive (for neurosurgery) archive of medical records, the main areas of our work are focused on the analysis of medical texts and images. Moreover, in this part of the chapter, a few words will explicitly address natural language processing. There is significantly less publications on using this artificial intelligence technology in neurosurgery than computer vision, so we believe our experience might be helpful.

One of our projects is devoted to length-of-stay prediction based on narrative text data from electronic health records [30, 31]. Our latest results (2021) indicate the possibility of building a machine prediction tool using modern neural language models based on narrative textual information with a median absolute error of 2 days, which is comparable to the doctor's prediction error [32].

Automatic medical text processing with supervised machine learning requires target variables (class labels) for each text. Typically, class labels are prepared by medical experts while reading the texts to extract specific information. We proposed an algorithm to extract information from unstructured medical texts semiautomatically, which dramatically gains the extraction process and makes it applicable to millions of records for a reasonable amount of time [33]. The use of the proposed algorithm in paresis detection in and pulmonary embolism extraction from medical records proved high performance with F1-score over 0.9 [34–36]. Currently we apply natural language processing to study complications in neurosurgery and proposed a data-driven definition for the term "complication" [37].

A few solutions to automatically track scientific literature topics of interest using machine learning with feature engineering based on small abstracts were proposed by our group [38, 39].

Our team is also involved in developing glioma image-based differentiating neural models, speech-to-text solutions, cortico-cortical evoked potential analysis, and other studies exploiting artificial intelligence in neurosurgery. However, we advocate for using medical texts in machine learning experiments as we see their potential for practice.

12 Conclusion

The use of artificial intelligence is becoming a trend in clinical neuroscience, including neurosurgery, in particular neurooncology. Artificial intelligence appeared to overcome the limitations of classic statistics when processing the multidimensional and unstructured data. Images and structured clinical data are the main sources for machine learning to date, while text mining is underexploited. The developers and researchers reasonably expect that the introduction of artificial intelligence will improve the quality and outcome of scientific research and clinical practice in neurosurgery.

The studies by the authors of the chapter are supported by Russian Foundation for Basic Research grant 18–29-01052.

References

1. Turing AM. Computing machinery and intelligence. Mind [Internet] 1950 LIX(236):433–60. [cited 2021 Oct 27]; Available from: https://academic.oup.com/mind/article/LIX/236/433/986238.
2. Danilov GV, Shifrin MA, Kotik KV, Ishankulov TA, Orlov YN, Kulikov AS, et al. Artificial intelligence in neurosurgery: a systematic review using topic modeling. Part i: major research areas. Sovrem Tehnol v Med. 2020;12(5)
3. Danilov G, Board Secretary S, Orlov Y, Kulikov A, Anesthesiologist S, Potapov A. Artificial intelligence Technologies in Neurosurgery: a systematic literature review using topic Modeling. Part II: research objectives and perspectives. Современные технологии в медицине. 2020;12(6):111–8. [cited 2021 Oct 28]. https://doi.org/10.17691/stm2020.12.6.12.
4. Louis DN, Ohgaki H, Otmar, Wiestler D, Cavenee WK, Burger PC, et al. The 2007 WHO classification of tumours of the central nervous system. Acta Neuropathol 2007;114:97–109.
5. Louis DN, Perry A, Reifenberger G, Von Deimling A, Figarella-Branger D, Webster OD. The 2016 World Health Organization classification of Tumors of the central nervous system: a summary. Acta Neuropathol. et al.;131:803–20.
6. Louis D, Perry A, Wesseling P, Brat D, Cree I, Figarella-Branger D, et al. The 2021 WHO classification of Tumors of the central nervous system: a summary. Neuro Oncol [Internet]. 2021;23(8):1231–51. [cited 2021 Oct 28]; Available from: https://pubmed.ncbi.nlm.nih.gov/34185076/
7. Lambin P, Rios-Velazquez E, Leijenaar R, Carvalho S, van Stiphout RGPM, Granton P, et al. Radiomics: extracting more information from medical images using advanced feature analysis. Eur J Cancer [Internet]. 2012;48(4):441. Available from: [cited 2021 Oct 28]; /pmc/articles/PMC4533986/
8. Tabatabaei M, Razaei A, Sarrami AH, Saadatpour Z, Singhal A, Sotoudeh H. Current status and quality of machine learning-based radiomics studies for glioma grading: a systematic review. Oncology [Internet]. 2021;99(7):433–43. [cited 2021 Oct 28]; Available from: https://www.karger.com/Article/FullText/515597
9. Ning Z, Luo J, Xiao Q, Cai L, Chen Y, Yu X, et al. Multi-modal magnetic resonance imaging-based grading analysis for gliomas by integrating radiomics and deep features. Ann Transl Med [Internet]. 2021 [cited 2021 May 11];9(4). Available from: https://doi.org/10.21037/atm-20-4076.
10. Danilov G, Korolev V, Shifrin M, Ilyushin E, Maloyan N, Saada D, et al. Noninvasive glioma grading with deep learning: a pilot study. In: Studies in health technology and informatics. IOS Press; 2021. [In press].
11. Jian A, Jang K, Manuguerra M, Liu S, Magnussen J, Di Ieva A, et al. Machine Learning for the Prediction of Molecular Markers in Glioma on Magnetic Resonance Imaging: A Systematic Review and Meta-Analysis. Neurosurgery [Internet]. 2021 Apr 7 [cited 2021 Apr 24]; Available from: https://academic.oup.com/neurosurgery/advance-article/doi/10.1093/neuros/nyab103/6214540
12. Krivov E, Kostjuchenko V, Dalechina A, Shirokikh B, Karchuk G, Denisenko A, et al. Tumor Delineation For Brain Radiosurgery by a ConvNet and Non-Uniform Patch Generation. Lect Notes Comput Sci (including Subser Lect Notes Artif Intell Lect Notes Bioinformatics) [Internet]. 2018 Aug 1 [cited 2021 Oct 29];11075 LNCS:122–9. Available from: https://arxiv.org/abs/1808.00244v1
13. D'Este S, Nielsen M, Hansen A. Visualizing glioma infiltration by the combination of multimodality imaging and artificial intelligence, a systematic review of the literature. Diagnostics (Basel, Switzerland) [Internet]. 2021 [cited 2021 Oct 29];11(4). Available from: https://pubmed.ncbi.nlm.nih.gov/33806195/
14. van Kempen E, Post M, Mannil M, Witkam R, Ter Laan M, Patel A, et al. Performance of machine learning algorithms for glioma segmentation of brain MRI: a systematic literature review and meta-analysis. Eur Radiol [Internet]. 2021; [cited 2021 Oct 29]; Available from: https://pubmed.ncbi.nlm.nih.gov/34019128/
15. Huang J, Shlobin NA, Lam SK, De Cuypere M. Artificial Intelligence Applications in Pediatric Brain Tumor Imaging: A Systematic Review. World Neurosurg [Internet]. 2021 Oct 11 [cited 2021 Oct 29]; Available from: https://linkinghub.elsevier.com/retrieve/pii/S1878875021015655
16. Kocher M, Ruge MI, Galldiks N, Lohmann P. Applications of radiomics and machine learning for radiotherapy of malignant brain tumors. Strahlentherapie Und Onkol [Internet]. 2020 Oct 1 [cited 2021 Oct 30];196(10):856. Available from: /pmc/articles/PMC7498494/.
17. Lohmann P, Elahmadawy M, Gutsche R, Werner J, Bauer E, Ceccon G, et al. FET PET Radiomics for Differentiating Pseudoprogression from Early Tumor Progression in Glioma Patients Post-Chemoradiation. Cancers (Basel) [Internet]. 2020 Dec 1 [cited 2021 Oct 31];12(12):1–17. Available from: https://pubmed.ncbi.nlm.nih.gov/33353180/
18. Lee J, Wang N, Turk S, Mohammed S, Lobo R, Kim J, et al. Discriminating pseudoprogression and true progression in diffuse infiltrating glioma using multi-parametric MRI data through deep learning. Sci Rep [Internet] 2020 Dec 1 [cited 2021 Oct

31];10(1). Available from: https://pubmed.ncbi.nlm.nih.gov/33230285/
19. Patel M, Zhan J, Natarajan K, Flintham R, Davies N, Sanghera P, et al. Machine learning-based radiomic evaluation of treatment response prediction in glioblastoma. Clin Radiol [Internet] 2021 Aug 1 [cited 2021 Oct 31];76(8):628.e17–628.e27. Available from: https://pubmed.ncbi.nlm.nih.gov/33941364/
20. Buchlak QD, Esmaili N, Leveque JC, Bennett C, Farrokhi F, Piccardi M. Machine learning applications to neuroimaging for glioma detection and classification: an artificial intelligence augmented systematic review. J Clin Neurosci. 89:177–98.
21. Tewarie I, Senders J, Kremer S, Devi S, Gormley W, Arnaout O, et al. Survival prediction of glioblastoma patients-are we there yet? A systematic review of prognostic modeling for glioblastoma and its clinical potential. Neurosurg Rev [Internet] 2021 Aug 1 [cited 2021 Oct 31];44(4):2047–2057. Available from: https://pubmed.ncbi.nlm.nih.gov/33156423/
22. Chaudhry F, Hunt R, Hariharan P, Anand S, Sanjay S, Kjoller E, et al. Machine Learning Applications in the Neuro ICU: A Solution to Big Data Mayhem? Front Neurol [Internet]. 2020 Oct 9 [cited 2021 Oct 31];11. Available from: https://pubmed.ncbi.nlm.nih.gov/33162926/
23. Danilov G, Kotik K, Negreeva A, Tsukanova T, Shifrin M, Zakharova N, et al. Classification of Intracranial Hemorrhage Subtypes Using Deep Learning on CT Scans. Stud Health Technol Inform [Internet]. 2020 [cited 2021 Oct 31];272:370–3. Available from: https://pubmed.ncbi.nlm.nih.gov/32604679/
24. Savin I, Ershova K, Kurdyumova N, Ershova O, Khomenko O, Danilov G, et al. Healthcare-associated ventriculitis and meningitis in a neuro-ICU: incidence and risk factors selected by machine learning approach. J Crit Care. 2018;45:95–104.
25. Kong X, Gong S, Su L, Howard N, Kong Y. Automatic detection of acromegaly from facial photographs using machine learning methods. EBioMedicine. 2018;27:94–102.
26. Johansson D, Ohlsson F, Krysl D, Rydenhag B, Czarnecki M, Gustafsson N, et al. Tonic-clonic seizure detection using accelerometry-based wearable sensors: a prospective, video-EEG controlled study. Seizure. 2019;65:48–54.
27. Livne M, Rieger J, Aydin OU, Taha AA, Akay EM, Kossen T, et al. A U-net deep learning framework for high performance vessel segmentation in patients with cerebrovascular disease. Front Neurosci. 2019;13:97.
28. Carson T, Ghoshal G, Cornwall G, Tobias R, Schwartz D, Foley K. Artificial Intelligence-enabled, Real-time Intraoperative Ultrasound Imaging of Neural Structures Within the Psoas: Validation in a Porcine Spine Model. Spine (Phila Pa 1976) [Internet]. 2021 Feb 1 [cited 2021 Oct 31];46(3):E146–52. Available from: https://pubmed.ncbi.nlm.nih.gov/33399436/
29. Hana T, Tanaka S, Nejo T, Takahashi S, Kitagawa Y, Koike T, et al. Mining-Guided Machine Learning Analyses Revealed the Latest Trends in Neuro-Oncology. Cancers (Basel). 2019 Feb;11(2).
30. Danilov G, Kotik K, Shifrin M, Strunina U, Pronkina T, Potapov A. Prediction of postoperative hospital stay with deep learning based on 101 654 operative reports in neurosurgery. Stud Health Technol Inform. 2019;258:125–9.
31. Danilov G, Kotik K, Shifrin M, Strunina U, Pronkina T, Potapov A. Predicting postoperative hospital stay in neurosurgery with recurrent neural networks based on operative reports. Stud Health Technol Inform [Internet]. 2020 Jun 16 [cited 2021 Oct 8];270:382–6. Available from: https://ebooks.iospress.nl/doi/10.3233/SHTI200187
32. Danilov G, Kotik K, Shevchenko E, Usachev D, Shifrin M, Strunina Y, et al. Length of stay prediction in neurosurgery with Russian GPT-3 language model compared to human expectations. In: Studies in health technology and informatics; 2021. In Press.
33. Danilov G, Shifrin M, Strunina U, Pronkina T, Potapov A. An information extraction algorithm for detecting adverse events in neurosurgery using documents written in a natural rich-in-morphology language. In: Studies in health technology and informatics, vol. 262; 2019.
34. Danilov G, Shifrin M, Strunina Y, Kotik K, Tsukanova T, Pronkina T, et al. Semiautomated approach for muscle weakness detection in clinical texts. In: Studies in health technology and informatics [internet]. IOS Press; 2020 [cited 2021 Feb 1]. p. 55–58. Available from: https://pubmed.ncbi.nlm.nih.gov/32604599/
35. Danilov G, Shifrin M, Strunina Y, Kotik K, Tsukanova T, Pronkina T, et al. Detection of muscle weakness in medical texts using natural language processing. In: Studies in health technology and informatics. IOS Press; 2020. p. 163–7.
36. Danilov G, Kosyrkova A, Shults M, Melchenko S, Tsukanova T, Shifrin M, et al. Inter-rater reliability of unstructured text labeling: artificially vs. naturally intelligent approaches. Stud Health Technol Inform [Internet]. 2021 May 27 [cited 2021 Sep 9];281:118–22. Available from: https://pubmed.ncbi.nlm.nih.gov/34042717/
37. Danilov G, Kotik K, Shifrin M, Strunina Y, Pronkina T, Tsukanova T, et al. A comparison of word embeddings to study complications in neurosurgery. In: Studies in health technology and informatics. In Press; 2021.
38. Danilov G, Ishankulov T, Orlov Y, Shifrin M, Kotik K, Potapov A. The classification of scientific literature for its topical tracking on a small human-prepared dataset. Stud Health Technol Inform [Internet]. 2020 [cited 2021 Sep 8];272:191–4. Available from: https://ebooks.iospress.nl/doi/10.3233/SHTI200526
39. Danilov G, Ishankulov T, Kotik K, Orlov Y, Shifrin M, Potapov A. The classification of short scientific texts using pretrained BERT model. Public Heal Informatics Proc MIE 2021 [Internet]. 2021:83–7. [cited 2021 Oct 31]; Available from: https://ebooks.iospress.nl/doi/10.3233/SHTI210125

MIX
Papier aus verantwortungsvollen Quellen
Paper from responsible sources
FSC® C105338

If you have any concerns about our products,
you can contact us on
ProductSafety@springernature.com

In case Publisher is established outside the EU,
the EU authorized representative is:
Springer Nature Customer Service Center GmbH
Europaplatz 3, 69115 Heidelberg, Germany

Printed by Libri Plureos GmbH
in Hamburg, Germany